ISBN 978-0-265-77794-7
PIBN 10968894

1 MONTH OF
FREE
READING

at

www.ForgottenBooks.com

By purchasing this book you are eligible for one month membership to ForgottenBooks.com, giving you unlimited access to our entire collection of over 1,000,000 titles via our web site and mobile apps.

To claim your free month visit:

www.forgottenbooks.com/free968894

English
Français
Deutsche
Italiano
Español
Português

www.forgottenbooks.com

Mythology Photography **Fiction**
Fishing Christianity **Art** Cooking
Essays Buddhism Freemasonry
Medicine **Biology** Music **Ancient**
Egypt Evolution Carpentry Physics
Dance Geology **Mathematics** Fitness
Shakespeare **Folklore** Yoga Marketing
Confidence Immortality Biographies
Poetry **Psychology** Witchcraft
Electronics Chemistry History **Law**
Accounting **Philosophy** Anthropology
Alchemy Drama Quantum Mechanics
Atheism Sexual Health **Ancient History**
Entrepreneurship Languages Sport
Paleontology Needlework Islam
Metaphysics Investment Archaeology
Parenting Statistics Criminology
Motivational

THE JOURNAL OF
XPERIMENTAL MEDICINE

EDITED BY

SIMON FLEXNER, M.D.

VOLUME TWENTY-SECOND

WITH EIGHTY-ONE PLATES AND NINETY-THREE

FIGURES IN THE TEXT

NEW YORK

THE ROCKEFELLER INSTITUTE FOR MEDICAL RESEARCH

1915

THE NEW ERA PRINTING COMPANY
41 NORTH QUEEN STREET
LANCASTER, PA.

CONTENTS.

No. 5, NOVEMBER 1, 1915.

No. 6, DECEMBER 1, 1915.

STUDIES ON THE CIRCULATION IN MAN.

XII. A Study of Inequalities in the Blood Flow in the Two Hands (or Feet) Due to Mechanical Causes (Embolism, Compression of Vessels, Etc.) or to Functional (Vasomotor) Causes, with a Discussion of the Criteria by Which the Conditions Are Discriminated.

By G. N. STEWART, M.D.

(From the H. K. Cushing Laboratory of Experimental Medicine of Western Reserve University, Cleveland.)

(Received for publication, March 17, 1915.)

In this paper a few typical cases are discussed to illustrate the criteria referred to in the title. Other cases presenting inequalities in the flow are included in Paper XIII of this series,[1] under unilateral peripheral neuritis and hemiplegia. Some have been published in previous papers, and of these one or two will be briefly alluded to. The flow observations are summarized in Table I.

A criterion which could be predicted *a priori* for a blood flow diminished by a mechanical obstruction would be the approximate constancy of the ratio of the flow in the obstructed part to that in the corresponding normal part in measurements made at such short intervals that collateral circulation could not be appreciably increased in the interval. If the obstruction is so great that the chief resistance of the vascular path is situated there, it is clear that a further criterion would be the lack of any marked response of the blood flow in the part to vasoconstrictor or vasodilator influences; *c. g.*, to reflex vasomotor effects from the contralateral part. It will, of course, depend upon the degree of the obstruction and the extent to which a collateral circulation has been opened up how great an effect on the flow will be produced through the vasomotors. Under no circumstances could we expect the ratio of the flow in the

---

[1] Stewart, G. N., *Arch. Int. Med.*, 1915, xvi (in press).

obstructed part to that in the normal part to be increased by conditions favoring vasodilatation acting equally on both parts or on the whole surface, for example, by an increase of the air temperature, whereas general cutaneous vasodilatation might very well alter the ratio to the disadvantage of the affected part. On the other hand, purely central changes, affecting only the driving power of the heart, would leave the ratio unchanged, however great the absolute changes in the flows might be.

These considerations have been verified in a number of cases. One of these has been already described.[2] The innominate and right common carotid arteries were ligated by my colleague, Dr. Carl Hamann, in a woman 68 years old. for Subclavian aneurysm. About a month after the operation the ratio of the flow in the right hand to that in the left was 1 : 3.54 and 1 : 3.48 on two successive days. The stability of the ratio, although the flow in both hands was somewhat greater at the first than at the second examination, is obvious. Also practically no vasodilatation was caused in the right hand when the left was immersed in warm water. The stability of the ratio is clearly, and the failure of the contralateral reflex to influence the flow is almost certainly, associated with the still very effective mechanical block on the arterial path of the limb.

Sixteen weeks later the collateral circulation had opened up so well that the ratio had increased to 1 : 1.3, and the vasomotor reflexes from the left to the right hand were now distinctly shown by the change in the blood flow.

A case with a different kind of mechanical block, namely, embolism (and thrombosis) in the left arm and the right leg, illustrates equally well the criterion of stability of the ratio of the flows. There was apparently at one time some temporary plugging of vessels in the right arm also, but this cleared up.

Costa B., a dyer, aged 47 years, was in Lakeside Hospital from Apr. 13, 1914, to May 9, 1914, suffering from rheumatic fever, and left with the physical signs of mitral stenosis and insufficiency. On July 31, 1914, he was readmitted to the hospital. That morning while at work he suddenly felt a severe pain in the right leg, worst in the groin. No pulsation could be detected in the right dorsalis pedis. The anterior tibial pulse could be felt, but was very feeble. No pulse in the right popliteal. Good pulsation in the left leg in all accessible arteries. No pulse in the left radial artery, but the right is strong, rhythmic, regular, and well sustained. The vessel wall is palpable, but not nodular. A strong pulse can be felt in the right brachial. Blood pressure, left arm, systolic 114, diastolic 85.

Sept. 26, 1914. The pulse is obliterated in the left radial. The left hand is cold and pains him. Sept. 30. He says there is tingling of the left hand and right foot. The left radial pulse is absent and never returned after this. The right radial pulse is diminished. Right brachial felt.

Oct. 3. 1914. He says that there is no discomfort except tingling in the right foot. Both extremities are warm. The right femoral pulse can be felt, but not the dorsalis pedis. No pulse in the left radial or brachial. Pulse in

[2] Stewart, *Arch. Int. Med.,* 1913, xii, 678.

right radial and brachial. On Oct. 5 it is stated that the right radial pulse could not be felt. On Oct. 8 the right radial pulse was again felt, and was always present thereafter. On Oct. 10 he complains of pain in the left shoulder and down the arm to the elbow. Physical signs of (compensated) mitral stenosis and insufficiency. The blood flow was examined on Oct. 26, 1914, at which time no improvement had occurred in the permeability of the vessels which could be detected·by palpation. The left foot is warmer than the right to the touch.

· *First Examination of Blood Flow.*—Costa B., Oct. 26, 1914. Hands in bath at 2.56 p.m., in calorimeters at 3.07, out of calorimeters at 3.28. As always, unless otherwise mentioned, the quantity of water in each hand calorimeter was 3,015 cc.

Time.	Temperature of			Time.	Temperature of		
	Calorimeters.		Room.		Calorimeters.		Room.
	Right.	Left.			Right.	Left.	
3.06½	31.63	31.53		3.18	31.72	31.435	
3.08	31.59	31.48		3.19	31.75	31.43	22.4
3.09	31.595	31.48	21.4	3.20	31.78	31.425	
3.10	31.61	31.475		3.21	31.80	31.42	
3.11	31.62	31.47	21.3	3.22	31.82	31.41	22.6
3.12	31.62	31.46		3.23	31.84	31.405	
3.13	31.63	31.46	21.4	3.25	31.895	31.40	
3.14	31.65	31.455		3.26	31.92	31.40	
3.15	31.67	31.45	21.7	3.27	31.945	31.395	
3.16	31.695	31.455		3.28	31.98	31.385	22.8
3.17	31.71	31.44	22.0	3.41	31.79	31.22	

Cooling of calorimeters in 13 minutes, right 0.19°, left 0.165°. Volume of right hand 473 cc., of left 441 cc. Water equivalent of calorimeters with contents, right 3,473, left 3,448. Pulse in carotid 102.

Feet in bath at 3.44½ p.m., in calorimeters at 3.55½, and out of calorimeters at 4.24. As always, unless otherwise mentioned, the quantity of water in each foot calorimeter was 2,775 cc.

Time.	Temperature of			Time.	Temperature of		
	Calorimeters.		Room.		Calorimeters.		Room.
	Right.	Left.			Right.	Left.	
3.55	32.17	32.19		4.14	31.69	31.95	22.0
3.57	32.03	32.06	21.9	4.16	31.68	31.97	
3.59	31.97	32.00		4.18	31.67	31.99	22.0
4.01	31.90	31.96	22.0	4.20	31.66	32.02	
4.03	31.85	31.93		4.22	31.65	32.04	22.05
4.05	31.78	31.91	22.0	4.24	31.64	32.06	
4.07	31.76	31.91		4.26	31.59	32.03	
4.09	31.73	31.91	22.0	4.34	31.40	31.84	22.1
4.12	31.70	31.93					

Cooling of foot calorimeters, 0.19° in 8 minutes for right and left. Volume of right foot 1,191 cc., of left 1,188 cc. Water equivalent of foot calorimeters with contents, right 3.856, left 3.854. Rectal temperature 37.45° C.

The blood flow in the hands for the last fifteen minutes in the calorimeters was 6.03 gm. per 100 cc. per minute for the right hand, and 1.28 gm. for the left, with room temperature 22° C. The ratio of the flow in the left foot to that in the right was 1:4.71. For the feet the flows were 1.25 gm. per 100 cc. per minute for the right, and 2.50 gm. for the left, with room temperature 22° C. calculated for the last 15 minutes in the calorimeters when the flows had become steady (ratio 1:2). The ratio of the sum of the foot flows to the sum of the hand flows was 1:1.94, and the ratio of the flow in the normal (left) foot to that in the normal (right) hand, 1:2.41, both within the normal range for the ratio of foot to hand flow.

The man was discharged on Nov. 20 from Lakeside Hospital, and was admitted at the City Hospital on Feb. 2, 1915. He complains of pain in the right leg from the groin down. When he walks a little, the pain gets worse. There are also shooting pains down the leg. On placing a constricting band on the left arm, the superficial veins filled slowly. When the band was put on the right leg, none of the superficial veins filled. The veins of the left leg and right arm filled rapidly (Feb. 16). On Feb. 9 the blood examination gave erythrocytes 5,100,000, leucocytes 6,400. Wassermann negative. Blood pressure (Feb. 9), systolic 95, diastolic 65; (Feb. 11) 110 and 75. On admission the boundaries of the cardiac dullness were the third rib, the right sternocostal margin, and 2 cm. to the left of the nipple line. The blood flow was examined on Feb. 24, 1915. At this time the grip of the left hand was strong, scarcely weaker than that of the right. The right foot and lower leg feel markedly cold to the touch, and there is no pulse in the dorsalis pedis or elsewhere in the leg. No pulse at the left wrist, and the left hand is colder to the touch than the right. There is no difference in the nails of the two hands. He complains of pains in the right leg and says that it feels tired all the time. The left hand does not now trouble him much. There is some wasting of the right leg.

Second Examination.—Feb. 24, 1915. Hands in bath at 3.12 p.m., in calorimeters at 3.22, and out of calorimeters at 3.35.

Time.	Temperature of			Time.	Temperature of		
	Calorimeters.		Room.		Calorimeters.		Room.
	Right.	Left.			Right.	Left.	
3.21	31.89	31.82		3.31	32.09	31.87	
3.23	31.85	31.80	23.8	3.32	32.12	31.88	23.8
3.24	31.88	31.82	23.8	3.33	32.16	31.89	
3.25	31.90	31.825		3.34	32.195	31.895	
3.26	31.93	31.83		3.35	32.26	31.90	23.9
3.27	31.97	31.84	23.8	3.55	32.06	31.67	
3.29	32.03	31.86	23.7				

Cooling of calorimeters in 20 minutes, right 0.20°, left 0.23°. Volume of right hand 498 cc., of left 452 cc. Water equivalent of calorimeters with contents, right 3,493, left 3,457. Pulse 81. The right radial pulse was surprisingly difficult to feel considering the fair blood flow.

The blood flow for the right hand at this examination was 6.96 gm. per 100 cc. per minute, and for the left hand 2.54 gm. (for 11 minutes in the calorimeters), with room temperature 23.8°. For the right hand this is practically the same as on Oct. 26, 1914, four months earlier, considering the higher room temperature at the second examination. In the left hand, however, the flow is twice as great as on the first occasion, so that the ratio of the hand flows is now 1 : 2.74, indicating a great improvement in the collateral circulation in the left hand. This is quite in accordance with the general condition of the hands. There is, of course, still a decided mechanical obstruction on the left side, and the new ratio of the hand flows is stable for successive measurements made at short intervals, as shown by the third examination.

Third Examination.—Feb. 26, 1915. The pulse rate was considerably greater than at the last examination, and the superficial veins of the hands were distinctly better filled. The veins of the right hand were fuller than those of the left, and the resistance to compression of the veins by the finger was greater in the right hand. The pulse was felt in the right brachial although not strongly, and was not felt in the left brachial. Feet in bath at 2.11 p.m., in calorimeters at 2.25½, out of calorimeters at 2.49. Pulse 102, counted in the carotid, as it was difficult to count it at the right wrist, although regular.

Time.	Temperature of			Time.	Temperature of		
	Calorimeters.		Room.		Calorimeters.		Room.
	Right.	Left.			Right.	Left.	
2.24	31.92	31.77		2.39	31.855	32.74	23.7
2.27	31.91	31.86	23.4	2.41	31.86	32.87	23.3
2.29	31.89	31.995	23.4	2.43	31.85	32.99	23.1
2.31	31.88	32.14	23.2	2.45	31.845	33.12	23.1
2.33	31.87	32.30	23.5	2.47	31.84	33.24	23.4
2.35	31.865	32.47	24.1	2.49	31.79	33.27	
2.37	31.86	32.61	24.1	3.18	31.42	32.79	

Cooling of foot calorimeters in 31 minutes, right 0.37°, left 0.48°. Volume of right foot 1,132 cc., of left 1,199 cc. Water equivalent of foot calorimeters with contents, right 3,812, left 3,862.

Hands in bath at 3.13½ p.m., in calorimeters at 3.23½, and out of calorimeters at 3.36.

Time.	Temperature of			Time.	Temperature of		
	Calorimeters.		Room.		Calorimeters.		Room.
	Right.	Left.			Right.	Left.	
3.23	31.48	31.34	23.5	3.31	31.80	31.46	24.3
3.25	31.48	31.36	23.3	3.32	31.87	31.48	24.2
3.26	31.51	31.37	23.1	3.33	31.90	31.49	
3.27	31.58	31.385	23.4	3.34	31.95	31.505	24.3
3.28	31.62	31.40		3.35	32.00	31.52	24.2
3.29	31.70	31.42	24.1	3.36	32.12	31.53	
3.30	31.73	31.43	24.3	3.52	31.96	31.37	

Cooling of hand calorimeters, 0.16° for right and left in 16 minutes. Volume of right hand 497 cc., of left hand 457 cc. Water equivalent of calorimeters with contents, right 3.492, left 3.460. Rectal temperature 37.85°.

The blood flow in the right hand on Feb. 26 was 9.98 gm., and in the left hand 3.70 gm. per 100 cc. per minute for the last 10 minutes in the calorimeters, with average room temperature 24° C. The flow in both hands is considerably greater than two days previously, corresponding to the increased pulse rate, but the ratio is exactly the same (1 : 2.70). The flow in the right foot was only 0.70 gm. per 100 cc. per minute even with the increase in the general circulation, a considerably smaller flow than at the first examination, four months previously, while the flow in the left foot was 6.50 gm., with room temperature 23.5° C., a great increase. The ratio between the flows in the two feet was 1 : 9.28, which shows a decided decrease in the circulation in the right foot. This agrees completely with the other signs of deterioration, and suggests the probability of impending gangrene. The flow, actually measured in the right foot, is probably still sufficient to nourish the tissues, since considerably smaller flows have not infrequently been met with in other conditions without gangrene. But should a further examination show that the diminution in the flow is still progressing, that of itself would be sufficient to indicate a grave prognosis. It is an interesting fact that the ratio of the sum of the foot flows to the sum of the hand flows (1 : 1.90) is precisely the same as four months ago, in spite of the great changes in the ratios between the two hand flows and between the two foot flows, and in the absolute amount of both foot and hand flows. This suggests that the blocking of the vascular path to one leg (doubtless the diminution in the flow extends to the whole of the right posterior extremity) is associated with a reciprocal dilatation of the path to the other leg, so that the normal partition of the blood between the legs and the rest of the body is scarcely disturbed. That is to say, the blood which normally finds its way through the two common iliacs seems eventually, when the main part of the path from one common iliac is blocked, still to find its way through the one which remains pervious, the normal limb making room, it may be supposed by vasodilatation, for an additional quantity of blood.

It would, of course, be rash to generalize from one or two measurements of this kind, but the result in this case is so precise that it needs an explanation. If we reflect that one important function of the circulation, and relatively more important in the limbs than elsewhere on account of the greater proportion of their surface to their mass, is the elimination of heat, it will not appear fantastic to suggest that when the blood flow through the skin of one leg is greatly interfered with, it might be advantageous for the flow through the skin of the other leg to be accelerated. The same consideration applies to the muscles in so far as interchange between them and the other tissues is of general utility in the metabolism. The regulation of the blood pressure may also be facilitated by such a redistribution of the blood, which, if it occurs, may be assumed to be brought about, mainly at any rate, through the vasomotor system. An occlusion coming on gradually and never becoming complete presents, of course, different conditions from a complete occlusion caused by ligation of large arteries or by amputation. One important difference is that the tissues fed by the partially occluded vessels continue to contribute to the total metabolism of the body and to the heat production, and must therefore to some extent influence the blood flow in the surfaces from which heat is lost and the organs concerned in excretion.

Note Added May 7, 1915.—Another examination of the blood flow in Costa B. was made on Apr. 7. The flow in the left hand was 3.43 gm. per 100 cc. per minute, and in the right hand 13.11 gm., with room temperature 24.5° C. (ratio 1 : 3.82). In the feet the flows were 1.51 gm. and 7.43 gm. for the right and left, respectively, with room temperature 23.3° C. (ratio 1 : 4.92). The ratio of the combined foot flows (per 100 cc. per minute) to the combined hand flows was 1 : 1.85; *i. e.,* practically the same as at the previous examinations. The man died on May 4. For 4 days before death the right leg and foot had been numb and discolored by subcutaneous hemorrhages. At the necropsy the right common iliac artery was found completely occluded by an old organized thrombus firmly adherent to the artery. A similar thrombus occupied the left subclavian artery. A fresh thrombus was found in the left external iliac artery. It was freely movable. There was marked mitral stenosis with hypertrophy of the left auricle, and the valve was covered with vegetations.

An apparent instance of a reciprocal relation of the lymph flow and blood flow in a part which leads to increased blood flow when the outflow of lymph is obstructed was observed in a case already reported.[3] The case was diagnosed as Hodgkin's disease. There was great and persistent edema in both legs and nowhere else in the body. The swelling remained quite unaltered during the five weeks of the man's stay in the hospital. It coincided with a relatively good blood flow in the feet, which was interpreted as indicating that the obstruction responsible for the edema was on the lymph path rather than on the venous path. The flow in the feet was particularly large in proportion to that in the hands (ratio of combined foot flows to combined hand flows 1 : 0.96). The sum of the foot flows per 100 cc. of part per minute actually exceeded slightly the sum of the hand flows, a very rare condition in our observations. It was suggested that far from being interfered with, the blood circulation in the edematous legs was accelerated through local vasodilatation.

It is known that the exchange between the blood in the capillaries and the tissues does not consist merely in the passage of materials out of the blood, but also in the passage of materials, including water, from the tissues into the blood. The intracellular liquids and tissue lymph are naturally in relation both with the blood and with the lymphatic lymph, and their normal composition is maintained by interchange with both. Is it not probable that when the lymphatic channels are blocked, and the elimination of waste products and the regulation of the quantity of tissue lymph by way of the lymphatics interfered with, the flow in the alternative channels, the blood capillaries, may be increased, so as to increase the excretion by way of the blood? The associated vasodilatation may very well be brought about by the accumulation of one or another of the waste products.

In the next case the question was raised at the clinical examination whether the circulatory changes observed in one hand could be at-

[3] Stewart, *Arch. Int. Med.,* 1913, xii. 678.

tributed to injury or irritation of the brachial plexus or some of its constituent cords. The blood flow measurements indicated clearly that the phenomena were due to mechanical obstruction on the arterial path of the limb, and further, a point of importance in connection with the question whether operative interference was advisable, that the blood flow in the affected hand was quite sufficient to nourish the part.

Walter L., aged 25 years, height 5 feet 10 inches, weight 210 pounds, a powerfully built man, well nourished. with a good deal of subcutaneous fat, was admitted to the City Hospital on Oct. 25' 1914, suffering from a gunshot wound. His partner who was shooting rats with a small rifle at the lunch hour accidentally shot him, the bullet (caliber No. 22) entering the right side of the chest close to the sternum and lodging near the right shoulder, as was afterwards shown by the x-ray. He drove home and then came at once to the hospital. The wound bled at first with a squirt, but soon stopped and has not bled much since. The other man was 5 feet 8 inches in height and the gun was pointed directly at the front of the patient and slightly upward. The general course of the bullet when it struck the patient was slightly upward, and very slightly to his right. On admission he complained of a continual dull ache localized under the armpit, and of pain when he breathed deeply. The fingers of the right hand are numb, but he can move them perfectly well. There is a tendency for the thumb and forefinger to flex themselves.

Nov. 1. He complains that the fingers of the right hand remain flexed rather than straight. and says that he must lay his hand upon something to keep the fingers from doubling up. It appears as if the median nerve were irritated. He can now move his right shoulder a little in all directions. There is a large hematoma in the axillary region, and the ecchymosis extends downwards to the waist line and also above the deltoid. On Nov. 4 he complains of severe pain in the deltoid region. On Nov. 5 the amplitude of the right radial pulse is less than that of the left. The pulse is regular. On Nov. 6 he still complains of his hand. The index and middle fingers now seem to be extended most of the time. On Nov. 11 there seems to be no pain in the arm, and the patient is allowed to be up. The index. middle finger, and, to a less extent. the thumb are numb. Tactile sensation and pain sensation are diminished on the palmar surface of these fingers and on the dorsal surface of their two last phalanges. The skin on the palmar surface over the median nerve distribution right up to the wrist is scaly, while the ulnar area of the palmar surface is normal. The finger nails on the right hand have grown less than on the left hand. The inner portion of the thenar eminence is atrophied. The grip between the little and ring fingers is less strong than in the left hand. The inequality in the radial pulses continues. The x-ray shows the bullet in the axilla about one inch below the lower lip of the glenoid.

The question was put by the surgeon in charge of the case, whether the diminished pulse could be attributed to injury of the brachial plexus or to pressure on it, and, in particular, whether the blood flow in the affected hand was so much diminished as to threaten gangrene and therefore to call for surgical interference.

First Examination of Blood Flow.—Walter L. Nov. 13, 1914. Hands in bath at 2.10 p.m., in calorimeters at 2.20¼. At 2.38 left hand put into water at 8.2° C. At 2.50 left hand put into water at 44°. At 3.04 right hand out of calorimeter.

| Time. | Temperature of | | | Time. | Temperature of | | Time. | Temperature of | | Notes. |
| | Calorimeters. | | Room. | | Right calorimeter. | Room. | | Right calorimeter. | Room. | |
	Right.	Left.								
2.20	31.06	31.04		2.39	31.44		2.57	31.725		
2.22	31.00	31.12		2.40	31.46	23.1	2.58	31.75		
2.23	31.02	31.18	23.3	2.41	31.47		2.59	31.775	23.3	
2.24	31.03	31.27	23.4	2.42	31.48	23.2	3.00	31.79		
2.25	31.05	31.35		2.43	31.49		3.01	31.82		
2.26	31.09	31.43	23.4	2.44	31.505	23.1	3.02	31.84	23.3	
2.27	31.12	31.52		2.45	31.52		3.03	31.89		
2.28	31.14	31.595		2.46	31.56	23.1	3.04	31.925	23.4	
2.29	31.16	31.67	23.4	2.47	31.595		3.19	31.715	23.3	Lt. 31.80
2.30	31.19	31.74		2.48	31.605	23.1				
2.31	31.22	31.81		2.49	31.615					
2.32	31.25	31.90	23.3	2.50	31.63					
2.33	31.28	31.98		2.51	31.64	23.3				
2.34	31.31	32.06	23.3	2.52	31.65					
2.35	31.33	32.13		2.53	31.66					
2.36	*31.36	32.20	23.3	2.54	31.68	23.3				
2.37	31.39	32.28	23.2	2.55	31.695					
2.38	31.42	32.345		2.56	31.71	23.3				

Cooling of calorimeters, right 0.21° in 15 minutes, left 0.545° in 41 minutes. Volume of right hand 503 cc., of left 464 cc. Water equivalent of calorimeters with contents, right 3,497, left 3,468. Rectal temperature 37.42° C.

For 15 minutes before the vasomotor reaction was tested the blood flows of the two hands came out 5.32 gm. per 100 cc. per minute for the right hand, and 14.76 gm. for the left hand, a ratio of 1 : 2.77, with room temperature 23.2°. When the left hand was immersed in cold water the flow in the right hand was diminished to 4.71 gm. per 100 cc. per minute for the first two minutes of the immersion, to 3.31 gm. for the next 3 minutes, and it rose only to 4.83 gm. per 100 cc. per minute during the remaining 7 minutes of the period. The vasomotor reflex elicited in the right hand by immersion of the left hand in cold water was accordingly small. When the left hand was immersed in warm water the flow in the right hand sank to 3.46 gm. per 100 cc. per minute for the first 3 minutes of the period. For the next 4 minutes it rose slightly (to 4.42 gm.), for the next 3 minutes to 5.33 gm., and for the remaining 4 minutes of the period to 7.32 gm. per 100 cc. per minute.

It is clear that the initial small flow in the right hand is not due to a vasoconstriction which can either be much increased by immersion of the contralateral hand in cold water, or much diminished by its immersion in warm water.   Even the maximum flow obtained under the influence of the reflex vasodilatation is scarcely half the normal flow in the left hand.   Such vasomotor reflex reactions are among the criteria of a circulation diminished by a mechanical block.   In the present instance the mechanical block is not extreme, as shown by the ratio of the flows, and therefore reflex vasomotor effects on the flow are still obtainable, although diminished.

There is in any event no probability that irritation of vasoconstrictors by pressure would cause such a great and permanent discrepancy between the blood flows in the two hands.   The conclusion was therefore drawn that the deficiency in the circulation of the right hand was due to pressure on the blood supply of the right arm either by the bullet itself, or by the hematoma, or by both.   As regards the question whether the flow in the right hand was dangerously small, it could be answered that it was not, and that although, of course, this matter should be tested from time to time, there was no reason to apprehend gangrene with a blood flow of this magnitude.

Five days later another observation was made on the blood flow with the view of determining whether the collateral circulation was increasing.   At the time of the second examination the right hand was in much the same condition as at the first examination, only the patient had observed that the finger nails on the right hand were now beginning to grow although not so fast as those on the left hand.

Second Examination.—Pulse 92. Hands in bath at 2.07 p.m., in calorimeters at 2.16½, and out of calorimeters at 2.33.

Time.	Calorimeters.		Room.	Time.	Calorimeters.		Room.
	Right.	Left.			Right.	Left.	
2.16	31.17	31.17		2.26	31.225	31.68	
2.18	31.13	31.22	23.7	2.27	31.24	31.73	22.9
2.19	31.13	31.285		2.28	31.25	31.78	
2.20	31.13	31.34	23.3	2.29	31.27	31.82	
2.21	31.17	31.41		2.30	31.28	31.87	
2.22	31.19	31.48		2.31	31.29	31.91	23.0
2.23	31.20	31.54	23.2	2.32	31.295	31.95	23.1
2.24	31.21	31.595		2.33	31.33	32.00	
2.25	31.22	31.63	23.05	2.42	31.22	31.86	

Cooling of calorimeters in 9 minutes, right 0.11°, left 0.14°. Volume of hands, right 492 cc., left 466 cc. Rectal temperature 37.22°. Water equivalent of calorimeters with contents, right 3,489, left 3,467.

The blood flows in both hands at this examination were somewhat less than on the previous occasion (3.93 gm. per 100 cc. per minute for the right hand, and 10.76 gm. for the left). The ratio, however, between the flows in the two hands was unchanged (1 : 2.74). The room temperature (23° C.) was almost the same as at the first examination. The fact that the ratio of the flows is practically identical with that obtained at the first examination of itself almost precludes the idea that the deficiency in the right hand is due to persistent irritation of vasoconstrictors. A stimulation of this kind could hardly remain constant for 5 days, while a mechanical block might well do so.

The patient was discharged from the hospital soon afterwards.

The blood flow in his hands was again examined on Mar. 8, 1915. Since Nov. 28, 1914, he has been at work and came to the hospital merely for the examination. He says that when he first resumed work, whenever he flexed the forearm on the elbow it was done with a jerk. He finds that the right hand is still weaker than the left and gets tired sooner. There is no pain in it and he uses it freely in his work. He has some little difficulty with the right hand in such movements as those concerned in buttoning his clothes. The right arm is not quite as strong as the left. There is now no pain at the shoulder and no trace of the injury can be detected except a small scar on the right side of the chest at the point of entrance of the bullet, which still remains in the body, no attempt having been made to extract it. The left radial pulse is distinctly stronger than the right. It was not possible to feel the right brachial artery, while the left brachial pulse was strong.

Third Examination.—Hands in bath at 2.50 p.m., in calorimeters at 3.00, and out of calorimeters at 3.15. Pulse 84.

Time.	Temperature of			Time.	Temperature of		
	Calorimeters.		Room.		Calorimeters.		Room.
	Right.	Left.			Right.	Left.	
2.59	32.03	32.10		3.09	32.26	32.59	23.9
3.01	32.01	32.15		3.10	32.29	32.64	23.9
3.02	32.03	32.20		3.11	32.32	32.695	
3.03	32.04	32.26	23.5	3.12	32.37	32.745	
3.04	32.06	32.31		3.13	32.41	32.795	
3.05	32.10	32.36	24.4	3.14	32.45	32.84	23.8
3.06	32.13	32.42		3.15	32.53	32.88	
3.07	32.18	32.48	24.5	3.33	32.35	32.68	
3.08	32.21	32.53	24.0				

Cooling of calorimeters in 18 minutes, right 0.18°, left 0 20°. Volume of right hand 481 cc., of left 493 cc. Rectal temperature 37.35° C. Water equivalent of calorimeters with contents, right 3,480, left 3,480.

The blood flow in the right hand at this last examination, 110 days after the previous one, was 9.30 gm. per 100 cc. per minute, and in the left hand 11.60

gm., with an average room temperature of 24° C. The sum of the flows in the two hands is only slightly greater than on Nov. 13, 1914 (20.9 gm. as against 20.08 gm.), but the distribution of the blood between the two hands is now very different, the ratio being only 1 : 1.24. Plainly a great improvement has occurred in the interval in the circulation of the right hand.

The volume measurement indicates a considerable increase in the left hand, probably due to work hypertrophy, as he still must use the left more than the right. There is a smaller diminution in the volume of the right hand due, it is to be supposed, either to absorption of a small amount of edema fluid or to slight atrophy. No edema of the hand was observed at any time, but there might have been some increase in tissue liquid not revealed by detectable swelling. The mechanical obstruction on the arterial path which was responsible for the diminished blood flow might, of course, have in some degree involved the venous outflow from the limb as well, although not to such an extent as to cause evident edema. That some pressure was exerted on nerves is indicated by the spasmodic contractions described, but there is no evidence of serious injury to nerves. The slight atrophy may well have been due simply to disuse.

Decided inequalities in the blood flow in the two hands (or feet) not associated with obvious functional or anatomical differences and not conforming to the criteria of inequalities due to a mechanical cause are sometimes observed in clinical cases, particularly, in my experience, in neuropathological conditions. The great characteristic of these inequalities of flow is their instability. Not only does the ratio of the flow in the two hands vary widely in examinations on successive days, without any apparent clinical change to account for the variation, but the ratio may one day be in favor of one hand and the next day in favor of the other, or the two flows, greatly different in amount at one examination, may be found equal at the next. A change in the external conditions (such as a sufficient increase or diminution of the air temperature) which are known to affect the vasomotor system is especially apt to alter or reverse the ratio.

For this reason it is suggested that some vasomotor peculiarity is responsible for the circulatory changes in these cases, and inequalities in the blood flow which possess the criterion mentioned are interpreted as depending on vasomotor rather than mechanical conditions. The results on a typical case (Thomas Q.) will be given.

The patient, a man 36 years of age, when he first came under observation was suffering from alcoholic neuritis affecting particularly the feet. There was no recognizable clinical difference between the two sides. Yet notable differences in the blood flow were made out, which, however, were not stable from day to day and were therefore interpreted as depending upon vasomotor differences on the two sides not so easily abolished or perhaps more easily produced than in normal persons by the preliminary bath and the subsequent long immersion in the calorimeters. About a year after the first series of examinations the patient was again seen in the hospital, this time suffering from (tubercular) pleurisy with febrile temperature in addition to the neuritis. The change in the clinical picture was very decided and so was the change in the results of blood flow measurements. The marked tendency to cutaneous vasoconstriction already found associated with fever[4] was reflected in a greatly diminished hand and foot flow.

Thomas Q., a railroad clerk, aged 36 years, was admitted to the City Hospital on May 24, 1912, suffering from chronic alcoholism, with alcoholic neuritis. He has been drinking for his entire life. Two years ago he had "alcoholic paralysis" and collapsed while he was walking along the street after being at a dance. His limbs gave way but he did not lose consciousness and resumed his work after four days. His legs swelled and became painful, so he stopped work and went to a hospital. For the past year he has had painful micturition, especially after a drinking bout. In 18 months his weight has declined from 204 to 147 pounds, and his strength has diminished. He has had night sweats at intervals and his appetite has become poor. He has been drinking heavily since March. The legs and feet are equally untrustworthy in walking. He feels no difference between the right and left. The feet feel rather cold to him and recently there has been a tendency for chilblains to form on them. He is nervous. There is tremor of the tongue and lips. The pupils react to light and accommodation. The knee jerk is absent on both sides. There is no special defect of sensation. There is pain on pressure over the nerve trunks of both lower extremities. The upper extremities show nothing special.

On May 29, 1912, the flow in his feet was measured. It came out 1.43 gm. per 100 cc. per minute for the right, and 2.44 gm. for the left (ratio 1 : 1.70), with the rather low room temperature of 21.5°. On May 31 the flow in the right foot was 3.39 gm. per 100 cc. per minute, and in the left foot 3 87 gm. (ratio 1 :1.14), with room temperature 25.4°. These flows are of the normal order of magnitude. Slight and transient reflex diminution of the flow in the left foot was caused by immersion of the right foot in cold water and a marked temporary diminution when the right foot was transferred to warm water. The subsequent increase of flow in the left foot, while the right continued in the warm water, only carried the flow slightly above the initial level. The much smaller flows in the feet on May 29 were probably due to the increased sensitiveness of these cases to vasoconstrictor stimulation occasioned by the considerably lower room temperature, for which there is other evidence. With the increased vasodilatation in both feet on May 31 the difference between them, if of vasomotor origin, would tend to become less.

Two examinations of the hands were made at this time. On June 4, with room temperature 25° C., the flow in the right hand was 8.51 gm. per 100 cc. per minute, in the left 13.20 gm. (ratio 1 : 1.55). The hands were not noticeably affected by neuritis at this time, but later on he returned to the hospital

[4] Stewart, *Jour. Exper. Med.*, 1913, xviii, 372.

with wrist drop in addition to foot drop. The difference in the flow in the two hands on subsequent examination was not found to be permanent, and it was therefore interpreted as due to a functional difference in vasomotor innervation. Immersion of the right hand in warm water caused, after a very slight and transient diminution in the flow (to 9.82 gm. for one minute), one of the greatest reflex increases witnessed in the whole series of observations; *viz.*, to 17.08 gm. per 100 cc. per minute. Immersion of the right hand in cold water was accompanied by a transient diminution in flow in the left hand (to 5.58 gm. per 100 cc. per minute for the first 3 minutes), the flow then increasing to 10.22 gm. per 100 cc. per minute for the remaining 7 minutes, during which the right hand continued in the cold water.

First Examination of Blood Flow.—Thomas Q. Examination of blood flow in the feet, May 31, 1912. Pulse 108. Feet in bath at 2.40½ p.m., in calorimeters at 2.55¼. 3,740 cc. of water in each calorimeter. At 3.13 p.m. the right foot was immersed in cold water at 9° C. He felt it very cold and complained much.

Time.	Temperature of Calorimeters.		Room	Time.	Temperature of Left calorimeter.	Room.	Notes.
	Right.	Left.					
2.54	31.42	31.40		3.22	32.25		
2.57	31.43	31.42	25.3	3.23	32.285	25.5	
2.59	31.50	31.50	25.3	3.24	32.31		
3.01	31.56	31.56		3.25	32.35		
3.03	31.61	31.63	25.3	3.26	32.40		At 3.26 rt. foot put in water at
3.05	31.675	31.70		3.27	32.41		43° C.
3.06	31.72	31.75		3.28	32.425		
3.07	31.75	31.79	25.4	3.29	32.45		
3.09	31.805	31.86		3.30	32.49	25.7	
3.11	31.88	31.93	25.45	3.31	32.525		
3.13	31.93	32.00		3.32	32.525		Stirring was brief and iusuf-
3.15		32.07		3.33	32.58		ficient.
3.17		32.11	25.5	3.34	32.60		Stirring insufficient.
3.18		32.14		3.35	32.635		Foot out of calorimeter.
3.19		32.165		3.44	32.525		Rt. is now at 31.595° C.
3.20		32.21					
3.21		32.23					

Cooling of calorimeters, right 0.335 in 31 minutes, left 0.11 in 9 minutes. Volume of right foot in calorimeter 1,311 cc., of left foot 1,290 cc. Water equivalent of calorimeters with contents, right 4,857, left 4,843. Rectal temperature 37.35°.

Second Examination.—Examination of flow in hands, June 4, 1912. Hands in bath at 2.23 p.m., in calorimeters at 2.33. At 2.47 p.m. the right hand was put into water at 8° C.; at 2.57 into water at 43°. At 3.07 p.m. the left hand was taken out of the calorimeter.

Cooling of calorimeters, right 0.33° in 39 minutes, left 0.13° in 9½ minutes. Volume of right hand 460 cc., of left hand 437 cc. Water equivalent of calorimeters with contents, right 3,463, left 3,445. Pulse 100. Rectal temperature 37.5.

| Time. | Temperature of Calorimeters. | | Room. | Time. | Temperature of | | Notes. |
	Right.	Left.			Left calorimeter.	Room.	
2.31½	31.57	31.51	24.8	2.52	32.53	24.95	
2.34	31.57	31.53		2.53	32.57		
2.35	31.63	31.62	25.2	2.54	32.605		
2.36	31.695	31.72		2.55	32.635		
2.37	31.74	31.78	25.0	2.56	32.66		
2.38	31.79	31.86		2.57	32.71		
2.39	31.83	31.92		2.58	32.745	25.1	
2.40	31.87	31.96		2.59	32.805		
2.41	31.895	32.02		3.00	32.86		
2.42	31.94	32.08		3.01	32.925	25.1	
2.43	31.995	32.15	25.0	3.02	32.995		
2.44	32.03	32.225		3.03	33.07		
2.45	32.08	32.29		3.04	33.135	25.15	
2.46	32.11	32.34		3.05	33.19		
2.47	32.14	32.38		3.06	33.25		
2.48		32.40		3.07	33.32	25.2	
2.49		32.41		3.16	33.19		Rt. 31.81
2.50		32.43					
2.51		32.48					

That the cause of the inequality between the flows in the two hands observed on June 4, 1912, was a transient functional difference is indicated by the result of the third examination on the following day (June 5). Here with a lower room temperature (22.4°) the flow was reduced in both hands, but far more in proportion in the left (7.31 gm. per 100 cc. per minute for the right, 7.35 gm. for the left hand). The excess in the flow in the left hand is, however, in reality somewhat greater than the numerical results indicate. For the left hand had suffered the loss of a portion of the thumb by amputation a while ago, and the ratio between its surface and its mass was therefore diminished. Also the man is right handed and in right handed persons the flow per 100 cc. of volume is usually somewhat greater than the left.

Third Examination.—June 5, 1912. He says he is feeling well and can walk fairly, although he soon gets tired. He was out on the hospital grounds yesterday and today. Hands in bath at 1.22½ p.m., in calorimeters at 1.32, out of calorimeters at 1.50. Pulse 110.

| Time. | Temperature of Calorimeters. | | Room. | Time. | Calorimeters. | | Room. |
	Right.	Left.			Right.	Left.	
1.31½	31.08	31.05		1.42	31.495	31.30	
1.33	31.07	31.03	22.2	1.43	31.52	31.415	22.4
1.34	31.11	31.07		1.44	31.575	31.45	
1.35	31.175	31.12	22.3	1.45	31.61	31.49	22.4
1.36	31.21	31.15		1.46	31.64	31.52	
1.37	31.27	31.20	22.4	1.47	31.60	31.55	
1.38	31.31	31.24		1.48	31.72	31.50	22.4
1.39	31.39	31.30		1.49	31.70	31.625	
1.40	31.425	31.335		1.50	31.785	31.635	
1.41	31.47	31.36		1.58	31.665	31.525	

Cooling of calorimeters in 8 minutes, right 0.12°, left 0.11°. Volume of right hand 468 cc., of left hand 436 cc. Water equivalent of calorimeters with contents, right 3,469. left 3,444. Rectal temperature 37.7°.

Thomas Q. was discharged from the hospital on June 29. 1912, and was re-admitted on Jan. 19. 1913. He has been drinking for 4 weeks and is decidedly worse than at the previous admission. He cannot walk at all and has wrist drop of both hands. He first noticed the wrist drop the day he came. In both extremities the extensor muscles are more involved than the flexors. The nerve trunks are tender, the reflexes absent. He was treated by hot baths and massage of the extensor muscles. He states that if he is made to sit around even a few minutes after the bath he gets a severe chill lasting 20 to 30 minutes, which does not occur if he is put straight to bed. He never has a chill except after the hot bath. This is of interest as it corresponds with the evidence of increased susceptibility to reflex vasoconstriction of the cutaneous blood vessels deduced from the blood flow examinations. The flow in the hands was examined on May 21 and again on May 24, 1913. and the flow in the feet on May 24. At this time he had some fever and on May 29 the physical signs of pleurisy were found. On June 15 and again on June 18, 1913, tubercle bacilli were found in the sputum, and he was sent to the sanitarium on June 24.

When examined on May 21, 1913, nearly a year after the first examination. the blood flow in the right hand was found to be only 4.49 gm. per 100 cc. of hand per minute, and in the left hand 3.06 gm. (ratio 1 : 1.46), with the relatively high room temperature of 26.4° C. These flows were for the last 12 minutes in the calorimeters. His rectal temperature was much above the normal (40.45° C.), and the tendency to vasoconstriction associated with fever is no doubt a factor in the small flow. This tendency is illustrated by the tardiness with which the rate at which the thermometers rose became steady. Three days later (on May 24, 1913) the flow in the hands was still smaller (2.31 gm. per 100 cc. per minute for the right hand and 2.26 gm. for the left hand, for 12 minutes in the calorimeters before the testing of the vasomotor reaction, with room temperature 23.2° C.). Calculated for the last 7 minutes of this period the flows were only 1.71 and 1.58 gm. per 100 cc. per minute for the right and left hands respectively, denoting a great tendency to the onset of vasoconstriction. A rapidly increasing vasoconstriction revealed by a decrease in the rate of heat loss to the calorimeters has been observed in other cases of fever; for instance, in pneumonia before the rapid rise of body temperature. In one case in which the condition had not as yet revealed itself clearly by the physical signs, this behavior of the hands in the calorimeters suggested the onset of fever, the temperature rose nearly 5° F. within eight hours, and the next day physical signs of pneumonia were present.

During immersion of the right hand in warm water, the flow in the left, far from being increased, was actually diminished (to 1.31 gm. per 100 cc. per minute), a further illustration of the increasing vasoconstriction, which cannot be overcome by the reflex vasodilatation normally associated with immersion of the contralateral hand in warm water. The patient felt worse than at the last examination.

The flow in the feet was also abnormally small (0.84 gm. per 100 cc. per minute for the right foot. and 0.87 gm. for the left for a period of 20 minutes in the calorimeters, with room temperature 23.3° C.). For the last 10 minutes of this period the flows were only 0.77 and 0.72 gm. per 100 cc. per minute for the right and left foot, respectively, showing that the vasoconstriction was increasing in the feet also with the duration of immersion. Naturally with a tendenc/ to vasoconstriction manifested so strongly in both hands and feet, the ratio of foot to hand flow (1 : 2.6) lies within the normal limits, whereas with a lesser degree or less general distribution of the vasoconstriction this ratio is apt to be markedly disturbed in favor of the hand flow.

Cases do occur, although in my experience quite rarely, in which without any obvious reason a marked inequality in the flow in the two hands is present and persists indefinitely, without reversal of the ratio. The ratio, however, is unstable, varying greatly from day to day, and this differentiates the condition from inequalities of mechanical origin and places it in the vasomotor group. This condition is illustrated in the case of William F.

William F., a laborer, aged 50 years, was admitted to the City Hospital on March 22, 1912, suffering from "combined system disease." His illness began last June with difficulty in walking, and this has become rapidly worse. Sometimes he almost falls. About three months before admission he had a severe nose-bleed lasting for three nights without known cause. For the past year he has suffered from severe frontal headache lasting from six to twenty-four hours. He also complains of pain in the pit of the stomach coming on after eating. No vomiting. He sweats much at night, and the sweats are followed by chills. The sweating is so copious that his gown is soaked in five minutes. He says he used to be "hot-blooded," wearing no overcoat in winter, but now he is always chilly. His feet get cold as ice in bed. His hands also get cold now, although not so readily as the feet. When he goes to bed at night his feet are apt to be swollen, but the swelling is all gone in the morning. He had chancre twenty years ago and it was not treated. He is married but never had any children. His wife had two or three miscarriages. He does not drink or chew tobacco. He walks with a spastic gait. Knee jerks exaggerated. and ankle clonus can be elicited on both sides. There is some toe drop on the left side. He says he has pain all the time in his "bones," in the legs and back. but not much pain in the arms. There seems to be some loss of sensation in the left leg, and he cannot always tell the difference between warmth and cold there. Pain perception is prompt, and vibration sense good. His power of localization is somewhat impaired. He can stand with his eyes closed. Both pupils react normally to light. No obvious wasting of the hands. The grip of both hands is fairly good, the right somewhat stronger than the left. He can not write now, although a well educated man. Blood count on Mar. 24. 1912. erythrocytes 5,480,000, leucocytes 7,400. Mar. 29. Wassermann test negative. Apr. 17, spinal fluid clear. Cell count 6 per cc. X-ray examination negative. Urine, nothing special. Heart examination negative. The blood flow in the hands was examined on Apr. 24, 1912, and that in the hands and feet the following day (Apr. 25).

First Examination of Blood Flow.—William F. Apr. 24, 1912. Hands in bath at 2.43 p.m., in calorimeters at 2.52. At 3.15 p m. left hand put into water at 44° C. Pulse 100. Mouth temperature 36.65°. At 3.29 left hand put in water at 8° C. At 3.41 left hand put in water at 43° C. At 3.48 hand out of calorimeter.

Time.	Temperature of Calorimeters.		Room.	Time.	Temperature of Right calorimeter.	Room.	Time.	Temperature of Right calorimeter.	Room.	Notes.
	Right.	Left.								
2.51	30.52	30.55		3.16	30.79	23.5	3.38	31.46		
2.53	30.47	30.50	22.3	3.17	30.80		3.39	31.49		
2.54	30.47	30.49		3.18	30.81		3.40	31.50		
2.55	30.47	30.49		3.19	30.84		3.41	31.505		
2.56	30.46	30.48		3.20	30.88		3.42	31.52		
2.57	30.46	30.47		3.21	30.915		3.43	31.525		
2.59	30.45	30.46		3.22	30.975	22.2	3.44	31.53	22.25	
3.00	30.45	30.455		3.23	31.02		3.45	31.54		
3.01	30.455	30.455	22.5	3.24	31.08		3.46	31.55		
3.03	30.48	30.45		3.25	31.105		3.47	31.56		
3.04	30.50	30.44		3.26	31.15		3.48	31.57		
3.05	30.52	30.435		3.27	31.195	22.0	3.56½	31.46		Lt. 30.04
3.06	30.54	30.445	22.5	3.28	31.24					
3.07	30.58	30.44		3.29	31.295					
3.08	30.60	40.44		3.30	31.305					
3.09	30.63	30.435		3.31	31.32					
3.10	30.67	30.435		3.32	31.335					
3.11	30.695	30.445		3.33	31.35	22.0				
3.12	30.705	30.445		3.34	31.365					
3.13	30.73	30.45		3.35	31.38					
3.14	30.75	30.455		3.36	31.40					
3.15	30.78	30.45		3.37	31.43					

Cooling of calorimeters, right 0.11° in 8½ minutes, left 0.41° in 41½ minutes. Volume of right hand, 517 cc., of left hand 523 cc. Water equivalent of calorimeters with contents, right 3,509. left 3,513. Rectal temperature 37.4° C.

On Apr. 24 the flow in the right hand came out 4.20 gm. per 100 cc. per minute; in the left, 1.11 gm. (ratio 1 : 3.78), with room temperature 22.5°, an average flow much below the normal. Immersion of the left hand in warm water caused a good contralateral vasomotor reflex, indeed considering the long duration of the vasodilatation, an exaggerated one. The vasodilatation was preceded in the normal way by a good vasoconstriction for the first 3 minutes of immersion of the left hand. The diminution of the flow in the right hand on immersion of the left in cold water was also prompt, substantial, and durable.

The next day the flow in the right hand was 5.83 gm., in the left 3.11 gm. (ratio 1 : 1.87), with room temperature 23.5°, still a marked preponderance in favor of the right hand.

In the feet the flow was remarkably small but the relative preponderance in the right foot was precisely the same as in the right hand, the ratio of the flow in the right hand to that in the right foot being 15 : 1. and the corresponding ratio for the left hand and foot also 15 : 1. Immersion of the left foot in warm water caused, if anything, a diminution of the flow in the right foot, but the change was insignificant.

The patient was discharged from the hospital at his own request on May 12, and readmitted on May 31, 1913. On June 6, 1912, the blood examination showed hemoglobin 85 per cent., leucocytes 11,800. He says his legs now get red if rubbed or scratched and feel "burning." This was not the case when he was in the hospital before. He describes the heat as coming "from the inside of the legs, from the bone." Neither the legs nor the feet feel warm to the observer's hand. The veins of the legs are larger than before. He does not sweat much now. The knee jerk is still exaggerated, and ankle clonus can be elicited. He is less able to walk than when previously in the hospital and must use a stick, which was not the case before. Romberg's sign is not present. He has had much trouble in urination for over a year, with burning pain in the penis. In the sitting position he can easily make water, but not standing up. The pulse in the left brachial is distinctly smaller than in the right. This is easier to make out than any difference between the two radials. He was discharged from the hospital "unimproved," Oct. 6, 1912. The blood flow in the hands was examined on June 11, and that in the feet on June 12, 1912.

Second Examination.—June 11, 1912. Hands in bath at 2.16 p.m., in calorimeters at 2.26¼, and out of calorimeters at 2.50.

Time.	Temperature of			Time.	Temperature of		
	Calorimeters.		Room.		Calorimeters.		Room.
	Right.	Left.			Right.	Left.	
2.25	31.50	31.40	24.5	2.39	31.65	31.43	
2.27	31.47	31.36		2.40	31.665	31.43	24.7
2.28	31.48	31.355	24.5	2.41	31.68	31.435	
2.29	31.49	31.355		2.42	31.695	31.44	
2.30	31.50	31.36	24.6	2.43	31.715	31.45	
2.31	31.515	31.365		2.44	31.73	31.455	
2.32	31.535	31.375		2.45	31.755	31.465	24.8
2.33	31.56	31.395		2.46	31.78	31.49	
2.34	31.595	31.40	24.8	2.47	31.80	31.51	
2.35	31.60	31.41		2.48	31.815	31.52	24.7
2.36	31.61	31.42		2.49	31.83	31.525	
2.37	31.625	31.425	24.7	2.50	31.85	31.53	
2.38	31.64	31.425		3.04	31.69	31.39	

Cooling of calorimeters in 14 minutes, right 0.16°, left 0.14°. Volume of right hand 493 cc., of left 499 cc. Water equivalent of calorimeters with contents, right 3.489, left 3.494. Rectal temperature 37.6°.

On June 11 the flow in the right hand was 4.40 gm., in the left 2.77 gm. (ratio 1 : 1.58) with room temperature 24.7° C. On June 12 the flows were 0.82 and 0.48 gm. in the right and left foot, respectively. Immersion of the right foot in warm water caused practically no change in the left foot. The foot flows are still abnormally small, although larger than on Apr. 25. It is to be remarked.

however, that the ratio between the flow in the right hand and that in the right foot (5.4 : 1) is again almost the same as the ratio between the flow in the left hand and that in the left foot (5.3 : 1). In other words, the relative preponderance of flow is almost the same in the right foot as in the right hand. The blood flow in the right hand and foot in this patient is therefore, it would appear, permanently and decidedly greater than in the left hand and foot, although at the time of examination no marked clinical difference in the condition of the two sides of the body could be detected. The variability in the ratio between the flows in the two hands from time to time, in the absence of material variations in the factors which determine general vasomotor changes (*e. g.,* decided changes in the external temperature), differentiates the condition from an inequality due to a mechanical cause and stamps it apparently as a functional peculiarity.

The curious fact that when the ratio between the flows in the two hands varies, the ratio between the flows in the two feet varies to the same amount, would seem to require for its explanation the assumption that the intensity of the vasomotor innervation of one-half of the body, or at least of the extremities on one side, is simultaneously affected and to the same degree with respect to the innervation of the other half. Changes in the bulb, for example in its circulation, affecting the "general vasomotor center," not necessarily exclusively on one side, but to a greater degree on one side than on the other, might be expected to produce such an effect. If the change responsible for the phenomenon were located below this level it would seem that the whole dorsolumbar region from which the vasoconstrictor outflow takes place would require to be subjected to it. Is there here, perhaps, an indication of a somewhat greater progress of the pathological change on one side than on the other, a difference which has not as yet otherwise revealed itself? If the primary lesion in this case is in the upper motor neurone, it is not unnatural to suppose that the vasomotor reflex arcs may maintain an even increased vasoconstrictor tone associated with a relatively small peripheral blood flow corresponding to the exaggeration of the skeletal reflexes and the spasticity of the skeletal muscles.

A marked variability in the ratio of the blood flows in the two hands (or feet) in our observations on normal persons has not been seen. But in one normal case a decided permanent difference in the flows in the two hands was made out, the ratio of the flows remaining practically constant in observations made at an interval of 3 days and varying surprisingly little even over long periods.

In John R., a normal man, at that time 20 years old, the flow on Mar. 22, 1913, in the right hand was 10.08 gm., and in the left 7.25 gm. per 100 cc. per minute, with room temperature 24.0° C. for a period of 13 minutes, the ratio of the flows being 1 : 1.39. On Mar. 25, 1913, the flows were 12.38 and 8.82 gm., respectively, for the right and left hands (ratio 1 : 1.40) with room temperature 24.8° C. for a period of 15 minutes. On Nov. 14, 1914, the flows were 14.85 gm. for the right hand and 10.07 gm. per 100 cc. per minute for the left hand for a period of 22 minutes, with room temperature 24.3° C. (ratio 1 : 1.47). There is no anatomical or functional difference between the two hands or arms, nor any history of injury which would explain such a difference in the blood flow. The stability of the ratio is strongly in favor of a mechanical explanation, a congenital difference in the cross-section of the two subclavians, for example, rather than an explanation based on a difference on the two sides in the permanent vasomotor tone or a difference in the reflex vasomotor reaction to the manipulations and external conditions connected with the measurements. No such difference was found in the foot flows, the ratio being 1 : 1.05 in the supine and 1 : 1.13 in the sitting position on Mar. 25, 1913, with room temperatures 25.1° and 24.6°, respectively. It was at first supposed that some as yet latent unilateral pathological process, for instance pulmonary tuberculosis, might be connected with the anomaly, but nothing has developed to justify that suggestion. The subject stated at the time of the last examination that the tendency to free bleeding from slight injuries and especially from the nose, from which he had suffered for 15 years, has now disappeared.[5]

John R. Nov. 14, 1914. Hands in bath at 11.42 a.m., in calorimeters at 11.55, out of calorimeters at 12.08 p.m.

Time.	Temperature of			Time.	Temperature of		
	Calorimeters.		Room.		Calorimeters.		Room.
	Right.	Left.			Right.	Left.	
11.54	32.12	32.22	24.1	12.08	32.67	32.51	
11.56	32.12	32.22		12.09	32.70	32.53	
11.57	32.16	32.24	24.2	12.10	32.75	32.555	24.3
11.58	32.20	32.255		12.11	32.79	32.58	
11.59	32.23	32.27		12.12	32.82	32.62	24.3
12.00	32.285	32.305	24.3	12.13	32.88	32.64	
12.01	32.32	32.32		12.14	32.91	32.66	
12.02	32.38	32.35		12.15	32.94	32.67	24.4
12.03	32.42	32.37		12.16	32.97	32.68	
12.04	32.48	32.40	24.3	12.17	32.995	32.69	
12.05	32.52	32.43		12.18	33.02	32.695	24.3
12.06	32.56	32.45		12.28	32.88	32.555	
12.07	32.61	32.48	24.3				

Cooling of calorimeters, right and left 0.14°, in 10 minutes. Volume of right hand 387 cc., of left hand 357 cc. Water equivalent of calorimeters with contents, right 3,405, left 3,381. Rectal temperature 36.64° C. Pulse 62. Blood pressure, right arm 98 (systolic), 84 (sudden change in sound), 76 (cessation of sound). Another observation 98, 83, 74. Left arm 97, 79, 65.

Note Added May 7, 1915.—On this day, which was close and muggy, the flow in the right hand was 18.52 gm. and in the left 14.01 gm. (ratio 1 : 1.32) with room temperature 25.1° C.

As a practical point of technique it may be suggested that when it is desired to test the stability of the ratio between the flows on the

[5] Stewart, *Jour. Exper. Med.*, 1913, xviii, 354.

two sides, in cases where the question arises whether the cause is a nervous or a mechanical one, this could perhaps often be quickly done by measuring the blood flow several times on the same day with considerably different room and calorimeter temperatures. If the inequality is of vasomotor origin, it may be expected to disappear or be greatly reduced in one or more of the sets of observations, whereas an inequality due to a permanent mechanical cause could not disappear.

SUMMARY.

1. In cases in which great inequalities in the blood flow in the two hands were produced by mechanical causes (ligation or compression of vessels, embolism), the stability of the ratio of the flows, in successive measurements at short intervals, was found to be characteristic. Over long intervals the opening up of collateral circulation or the progressive increase of the block (in a case of multiple embolism with thrombosis) was followed by changes in the ratio of the blood flows in the normal and the affected part. Another criteriou of these conditions was found to be that the inequality was not abolished by producing general vasomotor changes; *e. g.*, by altering the external temperature.

2. In certain cases inequalities in the blood flow in the two hands (or feet) were found which were not stable from day to day, and which could be abolished. reduced, increased, or reversed by alterations in the external conditions which bring about general vasomotor changes. These inequalities, not associated with clinically recognizable differences between the parts compared. were interpreted as due to unequal activity of the vasomotor mechanism on the two sides. The condition appeared to be most frequent in certain groups of neurological cases.

I wish to express my obligations to the staff of the City Hospital and to my colleagues at Lakeside Hospital for many courtesies.

TABLE I.

Date.	Case.	Age.	Pulse rate.	Blood pressure.	Room.	Art'l blood.	Right.	Left.	Right.	Left.	Right.	Left.	In min.	Right.	Left.	Right.	Left.	Notes.
							Temperature (C.) of Calorimeters.		Volume of part in cc.		Heat given off in gm.-cal.			Blood flow in gm. per min.		Flow per 100 cc. of part per min.		
Oct. 26, 1914	Costa B...	47 yrs.	102	114.85	22.0	36.95	31.81	31.42	473	441	1980	424	15	28.53	5.67	6.03	1.28	Hands } Embolism of lt. hand
Feb. 24, 1915			81	110.75	22.0	36.85	31.68	31.99	1191	1188	1040	1946	15	14.90	29.67	1.25	2.50	Feet } and rt. leg.
Feb. 26, 1915			102		23.8	37.05	32.07	31.86	498	452	1711	690	11	34.70	11.48	6.96	2.54	Hands.
					24.0	37.35	31.82	31.45	497	457	2469	900	10	49.60	16.95	9.98	3.70	Hands.
					23.5	37.25	31.87	32.71	1132	1199	694	5847	18	7.96	77.95	0.70	6.50	Feet.
Nov. 13, 1914	Walter L..	25 yrs.			23.3	36.92	31.22	31.77	503	464	2063	4769	15	26.80	68.59	5.32	14.76	Hands. Bullet in rt. shoulder.
					23.1		31.44				234		3	23.72		4.71		Lt. hand in ater at 8.2° C.
					23.2		31.48				245		3	16.68		3.31		Lt. and still in old water.
					23.1		31.56				821		7	24.31		4.83		Lt. and still in old water.
					23.3		31.65				248		3	17.42		3.46		Lt. hand in water at 44° C.
					23.3		31.69				420		4	22.26		4.42		Lt. hand still in warm ater.
					23.3		31.76				374		3	26.84		5.33		Lt. and still in warm ter.
					23.3		31.86				671		4	36.83		7.32		Lt. hand still in warm water.
Nov. 18, 1914			92	R. 115.80	2.30	36.72	31.22	31.67	492	466	1246	2964	13	19.36	50.16	3.93	10.76	Hds.
Mar. 8, 1915				L. 130.80	24.0	36.85	32.30	32.60	481	493	2015	2407	11	44.73	57.20	9.30	11.60	Hds.
May 29, 1912	Thomas Q.	36 yrs.	84		21.5	36.50	30.67	30.80	1265	1262	1708	2846	18	18.08	30.82	1.43	2.44	Feet. ic neuritis.
May 31, 1912			168		25.4	36.75	31.68	31.71	1311	1290	3250	3630	16	44.51	50.01	3.39	3.87	Feet.
					25.5			31.98				1119	6		43.44		3.36	Rt. fot in ater at 9° C.
					25.5			32.18				1501	7		52.13		4.04	Rt. ft still in old, ter.
					25.7			32.31				238	2		29.78		2.30	Rt. foot in water at 43° C.
					25.6			32.43				1405	7		51.62		4.00	Rt. fot still in warm, ter.
June 4, 1912			100		25.0	37.00	31.86	31.95	460	437	2355	3410	13	30.16	57.71	8.51	13.20	rds.
					25.0			32.40				303	3		24.30		5.58	Rt. hand in water at 8° C.
					25.0			32.57				1247	7		44.68		10.22	Rt. hand still in old water.
					25.1			32.73				165	1		42.93		9.82	Rt. hand in ter at 43° C.
					25.1			33.03				1210	9		74.66		17.08	Rt. hand still in warm ater.
June 5, 1912			110		22.4	37.2	31.43	31.33	468	430	3330	2879	17	34.23	32.05	7.31	7.35	Hands.

TABLE I.—Concluded.

Date.	Case.	Age.	Pulse rate.	Blood pressure.	Temperature (C.) of Room.	Art'l blood.	Calorimeters. Right.	Left.	Volume of part in cc. Right.	Left.	Heat given off in gm.-cal. Right.	Left.	In min.	Blood flow in gm. per min. Right.	Left.	Flow per 100 cc. of part per min. Right.	Left.	Notes.
May 21, 1913	Thomas Q.		140		26.4	39.95	31.43	31.35	400	379	1656	1077	12	17.99	11.59	4.49	3.06	Hands. Much worse. Has fever.
May 24, 1913			116	85.60	23.2	40.1	31.24	31.20	393	372	869	810	12	9.08	8.42	2.31	2.26	Hands.
					23.5	....	31.28	....			409	....	10	5.15	....	1.31	....	Lt. hand in water at 44° C.
					23.3	40.0	30.91	30.86	1063	1057	1460	1520	20	8.02	9.23	0.84	0.87	Feet.
Apr. 24, 1912	Wm. F.	50 yrs.	100		22.5	36.9	30.62	30.45	517	523	1719	474	14	21.72	5.83	4.20	1.11	Hands. Combined degeneration.
					23.5	....	30.80	....			228	....	3	13.84	....	2.67	....	Lt. hand in water at 44° C.
					22.1	....	31.05	....			2193	....	11	37.86	....	7.32	....	Lt. hand still in warm water.
					22.0	....	31.34	....			561	....	6	18.68	....	3.61	....	Lt. hand in water at 8° C.
					22.0	....	31.44	....			702	....	6	23.81	....	4.60	....	Lt. hand still in cold water.
					22.2	....	31.54	....			544	....	7	16.11	....	3.11	....	Lt. hand in water at 43.4° C.
Apr. 25, 1912			104		23.5	36.9	30.78	30.63	493	502	2058	1188	13	28.74	15.64	5.83	3.11	Hands.
					22.4	36.8	30.42	30.35	1370	1265	742	366	24	5.38	2.62	0.39	0.21	Feet.
					22.1	....	30.32	....			262	....	10	4.49	....		0.32	Lt. foot in water at 43° C.
June 11, 1912					24.7	37.1	31.76	31.48	493	499	1046	690	10	21.76	13.82	4.40	2.77	Hands.
June 12, 1912			108		23.4	36.85	30.82	30.73	1360	1317	855	492	14	11.25	6.38	0.82	0.48	Feet.
					23.6	....	....	30.68				369	10		6.64		0.50	Rt. foot in water at 43° C.
Nov. 14, 1914	John R.	22 yrs.	62	98.75	24.3	36.14	32.57	32.46	387	357	4065	2620	22	57.50	35.95	14.85	10.07	Hands.

HIBERNATION AND THE PITUITARY BODY.

By HARVEY CUSHING, M.D., and EMIL GOETSCH, M.D

(*From the Surgical Departments of the Harvard Medical School and the Peter Bent Brigham Hospital, Boston.*)

PLATES 1 TO 3.

(Received for publication, February 19, 1915.)

INTRODUCTION.

Not only has the phenomenon of sleep offered a problem which has not been solved to the satisfaction of physiologists,[1] but still less satisfactory have been the explanations accounting for the prolonged dormant periods to which certain species of animals are subject, and of which the hibernation of winter latitudes and æstivation of tropical and arid regions are the most notable examples. Marshall Hall[2] expressed the opinion nearly a century ago that natural sleep, hibernation, and the so called diurnation of animals with nocturnal habits, differed from one another merely in degree, and this idea has again been revived by various writers.

According to the prevailing views of hibernation,—the form of seasonal sleep to which chief attention has been paid,—the lethargy, which is the striking feature of the condition, is attributable to two extracorporeal factors, namely, low external temperature and diminished food supply.[3] However, such attempts as have been made to determine the influence of these factors under experimental conditions have given results which are not entirely convincing.

Of late years some attention has been paid to the central nervous system and the histological alterations which occur in the nerve cells

---

[1] Piéron, H., Le problème physiologique du sommeil, Paris, 1913.

[2] Hall, M., On Hybernation. *Phil. Tr. Roy. Soc.*, 1832. pt. 1, 335.

[3] A résumé of the subject with bibliography will be found in Merzbacher. L.. Allgemeine Physiologie des Winterschlafes, *Ergebn. d. Physiol.*, 1904, iii. pt. 2, 214.

of dormant animals,[4] but the larger number of the more recent studies upon hibernation deal with respiratory problems. These studies have shown (Nagai,[5] and others) that there is a marked decrease in the respiratory quotient during the sleeping period, while at the time of awakening, coincident with the rapid rise in body temperature, the output of carbon dioxide is greatly increased, presumably through combustion of stored glycogen and fat.[6] Valuable as these observations have been, they have not served to shed light on the underlying cause of the phenomenon of hibernation. They have merely added to our knowledge of the physiology of the state when once induced.

Though lethargy is the dominant manifestation of hibernation, it is important in our present connection to recall certain other factors of the condition, particularly the preliminary storage of fat and the subsequent evidences, during the period of somnolence, of retarded tissue combustion. One expression of this is the lowering of the body temperature—a matter which aroused the inquisitive mind of John Hunter, who, with Jenner, was perhaps the first to make accurate studies in this direction. Additional factors are the brachycardia, the slowed respiration with diminished output of carbon dioxide, the lowered blood pressure, the relative peristaltic inactivity, and the marked insensitivity to painful or emotional stimuli.

It is well known that secretory overactivity, or even the administration of extracts of certain of the glands of internal secretion, is capable of accelerating tissue metabolism with the production of symptoms the converse of those recounted above. The purpose of the present communication is to point out that a seasonal wave of physiological inactivity on the part of these glandular structures may well account for the phenomenon of hibernation.

[4] Zalla, M., Recherches expérimentales sur les modifications morphologiques des cellules nerveuses chez les animaux hibernants. (Résumé), *Arch. ital. de biol.*, 1910–11, liv, 116.

[5] Nagai, H., Der Stoffwechsel des Winterschlafers, *Ztschr. f. allg. Physiol.*, 1909. ix, 243.

[6] Pembrey, M. S., Further Observations upon the Respiratory Exchange and Temperature of Hibernating Mammals, *Jour. Physiol.*, 1903, xxix, 195.

The Somnolence of Experimental and Clinical Hypopituitarism.

Experimental Observations.—In the course of our earlier experiences with experimental hypophysectomy in the canine, it was observed[7] that after a total or nearly total surgical removal of the gland the animals, though they remained vigorous and active for a day or more, soon passed into a profoundly lethargic state, which often terminated within the next forty-eight hours in a fatal coma. In describing this state of so called cachexia hypophyseopriva, which had been previously observed by Vassale and Sacchi, by Paulesco, and some others, it was pointed out that a lowering of body temperature was one of the inaugural symptoms, and that in the fatal cases the temperature often fell to that of the surrounding medium, rectal registrations of 20° C. being not uncommon. It was noted, moreover, in the case of animals subjected to an incomplete though nearly total extirpation that a pronounced lethargy with coincident fall in body temperature was more apt to supervene in the winter months than after similar operations in warmer seasons. This we at first attributed to the fact that the animal's quarters were insufficiently heated, but we soon found that the fall in temperature was characteristic of the condition regardless of the surrounding temperature.

In addition to the persistent and often extreme lowering of body temperature, it was observed that respiration became shallow and slow, in some cases registering only three or four respirations a minute, that the pulse rate was usually retarded, that the blood pressure was very low, and that there was a peculiar insensitiveness to external painful stimuli, often so complete that the animal might not be aroused from its lethargy even by the taking of the blood pressure registration from the open femoral.

Though the main points which we wish to emphasize in connection with these experimental states of hypophysial insufficiency are the lethargy, the subnormal temperature, the slow pulse and respiration, the lowered blood pressure, and the insensitivity, following the total or nearly total deprivation of the gland, there are other matters

[7] Crowe, S. J., Cushing, H., and Homans, J., Experimental Hypophysectomy. *Bull. Johns Hopkins Hosp.*, 1910, xxi, 127.

of interest in our present connection. It was observed both by Crowe and Goetsch that the hearts of the animals which had been in a lethargic state for some hours would continue to beat, after the fashion of a batrachian heart, for an unusual length of time when removed from the body after death.[8] It was found that the subnormal body temperature could, with some difficulty, be raised approximately to the normal by the application of external heat (wrapping in electric pads, etc.),—an effect, however, which usually led to extreme panting and which appeared to accelerate rather than to postpone the fatal issue after a total hypophysectomy. It was also observed that the injection of pituitary extract was the most effective means of raising the body temperature with coincident improvement in the animal's condition; and by such injections threatened states of cachexia hypophyseopriva were seemingly forestalled in a number of animals in which at the operation a small fragment, sufficient to maintain life, had been left, though for a time it seemed to have lost its capacity to furnish the necessary secretion.

In the course of these earlier studies we made what proved to be, from a clinical standpoint, the most important of our observations; namely, that animals surviving after a considerable deprivation of the gland, though not sufficient to precipitate a fatal state of cachexia hypophyseopriva, tended to become adipose and to lose their sexual activity. This discovery furnished the first experimental explanation of the clinical syndrome of dystrophia adiposogenitalis (*typus Fröhlich*), by demonstrating that the condition was in all certainty due to hypophysial insufficiency.[9]

As a rule, these adipose animals were apt to be drowsy and apathetic, with a slowed pulse and a more or less persistently subnormal temperature, which could be raised by the administration of hypo-

[8] This we have since observed in the case of a hibernating woodchuck, which we have autopsied, cardiac pulsations having continued for half an hour.

It may be recalled in this connection that attention has been drawn by others to the resistance (*Ueberleben*) of the nervous system of the hibernating animal. Both Valentin and Schiff have shown that the nerve-muscle preparation of the hibernating marmot is almost as slow in losing its excitability as is that of the frog.

[9] Cushing, H., The Hypophysis Cerebri. Clinical Aspects of Hyperpituitarism and Hypopituitarism. *Jour. Am. Med. Assn.*, 1909, liii, 249.

physial extracts.[10] Thus, in addition to the symptoms in the acute cases of cachexia hypophyseopriva above recounted, a tendency to the deposition of fat was shown to occur in the more chronic states of secretory deprivation. All of these things, it will be seen, fall in with the view of a greatly lowered tissue oxidation or metabolism, and we came to recognize and to speak of the state as one akin to hibernation.[11]

Clinical Observations.—With the realization, through the experimental studies mentioned above, that the so called syndrome of Fröhlich is due to a deficient hypophysial secretion, it became apparent that other symptoms besides the adiposity and genital dystrophy of this clinical state were probably attributable to the same cause. Of these symptoms the drowsiness, which in extreme cases amounts to somnolence or lethargy, is the most important in our present connection. In many examples of the disease which we have observed, the condition has been so marked that the patients fall asleep during conversation or in the midst of an examination, and in a few instances somnolence has been so notably the most striking feature of the disorder that it has been the primary complaint.

Considerable attention was paid to this symptom in a volume[12] by one of us, dealing with the subject of pituitary disorders in general, and a number of clinical examples were given there, in one of which (Case XVI), a case of stationary hypophysial insufficiency, an extraordinary seasonal period of somnolence had recurred for several years.

The following case history will serve as a good illustration of the extreme degree of torpor which may accompany clinical hypopituitarism.

[10] The administration of glandular extracts to hypophysectomized animals or in clinical cases of hypophysial insufficiency tends to increase tissue oxidation. with a resultant rise in temperature. These thermic responses do not occur in controls.

[11] It may be added that studies by F. G. Benedict and J. Homans (The Metabolism of the Hypophysectomized Dog. *Jour. Med. Research.* 1911-12. xxv. 409) on the metabolism of an hypophysectomized dog have shown very much the same decrease in the respiratory quotient that has been observed in hibernating animals.

[12] Cushing, H., The Pituitary Body and Its Disorders. Philadelphia. 1912.

Interpeduncular Cystic Tumor of Pharyngeal Anlage. Characteristic Neigh-borhood Pressure Symptoms with Pronounced Glandular Manifestation of Hypophysial Insufficiency. Previous Diagnoses of Diabetes Insipidus, Dementia Paralytica, etc.—W. B., a business man, 48 years of age, entered the Johns Hopkins Hospital, Nov. 23, 1911, with the following symptoms notable in his complaint: constant headache, profound drowsiness, restricted vision, fainting spells, loss of potentio sexualis, great sensitiveness to cold, and polyuria.

His family record was remarkable merely from the standpoint of stature, several members exceeding 6 feet in height. His father was 6 feet tall and weighed over 200 pounds.

The story of his childhood and youth revealed nothing noteworthy, though he had not been particularly rugged and had been subject to headaches. He had married at the age of 30, and was the father of four healthy children.

Present Illness.—In 1905, six years before his admission, he received a severe blow in the face, which fractured his nose. He soon began to suffer from persistent headaches, and a noticeable loss of nervous energy subsequently became apparent. At about this time he became impotent.

In 1907 a diagnosis of diabetes insipidus was made, on the basis of a marked polyuria and polydipsia. In 1909 his vision began to fail. There was occasional diplopia, and a bitemporal hemianopsia was observed. At this time he had what was called a nervous breakdown.

With the further progress of his symptoms, a definite change occurred in his disposition. He neglected his business and financial affairs. He became forgetful, disoriented, and often more or less irresponsible. From the outset he was totally apathetic in regard to and seemingly oblivious of his illness. Osteopathy, a vigorous antiluetic régime, and other measures were tried without avail.

Drowsiness, first noticed a year previously, had become at the time of his admission the leading symptom. Not only did he pass most of the day in slumber, but on a number of occasions there was a wave of prolonged sleep lasting over several days, from which he could not be aroused. In two of these periods of deep somnolence, such a marked slowing of respiration occurred that recourse was had to artificial means for its continuance.

During his comparatively wakeful hours, he complained of extreme muscular enfeeblement and of chilliness even on the warmest days when others suffered from the heat. The skin was exceedingly dry; the hands and feet were always cold to the touch. For a long time the temperature had been persistently subnormal.

Physical Examination.—Briefly, this revealed a fair complexioned, well nourished man, 5 feet 9 inches in height, weighing 138 pounds, with thin hair and scant beard, and dry, parched skin. He was dull, apathetic, and drowsy, and unable to follow a consecutive train of thought. His speech was slow, drawling, and incoherent. The temperature (rectal) was 96.5° F., pulse 60, with a low blood pressure. The x-ray showed a sella of normal configuration. There was bitemporal hemianopsia with primary optic atrophy, which had progressed to near-blindness in the left eye. The Wassermann reaction was negative.

On Nov. 30, a week after his admission, he passed into a profoundly lethargic state, with a systolic blood pressure below 100, a subnormal temperature of 96° F., slow pulse, and feeble, shallow respiration of Cheyne-Stokes type.

Under the misapprehension that the symptoms were manifestations of pressure, a right subtemporal decompression was performed. No anesthetic was used, and there was apparently complete insensitivity of the tissues. No increase in intracranial tension was disclosed.

His condition remained unmodified, and in desperation two days later. he was given a subcutaneous injection of pituitary (anterior lobe) extract. This was followed by the characteristic thermic reaction which we have described, with a rise of temperature to 101° F. Four hours after the injection he roused from his lethargy, seemed fairly rational, and asked for nourishment. This was the turning point in his critical condition, and for the succeeding two weeks he was kept in a fairly normal state of activity by repeated injections of extracts.

On Dec. 15 a transphenoidal operation with sellar decompression was performed, in the hope that the supposedly compressed gland might thus be relieved, and for some weeks there was a most unexpected betterment in his condition, both physical and mental, so that he was up, dressing. and caring for himself, and taking exercise independently. During the succeeding month he remained for the most part perfectly rational and physically active. Then followed an occasional period of somnolence, and finally he lapsed into his former state of lethargy.

The feeding of glandular extract in large doses proved unavailing, and recourse was again had to intramuscular injections. Under these injections, given daily, he invariably improved, but their frequent repetition caused so much muscular soreness that they had to be discontinued.

On Feb. 9, 1912, under primary anesthesia the pituitary, gland of a still-born child was implanted in the subcortex at the seat of the decompression, and glandular administration was discontinued. There followed an unexpected and astonishing improvement in his condition. An analysis of his mental state at this time, made by Dr. Adolph Meyer, showed practically no deviation from the normal. His memory of past events was good and he had a peculiar subconscious recollection of his preceding months of somnolence.

He gained rapidly in strength, and on Mar. 10 he was discharged. apparently well, despite a tendency to nap in the afternoon. a continuance of polyuria, and unaltered neighborhood symptoms so far as the hemianopsia was concerned. He had been kept under observation during the full month in the expectation that the implanted gland might not actually have " taken."

On his departure he took a ten hour trip to his former home. attending personally to all the details of buying tickets. hotel registration, payment of bills, and so on. He met his many friends, remembered and kept appointments with promptitude, and went about independently.—indeed from a psychic standpoint seemed an absolutely normal individual and " like his old self."

More might be said concerning the return of somnolence. the patient's ultimate death, and the autopsy, which disclosed a characteristic infundibular cyst surmounting an almost completely destroyed hypophysis. but enough has been given to illustrate the hibernation-like symptoms which characterize some of these extreme states of hypopituitarism.

In all of our examples of pituitary disease associated with tumor or glandular enlargement we endeavor to distinguish clearly between the symptoms which are produced by pressure against neighboring structures and those due to a secretory disturbance on the part of the gland itself. Many of the symptoms which we are inclined to attribute to a primary hypophysial derangement, others have ascribed to an implication of hypothetical " nerve centers " at the base of the brain in the neighborhood of the gland. For example, in the case just cited it is true that the growth—a characteristic *Hypophysen-gangtumor* as described by Erdheim—compressed and distorted the structures in the interpeduncular space as well as the pituitary body itself; and were it not for the fact that quite comparable symptoms arise in cases of hypophysial disease in the absence of tumor, one might readily be led to accept the views of Erdheim, Aschner, Purves Stewart,[13] and others that some of the symptoms to which we have drawn attention are not evidences of dyspituitarism, but are due to pressure upon some predicated center in the tuber cinereum or elsewhere in the floor of the third ventricle. However, if the latter assumption were correct, the symptoms would hardly be improved by the administration of glandular extracts.

Without prolonging this report with further clinical illustrations, it is apparent that states of outspoken hypophysial insufficiency may be accompanied by symptoms comparable to those following the experimental deprivation of a large part of the gland in animals. The metabolic processes are at a low ebb, with a tendency in most individuals to the storage of fat, to an increased tolerance for carbohydrates, a subnormal temperature, slowed pulse and respiration, lowering of blood pressure, and often extreme somnolence.[14]

In addition to these symptoms, a notable expression of hypopituitarism is the more or less complete retardation of sexual development if the process antedates puberty, and the retrogressive changes

[13] Stewart, P., Four Cases of Tumour in the Region of the Hypophysis Cerebri, *Rev. Neurol. and Psychiat.*, 1909, vii, 225.

[14] One of us has commented elsewhere on the fact that the polyuria which often accompanies these states remains difficult of interpretation on the basis of glandular insufficiency (Cushing, H., Concerning Diabetes Insipidus and the Polyurias of Hypophysial Origin, *Boston Med. and Surg. Jour.*, 1913, clxviii, 901).

on the part of the reproductive organs and secondary sex characters if it affects adults. These disturbances are in all likelihood a consequence primarily of anterior lobe deficiency, as our experiments, to be reported under another title, would indicate.[15]

It can readily be seen that these experimental and clinical states possess many features in common with the condition of hibernation, —the deposition of fat, the somnolence, slow pulse and respiration, insensitivity, and subnormal temperature,—evidences, in brief, of a condition of lowered metabolic activity. We will return to the question of the activity of the reproductive glands.

Alterations in the Ductless Glands during Hibernation.

During the winter of 1912–13 we had the opportunity of studying the ductless glands of seven woodchucks[16] during and after the hibernating period. The carefully preserved glandular tissues of five of these animals (I, II, III, IV, and VII), which had hibernated in captivity, were forwarded to us through the kindness of Professor Sutherland Simpson, whose well known interest in the subject of hibernation has led to important studies,[17] but whose concern in this series lay in the direction of the nervous system. The two other animals (V and VI) were trapped in the open.

Professor Simpson's records state that his five woodchucks, all adults, fat and in good condition, were enclosed in their den without food, on December 9. One of the animals was sacrificed while dormant, on January 18, another on February 6, the third on February 22; the fourth was killed on March 15, soon after awaking, and the seventh on April 6, a month later. The fifth and sixth of the series were both caught for us in the open, on March 17 and 22, and killed

---

[15] We are coming to believe that not only is the factor of growth chiefly related to the pars anterior, but that this portion of the gland is more intimately associated with the activity or inactivity of the reproductive organs. We are only at the threshold of a satisfactory clinical differentiation between the constitutional effects of over- (perverted?) or undersecretion of the two lobes of the gland (Cushing, H., Concerning the Systematic Differentiation between the Two Lobes of the Pituitary Body, *Am. Jour. Med. Sc.*, 1913, cxlv, 313).

[16] The North American marmot (*Arctomys monax* or *Marmotta monax*).

[17] Simpson, S., The Food Factor in Hibernation (Preliminary Communication), *Proc. Soc. Exper. Biol. and Med.*, 1912, ix, 92; The Relation of External Temperature to Hibernation, *ibid.*, 1913, x, 180.

a few days later.   Thus three of the animals were killed while hibernating, three soon after awaking, and the last a few weeks later.

The tissues received from Professor Simpson had all been preserved in the same fashion.   The animals had been killed by cardiac transfixion and the organs immediately hardened *in situ* by aortic injections of saturated mercury bichloride in 10 per cent formalin. In the case of Woodchuck I the pituitary gland alone was received.

Woodchuck I.—Jan. 18, 1913. Animal found asleep. Rectal temperature 10.2° C. (50.3° F.). Hibernating for a period presumably of five or six weeks.

The block of tissue comprised a portion of the sphenoidal bone, the gland, and superimposed nervous tissues. The dural envelope and enclosed gland were carefully dissected away from the shallow sella turcica, mounted in paraffin, and serially sectioned in a sagittal direction.[18] The gland measured 3 mm. in its anteroposterior diameter.

Under the microscope the pars anterior shows a lack of the characteristic arrangement of the cells into cords and alveoli, though there are many well filled venous sinuses. The cells appear closely packed and possess deeply staining pycnotic nuclei, rich in chromatin and with but a small amount of surrounding, faintly pink staining, non-granular protoplasm. Particularly in the vicinity of the pars intermedia are the nuclei closely crowded together. Stained with hematoxylin and eosin the cells prove to be neutrophilic, none of them showing an especial affinity either for the acid or basic dye, such as characterizes the pars anterior cells under ordinary conditions.

There is a well formed cleft separating the pars anterior from the pars intermedia. The latter appears unmodified. The nuclei resemble those of the pars anterior, but they are surrounded by a larger amount of protoplasm with neutral staining affinities.

The pars nervosa also appears unmodified. The sections of the infundibular block show no apparent changes. The clusters of nerve cells of the tuber cinereum appear normal.

The finding of such definite histological changes in the pars anterior in this animal encouraged us to investigate the matter further, and when each of his remaining animals was killed, Professor Simpson was good enough to send us not only the pituitary body but fragments of the other ductless glands as well.

[18] In order to insure accuracy in the histological comparison between the glands of different individuals, it is essential not only that the sections be cut on the same plane, but that corresponding sections on this plane be compared. In this series of woodchucks, therefore, as in other animals and in man, median sagittal sections have been those of choice.

Owing to the cup-like shape of the enveloping anterior lobe these median sections show less of the pars anterior than do sections taken more laterally.

Woodchuck II.—This animal, a female, removed from the den and killed on Feb. 6, was still hibernating. Its rectal temperature was 14.8° C. The tissues were fixed as described, and the sphenoidal block, together with fragments of the other ductless glands, was sent to us.

Pituitary Body.—The gland was even smaller than that of Woodchuck I, measuring a scant 2.5 mm. in its anteroposterior diameter. In its histological appearance it corresponds precisely with that of the first animal, showing the same uniformly staining, closely packed pycnotic nuclei in the pars anterior (Figs. 1, 3, and 5).

Thyroid.—This shows no notable change. The acini are small, containing massive, deeply staining colloid, which fractures on section, making the tissue difficult to cut. The cells for the most part are flattened, though with cuboidal configuration around some of the acini. There is an abundance of intra-acinal fat, in which many clusters of hibernating gland cells appear.

Hibernating Gland.—The fragment of tissue marked "thymus" proves to be from the hibernating gland. The swollen cells hold fairly large, round nuclei and show an abundant granular protoplasm, free from vacuoles.

Adrenal.—The nuclei of both medulla and cortex appear somewhat shrunken, and more closely placed than normal, due to the relatively scant protoplasm. The nuclei of the medulla are somewhat irregular in shape.

Pancreas.—Here too the cells are moderately shrunken and the acini small. There are a few faintly staining secretion granules in the protoplasm (*Säure-violett* and *safranin O*). The nuclei are small and stain deeply. The islets are cellular, with scant protoplasm. There is an abundance of intra-acinal fat in the organ. The sections show a lymph gland with practically no eosinophils (compare No. VII).

Ovary.—There are a large number of primordial follicles and a few developing Graafian follicles, but none that are mature, and there are no corpora lutea. There is a small amount of interstitial tissue, but interstitial cells are not definitely recognized. The fimbriated extremity of the tube is lined by a low type of epithelium with very scant protoplasm.

The tissues of this animal (No. II) made it apparent, therefore, that hibernation is accompanied by histological changes in each member of the ductless gland series, though in none of them were the deviations from the normal as striking as in the pituitary body. The tissues of the third of the hibernating animals corresponded closely with the above, as follows.

Woodchuck III.—This animal, also a female, removed from the den while dormant and killed on Feb. 22, had a rectal temperature of 7° C. (44.6° F.). Presumably it had been hibernating for ten weeks, since about Dec. 9.

Pituitary Body.—The gland measured barely 3 mm. in its anteroposterior diameter. Histologically it corresponds in general with the glands of Nos. I and II. However, there is some tendency toward alveolar arrangement in the cells of the anterior lobe; they have somewhat more protoplasm and the nuclei are less pycnotic and are rounder and clearer than in Nos. I and II. A few of the cells show a slight affinity for acid stains.

Thyroid.—There is a large amount of intra-acinal fat. The acini themselves are similar to those described under No. II. The vesicles are possibly more distended with colloid, which stains uniformly pink. No cells of the hibernating gland type are observed.

Hibernating Gland.—This shows very little change from the hibernating gland of No. II, though the cells have begun to show some vacuolation.

Adrenal.—The gland is very small. The nuclei are shrunken, particularly those of the cortical cells, and compared with the awakening state (compare No. IV) the protoplasm is very scant.

Pancreas.—The acinus cells are distended and contain a large number of secretory granules. The nuclei are larger and more vesicular than in No. II, and occupy the base of the cells. The islets have a more open structure and are composed of loosely intertwined cell cords.

Ovary.—Much the same appearance as in No. II. A few follicles are somewhat more advanced, but none are mature, and there are no corpora lutea. The cells lining the fimbriated ends are taller and contain more protoplasm.

In the following example, the first of the awakened animals, a very obvious physiological reactivation was apparent, particularly in the anterior lobe of the hypophysis, but also in the cells of the adrenal. Unfortunately the sex of the animal was undetermined, as no block from the sexual glands was received.

Woodchuck IV.—This animal was found awake and active with a rectal temperature of 35° C., and was killed on Mar. 15. It had been incarcerated since Dec. 9 and had not had access to food during this period of ninety-six days. It had been observed on Mar. 8 and was then still asleep, so that its awakening must have been recent.

Pituitary Body.—The gland was macroscopically enlarged and swollen when compared with the glands of Nos. I, II, and III. It was thicker and measured 4 mm. in its anteroposterior diameter. Histologically the cells are swollen; their protoplasm is abundant, granular, and has begun to take differential eosinophilic and basophilic stains. A single mitotic figure was found.

The cleft contains colloid. The pars intermedia and pars nervosa are larger than in the earlier specimens; the cells are more active. The tissue of the posterior lobe is more distended and the nuclei are crowded farther apart, which makes the lobe itself appear less cellular.

Thyroid.—There is no particularly striking change, though the cells on the whole are more of the cuboidal type than in Nos. I and II. The intra-acinal tissue is composed of very scant connective tissue fibers, without fat tissue.

Hibernating Gland.—Sections were received from the three situations (cervical, thoracic, and subscapular), where the chief masses of this glandular structure occur.

The lobules appear smaller than those described under No. II, but there is no great change in the cells. The protoplasm is less abundant and stains less well, and many of the cells contain vacuoles of considerable size.

Adrenal.—Significant changes are apparent. The gland is larger than that of

No. II, and the cells of both medulla and cortex, but particularly those of the former appear swollen and show abundant protoplasm of granular structure. The nuclei are larger and more vesicular. The cortex shows more clearly than in No. III the normal characteristic division into zones.

Pancreas.—Very little change from No. III. The cells show a fair amount of protoplasm with abundant prozymogen granules at their base. No islets are seen on sections.

Ovaries or Testes.—Tissues not received.

Lymph Gland.—A single gland in one of the sections shows no eosinophilia.

The following animal (No. V) was trapped in the open in the neighborhood of Worcester, Mass., on March 17 and was forwarded to us for study. Its period of awakening may, therefore, have corresponded with that of the preceding animal (No. IV), though having been in natural surroundings it possibly became dormant earlier than December 9. It may also have had access to food during the winter, and the autopsy showed that it had fed on awaking and before capture, though subsequently it refused food. It may be noted that the winter of 1912–13 was not a very prolonged or severe one.

The animal was lively, fat, and appeared to be in good condition. It was sacrificed March 22, seven days later than No. IV.

Woodchuck V.—Mar. 22. An adult male, weighing 2,360 gm., very lively and fierce and apparently well nourished. Sacrificed by anesthetizing with ether and then bleeding from the carotid and femoral arteries. Immediate autopsy.

The abdominal cavity showed what appeared to be masses of adipose tissue in the mesentery and what was taken to be the omentum. This fat was white in appearance and of soft consistency. There was practically no subcutaneous fat. The cecum and large bowel contained food remnants, though the stomach and small intestine were empty.

Pituitary Body.—On opening the cranial cavity from above, the brain was elevated and the stalk of the hypophysis divided at the base of the third ventricle. The gland appeared to be considerably larger than those received from Professor Simpson, though the latter had not been seen in the fresh state. The posterior lobe was of a whitish appearance, very distinct from the dark pinkish pars anterior. Both lobes were flattened. The sella turcica was shallow. A block of tissue, including the sphenoidal bone and the floor of the sella, was cut out from the base of the skull and placed in Bensley fixative. The gland after fixation measured 4.5 mm. in its anteroposterior diameter, and 1.5 mm. in thickness.

Histological Appearance.—The cells of the pars anterior show the normal arrangement in cords; they are rich in protoplasm and have begun to show clearly their differential acid and basic staining reactions (Figs. 2, 4, and 6).

In the posterior lobe there is some increase in cellularity over the specimens previously studied.

Thyroid.—What was taken for the left thyroid, apparently with an accessory tag, was dissected out. It was not especially large and was of a faint pinkish yellow color. What appeared to be a parathyroid was evident at the upper pole and was left attached to the body of the gland. On account of bleeding from the right carotid, the tissues were suffused with blood and the right thyroid was not found. No thyroid tissue identified histologically.

Hibernating Gland and Thymus.—When the thorax was opened, in the position of the thymus there was found a very large, soft, brownish gland, of lobular structure, filling the upper mediastinum and embedded in a sort of adipose tissue. Nodules of similar tissue accompanied the intrathoracic vessels and extended downward along the course of the sympathetic trunk in the intercostal spaces as far as the diaphragm. Fragments of this tissue were preserved in various fixatives; the main bulk of the gland was preserved in formalin.

Histologically the gland differs greatly from the specimens previously studied. In many areas very little remains except the eccentric marginal nuclei of completely vacuolated cells. No thymic tissue identified in the sections.

Pancreas.—Diffuse lobules of the gland were found in a long process extending from the duodenum to the spleen, and small bunches of glandular tissue were seen in the mesentery extending downward almost to the kidneys. Pieces of the pancreas in the region of the splanchnics were taken out and placed in acetic-osmic-bichromate mixtures of 2 and 4 per cent osmic acid.

Histologically the gland is of normal appearance. The cells are large and filled with secretory granules.

Adrenals.—These appeared as small discrete glands of a greenish brown color. On section it was observed that a very large part of the cut surface was made up of pinkish white, soft, pulpy medulla. The cortex contained the pigment substance above described.

Histologically the medulla, when compared with the previous specimens, appears much enlarged in proportion to the cortex. Its cells are large, swollen, and compact. There is a much more distinct line of demarcation between the medulla and cortex, the lower zone of which appears flattened, as though the cells were compressed by the swollen medulla.

Testes.—The glands were large, and the structural elements were readily recognized. They measured about 1.75 cm. in length by 1 cm. in breadth. A surface scraping from the fresh section showed a few non-motile spermatozoa (?).

Histologically the gland shows an abundance of interstitial cells. The tubules contain a few spermatids, but no formed spermatozoa, nor are there any spermatozoa in the epididymis.

The next animal, a female, was likewise trapped in the neighborhood of Worcester, Mass., and was received a few days later than the preceding one.

Woodchuck VI.—An adult female, weighing 2,002 gm., active and seemingly well nourished. Rectal temperature 36.2°. Sacrificed on Mar. 27 by the same procedure as in No. V. The mammary glands were considerably hypertrophied and contained secretion.

Pituitary Body.—The gland appeared large, as in No. V. The two lobes were clearly distinguished. The sella was depressed, and its thin floor contained a small defect, as though produced by pressure atrophy.

Histologically a median section shows the pars anterior to be composed of large cells with differentially staining protoplasm, but with a preponderance of eosinophilic cells. The lobe is more vascular than those heretofore observed.

There is apparently some increase in the cellularity of the pars nervosa.

Thyroid.—What were taken for the thyroid lobes were removed, but the tissues proved to be symmetrically placed *lymph glands*. These show no excess of eosinophils (compare No. VII).

Hibernating Gland and Thymus.—Very much less in evidence than in No. VI. A small amount was apparent in the region of the thymus, with a few fine extensions downward on either side of the spinal column. Thymus not identified in gross.

Histologically the hibernating gland tissue shows greatly shrunken and vacuolated cells with small nuclei—a stage of absorption of the gland in advance of No. V. A small area of normal *thymus* appears on a few of the sections.

Pancreas.—Of normal appearance in gross.

Histologically the cells are larger than in the previous cases and are loaded with secretory granules. The islets present a normal appearance, with well developed cords and cells rich in protoplasm.

Adrenals.—In the fresh state, of a light green color, and somewhat larger than those of No. V.

On section the medulla appears less abundant in comparison with the cortex. When contrasted with the gland from a hibernating animal (*e. g.,* No. II), a very marked change is apparent, both in the size of the nuclei and amount of protoplasm. Next to the changes in the pituitary gland, the changes in the adrenal are the most striking.

Ovaries.—The right gland was considerably larger than the left. On section it was of a more pinkish color and contained what appeared to be numerous corpora lutea. The left ovary was whiter in appearance and showed one nodular elevation, apparently a corpus luteum.

Histologically the right gland is largely made up of several corpora lutea. with but little interstitial substance containing a few immature follicles.

Fallopian Tubes.—The uterus appeared to be thickened, edematous, and vascular. At uniform intervals there were small dilatations suggesting early pregnancy.

Histologically the tubes show a thick, vascular endometrium with high columnar epithelium.

Whether or not this animal was pregnant was not determined. The condition of the tubes and ovaries resembled closely that which will be described elsewhere as a consequence of pituitary feeding.

The following animal, the last of Professor Simpson's series. was found awake on March 15, when No. IV was sacrificed; but on March 22, when again observed, it was in a condition of semihibernation, with a rectal temperature of 15° C. It was observed again on March 29 and was found to be in the same dormant condition, with a rectal temperature of 12.7° C., and when replaced in the den, a dish of water and some cabbage were set beside it. On April 5 it was found to be awake and active, with a temperature of 38° C. (100.6° F.) and a body weight of 1,150 grams. A runway was attached to the den, so that the animal was free to move about, and it was abundantly supplied with food and water.

When observed on April 16 the rectal temperature was 35.8° C. (96.5° F.), body weight 1,500 grams. On May 5 rectal temperature 37.7° C. (100° F.), body weight 1,650 grams.

Woodchuck VII.—Female; killed May 5. Tissues from Professor Simpson were fixed as in the early series. It may be noted that it had been actively awake for a month and had put on weight. In comparison with our other animals it was small and may possibly have been younger.

Pituitary Body.—The gland, though somewhat smaller than the foregoing, shows characteristic anterior lobe differential staining reactions. The pars intermedia and pars nervosa are unchanged.

Thyroid.—The vesicles are surrounded by possibly a somewhat more cuboidal type of cell on the average than were those in No. II, though there appears to be very little difference from the winter state.

Hibernating Gland.—Here the changes are very marked from the winter state. The cells have lost their granular appearance and sharp outlines and are more ragged in form. The nuclei, from loss of protoplasm, appear more closely packed and smaller, with more dense chromatin. The cells are considerably vacuolated.

Pancreas.—No changes were observed. Special stains for zymogen granules, etc., were not employed.

Adrenals.—The cortex appears much thickened in comparison with the medulla. The cells of the medulla are large and active in appearance. A slight differentiation is evident between the zones of the cortex.

Ovary.—Gland apparently active, with increase of interstitial cells. Fully ripened, large Graafian follicles are present, and there is one large corpus luteum.

Lymph Gland.—A gland present in the tissues from the pancreas, and another present in the tissues from the hibernating gland in the neck, show an extraordinary preponderance of eosinophils.

Comment.—Though the observations on this series of seven animals are somewhat fragmentary,[19] and though more or less definite

[19] We had hoped to supplement these observations by the study of a series of animals during the winter of 1913–1914. We were able to secure for the purpose, however, only one woodchuck; but at the same time three hedgehogs and three raccoons were also interned. The winter was a particularly severe one and all four animals succumbed to the low temperature. Our provisions for

evidences of a pluriglandular insufficiency are apparent when the waking is contrasted with the dormant state, nevertheless the most striking histological changes occur in the anterior lobe of the pituitary gland. These changes, in brief, consist of a loss of the characteristic cellular topography of the pars anterior, and, in the case of the individual cells, of a shrinkage of both the nuclear and protoplasmic substance of the cells, with complete loss of the characteristic histological picture of the active gland,—namely, the differential staining qualities of the granular content to acid and basic dyes, —qualities which reappear at the end of the dormant period. It may be added that in only one of our specimens have we observed the karyokinetic figures which Gemelli (see below) has described as an especial feature of the awakening state. This may be for the reason that we have not happened upon the exact time when, after awakening, the cell division is particularly active. Significant of Gemelli's findings, however, is the fact that the writers, except in Case IV of this series, have never observed mitosis in any of the countless pituitary glands of the lower animals they have examined.

In view of the close interrelation of the pituitary body and the sexual glands it is unfortunate that we were deprived of the latter organs in Nos. I and IV, and that four of the remaining five animals were females. It is our presumption that very marked activation occurs in the reproductive organs when the animal emerges from the dormant state, and it is tempting to attribute this to the influence of the functionally reactivated pars anterior. This conjecture is supported by the experiences of one of us (Goetsch) with the

proper hibernation were doubtless inadequate. Food was supplied so that the animals might eat as they desired, and they all did so with more or less regularity. The woodchuck, which was a male, survived a particularly cold spell during which the other animals perished, but at no period had it definitely hibernated. On Jan. 12 it was fairly lively, with a rectal temperature of 32.5° C., weight 1,807 kilos. It was sacrificed. There was very little fat, but a large amount of hibernating gland. The pituitary body resembled the glands of Woodchuck III of the series. It was small and the cells were somewhat closely packed, and some of them took differential stains. The testes showed active cell division of spermatogenous cells, but there were no spermatozoa; the interstitial cells were abundant. There were no notable changes in the other glands. No changes of recognizable significance were made out from the tissues of the raccoons or hedgehogs, and no tissue resembling hibernating gland tissue was positively identified in any of them.

feeding of pars anterior extract and its effect in stimulating reproductive activity,[20] which will be the subject of a subsequent paper.

The Observations of Gemelli and Salmon.

As far as we are aware, only two writers, both of them Italians, have ap-
proached this subject in such wise that their views and observations are at all
pertinent to this discussion.

In 1905 a monograph on sleep was published by Salmon[21] in which the view
was advanced that physiological sleep is due to an internal secretion of the
pituitary gland. In later papers, published in French,[22] this view was still
further elaborated.

In 1906 an important paper appeared in Italian, written by Agostino Gemelli,[23]
in which the histological changes occurring in the pituitary gland of the hibernating marmot were fully described for the first, and as far as we can learn,
for the only time. In a second paper[24] he refutes the contentions of Salmon
and ascribes to the hypophysis in cooperation with the other ductless glands the
function of the neutralization of toxins.

It is our impression that the observations of these writers are in entire
accord with our own, but that their conclusions, particularly those of Salmon,
in so far as they were drawn from clinical observations, are based on wrong
premises.

Salmon advanced the opinion that the hypophysis is actually a center for
sleep, normal physiological sleep being explained on the basis of an internal
secretion from the gland which has some vasomotor or autotoxic effect on the
nervous system: in other words, that physiological sleep is essentially dependent upon the secretion of the pituitary gland.

As an argument favoring this view he gives a list of disorders in which
drowsiness is commonly observed, and in many of which there occurs some
hypophysial alteration. Strangely enough, however, he attributes drowsiness
to conditions of glandular hyperplasia, accompanied by hyperfunction, whereas
insomnia is ascribed to states of glandular atrophy with functional insufficiency.
This, it will be seen, is precisely the reverse of the view which is now commonly

[20] Certain clinical experiences are also in accord with this view: for the feeding of hypophysial extracts over prolonged periods to patients who are the
victims of hypopituitarism not infrequently arouses sexual activity after periods
of amenorrhea or impotence.

[21] Salmon, A., Sull' origine del sonno. Studio delle relazioni tra il sonno e
la funzione della glandula pituitaria, Florence, 1905.

[22] Salmon, A., Le Sommeil pathologique; l'hypersomnie, *Rev. de méd.,* 1910,
xxx, 765; La fonction du Sommeil. Physiologie, psychologie, pathologie, Paris,
1910.

[23] Gemelli, A., Su l'ipofisi delle marmotte durante il letargo e nella stagione
estiva, *Arch. p. le sc. med.,* 1906, xxx, 341.

[24] Gemelli, A., Nuove osservazioni su l'ipofisi delle marmotte durante il
letargo e nella stagione estiva; contributo alla fisiologia dell' ipofisi, *Biologica,*
1906, i, 130.

accepted. At the time of Salmon's publication, it may be recalled, there was no satisfactory differentiation of the clinical expressions of hyper- and hypopituitarism.

Gemelli was quick to see the falsity of Salmon's position, for he says: "If Salmon's hypothesis were correct we should expect to find an increase in the activity of the pituitary gland during hibernation, for only in this way could we confirm the view that physiological sleep is due essentially to the secretion of the glandular portion of the pituitary body. But we have really found the contrary to be the case—*viz.*, a lowering of the proper activity of the organ characterized by a decrease of the secretory cells. After the marmots awake we find a regeneration of the glandular tissue, with a great number of karyokinetic figures and an increase in the number of eosinophil cells. These observations point to an organic renovation of the gland and an increase in its activity, and therefore refute entirely Salmon's hypothesis."

If Salmon had reached the opposite conclusion, which assuredly he would have done had he been in possession of the facts before us today, namely, that drowsiness or somnolence is induced by a deficiency rather than an increase in hypophysial secretion, it is possible that Gemelli might have found reason to ascribe the somnolence of his marmots to the changes in the gland which he was the first to observe. As it was, Gemelli interpreted his findings as an argument in support of the view, which has become generally accepted, that the gland is of prime physiological importance and is not a rudimentary structure. With many other French and Italian investigators he contends that the ductless glands—adrenal cortex, thyroid, parathyroid, and hypophysis—are organs whose chief function is the neutralization of toxins, not only such as are furnished by organic, inorganic, and bacterial poisons, but those the products of metabolism as well. To quote from him: "During hibernation metabolism is decreased and therefore we have less toxic substances. When the animals awake in the spring the vegetative functions are resumed, and we therefore find a very great increase of toxic substances, corresponding to which the changes described in the pituitary gland occur." Gemelli's conclusions are as follows:

1. The hypophysis of the marmot follows the general law which governs the other organs during hibernation and the spring awakening.

2. The decrease in cyanophil cells during hibernation, the appearance of numerous karyokineses, and increase in cyanophil cells during the spring awakening, serve to confirm the hypothesis that the function of the glandular lobe of the hypophysis cooperates with the other ductless glands in neutralizing toxins.

3. The anterior part of the glandular lobe of the hypophysis cannot be regarded as a hypothetical center of physiological sleep.

Hibernation in General.

In physiological adaptation to periodic conditions unfavorable to life, of which a seasonal diminution of food supply is perhaps the most important element, certain animals have acquired a capacity for seasonal lethargy which appears to be but a form of prolonged diurnation, just as diurnation is an exaggeration of normal sleep. A low external temperature would appear to be merely a predisposing factor in hibernation, just as it is in the case of normal sleep, but from what we now know of æstivation the external temperature in itself can hardly be regarded as essential to the process.

Of the fur-bearing animals of the temperate latitudes the marmot, as far as is known, shows the phenomenon of hibernation in its most outspoken form. Winter sleep, however, is a peculiarity of other so-called poikilothermic animals, such as the dormouse, hedgehog, bat, and prairie-dog; and it is alleged that raccoons, skunks, badgers, bears, and even some squirrels and other animals, such as chipmunks and gophers, which store their winter provender, at least have the capacity of passing the months of scant food supply in a semihibernating condition. It is a physiological characteristic which has never been acquired by birds, due in all likelihood to their ability to migrate to more favorable surroundings.

Many of the animals which undergo a period of hibernation or semihibernation in their normal habitats when transplanted to other surroundings and furnished with food may not lapse into their wonted months of continuous sleep, but may remain more or less active; and this is true even of the typically hibernating marmot. It is conceivable that under these changed conditions they might grow out of the habit, though unquestionably they would still retain the capacity to hibernate were the original conditions resumed. It may be added, however, that certain animals when transplanted retain their habit of periodic somnolence. This is said to be true of the tenec of Madagascar, an animal belonging to the same order as the hedgehog. The tenec undergoes what is rare in mammals, namely, a period of summer sleep, and the specimens in captivity in the London Zoölogical Garden are said to have continued in their regular seasonal sleep even in the northern latitude.

The same physiological changes, though of brief duration, appear to take place in animals which have the habit of diurnation, a condition which apparently differs from hibernation only in its brevity. This is true of the bat, for example. Dormice, too, awake every twenty-four hours merely to feed; and the sleep of hedgehogs may last for two or three days at a time. It is alleged that during diurnation in these animals, respiration practically ceases and the temperature falls to about that of the surrounding medium and rises again with their awakening. Indeed, the fall in temperature which occurs during the natural sleep of all non-hibernating species[25] suggests that no sharp line of demarcation separates the two conditions.

Similar seasonal physiological rhythms may possibly affect other mammals besides those mentioned,[26] and attention was drawn, in the course of our canine hypophysial studies during three or four consecutive winters, to the fact that

[25] Simpson, S., and Galbraith, J. J., An Investigation into the Diurnal Variation of the Body Temperature of Nocturnal and Other Birds, and a Few Mammals, *Jour. Physiol.*, 1905–06, xxxiii, 225.

[26] Among cold-blooded animals, both vertebrate and invertebrate—snakes, lizards, tortoises, frogs, toads, snails, and so on—the phenomenon is of quite general occurrence. Hibernation in the carnivora is limited to the *Arctoidea*. The single canine example is apparently the raccoon-dog of Japan, which is said (Radde) to hibernate if food has been sufficiently plentiful to enable the animal to store sufficient fat.

in the midwinter months hemodynamic and other reactions to the injection of hypophysial extracts often did not occur or were far less pronounced than at other times of the year. It was suggested in explanation, (1) that a seasonal change in the animals experimented upon caused the administration of the extracts to fail to give the usual responses, or (2) that the glands of the animals from which the extracts had been prepared were in a condition of relative seasonal inactivity and that the extracts consequently were less active in their effect. It is possible that we may thus account for some of the discrepancies in the results obtained by various observers of the physiological reactions of prepared extracts.[27]

Even man appears to have a certain capacity for what may be called winter sleep, as has been shown by the experiences of those in the arctic regions; and it is possible that such a capacity might be further developed, as seems to be the case with the peasants in certain parts of Russia, who during the winter months, when there is a scarcity of food, pass weeks at a time in a somnolent state, arousing once a day for a scant meal.

Unquestionably the underlying factors in hibernation are extracorporeal ones relating to food supply, but it may be presumed that animals subjected to seasonal periods of food deprivation have acquired the capacity to survive, partly through the preliminary storage of tissue fuel and partly through the ability to conserve that fuel so that its combustion is extremely slow. Nothing would be so likely to favor these conditions as a physiological period of inactivity of those glands which are concerned with tissue metabolism, of which the pituitary body has been shown to be a striking example.

That deprivation of food and exposure to cold cannot be the only factors is shown by the fact that such an animal as the woodchuck in its natural surroundings often goes into winter quarters long before the onset of cold weather and emerges again at the vernal equinox, even though existing climatic conditions may be most unfavorable for the securing of food. Moreover, it has been found impossible to force hibernation by the withdrawal of food and exposure to cold out of season.

The occurrence of such a seasonal wave of inactivity on the part of certain

[27] R. Hunt (The Effects of a Restricted Diet and of Various Diets upon the Resistance of Animals to Certain Poisons, *Bull. Hyg. Lab., U. S. P. H..* 1910. No. 69) called attention to the great variation in the resistance of mice and guinea pigs to certain poisons and attributed them to seasonal variations in the activity of the thyroid glands. A. Seidell and F. Fenger (Seasonal Variation in the Composition of the Thyroid Gland, *Bull. Hyg. Lab., U. S. P. H..* 1914. No. 96) showed that there are, as a matter of fact, seasonal changes in the weight as well as in the iodin content of the thyroid glands of beef and sheep. G. B. Roth (Pituitary Standardization. A Comparison of the Physiological Activity of Some Commercial Pituitary Preparations, *Bull. Hyg. Lab., U. S. P. H..* 1914. No. 100) has called attention to the wide variability and the physiological activity of commercial pituitary extract, but has assigned no reasons for the differences.

Dr. Hunt expresses the belief that pituitary preparations vary on account of the difference in the methods of manufacture, rather than on account of the difference in the original gland.

members of the ductless gland series as our theory of hibernation predicates is no more remarkable than other periodic waves of inactivity or .activity in the case of other glands. This is particularly true of the glands directly concerned with reproduction,—glands which have been shown to have a very close functional interrelationship with the pituitary body.

SUMMARY.

A train of symptoms, coupled with retardation of tissue metabolism and with inactivity of the reproductive glands, not only accompanies states of experimentally induced hypophysial deficiency, but is equally characteristic of clinical states of hypopituitarism. The more notable of these symptoms are a tendency, in the chronic cases, toward an unusual deposition of fat, a lowering of body temperature, slowing of pulse and respiration, fall in blood pressure, and oftentimes a pronounced somnolence.

These symptoms bear a marked resemblance to the physiological phenomena accompanying the state of hibernation which have heretofore been unsatisfactorily ascribed solely to extracorporeal factors; namely, a seasonal deprivation of food and low temperature.

In a series of hibernating animals (woodchucks) it has been found that during the dormant period histological changes are apparent in many of the ductless glands. The most notable of these changes occur in the pituitary body, as previously observed by Gemelli. The gland not only diminishes in size, but the cells of the pars anterior in some animals at least completely lose their characteristic staining reactions to acid and basic dyes. At the end of the dormant period the gland swells, and as the cells enlarge they again acquire their differential affinity for acid, basic, and neutral stains, and at the same time karyokinetic figures may appear.

CONCLUSIONS.

On the basis of these observations hibernation may be ascribed to a seasonal physiological wave of pluriglandular inactivity. The essential rôle may perhaps be ascribed to the pituitary body, not only for the reason that the most striking histological changes appear in this structure, but also because deprivation of the secretion of this gland alone of the entire ductless gland series produces a group of symptoms comparable to those of hibernation.

FIG. 1

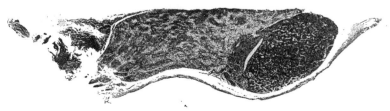

FIG. 2.

(Cushing and Goetsch: Hibernation and the Pituitary Body.)

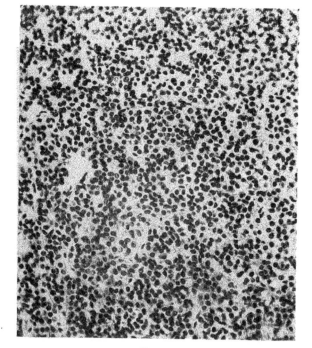

FIG. 3.

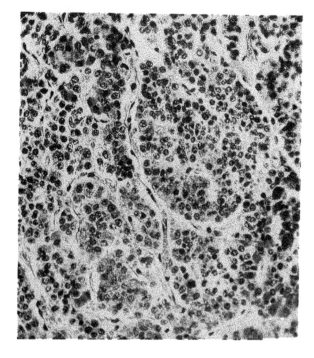

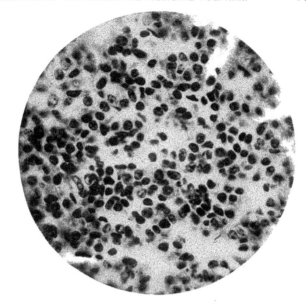

FIG. 5.

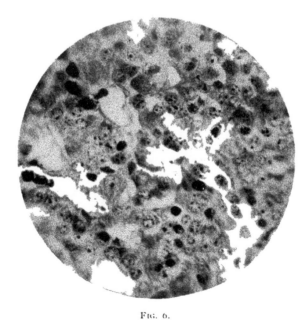

FIG. 6.

(Cushing and Goetsch: Hibernation and the Pituitary Body.)

EXPLANATION OF PLATES.

PLATE I.

FIG. I. Median section of the pituitary gland of an hibernating woodchuck, killed Feb. 15, 1913 (No. II). Pars anterior broken.

FIG. 2. Median section of the pituitary gland of a woodchuck (No. V), killed Mar. 22, 1913, a few days after awakening.

PLATE 2.

FIG. 3. Undifferentiated anterior lobe cells of the hibernating state (Woodchuck II), Feb. 15, 1913. Hematoxylin and eosin. × 300.

FIG. 4. Well differentiated anterior lobe cells of an animal on awakening (Woodchuck V), Mar. 22, 1913. Hematoxylin and eosin. × 300.

PLATE 3.

FIG. 5. Higher magnification of the cells of the hibernating state (Woodchuck II, Feb. 15, 1913). *Säureviolett* and *safranin O.*

FIG. 6. Higher magnification of the anterior lobe cells of the awakening state (Woodchuck V, Mar. 22, 1913). *Säureviolett* and *safranin O.*

LATE POISONING WITH CHLOROFORM AND OTHER ALKYL HALIDES IN RELATIONSHIP TO THE HALOGEN ACIDS FORMED BY THEIR CHEMICAL DISSOCIATION.

By EVARTS A. GRAHAM, M.D.

(*From the Otho S. A. Sprague Memorial Institute Laboratory of Clinical Research and the Department of Surgery, Rush Medical College, Chicago.*)

PLATES 4 AND 5.

(Received for publication, March 13, 1915.)

It is well known that the prolonged administration of chloroform may be followed by certain well marked morphological changes in the tissues, most conspicuous of which are edema, fat infiltration, multiple hemorrhages, and necrosis of the central portion of the liver lobule. These changes have been extensively studied; and excellent bibliographies may be found in articles by Bevan and Favill,[1] Wells,[2] Whipple and Sperry,[3] Howland and Richards,[4] and others. In this article, therefore, no detailed discussion of the changes will be given. However, in spite of the large amount of study that has been devoted to the lesions, there has apparently been little effort to analyze the factors involved in their production. At the present time there is no adequate explanation of how these changes are produced. A satisfactory explanation would be important not merely

---

[1] Bevan, A. D., and Favill, H. B., Acid Intoxication, and Late Poisonous Effects of Anesthetics. Hepatic Toxemia. Acute Fatty Degeneration of the Liver Following Chloroform and Ether Anesthesia. *Jour. Am. Med. Assn.*, 1905, xlv, 691.

[2] Wells, H. G., Chloroform Necrosis of the Liver, *Arch. Int. Med.*, 1908, i, 589.

[3] Whipple, G. H., and Sperry, J. A., Chloroform Poisoning. Liver Necrosis and Repair, *Bull. Johns Hopkins Hosp.*, 1909, xx, 278.

[4] Howland, J., and Richards, A. N., An Experimental Study of the Metabolism and Pathology of Delayed Chloroform Poisoning, *Jour. Exper. Med.*, 1909, xi, 344.

48

in connection with the action of chloroform itself but rather because of the light which it would throw on the nature of the fundamental processes involved, which without doubt are identical with those concerned in poisonings of the body with a large group of toxic substances, including other narcotics, arsenic, salvarsan, phosphorus, and probably most bacterial poisons. Evidence will be brought to show that in chloroform poisoning the liver necrosis is produced chiefly by the action of acid (largely probably by hydrochloric acid which is formed in the metabolic destruction of chloroform), and this ability to produce liver necrosis is a general property of alkyl halides, all of which probably yield halogen acids in their breakdown in the body. Reasons will also be brought in support of the view that the accompanying cloudy swelling, fat infiltration, hemorrhages, and edemas are also acid effects. These changes, however, are not limited to intoxications with substances which can split off mineral acids, but they may follow the administration of any substance that can cause tissue asphyxia with its attendant accumulation of organic acids.

The ease with which extensive liver necrosis is induced by a two or three hour narcosis with chloroform, and the failure to obtain it by a narcosis with ether of three or four times that duration, suggest strongly that this difference in the behavior of the two substances depends upon either a difference in molecular action or a difference in products formed during their breakdown.

When chemically pure chloroform is exposed to the action of sunlight at room temperature it is oxidized to phosgene and hydrochloric acid. The end reaction may be expressed thus:

$$CHCl_3 + O \rightleftarrows COCl_2 + HCl.$$

This oxidation occurs so easily that it has become necessary for manufacturers of chloroform for anesthetic purposes to add alcohol to it to prevent its decomposition.[5] Phosgene in the presence of

[5] Baskerville has shown that the alcohol protects the chloroform because it is more easily oxidized than chloroform, and hence when, as in a stoppered bottle, there is only a limited amount of oxygen present, the alcohol uses all the available oxygen for its own oxidation (Baskerville, C., and Hamor, W. A., The Chemistry of Anesthetics. IV. Chloroform, *Jour. Indust. and Eng. Chem.*, 1912, iv, 362).

water is decomposed into carbon dioxide and two molecules of hydrochloric acid. This hydrolysis may be expressed thus:

$$COCl_2 + HOH \rightleftharpoons CO_2 + 2HCl.$$

It is at once apparent, therefore, that if one molecule of chloroform is oxidized in the presence of water, three molecules of hydrochloric acid may be formed. The liberation of such a strong inorganic acid in the tissues could very easily produce necrosis and other acid effects; and since three molecules of the acid are formed from each one of the chloroform, it might easily be supposed that the oxidation of only a small amount of chloroform would suffice to produce a considerable effect on the tissues. Moreover, the well known facts that the liver is the organ which most strikingly manifests the chloroform necrosis, and that this organ is also the site of a most active metabolism, harmonize well with this hydrochloric acid theory.

With these facts and considerations as the basis for departure, the hypothesis was subjected to a series of experimental tests which may be outlined as follows.

1. A study was undertaken of the morphological changes induced by hydrochloric acid, with special reference to the liver. 2. Attempts were made to demonstrate free hydrochloric acid in the necrotic areas in the livers of animals poisoned with chloroform. 3. Observations were made on the relative necrosis-producing power of different chlorine substitution products of methane (*e. g.*, dichlormethane, chloroform, and tetrachlormethane), which on theoretical grounds could be considered to yield different amounts of hydrochloric acid in their breakdown. 4. The inhibiting effect of alkali was studied. 5. Attempts were made to produce the typical picture of chloroform poisoning by other alkyl halides of the same type as chloroform, *viz.*, bromoform ($CHBr_3$) and iodoform (CHI_3), which might be expected to give analogous products in their breakdown, with, however, the liberation of hydrobromic acid and hydriodic acid, respectively, instead of hydrochloric acid. 6. Experiments were made to ascertain whether or not morphological effects like those produced by chloroform can also be induced by alkyl halides in general, and whether these substances are decomposed in the body in such a way that the corresponding halogen acid is liberated.

7. A comparative study of the lesions produced by chloral hydrate and chloroform was made, since the former substance yields practically no hydrochloric acid in its metabolic breakdown, although it contains the same number of chlorine atoms as chloroform.

The Tissue Changes Induced by Administration of Hydrochloric Acid.

Oral and intraportal administrations of hydrochloric acid in suitable concentrations were followed by edema, hemorrhages, necrosis, and increased fat accumulation in the liver. When injected intraportally in dogs in relatively high concentrations (10 to 25 cubic centimeters of 0.5 (N/7) to 1 per cent (N/3.5) solutions), hydrochloric acid produced extensive liver necrosis involving large areas of many lobules about equally. When given in lower concentrations (*e. g.,* 0.37 per cent (N/10)), it produced edema and other degenerative changes as well as numerous subcapsular hemorrhages. The parenchymatous changes were most conspicuous at the periphery of each lobule. This was to be expected since, because of the arrangement of the blood supply, the periphery would be the first part of the lobule reached by the acid. The degenerative changes consisted of pycnosis and fragmentation of the nuclei with swelling of the cells. It was felt that these were precursors of necrosis. When several administrations were given by mouth to rabbits in concentrations of 1 (N/3.5) or 2 per cent (N/1.8) at intervals of from twelve to twenty-four hours, the animals usually died after the third or fourth administration. They showed extensive hemorrhage in the mucosa of the stomach and duodenum, large fatty livers, and swollen kidneys.

Microscopically, by the use of Sudan III, the fat in the liver was found chiefly around the central veins. In short, all of the marked morphological changes induced by chloroform have been seen to follow the administration of hydrochloric acid alone. They were, however, differently distributed. The liver necrosis appeared at the peripheries of the lobules instead of at the centers, as in chloroform poisoning, but this difference in location is not regarded as important, since a discussion of the site of the necrosis in chloroform poisoning involves the question of the site of the greater formation of hydrochloric acid together with that of the relative susceptibility of the central and peripheral parts of the lobule to its effects. These points will be further elaborated in the discussion.

The production of fatty changes in the liver by the use of hydrochloric acid is in harmony with a statement of Leathes,[6] that as an effect of "the action of mineral acids an active mobilization of the fat reserves may occur . . . and the liver cells are found loaded with fat of a low iodine value." Whether or not this mobilization of fat is to be considered as due to a primary effect of the acid or to the cellular asphyxia induced by the acid will not be discussed in this paper.

The extensive edema which follows the administration of hydrochloric acid confirms the well known observations of Fischer on the imbibition of water by tissue colloids under the influence of acids generally. The production of hemorrhages harmonizes with the frequency of their occurrence after tissue asphyxia, with its attendant formation of acid. Doyon[7] and his pupils first (1905-1906) called attention to the probability that fibrinogen is formed mainly in the liver and that interference with normal liver function induces a diminished coagulability of the blood, a conclusion which later was confirmed by Whipple and Hurwitz.[8] In a previous paper in which it was shown that the various hemorrhagic diseases of the newly born are probably expressions of an asphyxial process,[9] we stated that this hemorrhagic tendency might be due to a "more fundamental and wide-spread change, as a result of which not only fibrinogen, but innumerable other proteins tend to remain in solution or to pass into solution, with the result that apart from diminished blood coagulability there is a great reduction in the firmness of the vessel walls." Typical protocols follow.

Experiment 1.—Normal dog, weighing 5 kilos. 2.00 p. m. Under ether anesthesia abdomen opened and 2 cc. N/10 (0.37 per cent) hydrochloric acid injected into radicle of portal vein (branch of superior mesenteric). Marked dyspnea and muscular spasms occurred two minutes later. 2.30 p. m. 2 cc. again injected, followed by same symptoms. 2.45 p. m. Another injection of 2 cc. Few small hemorrhages noted on surface of liver. 3.00 p. m. Another injection of 2 cc. Hemorrhages more numerous; also appearing on wall of stomach. When liver is stroked with handle of scalpel a line is marked which immediately becomes dark red (as if hemorrhagic). 3.30 p. m. After another injection of 2 cc., marked respiratory spasm occurred with muscles of chest rigid. Artificial respiration necessary. Hemorrhages increasing. Stomach greatly distended. 4.00 p. m. Another injection of 2 cc. 5.00 p. m. Dog died. Stomach enormously distended with gas. Wall very hemorrhagic. About 10 cc. of unclotted blood in abdomen. On cutting the small vessels there is little tendency of the blood to coagulate. Several subcapsular hemorrhages on the surface of the liver were noted, varying from pin-head to 4 or 5 mm. in size. No gross increase in liver fat.

Experiment 2.—Dec. 23, 11.30 a. m. Two adult rabbits each given 50 cc. of

[6] Leathes, *J.* B., The Fats, London, 1910, 111.

[7] Doyon, M., Modifications de la coagulabilité du sang consécutives à la destruction du foie. Pathogénie des hémorragies symptomatiques des affections du foie, *Jour. de physiol. et de path. gén.,* 1905, vii, 639.

[8] Whipple, G. H., and Hurwitz, S. H., Fibrinogen of the Blood as Influenced by the Liver Necrosis of Chloroform Poisoning, *Jour. Exper. Med.,* 1911, xiii, 136.

[9] Graham, E. A., The Pathogenesis of the Hemorrhagic Diseases of the New-Born, *Jour. Exper. Med.,* 1912, xv, 326.

2 per cent hydrochloric acid by stomach tube. Marked dyspnea followed. The same amounts of acid were again given on Dec. 26, Jan. 4, Jan. 10, and Jan. 15. Jan. 16. One rabbit was found dead in the morning.

Autopsy.—The liver was large and yellowish, with very distinct lobular markings. Frozen sections stained with Sudan III showed very extensive fat accumulation, most marked around the central veins. Kidneys were swollen and pale brown. Microscopically they showed marked parenchymatous degenerative changes, especially marked in the convoluted tubules. A few casts were evident.

Jan. 16. 1.30 p. m. The other rabbit was given the same amount of acid again. Dead at 4.30 p. m.

Autopsy.—The liver was moderately fatty. The stomach and duodenum were markedly hemorrhagic; the duodenum was empty and firmly constricted throughout its whole length. The kidneys were swollen and pale.

Attempts to Recognize Hydrochloric Acid in the Necrotic Areas in the Liver.

The practical identification of free hydrochloric acid depends upon the recognition of (1) free hydrogen ions and (2) free chlorine ions. It is obvious then that efforts aiming at the direct proof of the presence of hydrochloric acid in the tissues are necessarily complicated by the fact that both ions are always present. Therefore, at best, only circumstantial evidence can be brought to show the presence of free hydrochloric acid. However, the following observations were made.

It is possible to show a high hydrogen ion content in the central, necrotic portions of the liver lobules if fresh unstained sections are treated with indicators which do not change when placed in contact with equally fresh normal tissues in the same way. In this work most of the observations were made with neutral red and phenylated Nile blue. In every case the reaction between the indicator and tissue was much more marked in the central (necrotic) portion than at the periphery of the lobule. In some instances the tint assumed by the neutral red was a deep rose, approximating that which is obtained with a concentration of acid represented by $H^{.}$ — 10^{-6}. Often the central portion was distinctly red, while the periphery was slightly yellowish. The Nile blue reactions were less satisfactory than those obtained with neutral red. If the area of necrosis was very marked, the response of the dye to the free hydrogen ion was shown by a definite blue color; but with moderate necrosis the color change often was insufficient to be satisfactory.

In order to minimize the formation of asphyxial organic acids during the time required to cut the sections and place them in contact with the indicators, the following method was used: Guinea pigs to which chloroform had been given two days previously for a period of four hours were again anesthetized with chloroform; and while unconscious the livers were removed. Sections were cut very quickly with a Valentine knife, washed in distilled water, and immersed in the dye on a slide. The whole process was sometimes done in less than one minute and always in less than two minutes. The neutral red was used as 1 per cent aqueous solutions; alcoholic phenylated Nile blue was used in different concentrations, but generally of 1 or 2 per cent strength.

In their work on the survival formation of lactic acid in amphibian muscle, Fletcher and Hopkins[10] have demonstrated that sarcolactic acid formation in frog muscles attains its maximum only after the lapse of hours, but in a much shorter time (thirty minutes) when the muscles, after removal from the frogs, are treated with chloroform vapor. In harmony with what is generally known concerning the action of narcotics during life there is no doubt that in chloroform poisoning the asphyxial formation of lactic acid is great. The question has therefore to be met whether the indicator changes described in the preceding paragraph are actually greater than might be explained by the presence of much asphyxial organic acid. This objection to interpreting the dye reaction as indicative of the presence of an inorganic acid in the necrotic areas is recognized and has not been removed. In fact when pieces of excised normal liver are kept for six to seven hours at 37° C. in order to insure their maximal content of survival acid, such tissues react to neutral red with about the same intensity as fresh chloroform livers. It is clear from this that the cell itself has the power to produce enough acid during life to effect the indicator changes described. The experiments are merely recorded in conjunction with other observations.

For the determination of an excess of chlorine ion in the necrotic areas sections of liver were cut in the same way as above and handled in general according to the method used by Macallum and Menten.[11] They were placed in N/10 silver nitrate, containing 1.5 per cent nitric acid, and kept in this bath, protected from the light, for from twelve to twenty-four hours. They were then mounted

[10] Fletcher, W. M., and Hopkins, F. G., Lactic Acid in Amphibian Muscle, *Jour. Physiol.*, 1906–07, xxxv, 247.

[11] Macallum, A. B., and Menten, M. L., On the Distribution of Chlorides in Nerve Cells and Fibres, *Proc. Roy. Soc., Series B*, 1906, lxxvii, 165.

in 50 per cent glycerol and exposed to the sunlight until the maximum effect was produced. The most pronounced blackening always occurred in the central necrotic portions of the lobules. The peripheries of the lobules were light brown instead of black. In control sections of normal livers the darkening amounted to only a brownish discoloration resembling the tint of the periphery of the lobule in a chloroform liver. This discoloration has been extensively discussed by Macallum and Menten, who consider that in the tissues it is a definite indication of the presence of chlorides.

Although an excess of free H^+ and Cl^- ions in the necrotic areas of the liver was clearly shown, nevertheless this finding alone cannot be interpreted as in any way indicative of the presence of free hydrochloric acid. The possibility of other sources of hydrogen ion has already been mentioned. Recently Fischer[12] has shown that protein gels (*i. e.*, fibrin) retain increased amounts of chlorides under the influence of acid. The excess of Cl^- ion in the necrotic areas, therefore, may represent only a greater accumulation or retention of neutral chlorides instead of hydrochloric acid.

As bearing on this point, however, it is of interest that the excretion of neutral inorganic chlorides in the urine is increased after the administration of chloroform, as has been shown by Zeller[13] and Kast.[14] Since hydrochloric acid formed in the tissues would doubtless be in part neutralized by metals in alkaline combination, this observation harmonizes with the present theory. Here also one must consider the possibility that an increased output of chlorides in the urine following administrations of chloroform may be due not so much to a splitting of the drug as to anomalies of metabolism and excretion set up in the process of narcosis. However, after the administration of iodoform there is, according to Mulzer,[15] an excretion in the urine of inorganic iodine indicative of a 60 per cent splitting of the iodoform. This could scarcely come from the tissues; and by analogy it would seem probable that the increased chloride excretion following chloroform administrations is due to

[12] Fischer, M. H., Relation between Chlorid Retention, Edema and "Acidos s." *Jour. Am. Med. Assn.*, 1915, lxiv, 325.

[13] Zeller, A., Ueber die Schicksale des Jodoforms und Chloroforms im Organismus. *Ztschr. f. physiol. Chem.*, 1883–84, viii, 70.

[14] Kast, A., Ueber Beziehungen der Chlorausscheidung zum Gesammtstoffwechsel, *Ztschr. f. physiol. Chem.*, 1888, xii, 267.

[15] Mulzer, P., Ueber das Verhalten des Jodoforms im Thierkörper. *Ztschr. f. exper. Path. u. Therap.*, 1905, i, 446.

decomposition of this drug. Moreover, after narcosis with chloral hydrate the urinary chlorides show no such increase as with chloroform.

Necrosis-Producing Power of Other Chlorine Substitution Products of Methane.

A study of other chlorine substitution products of methane should be of interest from the standpoints of determining (1) whether they all have the power of producing central liver necrosis, and (2) whether this property is proportional to the number of molecules of hydrochloric acid which could theoretically be derived from them. If, starting with methane, we should outline a series of its various chlorine substitution products, we might expect, according to the theory, that those which could give the largest amounts of hydrochloric acid in their breakdown would manifest the strongest tendency to produce the necrosis and other changes. Thus *a priori* we might expect the series to run in this order:

$$CH_3Cl < CH_2Cl_2 < CHCl_3 < CCl_4,$$

if all of these substances were equally broken down in the body. Experiments were carried out to determine this point. The last three of the series (*viz.*, CH_2Cl_2, $CHCl_3$, CCl_4) were selected as sufficient to test the tenability of this hypothesis. Reason for the assumption that tetrachlormethane might have greater necrosis-producing power than chloroform, and this in turn than dichlormethane lies in the fact that, in the ultimate breakdown of these substances outside of the body in the presence of water, four molecules of hydrochloric acid can be obtained from one of tetrachlormethane, three from chloroform, and two from dichlormethane. Thus Goldschmidt[16] has found the saponification of tetrachlormethane at 250° C. to take place as follows:

$$CCl_4 + H_2O \rightleftarrows COCl_2 + 2HCl$$

$$COCl_2 + H_2O \rightleftarrows CO_2 + 2HCl$$

[16] Goldschmidt, H., cited by Meyer, V., and Jacobson, P., Lehrbuch der organischen Chemie, 2d edition, Leipzig, 1907, i, pt. 2, 23.

The well known reactions by which three molecules of hydrochloric acid can be derived from chloroform, by simple oxidation in the presence of water, have already been discussed on pages 49 and 50. Concerning the transformation of dichlormethane, André[17] has shown that in five hours at 180° C. 73 per cent is decomposed as follows:

$$2CH_2Cl_2 + 2H_2O \rightleftarrows 3HCl + CH_2O_2 + CH_3Cl$$

$$CH_3Cl + H_2O \rightleftarrows HCl + CH_3OH$$

Thus from one molecule of dichlormethane two molecules of hydrochloric acid may be formed.

When these various substances were administered to animals by inhalation, it was found that not only did all three possess the power of producing central necrosis of the liver, but that this power was shown in greatest degree by tetrachlormethane, notwithstanding its higher boiling point, and least by dichlormethane, the most volatile of the three. In all respects the toxicity of tetrachlormethane was greatest, and that of dichlormethane least. A comparison of the minimum fatal doses of the three substances when given intravenously to rabbits again bore out the same relationship as by the inhalation method. The following table shows the minimum fatal doses:

Substance.	Minimum fatal doses per kilo of rabbit.	Minimum fatal doses expressed in gram-molecular concentrations.
CCl₄	0.053 gm.	0.000344
CHCl₃	0.085 gm.	0.000711
CH₂Cl₂	0.147 gm.	0.00161

The quantities injected were shaken up with enough water to make one cubic centimeter.

In other words, both the general toxicity of these substances as well as their power to produce the extensive morphological changes paralleled the amounts of hydrochloric acid which they can give in their respective breakdowns outside of the body. The toxicity of tetrachlormethane was so great that if it were administered by inhalation for two hours the animal almost invariably died within twenty-four hours. At autopsy numerous hemorrhages were pres-

[17] André, G., Action de l'eau et de l'ammoniaque sur le chlorure de méthylène, *Compt. rend. Acad. d. sc.*, 1886, cii, 1474.

ent with extensive visceral edema, fatty changes, and beginning central liver necrosis.   In a few instances the abdomen contained a large amount of free, unclotted blood.   It was only by giving the tetrachloride for about one hour or less that an animal could be expected to survive long enough to develop a well marked typical liver necrosis.   As has been said above concerning chloroform, the most extensive necrosis is seen about two days after the administration of tetrachlormethane.   A two hour narcosis with dichlormethane is usually not sufficient to induce an outspoken necrosis of the liver.   A comparison of the narcotic properties showed that dichlormethane is less powerful than chloroform.   With tetrachlormethane it was difficult to produce a quiet narcosis analogous to that accompanying the use of chloroform.   More or less severe muscular spasms, particularly of the extremities, occurred intermittently. This fact has already been·noted by von Ley.[18]

The experiments were conducted as follows: Each of a series of three guinea pigs of approximately the same weight was placed under a bell jar, opened at the top to admit air. Into each jar was dropped enough of one of the three substances to induce narcosis, which was then carefully maintained at as nearly as possible the same depth and for the same length of time. Seven sets of animals were used; and the duration of the narcosis was varied from one and one-half to four hours. A typical protocol follows.

Guinea pigs A, B, and C, weighing respectively 550, 530, and 505 gm., were each put under a bell jar, as described above. To A was given dichlormethane; to B, chloroform; and to C, tetrachlormethane. These substances were administered for two hours. On the morning of the second day C was moribund; then all three were killed by a sudden, overwhelming dose of chloroform. At autopsy C (tetrachlormethane) showed several pulmonary and subperitoneal hemorrhages; the liver was large and fatty; and the kidneys were swollen and gray. B (chloroform) showed no marked gross changes except a large, fatty liver. A (dichlormethane) had a moderate accumulation of fat in its liver; but otherwise there were no gross changes of importance. Microscopically the liver of C (tetrachlormethane) showed definite areas of necrosis which involved nearly the whole of the lobule in each case, but which nevertheless had apparently begun at the central portion of the lobule. This necrotic area of each lobule contained cells with fragmented nuclei; and there was a tendency for the whole area to stain intensely with eosin. There were no parenchymatous cells evident which contained normal-looking nuclei except at the periphery of the lobule. There was a large amount of infiltrated fat, as shown both by staining with Sudan III and by the presence of many fat vacuoles in sections prepared and stained in the ordinary way with hematoxylin and eosin. The cells

[18] von Ley, Inaugural Dissertation, Strassburg, 1889.

of the kidney were swollen and granular, and fat vacuoles could be made out here and there. Guinea pig B (chloroform) microscopically showed changes which differed from those of C only in degree. The liver contained areas of well marked central necrosis, which, however, involved only about one-fourth to one-third of the lobule. A (dichlormethane) showed no definite areas of liver necrosis at all comparable to the other guinea pigs. But about the central veins there were occasional necrotic cells and a conspicuous accumulation of fat. The typical necrosis, however, such as is seen after chloroform, was obtained with dichlormethane when it was administered to a guinea pig for from four to six hours.

The Inhibiting Effect of Alkali.

If the theory is correct that chloroform liver necrosis is an effect chiefly of acid, then it might be expected that the administration of an alkali simultaneously with the chloroform would inhibit, if not actually prevent, its occurrence. Such was found to be the case. When sodium carbonate was given intravenously in a proper concentration in a hypertonic solution of sodium chloride, the liver necrosis was either entirely prevented or greatly inhibited. It was of course difficult to know in any given case how much alkali to administer, as it was not possible to estimate how much acid was being formed. Moreover, it was obviously desirable to avoid an excess of alkali, since this in turn will give rise to some of the serious effects of an excess of acid, such as swelling of protein colloids, etc., as shown by Fischer.[19] The amount of alkali given in each experiment was therefore decided empirically; and consequently the degree of inhibition of the necrosis was subject to wide variations in the different experiments. In only one instance, however, and that in the first experiment, was there a failure to observe a definite diminution of necrosis when alkali in hypertonic sodium chloride solution was administered. In this one instance the dose of alkali was excessive (17 grams of $Na_2CO_3 \cdot 10H_2O$ in 1,700 cubic centimeters of 1.4 per cent sodium chloride solution to a dog weighing twenty-one kilos, intravenously). Because of the onset, on the following day, of extreme thirst and a severe hemorrhagic nephritis, with passing of urine thick with blood, it was felt that clearly an overdose of the solution had been given. In all of six subsequent experiments, in which a much smaller dose was given, there was less

[19] Fischer, M. H., Oedema, a Study of the Physiology and the Pathology of Water Absorption by the Living Organism, New York, 1910.

necrosis than in the control animals which did not receive alkali. In one case no necrosis at all occurred. Another interesting fact was that in all of these six experiments the alkali animals seemed less toxic, and at autopsy other changes characteristic of chloroform poisoning, as well as the liver necrosis, were less conspicuous than in the control animals. This beneficial effect of the alkali was particularly striking in the kidneys. In the control animals, which received no alkali, these organs were always enormously swollen and weighed much more than those of the alkali animals. The effect of the alkaline hypertonic salt solution in inhibiting the swelling of these organs, tends to support Fischer's views on the nature and origin of edema. In only three of the experiments was the amount of visible fat in the alkali livers conspicuously less than in the controls. In none of the experiments was there noted any particular influence of the alkali either in strengthening or weakening the narcotic power of chloroform.

The alkaline solution used was that employed by Fischer in his work on edema; *viz.*, distilled water 1.000 cubic centimeters; $Na_2CO_3 \cdot 10H_2O$, 10 grams; sodium chloride, 14 grams. The method of conducting the experiments was as follows: The animals were always run in pairs; and the duration of the anesthesia was the same for both animals, as was also the depth of the narcosis, as nearly as could be determined. To one animal was given the alkaline solution and to the other was given the same quantity of a 0.85 per cent sodium chloride solution. Two sets of guinea pigs and four pairs of dogs were used, in all twelve animals, exclusive of the first pair already mentioned, in which clearly an excessive dose of alkali was given. The solutions were always injected into the blood stream. With the guinea pigs the injections were made by means of a syringe into the heart; but with the dogs, they were given by means of a cannula into either the saphenous or femoral vein. The solutions were always warmed to body temperature previous to being injected; and the amounts were varied in different experiments. In general, the injections were made shortly after the animals had lost consciousness, and they were given slowly. Two days after the administration of the chloroform the animals were killed with chloroform and examined. Pieces of the various tissues were fixed in Zenker's fluid and 10 per cent formalin and were stained both with hematoxylin and eosin and with Sudan III for fat. Typical protocols follow.

Experiment 3.—Two adult dogs. A weighed 4.5 kilos; B, 5 kilos. Both dogs were given chloroform by inhalation for four and one-half hours. The chloroform used was Mallinckrodt's, "Purified for Anesthesia." As soon as both dogs had lost consciousness cannulas were inserted into the saphenous vein; and into A were injected 150 cc. of the alkaline solution already described, and into B, 150 cc. of 0.85 per cent sodium chloride solution. Both solutions were injected slowly over a space of one hour. Both dogs were still in a condition of deep narcosis when they were returned to their cages at the close of the experiment. On the next day Dog A (alkali) was lively and playful; but Dog B (control) was exceedingly drowsy and difficult to arouse. Water and meat were allowed freely. On the second day after the experiment both dogs were killed with chloroform and examined. The difference between the two dogs was very striking. Dog B (the control) had a very fatty, yellowish liver. There was a large subperitoneal hemorrhage on the liver; and there were numerous ecchymoses on the parietal peritoneum. The kidneys were swollen and gray in appearance; and together they weighed 68 gm. Dog A (alkali), on the contrary, showed no hemorrhages. There was little, if any, less fat in its liver than in B's, but the kidneys were practically normal in appearance and weighed only 43 gm. This marked difference in weight between the kidneys of the two animals is particularly striking since the weight of the dogs before the experiment was practically the same. Of still greater interest, however, is the difference in the microscopical findings in the two animals. The liver of B (control) shows areas of necrosis so extensive that practically the entire lobule is involved. There is little more than a fringe of cells about the periphery which have not lost their nuclei. The necrotic areas stain deeply with eosin; and the only nuclei that can be distinguished are those of the capillaries and not of the parenchymatous cells. There are numerous fat vacuoles, especially in the peripheral portions of the lobules. The liver of A (alkali), however, shows practically no necrosis. Only an occasional necrotic cell about the central veins is evident. The columns of cells stand out plainly, and their nuclei appear to be unchanged. There is a moderate amount of fat accumulation about the central veins. These differences are well shown in the accompanying illustrations (Figs. 1 to 4). The kidneys show also rather a marked difference, as might be expected from the striking difference in weight. The epithelium of the convoluted tubules in B's (control) kidney is swollen and granular, and it contains numerous fat vacuoles. There are occasional casts. A's (alkali) kidneys show no striking changes. The changes in the other organs were relatively slight in both dogs.

Experiment 4.—June 3, 1914. Two adult dogs: A, weighing 2.7 kilos: B, 2.5 kilos. Both dogs were given chloroform by inhalation for four and one-half hours. As soon as consciousness was lost A was given 80 cc. of the alkaline solution into the saphenous vein; and B was given the same amount of 0.85 sodium chloride solution. Two days later both dogs were killed with chloroform and examined. Both livers were very fatty, but microscopically B (control) showed much more necrosis than A. The pictures were very similar to those described in the previous experiment. The kidneys here again showed a striking difference. Those of B (control) weighed 35 gm., and those of A (alkali) only 26 gm., although A was the larger dog.

In the two experiments on guinea pigs the method consisted of putting two guinea pigs of approximately the same weight under a large bell jar and dropping chloroform into it through an opening at the top. In this way it was assured that both guinea pigs were breathing air with the same concentration of chloroform. After losing consciousness the alkaline solution was injected into the heart of one of the guinea pigs; and the other guinea pig was given 0.85 per cent

Sodium chloride solution in the same way. Both were then returned to the bell jar, and the anesthesia was continued.

Production of Typical Chloroform Liver by Iodoform and Bromoform.

Iodoform (CHI_3) and bromoform ($CHBr_3$) are so similar to chloroform ($CHCl_3$) in chemical structure as to suggest that they might have a similar power to produce the characteristic morphological changes of chloroform poisoning. Here, however, we should of course be dealing with an effect of hydriodic acid and hydrobromic acid, respectively, instead of hydrochloric acid. As a matter of fact, not only do both of these substances induce lesions which are in every way identical with chloroform effects, but it is also possible to obtain some evidence that in each case the respective halogen acid is produced in the body. It has been known for some time that the administration of iodoform in large quantities is frequently followed by visceral fatty changes and multiple hemorrhages.[20] In our experiments we found, in addition to these changes, a definite central liver necrosis in every way comparable to that produced by chloroform. Bromoform was found to produce identical changes. Evidence that here also we were dealing with an effect of acid was obtained by finding that the necrotic areas in the liver reacted to neutral red in the same manner as has been described already for chloroform. Attempts to identify iodine ions in these necrotic lesions failed, but iodine was found in large quantities in the urine. It is not altogether surprising that we failed to find it in the tissues, since it doubtless occurs in small quantities at most; and, furthermore, because of its tendency to combine with fats and protein, probably nearly all of the iodine would be present in an organic, and hence non-ionic, form. The methods used for its identification will be described below. They were based on the standard means of identification of inorganic iodine by use of dilute sulphuric acid, sodium nitrite, and starch paste.

It is especially interesting that inorganic iodides may be found in large quantities in the urine after the administration of iodoform;

[20] Cushny, A. R., A Textbook of Pharmacology and Therapeutics, 3d edition, Philadelphia, 1903, 522. Meyer, H. H., and Gottlieb, R., Pharmacology, Clinical and Experimental, translated by Halsey, J. T., Philadelphia, 1914, 519.

for this necessarily implies the previous formation of hydriodic acid, at least somewhere in the body. The finding of iodides is an old observation. Thus Binz[21] in 1877 attributed the toxic action of iodoform to iodine, because of the presence of inorganic iodine in the urine. Hogyes,[22] Harnack and Gründler,[23] and Mulzer[24] have all concurred in this observation. Mulzer's study is especially interesting. He states that most of the urinary iodine is inorganic, in the form of alkali iodides and iodates. Of these, the iodides always appear first. Only about 60 per cent of the calculated iodine can be found in the urine. The rest is eliminated into the sweat, hair, and intestines. Different structures of the body vary in their power to transform iodoform into inorganic iodine compounds. Muscle and liver hash seem to be the most powerful; and the muscle hash is a little more powerful than the liver. These findings are all in harmony with the idea which is being developed in this article; *viz.*, that the decomposition of this group of drugs in the body is associated with the formation of the respective halogen acid, in this case hydriodic acid. Furthermore, if we regard this decomposition as essentially dissociation in the presence of water, as we did in the case of chloroform, then we are not surprised that it occurs in greatest quantity in the liver and muscle tissue where metabolic activity is greatest. The apparent discrepancy between Mulzer's and our own results, in that he found inorganic iodine in the liver where we failed to find it, is easily explained by the fact that he subjected relatively enormous quantities of iodoform to the action of the tissue hash and so obtained a recognizable amount of inorganic iodine which was not bound to protein. Even then his yields of iodine were small.

In our experiments with bromoform no attempt was made to find bromine ions. After demonstrating the identity of the morpholog-

[21] Binz, C., Ueber Jodoform und über Jodsäure, *Arch. f. exper. Pa'h. u. Pharmakol.*, 1878, viii, 309.

[22] Hogyes E., Anmerkungen über die physiologische Wirkung des Jodoforms und über seine Umwandlung im Organismus. *Arch. f. exper. Path. u. Pharmakol.*, 1879, x, 228.

[23] Harnack, E., and Gründler, J., Ueber die Form der Jodausscheidung im Harn nach der Anwendung von Jodoform. *Berl. klin. Wchnschr.*, 1883, xx, 723.

[24] Mulzer, *loc. cit.*

ical findings obtained with chloroform, iodoform, and bromoform, and recognizing that chloroform and iodoform may form in the body hydrochloric acid and hydriodic acid, respectively, the parallelism with bromoform was so close that this detail was omitted as uuessential. Typical protocols follow.

Experiment 5.—Adult dog, weighing 2.7 kilos. Injected daily subcutaneously with 2 gm. of iodoform (Merck). Stirred up in 10 cc. of paraffin oil, for three days. On the third day when the injections were stopped, there was a strong reaction for inorganic iodine in the urine, as determined with starch paste. The dog had lost its appetite and had begun to look emaciated. On the sixth day (three days after stopping the injections) it was unable to stand, but, at intervals, while lying down, it would moan and howl feebly; and simultaneously there would occur frequent twitchings and cramp-like motions of the legs. At no time did the animal seem conscious of its surroundings. It refused meat and water; and between these times of activity it was drowsy and somewhat difficult to arouse. The strong reaction for iodine in the urine had persisted. The dog was killed with ether, but before death occurred a portion of the liver was removed and examined for free H⁺ ions in the manner already described in connection with chloroform. The reactions to the indicators were less intense than with chloroform. Other liver sections were then treated for various periods for from one to thirty minutes with a little dilute sulphuric acid and sodium nitrite, later removed from this bath, and placed on slides. To the sections on the slide thin starch paste was added, and they were then examined to see if the necrotic portions of the lobules became blue. No reaction for iodine was obtained in this way. Then after merely removing enough of the liver for proper histological examination, the remainder was ground in a meat grinder and treated with sulphuric acid of about M/5 strength, and a little sodium nitrite. The fluid was then poured off and treated with starch paste. No reaction occurred. The anatomical changes noted in the dog corresponded in every way to those so characteristic of chloroform poisoning. There was a number of ecchymoses on the parietal peritoneum and pleura. The liver was large, very fatty, and slightly yellowish. It contained a small subcapsular hemorrhage on its upper aspect. The kidneys were large and swollen, and the site of marked cloudy swelling. Microscopically the changes again were identical with those which follow chloroform. The liver contained areas of central necrosis involving about one-fourth of the lobule. There was much fat accumulation in the cells that were not destroyed. The kidneys showed numerous casts; and the tubular epithelium was swollen and granular.

Experiment 6.—Guinea pig, weighing 560 gm. Given daily for three days subcutaneous injections of 0.2 gm. iodoform (Merck) in paraffin oil. On the fourth day the guinea pig was killed and examined. The anatomical changes were essentially the same as those described in the preceding experiment, but not developed to quite so great a degree.

Bromoform was found to be much more toxic than chloroform; so it was impossible to administer it for two hours to an animal and have any assurance that it would live for forty-eight hours afterwards, and thus give time for the maximum development of the anatomical lesions. It was given by inhalation. In all, eight animals were used. A typical protocol is given.

Experiment 7.—A guinea pig weighing 672 gm. was placed under a bell jar and anesthetized by dropping bromoform through the opened top. It was kept in a state of deep narcosis for forty-five minutes and then returned to its cage and allowed to eat carrots. It remained lying on its side for nearly an hour after returning to its cage. Two days later it was killed by a blow on the head and examined. There were several small hemorrhages in the lungs. The liver was very fatty; and the kidneys were large and grayish. Microscopically the picture in the liver could not be distinguished from that of a case of chloroform poisoning. Around the central veins were small areas of necrotic cells without nuclei, constituting perhaps one-sixth or one-seventh of the lobule. There was a large amount of fat accumulation. Everywhere the changes were most conspicuous in the central portions of the lobules.

Liver Necrosis, Etc., as an Effect Common to Aliphatic Alkyl Halides.

The readiness with which central liver necrosis, fatty changes, hemorrhages, edema, etc., could be produced by methyl halogen compounds other than chloroform, immediately suggested the probability that ethyl and ethylene halides would react in the same way. Support for this idea was furnished also by the fact that these substances, like the methyl compounds, readily yield halogen acids outside the body. Niederist[25] has shown that at 100° C. ethyl iodide and water after fifteen hours yielded 98 per cent of calculated hydriodic acid, and that ethyl bromide after eighteen hours gave 94 per cent of calculated hydrobromic acid. Ethylene bromide after heating with water for fifty-two hours at 140° to 150° C. yielded 96 per cent of the theoretical hydrobromic acid.

$$\begin{array}{ccc} CH_2Br & H_2O & CH_2OH \\ | & + & \rightleftharpoons | \\ CH_2Br & H_2O & CH_2OH \end{array} + 2HBr$$

Butlerow[26] had previously found that, after heating ethyl chloride to 100° C. in a sealed tube for ninety-two hours, much hydrochloric acid and alcohol were present.

The following halogen substitution products of ethane have been

[25] Niederist, G., Ueber die Einwirkung von Wasser auf die Haloïdverbindungen der Alkoholradicale, *Ann. d. Chem.*, 1877, clxxxvi, 388.

[26] Butlerow, cited by Niederist, *loc. cit.*

tried in this work: ethyl chloride, ethyl bromide, ethyl iodide, and ethylene bromide.   Not only do all these substances produce morphological changes indistinguishable from those following chloroform, but also it is possible to obtain evidence that the respective halogen acid is liberated in each case.   This evidence is found in the fact that the neutral salts of the respective halogen acids have been found in the urine after the administration of ethyl bromide and iodide. That an acid effect is again concerned is shown by the fact that when sections of the liver are treated with neutral red, according to the method already given, the necrotic portions show a high hydrogen ion content.   Although apparently no measurements have been made of neutral chlorides in the urine after the administration of ethyl chloride, an increase might nevertheless be expected by analogy with the findings after the use of ethyl bromide and iodide.   Dreser[27] found inorganic bromine in the urine, both of man and experimental animals, after inhalations of ethyl bromide.   Inorganic iodine has been found in the liver of a guinea pig after the use of ethyl iodide by Loeb.[28]   The essential morphological changes under discussion in this article have been observed by previous workers to follow the use of ethyl chloride and ethyl bromide.   These have been rather extensively studied by Haslebacher,[29] who also found that the urine became strongly acid and then contained large amounts of casts and albumen, an observation which, in the case of ethyl bromide, had previously been made by Regli.[30]

In our experiments it was found that when given by inhalation the relative power of producing the morphological changes as well as the general toxicity of these substances was greatest in the case of the iodide and least with the chloride.   The most toxic of all was ethylene bromide.   This was so toxic that a narcosis of fifteen minutes' duration or longer was invariably followed within forty-eight hours by death.   In each instance Kahlbaum's preparations were used. Typical protocols follow.

[27] Dreser, H., Zur Pharmakologie des Bromäthyls, *Arch. f. exper. Path. u. Pharmakol.*, 1895, xxxvi, 285.

[28] Loeb, O., Die Jodverteilung nach Einfuhr verschiedener Jodverbindungen, *Arch. f. exper. Path. u. Pharmakol.*, 1906, lvi, 320.

[29] Haslebacher, A., Experimentelle Beobachtungen über die Nachwirkungen bei der Bromaethyl- und Chloraethylnarkose, Inaugural Dissertation, Bern, 1901.

[30] Regli, cited by Haslebacher, *loc. cit.*

Ethyl Bromide.—Adult guinea pig given ethyl bromide by inhalation for eighty minutes. Found dead about thirty hours after the administration. At autopsy the liver was fatty, and microscopically contained definite areas of central necrosis. There were extensive hemorrhages in the lungs. The kidneys were swollen and pale.

Ethyl Iodide.—Experiment performed as above, except that inhalation of ethyl iodide was continued for only forty-five minutes. The guinea pig was found dead twenty hours later. The liver was fatty, and microscopically there were areas of beginning central lobular necrosis. The kidneys were pale and apparently edematous. There were large hemorrhages in the lungs.

Ethyl Chloride.—After an inhalation of ethyl chloride for two hours, the guinea pig was killed two days later. It showed a moderately fatty liver with beginning central necrosis evident microscopically. There were no large hemorrhages in the lungs, and the kidneys were less pale and swollen.

Ethylene Bromide.—An adult guinea pig was given ethylene bromide by inhalation for twenty minutes. It was found dead forty hours later. Autopsy showed 2 or 3 cc. of a slightly blood stained fluid in the abdominal cavity. The liver was large and very fatty and showed beginning central necrosis. The kidneys were large and pale and the lungs had several large hemorrhages.

Lesions Produced by Chloral Hydrate Are Relatively Insignificant.

If the general theory is correct that the severe morphological changes induced by chloroform are due largely to hydrochloric acid liberated by its chemical dissociation in the tissues rather than merely to the fact that it contains three chlorine atoms, then we should expect that another narcotic substance, which would not yield hydrochloric acid in the body although it contained the same number of chlorine atoms, would not produce these changes. Chloral hydrate is a suitable substance with which to investigate this point, because it is eliminated almost completely as urochloralic (trichlorethylglucuronic) acid, and therefore is not capable of yielding appreciable amounts of hydrochloric acid. Only a very small portion of it is decomposed, with a resulting increase in urinary chlorides.[31] Experiments showed that guinea pigs, which had been profoundly narcotized with chloral hydrate for from fifteen to twenty hours, failed to show any liver necrosis, hemorrhages, or extensive edema. Only a very slight accumulation of fat occurred in the liver. This was demonstrable usually only by microscopic examination of sections stained for fat. Whipple[32] has stated that he also failed to produce liver necrosis with this agent.

[31] Meyer and Gottlieb, *loc. cit.*, p. 89.
[32] Personal communication.

These results seem to afford striking confirmation of the idea that the essential factor in the production of these severe lesions by alkyl halides is the halogen acid formed by decomposition rather than merely the halogen content of the molecule. A representative protocol follows.

Dec. 19. Three adult guinea pigs were used: A, weighing 615 gm.; B, 665 gm.; and C, 485 gm. At 12.30 each was given subcutaneously 0.2 gm. of chloral hydrate dissolved in 2 cc. of water. Fifteen minutes later Guinea Pigs A and B were lying on their sides, but still responsive to stimulation. At 3.00 p. m. there had still been no deep narcosis. Each guinea pig therefore was given 0.3 gm. At 4.30 each was given another injection of 0.1 gm. At 4.45 all three guinea pigs were in deep narcosis.

Dec. 20, 11.30 a. m. Each guinea pig was injected with 0.4 gm. An hour later all were in deep narcosis and were still so at 4.30 p. m.

Dec. 21. Guinea Pig B was found dead in the morning. Autopsy showed reddish purple liver with no appreciable fat accumulation and no noteworthy changes elsewhere. Microscopically there was found no appreciable accumulation of fat in the liver, no necrosis, and no marked changes in any of the viscera. At 11.30 a. m. Guinea Pigs A and C were given 0.5 gm. of chloral hydrate as before. At noon both animals were in deep narcosis. At 9.00 p. m. C died; A was still in profound narcosis. Autopsy on C showed slight fatty liver, but no hemorrhages or marked edema anywhere. Microscopically there was no necrosis in the liver.

Dec. 22. Guinea Pig A was found dead at 7.00 a. m. Aside from moderate postmortem decomposition there were no striking changes. Sections of the liver showed no central necrosis.

GENERAL DISCUSSION.

It is a striking fact that certain narcotic agents readily induce marked morphological changes, the most conspicuous of which are central necrosis of the liver lobules, fat infiltration, and a tendency to hemorrhage and edema. This property is particularly evident in those agents whose chemical structure places them in the group of alkyl halides. That this property is not necessarily connected with their ability to induce narcosis is shown by the fact that other narcotic agents (*e. g.,* ether and chloral hydrate), which do not belong to this general chemical group, fail to induce tissue changes which are at all commensurate with those following the administration of the alkyl halides. Moreover, that the mere presence of halogen atoms in the molecule is not the responsible factor is demonstrated by the fact that chloral hydrate ($CCl_3-CH{<}{OH \atop OH}$), which, like chloroform

($CHCl_3$), possesses three chlorine atoms, produces relatively insignificant morphological effects. Some other factor must therefore be responsible. Evidence has been submitted to show that an important factor is probably the halogen acid (hydrochloric, hydrobromic, or hydriodic acid) which is formed by chemical dissociation of the alkyl halides within the body. That these substances form their respective halogen acids in the body is shown by the occurrence in large quantity of the neutral salts of these acids in the urine. In this respect they differ from chloral hydrate, which is excreted mainly as urochloralic acid, and of which therefore only a small portion is decomposed to give neutral chlorides.

The idea of placing the chief responsibility for these severe tissue effects upon the halogen acids formed in the tissues is based upon a number of experimental findings. Pictures practically identical with that of late chloroform poisoning have been produced simply by the administration of hydrochloric acid in suitable concentrations, the only essential difference being that with the injection of the acid the liver necrosis was peripheral rather than central. That these substances form their respective halogen acids within the body is shown by the appearance of their neutral salts in the urine. Alkali, in suitable concentration and combined with hypertonic saline, prevented the liver necrosis and greatly inhibited the other tissue changes. In the series CH_2Cl_2, $CHCl_3$, and CCl_4, the tetrachloride was the most powerful and the bichloride the least, in their ability to induce the morphological changes. This comparison parallels the respective amounts of hydrochloric acid which these substances can yield in their breakdown outside the body; that is, CCl_4 can give four molecules of HCl; $CHCl_3$, three, and CH_2Cl_2, two. Of the ethyl compounds, the iodide was most toxic, the bromide less, and the chloride least of all. This relationship agrees with their relative chemical reactivities outside the body. The question of how the acid is formed will not be discussed extensively in this paper. Nef.[33] who has extensively investigated the nature of the chemical reactions of the alkyl halides, has submitted a large amount of evidence to show that the halogen acid is dissociated off, leaving a methylene

[33] Nef, J. U., Dissociationsvorgänge bei den Alkyläthern der Salpetersäure. der Schwefelsäure und der Halogenwasserstoffsäuren. *Ann. d. Chem.,* 1890. cccix, 126.

or bivalent carbon residue. In the case of chloroform the type of dissociation, according to him, is:

$$CHCl_3 \rightleftarrows CCl_2 + HCl;$$

and in the case of ethyl chloride it is:

$$C_2H_5Cl \rightleftarrows CH_3CH + HCl.$$

He has likewise produced evidence to show that in general the iodides are more dissociated than the bromides, and these in turn more than the chlorides. Our finding, therefore, that ethyl iodide produces the tissue changes more readily than the bromide and this in turn more readily than the chloride, conforms to Nef's idea of the readiness with which these substances can form respectively hydriodic, hydrobromic, and hydrochloric acids. The applicability of Nef's conceptions of dissociation and dynamic chemical equilibrium to problems of intermediate metabolism has already been under study for a number of years in this laboratory by Woodyatt, with particular reference to the chemical phenomena of diabetes.[34]

The tendency to ascribe the anatomical changes of chloroform to the production of phosgene ($COCl_2$), as has been done by Müller,[35] is probably an inadequate explanation, since this substance would almost certainly be quickly hydrolyzed to $2HCl$ and CO_2 in the body, and again we should be dealing with HCl as an important factor. Binz[36] has considered the liberation of the halogen itself (as molec-

[34] Woodyatt, R. T., Studies on the Theory of Diabetes. I. Sarcolactic Acid in Diabetic Muscle, *Jour. Biol. Chem.*, 1913, xiv, 441; Greer, J. R., Witzemann, E. J., and Woodyatt, R. T., Studies on the Theory of Diabetes. II. Glycid and Acetole in the Normal and Phlorhizinized Animal, *ibid.*, 1913–14, xvi, 455; Sansum, W. D., and Woodyatt, R. T., Studies on the Theory of Diabetes. III. Glycolic Aldehyde in Phlorhizinized Dogs, *ibid.*, 1914, xvii, 521; Woodyatt, R. T., Studies on the Theory of Diabetes. IV. The Parallelism between the Effects of the Pancreas and Those of Metallic Hydroxides on Sugars, *ibid.*, 1915, xx, 129; Sansum, W. D., and Woodyatt, R. T., Studies on the Theory of Diabetes. V. A Study of Narcotic Drugs in Phlorhizin Diabetes, *ibid.*, 1915, xxi, 1. Wells, H. G., Chemical Pathology, 2d edition, Philadelphia, 1914, 573.

[35] Müller, R., Ueber die Einwirkung des Phosgens auf den menschlichen und thierischen Körper, *Ztschr. f. exper. Path. u. Therap.*, 1911, ix, 103.

[36] Binz, C., *loc. cit.*

ular halogen) to be the chief toxic factor in all these drugs. There is, however, but little evidence in favor of such a conception; for at least outside the body most of the facts, as Nef has shown, point to a dissociation of a type to yield halogen acid instead of molecular halogen. Our experiments tend to show the existence of a similar type of dissociation within the body.

Fischler[37] has sought an explanation for the chloroform changes in an associated fat necrosis from injury to the pancreas which he noted in a number of dogs. This idea was based on the fact that central necrosis was produced in Eck fistula dogs without the use of chloroform but after intraperitoneal injections of trypsin and hydrazin sulphate, and after severe crushing of the pancreas. He considers some albumen-splitting substance, whose nature he does not discuss, as the responsible factor. Obviously such a suggestion does not explain the production of the changes. It is not surprising that such drastic measures resulted in severe morphological changes, of which one was central necrosis of the liver. It is interesting, however, that degenerative changes most conspicuous in the central part of the lobule (including even well marked necrosis), sometimes followed simply the establishment of the Eck fistula without other experimental procedures. In one experiment, in which he also ligated the hepatic artery, the central necrosis was marked.

Wells[38] has expressed the view that the changes in late chloroform poisoning arise because, although the oxidizing enzymes are suppressed, the autolytic enzymes are left free to digest the cell. There is no reasonable doubt that oxidations, as well as many other metabolic activities, are altered by chloroform and these other substances under discussion. It is likewise true that in many respects there is a similarity between the production of chloroform liver necrosis and the self-disintegration of tissue *in vitro* which is called autolysis. But the difficulty in such an interpretation lies in an inability to gain a definite conception of the nature of an autolytic enzyme and to distinguish its effects from those of acids formed in the tissues.

[37] Fischler, F., Ueber das Wesen der zentralen Läppchennekrose in der Leber und über die Rolle des Chloroforms bei dem sogenannten Narkosenspättod. *Mitt. a. d. Grenzgeb. d. Med. u. Chir.*, 1913, xxvi, 553.

[38] Wells, *Arch. Int. Med.*, loc. cit.

Particular care has been exercised all through this article to state that the halogen acids are suggested to be important factors rather than the only factors involved. Other acids must play a part; and possibly even other substances than acids are concerned. Since chloroform, like other narcotic agents, induces a severe tissue asphyxia, we are compelled to assume the presence of various organic acids, notably lactic acid; and lactic acid, as has been already shown by Fischer and others, is capable of producing extensive tissue alterations. The comparatively slight tissue changes which follow the use of ether and chloral hydrate, for reasons already stated, cannot be due to the liberation of a halogen acid in the tissues; but they may perhaps be attributable chiefly to the tissue asphyxia and resultant weak acid formation which they induce. Nor can a halogen acid be a responsible factor in poisoning with phosphorus, which produces morphological changes similar to those of chloroform, except that in the liver the most extensive alterations are at the periphery instead of the center of the lobule. The mechanism involved here is not clear. Cell asphyxia, however, without doubt occurs; and it is well known that lactic acid is formed in relatively large amounts,[39] which in itself could play an important part. In addition, the possibility should be considered of the formation in the tissue and action of some of the phosphoric acids which are known to occur so easily in the oxidation of phosphorus *in vitro*. As yet there is no direct evidence to support this view. Doubtless also other factors are more or less involved which concern physical alterations in the cell induced directly by the action of the phosphorus.

No detailed discussion will be taken up here concerning the question of whether these morphological changes (necrosis, fat infiltration, hemorrhages, and edema) are to be regarded as primarily acid effects or primarily asphyxial effects. It is of great interest that they can be readily obtained merely by the administration of such an acid as hydrochloric acid. But it is equally true that the administration of an acid also leads to the production of asphyxia. We are, therefore, confronted by the facts that acids give rise to asphyxia, and asphyxia in turn gives rise to acid production. It cannot be assumed that the changes are exclusively acid effects.

[39] Lusk, G., The Elements of the Science of Nutrition, 2d edition, Philadelphia and London, 1909, 304.

In this connection the location of the necrosis in the central part of the lobüle is of interest; for this is the region which is farthest removed from the oxygen supply and from compensatory influences carried in by the blood. It is, therefore, the part where, in general asphyxial conditions, asphyxial acids (lactic, etc.) would be formed in largest quantity. This is strikingly borne out in Fischler's experiments which showed that interference with blood supply, by ligation of the hepatic artery and production of Eck fistula, was followed by degenerative changes which were always most marked in the centers of the lobules. In the case of chloroform poisoning we might suppose that the central location of the necrosis is due to the fact that in that region we have the greatest total acid formed (*i. e.*, asphyxial acids in addition to hydrochloric acid). It is also farther removed from the neutralizing effect of alkalis brought in by the blood than is the periphery of the lobule.

In a previous article[40] it has been shown that newly born pups. which are relatively immune to the production of late chloroform poisoning, owe their resistance to their rich supply of glycogen. The nature of this protective action of glycogen is not clear. It is possible that the observation of Bechhold and Ziegler,[41] that glucose retards the diffusion of sodium chloride and some other substances in protein gels, bears on this phenomenon. In some work, as yet unpublished, we have observed that the presence of glucose in gelatin and agar gels markedly retards the diffusion of hydrochloric and other acids through them. Fischer and Sykes[42] have recently shown that glucose and other sugars inhibit the swelling of a colloid, like fibrin, in water.

SUMMARY.

The central lobular necrosis in the liver, which has been regarded by some writers as characteristic of late chloroform poisoning. has been produced experimentally with a number of other drugs. It

---

[40] Graham, E. A., The Resistance of Pups to Late Chloroform Poisoning in Its Relation to Liver Glycogen, *Jour. Exper. Med.*, 1915, xxi, 185.

[41] Bechhold, H , and Ziegler. J , Die Becinflussbarkeit der Diffusion in Galletten, *Ztschr. f. physikal. Chem.*, 1906, lvi, 105.

[42] Fischer. M. H., and Sykes, A., Ueber den Einfluss einiger Nichtelektrolyte auf die Quellung von Protein. *Kolloid-Ztschr.*, 1914. xiv. 215.

is, therefore, in no sense peculiar to chloroform poisoning. Substances which have been shown to produce a morphological picture indistinguishable from that of late chloroform poisoning are: (a) dichlor- and tetrachlormethane, (b) tribrom- and triiodomethane, (c) monochlor-, monobrom-, and monoiodoethane, also the dibromethane; that is, in general, the halogen substituted aliphatic hydrocarbons containing one or two carbon atoms. Presumably similar results might be obtained with the higher members of the same series.

The mechanism by which chloroform produces its characteristic tissue changes must accordingly be considered as a group reaction. Outside the body the similarities between the chemical behavior of different members of this group have been correlated by Nef on the basis of the type of dissociation which these substances undergo and the differences in their behavior on the basis of the differences of the degree to which such dissociations occur. According to the work of Nef, the group of substances under discussion has the property of dissociating to yield a halogen acid and an unsaturated alkylidene rest. Thus with chloroform the type of dissociation may be expressed thus:

$$CHCl_3 \rightleftarrows CCl_2 + HCl.$$

In this paper the view is developed that the changes characteristic of late poisonings with the above named group, namely edema, multiple hemorrhages, fat infiltration, and necrosis are ascribable (1) to acids and (2) to the fact that the amount of acid formed parallels the chemical dissociability of the drug outside of the body.

Favoring the view that acid is responsible for the changes are the following observations.

1. All the characteristic features of late chloroform poisoning have been produced merely by the administration of hydrochloric acid, except, however, for a different distribution of the liver necrosis.

2. The areas of central necrosis produced in the liver by the various substances under discussion give an acid reaction to neutral red.

3. Sodium carbonate in a hypertonic sodium chloride solution markedly inhibits the production of the lesions.

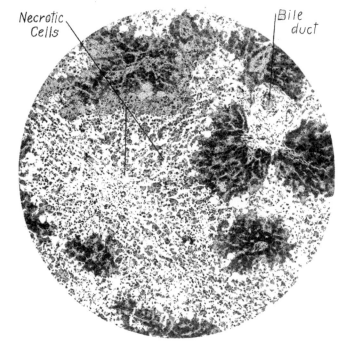

FIG. I.

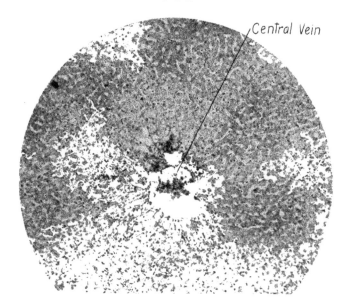

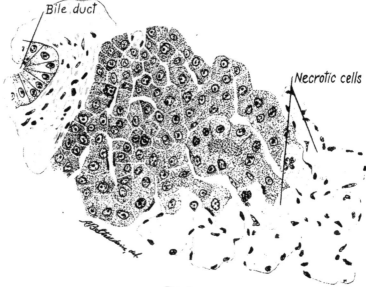

FIG. 3.

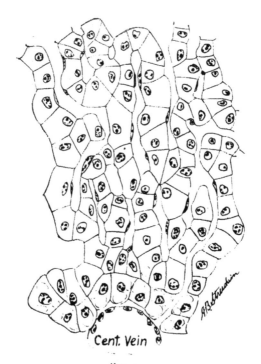

FIG. 4

In favor of the view that the respective halogen acids play an important part are the following.

1. After the administration of some of these drugs there has been noted an increase of the neutral salts of the halogen acids in the urine, a fact which indicates that the corresponding halogen acids must have been formed somewhere in the body.

2. The necrosis-producing powers of dichlormethane, chloroform, and tetrachlormethane parallel the amounts of hydrochloric acid which these substances theoretically can yield in their breakdown outside of the body. Likewise, the power to produce tissue changes exhibited by the ethyl compounds varies directly with the ease with which they·form their respective halogen acids *in vitro*.

3. Ether and chloral hydrate which do not yield halogen acid in their breakdown in the body likewise also do not produce necrosis. They induce only edema and fat infiltration to a less marked degree.

The suggestion is made that the halogen acid (hydrochloric, hydrobromic, or hydriodic acid), directly liberated in the process of dissociation, may be the important factor which makes the tissue changes seen in poisoning with chloroform and other alkyl halides so different from those following the administration of narcotic drugs of a different type.

EXPLANATION OF PLATES.

PLATE 4.

FIG. 1. Typical central liver necrosis produced by chloroform anesthesia for four and one-half hours in Dog B of Experiment 3 (page 61). Nearly all the lobule is affected. Microphotograph. × 83.

FIG. 2. Inhibition of necrosis by the intravenous injection of 150 cc. of Fischer's alkaline, hypertonic salt solution as shown in Dog A of Experiment 3 (page 61) which had received chloroform for the same length of time as Dog B (Fig. 1). There is practically no necrosis. Only a slight amount of fat accumulation in the cells about the central vein has occurred. Microphotograph. × 83.

PLATE 5.

FIG. 3. High power drawing of liver shown in Fig. 1. The necrotic cells in the central portion of the lobule stain intensely with eosin. The nuclei of the parenchymatous cells have disappeared; and only those of the capillaries are evident. × 356.

FIG. 4. High power drawing of the liver shown in Fig. 2. There is no appreciable necrosis; and only a moderate accumulation of fat has occurred in the parenchymatous cells around the central vein. × 356.

THE FORM OF THE EPITHELIAL CELLS IN CULTURES OF FROG SKIN, AND ITS RELATION TO THE CONSISTENCY OF THE MEDIUM.

By EDUARD UHLENHUTH, Ph.D.

(*From the Laboratories of The Rockefeller Institute for Medical Research.*)

PLATES 6 TO 21.

(Received for publication, May 12, 1915.)

INTRODUCTION.

In a previous publication (1) the author has shown that the polygonal epithelial cells of frog skin assume a fusiform or thread-like shape ("spindle cells" and "thread-like columnar cells") shortly after they have wandered out into the medium. In this form they closely resemble connective tissue cells.

Champy (2) found in the case of the explanted kidney that both connective tissue and epithelial cells gradually took on an indefinite character, and he observed the same phenomenon in the explanted parotis, submaxillaries, and thyroids. It is of especial importance to note that the indefinite tissue of these organs is absolutely identical, and that, for example, the tissue derived from the kidney cannot be distinguished from that from the parotis. He calls this phenomenon "dedifferentiation," and attributes it to loss of function. It seems to us, however, that but little is gained by this explanation, for we possess no information regarding the functions of these organs in the embryonic stage and are unaware whether, for instance, the parotis or submaxillary functionates at all in the fetal organism. The point to be emphasized is rather the fact that organs as different as the kidney, salivary glands, and thyroids develop identical tissues as soon as the same chemicophysical conditions are brought to bear on the respective organs. This result signifies that the factors responsible for this equalization are present in the culture medium. Thus, in view of the fact that we can modify at will the character of the medium, we are supplied with a definite means of further analyzing these phenomena.

It stands to reason that the above mentioned changes occurring in the cells are the result not of one factor, but of a number of different chemical and physical factors. In this paper, however, we intend to confine ourselves to the description of but one of these factors and to show in what way the consistency of the medium influences the form of the cell.

It is an easy matter to vary the firmness of the medium, which consisted of frog plasma and frog muscle extract obtained from the same species (*Rana pipiens*) that supplied the fragments of skin. The plasma of all individuals of the same species does not coagulate uniformly well, and it was soon apparent that this fact greatly influences the form which is subsequently assumed by the cells. Furthermore, by the addition of chicken plasma the frog medium can usually be made to gain a semi-firm consistency, and by the combined addition of chicken plasma and embryonic chicken extract the medium becomes very firm.

In previous experiments (1) we carried out prolonged observations on cultures that were cultivated in a pure frog medium (frog plasma + frog muscle extract), and found that the rim of tissue which formed around the explanted fragment of skin began by assuming the form of a compact membrane consisting of polygonal cells. Next the cells situated at the edge of the membrane began to project out into the medium, at the same time taking on an elongated spindle-shaped form (spindle cells). Many of these cells became detached from the membrane and wandered out into the plasma. By this means the edge of the membrane assumed a ragged appearance. Finally spaces occurred in the membrane which were bridged over by very elongated cells. often thread-like in appearance (thread-like columnar cells). As a result of successive processes the entire membrane gradually became divided up into long cells of the connective tissue type which strayed out into the medium.

It has now been shown that this change in the form of the tissue membrane and of its cells results from the gradual softening of the medium, and we are thus able, by the employment of suitable media. to produce artificially the individual stages of this process. accompanied by the characteristic form of the cells. The shape of the cells can be maintained permanently polygonal or fusiform. according as the medium selected is firm or soft.

As in our previous experiments, the material chosen was obtained from the dorsal skin of the adult leopard frog (*Rana pipiens*), which was cultivated in small fragments (approximately two square millimeters). The following are the results obtained with 179 cultures, and they are in accordance with the findings and records of the 480 cultures of our previous work.

EXPERIMENTAL.

Experiments with a Firm Medium.—We shall first describe the experiments made with a firm medium. A medium was considered firm when, in the process of changing, the culture could be cut out of the old medium enclosed in a firm block of medium. The firmness of these blocks was so great that they failed to bend over the edges of the cataract knife, even when they were considerably broader than the knife itself. We usually obtained such a medium by adding to two drops of frog plasma (F.Pl.)[1] one drop of frog muscle extract (F.E.), one drop of chicken plasma (Ch.Pl.), and one drop of embryonic chicken extract (Ch.E.).

Fig. 1 shows a culture (B 174) in the firm medium (F.Pl. + F.E. + Ch.Pl. + Ch.E.); photographed about twenty hours after explantation. A considerable rim of tissue has formed around the old fragment. It consists of strictly polygonal cells, and its line of demarcation towards the medium is uniform or only slightly irregular. The cells are closely pressed together. A vertical section of these cultures shows that the rim of tissue usually consists of from three to four layers of cells at the center, and of two layers at the periphery, and that the cells are polyhedral in shape, as shown in Fig. 2. In Fig. 3 are seen from above some of these polyhedral cells taken from a culture fixed *in toto* and stained with Giemsa (B 169); it was prepared in the same medium as culture B 174 in Fig. 1. Thus viewed we clearly see the polyhedral shape of the cells.

Fig. 4 (Culture B 114) represents a stained culture after a six days' sojourn in a firm medium. The culture was prepared on Dec. 15 and the following day showed but a relatively small rim of cells. On Dec. 17 the fragment of skin, together with the cell membrane, was placed in a fresh medium. On Dec. 19 the rim of cells had increased in size while remaining compact. It is noteworthy that no migration of cells had taken place. On the same day the old medium, which had remained completely firm, was exchanged for a fresh medium, into which was transferred the entire membrane of tissue. On the morning of Dec. 31 the medium showed signs of liquefying on one side. One protuberance into the medium had already formed, as well as a small cavity, which latter, however, was subsequently destroyed by fixation (to the left of the figure). After six days the culture was fixed. The membrane consisted exclusively of polygonal cells (Fig. 4), with the exception of that portion where the medium had begun to liquefy. Fig. 5 represents a part of the epithelial rim of Culture B 114 mag-

[1] We shall abbreviate as follows: frog plasma, F.Pl.; frog muscle extract, F.E.; chicken plasma, Ch.Pl.; chicken embryonic extract, Ch.E.

nified 1,000 times. The polygonal shape of the cells is distinctly seen. It should be noted that the tissue membrane of this culture (as also of all others subjected to the same conditions) maintained its compact form for six days without becoming loosened up, and that even the outermost cells of the rim, shown in Fig. 5, remained polyhedral in shape.

Fig. 6 was taken from a living culture (B 103) ten days after explantation. The culture was started on Sept. 12, 1914, and at first showed but slight signs of growth. The epithelium did not attain any considerable dimensions until Dec. 13, at which time it was quite thick. On Dec. 14 the medium was changed for the third time, whereupon it was found to be of exceedingly firm consistency. Thus it was possible to transfer the large tissue membrane without injury into the new medium. On Dec. 15 the rim of tissue was not as compact as in similar cultures which had been maintained under the same conditions. When the medium was changed on the same day it was also found to be less firm than that of the other cultures. Nevertheless, it was possible to cut out a block sufficiently firm to permit of the transfer of the entire rim of cells into the new medium. The cells of the outermost edges, however, which were the result of the growth from Dec. 14 to 15, tended to become massed together. On the 16th additional increase in the size of the cell membrane could be seen and new cells had been added. No migration of cells had taken place, however, and the membrane was quite compact. On Dec. 17 the medium was again changed. On Dec. 19 enlargement of the membrane could again be seen in certain sections, but the form of the membrane remained uniform and was unindented. No cell migration had taken place. At this stage the culture was photographed (Fig. 6). To the left of the figure can plainly be seen the thin layers of cells which sprouted out on the day preceding that on which the photograph was taken. They are separated from the remaining tissue by a lighter band of tissue. No detached cells can be seen around the culture. The latter, therefore, maintained for ten days consecutively a compact rim of tissue, which grew almost continuously and failed to undergo loosening up. The cells remained permanently polygonal in shape.

As long as the medium remained firm the same conditions were observed in all the other cultures subjected to the above treatment.

Experiments with a Medium of Semi-Firm Consistency.—A medium of semi-firm consistency was supplied by the addition of chicken plasma to the frog plasma and frog muscle extract (2 drops of F.Pl. + 1 drop of F.E. — 1 drop of Ch.Pl.). With a medium of this consistency it is still possible to remove a clean-cut block from the drop of plasma, but these blocks are so soft that if they extend beyond the edge of the knife their sides double over and adhere to the under surface of the knife. In order, therefore, to transfer into a fresh medium large blocks of this consistency with the enclosed culture every precaution must be used to avert this doubling over.

Moreover, it should be stated that owing to the extreme fluctuations in the consistency of the above mixture (F.Pl. + F.E. + Ch.Pl.) the results here obtained were by no means as uniform as those secured with the mixture composed of F.Pl. + F.E. + Ch.Pl. + Ch.E. Sometimes, for instance, a very firm medium is produced, at other times a soft one, such as results from the use of a pure frog medium. But it is easy in each case to ascertain the degree of firmness. Here, however, we shall mention only the behavior of the cultures kept in a semi-firm medium; for their behavior when in a firm or soft chicken plasma mixture (F.Pl. + F.E. + Ch.Pl.) is the same as when kept in a firm medium (F.Pl. + F.E. + Ch.Pl. + Ch.E.) or in a pure frog medium (F.Pl. + F.E.), respectively.

Fig. 7 represents a culture about twenty hours old (B 86), cultivated in a mixture of F.Pl. + F.E. + Ch.Pl. The rather broad rim of cells is seen to be very dissimilar from the tissue membranes obtained in a firm medium. The central zone of cells cannot be clearly made out in the photograph, owing to too intensive staining. Nearly all the peripheral cells have assumed a spindle-shaped form. At several places sharp projections of cells can be seen jutting out into the medium, and at their extremities the typical fusiform cells are beginning to stray out individually into the medium. This process of migration is seen to be specially active at the projection seen in the right-hand upper corner of the photograph, as a result of which a small space has been formed in the membrane. The rim of tissue has consequently assumed an irregular, serrated appearance, and the medium in the neighborhood of the culture is beginning to fill up with detached fusiform cells that have strayed out, as can be clearly discerned in the figure.

The above characteristics are seen still more plainly in a three day old culture (B 7), which is shown under higher magnification in Fig. 8. The edges of the membrane project into the medium on all sides in the form of sharp tongues, the cells of which are elongated examples of the spindle form. An enormous number of cells have wandered out into the medium, and consequent upon this migration and upon the advance of the border cells, it is seen that behind the edge spaces of varying size are being formed in the membrane. These spaces, however, contain no liquid.

Figs. 9 a and 9 b are drawings of living cultures made without a camera lucida. They represent the process involved in the projecting out of the cells at the edge of the membrane and their ultimate detachment and migration (page 86).

It is possible to carry on for a prolonged period of time observations with chicken plasma cultures in the same manner as previously described in connection with the cultures grown in a firm medium, and furthermore to transfer them practically uninjured into fresh media. Cultures kept in a plasma of semi-firm consistency maintain permanently the characteristics described above, in contradistinction to those maintained in a firm medium, in which the cell membrane is permanently compact and the form of the cells unvaryingly polygonal, while the latter usually refrain from wandering off into the medium. Culture B 78 was observed for a period of twenty-two days. Not until the sixth day had a fairly large rim of epithelium formed; later, however, it became greatly reduced in dimensions, owing in part to considerable cell migration, and in part to losses incidental to change of medium. After fourteen days, as a result of this, only a small rim of cells remained, and some of the peripheral cells had become spherical in form, owing to insufficient nourishment, a circumstance which is apt to occur in any medium. On the fifteenth day, however, the peripheral cells recovered, and began to form a new thin membrane, the edges of which showed the usual sharp indentations, and consisted of greatly elongated cells. Two subsequent transfers again occasioned a reduction of the membrane, and this was followed by renewed active growth of the fusiform cells accompanied by the usual migration of cells. The same process was repeated after another change of medium, at which stage the culture was photographed. Fig. 10 represents Culture B 78 twenty-one days after explantation. The central parts of the membrane consist chiefly of polygonal cells; the edges, however, show sharp and multiform indentations, and their cells, even those remote from the edge, are spindle-shaped. Cell migration can likewise be clearly seen.

After twenty-two days Culture B 78 was fixed and stained with Giemsa, in which condition it appears in Fig. 11. The membrane of cells is seen to be even more loosened up than on the preceding day, and long tissue processes, consisting of elongated spindle-shaped cells, are seen to project into the medium. The migration of cells has assumed exceptionally large proportions; the detached cells are for the most part fusiform in shape, but some are round.

Fig. 12 shows individual spindle-shaped cells, drawn under a high magnification. In Fig. 12 a we see the transformation of typical polygonal cells of the membrane into fusiform cells. Figs. 12 b and c represent typical examples of the isolated fusiform cells.

Experiments with a Soft Medium.—The phenomena which we have observed in the medium of semi-firm consistency occupy a position midway between those occurring in the firm and in the soft media. We shall now consider the latter. A soft medium is obtained by employing a pure frog medium (F.Pl. + F.E.). It is impossible to cut out little square blocks of this medium: on the contrary, these blocks invariably assume a spherical form, irrespective of the shape in which they have been cut out, and they furthermore have a tendency to adhere to the knife. Soft media can be drawn out in threads (isolation of the fibrin); they contain much liquid which is accumulated in the minute vacuoles in the network formed by the fibrin. These media rapidly succumb to complete liquefaction, whereupon a number of particular phenomena take place, which we must consider in connection with the rest, as they usually occur coincidently.

Fig. 13 was photographed from a living culture (B 119), about twenty hours after explantation of the fragment. We are unable to determine whether in this culture a "secondary epithelial rim," as we termed in our first communication the compact rim of epithelium, preceded the loosening up stage, illustrated in this figure; the former process may have developed over night. Nevertheless, in cultures consisting exclusively of frog medium the cells frequently begin to leave the fragment immediately in the form of a star-shaped membrane, with very ill defined edges; for these pure frog media usually become soft very soon, sometimes from the very outset. Fig. 13 gives a good impression of the condition of the cell membrane in a soft medium. Almost the entire rim of epithelium is drawn out in the form of long chains of cells and bands of tissue. Even the compact parts of the tissue membrane consist exclusively of elongated cells. On all sides large spaces are seen in the tissue membrane, lined by highly delicate, thread-like cells. These spaces contain no liquid, but soft medium. The whole form of the rim of tissue clearly indicates that the tissue is in a condition of active movement, as can also clearly be seen in living cultures under the microscope. The chief characteristic of this particular culture is the enormous migration of cells. The entire drop of plasma is filled with wandering cells, which are partly fusiform in shape, partly round. In this culture whole fragments of tissue strayed out into the plasma. We were able to certify that they moved in a soft, not liquid substrate, by shaking the culture, a test invariably applied by us; for fragments floating in a liquid move from side to side, which did not occur in this case.

Fig. 14 shows Culture B 121 (stained with Giemsa), six days old, which was kept under parallel observation with the above mentioned Culture B 119 (Fig. 13). The culture was started on Dec. 15, and on Dec. 16 already possessed a fairly extensive star-shaped membrane of tissue, which projected out into the medium in long tissue processes, as is the rule in a soft frog medium culture. On Dec. 17 the medium was renewed, and on Dec. 19 the membrane of cells had attained considerable proportions. It was characterized by extremely long projections of tissue, as well as by bridges of thread-like columnar cells, and exceptionally active cell migration. The medium was renewed the same day, and on Dec. 21, six days after explantation, the culture was fixed, stained, and photographed (Fig. 14). The same characteristics can be noted as in Fig. 13, but more strongly emphasized.

Fig. 15 is a four day old culture in a pure frog medium photographed while living (B 117). The medium has partly liquefied and the resulting liquid has collected first of all in the holes in the tissue membrane caused by the migration of the peripheral cells. In such cases we are unable to decide whether these holes are primarily formed through the wandering out of the cells, as occurs in non-liquid cultures, or whether the membrane becomes torn by the vacuoles of liquefaction which are in process of being formed. The vacuoles, which at first are small, soon increase in size, the bridges of cells which separate them at the same time becoming thinner and thinner until at length they break away. By this means one or more larger vacuoles are formed, which in turn force the remaining part of the tissue to give way at the sides. Finally the cells are left in the form of a narrow wreath of cells encircling the large vacuole. In Fig. 15 can be seen the delicate, thread-like columnar cells which connect the ring of cells with the old fragment. At one place in this culture a single fragment of compact, but very thin tissue membrane yet remains; here the plasma had remained very firm, so much so that it was possible to cut out a block containing the membrane and transfer it into a new medium.

At this point it may be well to consider the behavior of the uppermost layers of epithelial cells. In Culture B 117 the same have become disconnected and are seen to lie over the old fragment and over a part of the cell membrane in the form of a thin membrane (Fig. 15) which possesses sharply defined edges while its cells have remained inactive. This phenomenon will be discussed in detail in a subsequent communication.

Cultures which are preserved in a pure frog medium, or in soft or liquid media generally, can only in rare cases, favored by every precaution, be maintained for any length of time as actively growing tissue membranes; usually the cells migrate in such large numbers that in a short time nothing remains of the original membrane of cells. But it occasionally happens that a portion of the epithelium, instead of migrating, grows around the connective tissue either partially or completely, thus retaining its epithelial character. This particular phenomenon, however, which we have mentioned elsewhere (1), does not concern us here, and will be reserved for future consideration. It behooves us, rather, to follow the fate of certain cultures which when cultivated in a soft frog medium retained their membrane in a growing condition for a considerable length of time.

Culture B 77 was started on Nov. 28, and the tissue membrane took a fairly long time to develop. The records fail to show clearly at what time this took place. On Dec. 13, fifteen days after explantation, the membrane of cells still showed signs of growth at its edge, and was transferred on Dec. 14 to the new medium in an incomplete condition. On the following day, however, there were signs of renewed growth. On the same day the culture was again supplied with medium, and on Dec. 16 fresh growth of spindle-shaped cells was noted. At this stage the culture was photographed (Fig. 16), eighteen days after explantation. The cell membrane is seen to be exceedingly delicate and much thinner than that of a twenty-one day old culture grown in a medium of semi-firm consistency (Fig. 10). The tissue protuberances are much longer and narrower, and some of the cells are as fine as threads. The migration of cells is very active.

Similar conditions obtained in the case of Culture B 81, which was likewise started on Nov. 28 in pure frog medium. By Dec. 2 a narrow rim of epithelium with long protuberances and columns of thread-like cells had formed. On Dec. 13, one day after the medium had been changed for the eighth time, long fusiform cells still projected from the rim of membrane, and on Dec. 14 the membrane had again greatly increased in size. On Dec. 16, after two more changes of medium, the rim of tissue was still growing and projecting long processes into the medium. By Dec. 19 the entire membrane had become loosened up and an enormous number of bipolar fusiform and round cells could be seen in the plasma which had partially liquefied. The cells had strayed to the very limits of the drop of plasma. Fig. 17 shows the living culture photographed on the same day (twenty-one days after explantation). The spherical zone of liquefaction can be plainly seen, containing innumerable fusiform and round cells.

The behavior described above of the membranes of cells in connection with the old tissue fragment, is likewise characteristic of individual, isolated fragments of membrane when grown in daughter cultures. Fig. 18 shows an isolated fragment of membrane (Culture B 100 a) in a firm medium (F.Pl. — F.E. + Ch.Pl. + Ch.E.), and Fig. 19 represents a fragment of a tissue membrane (Culture B 119 a) in a soft medium (F.Pl. + F.E.). The former was taken from Culture 100, the latter from Culture 119. Culture B 100 a produced two days after isolation a fairly thick membrane, consisting of polygonal cells. whereas in Culture B 119 a twenty hours after isolation the elongated fusiform cells had formed a connective tissue-like network of cells. The thread-like columnar cells had developed exceptionally well in this culture, and can be seen in great numbers in Fig. 19.

At this stage it might be asked whether the above mentioned variations which we have ascribed to differences in the consistency of the medium, should not rather be attributed to the chemical differences prevailing in the three kinds of medium, in view of the fact that a firm consistency was secured by the addition to the pure frog medium of chicken plasma and chicken embryonic extract, which thus at the same time enriched the culture by the addition of the specific chemical materials characteristic of the body juices of a foreign species.

The extended period of growth of the said cultures is a sufficient indication that these foreign substances exert no injurious effect on them. For reasons not yet known it occasionally occurs that in spite of the addition of these foreign substances, the medium yet fails to solidify but remains soft, and in such cases the form of the membrane and cells is seen to be identical with those observed in a pure frog medium. On the other hand, it occasionally happens that a pure frog medium or one to which chicken plasma alone has been added, remains unusually firm; the result in such cases will be the formation of great compact rims of tissue and polygonal cells, such as are usually seen only in a culture composed of F.Pl. + F.E. + Ch.Pl. + Ch.E. The following illustrations prove this point. Fig. 20 shows a culture which was kept for about twenty hours in a pure frog medium. The rim of tissue is uncommonly thick and consists chiefly of polygonal cells; only in the left-hand top corner have the cells begun to assume a spindle form. Fig. 21 is a two day old culture (B 13), which was kept in a mixture of F.Pl. + F.E. + Ch.Pl. In this case the medium was exceptionally firm, the epithelium as a result thereof exceedingly compact, and the cells polygonal in shape. Fig. 22, on the other hand, shows culture B 100 which was grown in a mixture of F.Pl. + F.E. + Ch.Pl. + Ch.E., and which for a period of ten days possessed an exceedingly compact rim of tissue, consisting of polygonal cells. Here cell migration failed to occur. On the tenth day the culture was transferred into a new medium, which in spite of being composed of the customary mixture failed to coagulate entirely. The membrane thereupon immediately assumed the form characteristic of one grown in a semi-firm medium; viz., in a combination of F.Pl. + F.E. + Ch.Pl. At the same time the cells became fusiform in shape and began to migrate.

Consistency of the Medium and Form of the Cells.

We have shown that the shape of the cells is dependent upon the consistency of the surrounding medium. We have also shown that the fusiform and thread-like epithelial cells of the connective tissue type, described in a former publication (1), arise only in media of a semi-firm or soft consistency, whereas epithelial cells grown in a firm medium remain permanently polyhedral. These observations may be considered as a distinct advance, as they furnish information regarding the causes governing the shape of the cells. Nevertheless, we can not yet arrive at a complete understanding of this important morphological phenomenon, for we are confronted by the question as to how and why the shape of the cells is affected by change in the consistency of the medium. Neither our own experiments nor those of others at present offer a satisfactory explanation; but it is important definitely to realize along what lines among the knowledge and facts at our disposal we should search for a solution of the problem.

Our point of departure is a two-fold phenomenon: first the increased cell migration in a soft medium; and second the difference in the shape of the cells consequent upon the varying firmness of the medium.

Burrows was the first to consider in an exhaustive manner the effect of the dilution of plasma on the growth of tissue. He found that the cells move more rapidly and actively in plasma which has been diluted with tissue juices than in undiluted plasma, and he attributes this fact chiefly to the reduction in the fibrin content of the medium, consequent upon this dilution (3), and secondly, to the reduced density of the medium (4). From his demonstrations, however, we fail to learn that by diluting the plasma the fibrin content was diminished. for, according to Burrows, upon pressing out the liquid portion of the medium a firm block of fibrin again remained, which although reduced in thickness had nevertheless preserved its consistency.

Be that as it may, the problem of a reduction in the content of fibrin is not involved in our method. for in the media which contained relatively the largest quantity of fibrin (F.Pl. + F.E. + Ch.Pl.) the migration of cells was considerable, whereas in the media which contained less fibrin but more muscle extract (F.Pl. + F.E. + Ch.Pl. — Ch.E.) no migration whatever took place. As a matter of fact the relation between the cell migration and fibrin content was exactly the reverse from what would have been expected according to Burrows. With respect to density, however, our findings agree with those of Burrows. The maximum density coincides with a complete lack of isolated migrating cells, whereas in a pure frog medium, i. e., one possessing a minimum density, the cells become detached from the tissue membrane and stray out into the medium in enormous quantities.

The form of the cells is closely connected with their movement; both phenomena originate from the same cause. By careful investigation of the above facts it will remain evident that neither can be ascribed to " activity " on the part of the cells, but are purely passive phenomena. The cell is impelled to move, and it receives its form by the same processes which necessitate movement, for which reason these two phenomena must be studied in combination.

We shall consider the various forms of cells in succession, beginning with the fusiform cells at the phase of their evolution from the polygonal type. This can be done without difficulty by selecting specimens whose medium is undergoing a change from the firm to the soft stage. Here we at once remark that the spindle-shaped cells are invariably formed from the border cells which are surrounded on three sides by the soft medium; they never arise from the polyhedral cells situated in the more central parts of the membrane where the latter is three to four layers thick. The border cells, as well as all the other cells constituting the membrane, possess a clearly differentiated ecto- and endoplasm. The latter is relatively dense and foamy in structure, containing vacuoles, granulations. and the nucleus. The ectoplasm, on the other hand, is homogeneous, and only plainly visible when the light in the microscope is partly shut off (Figs. 9 and 9 a). The above facts can be clearly distinguished in preparations *in toto,* which have been fixed with formol and stained with Giemsa (Fig. 12). The transition from the polygonal to the fusiform shape of the cells is a gradual process. First of all the ectoplasm projects from the edge of the membrane into the medium in the form of one or more exceedingly delicate but broad processes. This results at first in the cell's assuming a flask-shaped form (Fig. 9 a Ia), and moving out into the medium. Many of the cells, however, do not become immediately disconnected upon starting this movement, but remain attached on one side to several of the back-lying cells of the membrane, which often results in the posterior end of the projecting cell becoming long drawn out (Fig. 12 a, Fig. 9 b Ia). As soon as this connection is severed the posterior side of the cell usually contracts immediately, sometimes also the anterior end (Fig. 9 a III, 9 b IIa), after which the ectoplasm gradually disperses into the medium again from the anterior end of the cell, the posterior end again becoming drawn out into a long projection (Fig. 9 a IV and Fig. 9 b III a). By means of these processes uni-, bi-, and tripolar fusiform cells are formed. This migration of the peripheral cells causes the cells situated nearer the center of the membrane to be brought into contact with the medium, whereupon they undergo the same changes as the cells which first migrated. From the circumstance that the migrating cells fail to become dissociated immediately from the remaining cells of the membrane the latter similarly become drawn out, and thus chains or triangular processes are formed composed of elongated cells.

If we base our conclusions on the fact that the dispersion of the peripheral cells. which initiates the change in form, as well as the movements of the polyhedral cells, occurs only in comparatively soft, but never in firm media, the fact is thus primarily established that the consistency of the medium is highly instrumental in causing these phenomena. Thus, there must exist a definite relation between the consistency of the medium and the consistency of the cytoplasm in order that the movement of the cells may be initiated. Additional light is thrown on this problem by a consideration of the observations which led Burrows to the formulation of his well thought out theory of cellular movement; for our own experiments fail to account for the dissociation from the tissue of the first migrating cells and the subsequent temporary contraction of their posterior end.

Burrows has proved that the mechanical laws underlying the movements of the cell are identical with Jacques Loeb's tropisms. According to Burrows, cells which are surrounded by a medium containing a uniform distribution of metabolic products are invariably spherical in shape in consequence of uniform contraction of the cytoplasm on all sides of the periphery of the cell. A culture, however, which has been freshly supplied with medium, does not possess this uniform distribution of metabolic products, for these are found in greatest concentration in the proximity of the original fragment, and in least concentration near the edge of the drop of medium. Thus, only the posterior end of the cell, *viz.*, that adjoining the old fragment of tissue, will become contracted. By means of repeated contraction of the posterior end and flowing out of the anterior end the cell is induced to move away from the old fragment.

This brilliant theory which is based on a series of painstaking observations lacks, however, one point; it fails to explain the extension of the anterior end of the cell. Our observations, therefore, here serve as a supplement to those of Burrows and are as follows: The cell plasma in flowing out simply obeys the law of gravity, and this movement is initiated at the moment when the relation between the consistency of the medium and the consistency of the cell plasma has attained a certain value, not yet ascertained, brought about by the reduction of the firmness of the medium.

This explanation likewise accounts for the shape of the fusiform cells; that is, of the unipolar ones. But even the bipolar form assumed by the migrating cells might be explained thus. We have seen that the peripheral cells, the free ends of which flow out into the medium, remain connected with the back-lying cells for some time longer, causing a passive elongation of the posterior end to take place. Later this connection is severed and the posterior end of the cell again becomes first contracted and then elongated. It may safely be assumed that the part that was in the first instance played by one of the cells of the inner rows is now assumed by a fibrin-thread or by other fragments of tissue lying in the medium, to which the posterior end of the cell adheres. Thus, contraction can only take place at the moment that the tension, which is threatening the rupture of the progressively thinner posterior end, has attained a certain maximum, and is added to the contracting force resulting from the products of metabolism. The cell will, therefore, be compelled to continue its advance, assuming at times a bipolar and at times a unipolar form.

From the above it is primarily evident that an isolated cell must be invariably accompanied by the occurrence of the fusiform shape, unless the conditions become changed in such a way as to cause the cell to become round instead (page 89). As the isolation of cells is possible only in relatively soft media (in accordance with the law of gravitation), we are unable to find examples of fusiform cells occurring in firm media. It is easy to account for the polyhedral form of the cells in a firm medium which are united to form a membrane and which, as in the case of the flattened form assumed by the cells attached to a firm base, must be a result of surface tension. If in addition the cells are made to adhere to a firm base they become flat instead of polygonal in shape. For this reason the cells situated over the connective tissue of the explanted fragment are always flat, whereas the cells which are distributed on all sides in the form of a compact membrane and are uniformly surrounded by the firm medium, take on a polygonal shape, even if they were originally flat, as happens in the case of the cells of the middle epithelial layers. The cell retains this polyhedral shape until the firmness of the medium is reduced to the point at which it is possible for the border cells to flow out.

Thus we see that the cells' adoption of the polyhedral and fusiform shapes is not an active process, and not confined to a definite class of cells; for the epithelial cells can also become passively fusiform, in which condition it is hard to distinguish them from connective tissue cells. There is no doubt that the thread-like columnar cells are likewise passively transformed epithelial cells; this change can at any time be observed in living cultures. They originate through tension from the polyhedral cells and are always seen to occur at places where the membrane has become torn and holes are formed in the tissue membrane. This, however, only occurs in semifirm and soft media. In both media the peripheral cells wander more rapidly than the other cells, because a greater portion of their surface is in contact with the medium. By this means continuous tension is exerted on the central parts of the membrane, and the outer parts are thus gradually disconnected from the inner. The cells which are. slow in giving up their connection become exceedingly drawn out before breaking off, as occurs in the case of the posterior end of the fusiform cells, and they thus bridge over the holes which have been formed. To this must be added in soft media the formation of vacuoles, by which the same result is effected.

We must now endeavor to account for the form of the round cells, of which mention has repeatedly been made. Round cells are found in all the media employed. The liquid medium contains them to the exclusion of all other forms (Fig. 17) whereas they are but rarely found in a firm medium and only under special conditions. In media which have liquefied but a few hours after the preparation of the culture, membrane formation never occurs, but all the cells rapidly become converted to the round type, and the entire rim of cells, even if when originally cultivated in a firm medium it consisted of a membrane of tissue, is then seen to be composed of loose round cells, irrespective of the frequency of the change of medium. In many cases the fusiform cells, after wandering about for a long time in an isolated state, become surrounded with a zone of liquid and then likewise become round in shape.

The round cells (Fig. 23) are distinguished from the other varieties by the fact that their cytoplasm almost always assumes a purple color in cultures stained with Giemsa. which suggests a difference in the chemical constitution of the round cell. Many of them possess a normal nucleus, some a pycnotic one (Fig. 23 a), and others no nucleus at all. Even cultures which have undergone complete loosening up and are surrounded by a fluid medium are still seen to contain a considerable number of normal round cells. Thus. the round

cell seen in Fig. 23 c originated in a culture (B 81) which had undergone complete loosening up (Fig. 17).

It is seen that the round cells are yet alive from the circumstance
that many of them resume the fusiform shape as soon as they are
placed in an appropriate medium. This was seen to occur in living
cultures observed under the above conditions. Moreover, it is also
possible in living cultures to witness the conversion of fusiform into
round cells. Fig. 24 shows a cell of a twenty-two day old culture in
the process of transformation from the fusiform to the round shape;
it is moving around in the medium. Here a dark, yellowish endoplasm, filled with vacuoles and granulations, can be distinguished
from a light, homogeneous ectoplasm. The latter is in constant wave-
like motion. The changes in shape are continuous and very rapid in
comparison with those of the fusiform cells. This particular cell
wandered about in every direction, alternately approaching and leaving the fragment, but never continuing for long in the same direction.

Burrows found that the cells assumed a round form when they
were placed into a medium containing evenly distributed products
of metabolism. Nevertheless, the facts appear to us to argue that
in addition to the chemical constitution of the medium account should
also be taken of the consistency of the latter, in view of the fact that
in liquid media round cells only were seen, and that the spindle-shaped
cells in the non-liquid media always become spherical when surrounded by a small quantity of liquid. It may be suggested that the
relation between the chemical constitution of the medium and that
of the plasma on the one hand, and the relation between the consistency of the medium and that of the plasma on the other
hand are also instrumental in determining the form of the
cell. The point whether the round cells are normal or degenerate
must depend on the length of time that they have ceased
movement. Round cells which have been stationary for a long
period of time doubtless degenerate as a result of the progressively deteriorating quality of the surrounding medium consequent
upon metabolism. These are the round cells in which the nucleus is
either pycnotic or entirely absent. At all events, it would be erroneous to designate the round cell *per se* as a degenerate form.

In this section we have seen that the shape of the cell is not an active manifestation on the part of the cell, nor even connected with the cellular functions or with the origin of the cell (connective tissue or epithelium). The form of the cell is in no way connected with the morphologic character possessed by it while in the organism, which it introduces into the culture; on the contrary, the flatly polygonal epithelial cell can be made to assume a polyhedral, fusiform, or round shape, if the relation between the chemical constitution and the consistency of the medium on the one hand, and of the cell plasma on the other, be varied accordingly. Certain forms, such as the thread-like form of the columnar cells and of the posterior end of the bipolar spindle-shaped cells, are produced directly through tension.

We wish to mention that the method used by us possesses many disadvantages, which, however, we intend to rectify in our later experiments. It is above all necessary to utilize a medium which is either less exposed to a spontaneous variation of its consistency, or one at least in which this factor can be strictly controlled. In our subsequent investigations we shall endeavor to utilize media. the consistency and firmness of which may be accurately expressed in mathematical terms. Owing to the fact that our investigations bearing on this point are not yet sufficiently advanced, they can not here be taken into consideration.

Form and Function of the Cells.

If in our work with cultures we are able to observe how through external agents a change of form is imposed upon the epithelial cells, rendering them very similar to the connective tissue cells. it is but reasonable that these observations should be borne in mind when we return to an examination of the cells in the organism itself.

Hitherto the relations existing in the organism have not been investigated from this point of view; on the contrary, the study of the shape of cells and organs has been conducted from a standpoint according to which they are considered as an expression of cellular activity and as a consequence of cellular function. Moreover. the study of these relations in the organism itself, with exact methods subject to the control of the observer. is a problem much more difficult to compass.

Nevertheless, we believe that by observations allied to those above described we shall be led to a clearer understanding of the morphological character of the cell than is possible by the adoption of other methods which proceed from the theory of functional adaptation.

The phenomena observed in the cultures likewise obtain in the epithelia in the organism. For example, frog skin epithelium consists of a basal layer of cuboidal cells and of several layers of flatly polygonal cells. The situation of the former is similar to that of our polyhedral membrane cells; they are enclosed between two comparatively firm media, the connective tissue and the more or less tense flat epithelial cells. It is, therefore, not surprising that they possess a similar form. It may be that the remaining flat cells of the epithelium owe their form to the tension there prevailing, a condition which can not be similarly produced in the cultures by artificial means. The flat cells of the cultures always originate at places where the cells lie between two media of different firmness; in such cases they appear to be pressed flat to the firmer of the two media by means of adhesive forces. Flat cells of this type either adhere to the coverglass (Harrison, Lambert, etc.), or, as observed in our own cultures, to the connective tissue. All the cells which remain attached to the connective tissue become flatly polygonal in shape. Nevertheless, in the organism cases are also seen to occur in which the epithelium is not under tension on all sides, but where the cells stray out freely while adhering to the connective tissue-like base. In such cases all the epithelial cells become flat in form, including those which were previously cuboidal in shape. This condition is always found in regenerating epithelial wounds (5). Here many epithelial cells also assume a spindle form, as they wander out from the edges of the wound into the wound scab, where they find similar conditions to those prevailing in the plasma medium. An abundant exudate in the wound, containing serum and fibrin, may furnish a medium more or less soft and conducing to this change. Leo Loeb (6) observed such spindle-shaped epithelial cells in the skin in process of regeneration in rabbits and guinea pigs.

The significance of these observations is apparent when they are compared with Champy's interpretation of his important observations made in connection with tissue cultures. Champy (2) states that the cells of some organs when cultivated in blood plasma undergo certain changes which he calls "dedifferentiation." In cultures of the kidney, submaxillary, parotis, and thyroid from the fetal rabbit the cells of the glandular tubes soon begin to multiply actively through mitotic division. The minute canals sprouting from the cut tubules at first retain their epithelial character, but they immediately forfeit their character of gland cells. Finally, the basal membrane of these canals bursts and the epithelial cells mingle with the connective tissue cells of the interstitial tissue. At this stage they also lose their epithelial character and in conjunction with the connective tissue cells form a new tissue which is completely indefinite in character. Moreover, the interesting point in this connection is the fact that the indefinite tissue produced by all four organs is identical.

Such a process of "dedifferentiation," however, occurs only in organisms the cells of which undergo mitotic division. Therefore, it fails to occur in the organs of adult animals, which remain inactive outside of the organism. In the case of explants of cartilage, however, although mitotic division occurs, dedifferentiation fails to take place. The cells simply revert to a primitive condition, while at the same time they remain differentiated as cartilage cells.

Thus we find, as Champy himself realized, that for the "preservation of functional structure" function itself is by no means indispensable. For there are many organs the cells of which. failing to undergo mitotic division, preserve their so called "functional differentiation" when explanted. In this category we must include above all certain adult organs in which loss of functional structure would be expected to take place much earlier than in embryonal organs; that is, if the declarations of those who support the theory of functional adaptation be correct.[2] The cartilage. moreover, in spite of mitotic division, remains permanently cartilage. although of a primitive form, and in this case no atrophy from inactivity has occurred, as would have been a necessary presupposition in a case of functional adaptation. The case under consideration is a parallel of that of the transplanted salamander eyes. In these cases it was shown that although the animals were kept in the dark and had had their optic nerves cut. thus depriving the eyes of function and functional stimulus, these organs nevertheless failed to lose their functional structure (7). Thus far do the theories of Champy coincide with the established facts.

But he next concludes that the functional structure fails to arise in the newly formed cells owing to the culture's lack of functional stimulus. He bases his

[2] Roux' "functional period." which is supposed only to commence at an advanced, or at least, postembryonal stage.

theory on the following consideration. Before undergoing mitosis a cell first discards its differentiation. As the newly formed cells in the body are immediately subjected afresh to the influence of functional stimulus, they at once resume their functional structure: but in cultures which are devoid of all functional stimuli the reestablishment of the functional structure remains absent. Champy sets up this hypothesis immediately after observing that cartilage fails to dedifferentiate, although when cultivated its cells undergo mitotic division.

As a matter of fact the problem at issue in Champy's experiments must be looked for elsewhere. If every cell when about to undergo division renounces its functional structure and becomes embryonal, as has often been observed to be the case and as Champy himself admits, it is evident that the cells of the culture, even when exposed to functional stimuli, can never take on their original functional structure, as they continually proceed from division to division. It is therefore incorrect to assert that the absence of functional structure is to be ascribed to absence of function. The problematic point is rather the fact that the cells in the cultures never have a chance of developing. Were it possible for them to do so they would simply develop into their normal, inherited structures, among which are included the majority of characters which, according to Champy, are functional structures. That part of the structure, however, which owes nothing to the factor of inheritance is seen to arise passively in the cell, even in the absence of functional stimulus, as soon as any forces are brought to bear upon it (8).

We should like to recall the interesting experiments of Burrows in connection with heart muscle cells, as bearing on the question of the relation between development and differentiation in the cultivated cells. These heart muscle cells can be artificially reduced to a condition at which cell division takes place, at which stage they also undergo "dedifferentiation." If, however, the cell division is intercepted the cell is able to develop, and upon reaching the end of its development it begins to pulsate normally, independently of the operation of any specific functional stimuli.

It is, however, incomprehensible how, for instance, "lack of function" in the culture can be designated as the cause of the difference existing between the cultivated and the fetal parotic cells, in view of the fact that this lack of function exists in the body itself. For at the embryonic stage the parotid of the rabbit has not yet begun to functionate.

Finally, we must again revert to Champy's observations on the explanted cartilage.

Thoma (9) has repeatedly attempted to replace Roux' doctrine of functional adaptation by a new theory, which he designates as histomechanism and histochemistry. According to Thoma's histomechanical laws, cartilage can be formed from connective tissue only when the forces of pressure and tension have reached a certain maximum; as soon as they fall below this degree the cartilage is unable to subsist and reverts to connective tissue.

As has been recognized by Roux, this theory is fundamentally identical with the theory of functional adaptation; the difference is only one of name, as was shown in the case of the cartilage cultures. For it was seen that although the conditions of pressure prevailing in the plasma cultures differ essentially from those found in the body, the cartilage nevertheless continues to exist as such. Its cells by no means revert to the connective tissue type of cell, but they merely assume a more primitive character as a consequence of the mitotic divisions to which they are subjected. Here, therefore, the theory of functional stimulus is again at fault.

In conclusion we should like to call attention to another hypothesis of recent origin, by which it is sought to explain the variations in the morphological characters of the cultivated cells. We refer to Champy's so called theory of "inhibition" (10), a hypothesis no less obscure than that of functional adaptation.

Champy noticed that a tissue fails to dedifferentiate if it is explanted together with another "antagonistic" tissue, although when cultivated alone it rapidly undergoes dedifferentiation. When connective tissue and epithelium are cultivated together, the latter not only fails to dedifferentiate, but as in the cases of cicatrization of wounds, it grows around the connective tissue and remains a typical epithelium

This observation of Champy's is accurate, and we have already reported similar findings in a previous communication (1). By the above described process we even succeeded in obtaining minute epithelial organisms, filled with connective tissue. But upon making an extensive examination of 600 cases we discovered that this result was but rarely obtained, and only in cases where the connective tissue was considerably firmer than the surrounding medium. In such cases the cells are pressed on to the connective tissue by means of simple adhesion, just as under certain conditions the cells in a firm medium may be pressed on to the still firmer cover-glass. The fact that this results in their assuming a flat polygonal shape can certainly not be attributed to any inhibiting influences, but simply to the force by which they are being pressed against the base. Moreover, the statement that the epithelial cells always dedifferentiate as soon as they wander out into the medium is incorrect, for in firm media in which the epithelial cells migrate into the medium, they remain united in the form of an epithelial membrane and also preserve the typical form of the epithelial cell. The fact that the cells which in the body were flat in shape under these circumstances also assume a polyhedral form is to be ascribed to purely physical influences. Champy himself has shown that the cells remain united in the form of a membrane; but he believes that these cells are not epithelial, because the mitotic spindles are situated parallel instead of vertical to the plane of extension of the membrane. This latter fact we have observed to be correct (Fig. 5); nevertheless, it would be a mistake to assert that such a phenomenon never occurs in true epithelia, nor in the organism itself. Fig. 25 shows the regenerating epithelium in the tail of a tadpole (*Rana pipiens*). Such a regenerating epithelium shows many examples of mitosis in which the spindle is parallel to the surface of the cutis; we have one such example in our figure. No one would venture to state that regenerating epithelium is not true epithelium, and for that reason we must consider Champy's judgment to be inaccurate. The epithelium loses its inherent characteristics only when the cells are placed in a soft or semi-firm medium (according to our relative nomenclature), where the cytoplasm begins to disperse and the cells, now fusiform in shape, begin to separate, owing to the movements which are being initiated. Then only does it become hard to distinguish them from the ordinary connective tissue type of the cultures. Thus, the above processes are by no means caused by the suspension of some inhibiting influence, but simply through the physical changes in and around the cell.

The technique of tissue cultivation permits us to study the elementary parts of organs (the cells) with great exactitude. The more we are enabled, by means of our modern methods of technique, to pursue this detailed study, the more are we able to dispense with the theories of "inhibiting influences" and "functional stimuli." By means of the above experiments in which we have shown that the form of the cell is dependent on the relation existing between the consistency of the medium and that of the cell plasma, we believe to have added some small contribution toward the attainment of this end. Changes in the consistency of these two substances, such as can be introduced into the culture by means of chemical procedures, result in a variation in the shape of the cell. Two additional factors of importance are, moreover, surface tension and adhesion, as was earlier shown by other investigators. But the effect produced by these latter must similarly depend upon the relative consistency of the medium and of the cell plasma. The next point of importance will be to establish the mathematical values of this relation, as applied to the various forms of cell. It is our intention to consider this aspect of the problem in subsequent investigations.

SUMMARY.

1. Fragments of skin from the leopard frog (*Rana pipiens*) were cultivated in media of varying consistency. A mixture of frog plasma, frog muscle extract, chicken plasma, and chicken embryo extract usually produced a very firm medium; a mixture of frog plasma, frog muscle extract, and chicken plasma, one less firm (semi-firm); and a mixture of frog plasma and frog muscle extract a medium of a consistency varying from soft to liquid.

2. (*a*) In a firm medium the cells which migrate into the medium are polyhedral (polygonal when seen from above) in form, which shape they retain permanently. They remain united in a compact membrane, the central parts of which consist of several (three to four) layers of cells. Migration of isolated cells into the medium does not take place.

(*b*) In the semi-firm media the cells situated at the edge of the membrane become fusiform in shape, gradually detach themselves from the membrane, and stray out individually into the medium This causes the membrane to become loose in character, and to contain holes, while its edges at the same time become very irregular and send out pointed projections.

(*c*) In a soft medium the cells are fusiform or thread-like in shape. The migration of isolated cells is much more pronounced than in the semi-firm media, as a result of which the membrane undergoes constant and rapid loosening up. By this means whole portions of the membrane become detached and their separate parts are at first united by the thread-like columnar cells, which become drawn out in the form of long threads upon the separation of the individual sections of the membrane. The loosening up of the membrane is further assisted by liquefaction and the consequent formation of vacuoles; the latter process likewise results in the formation of thread-like columnar cells.

(*d*) Liquid media contain only round cells.

3. This serves to explain numerous internal processes of the organism, especially certain changes of form observed by Leo Loeb in transplantations of wound scabs and of skin; the conditions artificially produced by Leo Loeb must have effected a change in the consistency of the medium.

4. It has been shown that it is unnecessary, for a satisfactory explanation of the above findings, to have recourse to the theories of "functional stimulus" or "inhibiting influences," by means of which Champy wished to account for the variations in the morphological character of the cells.

I wish here to acknowledge my indebtedness to Mr. Louis Schmidt, for the care with which he has made the photographic reproductions of my cultures. Mr. Schmidt has succeeded in working out a new method for counteracting the destructive effects of the condensation of water which is apt to occur in the cultures when exposed to the rays of the arc light after having been previously kept at a low temperature ($21°$ to $23°$ C.).

BIBLIOGRAPHY.

1. Uhlenhuth, E., Cultivation of the Skin Epithelium of the Adult Frog, *Rana pipiens, Jour. Exper. Med.*, 1914, xx, 614.
2. Champy, C., Dédifférentiation des tissus cultivés en debors de l'organisme, *Bibliog. anat.*, 1913, xxiii, 184.
3. Burrows, M. T., The Tissue Culture as a Physiological Method, *Tr. Cong. Am. Phys. and Surg.*, 1913, ix, 77.

4. Burrows, Grafting of Normal Tissues as Dependent on Zoological or Individual Affinity: Autoplastic, Isoplastic, Heteroplastic. Tissue Culture *in Vitro, Transactions of the XVIIth International Congress of Medicine, Section III,* London, 1913, 217.
5. See, for example, Oppel, A., Demonstration der Epithelbewegung im Explantat von Froschlarven, *Anat. Anz.*, 1913–14, xlv, 173.
6. Loeb, L., Über die Entstehung von Bindegewebe, Leucocyten und roten Blutkörperchen aus Epithel, und über eine Methode, isolierte Gewebsteile zu züchten, Chicago, 1897.
7. Uhlenhuth, Die Transplantation des Amphibienauges, *Arch. f. Entwcklngsmechn. d. Organ.*, 1912, xxxiii, 723; Are Function and Functional Stimulus Factors in Preserving Morphological Structures?, *Biol. Bull.*, 1915, xxix (in press).
8. Apropos of the non-specificity of what has often been designated "functional stimulus," see Loeb, J., Aphorismen zur Vererbungslehre, *Monist. Jahrh.*, 1912, No. i, 6. Uhlenhuth, *loc. cit.* With reference to the value of the term "stimulus," see Loeb, J., Die chemische Entwicklungserregung des tierischen Eies, Berlin, 1909, 10.[1]
9. Thoma, R., Synostosis suturae sagittalis cranii. Ein Beitrag zur Histomechanik des Skeletts und zur Lehre von dem interstitiellen Knochenwachstum, *Virchows Arch. f. path. Anat.*, 1907, clxxxviii, 248; Über die Histomechanik des Gefässkanals und die Pathogenese der Angiosklerose, *ibid.*, 1911, cciv, 1; Anpassungslehre, Histomechanik und Histochemie. Mit Bemerkungen über die Entwicklung und Formgestaltung der Gelenke. *ibid.*, 1912, ccvii, 257; Anpassungslehre, Histomechanik und Histochemie. *ibid.*, 1912, ccx, 1.
10. Champy, La présence d'un tissu antagoniste maintient la différenciation d'un tissu cultivé en dehors de l'organisme, *Compt. rend. Soc. de biol.*, 1914, lxxvi, 31.

EXPLANATION OF PLATES.

PLATE 6.

Fig. 1. Frog skin, cultivated in a firm medium (F.Pl. + F.E. + Ch.Pl. + Ch.E.); about twenty hours after explantation. A dense membrane of polygonal epithelial cells surrounds the old fragment. A small vacuole still exists on the right side only, which has caused some cells to become elongated. Culture B 174; Giemsa stain; magnification X 30.

Fig. 2. Frog skin, cultivated in a semi-firm medium (F.Pl. + F.E. + Ch.Pl); three days after explantation. Part of the cell membrane, fairly remote from the border of the membrane, where the latter still consists of two to three layers of cells. As shown by this cross-section, they are cuboidal in shape; as shown in Fig. 8, seen from above, they are polygonal. Therefore their actual shape is polyhedral. Camera lucida drawing; Culture 7. Cross-sections stained with Giemsa solution after fixation in potassium bichromate, 70 per cent; glacial acetic acid, 20 per cent; formalin, 10 per cent; magnification, Leitz oc. 3; obj. 1 12, oil immersion; tube 170.

Fig. 3. Frog skin, cultivated in a firm medium (F.Pl. + F.E. + Ch.Pl +

Ch.E.) ; about twenty hours after explantation. A piece of the border of the cell membrane, seen from above, where the cells have a polygonal shape. The spaces between the cells are artificially produced. Camera lucida drawing. Culture B 169. *In toto* preparation. Stained with Giemsa solution after fixation in formalin 2 per cent; magnification. Leitz oc. 3; obj. 6; tube 160.

PLATE 7.

FIG. 4. Frog skin, cultivated in a firm medium (F.Pl. + F.E. + Ch.Pl. + Ch.E.) ; six days after explantation. A dense membrane of polygonal cells (epithelial) surrounds the original fragment. No cells migrate out of the membrane into the medium. Culture B 114; Giemsa stain; magnification × 30.

FIG. 5. Border of the cell membrane of the same culture as seen in Fig. 4. highly magnified. The shape of the cells is shown in a firm medium; they are polygonal, even those situated at the edge of the membrane. Two of the cells are undergoing mitotic division. Culture B 114; Giemsa stain; magnification × 1,000.

PLATE 8.

FIG. 6. Frog skin, cultivated in a firm medium (F.Pl. + F.E. + Ch.Pl. + Ch.E.) ; ten days after explantation. A dense membrane of polygonal epithelium-like cells surrounds the original fragment. The border of the membrane is uniform, and even the border cells are polygonal in shape. No cells are migrating from the membrane into the medium. Culture B 103; photographed while alive; magnification × 30.

FIG. 7. Frog skin, cultivated in a semi-firm medium (F.Pl. + F.E. + Ch.Pl.) ; about twenty hours after explantation. The cell membrane is dense only in the center; at the edge of the fragment it is loose, and here large pointed tongues of tissue project out into the medium. At the tips many spindle-shaped cells are in the act of becoming detached from the membrane and migrating into the plasma. Most of the cells situated at the edge have changed from being polygonal to fusiform. Therefore, the edges of the membrane are serrated, the medium being filled with free spindle-shaped cells. Culture B 86; Giemsa stain; magnification × 30.

PLATE 9.

FIG. 8. Frog skin, cultivated in a semi-firm medium (F.Pl. + F.E. + Ch.Pl.) : three days after explantation. The same characteristics are seen as in Culture B 86 (Fig. 7), but more pronounced. There are many holes in the membrane at the back of the projections, produced by the forward motion of the cells into the medium, and they are not filled with fluid. Culture B 7; photographed while alive; magnification × 60.

FIG. 9. The polygonal cells at the edge of the cell membrane have assumed a spindle shape and become isolated from the membrane when cultivated in a soft medium. Drawing, made with a magnification. Leitz oc. 3; obj. 6; but without camera lucida.

FIG. 9 a. Culture A 406 c. cultivated in F.Pl. + F.E., six days after explantation. I. The homogeneous ectoplasm flows out into the medium. II. Cells a and b assume a flask-like shape. III. Cell a has become isolated and contracted. IV. Cell a resumes the fusiform shape.

Fig. 9 b. Culture B 9, cultivated in F.Pl. + F.E. + Ch.Pl., four days after explantation. I. Cell a moves out into the medium, but remains connected by a thin plasma bridge with the other cells. II. Cell a assumes a round shape. III. Cell a resumes the spindle shape.

PLATE 10.

Fig. 10. Frog skin, cultivated in a semi-firm medium (F.Pl. + F.E. + Ch.Pl.) ; twenty-one days after explantation. Only the most central zone of the membrane is compact, the borders are drawn out into long cloven tissue tongues, giving the culture a star shape. The migration of isolated spindle cells is considerable, and more extensive than in Figs. 7 and 8. Culture B 78; photographed while alive; magnification × 30.

PLATE 11.

Fig. 11. The same culture as in Fig. 10, but twenty-two days after explantation. The cell membrane has further developed in size; there are long tissue projections and large holes in the cell membrane, some of which have been filled up with a softer medium. The hole in the upper part of the figure especially is bridged over by thread-like columnar cells. There are large numbers of isolated, migrating cells around the cell membrane, which spread out far into the medium. Culture B 78; Giemsa stain; magnification × 30.

PLATE 12.

Fig. 12. Spindle-shaped cells of epithelial origin in a culture of frog skin, cultivated in a semi-firm medium (F.Pl. + F.E. + Ch.Pl.) for twenty-two days (same culture as in Fig. 11) ; the cells are migrating from the border of the tissue projections into the medium.

Fig. 12 a. Tip of a tongue of cells, projecting from the membrane into the medium, where the cytoplasm of the originally polygonal cells is seen flowing out into the medium, by means of which process the cells are assuming a spindle-like shape. The nuclei are also stretched out and elongated, and have a clear nuclear plasma and in addition one or two nucleoli, but no chromatin net, as is always the case in the polygonal cells, when fixed in formalin : the cytoplasm is more or less foamy in structure.

PLATE 13.

Fig. 12 b. Two spindle cells shortly after isolation from the membrane, still connected with each other. Both possess some vacuoles in the cytoplasm, one of which is distinctly differentiated into a clear ectoplasm and a foamy endoplasm.

Fig. 12 c. Isolated and moving spindle cell. The anterior end, directed to the external edge of the drop of medium, is stretched out into a long, thin process, at the end of which a slight swelling has developed.

PLATE 14.

FIG. 13. Frog skin, cultivated in a soft medium (F.Pl. + F.E.) ; about twenty hours after explantation.   Nearly the whole membrane is migrating away from the old fragment into the medium, in which we see little pieces of tissue moving freely. • Holes of various sizes have developed in consequence of the migration of the border cells, the external parts of the membrane being connected with the internal ones by means of the thread-like columnar cells, which are long drawn out. The holes are not filled with a medium of different consistency from that prevailing outside, both being soft but not liquid. Nearly all the cells of the membrane, including the internal ones, are elongated ; the cells at the edge are spindle-like.   Culture B.119 ; photographed while alive ; magnification × 30.

PLATE 15.

FIG. 14. Frog skin, cultivated in a soft medium (F.Pl. + F.E.) ; six days after explantation. Similar to Fig. 13. Isolated spindle-shaped cells and thread-like columnar cells can plainly be seen. Culture B 121 ; stained with Giemsa solution after fixation in formalin 2 per cent ; magnification × 30.

PLATE 16.

FIG. 15. Frog skin, cultivated in a medium which is partly firm, and partly in the act of liquefying (F.Pl. + F.E.) ; four days after explantation. Large vacuoles are causing the cells to become stretched out, thus producing thread-like columnar cells. In the firm part of the medium a compact cell membrane is seen. The upper layers are separated from the fragment and form a thin membrane, which covers a part of the old fragment and part of the compact cell membrane. No change has been going on in these upper layers, the cells of which are neither growing nor moving. Culture B 117 ; photographed while alive : magnification × 30.

FIG. 16. Frog skin, cultivated in a soft medium (F.Pl. + F.E.) : eighteen days after explantation. Only a very small, thin membrane remains around the old fragment. This membrane is extended into several tissue projections, from which the spindle-shaped cells are becoming separated and are migrating into the medium. The medium contains numbers of isolated cells, which are spindle-like or round. Culture B 77 ; photographed while alive ; magnification × 30.

PLATE 17.

FIG. 17. Frog skin, cultivated in a soft medium (F.Pl. + F.E.), which had almost completely liquefied when the photograph was taken twenty-one days after explantation. The old fragment has become a globular ball. The cell membrane has entirely loosened up ; in its place there are only isolated cells, most of which are round, but some are spindle-shaped. Culture B 81 ; photographed while alive ; magnification × 30.

PLATE 18.

FIG. 18. A small fragment of a cell membrane, grown out of a piece of frog skin in a firm medium, isolated five days after explantation and kept for two days in a firm medium (F.Pl. + F.E. + Ch.Pl. + Ch.E.), at which time the picture was taken. A dense, compact membrane has grown out of the original fragment of membrane, consisting of polygonal cells. No migration of isolated cells into the medium is taking place. Culture B 100 a; photographed while alive; magnification $\times$ 30.

Fig. 19. A fragment of a cell membrane, grown out of a piece of frog skin in a soft medium, isolated four days after explantation, and kept in a soft medium (F.Pl. + F.E.) for about twenty hours, at which time the picture was taken. A network of elongated, spindle-shaped, and thread-like cells has developed from it, which greatly resembles connective tissue. The thread-like columnar cells are clearly seen in this photograph. Culture B 119 a; stained with Giemsa solution after fixation in 2 per cent formalin; magnification $\times$ 30.

PLATE 19.

Fig. 20. Frog skin, cultivated in an unusually firm F.Pl. + F.E. medium. It possesses the same characteristics as a culture in a firm F.Pl. + F.E. + Ch.Pl. + Ch.E. medium (compare Figs. 1, 4, and 6). Culture A 157; about twenty hours after explantation; Giemsa stain, after fixation in 2 per cent formalin; magnification $\times$ 30.

Fig. 21. Frog skin, cultivated in an unusually firm F.Pl. + F.E. + Ch.Pl. medium; two days after explantation; the cell membrane and its cells, grown out of the old fragment, have the same characters as in a firm (F.Pl. + F.E. + Ch.Pl. + Ch.E.) medium (compare with Figs. 1, 4, and 6). Culture B 13; photographed while alive; magnification $\times$ 30.

. PLATE 20.

Fig. 22. Frog skin, which was cultivated in a firm medium (F.Pl. + F.E. + Ch.Pl. + Ch.E.); the membrane had for ten days the compact character peculiar to a firm medium. On the eleventh day the fresh medium, although of the same composition, had become softer and the membrane and the cells at once assumed the characteristics of cultures kept in semi-firm media of the composition of F.Pl. + F.E. + Ch.Pl., as is shown in this photograph (compare Figs. 7, 8, and 11). Culture B 100; eleven days after explantation; stained with Giemsa solution after fixation in 2 per cent formalin; magnification $\times$ 30.

PLATE 21.

Fig. 23. Round cells. Fig. 23 a and b, from Culture B 81, which was placed in a liquid medium (compare Fig. 17), consisting of F.Pl. and F.E.; Fig. 23 c, from a semi-firm medium, consisting of F.Pl. + F.E. + Ch.Pl. — Ch.E.; fixed in formalin 2 per cent; Giemsa stain; camera lucida drawing; magnification. Leitz oc. 3; obj. 1/12, oil immersion; tube 170.

Fig. 24. Various stages of the movement of a formerly spindle-shaped cell, which is in process of becoming round; a twenty-two day old culture of frog skin, kept in a medium which is becoming liquid (F.Pl. + F.E. + Ch.Pl.). Culture B 45; drawing made from the living specimen; magnification, Leitz oc. 3; obj. 6; tube 160; schematic.

FIG. 25.   Skin epithelium of a regenerating tail of the tadpole of *Rana pipiens*, showing a mitosis, the spindle of which is parallel with the surface of the connective tissue.   Taken one day after the tip of the tail had been cut; fixed in potassium bichromate, 70 per cent; glacial acetic acid, 20 per cent; formalin, 10 per cent; stained with iron hematoxylin.   Camera lucida drawing; half size; magnification, Leitz oc. 3; obj. 6; tube 160.

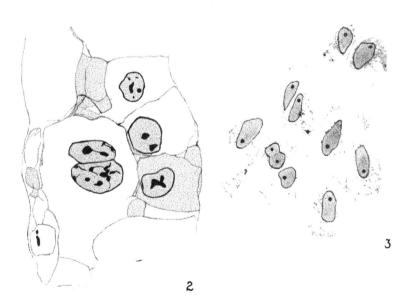

3

2

(Uhlenhuth: Epithelial Cells in Cultures of Frog Skin.)

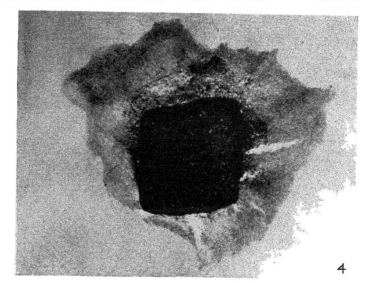

4

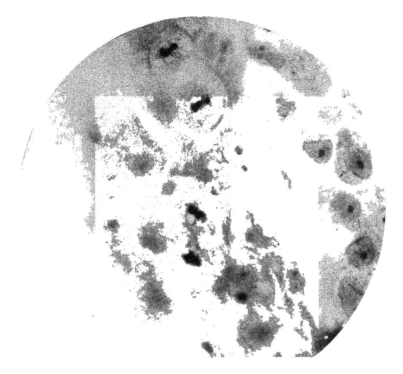

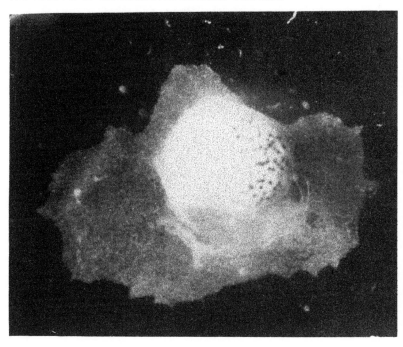

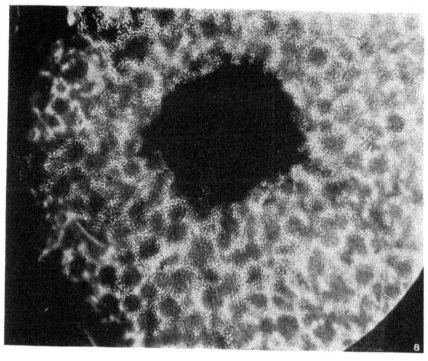

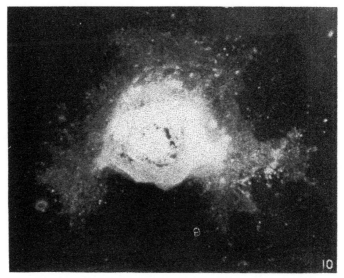

(Uhlenhuth: Epithelial Cells in Cultures of Frog Skin.)

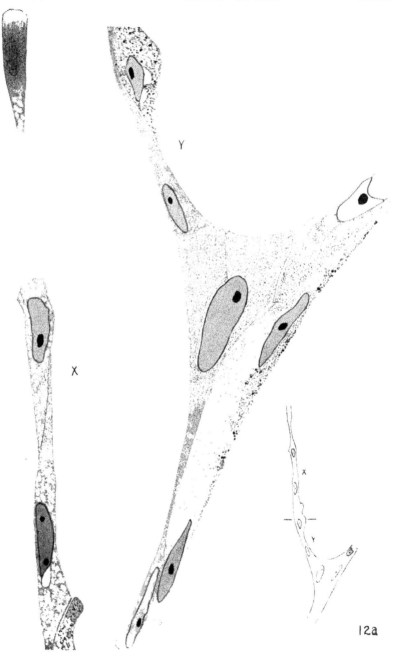

12a

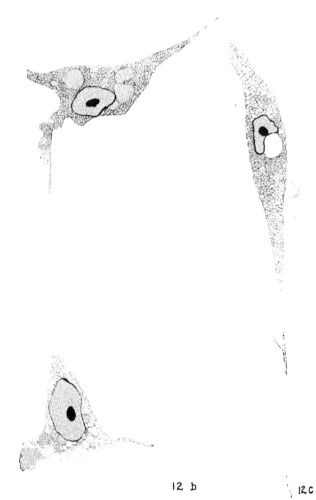

12 b   12 c

(Uhlenhuth: Epithelial Cells in Cultures of Frog Skin.)

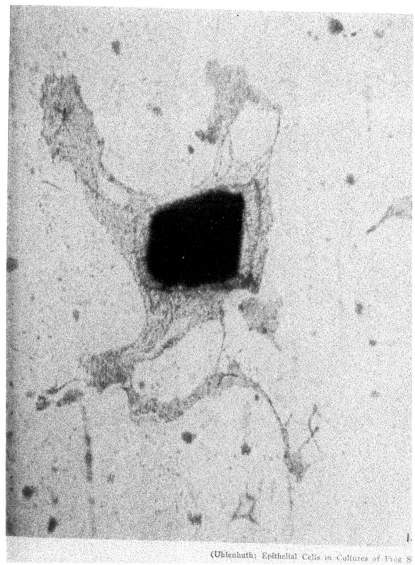

(Uhlenhuth: Epithelial Cells in Cultures of Frog S

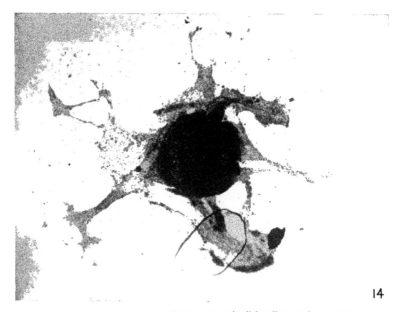

14

(Uhlenhuth: Epithelial Cells in Cultures of Frog Skin.)

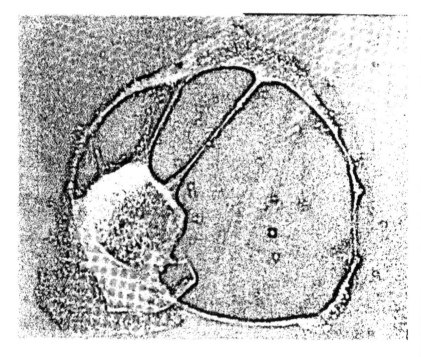

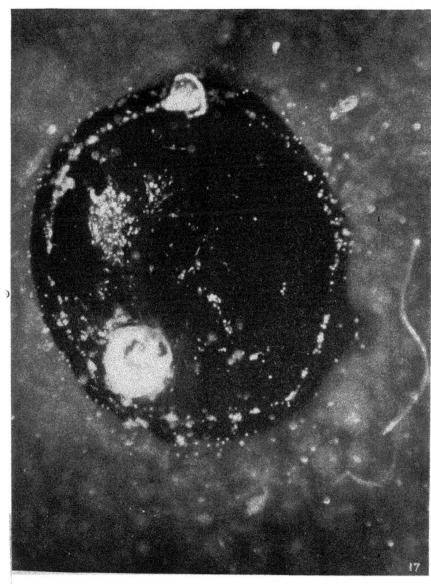

(Uhlenhuth: Epithelial Cells in Cultures of Frog Skin.)

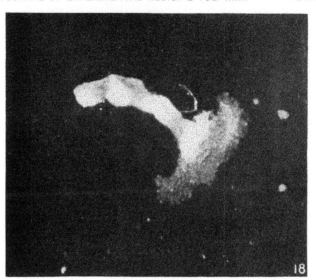

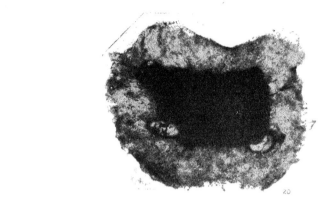

(Uhlenhuth: Epithelial Cells in Cultures of   Skin

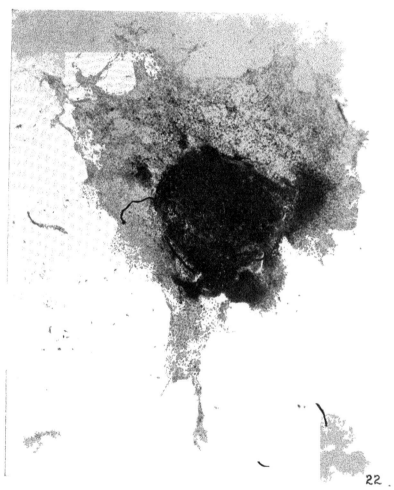

22

(Uhlenhuth: Epithelial Cells in Cultures of Frog Skin.)

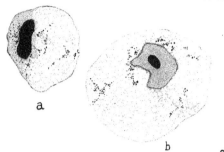

a

b   23

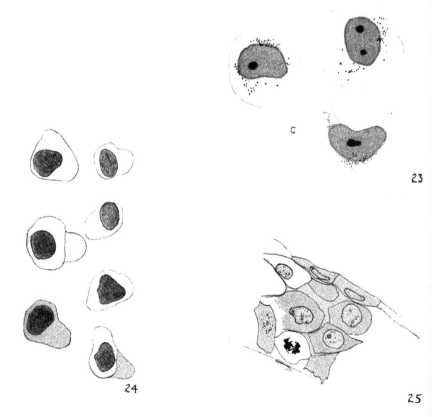

c

23

24

25

THE OCCURRENCE OF CARRIERS OF DISEASE–PRODUCING TYPES OF PNEUMOCOCCUS.

By A. R. DOCHEZ, M.D., and O. T. AVERY, M.D.

(*From the Hospital of The Rockefeller Institute for Medical Research.*)

(Received for publication, May 20, 1915.)

In previous papers[1] we have discussed the varieties of pneumococcus responsible for the production of lobar pneumonia and the differences that exist between such organisms and those found in the mouths of healthy individuals. The pneumococci obtained from persons suffering from lobar pneumonia have been divided into four groups. We have pointed out that the first three groups are comprised of disease-producing races, and are responsible for about 75 per cent of all cases of lobar pneumonia. Approximately 25 per cent of cases of pneumonia are due to the fourth group. The pneumococci of this group cannot readily be distinguished from those dwelling in the normal human mouth. Disease caused by the highly parasitic types is usually much more severe than that occasioned by the strains which are indistinguishable from the sputum pneumococci. Inasmuch as the highly virulent forms are always associated with disease, and only occur in the mouths of healthy individuals under the special conditions which are reported in this paper, the evidence is strong that in the spread of lobar pneumonia the disease in a majority of instances is transmitted from one individual to another. Infectious diseases usually spread by immediate contact, through the intermediation of a temporary host, or by the agency of a healthy carrier. The importance of these modes of transmission varies with different diseases and in many instances more than one of these mechanisms may be involved. In all likelihood the healthy carrier may be a greater menace to the health of the community than the

[1] Dochez, A. R., and Gillespie, L. J., *Jour. Am. Med. Assn.*, 1913, lxi, 727. Dochez, A. R., and Avery, O. T., *Jour. Exper. Med.*, 1915, xxi, 114.

infected individual, largely because of failure to recognize the carrier condition. In the epidemiology of certain diseases, notably typhoid and epidemic cerebrospinal meningitis, the importance of the carrier state is well known because of the definite tracing of foci of disease to such a condition. The more readily a disease is transmitted from one individual to another, the greater becomes the number of instances at any one time, and consequently the more vigorous the search for the source of infection. The sporadic occurrence of lobar pneumonia, combined with the inability to distinguish disease-producing types of pneumococcus from those habitually living in normal mouths, has probably been responsible for the failure to establish a well defined epidemiology for this disease.

It is not our purpose in this paper to discuss the importance of immediate contact as a cause of the spread of pneumonia, for, although instances have occurred in which two closely associated individuals, such as husband and wife, have been infected within a short period with the same type of pneumococcus, cases of pneumonia usually develop at such wide intervals of space and time that direct contact relationships are obscured. Inasmuch as fully 75 per cent of all cases of lobar pneumonia are caused by peculiarly distinct races of pneumococci, not occurring in the normal mouth, then, as we have previously assumed, these instances of the disease must be due to contact infection, either direct or indirect, and some mechanism must exist by means of which the etiological agent is transmitted from one individual to another. In the present study we shall show that persons closely associated with individuals suffering from pneumonia in a large percentage of instances harbor in their mouths pneumococci of the same type as those causing the disease, that such organisms are not found in the mouths of normal individuals not exposed to pneumonia, and that the considerable period of time during which these organisms are carried may in part account for the sporadic occurrence of individual cases of pneumonia. An additional means of transmission exists in the fact that the recovered case also may carry the pneumococcus responsible for his disease during a relatively long period of time.

In the present paper are presented the studies of the pneumococci obtained from the mouths of persons associated with cases of pneumonia, of the type of pneumococcus encountered in the mouths of normal unexposed individuals, and of the period of time during which convalescents harbor the disease-producing organisms. For

TABLE I.

Incidence of Carrier Condition in Healthy Individuals in Contact with Lobar Pneumonia.

Case No.	Type of infecting pneumococcus.	Relationship of associates.	Type found in associates.	Duration of period of carrying
2276	Type I	Wife	Type I	10 days +
2195	Type I	Husband	Type I	24 days.
2203	Type I	Wife	Type IV	
		Mother	Type IV	
2237	Type I	Nurse	Type I	Undetermined.
W	Type I	Wife	Type IV	
B	Type I	Wife	No pneumococcus	
		Daughter	Type IV	
2269	Type I	Wife	Type IV	
2286	Type II	Mother	Type IV	
		Father	Type IV	
2299	Type II	Sister	No pneumococcus	
2301	Type II	Wife	No pneumococcus	
2294	Type II	Wife	No pneumococcus	
2309	Type II	Mother	Type IV	
2314	Type II	Wife	Type II	Undetermined.
2335	Type II	Mother	No pneumococcus	
		Sister	No pneumococcus	
2326	Type II	Wife	No pneumococcus	
2330	Type II	Daughter	Type II	Undetermined.
2344	Type II	Sister	No pneumococcus	
1835	Type II	Wife	Type II	9 days +
2174	Type II	Son	Type IV	
2175	Type II	Mother	Type II	27 days +
		Sister	No pneumococcus	
		Nurse	Type II	Developed pneumonia, Type II.
2202	Type II	Mother	Type II	45 days.
2199	Type II	Husband	Type IV	
2298	Type II	Sister	Type IV	
2226	Type II	Wife	Type IV	
2245	Type II	Mother	Type IV	
		Brother	Type II	39 days +
2230	Type II	Wife	No pneumococcus	
2247	Type II	Daughter	Type II	Undetermined.
		Wife	No pneumococcus	
		Son	Type IV	
2266	Type II	Wife	No pneumococcus	
		Brother-in-law	No pneumococcus	
2270	Type II	Mother	Type II	7 days +
		Sister	No pneumococcus	
2265	Type II	Physician	Type II	21 days.

Summary.

TABLE I.—*Concluded.*

Type of pneumococcus.	No. of cases.	Positive contacts.	Per cent.
Type I.................	8	3	37.5
Type II.'..............	24	10	41.7
Total.................	32	13	40.6

Approximate duration of carrier state 23 days.

the sake of clearness, an explanation of our classification of pneumococci is given. The organisms have been placed into four groups, numbered from I to IV. Groups I, II, and III are found only in association with disease and are distinctly parasitic in type. Members of Groups I, II, and III are recognized by their immune reactions which are identical within the respective group. Group IV consists of a heterogeneous series of strains which are not related antigenically, and which cause a minority of cases of pneumonia, and from which the pneumococcus occurring in the normal mouth is indistinguishable. In the Tables I to IV the pneumococci studied are classified according to this numerical grouping.

TABLE II.

Type of Pneumococcus Isolated from Sputum of Normal Individuals.

Pneumococcus.		Incidence.	Per cent.
Contacts {	Type I	3	2.6
	Type II	8	7.0
	Type IV	55	48.6
	No pneumococcus	47	41.6
	Total	113	
	Pneumococcus present	66	58.4
	Pneumococcus absent	47	41.6

TABLE III.

Type of Pneumococcus Isolated from Individuals with Lobar Pneumonia.

Pneumococcus.	No. of cases.	Per cent.
Type I	78	34.97
Type II	75	33.63
Type III	22	9.86
Type IV	48	21.52
Total No. of cases	223	

TABLE IV.

Persistence of Disease-Producing Type of Pneumococcus during Convalescence.

Case No.	Type of pneumococcus during height of disease.	Type of pneumococcus after recovery.
1654	Type I	60 days, Type I.
		65 days, Type IV.
1751	Type I	59 days, streptococcus.
		73 days, streptococcus.
1775	Type I	30 days, Type IV.
1867	Type I	90 days, Type I.
1828	Type I	33 days, Type IV.
1792	Type I	30 days, Type IV.
2167	Type I	12 days, Type IV.
2168	Type I	45 days, Type I.
2191	Type I	15 days, Type IV.
2195	Type I	28 days, Type I.
		49 days, no pneumococcus.
2203	Type I	12 days, no pneumococcus.
2237	Type I	13 days, Type I.
W	Type I	15 days, Type I.
		23 days, Type IV.
B	Type I	14 days, no pneumococcus.
2250	Type I	10 days, Type I.
2269	Type I	13 days, no pneumococcus.
2284	Type I	15 days, Type I.
		29 days, no pneumococcus.
2267	Type I	15 days, no pneumococcus.
2276	Type I	15 days, Type I.
1679	Type II	30 days, streptococcus.
		48 days, Type IV.
		108 days, Type IV.
1753	Type II	60 days, Type IV.
1763	Type II	34 days, streptococcus.
		40 days, Type IV.
1761	Type II	47 days, streptococcus.
		78 days, Type IV.
1825	Type II	53 days, Type II.
1786	Type II	20 days, Type IV.
		25 days, Type IV.
1820	Type II	30 days, Type II.
1827	Type II	14 days, Type IV.
1880	Type II	63 days, Type II.
1950	Type II	21 days, Type IV.
1969	Type II	24 days, Type IV.
2174	Type II	15 days, streptococcus.
2175	Type II	17 days, no pneumococcus.
2212	Type II	12 days, Type II.
		17 days, no pneumococcus.

TABLE IV.—*Concluded.*

Case No.	Type of pnemococcus during height of disease.	Type of pneumococcus after recovery.
2202	Type II	37 days, Type II.
		43 days, Type IV.
2199	Type II	32 days, no pneumococcus.
2226	Type II	29 days, Type II.
2245	Type II	16 days, Type IV.
2266	Type II	15 days, Type II.
2270	Type II	37 days, Type II.
2286	Type II	10 days, Type II.
		25 days, no pneumococcus.
2292	Type II	29 days, Type IV.
2296	Type II	19 days, no pneumococcus.
1743	Type III	13 days, Type III.
		73 days, Type IV.
2185	Type III	16 days, Type III.
		34 days, no pneumococcus.
2249	Type III	14 days, Type III.

In the above four tables are given the main facts upon which we base our assumption that in the majority of cases lobar pneumonia is a disease the continued wide-spread incidence of which is dependent upon communication of infection from one individual to another. Table I establishes beyond doubt the existence of healthy carriers of the disease-producing types of pneumococcus. The study of the carrier state was limited to the investigation of infection with pneumococcus Types I and II because of the relative ease with which these organisms can be distinguished from other types of pneumococcus. Out of a total of thirty-two cases studied, at least one carrier of infection among the patient's associates was found in thirteen instances, 40.6 per cent. Types I and II show approximately the same percentage incidence of the carrier condition, and in every instance the pneumococcus isolated corresponds in type with that of the infected individual. The approximate duration of the carrier state has been twenty-three days, which is probably somewhat shorter than would be found had it been possible to retain under observation all carriers until the disease-producing type of pneumococcus had disappeared from the mouth flora. Study of Table I shows that positive carriers are more commonly observed among females than among males, a fact that is probably accounted for by the more frequent service of the former in a nursing capacity, thus entailing more intimate association with the sick.

Tables II, III, and IV develop somewhat further, points brought out in a previous communication. In Table II is shown the incidence of pneumococcus in the mouth flora of normal individuals and the classification of such pneumococci. All normal individuals studied are given in this table, whether in contact with cases of lobar pneumonia or not. Pneumococcus was found in 58.4 per cent of all instances and was absent in 41.6 per cent. Of the cases in which pneumococcus has been found, 48.6 per cent of the organisms have been of the sputum type and 9.6 per cent have been of the disease-producing type. All the latter have occurred in individuals intimately associated with cases of lobar pneumonia, and their presence is dependent upon this association. Determination of contact carriers of pneumococci belonging to Group III, *Pneumococcus mucosus,* presents certain difficulties. Only recently have we been able to obtain an immune serum effective against organisms of this group. Further development of the study of the mucous group is showing that certain organisms resembling in their cultural reactions *Pneumococcus mucosus,* are, in reality, mucous types of streptococcus. The latter varieties are frequently found in normal sputum, and have led to some confusion. With the working out of the serological reactions of the true *Pneumococcus mucosus,* evidence is accumulating that this organism is as strictly pathogenic in type, and has quite as specific immunological characteristics as pneumococci belonging to Groups I and II.

For comparison with the statistics given in Table II, which illustrates the prevalence of the mouth type of pneumococcus in normal individuals, Table III is added to show the percentage incidence of the fixed types of pneumococci observed in individuals suffering from lobar pneumonia. In this table the highly pathogenic types are dominant, being responsible for 77.2 per cent of all infections studied. This fact is convincing evidence that specific types of pneumococci are mainly responsible for the production of lobar pneumonia. The dominance of these organisms in disease is in striking contrast to the high percentage incidence of the sputum type of pneumococcus in the mouth flora of normal individuals.

Table IV shows the length of time during which recovered cases of pneumonia harbor the organism responsible for their disease. The period of carrying is measured from the date of onset of the pneumonia. The shortest time in which the disease-producing pneumococcus has disappeared from the mouth has been twelve days, and the longest duration of carrying has been ninety days. These results show that pathogenic types of pneumococcus persist in the mouths of individuals recovering from lobar pneumonia for a variable period of time. Because of wide intervals between observations, the average duration of this condition can only be approximately determined, and has been found to be about twenty-eight days.

The studies detailed in this paper show that there are two sources of danger in the spread of pneumococcus infection. One lies in the occurrence of healthy carriers of disease-producing pneumococci among individuals associated with cases of pneumonia, and the other in the fact that patients recovering from the disease harbor the responsible organism for a considerable length of time. The actual tracing of cases of pneumonia to examples of the carrier state is difficult, but even the small amount of effort that has so far been devoted to this side of the study has brought to light an occasioual suggestive fact.

SUMMARY.

Lobar pneumonia in 75 per cent of instances is due to specific types of pneumococci possessed of a high degree of pathogenicity. Although pneumococci occur in the mouths of 60 per cent of normal individuals, such organisms are readily distinguishable from the highly parasitic types of pneumococcus responsible for the severe forms of lobar pneumonia, a convincing proof that infection in this disease is, in the majority of instances, not autogenic in nature, but is derived from some extraneous source. In a high percentage of instances healthy persons intimately associated with cases of lobar pneumonia harbor the disease-producing types of pneumococcus. In every such instance the pneumococcus isolated has corresponded in type with that of the infected individual. Convalescents from pneumonia carry for a considerable length of time the type of pneumococcus with which they have been infected. The existence of the carrier state among healthy persons and among those recently recovered from pneumonia establishes a basis for understanding the mechanism by means of which lobar pneumonia spreads and maintains its high incidence from year to year.

FURTHER EXPERIMENTS ON THE EFFECTS OF LONG CONTINUED INTRAPERITONEAL INJECTIONS OF PROTEINS.[*]

By PAUL G. WOOLLEY, M.D., AMIE DeMAR, and DAISY CLARK.

(*From the Pathologic Institute of the Cincinnati General Hospital and the Department of Pathology of the University of Cincinnati, Cincinnati.*)

(Received for publication, April 12, 1915.)

In a previous article[1] we said: "Albumose (Witte's peptone) introduced parenterally into the guinea pig has very little, if any, harmful effect unless the oxidative powers of the organism are below normal. In view of the results of Longcope's experiments, as compared with ours, it seems possible that the more complex proteins will produce effects in the absence of decreased oxidation which the less complex ones will not produce under similar circumstances." These remarks were based upon a short series of experiments in which seven guinea pigs were given intraperitoneal injections of 1 per cent solution of Witte's peptone. Since the publication of this series we have continued the work, as will be detailed hereafter.

The different experiments are grouped as follows: 1. Guinea pigs which were given only Witte's peptone (Nos. 1, 2, 3, 4, 5, 18, 19). 2. Guinea pigs which received Witte's peptone, followed by chloroform anesthesia for fifteen minutes (Nos. 11, 12, 21, 22). 3. Guinea pigs which received chloroform only (Nos. 24, 27, 29). 4. Controls (Nos. 20, 23). 5. White rats which received intraperitoneal injections of a solution of albumen (Nos. 12, 13). 6. White rats which received intraperitoneal injections of a solution of albumen together with subcutaneoüs injections of chloroform in oil (Nos. 21, 22). 7. White rats which received intraperitoneal injections of casein (Nos. 14, 17). 8. White rat which received intraperitoneal injections of casein together with subcutaneous injec-

---

[1] Woolley, P. G., DeMar, A., and Clark, D., *Science*, 1914, N.S., xl, 789.

tions of chloroform in oil (No. 16).´ 9. Controls (Nos. 1, 2, 3, 4, 5, 6, 7, 8, 9, and 10).

In the protocols the expression albumose solution means one prepared according to the following formula and then sterilized:

Witte's peptone	1.0 gm.
Sodium chloride	0.5 gm.
Distilled water	100.0 gm.

The albumen solutions were made by dissolving 10 mg. of desiccated egg-albumen in 2 cc. of distilled water. The solutions were freshly prepared before each inoculation.

The chloroform solution was a 2 per cent solution in sterile olive oil.

The casein was given in the form of a solution made by dissolving 1 gm. of casein in 100 cc. of normal sodium hydrate and then carefully neutralizing (to phenolphthalein)´ with normal hydrochloric acid.

Group I.

Guinea Pig 1.—Weight 400 gm. This animal received 17 daily injections each of 1.5 cc. of the peptone solution, a total of 25.5 cc., or 0.255 gm. of albumose. It died suddenly on the day following the last injection. The cause of death was not discovered. The postmortem examination was done while the animal was still warm, and showed no other changes than a slight mediastinal edema, moderate hyperplasia of the lymph nodes, and congestion of the lungs, liver, spleen, and kidneys. Microscopic examination of the tissues showed edema and congestion with occasional small hemorrhages in the kidneys, with a few areas of small round cell infiltration, enormous congestion of the adrenals, edema and focal necroses of the thymus, and hyperplastic changes associated with congestion in the lymph glands and spleen. The spleen was more than normally pigmented.

Guinea Pig 2.—Weight about 350 gm. This animal received 57 daily injections each of 1.5 cc. of the peptone solution, a total of 85.5 cc, or 0.855 gm. of albumose. During the period of treatment it gave no sign of any untoward effects of the treatment. It ate well, lost no weight, and was finally chloroformed 72 hours after the last injection. The postmortem examination was done while the body was still warm. The organs showed no abnormal macroscopic or microscopic lesions, other than a moderate, generalized congestion associated with a very moderate edema of the parenchymatous organs. This, however, was no more than is usual after chloroform anesthesia.

Guinea Pig 3.—Weight about 400 gm. This animal received 30 daily intraperitoneal injections each of 1.5 cc. of the peptone solution, a total of 5 cc., or 0.45 gm. of albumose. It was killed with chloroform. The postmortem examination showed only a very moderate congestion and edema of the liver, spleen, kidneys, and adrenals, and a slight hyperplasia of the mesenteric lymph glands. Microscopic examination showed nothing abnormal except perhaps a slight degree of hyperplasia of the mesenteric lymph glands.

Guinea Pig 4.—Weight about 400 gm. This animal received 5 cc. of the peptone solution each day for 7 days, a total of 35 cc., or 0.35 gm. of albumose. It was killed with chloroform. At autopsy nothing abnormal was found. Microscopic examination was also negative.

Guinea Pig 5.—Weight about 350 gm. This animal was treated in the same manner as No. 4 for a period of 20 days, during which time it received a total of 100 cc. of the peptone solution, or 1 gm. of albumose. It was chloroformed and autopsied. During the period of treatment it lost 87 gm. in weight. At autopsy nothing noticeable was found except a partially healed meager exudate on the surface of the spleen. The peritoneal cavity contained 2 cc. of a clear fluid. Microscopic examination of the tissues showed no lesions except in the case of the spleen, in which there was an increased amount of pigment and a moderate hypertrophy of the Malpighian follicles. The capsule was thickened.

Guinea Pig 18.—Weight 470 gm. This animal received 5 cc. of the peptone solution, intraperitoneally, each day for 42 days, a total of 210 cc., or 2.1 gm. of albumose. At the time of death it weighed 485 gm., a gain of 15 gm. It was killed with chloroform. The organs showed but little change. The spleen was congested; the liver and kidneys showed occasional foci of small round cells. The peritoneal cavity contained no free fluid.

Guinea Pig 19.—Weight 452 gm. This animal received exactly the same treatment as Guinea Pig 18. At the end of the experiment the weight had dropped to 420 gm., a loss of 32 gm. The organs showed more noticeable microscopic changes than those of Guinea Pig 18. In the kidney the small round cell infiltration was rather diffuse,—the foci of accumulation were more numerous. The liver was unusually congested. The spleen was diffusely hyperplastic. The adrenal exhibited small round cell infiltration in foci in the medulla, and the cortex showed an unusual amount of lipoid metamorphosis. In the medulla there was an increase of hyaline material.

Remarks.—This group of experiments seems to show that comparatively large doses of albumose introduced in small doses over a considerable period of time have practically no harmful effect upon the organs of the body, unless it may be that the foci of small round cells are due to the treatment. It is possible that were the experiments continued over very long periods of time these foci might become fibroid.

Group II.

Guinea Pig 12.—Weight 412 gm. This animal received 50 intraperitoneal injections of 5 cc. of the peptone solution in the course of two months, a total of 250 cc., or 2.5 gm. of albumose. After each injection it was submitted to deep chloroform anesthesia for 15 minutes. After 25 treatments the weight had increased to 455 gm. At the end of the treatments the weight was 485 gm. The postmortem examination revealed nothing macroscopically abnormal, and physically the animal seemed to be in good condition. Histological examination showed that there was a certain amount of anatomic modification of the tissues of some of the organs. The report was as follows: The kidney shows a well marked edema and cloudy swelling. The glomerular spaces are dilated and the tufts compressed, and in the spaces there is considerable coagulated albuminous material. About the glomeruli there are frequent small accumulations of small round cells, and in the outer layers of the cortex there are occasional lines of interstitial fibrosis. The whole organ showed congestion. The liver showed a very well developed edema, to the extent that in many areas the cells show what seem to be hydropic changes. With this is associated congestion and very moderate interstitial fibrosis as exemplified in the occasional collections of small round cells in the perilobular connective tissues. The spleen shows enormous hyperplasia of the Malpighian follicles together with some increased pigmentation. Within the corpuscles there is evidence of cellular fragmentation. The adrenals show a few collections of formative cells in both medulla and cortex, chiefly in the latter. The other organs revealed nothing remarkable.

Guinea Pig 11.—Weight 445 gm. This animal was treated in exactly the same way as No. 12. After 25 treatments it weighed 482 gm., and at the end of the experiment 565 gm. The report of the histologic examination stated that the changes were similar to those found in No. 12, except that there were a few retention cysts in the kidneys and that there was nothing of note in the adrenals except an intense congestion.[2]

Guinea Pig 21.—Weight 365 gm. This animal received 40 daily injections (intraperitoneal) each of 5 cc. of the albumose solution, and each injection was followed by 15 minutes of deep chloroform anesthesia. At the end of the experiment the weight had fallen to 340 gm., a loss of 25 gm. It was killed with chloroform. The cellular and organic changes were practically the same as those exhibited by Guinea Pig 19, though somewhat less in degree.

Guinea Pig 22.—Weight 340 gm. This animal was treated in exactly the same manner as Guinea Pig 21. At the end of the experiment it weighed 360 gm., a gain of 20 gm. The sections made from the organs showed changes similar to those described in Guinea Pig 18.

Remarks.—It was the results obtained in the first two animals in this group which, compared with those of Group I, called forth the remarks quoted in the first paragraph of this report. Chloroform undoubtedly reduces the oxidations of the body, and when administered at the same time with other substances may lead to the production of grave changes in the tissues, as Opie[3] has shown. It

[2] The histologic examinations in these cases were made by Dr. T. H. Kelly, who had no knowledge of the experimental procedures used in the individual cases. They were subsequently verified by one of us.

[3] Opie, E. L., *Tr. Assn. Am. Phys.*, 1910, xxv, 140.

is, however, to be remarked at this place that the results in the last two animals do not support the conclusion that because chloroform may do these things, it must do them. What conditions were present in Animals 11 and 12 to cause them to react differently, we cannot tell. They were from a different lot of animals from Nos. 21 and 22; they were fed more liberally with fresh vegetables than the latter, because they were procured in the late summer. But, on the other hand, they were of the same lot as Nos. 1, 2, 3, 4, and 5. Nos. 21 and 22 belonged in the same lot as 18 and 19. Still, perhaps the effect is dietary.

Group III.

Guinea Pig 24.—Weight 340 gm. This animal was treated on successive days by deep anesthetization for 15 minutes. At the end of the experiment it had lost 55 gm. It was killed by complete anesthetization. The tissues showed no changes other than congestion.

Guinea Pig 27.—Weight 616 gm. This animal was treated in exactly the same manner as was Guinea Pig 24, except that the number of anesthetizations was 9. After the 9th treatment it weighed 620 gm. The organs were similar to those of Guinea Pig 24.

Guinea Pig 29.—Weight 358 gm. This animal was treated like the two preceding, except that it received 40 treatments. At the end of the experiment it weighed 435 gm., a gain of 77 gm. The organs showed no macroscopic abuormality. Sections showed intense fatty metamorphosis of the liver and little else.

Remarks.—Apparently chloroform alone produces no more change, so far as interstitial changes are concerned, than the combination of chloroform and albumose. It does reduce oxidations and therefore tends to bring about an increase in weight, largely perhaps because of a decrease in oxidation of the fats.

Group IV.

Guinea Pigs 20 and 23.—These were normal controls, the organs of which were used for comparison. In each there were occasional fibrotic glomeruli in the kidney and occasional foci of small round cell infiltration.

Remarks.—The preceding group of experiments brings out no revolutionary fact. They merely indicate that while albumose administered, as we have done it, to normal guinea pigs produces, as a rule, no obvious symptoms or histologic changes, even when chloroform is given at the same time, yet occasionally early sclerotic changes appear.

Group V.

White Rat 12.—Weight 286 gm. This animal received 25 injections on successive days, each of 5 mg. of albumen. At the end of the experiment it weighed 305 gm. Postmortem examination showed no gross abnormality. Sections showed excessive edema and cloudy swelling, and in some of the glomerular spaces small mounts of coagulated albumen.

White Rat 13.—Weight 225 gm. This animal was treated in the same manner as Rat 12, except that it was given 44 injections, a total of 220 mg. of albumen. At the end of the experiment the weight had fallen 40 gm. The animal was killed by a blow on the neck.

The organs were apparently normal, macroscopically. The sections of the organs showed a considerable degree of cloudy swelling, especially of the convoluted tubules. In places the appearances were almost those of necrosis. Otherwise there were no microscopic changes, except a small granulomatous area in the liver.

Remarks.—The only essential difference in the tissues of Rats 12 and 13 lay in the presence of albumen in the glomerular spaces in No. 12, which was killed with chloroform. In No. 13 the marked cloudy swelling was perhaps due to whatever produced the grauuloma.

Group VI.

White Rat 21.—Weight 282 gm. This animal received 40 daily injections of 5 mg. of albumen, intraperitoneally, and subcutaneous doses of 0.5 cc. of a 2 per cent chloroform solution in oil. At the end of the series it had lost 12 gm. in weight. Neither the organs nor the sections showed anything remarkable.

White Rat 22.—Weight 273 gm. This animal was treated for 7 days. like Rat 21. It lost no weight. The organs showed merely edema.

Group VII.

White Rat 14.—Weight 163 gm. This animal received 25 injections, on successive days, of 10 mg. of casein. At the end of the experiment it weighe l 190 gm. Nothing anatomic or histologic seemed to result from the treatment.

White Rat 17.—Weight 290 gm. This animal received daily intraperitoneal injections of 10 mg. of casein for 17 days, followed by 23 similar injections each of 70 mg. of casein. At the end of the series it had lost 20 gm. There were no essential changes observed, macroscopically or microscopically, in the organs or tissues.

Group VIII.

White Rat 16.—Weight 310 gm. This animal was treated exactly like Rat 17, except during the last 23 days, when it was receiving 20 mg. of casein daily.

It also was given 0.5 cc. of the chloroform oil solution daily. At postmortem examination there was no evidence of gross changes, and sections showed nothing more than marked congestion and cloudy swelling.

Group IX.

White Rats 1, 2, 3, 4, 5, 6, 7, 8, 9, and 10.—These animals of various weights and ages furnished material for comparison.

Rat 1 had been fed on a strictly meat diet for several weeks.

Rat 2 had been fed on grain and vegetables for several weeks.

Rats 4 and 5 had been injected with trypan blue.

Rats 6, 7, 8, 9, and 10 had been given a single injection of uranium acetate, and had been killed after 24 to 48 hours.

DISCUSSION.

As stated in our earlier report, it was our object in carrying out this series of experiments to determine, if possible, the organic effects of certain substances of protein nature which might be produced within the tissues, or which might be absorbed from the intestinal tract.

Longcope[4] has said that parenteral digestion of egg-albumen may (under certain circumstances) seemingly produce organic renal and hepatic changes. This may be taken to mean that splitting of the whole protein (egg-albumen) leads to the production of these phenomena; that the effects are the results of the irritant action of substances set free during splitting; or that moieties of the protein molecule, by making abnormal combinations, may act as irritants or in some other way embarrass the functional cells of the tissues and at the same time stimulate the overgrowth of fibrous tissue.

Longcope produced the effects which he reported, not by the mere administration of albumen, but by so timing the injections that a mild state of anaphylaxis resulted. In our series we have not done this, but have endeavored to discover whether the substances which we have used would produce similar effects in the absence of anaphylaxis, and whether, at the same time, the addition of materials (such as chloroform) which hinder the oxidations in the cells, would modify or intensify the reactions. In a former series of experiments Woolley and Newburgh[5] showed that indol and tyrosine pro-

---

[4] Longcope. W. T.. *Jour. Exper. Med.,* 1913, xviii, 678.

[5] Woolley, P. G., and Newburgh, L. H., *Jour. Am. Med. Assn.,* 1911, lvi, 1796. Newburgh and Woolley, *Lancet-Clinic,* 1912, N.S., lxviii, 404.

duced no remarkable changes. In this series we have shown that egg-albumen, casein, and albumen give practically the same negative results whether or not chloroform is administered with them.

In all our experiments we have used comparatively small doses (at any rate they were not massive); because if, as has been suspected, such anatomic conditions as chronic interstitial nephritis are the result of absorption from the gastro-intestinal tract, then the quantities of materials absorbed are probably small, and the process covers a considerable period of time. Perhaps we have not continued our work over a long enough period. However that may be, there is evidence that the living proteins in the form of bacteria are far more potent than the non-living, and that even the less pathogenic ones may produce greater damage in the length of time covered in our experiments than has been occasioned by the subs. inces we have used.[6] These substances (microorganisms) are entering the system continuously, and since we know that they produce, often rapidly, anatomic changes, a series of experiments such as ours is chiefly of value in proving the relative lack of danger from absorption of the substances which we have employed,—at least in normal animals.

SUMMARY.

We have attempted to discover whether or not certain protein materials, such as albumose, casein, and albumen, when introduced parenterally (peritoneally) into experimental animals, are able to produce organic lesions. These proteins were used alone or in combination with chloroform, which was administered in oil and as an anesthetic. Our results were negative.

[6] Opie, *loc. cit.*

THE HEART MUSCLE IN PNEUMONIA.[1]

By L. H. NEWBURGH, M.D., and W. T. PORTER, M.D.

(*From the Laboratory of Comparative Physiology and the Department of Medicine of the Harvard Medical School, Boston.*)

PLATE 22.

(Received for publication, April 24, 1915.)

It is very generally believed that the heart muscle is seriously injured in pneumonia and that heart failure from this source is a frequent cause of death in this infection.

The experiments presented in this communication show that the cardiac ventricle from dogs that have died from pneumonia contracts as well as the ventricle from healthy dogs, provided the pneumonic muscle is fed with normal blood. When a normal ventricle is fed with pneumonic blood, the contractions are much impaired. If, however, the ventricle from a dog with pneumonia is fed with pneumonic blood, the contractions are almost normal in extent and may be normal in duration.

Thus in pneumonia the heart muscle is essentially normal, whereas the pneumonic blood is distinctly poisonous to heart muscle suddenly fed with it. In the body, during the gradual course of the disease, the blood is progressively affected and the heart muscle gradually adjusts itself to the poison, with striking success.

Method.

The experiments consist of four series of ten dogs each. In the first, the normal ventricle was fed with normal blood; in the second, the pneumonic ventricle was fed with normal blood; in the third, the normal ventricle was fed with pneumonic blood; in the fourth, the pneumonic ventricle was fed with pneumonic blood.

[1] Aided by a grant from the Dalton Scholarship of the Massachusetts General Hospital. A preliminary abstract of the conclusions reached in this investigation was published in the *Boston Med. and Surg. Jour.*, 1915, clxxii, 718.

123

The contractions of the heart muscle were recorded by a method devised by Porter in 1897.[2] The dog is anesthetized with ether (no morphin), bled from the carotid artery, the blood defibrinated and filtered through glass-wool. Meanwhile, warm normal saline solution[3] is allowed to flow into the crural vein. After a short interval, the dog is bled again from the carotid artery and the blood defibrinated as before. The heart is now rapidly removed and placed still beating in a beaker filled with warm saline solution. Often the beats are so vigorous that the heart with each ventricular systole springs more than an inch from the bottom of the beaker. Thus the organ is self-cleansed from blood. A cannula is now tied into the branch of the left coronary artery supplying the area the contractions of which are to be studied, and the part of the left ventricular wall supplied by the artery is cut out. The cannula bearing the attached ventricular segment is filled with defibrinated blood and joined to a glass tube containing defibrinated blood at a temperature of 23° to 25° C. and at a pressure of 56 mm. of mercury. This tube is surrounded with a larger glass tube through which warm water is circulated to keep the blood at the desired temperature. An adjustable clamp supports the coronary cannula and thus the attached heart muscle in a suitable position. A bent hook is passed through the lower end of the muscle and attached to a light lever magnifying five times. The contractions are recorded on a kymograph moving 40 mm. per hour (Fig. 1). The defibrinated blood in all our experiments was diluted with twice its bulk of normal saline solution.

The work done by the ventricular muscle was judged (1) by the length of the period during which the heart contracted, and (2) by the total area of the contractions. The curves, illustrated by Fig. 1, were laid upon a glass plate illuminated from below with electricity. Over the curve was placed a card and the total area of the curve was carefully traced upon this card. The area was then cut out with scissors, placed with the other contraction areas

[2] Porter, W. T., *Jour. Boston Soc. Med. Sc.*, 1897, i, 15; *Jour. Exper. Med.*, 1897, ii, 391; *Am. Jour. Physiol.*, 1898, i, 514–515.

[3] Sodium chloride, 9.00 gm.; calcium chloride, 0.26 gm.; potassium chloride, 0.10 gm. in 1,000 cc. of water.

in its own group and the entire group weighed in a good balance. The cards used were of uniform thickness. The small errors made in tracing the areas and cutting them out were "accidental errors"; those falling above the true value were compensated by those falling below it. Ventricular strips from large hearts sometimes give higher contractions than the strips from small hearts, thus influencing the contraction area. As there were ten hearts in each group, compensation took place here also, and there is no reason to doubt that for the purpose of comparing contraction areas in groups of ten the method is substantially accurate in a problem in which only marked similarities and differences are of value.

When a kymograph moves so slowly that the contractions of the slowest heart in a series are fused, the contractions of a more rapid heart will overlap. In such a case, the fused curve from the more frequently contracting heart will contain more contractions, and should thus express more work than the fused curve of equal height from the less frequently contracting heart. But under these circumstances, the more frequently contracting heart will conceal its extra work and the two curves, although equal in area, will not be equal in work done. In our experiments, this error does not affect the main conclusion, because the great majority of the ventricles beat at nearly the same rate, about 55 per minute. The relatively infrequent variations were chiefly in Series III, in which normal ventricles were fed with pneumonic blood, which seemed to make some of them more irritable. Such ventricles beat more rapidly than 55; their contractions, however, were increased in force as well as in frequency, giving a higher contraction area and thus compensating in part for overlapping. At the most, the error from occasional overlapping disappears in comparison with the wide difference in performance between the normal ventricle fed with normal blood and the normal ventricle fed with pneumonic blood.

The organism employed was the *Bacillus pneumoniæ* (Friedländer) obtained from the stock culture in the Bacteriological Laboratory of the Harvard Medical School through the kindness of Dr. Sisson. It was passed through three guinea pigs to increase its virulence to such a degree that 1 cc. of a broth culture injected into the peritoneum killed a guinea pig in twelve hours. Its virulence was kept at this point by occasional passage through additional guinea pigs. Broth cultures incubated from eighteen to twenty-four hours were used for injection into the trachea. The dose was 5 to 7 cc. per kilo.

The culture was administered as follows:[4] The dog was given subcutaneously from 0.5 to 1.0 cc. of a 3 per cent solution of morphin sulphate so that it might not cough up the culture. From thirty to sixty minutes later, the dog was placed on the operating table, etherized, the jaws held open by an assistant, the tongue drawn forward and upward, the epiglottis drawn forward, and a tube passed through the glottis into a bronchus, usually the right bronchus. The culture was then forced through the tube with a syringe, and the dog was taken from the table and placed in a sick room to await the disease. Many of the dogs were at the point of death as early as eighteen hours after the inoculation.

In six of the ten experiments in which the pneumonic heart was fed with normal blood, the dogs were allowed to die of the disease, to forestall critics who might say that the dogs could not have been very ill. In one of the remaining four, the temperature had fallen to 38° C., and the respiration was very rapid and labored; the second of the four was entirely insensible to pain; the temperature of the third had fallen from 40° to 36° C.; and the fourth was in coma with a temperature of 32° C.

In the ten experiments in which the normal ventricle was fed with pneumonic blood, the normal ventricle was taken from ten healthy dogs and the pneumonic blood from ten dogs in that stage of the disease that just precedes death. The temperature of these animals had begun the fatal descent which in dogs with pneumonia is the precursor of the end. It had been 40° to 41° C., and the operation was performed when the temperature had fallen to 38° or 37°, sometimes as low as 34°. Five of the dogs were in complete coma; these dogs were entirely insensitive to pain.

In the ten experiments in which the pneumonic heart was fed

[4] Lamar, R. V., and Meltzer, S. J., *Jour. Exper. Med.*, 1912, xv, 133.

with pneumonic blood, the blood of each animal was used for its own heart preparation. In nine of these animals, the fatal descent in temperature had begun. Six of them were entirely comatose. The tenth dog in this series was not used until the respiration had ceased and the pulse was absent from the exposed carotid artery. The dog was thereupon revived with normal saline injections and artificial respiration until the blood had been withdrawn and the heart extracted. In every experiment upon a pneumonia dog, an autopsy revealed an extensive consolidation of the lung.

Observations.

TABLE I.

The Average Duration of Contraction and the Total Weights of the Contraction Areas in Four Series of Ten Dogs Each.

Preparation.	Average duration of contractions.	Total weights* of contraction areas.
	min.	*gm.*
Series I. Normal ventricle fed with normal blood...	181	8.84
Series II. Pneumonic ventricle fed with normal blood	187†	8.46
Series III. Normal ventricle fed with pneumonic blood	70	3.40
Series IV. Pneumonic ventricle fed with pneumonic blood	176	6.46

* Each contraction area was cut out of a thin card. These cards were of uniform thickness, and 3.5 sq. cm. of card weighed 0.1 gm.

† By a clerical error, this figure was given as 200 minutes in the abstract published in the *Boston Med. and Surg. Jour.* (Newburgh and Porter, *loc. cit.*).

The forty experiments on which our conclusions are based are presented in Table I. It is at once evident that the results in Series I and II are identical. The pneumonic ventricle fed with normal blood contracts as long and as vigorously as the normal ventricle. In Series III the normal ventricle is fed with pneumonic blood, under which condition the duration of contraction and the contraction area are little more than one-third the normal value. In Series IV, in which the pneumonic heart is fed with pneumonic blood to which it had been exposed during the course of the disease, the duration of contraction is but 4 per cent, and the area of contraction 27 per cent less than normal.

CONCLUSIONS.

1. The heart muscle is not functionally impaired in pneumonia, since the pneumonic ventricle beats normally as soon as its food is normal.

2. Pneumonic blood, suddenly fed to normal heart muscle, lowers its efficiency, lessening the duration and the area of contraction.

3. The heart muscle in pneumonia, exposed gradually to the action of the poison, largely adjusts itself to its poisoned food.

EXPLANATION OF PLATE 22.

FIG. 1. Original size. Experiment of May 17, 1915. Contractions of the part of the left ventricle supplied by the first portion of the circumflex branch of the left coronary artery. The preparation was perfused with defibrinated blood mixed with two parts of normal saline solution. The blood was from a healthy dog. The lever magnified five times. May 16, 3 p. m., 45 cc. of a broth culture of Friedländer's bacillus were injected into a bronchus. May 17, 9 a. m., the temperature was 41° C. At 6 p. m. the temperature was 38°, the dog was completely insensible, and no ether was required during the operation.

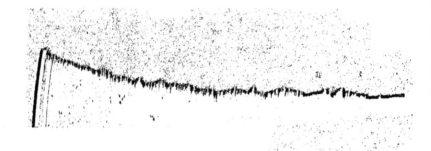

6.55 p. m.   9 45

FIG. 1.

(Newburgh and Porter: The Heart Muscle in Pneum

SERUM FERMENTS AND ANTIFERMENT AFTER FEEDING.

Studies on Ferment Action. XXI.

By JAMES W. JOBLING, M.D., WILLIAM PETERSEN, M.D., and
A. A. EGGSTEIN, M.D.

(*From the Department of Pathology, Medical Department, Vanderbilt University, Nashville.*)

(Received for publication, April 13, 1915.)

Considerable progress in our understanding of the metabolism of protein digestion has been made possible in recent years due largely to microchemical investigations carried out by Van Slyke (1) and by Folin(2), and by means of the ingenious vividiffusion apparatus of Abel, Rowntree, and Turner (3).

There has been, however, no corresponding investigation of the serum ferments which might be involved during the process, possibly because of the lack of a suitable quantitative method of estimating the amount of protease in the serum. Neither the optical nor the dialysis method of Abderhalden are of value for such work, for apart from several objections of a theoretical nature, which we have reviewed in a former paper (4), they are inconveniently cumbersome, and the dialysis method particularly is open to the serious objection that it is not the introduced protein (placenta, tumor tissue, etc.) which is digested, but that the ferment acts on the serum proteins. The introduced substrate therefore adds an unnecessary complication and confusion.

Method of Determining Protease Action.

The method which we have used is based on the following considerations. When serum antiferment is removed by means of lipoid solvents, chloroform, ether, etc., whatever protease is present can become active and digest the serum proteins (5). If, therefore, the amount of non-coagulable nitrogen is determined both before and after incubation, the difference in amount will be an index of protease action. The nitrogen determinations can be made with great facility by the Folin or Van Slyke method. We have noticed, however, that under chloroform the increase in amino-acids need not be in proportion to the amount of non-coagulable nitrogen found, the chloroform evidently offering no insult to the true proteases, while the ereptase action may be retarded.

The normal activity of protease has been assumed to be lytic, while the synthetic action of the ferment has not been demonstrated. Few observations have been made in this direction, the idea being held that the synthetic process is more intimately connected with the function of the individual cell. Theoretically there is, of course, no reason why proteosynthesis should not occur in the serum; indeed, it seems rational that such ferment activity must be functionating when the serum proteins are formed.

The technique used is as follows: The clear hemoglobin-free serum is measured with an accurate 1 cc. pipette into a rather wide test-tube (about 18 mm.). To the tube 0.5 to 0.75 cc. of chloroform is added and the tube is sharply shaken, at intervals, until a milky emulsion is formed. We prefer chloroform because the emulsion is more stable than with ether or other lipoid solvents. A control tube is inactivated at 60° C. for 30 minutes and a drop of toluol is then added in place of chloroform. Both tubes are then incubated over night (15 to 16 hours at 37°). In the morning about 1 cc. of a mixture of 10 per cent acetic acid plus 20 per cent salt solution is added, and the tubes are then gently warmed in a water bath until the chloroform has been evaporated. About 2 or 3 cc. of distilled water are then added slowly and the tubes boiled for at least 10 minutes. The coagulated protein is filtered off by means of hard filter papers, previously moistened, the filtrate being permitted to filter directly into the large tubes used for oxidizing. The tubes are then oxidized and Nesslerized according to the usual Folin method, the readings being made against varying dilutions of the 1 mg. standard, so that test readings are made against a standard of apparently equal color.

Whenever possible we made duplicate determinations of each serum to obviate any error. When properly made the difference in any two readings should not exceed a few hundredths of a mg. per cc. of serum, a degree of accuracy sufficient for the work. If, for example, the total non-coagulable nitrogen per cc. of serum is proved to be 0.32, and after incubation, has increased to 0.45 mg., the protease activity can be expressed as 0.13 mg. per cc., and has been so designated in the accompanying text-figures. The proteins are not precipitated by alcohol, as in the usual Folin method, because we wish to determine the increase of higher splitting products of the proteins as well as of the amino-acids. By precipitating with alcohol the higher split products would, of course, be retained on the filter. The non-coagulable nitrogen which we determine is therefore not quite analogous to the non-protein nitrogen determined by the Folin method.

Method of Determining Serum Esterase.

Serum esterase has been determined as follows: To 1 cc. of the serum, 1 cc. of neutral, redistilled ethyl butyrate and 0.5 cc. toluol are added, the volume being brought to 10 cc. with physiological salt solution. The flasks are then shaken 100 times and incubated for 4 hours. 25 cc. of neutral 95 per cent alcohol are then added to each flask and the acidity which has developed is titrated with N/50 sodium hydrate (alcoholic) to a faint pink with phenolphthalein. After deducting the proper controls, *i. e.*, serum alone, ethyl butyrate alone. etc., the esterase index is expressed in terms of cc. of N/100 sodium hydrate used to neutralize the acidity developed by 1 cc. of serum from 1 cc. of ethyl butyrate.

The antiferment has been determined by the method described in our previous papers (5).

Healthy adult dogs have been used exclusively. Blood was taken from the ear veins before feeding, and the animals were then fed with chopped boiled meat and bread. They were permitted to eat as much as 'desired. No water was given until the following day. Blood samples were usually taken at noon, at 2 and 4 p. m., and the following morning.

EXPERIMENTAL.

Dog 22.—(Text-fig. 1.) Fed at 9.30 a. m. Bled at 8.30 a. m., noon, 2 and 4 p. m., 9 p. m., and the following morning. It will be noted from the chart that the antiferment index dropped rather sharply, rose again by 2 p. m., and remained fairly constant after that time. The non-coagulable nitrogen increased from 0.45 mg. per cc. to 0.5 mg. at noon and at 2 p. m., after which time it continued to decline.

The protease action which was 0.35 mg. per cc. before the meal had decreased to 0.25 mg. at noon, and at both 2 and 4 p. m. was negative. The serum on incubation at this time showed an actual decrease in non-coagulable nitrogen. The serum taken at 9.00 p. m., and that of the morning showed almost the original protease index. Lipase was not determined.

Dog 17.—(Text-fig. 2.)   Fed at 9.45 a. m.   Bled at 9.00 a. m., 2 and 4 p. m., and the following morning.   In this animal there was a well marked rise in the antiferment index, which persisted till the following morning.   The non-coagulable nitrogen showed a striking increase, reaching more than twice the original amount at 4 p. m.   In view of this result we repeated the experiment at a later date on the same animal with quite similar results.   Thinking that possibly a

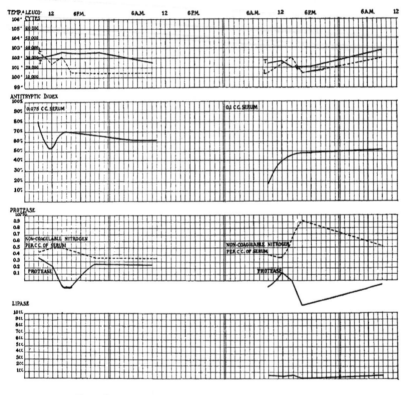

TEXT-FIG. 1.   TEXT-FIG. 2.

TEXT-FIGS. 1 and 2.   Serum ferments and antiferment index after feeding.

kidney lesion might be at fault, in view of the repeated observation of renal pathology reported in dogs, the urine was tested for albumin and casts; and the kidneys were examined histologically with entirely negative results.

The serum showed an increase of protease at noon, after which no protease action was found.   The blood drawn at 4 p. m., and containing 0.9 mg. of non-coagulable nitrogen per cc., when incubated with chloroform, showed a decrease to 0.5 mg. per cc.; *i. e.*, a loss of 0.4 mg. per cc. The lipase showed a slight decrease.

Dog 26.—(Text-fig. 3.) Fed at 9.30 a. m.; relatively small feeding. Bled at 9 a. m. and 1.30 p. m. Killed at 4 p. m.

The antiferment index remained constant for the first two samples and showed a slight decline at 4 p. m. The lipase remained constant throughout. The non-coagulable nitrogen showed a marked increase, as will be noted in Table I.

TABLE I.

Serum.	Non-coagulable nitrogen per cc.	After incubating under chloroform 16 hrs.	Increase of non-coagulable nitrogen.	After incubating under toluol.	Decrease in non-coagulable nitrogen.
	mg.	*mg.*	*mg.*	*mg.*	*mg.*
9 a. m.	0.23	0.72	0.49	0.23	0
	0.22				
1.30 p. m.	0.31	0.66	0.35	0.27	−.04
	0.31				
4 p. m.	0.36	0.70	0.31	0.29	−.10
	0.39				

The protease action decreased progressively, although it did not fall to quite so low a level as in other experiments.

When the serum, instead of being inactivated at 60°, was permitted to remain under toluol (0.5 cc.), it will be noted that for the first sample there was no change, while in both the other specimens of serum a distinct decrease in non-coagulable nitrogen occurred.

Estimation of Proteoses.

In order to determine the nature of the increase in the non-coagulable nitrogen, which was not due to an increase in urea, we estimated the amount of proteoses as follows:

To 10 cc. of serum, acetic acid and salt were added and the coagulable proteins removed by filtration after acidifying and boiling for 30 minutes. The clear fluid was now neutralized and permitted to stand until whatever acid albuminates were present had been precipitated. It was again filtered and the material placed in a dialyzing bag under toluol. After 24 to 48 hours. depending upon the rate of dialysis, and having been freed from all urea and other diffusible nitrogenous substances. nitrogen determinations were made on the material remaining in the dialyzing membranes.

In place of dialyzing we have made direct determinations by precipitating all the proteoses by saturating the boiling filtrate with sodium sulphate. The precipitate collected on the filter while hot is washed with a saturated solution of sodium sulphate and is then washed through the filter with water made slightly alkaline with sodium carbonate. Nitrogen determinations are made on the proteoses so dissolved. As a rule, the figures obtained after dialyzing are slightly higher than those obtained by this method.

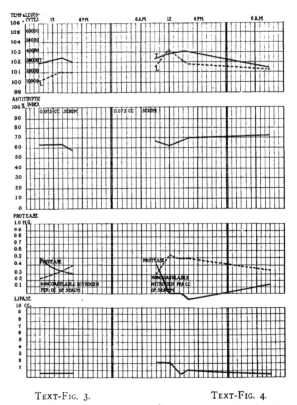

TEXT-FIG. 3.   TEXT-FIG. 4.

TEXT-FIG. 3. Serum ferments and antiferment index after feeding.
TEXT-FIG. 4. Serum ferments and antiferment index after feeding in dog with pancreatic insufficiency.

In the following experiments somewhat larger amounts of blood were withdrawn by means of the suction pump from the ear veins (about 60 cc. of blood) before and after feeding. The animal was usually anesthetized after about six or seven hours and blood samples were removed from various parts of the circulation. In these experiments it frequently happens that the terminal blood so collected may contain proteases unless great care is taken to prevent handling of the intestine and any unnecessary trauma, for blood taken during such conditions, as well as agonal blood under any circumstances, always contains proteases. This is shown in the following experiment.

Dog 42.—Small feeding at 9.00 a. m. Bled before feeding and at 1.15 p. m. Killed at 3.30 p. m. (Table II).

TABLE II.

Serum.		Non-coagulable nitrogen per cc.	After incubating under chloroform.	Increase in non-coagulable nitrogen.	Non-dialyzable nitrogen per 5 cc. of serum.
		mg.	*mg.*	*mg.*	*mg.*
9 a. m.		0.32	0.32	0	0.8
1.15 p. m.		0.3	0.31	0.01	0.71
3.30 p. m.	Arterial	0.21	0.5	0.29	0.44
	Inferior vena cava	0.22	0.5	0.28	
	Portal	0.22	0.45	0.23	0.45
	Hepatic	0.22	0.35	0.13	

This experiment differs from the others in that there is a decrease in non-coagulable nitrogen, made evident also by the decrease in the non-dialyzable nitrogen, as shown in the last column. The animal, furthermore, was without protease in the serum at the beginning of the experiment, but showed the presence of the ferment when killed. Inasmuch as the ferment was found in all the samples and not only

TABLE III.

Serum.		Non-coagulable nitrogen per cc.	Amino nitrogen in 2 cc. of serum (direct determination).	Increase in non-coagulable nitrogen after incubating under chloroform.
Dog 50		*mg.*	*mg.*	*mg.*
9 a. m.		0.34	1.05	
Noon		0.34	0.96	
3.00 p. m.	Arterial	0.35	0.99	
	Portal	0.45	1.35	
	Hepatic	0.25		
Dog 41			Amino nitrogen per 5 cc. of serum. Protein precipitated by alcohol	
9 a. m.		0.25	0.23	0.55
3.00 p. m.	Arterial	0.34	0.215	0.43
	Portal	0.45	0.283	0.55
	Hepatic	0.38		0.45

in the portal blood, in which the concentration, as the following experiment will show, is usually greatest, we believe that the ferment present was due to trauma incident to the collection of the samples and possibly to an agonal condition.

The non-coagulable nitrogen is usually greatest in the portal blood, as is shown in Table III, taken from specimens of two dogs.

It is evident that only part of the increase in non-coagulable nitrogen is due to an increase in amino-acids.

The following experiment (Table IV) shows that there is no increase in the higher split products (primary and secondary proteoses) in the serum.

Dog 52.—No food for 24 hours. Large feeding of meat and milk at 8.30 a. m.

TABLE IV.

Serum.	Non-coagulable nitrogen per cc.	Incubated under chloroform.	Increase of non-coagulable nitrogen under chloroform.	Incubated under toluol.	Decrease in non-coagulable nitrogen.	Proteoses per 1 cc. of serum.	Lipase per cc.
	mg.	*mg.*	*mg.*	*mg.*	*mg.*	*mg.*	
8.30 a. m........	0.31	0.50	+0.19	0.25	0.06	0.17	0.0
12 noon........	0.38	0.30	−0.08	0.3	0.08	0.15	0.3
3.00 p. m. ⎰ Arterial....	0.45	0.43	−0.02	0.38	0.07	0.09	0.5
Inferior vena cava	0.37	0.34	−0.03	0.34	0.03		0.3
Hepatic....	0.41	0.34	−0.07	0.3	0.11		0.0
Portal.....	0.45	0.51	+0.06	0.25	0.2	0.11	

As will be noted, there is a distinct increase in non-coagulable nitrogen after the feeding, while the serum protease, originally 0.19 per cc., drops in all the samples, although the portal blood still shows some protease action. When preserved under toluol this serum shows the greatest decrease in non-coagulable nitrogen. The proteoses, estimated by the direct precipitation with sodium sulphate, are decreased throughout. It is interesting to note that the hepatic blood contained no lipase.

The increase in amino-acids after feeding has previously been noted, and Van Slyke, Cullen, and McLean (6) have recently demonstrated an increase in urea in the blood after feeding. We are rather inclined to assume that some of the increase in non-coagulable nitrogen must be due to the lower dialyzable split products other than amino-acids. In certain pathological conditions Pribram (7) has noted an increase in non-dialyzable protein split products; this is especially true during anaphylactic shock, as we shall show in a later paper.

Occlusion of the Pancreatic Duct.

Dog 26.—(Text-fig. 4.) The animal was operated on Feb. 11, 1915, the pancreatic duct ligated, and bile salts were injected into the pancreatic duct. The animal recovered from the acute pancreatitis so produced. On Feb. 18 it was used for the feeding experiment, and was killed the following day. The pan-

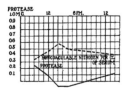

TEXT-FIG. 5. Average change in non-coagulable nitrogen and serum protease after feeding.

creatic duct was found completely occluded and the pancreatic tissue showed marked evidences of a chronic inflammation. The animal was fed at 9.30 a. m., and bled at 9.00 a. m., noon, 2 and 4 p. m., and the following morning. The animal ate a very large meal and showed obvious evidence of discomfort and distress after the feeding.

The antiferment index showed a slight drop and a following recovery. The esterase showed rather a marked fall. The relation of the non-coagulable nitrogen and the protease are typical and similar to those of normal animals.

The Non-Coagulable Nitrogen of the Serum.

A composite chart (Text-fig. 5) made from eight feeding experiments shows a very distinct increase in the non-coagulable nitrogen of the serum, reaching a maximum about five to six hours after the feeding. Holweg (8) has recently shown that this increase is due partially to an increase in urea, while the change in amino-acid and albumoses is not constant. Abderhalden and numerous other workers with the dialysis method have called attention to the fact that for the Abderhalden test the serum should be drawn before a meal in order to overcome the presence of split products and non-specific ferments in the blood stream. In several previous papers Abderhalden had denied the presence of split products of proteins in the serum. The increase in amino-acids as determined by the Van Slyke apparatus is very slight, except in the portal blood, but nevertheless we have found it constant. On the other hand, there is no increase in proteoses. Whether or not there is an increase in peptone we have not determined.

Serum Protease.

The protease curve in Text-fig. 5 indicates a marked fall in the amount of ferment present in the serum after feeding. In view of the fact that proteolytic ferments have been demonstrated in the urine after feeding, the tryptase reaching its maximum in from six to seven hours, these results were quite contrary to our expectations. Inasmuch as the serum reaction under chloroform is slightly acid, we have several times altered the reaction to one slightly alkaline, but have not been able to show an increased proteolysis in this way. It is to be remembered, however, that we are here dealing with true protease and not an ereptase, the latter being the ferment most frequently tested for in the urine.

While this decrease in protease has usually been noted in the peripheral circulation, the portal blood may at the same time show an increase in ferment, as is illustrated in the following experiment (Table V).

Dog 34.—Fed at 8.30 a. m. Bled at 8.45 a. m. and 3.30 p. m.

TABLE V.
Total Non-Coagulable Nitrogen per Cc.

	mg.	
Before feeding	0.22	(Peripheral.)
After feeding	0.32	(Peripheral and portal.)

Protease. Increase of Non-Coagulable Nitrogen on Standing.

Before feeding	0.22	(Peripheral.)
After feeding	0.18	(Peripheral.)
After feeding	0.42	(Portal.)

The increased amount of ferment in the portal blood evidently does not pass the liver. It is possible that in the portal blood a large amount of split products of proteins are present which might offer an available substrate for ereptase rather than tryptase action. The absorption of free tryptic ferment from the normal intestinal tract must be small in amount, for the ferment is rapidly bound by the proteins present, so that even in the actively digesting intestine the amount of free ferment is not large.

Whether the decrease of non-coagulable nitrogen noted on incubation in various of these experiments depends on any actual synthesis to coagulable proteins or upon purely experimental manipulation will be discussed fully in a later paper.

Serum Antiferment.

The average antiferment index following feeding is shown in Text-fig. 6. The increase, previously noted, being coincident with an increase in split products in the serum was supposed to lend support to the view of Rosenthal that the antiferment was due to protein split products.

This increase is not uniform in all animals. While in some there is a very sharp rise, others may remain unaltered or even show a

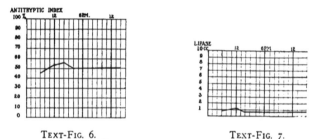

TEXT-FIG. 6.   TEXT-FIG. 7.

TEXT-FIG. 6. Average change in antiferment titer after feeding.
TEXT-FIG. 7. Average change in lipase after feeding.

decrease. The increase is probably due to an increase in unsaturated lipoids, a subject which we have previously discussed (5).

Serum Esterase.

The increase in serum esterase is very slight in amount, the average for the experiments (Text-fig. 7) rising from 0.9 cc. N 100 sodium hydrate to 1.2 cc. after three hours. When samples are collected at various parts of the circulation, the blood from the hepatic veins usually contains less ferment than from other sources, indicating that the source of the ferment is probably not in the liver.

CONCLUSIONS.

1. After feeding, an increase in non-coagulable nitrogen of the serum can be determined, reaching a maximum in about six hours.

2. This increase is greatest in the portal blood and is partially due to an increase in amino-acids. There is no increase in proteoses.

3. There is usually a progressive decrease in serum protease, reaching a minimum after from five to seven hours.

4. The portal blood may show an unaltered or an increased amount of protease.

5. The serum antiferment shows a slight increase, but is subject to considerable fluctuation.

6. The serum lipase (esterase) shows a slight increase, reaching a maximum after three hours. The hepatic blood usually contains the lowest concentration of lipase.

BIBLIOGRAPHY.

1. Van Slyke, D. D., *Jour. Biol. Chem.*, 1912, xii, 275.
2. Folin, O., and Denis, W., *Jour. Biol. Chem.*, 1912, xi, 527.
3. Abel, J. J., Rowntree, L. G., and Turner, B. B., *Jour. Pharmacol. and Exper. Therap.*, 1913–14, v, 275.
4. Jobling, J. W., Eggstein, A. A., and Petersen, W., *Jour. Exper. Med.*, 1915, xxi, 239.
5. Jobling, J. W., and Petersen, W., *Jour. Exper. Med.*, 1914, xix, 459, 480.
6. Van Slyke, D. D., Cullen, G. E., and McLean, F. C., *Proc. Soc. Exper. Biol. and Med.*, 1914–15, xii, 93.
7. Pribram, H., *Zentralbl. f. inn. Med.*, 1914, xxxv, 153.
8. Holweg, H., *Med. Klin.*, 1915, xi, 331.

SERUM FERMENTS AND ANTIFERMENT DURING TRYPSIN SHOCK.

Studies on Ferment Action. XXII.

By JAMES W. JOBLING, M.D., WILLIAM PETERSEN, M.D., and
A. A. EGGSTEIN, M.D.

(*From the Department of Pathology, Medical Department, Vanderbilt University, Nashville.*)

(Received for publication, April 19, 1915.)

In an extended series of papers Kirchheim (1) has studied the question of the toxicity of trypsin in both its local and general effects. The resistance of living tissue to the local effect of trypsin has also been studied by Langenskiöld (2) and by Marie and Villandre (3). Kirchheim called attention to the similarity of trypsin intoxication to anaphylactic and peptone shock. He determined that the toxicity was destroyed when the ferment was inactivated by heat and that the fresh pancreatic secretion was not toxic unless activated by enterokinase. He concluded therefrom that the toxicity was not due to admixed protein split products. In order to determine whether the toxicity depended upon the effect of the ferment directly on the living cell, or whether split products were first formed from soluble proteins, in this way leading to an indirect intoxication, Kirchheim tried the effect of the ferment directly on spermatozoa. Since he found, however, that the spermatozoa were injured neither by the ferment nor by split products produced by the action of the ferment, Kirchheim drew no definite conclusion. In his work on the serum antiferment Kirchheim showed that both theories held (split products in the serum, and true antibody formation) were erroneous. Incidentally he noted that chloroform rendered the serum albumin more digestible by trypsin.

In view of the similarity of trypsin intoxication to anaphylactic and peptone shock we have undertaken a series of experiments to determine the effect of the ferment when injected into the blood stream of dogs. The trypsin used was either commercial pancreatin (for intestinal injection) or purified according to the method previously described (4). The latter ferment was very active and when dried retained its strength unimpaired. The serum ferments were titrated according to the method described fully in a previous paper

141

(5). Dogs of medium weight (five to nine kilos) were used throughout.

EXPERIMENTAL.

Dog 11.—(Text-fig. 1.) Weight 5.5 kilos. 0.27 gm. of purified trypsin was dissolved in 5 cc. of normal saline and injected intravenously at noon. The animal became ill immediately, cried with pain, and was nauseated. Bled after 10 minutes, at 2 and 4 p. m., and the following morning. As will be noted from Text-fig. 1 there was an immediate rise in serum protease, the blood taken 10 minutes after the injection digesting slightly more than 0.6 mg. of nitrogen

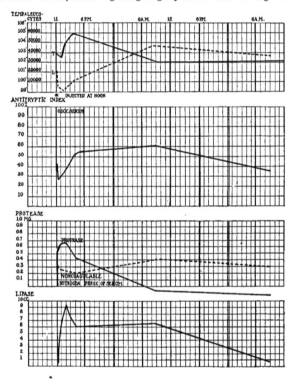

TEXT-FIG. 1. Effect of trypsin injection on the serum ferments and antiferment titer.

from serum proteins per cc. The maximum effect was noted after 2 hours, followed by a gradual decline. The non-coagulable nitrogen showed a gradual decrease, with an increase after 24 hours. The antiferment index fell immediately, recovered gradually, and increased after 24 hours, after which the original titer was reached.

The lipase curve shows a striking rise, this being progressive for 2 hours, after which the amount is lessened, but remains at a high level during the following day.

There was a marked leucopenia immediately following the injection, followed by a leucocytosis for the next 48 hours. The maximum rise in temperature was noted after 4 hours.

A similar experiment is shown in Text-fig. 2.

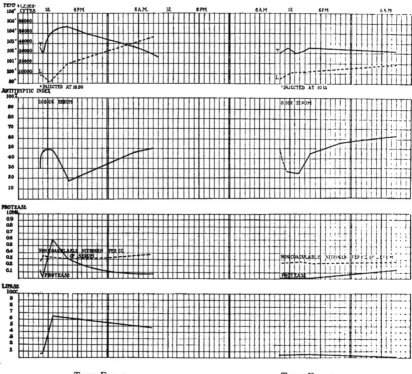

TEXT-FIG. 2.   TEXT-FIG. 2 a.

TEXT-FIG. 2. Effect of trypsin injection on the serum ferments and antiferment titer.

TEXT-FIG. 2 a. Effect of subcutaneous injection of trypsin on the serum ferments and antiferment titer.

Dog 14.—(Text-fig. 2.) Weight 9.4 kilos. Injected 0.1 gm. of very active trypsin at 10.55 a. m. Bled at 11 a. m., 1 and 4 p. m., and the following morning. In this experiment the increase in protease was not noted immediately after injection, but only at the time of the second bleeding, when the maximum was reached. The lipase, too, showed no immediate change, but remained constant, showing a marked increase at 1 p. m. The antiferment, instead of showing an immediate decline, rose at first, and the fall was noted only during the afternoon. The temperature and leucocyte count were similar to those in the previous animal.

Text-fig. 3 illustrates a similar experiment.

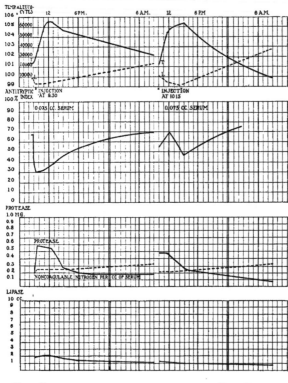

TEXT-FIG. 3.   TEXT-FIG. 3 a.

TEXT-FIG. 3.   Effect of trypsin injection on serum ferments and antiferment titer.

TEXT-FIG. 3 a.   Effect of inactivated trypsin injection.

Dog 35.—(Text-fig. 3.)   Weight 6 kilos.   0.2 gm. of an active trypsin preparation was given intravenously at 9.30 a. m.   The animal became quite ill, the temperature reaching 105° F.   The rise in the protease in the serum was immediate. The antiferment dropped quite markedly. The animal showed no rise in lipase. The non-coagulable nitrogen rose almost immediately and remained high for 24 hours.

Inactive Trypsin.

According to Kirchheim, his inactive trypsin preparations were non-toxic in the animals which he used (guinea pigs and rabbits). In our experiments we have not been able to confirm this finding. Inasmuch as by this method the question as to whether the ferment is toxic because of its ferment property, or because of its inherent toxicity as a proteose, is to be decided, we have repeated our experiments under various conditions and with several different trypsin preparations. We have without exception found the inactivated ferment toxic for dogs, as indicated by the resulting malaise and the effect on temperature and leucocyte count. The effect on the ferments of the serum is shown in the accompanying charts.[1]

Dog 35.—(Text-fig. 3 a.) The same trypsin preparation was used as before, but inactivated by boiling for 10 minutes after alkalinizing with sodium carbonate. 0.2 gm. was given intravenously at 10.15 a. m. The animal became ill and showed a temperature and leucocyte count similar to that following the active trypsin injection. The antiferment showed a slight rise followed by a drop, with recovery the following day. The protease, which was high at the beginning of the experiment, showed a progressive decline. The non-coagulable nitrogen increased after 24 hours, as with the active preparation. The lipase showed practically no change.

Dog 13.—(Text-fig. 4.) Weight 5.7 kilos. 0.25 gm. of inactivated trypsin was injected at 10.20 a. m. This animal responded with a picture of intoxication similar to that produced by the active preparation. The temperature and leucocyte curve are typical. There was a distinct fall in the antiferment, followed by a gradual recovery. There was a moderate rise in the serum lipase. The serum protease showed a slight increase, with a following decline to zero the next day. The non-coagulable nitrogen, with the exception of a slight initial rise, remained unchanged.

Subcutaneous Injections.

If the dose, instead of being brought directly into the blood stream. is injected subcutaneously, there are apparent no toxic effects as far as the general condition of the animal is concerned. Such an experiment is as follows:

[1] F. Ishiwara (abstracted in *Sei-I-Kwai Med. Jour.*, 1915. xxxiv. 19) has recently noted the toxicity of inactivated trypsin and considers it due to the presence of diamino-acids.

Dog 14.—(Text-fig. 2 a.) 0.1 gm. of active trypsin was injected subcutaneously at 10.15 a. m. (the intravenous dose is shown in Text-fig. 2). The temperature of the animal remained practically unaltered; the leucocyte count increased gradually. The antiferment curve showed rather a marked fall, with

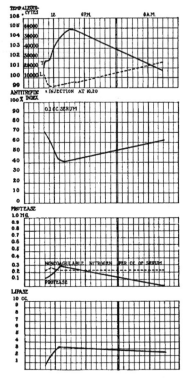

TEXT-FIG. 4. Effect of inactivated trypsin on serum ferments and antiferment titer.

an increase the following day. The protease was unaltered until the following morning when a slight increase was noted. The lipase showed a slight decrease.

Gastric Absorption.

When pancreatin is introduced into the stomach there is practically no increase in the ferments of the serum, the introduced ferment being probably destroyed before the duodenal tract is reached. The following experiment is illustrative.

Dog 14.—(Text-fig. 5.) 1 gm. of commercial pancreatin was dissolved in 25 cc. of water and injected into the stomach at 9.30 a. m. Bled at 10.00 and 11.30 a. m., 1.30 and 4.00 p. m., and the following morning. There resulted a gradual rise in temperature and leucocyte count, and a rather marked fall in antiferment, with rapid recovery, was noted. The lipase remained constant. The non-coagulable nitrogen showed first a slight decrease, followed by a rise in the afternoon. The protease, which fell to zero after injection, reached the original titer in the afternoon, but declined the following morning.

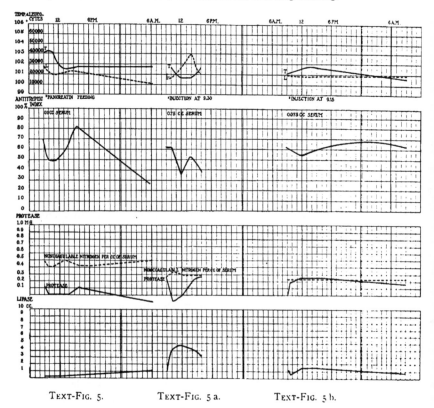

TEXT-FIG. 5.　　　TEXT-FIG. 5 a.　　　TEXT-FIG. 5 b.

TEXT-FIG. 5. Effect of gastric absorption of pancreatin on serum ferments and antiferment titer.

TEXT-FIG. 5 a. Effect of intestinal absorption of trypsin on serum ferments and antiferment titer.

TEXT-FIG. 5 b. Effect of intestinal absorption of pancreatin on serum ferments and antiferment titer.

Intestinal Absorption.

In order to determine whether a toxic effect of the ferment could result from intestinal absorption the following experiment was undertaken in which the dissolved ferment was injected directly into a loop of the small intestines.

Dog 24.—(Text-fig. 5 a.) The animal was anesthetized and a loop of small bowel isolated through a small laparotomy wound under strict aseptic precautions; 0.5 gm. of a very active trypsin preparation was injected. First blood taken during anesthesia. Bled at 10.00 a. m., noon, 2 and 4 p. m., when the animal was killed. Text-fig. 5 a gives the ferment picture in detail. Contrary to the direct intravenous injection the dog showed a marked leucocytosis and only a slight fall in temperature. The antiferment showed the usual fall with partial recovery and secondary drop. The lipase curve showed the typical rise indicative of the intoxication. The serum protease, instead of a rise, showed a distinct fall, a recovery to the normal titer occurring in the afternoon.

In place of the very active and toxic preparation the following experiment was made with pancreatin (commercial).

Dog 40.—(Text-fig. 5 b.) Similar to the above; 0.5 gm. of commercial pancreatin was injected into a loop of the small bowel at 9.15 a. m. Bled at 9.30 a. m., noon, and 3.00 p. m. In this experiment there was no evidence of an intoxication in so far as the temperature and leucocyte count are concerned. There was practically no change in the antiferment and non-coagulable nitrogen of the serum; the protease showed an immediate increase, while the lipase showed at first a fall and then a slight rise in titer.

DISCUSSION.

The experiments have demonstrated that when active trypsin solutions are injected into the blood stream an intoxication results, manifested by marked gastro-intestinal irritation, a rise of temperature, with a primary leucopenia, followed by a leucocytosis. There is usually an immediate rise in serum protease and serum esterase, together with a lengthening of the coagulation time, which in some instances may lead to a complete inhibition of coagulation. In many ways the picture is not dissimilar to anaphylactic shock, which we shall discuss in a later paper. In the latter condition there is associated, however, a marked increase in the non-coagulable nitrogen of the serum, representing protein split products, and in so far differing from the effect with trypsin.

The antiferment change usually consists of an immediate drop, with recovery following in a few hours. This lowering of the antitryptic titer is probably not due to a saturation of the antiferment by the injected trypsin, for the extent of the fall is not related to the amount of ferment injected nor to its activity. It is more probably the expression of a colloidal change in the serum whereby the lipoids are changed from a greater to a less disperse state. This would seem reasonable in view of the fact that a similar change occurs in anaphylactic and peptone shock and following the injection of various other substances,—bacteria, serum, etc. There is, of course, the possibility of a rapid oxidation occurring, with a resulting lowering of the antiferment strength because of an alteration of the unsaturated carbon bonds. A simple physical change seems, however, to account more readily for the rapidity of the various fluctuations that occur.

The question as to the toxicity of trypsin resolves itself into an endeavor to decide whether free protease activity in the blood stream is noxious; whether the trypsin molecule apart from its tryptic activity is toxic; whether due to admixed foreign toxic material (protein split products); or whether the ferment free in the serum is able to split non-toxic body proteins to toxic products and so induce an indirect toxic effect.

We have been able to demonstrate a considerable increase in serum protease immediately following the injection, an increase maintained for two or more hours and then rather rapidly disappearing. It would seem reasonable that this represents the injected ferment, were it not for the fact that the ferment activity so demonstrable is not active in an alkaline medium, but only in a neutral or acid reaction; that the increase may be progressive; that in one experiment the ferment was not demonstrable in blood drawn five minutes after the injection, but became apparent later. Furthermore, a similar, if not so marked an increase in serum protease may occur during peptone and anaphylactic shock. If we regard the ferment rise as a mobilization of ferments from the animal organism, the explanation that the lowering of the antiferment titer is due to a saturation with ferment finds a stronger basis, for the amount of ferment so liberated need have no relation to the amount and strength of the ferment injected. On the other hand, the inactivated preparations, while toxic, usually do not produce the increase in serum protease, although the antiferment may show marked changes.

Our experiments differ from those of Kirchheim in that we have found the inactive trypsin solution toxic for dogs; Kirchheim found that his inactivated preparations were non-toxic for guinea pigs and rabbits. We are not able to account for this discrepancy. As evideuces of the toxic effect we have observed the general conditions of the animals, together with the temperature and leucocyte count, and in no instance have we noted any great difference between the active and inactive preparations in these respects. From our results we are inclined to the view that the toxicity does not depend on the activity of the ferment, but rather on the ferment molecule itself. While it is true that the more active trypsin preparations are most toxic, the increase may in part be due to the elimination of protective colloids during the course of purification.

The rise in lipase noted after the injections is not peculiar to the trypsin intoxication, but occurs during similar intoxications from other causes (anaphylaxis, peptone, etc.).

As might be anticipated, the experiments with the feeding or subcutaneous injections of trypsin lead to no appreciable increase in the amount of serum protease demonstrable, nor were there any definite effects of intoxication from the amounts used.

On the other hand, when the ferment is brought directly into the small bowel, an influence on the serum protease is apparent. This may result in a simple increase when a large amount of a relatively weak trypsin preparation is used, without evidences of an intoxication. This would in a way simulate the normal condition during digestion, except that the ferment finds no substrate to which it can become attached. When, however, a very active preparation is injected, an intoxication may result. Instead of the blood showing an increase of protease in this case, an actual decrease may occasionally be observed, as shown in Text-fig. 5 a. When the liver of such an animal is examined histologically, profound fatty changes are observed. It is possible that the ferment, when absorbed, does not pass the liver into the general circulation. This would correspond to the condition which we have noted after feeding, in which the portal blood may contain considerable protease, while the peripheral blood may contain much less or none at all.

The possibility that intoxication may occur from the absorption of large amounts of free tryptic ferment from the bowel under pathological conditions must be considered. This would be more apt to occur if the intestinal tract were empty, so that the secreted ferment would remain unbound by proteins.

As will be noted from the text-figures, there is no fixed relation between the number of leucocytes and the ferments or antiferment, thus corroborating the recent finding of Rosenow and Färber (6), who noted that the leucocyte count and the destruction of leucocytes had no constant effect on the antiferment titer. Caro (7) noted that the number of lymphocytes (which are supposed to contain a lipolytic ferment) bore no relation to the lipolytic activity of the serum.

The questions which arise in view of the marked increase in both protease and lipase during these intoxications (inclusive of anaphylactic shock, peptone poisoning, etc.) concern their origin and effect on the organism. We are inclined to the view that they are mobilized from the tissue cells in general, and not from any specific organ. It seems certain at least that they do not come from the liver, for the hepatic blood usually contains the minimal concentration of the ferments under discussion.

In order to discuss their effect we must keep in mind certain fundamental factors concerned in protein intoxication. The true proteases act only on native proteins, which are split into certain definite components probably preexisting in the protein molecule (Levene). These may be toxic, certain of the proteoses being especially toxic for guinea pigs (8, 9), others, as far down as the peptones, being even more markedly toxic for dogs.

The proteases do not act in the presence of the antiferment. Inasmuch as the serum always contains an antiferment, protease action is normally held completely in abeyance. Therefore under normal conditions, even with proteases present in the serum, toxic split products can not be formed from the serum proteins. This protection is not necessarily intracellular.

Any condition which tends to remove the antiferment protection may result in splitting and a consequent intoxication. In this way the so called anaphylatoxins are formed when the antiferment is adsorbed by the various agents used.—bacteria, agar, kaolin, etc.

The released serum protease then can simply split the serum protein (10). In a similar manner when the antiferment is removed by lipoidal solvents the serum becomes toxic (11); and the splitting which takes place in the Abderhalden test has its basis along similar lines (12).

It is possible that a lowering of the antiferment titer may have some part in producing an intoxication. Such a lowering takes place not only following trypsin injections, but following many other conditions, some of which have been noted by Pfeiffer and Jarisch (13). The toxicity of kaolin (14) and certain colloids can be explained in this way. Conversely the raising of the antiferment titer may prevent an intoxication if that intoxication depends upon the formation of toxic split products and not upon the introduction of preformed toxic substances. We have previously called attention to such conditions (15).

We may sharply differentiate the true serum protease from the ereptase action of the serum, which is solely concerned with the split products of the protein molecule (casein being an exception to the general rule that the native proteins are not split by this ferment). The ereptase is not influenced by the antiferment, so that under normal serum conditions it is only the ereptase which can be active. It is this ferment which is chiefly responsible for the changes occurring in the Abderhalden reaction. The ereptase may be responsible as an intoxicating, as well as a detoxicating agent. Thus a serum may contain non-toxic proteoses which may be rapidly split to toxic peptones by the ferment. Later, however, the serum ferment may split these same toxic substances to amino-acids and thus cause a complete detoxication.

The ereptase must not be supposed to be only lytic in action. There is considerable evidence, which we shall discuss in a later paper, that the ferment is actually synthetic under certain conditions. From these conditions it would seem probable that the increase in protease and ereptase which occurs during the various so called protein intoxications may be of considerable aid in the process of detoxication.

CONCLUSIONS.

1. The intravenous injection of trypsin in dogs results in a shock similar in many respects to anaphylactic and peptone shock.

2. The injection is followed by a marked rise of serum protease and lipase.

3. The antiferment usually shows a distinct drop in titer, with a recovery following in from four to twenty-four hours.

4. The non-coagulable nitrogen shows no constant alteration, but is never greatly changed in amount.

5. Inactivated preparations were in some respects followed by symptoms similar to those following the injection of the active preparation.

6. Subcutaneous and gastric absorption was practically without effect.

7. Intestinal absorption was followed by an increase in serum protease without evidence of intoxication, or by typical symptoms of acute poisoning.

8. The leucocyte curve bears no constant relation to the serum protease or lipase.

BIBLIOGRAPHY.

1. Kirchheim, L., *Arch. f. exper. Path. u. Pharmakol.*, 1911, lxvi, 352; 1912–13, lxxi, 1; 1913, lxxiii, 139; lxxiv, 374. Kirchheim, L., and Reinicke, H., *ibid.*, 1914, lxxvii, 412. Kirchheim, L., and Böttner, A., *ibid.*, 1914, lxxviii, 99.
2. Langenskiöld, F., *Skand. Arch. f. Physiol.*, 1914, xxxi, 1.
3. Marie, P.-L., and Villandre, C., *Jour. de physiol. et de path. gén.*, 1913, xv, 602.
4. Jobling, J. W., and Petersen, W., *Jour. Exper. Med.*, 1914, xix, 239.
5. Jobling, J. W., Eggstein, A. A., and Petersen, W., *Jour. Exper. Med.*, 1915, xxi, 239.
6. Rosenow, G., and Färber, G., *Ztschr. f. d. ges. exper. Med.*, 1914, iii, 377.
7. Caro, L., *Ztschr. f. klin. Med.*, 1913, lxxviii, 286.
8. Jobling, J. W., and Strouse, S., *Jour. Exper. Med.*, 1913, xviii, 591.
9. Zunz, E., and György, P., *Ztschr. f. Immunitätsforsch., Orig.*, 1914, xxiii, 296.
10. Jobling and Petersen, *Jour. Exper. Med.*, 1914, xx, 37.
11. Jobling and Petersen, *ibid.*, 1914, xix, 480.
12. Jobling, Eggstein, and Petersen, *loc. cit.*
13. Pfeiffer, H., and Jarisch, A., *Ztschr. f. Immunitätsforsch., Orig.*, 1912–13, xvi, 38.
14. Friedberger, E., and Tsuneoka, R., *Ztschr. f. Immunitätsforsch., Orig.*, 1913–14, xx, 405.
15. Jobling and Petersen, *Jour. Exper. Med.*, 1914, xx, 468.

THE IMMUNOLOGICAL RELATIONS OF THE ROUS CHICKEN SARCOMA.[1]

By WILLIAM H. WOGLOM, M.D.

(*From Columbia University, George Crocker Special Research Fund, New York.*)

(Received for publication, April 22, 1915.)

The tumor[2] used throughout these experiments was the first chicken sarcoma reported by Dr. Peyton Rous,[3] to whose courtesy the laboratory of the Crocker Fund is indebted for the fowl from which the transplants were made.

This growth, according to Rous's description, is composed of loose bundles of spindle cells, crossing in every direction and separated from the smaller blood vessels only by endothelium. Intercellular fibrils can be demonstrated with Mallory's phosphotungstic acid stain, though they are rare in the more cellular portions of the tumor. Areas of necrosis are present, dependent, in general, upon insufficient vascularization. The sarcoma shows a marked tendency to invade the surrounding tissues; furthermore, it metastasizes, generally by way of the blood stream and most commonly in the lungs, although secondary nodules in the heart, liver, and spleen are not rare. It possesses, accordingly, many of the characteristics of the transplantable tumors of the mouse and rat, but differs fundamentally from these in being transmissible in the form of a Berkefeld filtrate or of dried tissue.

When the present investigation was started, the immune reactions associated with this growth had not been fully investigated;

---

[1] Read before the American Association for Cancer Research, St. Louis, April 1, 1915.

[2] The employment of such terms as "sarcoma" and "tumor" throughout this paper is to be looked upon rather as a concession to convenience than as indicating the possession of any definite idea regarding the nature of the material in question.

[3] Rous, P., *Jour. Exper. Med.*, 1910, xii, 696.

thus, it was not known whether fowls can be rendered resistant to its inoculation by previous treatment with fowl tissue, after the manner in which mice can be made refractory to mouse tumors with the normal tissues of their species.

The injection of 0.05 gm., or even less, of mouse spleen, kidney, embryo, or blood corpuscles will confer a resistance to the subsequent implantation of mouse tumor in from 70 to 100 per cent of treated animals, which sets in by the third day, reaches its height about the tenth, and persists for approximately three months.

It has been suggested by Pitzman[4] that the refractory condition so evoked is due solely to a bacterial infection set up at the time when the immunizing material is introduced. If this were true, it should be possible to elicit resistance by preliminary treatment with the tissues of animals other than the mouse; the majority of observers, however, deny that this can be accomplished. In serious conflict with such an hypothesis, furthermore, is the observation of Woglom[5] that the highest degree of resistance is procured by treatment with embryo skin, although, as both aerobic and anaerobic cultures made in this laboratory show, this is the tissue, of all those used to induce the refractory condition, which is certain to be sterile. It is highly probable, therefore, that the presence of an artificial immunity to tumor implantation represents the completion of a specific reaction.

A preliminary communication[6] described unsuccessful attempts to duplicate this reaction in fowls by treatment with ten day chick embryos, five to forty days before tumor inoculation. The small size of some of the growths in fowls thus injected thirty-two or forty days before introduction of the sarcoma suggested the possibility that a much longer time might be required for the development of complete immunity than the ten days necessary in mice. Hence, the period was extended to 100 days.

Details of individual experiments are to be found in the accompanying table.

The fowls which it was sought to immunize were injected in the

[4] Pitzman, M., *Ztschr. f. Krebsforsch*, 1914, xiv, 57.
[5] Woglom, W. H., *Jour. Exper. Med.*, 1912, xvi, 629.
[6] Woglom, *Proc. N. Y. Path. Soc.*, 1914, N. S., xiv, 202.

left breast with from 1 to 10 cc. of fresh hashed chicken embryo, and at periods varying from 5 to 100 days afterward were inoculated with intact grafts (0.02 gm.) of tumor in the right breast,

TABLE I.

Experiment No.	Interval between treatment and tumor inoculation.	Dose of embryo emulsion.	No. of chickens.	Tumors.	No tumors.
	days	*cc.*			
I	5	5.0	10 controls*	10	0
			8 treated	5	3
II	5	10.0	6 controls	6	0
			6 treated	6	0
III	10	1.0	15 controls	15	0
			17 treated	16	1
IV	12	5.0	11 controls	11	0
			10 treated	10	0
V	14	2.0	5 controls	5	0
			9 treated	9	0
VI	14	4.0	2 controls	2	0
			5 treated	5	0
VII	25	5.0	10 controls*	10	0
			11 treated	10	1
VIII	28	2.0	3 controls	3	0
			3 treated	3	0
IX	32	5.0	5 controls	5	0
			3 treated**	2	1
X	40	5.0	11 controls	11	0
			7 treated**	7	0
XI	70	5.0	22 controls	21	1
			20 treated	18	2
XII	100	5.0	16 controls	15	1
			18 treated	18	0

* Same controls used for both experiments.
** More small tumors than among controls.

together with an equal number of normal controls. Three weeks after implantation of the tumor the fowls were autopsied.

117 treated and 106 control fowls lived long enough to come to autopsy. Of the treated, 109 (93 per cent) proved receptive for the tumor, and among the controls 104 (98 per cent) developed growths.

The difference between the treated fowls and their controls is slight enough to warrant the statement that immunity to the sarcoma in question can not be produced by preliminary injection with chicken embryo in the amounts administered. The number of tumors in each group is approximately the same, and, with the exception of Experiments IX and X, the growths in the treated fowls were fully as large as those in the controls.

Although three treated chickens out of eight in Experiment I did not develop tumors, a repetition of the experiment with double the amount of embryo emulsion gave a clean cut result, none of the treated chickens being found resistant. These, as well as the other five instances in which fowls previously injected with embryonic material failed to develop tumors, are, therefore, referable in all probability to natural resistance rather than to an artificial immunity consequent upon the preliminary treatment, a view strengthened by the fact that Rous and Murphy[7] also have recently recorded their failure to immunize against this growth with normal tissue. The immune fowls in the treated series are partially offset, moreover, by two controls in which the grafts failed to proliferate.

The absence of any resistance 70 and 100 days after attempted immunization completely nullifies the suggestion contained in a preceding paragraph, that a period much longer than the 10 days requisite in mice might be necessary for the development of complete immunity in the chicken.

The failure to produce resistance can not be ascribed to insufficient dosage. A mouse weighing 15 grams can be made refractory to transplantable tumors by preliminary injection with 0.05 cc. of normal mouse tissue, an amount representing about 1/300 of its body weight, and 5 cc. of chicken embryo, when injected into a fowl weighing 1,500 grams, is roughly equivalent to this quantity.

The absence of immunity is not an unanswerable argument against the neoplastic nature of this tumor; for, in the first place. it is not known that the immunological reactions characteristic of the mouse have their counterpart in the chicken, and, secondly, a tumor is occasionally found, even in the mouse, against which no resistance can be produced. At most it can be said only that the outcome of these experiments is a warning against the unreserved acceptance of this growth, at present, as a true tumor.

CONCLUSION.

The injection of chicken embryo, in amounts of from 1 to 10 cc., confers no resistance against the Rous chicken sarcoma, when this is inoculated from 5 to 100 days after the preliminary treatment.

[7] Rous, P., and Murphy, Jas. B., *Jour. Exper. Med.*, 1914, xx, 419.

CASTRATION EXERTS NO INFLUENCE UPON THE GROWTH OF TRANSPLANTED OR SPONTANEOUS TUMORS IN MICE AND RATS.

By JOSÉ S. HILARIO, M.D.

(*From Columbia University, George Crocker Special Research Fund, New York.*)

(Received for publication. May 18, 1915.)

Graf[1] was the first to seek some relationship between the internal secretion of the sexual organs and the development of transplanted tumors. The sexual glands were removed from 58 male and 60 female mice, which were afterwards inoculated with a tumor having an inoculation percentage of from 80 to 100, 40 male and 31 female unoperated mice serving as controls. He concluded that castration exerts no essential influence either upon the percentage of takes or upon the rate of growth. A few years later, Almagià[2] described five experiments, comprising 30 castrated and 50 normal mice inoculated with a tumor characterized by slow growth and a comparatively low inoculation percentage. Among the 50 normal animals, 42 had tumors and 10 had not, while in the castrated group only 6 were positive, 24 having no tumor. It was observed, furthermore, that the injection of testicular extract into tumor-bearing mice exaggerated the development of their growths, and that castrates which had proved refractory to inoculation could be restored to a condition of receptivity by the introduction of this material. He concluded, therefore, that the development of a tumor is stimulated, either directly or indirectly, by the sexual glands, and inhibited by their absence.

Rohdenburg, Bullock, and Johnson,[3] like Graf, expressed the opinion that castration exerts no effect. or. at the most. a very slight one, upon either the inoculation percentage or the rate of growth.

An inference exactly contrary to that of Almagià has recently been recorded by Sweet, Corson-White. and Saxon,[4] whose material consisted of 73 castrated mice and 14 castrated rats. compared with 73 normal mice and 15 normal rats. These authors used a tumor with but moderate powers of adaptation and proliferation, since such a growth would be more sensitive to any slight degree of immunity present than would one with a very high inoculation percentage.

[1] Graf. R., *Centralbl. f. Path.*, 1909, xx, 783.

[2] Almagià. M., *Bull. d. r. Accad. di med. di Roma*, 1912, xxi. 102.

[3] Rohdenburg, G. L., Bullock, F. D., and Johnson, P. J., Studies in Cancer and Allied Subjects. New York. 1913. iii. 87

[4] Sweet. J. E., Corson-White, E. P., and Saxon. G. J., *Jour. Biol. Chem.*, 1913. xv. 181.

They asserted that receptivity for such a tumor is increased by castration, and that the proliferative power of the neoplasm is augmented.

Since each of the three possibilities—(*a*) that castration increases tumor growth, (*b*) that castration retards tumor growth, (*c*) that castration exerts no effect upon tumor growth—has been advocated, it is necessary that the experiment be repeated on a larger number of animals and with tumors of various types.

The following experiments, amounting to nine in number, include in the final reckoning 256 male animals, of which 88 were castrated mice, 35 were castrated rats, 103 were normal mice, and 30 were normal rats. The animals in every experiment except VI and IX were of approximately equal weight and age, and all those in any one experiment were obtained from the same dealer. Castration was performed under ether anesthesia, through an incision in the median line of the scrotum, care being taken not to crush the glands in the forceps so that complete removal of all testicular tissue might be assured. The wound was closed by a continuous catgut suture.

From ten to twenty-six days after castration the animals were inoculated in the right axilla, seven series with carcinoma, and two with sarcoma. The castrated animals and their controls were inoculated at one sitting from the same tumor and with grafts of similar size, and in every instance the castrates and their controls were kept in adjoining boxes and on the same diet. Details regarding tumor dose, weight of animals, etc., are to be found under Text-figs. 1 to 10, in which the tumors are reproduced in silhouette. their actual size being readily calculable from the 10 cm. scale attached. The first charting was made ten days after inoculation. subsequent ones at weekly intervals; a † sign indicates the death of an animal. The results were read at a time when the largest number of animals could be included. More exact information on this point will be found in the detailed discussion of the charts.

Growths of various kinds were employed to test the presence or absence of resistance. Thus, among the mouse carcinomata. Crocker Fund Tumors Nos. 11, 15, and 46 are adenocarcinomata with an inoculation percentage of from 50 to 75, which grow slowly and occasionally retrogress, while Tumor 180, a carcinoma of the solid type, grows in every mouse with extreme rapidity and never recedes. The Ehrlich mouse sarcoma, with an inoculation percentage of from 50 to 100, grows quickly and sometimes regresses. As for rat tumors, the Flexner-Jobling adenocarcinoma grows with moderate speed in from 50 to 100 per cent of inoculated rats, receding rather frequently; the Jensen sarcoma grows more rapidly, gives an inoculation percentage of from 50 to 100 and often retrogresses.

With this preliminary description of experimental material, the text-figures may be introduced.

Experiment I.—(Text-fig. 1.) The result was read at the fourth charting, in order to include Nos. 6, 24, 26, 31, and 32, which died before the fifth. The control group contains 17 positives (85 per cent) and 3 negatives, the castrates 8 positives (67 per cent) and 4 negatives. The rate of growth is the same in both series. In the control group 2 tumors (7 and 18) receded, and in the castrates none.

Experiment II.—(Text-fig. 2.) Here the computation was made at the third charting to include Nos. 2, 9, 13, and 32, No. 15 being counted among the positives since it developed a tumor before the next charting. Hence there are 15 positives (75 per cent) and 5 negatives in the control, and 13 positives (62 per cent) and 8 negatives among the castrated mice. There were 10 cases of involution among the control animals (Nos. 3, 6, 7, 10, 11, 12, 17, 18, 19, 20) and 2 (Nos. 29, 36) in the castrated. Growth was more vigorous in the castrates, but this circumstance cannot be regarded as significant because this is the only experiment in which such a phenomenon was observed, and, secondly, because the rate of growth in the castrated animals is characteristic of Tumor 46 under normal conditions. It would, therefore, be more accurate to say that proliferation was diminished in the control groups, by some factor unconnected with the experiment. The high mortality suggests an infection, more severe, perhaps, in the controls, since it is known that tumors do not grow so well in sick animals.

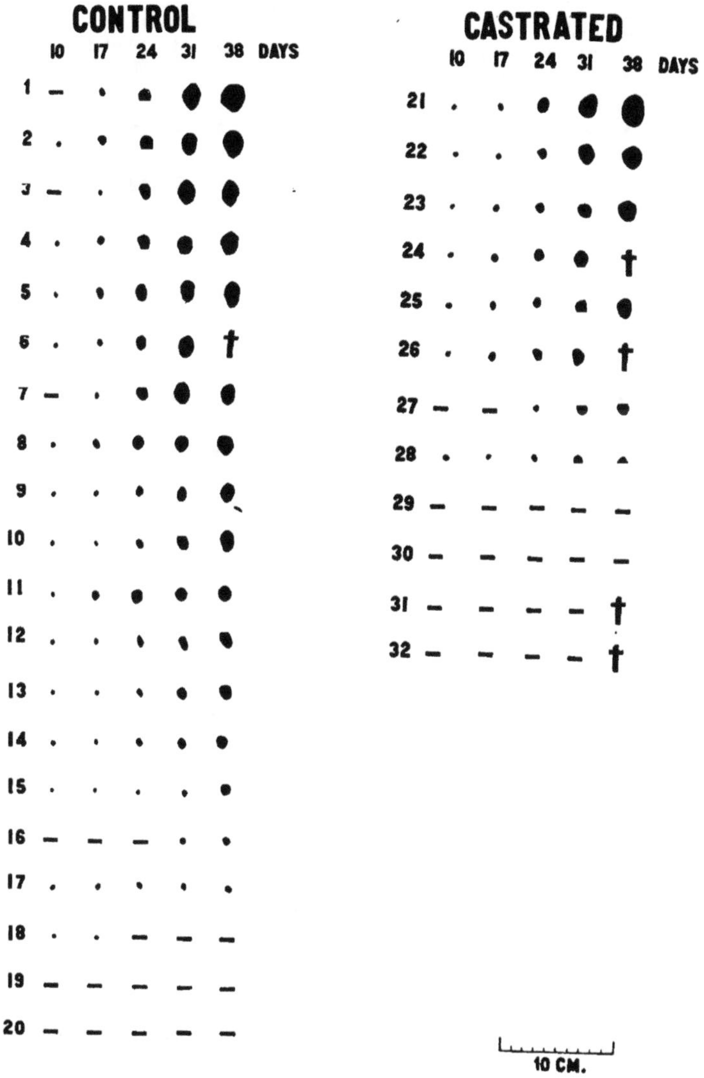

CONTROL

	10	17	24	31	38 DAYS
1	—	·	▲	●	●
2	·	●	▲	●	●
3	—	·	●	●	●
4	·	·	●	●	●
5	·	●	●	●	●
6	·	·	●	●	†
7	—	·	●	●	●
8	·	●	●	●	●
9	·	·	●	●	●
10	·	·	●	●	●
11	·	●	●	●	●
12	·	·	●	●	●
13	·	·	●	●	●
14	·	·	●	●	●
15	·	·	·	●	●
16	—	—	—	·	·
17	·	●	·	·	·
18	·	·	—	—	—
19	—	—	—	—	—
20	—	—	—	—	—

CASTRATED

	10	17	24	31	38 DAYS
21	·	·	●	●	●
22	·	·	●	●	●
23	·	·	●	●	●
24	·	·	●	●	†
25	·	·	·	▲	●
26	·	·	●	●	†
27	—	—	·	●	●
28	·	·	·	▲	▲
29	—	—	—	—	—
30	—	—	—	—	—
31	—	—	—	—	†
32	—	—	—	—	†

L┈┈┈┈┈┈┈┘
10 CM.

TEXT-FIG. I. Experiment I. $\frac{11Q}{20}$. Nos. 1-20, normal male controls (average weight 14.8 gm.). Nos. 21-32, castrated Dec. 12, 1914 (average weight 15.4 gm.). All mice inoculated Dec. 28, 1914, in the right axilla with 0.01 gm. by the needle method.

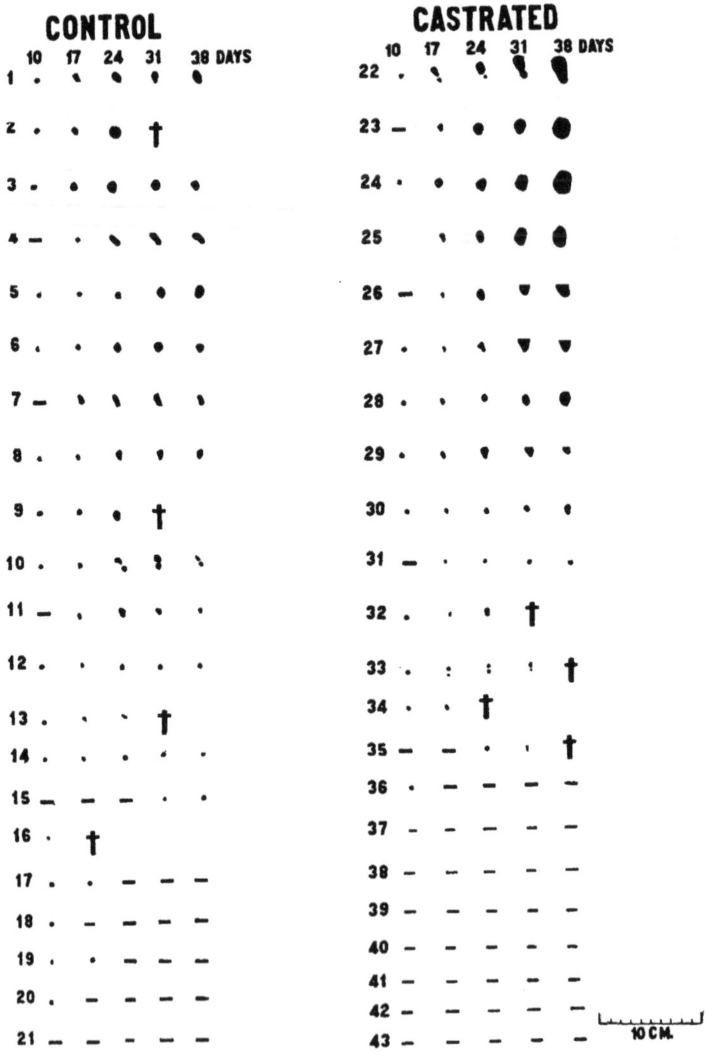

TEXT-FIG. 2. Experiment II. $\frac{46}{8D}$. Nos. 1–21, normal male controls (average weight 17.5 gm.). Nos. 22–43, castrated Dec. 1, 1914 (average weight 17.3 gm.). All mice inoculated Dec. 15, 1914, in the right axilla with 0.01 gm. by the needle method.

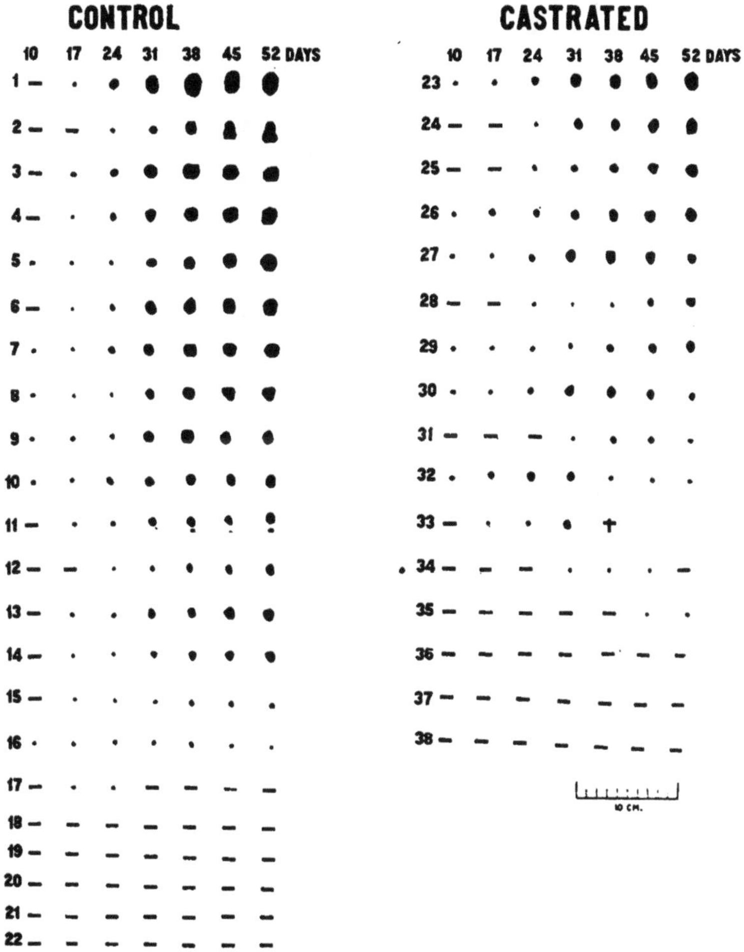

TEXT-FIG. 3. Experiment III. $\frac{15}{8D}$. Nos. 1-22, normal male controls (average weight 17.4 gm.). Nos. 23-38, castrated Dec. 10, 1914 (average weight 17.5 gm.). All mice inoculated Dec. 28, 1914, in the right axilla with 0.01 gm. by the needle method.

Experiment III.—(Text-fig. 3.) The calculation in this case was made at the seventh charting. 16 controls (73 per cent) were positive and 6 negative, while in the castrates 11 (73 per cent) were positive and 4 negative. There were 3 cases of involution (Nos. 9, 16, 17) among the control mice, and 4 among the castrated (Nos. 27, 30, 32, 34). The rate of growth was slightly decreased in the castrated animals.

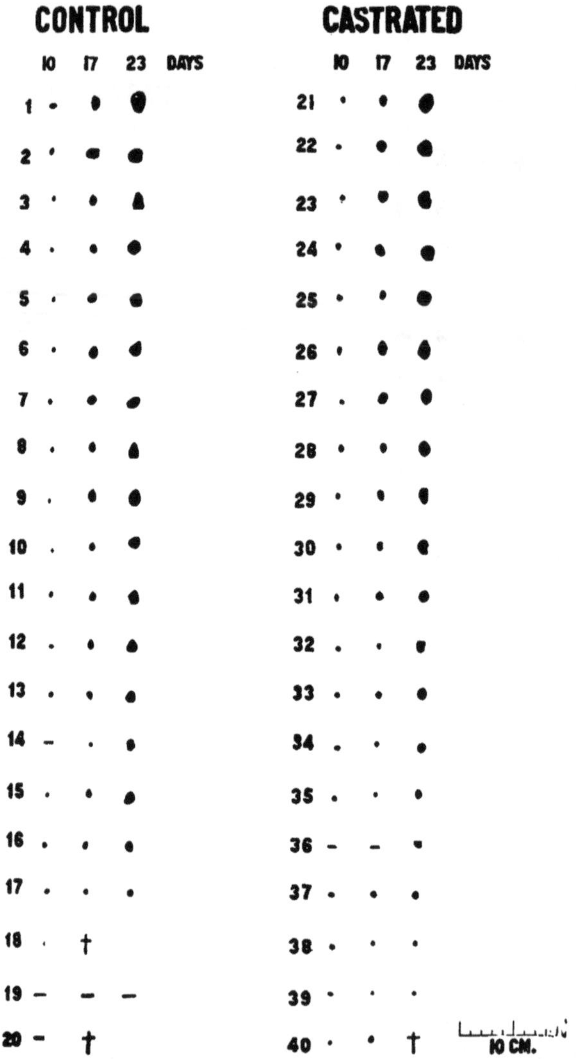

TEXT-FIG. 4. Experiment IV. $\frac{180}{3D}$. Nos. 1–20, normal male controls (average weight 16.4 gm.). Nos. 21–40, castrated Nov. 28, 1914 (average weight 15.7 gm.). All mice inoculated Dec. 8, 1914, in the right axilla with 0.01 gm. by the needle method.

Experiment IV.—(Text-fig. 4.) At the third charting, 17 controls (94 per cent) were positive and 1 was negative. Of the castrates 19 (100 per cent) were positive and none were negative. The rate of growth was the same in both groups, and there was no instance of regression.

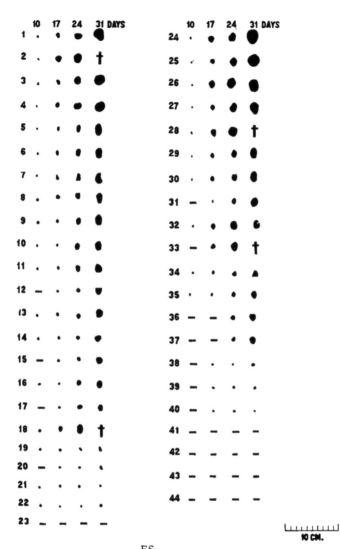

	10	17	24	31 DAYS		10	17	24	31 DAYS
1	.	•	●	●	24	.	•	●	●
2	.	•	●	†	25	،	•	●	●
3	.	،	●	●	26	.	●	●	●
4	.	•	●	●	27	.	•	●	●
5	.	•	●	●	28	،	●	●	†
6	.	•	●	●	29	.	•	●	●
7	.	،	●	●	30	.	•	●	●
8	.	•	●	●	31	—	•	●	●
9	.	•	●	●	32	.	•	●	●
10	.	•	●	●	33	—	•	●	†
11	.	•	●	●	34	.	•	●	●
12	—	•	●	●	35	.	•	•	●
13	.	•	•	●	36	—	—	●	●
14	،	•	•	●	37	—	—	•	●
15	—	•	●	●	38	—	.	.	•
16	.	•	●	●	39	—	•	•	•
17	—	•	●	●	40	—	.	.	.
18	.	•	●	†	41	—	—	—	—
19	.	•	،	،	42	—	—	—	—
20	—	•	.	،	43	—	—	—	—
21	.	•	•	.	44	—	—	—	—
22	.	.	•	•					
23	—	—	—	—					

10 CM.

TEXT-FIG. 5. Experiment V. $\frac{ES}{20D}$. Nos. 1-23, normal male controls (average weight 17.5 gm.). Nos. 24-44, castrated Dec. 9, 1914 (average weight 15.5 gm.). All mice inoculated Dec. 24, 1914, in the right axilla with 0.01 gm. by the needle method.

Experiment V.—(Text-fig. 5.) The reading in this case was made at the third charting, to include Nos. 2, 18, 28, and 33. In the controls there were 22 positives (96 per cent) and 1 negative, and among the castrates 17 positives (81 per cent) and 4 negatives. The growth rate was similar for the two groups, and no tumors regressed.

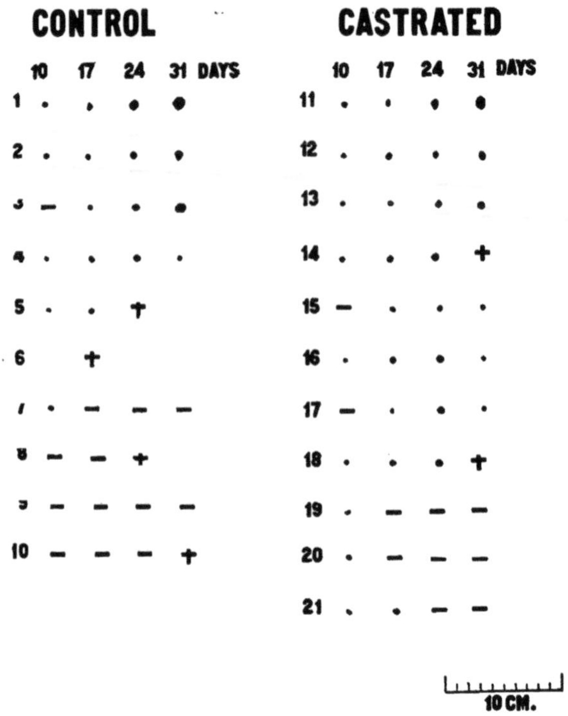

CONTROL   CASTRATED

TEXT-FIG. 6. Experiment VI. $\frac{FRC}{13S}$. Nos. 1–10, normal male controls (average weight 85.0 gm.). Nos. 11–21, castrated Dec. 14, 1914 (average weight 51.7 gm.). All rats inoculated Jan. 9, 1915, in the right axilla with 0.02 gm. by the needle method.

Experiment VI.—(Text-fig. 6.) This experiment offers another example of retarded growth in sick animals, except that, in contrast to Experiment II, both groups are about equally affected. To include this series, it is necessary to read the result at the second charting, where 5 of the controls (55 per cent) were positive and 4 negative. Among the castrates, 9 (82 per cent) were positive and 2 negative. The rate of growth was approximately the same in both groups. Two minute tumors regressed in the controls (Nos. 4, 7) and 6 in the castrates (Nos. 15, 16, 17, 19, 20, 21).

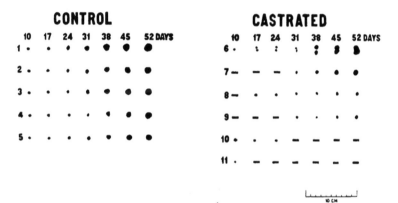

TEXT-FIG. 7. Experiment VII. $\dfrac{\text{FRC}}{13D}$. Nos. 1–5, normal male controls (average weight 50.0 gm.). Nos. 6–11, castrated Dec. 3, 1914 (average weight 48.3 gm.). All rats inoculated Dec. 14, 1914, in the right axilla with 0.02 gm. by the needle method.

Experiment VII.—(Text-fig. 7.) At the seventh charting, 5 controls (100 per cent) were positive and none negative, while 4 (67 per cent) of the castrates were positive and 2 negative. The rate of growth was somewhat diminished in the castrated rats, only 2 small tumors having regressed (Nos. 10 and 11).

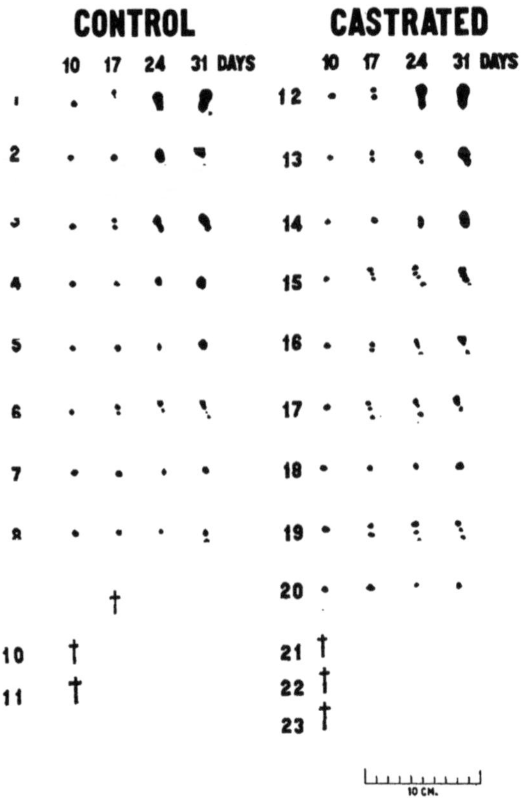

TEXT-FIG. 8. Experiment VIII. $\frac{FRC}{12U}$. Nos. 1-11, normal male controls (average weight 44.5 gm.). Nos. 12-23, castrated Dec. 4, 1914 (average weight 50.0 gm.). All rats inoculated Dec. 18, 1914, in the right axilla with 0.02 gm. by the needle method.

Experiment VIII.—(Text-fig. 8.) Among the controls 8 (100 per cent) were positive at the fourth charting, and none negative, while among the castrated rats 9 (100 per cent) were positive and none negative. The rate of growth was similar in both groups and none of the tumors regressed.

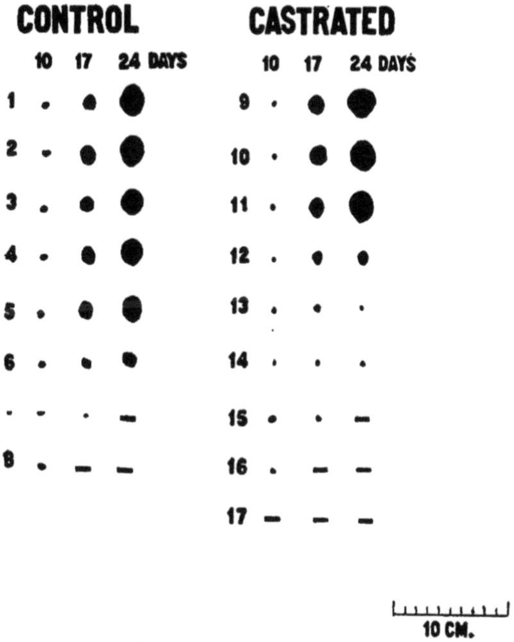

TEXT-FIG. 9. Experiment IX. $\frac{JRS}{17B}$. Nos. 1–8, normal male controls (average weight 39.0 gm.). Nos. 9–17, castrated Dec. 15, 1914 (average weight 72.7 gm.). All rats inoculated Dec. 29, 1914, in the right axilla with 0.02 gm. by the needle method.

Experiment IX.—(Text-fig. 9.) At the third charting, 6 controls (75 per cent) were positive and 2 negative, while of the castrates 6 (67 per cent) were positive and 3 negative. There were more large tumors in the control than in the castrated rats. Two small tumors (Nos. 7, 8) receded in the controls and 3 (Nos. 13, 15, 16) in the castrates.

Summing up, it appears that with the exception of Experiments VI and VII, where it was 27 and 33 per cent, there was a difference of less than 20 per cent between the castrated animals and their controls, a disparity which does not exceed the rather wide margin which must be allowed for experimental error in the biological investigation of cancer.

When the figures are set in two parallel columns (Table I) with their differences between, it is seen that in two experiments (III and VIII) the inoculation percentages were identical, in two (IV

TABLE I.

Experiment No.	Controls.	Difference.	Castrated.	Experiment No.	Controls.	Difference.	Castrated.
I......	85	18	67	VI	55	27	82
II......	75	13	62	VII	100	33	67
III......	73	0	73	VIII	100	0	100
IV......	94	6	100	IX	75	8	67
V......	96	15	81				

and VI) this percentage was larger in the castrated group, and that in five of the experiments (I, II, V, VII, and IX) it was larger in the controls. Although this might appear to indicate that castration decreases the receptivity for a transplanted tumor, it must be recollected that in two experiments (II and VI) the animals were probably sick, and that in one (VII) the number is very small, and it must be noted, also, that it is in two of these three experiments (VI and VII) that the greatest differences between the control and the castrated animals are found (27 and 33 per cent, respectively). In the other six experiments the difference lies between 0 and 18 per cent, averaging but 8 per cent.

When the single experiments are gathered into two groups, one including mice and the other rats, it is shown that among 88 castrated mice 68 (77.3 per cent) had tumors and 20 (22.7 per cent) had none, while among 103 normal controls 87 (84.5 per cent) had growths and 16 (15.5 per cent) had not. Of 35 castrated rats 28 (80 per cent) developed tumors and 7 (20 per cent) proved refractory, the group of 30 normal controls containing 24 (80 per cent) with, and 6 (20 per cent) without tumors. The total inocu-

lation percentage in the castrated rats and their controls is therefore identical, and, in the case of the mice, is nearly so, since the
7 per cent decrease is so slight that it may be disregarded. Individual disparities in the various experiments thus vanish when all
the groups are added together, and it is quite possible that the
slight difference in favor of the controls might similarly disappear
if a still larger number of experiments were available.

Nor is there any distinct evidence that, once the graft has obtained a foothold, the conditions governing its further development
are either more or less favorable in the castrated series than in the
controls. In five experiments (I, IV, V, VI, VIII) the growth
rate is identical in the two groups; in one (II) it is better in the
castrates, possibly by reason of a source of error already pointed
out; in three (III, VII, IX) it is more rapid in the controls. In
the control series seventeen tumors receded wholly or partially, and
the same number in the castrated animals.

The fact that the growth rate of transplanted tumors in normal
control mice and rats is occasionally a little in excess of that in
castrates, coupled with the observation that the inoculation percentage is a trifle higher in normal animals, might seem to indicate
some slight decrease in receptivity after castration. In both càses,
however, the difference is so small that it may justly be referred
to an outside factor such as minor nutritional disturbances or some
mild and unavoidable infection at the time of operation. Certainly if castration exerts a specific effect upon tumor growth, its
results should be more evident.

Almagià described a higher degree of cachexia among the controls than among the castrates, a statement which the experiments
now under consideration do not support, since there was no sign of
such a condition in either of the two groups. This corresponds
with the experience of the majority of observers, that cachexia
does not appear until the tumor has ulcerated, and that it is then to
be attributed to infection.

While it is well known from the work of Beatson, Abbe, and
others, that oophorectomy in young women may check temporarily
the growth of a carcinoma of the breast and even induce the disappearance of small metastatic nodules in the skin, yet it is equally

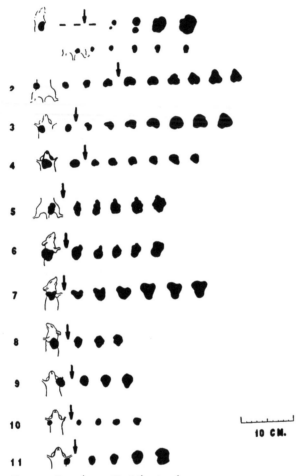

TEXT-FIG. 10. Experiment X. Mice bearing spontaneous mammary adeno-carcinomata, and castrated at the periods shown by. the arrows. The tumors were charted weekly. The removal of the ovaries exerted no influence upon the growth of the neoplasms.

Experiment X.—(Text-fig. 10.) From these 11 mice, old females with spontaneous carcinoma of the mamma, the ovaries were removed at periods designated by the arrows, the tumors being charted at weekly intervals there-after. No. 1 was castrated in order to observe the effect upon the inguinal growth; in the meantime, recurrence took place in the axilla from a tumor re-moved about 4 weeks previously. .The decrease in the dimensions of the growth in No. 6 was brought about by a hemorrhage. The other mice require no com-ment; it is perfectly evident that the removal of the ovaries had no effect on the activity of the tumor in any of the cases.

well known that none of these cases .are permanently benefitted. The period of influence is generally not more than six months. That 'no effect followed removal of the ovaries in this experiment may be due to the fact that the animals used had passed the age when these organs have any marked influence upon the blood supply of the breast.

CONCLUSIONS.

Castration neither increases nor decreases the inoculation percentage of transplantable carcinomata and sarcomata of the mouse and rat, nor does it either stimulate or retard their proliferation; it exerts not the slightest effect upon the growth of spontaneous carcinomata of the mouse.

The writer wishes to express his obligation to Professor Francis C. Wood, Director of the George Crocker Special Research Fund, for the privilege of working in the laboratory of the Fund, and to Drs. William H. Woglom and Frederick Prime, Jr., for many helpful suggestions during the prosecution of these experiments.

INFLAMMATORY REACTIONS IN RABBITS WITH A SEVERE LEUCOPENIA.[1]

By W. E. CAMP, M.A., and E. A. BAUMGARTNER, Ph.D.

(From the Department of Pathology of the University of Minnesota, Minneapolis.)

PLATES 23 AND 24.

(Received for publication, April 19, 1915.)

The object of this investigation is to study the local reaction to injury in animals in which a severe leucopenia has been produced experimentally. Inasmuch as the leucocytes are believed to play a very important part in the inflammatory process, it occurred to us that it would be of interest to study the inflammatory reaction under conditions in which the leucocytes could take part only to a very limited extent. The leucopenia was produced by subcutaneous injections of benzol after the method employed by Selling.

Benzol was first used experimentally as a leucotoxin by Selling.[2] It was mixed with an equal volume of olive oil, and given daily, by subcutaneous injection, in doses of 2 cc. of the mixture per kilo of body weight of the animal. In every case there was a marked decrease of leucocytes. After eight or ten daily injections the leucocyte count was usually between 500 and 1,000 per cmm. In one case the count was only 200. The decrease of the polymorphonuclear leucocytes was somewhat greater than that of the lymphocytes. When the leucocyte count was very low there was always a pronounced relative lymphocytosis.

Selling found almost complete aplasia of the bone marrow in cases in which the leucocytes were greatly reduced, There was considerable destruction of the lymphoid tissue of the lymph nodes, but the injury of these organs was not nearly so severe as that of the bone marrow. He did not find an increase of leucocytes in the capillaries of the viscera. Benzol undoubtedly destroys the leucocyte-forming tissue. Selling believes that it also kills the leucocytes in the circulating blood, since their decrease is too rapid to be accounted for by natural death.

[1] This study was undertaken at the suggestion of Dr. E. T. Bell, under whose supervision the work was conducted. We wish to thank Dr. Bell for his valuable aid and criticisms.

[2] Selling, L., Benzol als Leukotoxin, *Beitr. z. path. Anat. u. z. allg. Path.,* 1911, li, 576.

174

Pappenheim[3] also studied the effects of the administration of benzol to rabbits. There was wide-spread necrosis in the liver and severe parenchymal injury of the kidney in his experiments. The leucocytes disappeared almost completely from the peripheral blood, but in some of the cases they were still very numerous in the widened capillaries of the liver, lung, spleen, and kidney. Pappenheim believes that the actual decrease of leucocytes is therefore not so great as one would think from an examination of the peripheral blood.

Pappenheim and Plesch[4] produced leucocyte-free rabbits by a single intravenous injection of thorium X. The animals lived 3 or 4 days after the injection. At the end of 48 hours they were usually leucocyte-free. The bone marrow showed only a very few leucocytes. These were pseudo-eosinophils and were scattered among the red blood cells in between the fat cells. The spleen and lymph nodes showed atrophy of the follicles, and increase of the connective tissue cells. The liver showed congestion of the sinusoids around the central veins, and necrosis of the parenchymal cells in these areas. Small areas of hemorrhage were found scattered throughout the spleen and liver.

In leucocyte-free animals, vitally stained with lithium carmine, Pappenheim and Fukushi[5] produced a lymphoid cell exudate by injection of tuberculin into the peritoneal cavity. The exudate cells were of the large and small lymphoid type and contained carmine granules. The authors concluded that these cells were of connective tissue origin.

Lippmann and Plesch[6] injected cultures of the bacillus of swine erysipelas into the pleural and peritoneal cavities of animals made leucocyte-free by administration of thorium X. In the pleural cavity they obtained only a mononuclear exudate, and hold, from the morphology of the cells, that they are derivatives of the endothelium of the pleura. In the peritoneal cavity they obtained not only the mononuclear cells such as were found in the pleural cavity, but also some polynuclear cells. Lippmann and Plesch also found that aleuronat-bouillon suspension when injected into the muscles of an animal made leucocyte-free by thorium X, produced no cellular exudate. In the normal animals this irritant gave a pure polymorphonuclear exudate.

Rosenow[7] produced leucocyte-free dogs that lived from 20 to 48 hours, by

[3] Pappenheim, A.; Zur Benzolbehandlung der Leukämie und sonstiger Blutkrankheiten, *Wien. klin. Wchnschr.*, 1913, xxvi, 48; Experimentelle Beiträge zur neueren Leukämietherapie, *Ztschr. f. exper. Path. u. Therap.*, 1914, xv, 39.

[4] Pappenheim, A., and Plesch, J., Experimentelle und histologische Untersuchungen zur Erforschung der Wirkung des Thorium X auf den thierischen Organismus, *Ztschr. f. exper. Path. u. Therap.*, 1912-13, xii, 95.

[5] Pappenheim, A., and Fukushi, M., Neue Exsudatstudien und weitere Ausführungen über die Natur der lymphoiden peritonealen Entzündungszellen. *Folia Haematol.*, 1913-14, xvii, 257; abstracted in *Centralbl. f. allg. Path. u. path. Anat.*, 1914, xxv, 498.

[6] Lippmann and Plesch, J., Studien am aleukozytären Tier: ueber die Genese der "Lymphozyten" in den Exsudaten seröser Höhlen, *Deutsch. med. Wchnschr.*, 1913, xxxix, 1395.

[7] Rosenow, G., Studien über Entzündung beim leukocytenfreien Tier, *Ztschr. f. d. ges. exper. Med.*, 1914, iii, 42; abstracted in *Centralbl. f. allg. Path. u. path. Anat.*, 1914, xxv, 499.

means of a single intravenous injection of 1.0 to 2.0 mg. of radium bromide. A copper wire was passed through the anterior chamber of the eye. The exudate consisted usually of fibrin masses, red blood cells, and iris stroma cells. In one case he obtained in the exudate large mononuclear cells which he believes are desquamated degenerating endothelial cells or cells of the iris stroma. Intracorneal injection of staphylococci produced an exudate of small lymphoid-like cells. Transition forms from endothelium to leucocytes were not observed. The author believes that these experiments tend to support the conception that the exudate cells in the first stages of an acute inflammation are extravasated leucocytes.

Winternitz and Hirschfelder[8] produced pneumonia in rabbits in which the leucocyte count had been greatly reduced by injections of benzol. The pneumonia was produced by intratracheal inoculation with 4 or 5 cc. of a culture of virulent pneumococci. The leucocyte count at the time of the inoculation with pneumococci varied from 280 to 880 in six of the animals. The resistance of the leucopenic animals to the infection was markedly reduced. The average life of the control animals after the inoculation was 61 hours; while the average life of the leucopenic animals was 20 hours. The gross appearances of the pneumonic lungs in the leucopenic animals were the same as in the controls. "In the most highly aplastic animals, . . . the pneumonic exudate . . . contained the usual number of red blood cells and the usual quantities of fibrin, but only occasionally polymorphonuclear leucocytes or undifferentiated mononuclear cells."[9]

Materials and Methods.—In our experiments adult rabbits were injected subcutaneously with a mixture composed of equal parts of benzol and olive oil. The injections were given daily, or sometimes twice a day, where it seemed necessary in order to keep the leucocyte count low. Usually 2 cc. of the mixture (1 cc. of pure benzol) per kilo of body weight were given daily until the leucocyte count fell below 1,000; then smaller doses were administered. When doses smaller than 2 cc. of the mixture were given at the beginning of the experiment, the animals showed a marked loss of weight without pronounced leucopenia. With larger doses the loss in weight was not so marked. and the leucocyte decrease was more regular and rapid. The best results in our experiments were obtained by giving two or three initial doses of about 2 cc. of the mixture per kilo and increasing this 0.5 to 1 cc. in the succeeding doses until the count was

[8] Winternitz, M. C., and Hirschfelder, A. D., Studies upon Experimental Pneumonia in Rabbits. Parts I to III, *Jour. Exper. Med.*, 1913, xvii, 657. Hirschfelder and Winternitz, Studies upon Experimental Pneumonia in Rabbits. IV. Is There a Parallelism between the Trypanocidal and Pneumococcicidal Action of Drugs, *ibid.*, p. 666.

[9] Winternitz and Hirschfelder, *loc. cit.*, p. 661.

below 1,000. Then small doses were given to keep the count low without causing death.

The following methods were used to produce inflammation, more than one method being often employed in the same animal: (1) Croton oil was rubbed into a deep scratch on the inner surface of the ear. (2) Intramuscular injections of an aqueous suspension of carmine were given in the lumbar region. (3) Half of the ear was immersed in water at about 55° C. for three minutes. In all cases controls were made on normal rabbits.

Both the microscopic and gross changes were observed. The tissues were fixed in Zenker and alcohol-formalin, and sectioned in paraffin.

EXPERIMENTAL.

The decrease of the leucocyte count in our experiments, as shown by Text-figs. 1 to 11, corresponds to that obtained by Selling. The initial rise in the number of leucocytes per cmm., as recorded in one-third of his cases, does not occur in any of our experiments. Secondary rises in the count usually occurred when a dose was omitted. As shown by the curves, the leucocyte count dropped rapidly to 1,000, or even to 500 per cmm., and remained at about this level except in a few cases where the daily dose was omitted.

Experiment 1.—Rabbit A (Text-fig. 1). In this, as in all the text-figures, the leucocyte count is shown by a curve. The dosage of the benzol-olive oil mixture is shown in the figures. The injections were usually made immediately after the leucocytes were counted. Reference to the curve will also show the time of application of the irritant. Croton oil was applied to the left ear, Apr. 4, 2 p. m. Death 82 hours later. No gross changes were noted until Apr. 6, 2 p. m., when the ear became slightly red and remained so until death On microscopic examination the surface epithelium is seen to be necrotic in places. Bacteria have invaded the tissues in some places where the surface epithelium is absent. No leucocytes are present. There is no edema.

Croton oil was applied to the right ear, Apr. 6, 2 p. m. Duration 35 hours. No gross changes developed. On microscopic examination free bacteria are seen deep in the tissues where the epithelium was scratched away. There is some edema of the connective tissue. No leucocytes are present.

Carmine was injected into the left lumbar muscles, Apr. 5, 2 p. m. Death 50 hours later. There is no phagocytosis of the carmine granules. No leucocytes are present. There is no reaction on the part of the connective tissue cells.

Carmine was injected into the right lumbar muscles, Apr. 6, 2 p. m. Death 35 hours later. In some of the sections a few polymorphonuclear leucocytes are found in contact with one small area of carmine. Two or three of the cells contain carmine granules.

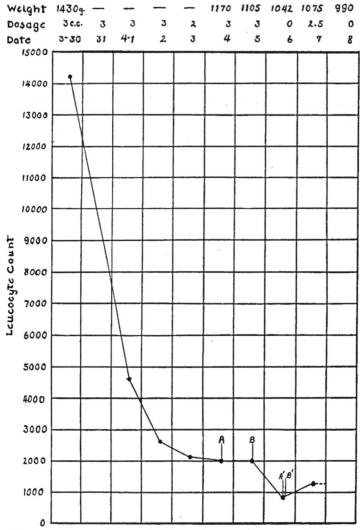

Weight	1430g.	—	—	—	—	1170	1105	1042	1075	990
Dosage	3 c.c.	3	3	3	2	3	3	0	2.5	0
Date	3-30	31	4-1	2	3	4	5	6	7	8

TEXT-FIG. I. Experiment I, Rabbit A. At the point marked A, croton oil was applied to the left ear. At B, carmine was injected into the left lumbar muscles. At A', croton oil was applied to the right ear. At B', carmine was injected into the right lumbar muscles.

The solid line represents the time the animal was known to be living. The broken line represents the estimated life after the last observation.

Experiment 2.—Rabbit B (Text-fig. 2). Apr. 6, 2 p. m., croton oil was applied to the left ear. Death 69 hours later. It is to be noted that the count was 3,800 at the end of the experiment. On microscopic examination the surface epithelium is absent over a small area. There is extreme congestion and rather marked edema. A few lymphocytes and large numbers of free bacteria are distributed through the edematous tissues.

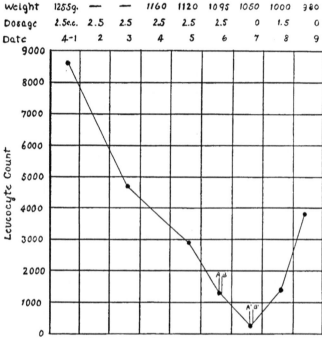

TEXT-FIG. 2. Experiment 2, Rabbit B. At A, croton oil was applied to the left ear. At B, carmine was injected into the left lumbar muscles. At A', croton oil was applied to the right ear. At B', carmine was injected into the right lumbar muscles. Note the abrupt rise after omission of a single dose of the benzol.

Croton oil was applied to the right ear on Apr. 7, 2 p. m. Death 45 hours later. No gross changes appeared. Microscopic examination was not made.

Carmine was injected into the left lumbar muscles, Apr. 6, 2 p. m. Death 69 hours later. At the site of the injection is a mass of necrotic muscle fibers surrounded by a zone of polymorphonuclear leucocytes. There are large numbers of free bacteria among the necrotic muscle fibers.

Carmine was injected into the right lumbar muscles, Apr. 7, 2 p. m. Death 45 hours later. There are a few leucocytes, mostly polymorphonuclear, on the edges of the carmine mass. Some of these contain carmine granules.

Experiment 3.—Rabbit C (Text-fig. 3). Croton oil was applied to the left ear, Apr. 6, 2 p. m. Death 59 hours later. The ear became red about 48 hours later and remained so until death. It was not swollen. Microscopically a mass of bacteria is seen in the torn epithelium where the croton oil was applied. The

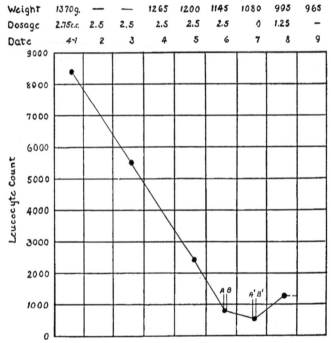

Weight	1370g.	—	—	1265	1200	1145	1080	995	965
Dosage	2.75c.c.	2.5	2.5	2.5	2.5	2.5	0	1.25	—
Date	4-1	2	3	4	5	6	7	8	9

Text-Fig. 3. Experiment 3, Rabbit C. At A, croton oil was applied to the left ear. At B, carmine was injected into the left lumbar muscles. At A', croton oil was applied to the right ear. At B', carmine was injected into the right lumbar muscles.

blood vessels are moderately congested. There is no edema. No leucocytes are present in the tissues.

Croton oil was applied to the right ear, Apr. 7, 2 p. m. Death 35 hours later. No gross changes developed. Microscopically there is some necrosis of the epithelium. No leucocytes are present. There is no edema.

Carmine injection in left lumbar muscle, Apr. 6, 2 p. m. Death 59 hours later. No leucocytes are present. There is no connective tissue reaction (Fig. 6).

Carmine injection in right lumbar muscle, Apr. 7, 2 p. m. Death 35 hours later. No leucocytes present. No connective tissue reaction.

Experiment 4.—Rabbit D (Text-fig. 4). Croton oil was applied to the inner surface of the left ear, May 30, 2 p. m.; 24 hours later the area was only slightly reddened, in which condition it was found at autopsy. There was a slight swelling of the ear, and considerable necrosis where the croton oil was applied. Death occurred the second night. Estimated time of action of croton oil 36

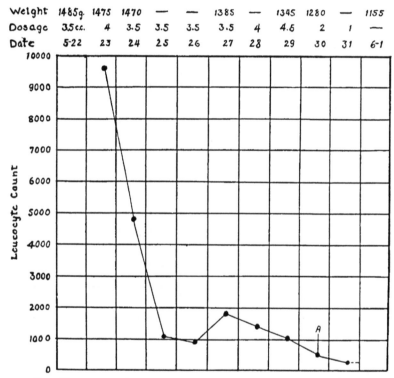

Weight	1485g	1475	1470	—	—	1385	—	1345	1280	—	1155
Dosage	3.5cc.	4	3.5	3.5	3.5	3.5	4	4.5	2	1	—
Date	5-22	23	24	25	26	27	28	29	30	31	6-1

TEXT-FIG. 4. Experiment 4. Rabbit D. At A, croton oil was applied to the left ear.

hours. Microscopically there is complete loss of surface epithelium over large areas on the inner surface of the ear. Some of the subepithelial connective tissue is necrotic. There is moderate edema. Numerous large clumps of bacteria are seen in the connective tissue. There are no leucocytes present.

Experiment 5.—Rabbit E (Text-fig. 5). Croton oil was applied to the left ear, June 1, 1 p. m.; 24 hours later the ear was only slightly congested, but markedly thickened. It remained in this condition until death, which occurred 45 hours after application of the irritant. Microscopic examination shows extensive necrosis of the epithelium at the line of application of the croton oil. Where the epithelium is absent masses of bacteria are distributed through the adjacent connective tissue. There is a very marked edema, especially of the tissue external to the cartilage. No leucocytes are present (Figs. 2 and 3).

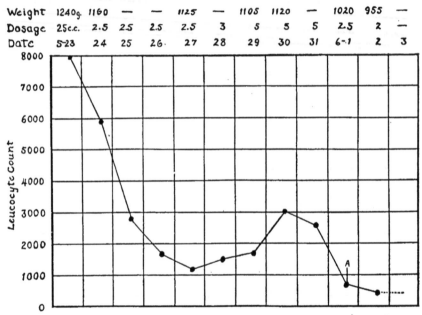

TEXT-FIG. 5. Experiment 5, Rabbit E. At A, croton oil was applied to the left ear.

Experiment 6.—Rabbit F (Text-fig. 6). On Apr. 21, 2 p. m., the right ear was immersed in water at 55° C. for 2.5 minutes. The ear immediately became congested and slightly thickened; 24 hours later it was very much thickened. Two or three large blebs were present. Death occurred about 35 hours after application of the heat. Microscopically edema is very marked. Some masses of fibrin are seen. No leucocytes are present. In a few places there is a little necrosis of the surface epithelium (Fig. 4).

On Apr. 22, 2 p. m., the left ear was immersed in water at 55° C. for 3 minutes. Only a slight reddening resulted. At 5.30 p. m. it was immersed in water at 54° C. for 3 minutes. The ear became slightly reddened immediately. The rabbit lived about 11 hours after the first application of heat. At autopsy the ear was found very much thickened and showed blebs similar to those on the right. Microscopically there is a very marked edema. No leucocytes are present. There is considerable necrosis of the surface epithelium.

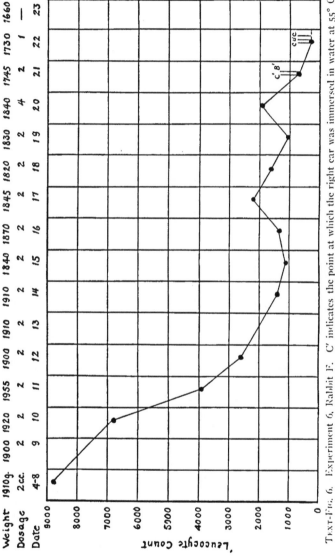

TEXT-FIG. 6. Experiment 6, Rabbit F. C indicates the point at which the right ear was immersed in water at 55° C. for 2½ minutes. At B, carmine was injected into the right lumbar muscles. At C, left ear in water at 55° C. for 3 minutes, which was repeated. At B, carmine was injected into the left lumbar muscles.

Carmine injection in right lumbar muscles, Apr. 21. Death about 35 hours later. Microscopic examination shows enormous numbers of bacteria distributed through the carmine masses and through the surrounding muscle tissue. No leucocytes are present. There is no reaction on the part of the connective tissue cells (Fig. 7).

Carmine injection in left lumbar region. Death about 11 hours later. On

Weight	1710g	1630	1550	1515	1435	1375	1360	1310	1240	1200
Dosage	25c.c.	2.5	2.5	2.5	2.5	2.5	2.5	25	1.25	1.25
Date	4-11	12	13	14	15	16	17	18	19	20

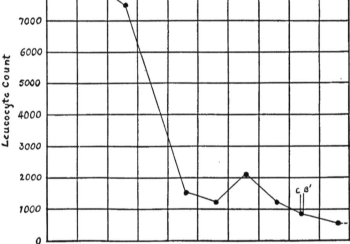

TEXT-FIG. 7. Experiment 7, Rabbit G. At C, left ear in water at 50° C. for 3 minutes. At B', carmine was injected into the right lumbar muscles.

microscopic examination a few bacteria are seen around the carmine mass. No leucocytes are present.

Experiment 7.—Rabbit G (Text-fig. 7). On Apr. 19, 10.40 a. m., the left ear was immersed in water at 50° C. for 3 minutes. 10 minutes afterwards no gross changes were to be seen. It was then immersed in water at 55° C. for 1 minute. After 10 minutes the ear became reddish, thick, and heavy. 30 hours later there was marked swelling and congestion. On microscopic examination there is marked necrosis of the surface epithelium. Edema is pronounced. No leucocytes are present. The blood vessels are congested.

Carmine injection in right lumbar muscles, Apr. 19, 10.40 a. m. Died 31 hours later. On microscopic examination large numbers of bacteria are seen around the carmine mass. No leucocytes are present. There is no reaction on the part of the connective tissue. There is considerable necrosis of the muscle fibers.

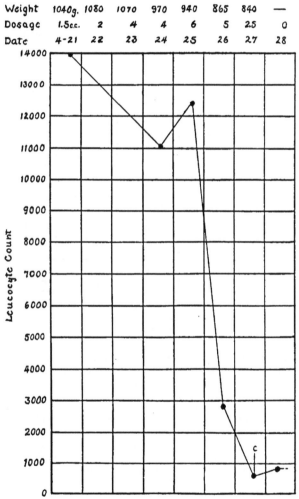

TEXT-FIG. 8.   Experiment 8, Rabbit H.   At C, left ear in water at 55° C. for 3 minutes.

Experiment 8.—Rabbit H (Text-fig. 8). Apr. 27, the left ear was immersed in water at 55° C. for 3 minutes. Immediately afterwards the ear became greatly congested and swollen, in which condition it remained until death 26 hours later. On microscopic examination there is congestion of the blood vessels and marked edema. In a few places the surface epithelium is partly necrotic. No leucocytes are present.

Experiment 9.—Rabbit I (Text-fig. 9). May 3, at 12.45 p. m., the left ear was immersed in water at 55° C. for 3 minutes. 20 minutes later, the ear was

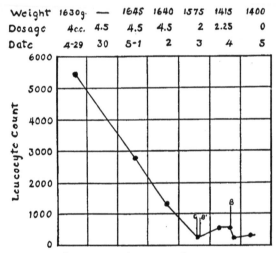

Weight	1630g.	—	1645	1640	1575	1415	1400
Dosage	4cc.	4.5	4.5	4.5	2	2.25	0
Date	4-29	30	5-1	2	3	4	5

TEXT-FIG. 9. Experiment 9, Rabbit I. At C, left ear in water at 55° C. for 3 minutes. At B', carmine was injected into the right lumbar muscles. At B, carmine was injected into the left lumbar muscles.

somewhat thicker, but was not hot or reddened. The following day some congestion and thickening were noticed. Death 44 hours after the application of heat. On microscopic examination a few areas of partly necrotic surface epithelium are seen. Moderate edema is also present. A few mononuclear leucocytes are found in the connective tissue.

Carmine injection in right lumbar muscles, 44 hours before death. On microscopic examination an occasional polymorphonuclear leucocyte is seen (5 or 6 to a section). There is no definite evidence of phagocytosis. Large numbers of bacteria are present. There is no connective tissue reaction.

Carmine injection in left lumbar muscles, 20 hours before death. Microscopic examination shows about the same as the preceding, except that there are only a few bacteria present.

Experiment 10.—Rabbit J (Text-fig. 10). May 7, 11.40 a. m., the left ear was immersed in water at 55° C. for 3 minutes; 24 hours later the ear was thick,

but not warm or reddened. Death 46 hours after application of heat. Microscopic examination shows some necrosis of the surface epithelium. No leucocytes are present. There is moderate edema.

Carmine injection in right lumbar muscles, 46 hours before death. Micro-

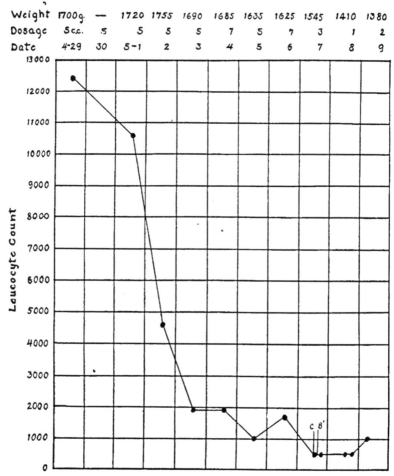

Weight	1700g.	—	1720	1755	1690	1685	1635	1625	1545	1410	1380
Dosage	5 c.c.	5	5	5	5	7	5	7	3	1	2
Date	4-29	30	5-1	2	3	4	5	6	7	8	9

TEXT-FIG. 10. Experiment 10, Rabbit *J.* At C, left ear in water at 55° C.
for 3 minutes. At B', carmine was injected into the right lumbar muscles.

scopic examination shows a considerable number of polymorphonuclear leucocytes in relation to the carmine masses in a few places; but only rarely does a

leucocyte-contain carmine granules. A large number of bacteria are distributed through the muscle. There is rather extensive necrosis of muscle around the carmine. The connective tissues show no reaction.

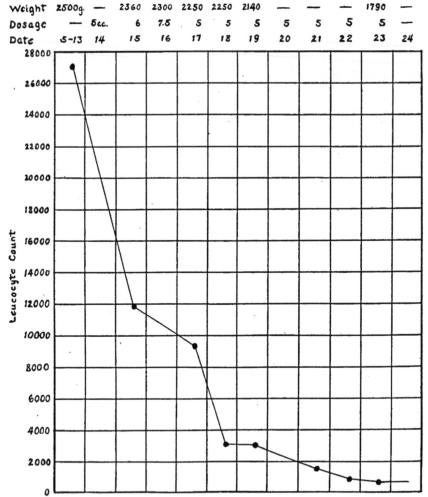

Weight	2500g.	—	2360	2300	2250	2250	2140	—	—	—	1790	—
Dosage		—	6cc.	6	7.5	5	5	5	5	5	5	5	—
Date	5-13	14	15	16	17	18	19	20	21	22	23	24

TEXT-FIG. 11. Experiment 11, Rabbit K. Abscess rabbit; died with 600 leucocytes per cmm.

Experiment 11.—Rabbit K (Text-fig. 11). This rabbit had a large abscess in the neck about 4 cm. in diameter. Benzol was given to see whether it would destroy leucocytes in an abscess as well as those in the bone marrow and circulating blood. The animal died when the leucocyte count was 600. The abscess did not vary in size during the experiment. Microscopically the abscess shows a thick connective tissue wall, inside of which it is composed of polymorphonuclear leucocytes. The outer layers at least show no marked retrogressive changes. There were also small abscesses in the liver composed of polymorphonuclear leucocytes. Benzol does not, therefore, destroy leucocytes in an abscess.

In several of our experiments the bone marrow was examined microscopically. We found a complete, or almost complete, disappearance of leucocytes, as described by Selling.

The liver was examined microscopically in a few of the rabbits. Small focal necroses were found. Some of the liver cells contained fat droplets. Only an occasional leucocyte was to be found in the capillaries. We never found accumulations of leucocytes as described by Pappenheim. The leucocytes were no more numerous than the count of the peripheral blood would lead us to expect.

DISCUSSION.

Croton Oil.—When croton oil is rubbed into a needle scratch on the ear of a normal rabbit, striking gross changes are always visible before the end of the first twenty-four hours. These are usually most pronounced at the end of forty-eight hours. The ear becomes hot, reddened, and swollen. Microscopically the ear shows considerable edema and the connective tissues are densely infiltrated with polymorphonuclear leucocytes (Fig. 1).

In some of the rabbits with severe leucopenia no gross changes appeared in the ear after the application of croton oil, although the animals lived one to three days. In one instance slight congestion and swelling were present at the end of twenty-four hours. In another there was very marked congestion and swelling at the end of twenty-four hours. On microscopic examination, no leucocytes were present in any of the five experiments, except Experiment 2, which showed a few lymphocytes. In this animal, however, the leucocyte count rose to 3,800 on the last day. Edema was very pronounced in one experiment (Fig. 4). In every case masses of bacteria were found invading the tissues where the epithelium had been torn off by the needle. In many instances they were widely scattered through the connective tissue. The bacteria were evidently growing rapidly and meeting little or no resistance from the tissues. A comparison of Figs. 1 and 2 will show that the ab-

sence of leucocytes is the most essential histological difference between the inflammatory reaction of a normal rabbit and one with
severe leucopenia. Evidently congestion of blood vessels and
edema may occur in the absence of leucocytes, but in many cases
there is no reaction whatever in the leucopenic animal. There are
apparently no protective bodies to prevent the entrance and growth
of bacteria in the leucopenic rabbit.

Application of Heat.—When the ear of a normal rabbit is injured by an exposure in water at 55° C. for three minutes, it rapidly
becomes congested, and greatly swollen. Microscopically one sees
marked edema and congestion. The surface epithelium is necrotic
in many places. A considerable number of leucocytes are present
near the surface epithelium, but they are much less numerous than
in croton oil ears.

In the leucopenic rabbit the application of heat usually causes
immediate congestion and swelling which may become very marked
before the end of the first twenty-four hours. Sometimes, however, the gross reaction is slow to appear and very slight. Microscopically no leucocytes were present in any case, except Experiment
9 which showed only a few mononuclear forms. Edema is usually
very pronounced, and there is often necrosis of the surface epithelium (Fig. 4). In the leucopenic rabbit the reaction to heat
differs from the reaction in the normal animal only in the absence
of leucocytes.

Carmine.—When an aqueous suspension of carmine is injected
into the lumbar muscles of a normal rabbit the carmine mass soon
becomes surrounded by leucocytes which are chiefly of the polymorphonuclear type (Fig. 5). The leucocytes ingest the carmine
granules, and before the end of the second day a proliferation of the
adjacent connective tissue begins (Fig. 8). In the leucopenic animals
no leucocytes were found around the carmine masses except in Experiments 2 and 10. In Experiment 2 there was a terminal rise of
the leucocyte count to 3,800, which evidently accounts for the leucocytic exudate. In Experiment 10 there was a terminal rise of
leucocyte count to 1,000. This was the only instance, however,
in which a leucocytic exudate appeared when the count was as low

as 1,000. There was no connective tissue proliferation around the carmine mass in any case (Figs. 6 and 7).

Reaction to Bacteria.—Since the carmine suspension injected into the muscles was not sterilized, some bacteria were evidently introduced along with the carmine. Microscopic sections in nearly every case showed masses of bacteria through the carmine and the surrounding tissues. The large numbers of bacteria present, and their wide distribution through the tissues show clearly that there is little or no resistance to their growth. Normal animals injected in the same way never showed any bacteria around the carmine.

The croton oil ears of the leucopenic rabbits showed the same profuse bacterial growth through the connective tissues as was seen in the carmine experiments. The distribution of the bacteria in these cases indicates that they entered the tissue where the epithelium was scratched away for the application of the croton oil. The absence of bacteria in the heated ears is presumably due to the fact that the surface epithelium was not broken.

Our experiments show that in almost every instance there is no leucocytic exudate in rabbits with a severe leucopenia. Lippmann and Plesch produced a lymphocytic exudate in the pleura and peritoneal cavities of leucocyte-free rabbits with cultures of the bacillus of swine erysipelas, but the same irritant did not produce any leucocytic exudate when injected into the muscles. Pappenheim and Fukushi, using tuberculin, produced a lymphocytic exudate in the peritoneal cavity.

Our results are not at variance with those of the investigators cited above, since we applied the irritants to the muscle and subcutaneous tissues only.

The evident absence of antibacterial bodies in rabbits with a severe leucopenia may not be due entirely to the destruction of the leucocytes, since benzol probably has a general cytotoxic, as well as a leucotoxic action.

SUMMARY.

Congestion of blood vessels and marked edema may occur in rabbits with a severe leucopenia. These two phenomena are, therefore, independent of the leucocytes.

When the leucocyte count is below 1,000, croton oil and heat produce no leucocytic exudate in the tissues of the ear, and carmine produces no exudate in the muscles.

In rabbits with severe leucopenia there are apparently no antibacterial bodies present.

Benzol does not destroy the leucocytes in an abscess.

EXPLANATION OF PLATES.

PLATE 23.

FIG. 1. Microphotograph of a section of the ear of Rabbit L, control. Magnified 29 diameters. Croton oil was applied 45 hours previously. This figure shows the enormous numbers of polymorphonuclear leucocytes due to the croton oil irritation in a normal rabbit.

FIG. 2. Microphotograph of the section of the ear of Rabbit E, Experiment 5. Magnified 29 diameters. Croton oil was applied 48 hours previously. The needle scratch, shown on the right of the figure, is covered with bacteria which are also invading the subepithelial tissues. No leucocytic exudate is present.

FIG. 3. Same as Fig. 2. Magnified 29 diameters. Taken a little nearer the tip of the ear to show the marked edema.

FIG. 4. Microphotograph of a section of the right ear of Rabbit F, Experiment 6. Magnified 14 diameters. 35 hours previously the ear was immersed in water at 55° C. for 2.5 minutes. Note the marked edema and complete absence of cellular reaction.

PLATE 24.

FIG. 5. Microphotograph of a section of lumbar muscles of Rabbit M, control. Magnified 387 diameters. Injection of carmine 30 hours previously. This figure shows the polymorphonuclear reaction due to carmine irritation in a normal rabbit.

FIG. 6. Microphotograph of a section of the left lumbar muscles of Rabbit C, Experiment 3. Magnified 387 diameters. Injection of carmine 59 hours previously. There is marked edema, but complete absence of leucocytes.

FIG. 7. Microphotograph of a section of the right lumbar muscles of Rabbit F, Experiment 6. Magnified 387 diameters. Injection of carmine 35 hours previously. Note the enormous numbers of bacteria and the complete absence of leucocytic exudate.

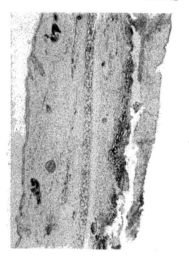

FIG. 1.

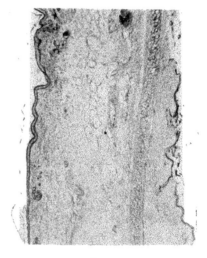

FIG. 2.

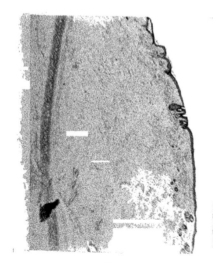

FIG. 3.

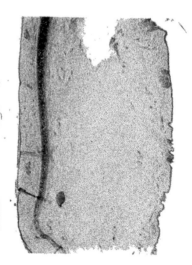

FIG. 4.

(Camp and Baumgartner: Inflammatory Reactions in Rabbits.)

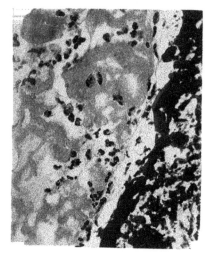

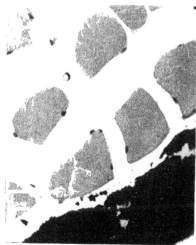

FIG. 5.   FIG. 6.

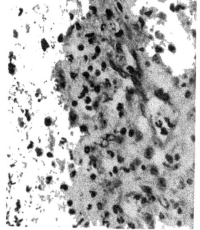

FIG. 7   FIG. 8.

(Camp and Baumgartner: Inflammatory Reactions in Rabbits.)

Fig. 8. Microphotograph of a section of the muscles of Rabbit N, control. Magnified 387 diameters. Injection of carmine 70 hours previously. This figure shows the reaction of the connective tissue cells to carmine irritation in the normal rabbit.

THE ARTIFICIAL PRODUCTION IN MAMMALIAN PLASMA OF SUBSTANCES INHIBITORY TO THE GROWTH OF CELLS.

By ALBERT J. WALTON, M.S.

(*From the Bacteriological Laboratory of the London Hospital, London.*)

PLATES 25 AND 26.

(Received for publication, April 6, 1915.)

The fact that the growth of mammalian tissue varies with the plasmatic medium and that this variation is probably due to the presence of inhibiting and stimulating substances has been shown in a previous communication.[1] It has also been shown that the extracts of various tissues when added to the plasma have a more or less specific action on the growth of the tissue.[2] This evidence is suggestive that an immunity to the growth of individual cells might be artificially acquired by an animal.

Many workers have attempted to obtain specific cytotoxins and cytolysins by injecting animals with certain cells and determining whether antibodies were found in the serum. In determining whether such bodies were present or not the method generally adopted was to inject the serum of the animal thought to be immune into another animal of the same species as that from which the immunizing agent had been obtained. Cytolysins were in many cases found to be present, but they were not specific. Lambert[3] after considering fully the published results observed by many investigators on the above lines gives the results of his experiments. He investigated the nature of the growth of cells in the plasma of immunized animals and was able to determine that immune bodies were present but were not specific. In his experiments guinea pigs were

[1] Walton, A. J., *Proc. Roy. Soc., Series B.*, 1913–14, lxxxvii, 452.
[2] Walton, *Jour. Exper. Med.*, 1914, xx, 554.
[3] Lambert, R. A., *Jour. Exper. Med.*, 1914, xix, 277.

194

immunized to rat sarcoma and the skin of rat embryos. The important question as to whether one animal can be immunized to the tissue of another animal of the same species has not been investigated by the method of growth *in vitro*. A consideration of this question would appear to be of importance as throwing light upon the correlation of growth of various tissues in normal and abnormal states of the body. In conducting such a series of experiments it is also of importance to determine what are the changes taking place in the cells injected. Do they act as tissue grafts, or are they destroyed by the host into which they are placed?

The present experiments were carried out with a view of determining whether such an immunity was specific to the one type of cell. A series of ten experiments which included the making of 396 cultures was employed.

Technique.

An adult rabbit was killed and the testicles and portions of the liver were removed. Each tissue was divided as finely as possible and made into a thick emulsion which could just pass through a large hypodermic needle by mixing with it a small quantity of Ringer's fluid. On the same or the subsequent day five or six cubic centimeters of testicular emulsion were injected into the peritoneal cavity of an adult rabbit, a corresponding amount of liver emulsion being injected into the peritoneal cavity of a second rabbit. The process was repeated twice at intervals of seven days, so that one animal received three doses of testicular emulsion and the other three doses of liver emulsion. Portions of liver and testicle were cultivated in the plasmata of these animals and in the plasma of a control animal at various intervals. The injection of this quantity of emulsion was in no case followed by any injurious results to the animal experimented upon.

Experiment 1.—Two adult male rabbits, A and B, were injected. Rabbit A receiving three intraperitoneal injections of 5 cc. each of liver emulsion, and Rabbit B three injections of 5 cc. each of testicle emulsion, an interval of one week elapsing between each injection.

Seven days after the last injection Rabbit C was anesthetized and blood removed from the carotid artery.[4] This animal was kept under ether while blood

[4] All the experiments were done under ether anesthesia.

was removed by puncture from Animals A and B. The three bloods were centrifugalized in ice. Portions of liver and testicle were removed meanwhile from Animal C and placed in Ringer's fluid. Cultures of both tissues were now made by the Carrel technique in each of the three plasmata.

Results.—Good growth occurred with both tissues in the control plasma of Animal C.

In the plasma of Animal A (immune to liver) there was fair growth in all the specimens of testicle, but this growth was always considerably less than in the controls. Of the liver cultures not one showed any growth, and the specimens after fixation stained very faintly or not at all, showing that the cells were dead (Text-fig. 1). In the plasma of Animal B (immune to testicle) there was only slight growth of the cultures of testicle, the growth being considerably less than in the cases of those pieces of tissue grown in the plasma of Animal A, and very much less than in the controls of the plasma of Animal C. The cultures of liver showed in every case no trace of growth (Text-fig. 2).

This experiment would appear to show that injections of liver or testicle confer an immunity which is more marked in both cases to liver than to testicle.

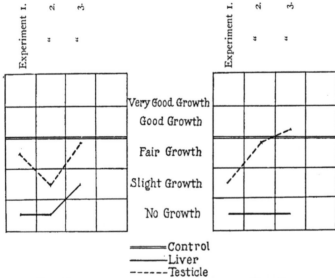

Control
Liver
Testicle

TEXT-FIG. 1.   TEXT-FIG. 2.

TEXT-FIG. 1. Animal A. Immunized with liver emulsion.
TEXT-FIG. 2. Animal B. Immunized with testicular emulsion.

Experiment 2.—The above experiment was repeated with the same two immune animals fourteen days after they had received the last injection, another control animal, C, being used.

Results.—In the control plasma of C there was good growth of both tissues. In Animal A (immune to liver) there was only slight growth in the cultures of testicle and no growth in the cultures of liver. In Animal B there was marked growth in all the cultures of testicle, this being nearly as good as in the controls. The cultures of liver showed no growth (Text-figs. 1 and 2).

Apparently, therefore, as regards the testicle there was an increased immunity in the animal injected with liver emulsion, but in the animal injected with testicle emulsion the immunity was passing off so that growth was nearly as good as in the controls. Both animals still showed, however, a definite immunity to the growth of liver.

Experiment 3.—The above experiment was repeated with the same immune animals twenty-six days after they had received the last injection, another control animal, C, being used.

Results.—Good growth occurred with both tissues in the control plasma of Animal C. In the plasma of Animal A there was good growth in all the cultures of testicle. The cultures of liver showed growth in four of six specimens, but the growth was only slight and was always less than in the case of the controls. In the plasma of Animal B there was again extensive growth of all the cultures of testicle, this being definitely greater than in the controls. The cultures of liver showed no growth (Text-figs. 1 and 2).

Thus after twenty-six days the immunity to the growth of testicle was wearing off in the animal injected with liver, while in the animal injected with testicle an anti-immune body appeared to have been formed. As regards the immunity to liver this was beginning to pass off with the animal injected with liver, but was still present in the animal injected with testicle.

These experiments seem to show that a certain immunity was produced to the growth of tissue by the injection of cellular emulsions, that this immunity was not definitely specific, although it was always better to liver than to testicle, and that it lasted but a short time.

In order to confirm these results further experiments were carried out, the cultures in this group being made after each injection and continued at weekly intervals after the last injection until the results were found to coincide with those obtained in a normal animal.

Experiment 4.—An adult male rabbit, D, was injected with 5 cc. of liver emulsion, and an adult female rabbit, E, with 5 cc. of testicle emulsion. Seven days later portions of liver and testicle were removed from a control animal, C, and cultivated in the plasmata of the three animals. As soon as the experiment was completed Animals D and E received a second intraperitoneal injection of liver or testicle emulsion.

Results.—There was good growth of both tissues in all specimens in the plasma of the control animal, C. In the plasma of Animal D there was no growth of any of the specimens of either liver or testicle. In the plasma of Animal E there was slight growth of liver and fair growth of testicle in all specimens (Text-fig. 3).

Hence it appeared that the animal injected with liver had already a good immunity to both liver and testicle, while that injected with testicle had a fair immunity to liver and only a slight immunity to testicle.

Five days after the above experiment Animal D died from the effects of severe bites inflicted by another rabbit. An adult female rabbit, G, was therefore taken for further injections with liver extract.

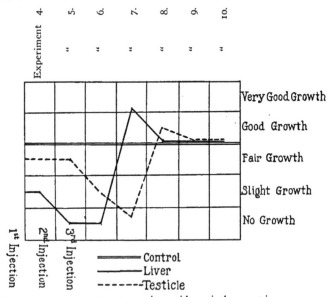

TEXT-FIG. 3.   Animal E.   Immunized with testicular emulsion.

Experiment 5.—Cultures were made in the plasma of Animal E seven days after the second injection, in that of Animal G before injection, and in that of Animal C as a control.

Results.—There was good growth of both liver and testicle in the control animal, C. In the plasma of Animal G before an injection there was again, as was to be expected, good growth of both tissues (Text-fig. 4). In the plasma of Animal E there was no growth of the liver tissue and only fair growth of the testicular tissue (Text-fig. 3).

It would appear therefore that after the second injection immunity to the growth of liver had definitely increased, but that to testicle was still slight.

Experiment 6.—Cultures were made seven days later in the plasma of Animal E, *i. e.,* seven days after the third injection of testicle emulsion, and in Animal G seven days after the first injection of liver emulsion, also in the plasma of the control animal, C.

Results.—There was good growth in both tissues of the plasma of the control animal, C. In Animal G there was fair growth of all the liver cultures, this being, however, distinctly less than in the control (Fig. 1), but there was still good growth of the testicle cultures (Text-fig. 4). In Animal E there was no growth of any of the specimens of liver and only slight growth in the specimens of testicle (Text-fig. 3).

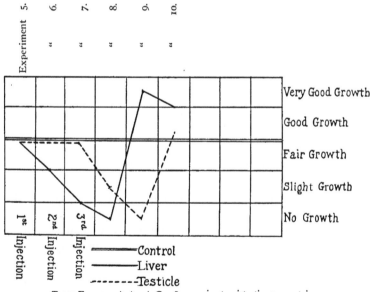

TEXT-FIG. 4. Animal G. Immunized with liver emulsion.

In this case the immunity in the animal injected with testicle was still very good for the liver and had greatly increased for the testicle.   In the case of the animal injected with liver there was already some immunity to the growth of liver and but little, if any, to the growth of testicle.

Experiment 7.—The above experiments were repeated eight days later with another control animal; that is, fifteen days after the third injection of testicle emulsion in Animal E, and eight days after the second injection of liver emulsion in Animal G.

Results.—There was again good growth in the controls.   In the plasma of Animal G there was only slight growth of the liver specimens, but the specimens of testicle still showed good growth (Text-fig. 4).   In the plasma of Animal E there was good growth of all the cultures of liver; in fact this growth was considerably better than in the controls, but with all the cultures of testicle there was only slight growth, several of the specimens showing no growth at all (Text-fig. 3).

In this experiment there was a distinct change in the plasma of the animal injected with testicle.   The immunity to the liver had entirely disappeared and was, in fact, succeeded by an anti-immunity. There was, however, still a very distinct immunity to the growth of testicular tissue.   The animal inoculated with liver injections showed an increasing immunity to the growth of liver, but had not yet developed any to the growth of testicle.

Experiment 8.—The above experiments were repeated seven days later; that is, twenty-two days after the third injection of testicle into Animal E, and seven days after the third injection of liver into Animal G.

Results.—The growth of the controls was again good in all cases.   In the plasma of Animal G there was no growth in the specimens of liver (Fig. 2), and only slight growth in the specimens of testicle (Text-fig. 4).   In the plasma of Animal E there was good growth in all the specimens of liver, but it was now no better than in the controls.   In the case of the testicle the growth was distinctly better than in the case of the controls (Text-fig. 3).

At this period, then, the animal injected with testicle had lost all its immunity to liver and the immunity to the testicle had disappeared to be followed by an anti-immunity.   The animal injected with liver had, on the other hand, an increasing immunity to liver and was now developing an immunity to the growth of testicle.

Experiment 9.—The above experiment was repeated seven days later; that is, twenty-nine days after the third injection of testicle in Animal E, and fourteen days after the third injection of liver in Animal G.

Results.—The controls as usual grew well.   In the plasma of Animal G there was now very marked growth of liver, this being in all cases more extensive than in the controls (Figs. 3 and 4). The specimens of testicle, however, showed only very slight growth in a few specimens (Text-fig. 4). In the plasma of Animal E the growth of both tissues was similar to that of the controls (Text-fig. 3).

At this stage, therefore, Animal E had lost all its immunity. In Animal G the immunity to liver had disappeared and this, as had been the case in Animal E, was followed by an anti-immunity. There was, however, a very definite immunity to the growth of testicle.

Experiment 10.—The above experiments were repeated seven days later; that is, thirty-six days after the third injection of testicle into Animal E, and twenty-one days after the third injection of liver in Animal G.

Results.—Both tissues grew well in the control plasma. In the plasma of Animal G there was still extensive growth of liver, especially as regards the parenchymatous cells, these being much more marked than in the controls. In the case of the specimens of testicle there was now good growth which was slightly better than the controls (Text-fig. 4). In the plasma of Animal E the growth of both tissues was good and was equal to that of the controls.

At this stage both animals had lost their immunity, but whereas Animal G was in a stage of anti-immunity to both tissues, so that they grew better in its plasma than in the controls, yet in the case of Animal E the anti-immunity was also lost and the plasma was of the same value as a medium as that of an uninjected animal. The results of these experiments are well shown in Text-figs. 3 and 4, and it is of interest to note that in both cases, although one was injected with liver and the other with testicle, an immunity to the growth of the liver was more rapidly developed, being followed at a later interval by an immunity to the growth of testicle. In both cases also this immunity was followed by an anti-immunity which was again rapidly lost, so that the plasma in this respect resembled that of a normal animal. The curves show that the extent of growth in the two animals, although varying in degree, simulate one another to a remarkable extent and confirm the results obtained in Animals A and B in so far that they show that an immunity was more rapidly obtained to the growth of liver, but was followed at a slightly later date by an immunity to the growth of testicle. The fact that the immunity to liver and testicular growth differs in point of time shows that to a certain extent the inhibition to this growth is specific, but the experiments also clearly show that it is not specific in the sense that the injection of liver limits the growth of liver tissue only, whilst the injection of testicle limits only that of testicular tissue. For in both Animals E and G it was seen that the immunity to liver was more rapidly developed, although one was injected with liver and the other with testicle, the same result being also obtained in Animals A and B. The fact, however, that the growth of each tissue is independently controlled would give rise to the belief that in the future it may be possible to determine a method of controlling the growth of one and only one tissue.

After these experiments were completed the immunized animals were killed and it was found that at the site of injection the emulsified tissue formed an encapsulated mass which on section was caseous and necrotic. Microscopical sections showed, as a general rule, necrotic cellular masses which stained poorly. These masses were surrounded by large numbers of round cells and commencing connective tissue formation. In no case was there any evidence of the injected cells having grown in the host. They were in all cases destroyed and surrounded by inflammatory exudates. In both animals sections were made of the liver and it was found that the cells resembled exactly those of a normal animal. It was impossible to say that they showed any degeneration or necrotic changes as the result of the injection of tissue emulsion.

CONCLUSIONS.

Active immunity to the growth of tissue may be obtained by intraperitoneal injections of tissue emulsions.

This immunity is of short duration and is followed by an anti-immunity.

The immunity varies for individual tissues both in extent and in the time of onset.

The immunity is not specific in so far that it is not more marked to the tissue similar to that forming the emulsion.

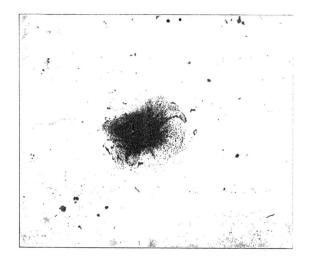

FIG. 1.

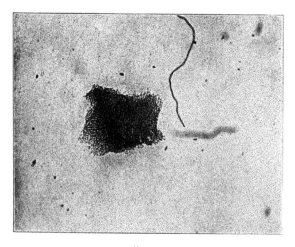

FIG. 2.

(Walton: Artificial Production of Inhibitory Substances.)

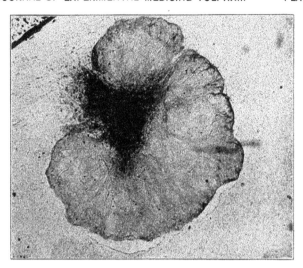

FIG. 3.

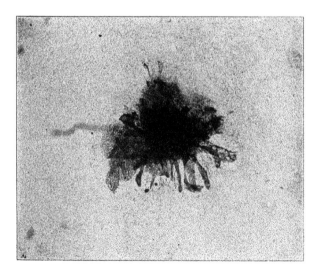

FIG. 4.

(Walton: Artificial Production of Inhibitory Substances.)

EXPLANATION OF PLATES.

THE LYMPHOCYTE IN NATURAL AND INDUCED RESISTANCE TO TRANSPLANTED CANCER.

II. Studies in Lymphoid Activity.

By JAMES B. MURPHY, M.D., and JOHN J. MORTON, M.D.

(From the Laboratories of The Rockefeller Institute for Medical Research.)

(Received for publication, May 28, 1915.)

In former communications, we have shown that the resistance to heteroplastic tissue grafts apparently depends on the activity of the lymphocyte. The chick embryo, which we found[1] to be devoid of any resisting power to the growth of grafted tissues from a foreign species, could be rendered as refractory as the adult animal, by the introduction of a bit of adult homologous tissue rich in lymphoid elements, such as spleen or bone marrow.[2] Furthermore, an adult animal deprived of the major portion of its lymphoid system by means of repeated small doses of x-ray, no longer resists the growth of heterologous tissues. The tissue cells from a foreign species will grow actively till such a time as the depleted lymphoid system of the animal is well advanced in regeneration.[3]

Histologically, there is a striking resemblance between the appearances presented by the failing heteroplastic graft in the foreign host and the cancer graft in the so called immune animal. The phenomenon is the same, whether the state of resistance to the cancer is one possessed naturally by the animal, a condition termed natural immunity, or whether it is of the nature of induced immunity developed by previous treatment of the animal. The details of the various stages of the processes as observed under the microscope need not be gone into here.[4] It is only necessary to say that the

---

[1] Murphy, Jas. B., *Jour. Exper. Med.*, 1913, xvii, 482.

[2] Murphy, Jas. B., *ibid.*, 1914, xix, 513.

[3] Murphy, Jas. B., *Jour. Am. Med. Assn.*, 1914, lxii, 1459.

[4] For the literature see Da Fano, C., *Ztschr. f. Immunitätsforsch., Orig.*, 1910, v, 1.

constant and undisputed fact is that in both the heteroplastic tissue graft and the homologous cancer graft in an immune animal, there is a pronounced collection of small round cells at the edges of the grafts and densely infiltrating the tissues about.[5] The lymphoid elements appear quite early in the process and last till it is completed. It is notable that these elements are absent about the growing cancer graft in the highly susceptible animal.

With the above fact at hand it would seem, arguing from analogy, that the lymphocytes might play an as equally important part in immunity to transplanted cancer as they appear to play in the resistance phenomenon to heteroplastic tissue grafts. With this problem in mind the following experiments were planned.

Resistance to Transplanted Cancer.

It is unnecessary to go into the details and theories of the various types of resistance to the transplanted cancer. This has been done at length in numerous communications on the subject.[6] The essential facts are that mice may be rendered relatively immune for a period by giving a subcutaneous or intraperitoneal injection of a certain amount of homologous living tissues at least ten days before inoculating the cancer graft. Various tissues have served, such as hashed embryo, embryo skin, spleen, and red blood cells, for this purpose. The efficiency of the protection varies considerably, depending on the type and amount of tissue used, as well as on the growth power of the cancer inoculated. A certain proportion of mice inoculated with cancer show a natural refractory state which varies in different families of mice and to an extent according to the virulence of the tumor. Sometimes this natural resistance is so effective that the cancer graft does not become established, while in the less resistant animal the introduced tumor may grow for a time and only later be overcome and absorbed. These are the types of immunity which we have elected to study from the point of view of their general lymphoid reaction.

[5] For general review and literature on the subject see Woglom. W. II.. The Study of Experimental Cancer, A Review, Studies in Cancer and Allied Subjects, New York, 1913, i, 128.

[6] Woglom, *loc. cit.*

In all we have made detailed studies of the blood in over eighty mice before and at intervals after inoculation with cancer in both immune and susceptible animals. Only one of these experiments will be reported as the others gave practically identical results.

Experiment I.—20 white mice of about the same age and size were selected. A white blood count and a differential count were made to establish the normal in each animal. 10 of the mice were given a subcutaneous injection of 0.3 cc. of defibrinated mouse blood. 10 days later all 20 animals were inoculated with equal sized grafts of a transplantable mammary carcinoma of the mouse.[7] Blood examinations were made after 24 hours and at intervals for something over 50 days. The grafts were measured and recorded weekly. The protection in the immunized animals of this series was 100 per cent. Among the controls, 60 per cent grew their grafts and 40 per cent were naturally immune. From the white count and differential, we have estimated the actual numbers of the various types of white cells present.

Typical examples of each group are shown in Text-fig. 1, A being an immunized animal, B a control animal with a growing tumor, and C a naturally immune animal. In these counts we have grouped together the large and small lymphocytes; and have not differentiated the types of polynuclear cells. The changes that have taken place have been, however, almost entirely in the small lymphocyte group. In Curve A, here given, no blood examinations were made between the primary count and one twenty-four hours after the cancer inoculation, but a study of a large series of animals has shown no variation in the count before the introduction of the cancer graft. It is assumed, therefore, in this case that the normal level is maintained.

The immediate increase in the circulating lymphocytes present in the immunized animal is a constant finding and in a large proportion of the cases has risen to 100 per cent or more above the normal level during the first twenty-four hours. This is well shown by the composite curve made from the average of all of this series (Text-fig. 2 A). In Curve C of Text-fig. 1 we show a typical example of a naturally immune animal. Here the lymphoid increase is delayed, and is preceded by a slight fall. The composite curve (Text-fig. 2 C), however, shows that as an average there is slight if any change during the first twenty-four hours, but the reaction develops

---

[7] This tumor was propagated by Dr. Woglom and kindly given us by him.

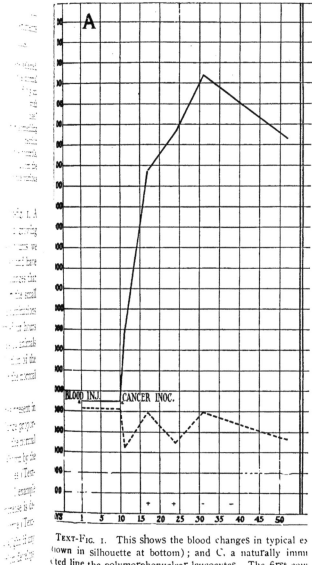

TEXT-FIG. 1. This shows the blood changes in typical e>
(own in silhouette at bottom); and C. a naturally imm
(ted line the polymorphonuclear leucocytes. The first cou:
(cancer inoculation.

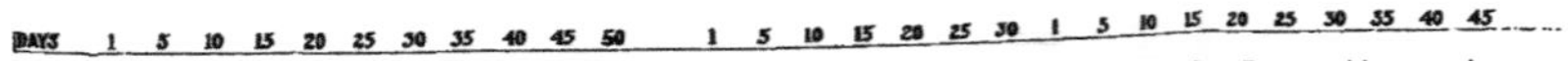

DAYS | 1 5 10 15 20 25 30 35 40 45 50 | 1 5 10 15 20 25 30 | 1 5 10 15 20 25 30 35 40 45 ...

TEXT-FIG. 1. This shows the blood changes in typical examples of A, an animal with induced immunity; B, one with a growing cancer (shown in silhouette at bottom); and C, a naturally immune animal. The solid line represents the actual number of lymphocytes, and the dotted line the polymorphonuclear leucocytes. The first count in Curve A was made before the immunizing injection, and in the others before the cancer inoculation.

(Murphy and Morton: Lymphoid Activity.)

274

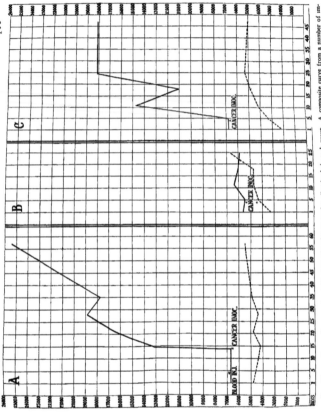

TEXT-FIG. 2. Composite curves formed by averaging the counts for all animals in each group. A, composite curve from a number of immunized animals. B, composite curve from a number of susceptible animals with growing cancers. C, composite curve from a number of animals, naturally immune to transplanted cancer. The solid line represents the actual number of lymphocytes, and the dotted line the polymorphonuclear leucocytes.

(Murphy and Morton: Lymphoid Activity.)

sometime during the first week and continues to rise with some variation to a level between 100 and 200 per cent above normal. These animals show, as a rule, some growth of the cancer during the first week but after this, retrogression sets in, continuing to complete absorption. In the susceptible animal (Text-fig. 1 B) there is a decrease in the lymphocytes and a tendency for the polynuclear cells to increase slightly. This particular animal offered little resistance apparently, as the rate of growth of the tumor, shown at the bottom of the text-figure in silhouette, would indicate. This is not uniformly so, however, as is demonstrated in the composite curve (Text-fig. 2 B) taken from a number of animals with growing tumors. Here is shown a slight reaction at the end of the first week which, however, subsides as the tumor increases in size. In one animal in the series, there was a fairly marked reaction of the lymphocytes in the second week, with a marked retardation in the growth of the tumor. The cancer later broke through the resistance which was accompanied by a corresponding drop in the mononuclear elements of the blood.

This series of experiments shows that the resistant state to transplanted cancer, whether of the natural or induced variety, is accompanied by a marked lymphocytosis. In the highly susceptible animals this phenomenon is absent, while in the less resistant ones, where the tumor is retarded but not checked absolutely, there is a slight but apparently inadequate reaction of these cells. Whether the immunity is dependent on the lymphoid increase or not is a different matter. If the reaction is a necessary part of the immunity process, the destruction of the lymphocytes in the immunized animals should be accompanied by a loss of the immunity.

The Effect of Lymphoid Destruction on the Induced Immunities to Transplanted Cancer.

Heineke has shown that x-ray effect is manifest first in its destruction of the lymphocyte.[8] We have found that repeated small doses of x-ray[9] destroy the major portion of the lymphoid system of the mouse without causing apparent injury to other tissues or pro-

[8] Heineke, H., *Mitt. a. d. Grenzgeb. d. Med. u. Chir.*, 1905, xiv, 21.

[9] Murphy, Jas. B., and Ellis, A. W. M., *Jour. Exper. Med.*, 1914, xx, 397.

ducing detrimental effect to the general health of the animal.   This method offers a means of testing out the importance of the lymphocytic reaction in the immune states to cancer.   We have found it necessary to regulate the dose very carefully, as too large or too frequent doses will cause such a general disturbance in the metabolism that tumors after they develop will grow slowly.   Likewise, too small an amount of x-ray will only partially bring about the desired effect.   The following experiment is one of four which have given uniform results.

Experiment II.—30 adult mice of about the same age and size were selected. 20 of these were given 0.3 cc. of defibrinated mouse blood, and 10 were put aside for controls. 10 of the immunized animals were subjected to small doses of x-ray for 7 consecutive days by means of the Coolidge tube, 10 milliamperes, 3 inch spark gap, and exposure of 1 to 2 minutes. Blood examination showed this to be sufficient to reduce markedly the circulating lymphocyte, leaving the general health of the animal unaffected. 10 days after the blood injection, all 30 animals were inoculated with a fragment of a transplantable mouse cancer. Text-fig. 3 gives the rate of growth of cancer in the various groups. The x-rayed immunized group shows the same number of takes as does the normal series, but the tumors in the x-rayed animals grew more rapidly.

The experiment would seem to indicate that the destruction of the lymphocytes between the period of immunizing injection and cancer inoculation by preventing the usual lymphoid reaction, suffices to abolish the immunity which would otherwise be present.

The Effect of Lymphoid Destruction on Natural Immunity to Transplanted Cancer.

The above experiments have shown that the lymphoid reaction is the same in the naturally immune animals and in the animals with induced immunity.   We would expect, therefore, that treatment of normal animals with x-ray before inoculation with cancer should greatly increase the number of takes.

Experiment III. Series A.—20 rats of the same age and size were selected. 10 of them were given 12 doses of x-ray on consecutive days in amounts sufficient to reduce greatly the lymphoid elements. All the 20 animals were then given grafts of the Jensen rat sarcoma. Of the untreated series, only 2 developed tumors and 1 of these later retrogressed, while in the 8 surviving animals in the x-rayed lot, all developed large tumors which eventually caused death.

Result.—Normal series 22 per cent takes; x-rayed animals 100 per cent takes.

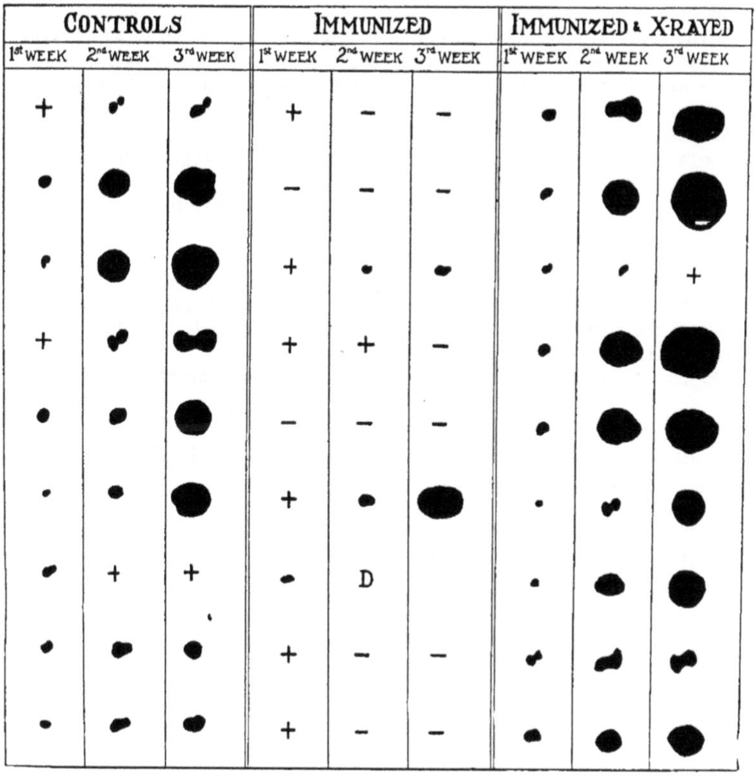

CONTROLS			IMMUNIZED			IMMUNIZED & X-RAYED		
1st WEEK	2nd WEEK	3rd WEEK	1st WEEK	2nd WEEK	3rd WEEK	1st WEEK	2nd WEEK	3rd WEEK
+			+	−	−			
			−	−	−			
			+					+
+			+	+	−			
			−	−	−			
			+					
	+	+		D				
			+	−	−			
			+	−	−			

TEXT-FIG. 3. This chart shows the effect of x-ray on induced immunity. The silhouettes show the rate of growth of the inoculated cancer for 3 consecutive weeks after inoculation. The first group are the normal animals: the second, the immunized animals; and the third, animals that were immunized and given a series of small exposures to x-ray between the immunizing dose and the cancer inoculation.

Series B.—20 rats of the same lot were selected, and 10 of these were given the same amount of exposures to x-ray as in Series A. All were inoculated with equal sized grafts of the Jensen rat sarcoma. 8 of the normal animals survived, 3 of which number developed tumors, 1 retrogressing, however, after a period of growth. Among the x-rayed animals 6 lived and all of these developed large tumors, but 1 retrogressed after a period.

Result.—Normal animals 37 per cent takes; x-rayed animals 100 per cent takes.

Series C.—20 rats were selected and 10 of these were given x-ray exposures of 4 minutes' duration on 5 consecutive days. A white blood count and a differential were done before and 3 days after the x-ray treatment was discontinued. The circulating lymphocytes fell during this time on an average to 26 per cent of their former number, while the polymorphonuclear cells remained practically at the same level in actual numbers. All the 20 animals were inoculated with grafts of the Flexner-Jobling carcinoma of the rat. In the normal series only 2 animals developed tumors and 1 of these was of very slow growth. Among the x-rayed animals, all developed tumors which continued to grow till the death of the animals.

Result.—Normal animals 25 per cent takes; x-rayed animals 100 per cent takes.

Series D.—20 mice were used in this experiment, 10 being subjected to 1 minute exposures of x-ray on 10 consecutive days. All the animals were then inoculated with grafts of a transplantable mouse carcinoma. Among the normal animals, 4 out of the 9 surviving developed tumors which progressed till the death of the animal, while the other 5 recovered completely. In the x-rayed series 7 out of 9 surviving mice developed tumors.

Result.—Normal animals 44 per cent takes; x-rayed animals 77 per cent takes.

It will be noted that tumors have been used in this experiment at periods when they were giving a low percentage of takes, so as to make the test more rigid. We have also by this same method successfully transplanted spontaneous tumors to x-rayed animals when like grafts in normal animals failed to grow. They act, however, much as the heteroplastic tissues,[10] retrogressing as the lymphoid tissue regenerates. With more carefully regulated dosage of x-ray it seems likely that uniform takes could be obtained in grafts from the spontaneous as well as from the transplantable tumors.

DISCUSSION.

The facts as they stand from these experiments are (1) that a marked lymphocytosis arises after inoculation with cancer in both the naturally immune animals and in the animals with induced immunity; and (2) when the lymphocytosis is prevented by a previous exposure of the animal to x-ray the resistant state is abolished. All the evidence points, therefore, towards the lymphocyte as a necessary factor in the immunity processes, and the finding offers a probable explanation of the results of Apolant who found that it was more difficult to render splenectomized animals immune to cancer than intact ones.[11]

[10] Murphy, Jas. B., *loc. cit.*

[11] Apolant, H., *Ztschr. f. Immunitätsforsch., Orig.,* 1913, xvii, 219.

Moreover, it has been noted that splenectomized animals exhibit less resistance to the growth of inoculated cancer than do intact ones giving both a higher percentage of takes and more rapidly growing tumors. This fact had been interpreted as the result of removal of one of the chief organs for antibody formation. But as no circulating antibodies have ever been demonstrated for cancer, it would seem more probable in the light of our work that the lymphocytic reaction has been interfered with by the removal of one of the principal lymphoid organs. We are not prepared, however, at the present time to discuss the mechanism of the lymphoid action in cancer immunity, or to do more than state that among the several possibilities that present themselves the data at hand are insufficient to establish one agency above all others.

SUMMARY.

The refractory state to transplanted cancer, induced by the subcutaneous injection of defibrinated blood, is accompanied in every case by a definite lymphoid crisis in the blood. The rise of lymphocytes is not present during the interval between the immunizing injection and the cancer inoculation but comes on sharply within twenty-four hours of the introduction of the cancer graft. In control animals where the graft leads to a definite take there is no such lymphoid response, but in instances of natural immunity the phenomenon is similar to that seen in artificially induced immunity, though the period of rise is often delayed for several days or a week.

The lymphoid crisis is not merely an accompanying factor in the immune period; it is essential to the process. This is demonstrated by the fact that destruction of the lymphocytes by x-ray is accompanied by the loss of natural or induced resistance to the growth of inoculated cancer.

The polymorphonuclear cells show a tendency to increase in the animals with growing tumors, but further study will be necessary before any conclusions regarding them can be drawn.

THE NUMERICAL LAWS GOVERNING THE RATE OF EXCRETION OF UREA AND CHLORIDES IN MAN.[1]

I. An Index of Urea Excretion and the Normal Excretion of Urea and Chlorides.

By FRANKLIN C. McLEAN, M.D.

(*From the Hospital of The Rockefeller Institute for Medical Research.*)

Plates 27 and 28.

(Received for publication, April 7, 1915.)

That excretion of at least certain substances is carried on by the kidneys according to definite laws, capable of numerical expression, was shown first for urea by Ambard, who formulated these laws in a form known as Ambard's coefficient, and demonstrated the relative constancy of the formula (1). Similar laws were formulated by Ambard and Weill (2) for the excretion of chlorides. In a later book Ambard (3) reviews the results obtained by other workers in the field, and applies similar laws to the excretion of water and of glucose.

Ambard's work has attracted attention among French clinicians, who have used his coefficient as a means of estimating renal function. His conclusions have been criticized by writers in this country on the ground that he used the inaccurate hypobromite method for determination of urea in both blood and urine, but his critics have failed to repeat his observations with more accurate methods.

Using the methods devised in Folin's laboratory for the estimation of urea in blood and urine we have previously (4) shown the relative constancy of the Ambard formula for urea in normal individuals. We now present the results of a much larger series of

[1] Preliminary reports of the work here presented have appeared in *Am. Jour. Physiol.*, 1914-15, xxxvi, 357; *Proc. Soc. Exper. Biol. and Med.*, 1915, xii, 164; *Med. Record*, 1915, lxxxvii, 624.

observations on urea and sodium chloride excretion in individuals with normal excretion.

Ambard's Laws.

Ambard's first experiments were with dogs. He found that when the concentration of urea in the urine is constant the rate of excretion varies directly as the square of the concentration of urea in the blood. His first law thus formulated is:

$$\frac{(\text{Urea in blood})^2}{\text{Rate of excretion}} = \text{Constant, or} \quad \frac{\text{Urea in blood}}{\sqrt{\text{Excretion per unit of time}}} = \text{Constant}$$

(when concentration in urine is constant).

Secondly, he found, by comparing different experiments in which the concentration of urea in the blood was the same, that the rate of excretion then varied inversely as the square root of the concentration in the urine; *i. e.,*

$$\frac{\text{Rate of excretion I}}{\text{Rate of excretion II}} = \frac{\sqrt{\text{Concentration II}}}{\sqrt{\text{Concentration I}}}.$$

This may also be expressed

$$\text{Rate I} \sqrt{\text{Concentration I}} = \text{Rate II} \sqrt{\text{Concentration II}}$$

or, in other words, when the blood urea remains constant the rate times the square root of the concentration in the urine remains constant. Introducing this factor into the first law we have

$$\frac{\text{Urea in blood}}{\sqrt{\text{Rate of excretion}} \sqrt{\text{Concentration in urine}}} = \text{Constant.}$$

This was shown to hold true for the same animal or for the same individual, or for animals or individuals of the same or nearly the same weight. For individuals of different weight it became necessary to introduce a further modification, in order to make all individuals comparable. This modification is based on the assumption, supported by the results of experiment, that, other factors remaining constant, the rate of excretion varies directly with the weight of the individual. One assumes that the amount of active kidney tissue and the circulation of blood through the kidneys vary directly with the weight. This may be expressed as

$$\frac{\text{Rate}}{\text{Weight}} = \text{Constant (other factors remaining constant).}$$

Introducing this correction into the previous formula, we have

$$\frac{\text{Concentration of urea in blood}}{\sqrt{\dfrac{\text{Rate}}{\text{Weight}}} \sqrt{\text{Concentration in urine}}} = \text{Constant.}$$

This formula expresses all of Ambard's laws in the simplest form. In order to

Standardize the formula for use in human subjects, it was expressed by Ambard as follows:

$$\frac{Ur}{\sqrt{D \times \frac{70}{Wt}} \times \sqrt{\frac{C}{25}}} = K \text{ (constant)}.$$

Ur = gm. of urea per liter of blood.
D = gm. of urea excreted per 24 hours.
Wt = weight of individual in kilos.
C = gm. of urea per liter of urine.

This formula is known as Ambard's coefficient, and the value obtained for K by Ambard and Weill in normal human subjects ranged between 0.060 and 0.070. We have found the normal variations to be somewhat greater, and have found the usual normal to be about 0.080. The difference in the value of the constant is probably due to the more accurate method used for determination of urea. In Table I will be found the results of a larger series of observations in individuals with normal excretion.

The introduction of a standard weight of 70 kilos, a standard concentration of 25 gm. per liter, and the expression of rate of excretion as the rate per 24 hours are purely arbitrary, and can not affect the general relationship between the variable factors. When these arbitrary factors are kept constant, they, in conjunction with the constant relationship between the four variable factors, tend to keep K constant, and the actual numerical value of K is determined by these arbitrary additions to the formula. All the observations of the French school are based on the formula as thus stated.

Ambard and Weill also applied laws to the excretion of sodium chloride in human subjects. They found that the same general laws were applicable, with the important exception that, while excretion of urea occurs no matter how low its concentration in the blood falls, there is a threshold for chloride excretion, and when the concentration in the plasma falls below this threshold value, excretion of chloride practically ceases. In view of the fact that there is a wide difference in chloride content of the corpuscles and plasma, plasma alone, as the fluid part of the blood, has been studied. Ambard and Weill, partly by direct experiment and partly by plotting curves, established the normal threshold value for sodium chloride as 5.62 gm. per liter of plasma. Therefore the sodium chloride above 5.62 gm. per liter determines the rate of excretion, and the law may be expressed as for urea,

$$\frac{\text{Excess NaCl over 5.62 gm. per liter of plasma}}{\sqrt{\frac{\text{NaCl in 24 hrs.}}{\text{Wt. in kilos}}} \sqrt{\text{NaCl per liter of urine}}} = \text{Constant.}$$

For practical use it appears best to calculate the plasma sodium chloride from the rate of excretion, and to compare the calculated concentration with that actually found. The formula, as derived with the use of values actually found for the constant in the above formula, reads

$$\text{Plasma NaCl} = 5.62 + \sqrt{\frac{D \times \frac{70}{Wt} \sqrt{\frac{C}{14}}}{79.33}}.$$

(The symbols have the same meaning as in the urea formulas.)

This, in its simplest form, reads

$$\text{Plasma NaCl} = 5.62 + \sqrt{\frac{\text{Gm. NaCl per 24 hrs. } \sqrt{\text{Gm. NaCl per liter of urine}}}{4.23 \times \text{Wt. in kilos}}}.$$

The constancy of this formula depends on two factors: (1) the constancy of the threshold, and (2) the constancy of the rate of excretion of sodium chloride above the threshold. In their original contribution Ambard and Weill (1, 2) believed that the threshold was quite constant in normal individuals. Later they recognized some variation in the threshold in normal individuals, though this variation seems to be relatively slight. Assuming that the laws for rate of excretion of sodium chloride over the threshold remain constant in normal individuals, one may calculate the threshold by subtracting the calculated excess from the sodium chloride actually found in the plasma, and our figures for the variations in the threshold are based on the following formula:

$$\text{Threshold} = \text{Plasma NaCl} - \sqrt{\frac{D\sqrt{C}}{4.23\,\text{Wt}}}.$$

This formula is subject to error if the rate of excretion over the threshold varies. We have not found that the urea excretion gives any basis for estimating the rate of sodium chloride excretion, and accordingly have not used Ambard's combined formula for calculating the threshold of sodium chloride excretion. Our figures on the threshold are only to be regarded as approximate, since we have so far no means of recognizing variations in rate of excretion over the threshold.

The principles of the laws of urea and chloride excretion may be illustrated in a very simple way. If we imagine a vessel into which water flows at a constant rate, escaping through an outlet at the bottom, the water will seek the level in the vessel at which the pressure is such that the rate of outflow is exactly equal to the intake. If we then increase or decrease the rate of inflow, the level will change to meet the new conditions. The change in level of the fluid in the vessel may be regarded as a compensatory change. Under physiological conditions fluctuations in the level of blood urea compensate for changes in the rapidity of formation of urea, and changes in the level of chloride in the plasma compensate for fluctuations in chloride intake. Under pathological conditions changes in the level of urea and sodium chloride in the blood also occur to compensate for changes in the outlet, in the form of diseased kidneys. In the case of chlorides the outlet of the vessel must be considered as being at some distance from the bottom, and in such a case only the level of fluid above the outlet would play any part in determining the rate of outflow, which would cease when the level of fluid fell to the level of the outlet. Similarly, only the chloride above the threshold determines the rate of excretion, which practically ceases when the threshold value is reached.

An Index of Urea Excretion.

As Ambard's coefficient expresses changes in urea excretion by variations in the value of K, these variations must be expressed on an arbitrary scale. Values for K increase directly with increase in Ur, other factors remaining the same. Changes in K then reflect changes in the blood urea, which changes occur as the square root of changes in rate of excretion. In order to express the changes in rate of excretion in a manner mathematically correct and based on a scale of 100 for the sake of comparison, we have used a formula adapted from the laws of Ambard, which we have called an index of urea excretion. An index of 100, corresponding to a value for Ambard's coefficient of 0.080, is the standard normal index, and variations are expressed directly in terms of the normal. Thus, an index of 50 indicates a rate of excretion 50 per cent of normal under the conditions of concentration in the blood and urine. The index is based on a standard normal Ambard's coefficient of 0.080. The actual normal variations in the index are shown in Table I. The derivation of the index is as follows:

$$\text{Index} = \frac{\text{(Rate of excretion found)}}{\text{(Standard normal rate)}} \times 100$$

(under the same conditions of weight, and concentration in blood and urine).

From the laws of Ambard:

1) $$K = \frac{Ur}{\sqrt{\text{Rate}}}, \quad \therefore \text{Rate} = \left(\frac{Ur}{K}\right)^2.$$

Similarly 2) $$0.080 = \frac{Ur}{\sqrt{\text{Normal rate}}}, \quad \therefore \text{Normal rate} = \left(\frac{Ur}{0.080}\right)^2.$$

Therefore 3) $$\frac{\text{Rate}}{\text{Normal rate}} = \frac{\left(\frac{Ur}{K}\right)^2}{\left(\frac{Ur}{0.080}\right)^2} = \left(\frac{0.080}{K}\right)^2,$$

and 4) $$\text{Index} = 100 \times \left(\frac{0.080}{K}\right)^2 = \left(\frac{0.80}{K}\right)^2.$$

Substituting for K (Ambard's coefficient) and simplifying,

$$\text{Index} = \frac{\text{Gm. urea per 24 hrs.} \sqrt{\text{Gm. urea per liter of urine}} \times 8.96}{\text{Wt. in kilos} \times (\text{Gm. urea per liter of blood})^2}.$$

When $K = 0.080$, the standard normal, I (index of urea excretion) $= 100$.

In this form the index offers a means of measuring the rate of excretion, under the conditions found at any given time, directly in terms of the normal, and does not require the use of an empirical scale for comparison of pathological cases with the normal.

Normal Excretion of Urea and Sodium Chloride.
Methods for Observation.

Collection of Specimens.—Short periods are preferable for observation. By collecting the urine over a given period and withdrawing blood at the middle of the period, the blood sample may be assumed to represent the average for the period. If no food or water is taken during the period, and the period is not too soon after a heavy meal, the rate of excretion during the period will remain practically constant. We have taken, as a rule, a period of 72 minutes, during either the forenoon or afternoon. One-half hour before the period the subject drinks 150 to 200 cc. of water, and takes no more fluid or food until the observation period is ended. At the beginning of the period the bladder is emptied. 36 minutes later about 10 cc. of blood are taken from an arm vein into a dry tube containing about 100 mg. of powdered potassium oxalate, and mixed with the oxalate to prevent clotting. At the end of 72 minutes after the bladder was first emptied, the specimen of urine, representing the total amount secreted during the 72 minute period, is collected, carefully measured, and used for analysis. A 72 minute period is chosen since it is one-twentieth of 24 hours, and the calculation to 24 hours is made somewhat easier. It should be remembered that the expression of rate of excretion on the basis of 24 hours need bear no relation to the amount actually excreted in 24 hours. The rate is actually determined for the shorter period, and calculated to 24 hours as a standard period on which to base all results.

Methods for Analysis.—Very accurate analyses are necessary, when one desires to determine a quantitative relationship such as is here described. The urea content of whole blood[2] and urine are at present determined by the use of urease, the specific enzyme of soy bean adapted to quantitative determinations of urea by Marshall (5). We have used the permanent preparation of urease described by Van Slyke and Cullen, and the procedure advised by them (6). This method is capable of great refinement, and gives very accurate results when carefully controlled. Urine determinations are always

[2] The urea content of whole blood is slightly lower than that of plasma. Whole blood is always used.

accompanied by determinations of preformed ammonia, and the figure for urea plus ammonia is thus corrected. Duplicate analyses of blood are of advantage when a sufficient amount of blood is at hand, but with careful technique are hardly necessary. Duplicate analyses on urine are unnecessary. One must carefully control the activity of the enzyme used, by the method for standardizing it detailed by Van Slyke and Cullen; since if the activity is below the specified standard, the enzyme may fail to decompose all urea present in the time allowed. Reagents contaminated with ammonia must, of course, be avoided.

After removing the portion of whole blood for urea determination, the remainder is centrifugalized at high speed to throw down all corpuscles, and the plasma is pipetted off.[3] The total chlorides of both plasma and urine are then determined and calculated as sodium chloride. For plasma we have used the method recently described by the author and Van Slyke (7). By this method accuracy within 1 per cent may be obtained with 2 cc. of plasma, and duplicate analyses are rarely necessary. When the plasma is allowed to stand in contact with the cells there is a slight tendency for plasma sodium chloride to diffuse into the cells. As will be seen by the following figures, this change occurs very slowly and it is necessary only to centrifugalize within two to three hours to avoid this danger. We have usually centrifugalized within an hour after the blood was drawn.

Effect of the Time of Separation of Plasma from Cells on the Chloride Content of Plasma.

	Plasma separated from red cells by centrifuge.		
	At once. NaCl per liter.	After	NaCl per liter.
	gm.	hrs.	gm.
Sample I.............	6.11	3	6.09
" II.............	5.95	3	5.97
" III.............	5.99	4	5.99
" IV.............	5.89	4	5.88
" V.............	6.13	24	6.07

Urine chlorides are determined by a modified Volhard titration. Titration of the excess silver with ammonium sulphocyanate is per-

[3] Serum of clotted blood is never used.

formed in the presence of the silver chloride precipitate and of 5 per cent nitric acid, after five minutes have been allowed for complete precipitation of the chloride. With this method removal of the proteins is unnecessary, as has been shown by Bayne-Jones (8), and results are uniformly accurate to within 0.2 of a gram per liter, which is sufficient for the present purposes.

From the urea and chloride determinations the values obtained are substituted in the proper formulas as described.

The administration of water before the period is in order to prevent apparent retention due to dehydration of the organism. If a fair amount of urine is thus obtained one need never obtain results in normal individuals simulating those in subjects actually retaining urea. Apparently sodium chloride is still less dependent on water intake than urea. Diet, especially as regards chloride and nitrogen intake, is unimportant from the standpoint of the observations, as the formulas are independent of the intake. It is therefore unnecessary to put an individual on a standard weighed diet in order to obtain comparable observations.

Calculation.—The substitution of values found by analysis in the formulas and the calculation of the formulas is in itself a considerable task if the ordinary arithmetical processes are used. Logarithms are of advantage, but they are also laborious. To simplify the process of calculation, and thereby reduce both the labor and the chances of error, a slide-rule has been adapted to the formulas. By the use of this device it is not even necessary to remember the formulas; the whole calculation becomes a matter of only a few seconds, and is purely mechanical. Figs. 1 and 2 show the rule as adapted to the calculation of both urea and sodium chloride formulas. It is the usual form of ten inch slide-rule, with the addition of certain scales and indices. The manipulation is quite simple and rapid, requiring no knowledge of the mathematical principles involved in the formulas.[4]

---

[4] The rule, with directions for use, may be obtained from Keuffel and Esser Co., 127 Fulton Street, New York.

The same rule is adaptable to all problems of multiplication and division, and is of great service in all laboratory calculations.

The Excretion of Urea.

Table I includes all observations made on urea excretion in normal individuals, including those previously reported by McLean and Selling (4). This table also includes observations made in a few hospital patients with normal excretion and selected for this purpose. It does not include a much larger series of observations made in hospital patients in whom the urea figures fell within normal limits, but who could not be properly classified as patients with normal kidneys. The patients utilized were, for the most part, cases with heart lesions, fully compensated, without signs of renal involvement, and, except where specified, not under the influence of drugs. These cases, in conjunction with those strictly normal, furnish a basis of comparison for study of patients with abnormal excretion.

Discussion of Table I.—As will be seen, we have calculated in each case, in addition to the index of urea excretion, the two formulas

$$\sqrt{\frac{D}{Wt}}\,\sqrt{C} \quad \text{and} \quad \frac{Ur}{\sqrt{\frac{D}{Wt}}\,\sqrt{C}}.$$

As stated above, the latter formula expresses the laws of Ambard in the simplest form, without distortion by addition of the constants used in Ambard's coefficient; and we have used this formula as a test of the validity of the laws and the constancy of the normal relationship. A great part of the apparently considerable variation in the index is due to the construction of the formula, in that the index varies directly with one factor in the law; *i. e.*, with the rate of excretion. To show the actual deviation from the law we have adopted 0.30 as a normal value for

$$\frac{Ur}{\sqrt{\frac{D}{Wt}}\,\sqrt{C}}$$

corresponding to an index of 100, and have tabulated the percentage variation from the law as expressed in this way. Deviations where the rate of excretion is greater than that corresponding to the standard value of 0.30 are indicated by +, those less by —.

TABLE I.*

The Relation of the Rate of Urea Excretion to Concentration in Blood in 107 Observations in Individuals with Normal Excretion, Arranged according to Concentration of Urea in the Blood.

Index of urea excretion calculated from the formula,

$$\text{Index (I)} = \frac{\text{Gm. per 24 hrs. } \sqrt{\text{Gm. per liter}} \times 8.96}{\text{Wt. in kilos} \times (\text{Blood urea})^2}.$$

No.	Subject.	Date.	Weight.	Urine per 24 hrs.	Urea. Gm. per liter of blood Ur.	Urea. Gm. per liter of urine C.	Urea. Gm. per 24 hrs. D.	Urea. Index of excretion I.	$\sqrt{\frac{D}{Wt} \times \sqrt{C}}$	$\sqrt{\frac{D}{Wt} \times \sqrt{C}}{Ur}$	Deviation from standard normal.	Remarks.
			kilos	cc.							per cent	
1	V.S....	May 5	72.0	2,300	0.191	7.85	18.0	170	0.84	0.23	+23	
2	F.C.M...	" 5	80.0	1,760	0.206	13.1	23.0	220	1.02	0.21	+30	
3	F.C.M...	" 5	80.0	2,560	0.206	9.5	24.3	196	0.97	0.21	+30	
4	A.R.D...	" 5	67.0	1,560	0.206	10.6	16.6	170	0.90	0.23	+23	
5	F.C.M...	" 17	81.6	2,800	0.211	10.1	28.2	220	1.05	0.20	+33	
6	E.S....	" 19	86.0	1,125	0.212	19.75	22.2	228	1.07	0.20	+33	
7	F.C.M...	Apr. 7	77.2	860	0.214	21.0	18.0	108	1.03	0.21	+30	
8	E.S....	May 19	86.0	1,295	0.220	14.4	18.7	153	0.91	0.24	+20	
9	F.C.M...	Jan. 20	80.0	1,073	0.221	18.14	19.45	190	1.02	0.22	+27	
10	V.S....	May 5	72.0	5,100	0.224	3.82	19.5	95	0.73	0.31	− 3	
11	D.A.P...	Apr. 8	60.0	1,900	0.225	8.35	15.9	136	0.88	0.25	+16	
12	V.S....	May 18	72.0	736	0.228	16.8	12.4	122	0.84	0.27	+10	
13	V.S....	" 18	72.0	600	0.232	13.62	8.17	70	0.65	0.36	−20	
14	2,313....	Jan. 22	67.0	1,150	0.232	11.5	13.2	110	0.82	0.28	+ 7	Neurasthenia.
15	R.A.S.	Apr. 10	70.0	1,150	0.235	21.0	24.2	255	1.26	0.19	+37	
16	F.C.M...	May 4	80.0	3,400	0.254	9.2	31.2	164	1.09	0.23	+23	
17	F.C.M...	" 6	77.2	4,248	0.256	8.0	31.2	157	1.07	0.24	+20	
18	A.M.C...	" 19	68.2	1,125	0.261	11.1	28.7	175	1.18	0.22	+27	
19	2,333....	Apr. 5	72.2	1,120	0.266	16.4	18.4	130	1.01	0.26	+13	Aortic regurgitation.
20	H.T.C...	Mar. 19	65.0	1,282	0.266	17.94	23.0	190	1.21	0.22	+27	
21	2,333....	" 1	68.4	920	0.276	14.9	13.7	91	0.88	0.31	− 3	
22	A.M.C...	May 19	68.2	4,850	0.280	6.15	29.8	123	1.04	0.26	+13	
23	2,316....	Feb. 16	73.2	1,000	0.286	16.42	16.42	100	0.95	0.30	0	Aortic regurgitation.
24	2,321....	Apr. 5	60.2	3,840	0.286	7.2	27.7	134	1.11	0.26	+13	Heart block.
25	2,333....	" 9	73.0	4,800	0.286	7.85	37.6	159	1.20	0.24	+20	
26	F.C.M...	Mar. 18	80.0	1,380	0.286	19.15	26.5	159	1.20	0.24	+20	
27	2,321....	Apr. 1	60.0	1,380	0.296	15.1	20.8	138	1.16	0.26	+13	Heart block.
28	2,365....	May 5	59.8	1,900	0.296	10 9	20.6	116	1.07	0.28	+ 7	Aortic regurgitation.
29	H.K.A...	Apr. 8	60.0	700	0.298	21.0	14.7	114	1.06	0.28	+ 7	

* Those with initials italicized were healthy normal adults. The numbers are hospital records.

TABLE I.—*Continued.*

No.	Subject.	Date.	Weight.	Urine per 24 hrs.	Gm per liter of blood Ur.	Gm. per liter of urine C.	Gm. per 24 hrs. D.	Index of excretion I.	$\sqrt{\frac{D}{Wt}} \times \sqrt{C}$	$\frac{Ur}{C}$	$\sqrt{\frac{D}{Wt}} \times \sqrt{C}$	Deviation from standard normal.	Remarks.
			kilos	cc.								per cent	
30	F.C.M...	Oct. 6	78.0	1,120	0.298	22.3	25.0	153	1.23	0.24		+20	
31	F.C.M...	June 1	77.2	720	0.299	20.7	14.7	87	0.93	0.32		- 7	
32	2,365....	May 4	59.8	1,460	0.301	12.1	17.7	102	1.02	0.30		0	
33	2,365....	" 27	65.2	1,940	0.304	14.3	27.7	155	1.26	0.24		+20	
34	A.T.D...	" 29	80.0	1,580	0.305	19.5	30.8	164	1.38	0.24		+20	
35	2,313	Feb. 18	66.2	2,050	0.309	9.85	20.2	89	0.98	0.31		- 3	
36	2,333	Apr. 30	72.2	1,860	0.316	18.0	33.5	176	1.40	0.23		+23	
37	E.S.....	Mar. 19	90.0	730	0.316	28.0	20.5	110	1.10	0.129		+ 3	
38	2,247	Dec. 21	62.2	1,343	0.316	13.25	17.8	93	1.02	0.31		- 3	Multiple sclerosis.
39	2,183	Nov. 17	67.1	1,690	0.316	13.47	22.75	112	1.12	0.28		+ 7	Chronic arthritis.
40	H.K.A...	June 5	60.0	2,400	0.321	8.2	20.9	87	1.00	0.32		- 7	
41	2,333....	Apr. 20	73.6	1,240	0.321	18.85	23.35	120	1.17	0.27		+10	
42	L.S.....	" 7	55.0	1,330	0.321	13.3	17.55	100	1.07	0.30		0	
43	R.A.S...	May 5	70.0	3,720	0.321	8.1	29.3	104	1.09	0.29		+ 3	
44	F.C.M...	" 21	77.2	816	0.321	24.8	20.1	112	1.14	0.28		+ 7	
45	F.C.M...	" 11	77.2	2,000	0.321	13.9	27.8	118	1.16	0.28		+ 7	
46	2,334. ...	Dec. 17	37.2	1,380	0.324	10.7	14.8	112	1.14	0.28		+ 7	Mitral stenosis.
47	2,365....	May 26	65.2	4,100	0.324	7.3	30.0	106	1.11	0.29		+ 3	
48	P.......	" 18	91.0	1,280	0.325	20.0	25.6	108	1.12	0.29		+ 3	
49	2,333	Apr. 16	73.0	1,440	0.326	22.2	32.0	174	1.43	0.23		+23	Exophthalmic goitre.
50	2,364....	" 20	46.6	1,285	0.326	16.85	21.6	160	1.38	0.24		+20	
51	2,236	Jan. 4	71.6	1,180	0.326	17.33	20.45	100	1.09	0.30		0	Aortic stenosis.
52	2,365....	May 13	61.0	1,700	0.326	11.7	19.9	94	1.06	0.31		- 3	Aortic regurgitation.
53	2,316	Jan. 27	71.0	1,620	0.328	15.7	25.5	110	1.20	0.27		+10	
54	2,365....	Mar. 8	59.0	462	0.330	27.0	12.5	91	1.06	0.31		- 3	
55	2,316....	Apr. 1	76.2	900	0.336	20.6	18.4	87	1.05	0.32		- 7	
56	2,333	Mar. 8	69.0	1,760	0.336	17.45	30.7	147	1.36	0.25		+16	
57	2,098	Dec. 1	71.0	870	0.336	20.4	17.75	90	1.06	0.32		- 7	Auricular fibrillation.
58	F.C.M...	Apr. 28	80.0	1,240	0.341	18.7	23.2	96	1.12	0.30		0	
59	C.A.L...	May 12	84.0	1,000	0.342	19.6	19.6	80	1.02	0.34		-13	
60	A.A.G...	" 15	64.0	3,500	0.342	11.9	41.9	172	1.50	0.23		+23	
61	P.......	" 18	91.0	1,300	0.343	17.4	22.7	80	1.02	0.34		-13	
62	2,316	Mar. 25	77.0	1,880	0.346	14.6	27.5	102	1.17	0.30		0	
63	2,316	Apr. 9	76.8	2,240	0.346	14.9	33.3	126	1.30	0.27		+10	
64	2,333	Mar. 29	71.0	2,080	0.346	15.8	32.9	138	1.36	0.25		+16	
65	2,375....	Nov. 3	36.2	680	0.350	17.15	11.67	98	1.16	0.30		0	Mitral stenosis.
66	L.S.....	May 19	55.0	1,056	0.353	24.8	26.3	171	1.54	0.23		+23	
67	2,316....	Mar. 8	75.2	3,280	0.356	8.88	29.1	84	1.07	0.33		-10	
68	2,316 ...	" 29	76.8	1,510	0.356	18.05	27.3	108	1.23	0.29		+ 3	

TABLE I.—Concluded.

| No. | Subject | Date | Weight | Urine per 24 hrs. | Urea. | | | | $\sqrt{\frac{D}{Wt}} \times \sqrt{C}$ | $\frac{Ur}{\sqrt{\frac{D}{Wt}} \times \sqrt{C}}$ | Deviation from standard normal. | Remarks. |
					Gm. per liter of blood Ur.	Gm. per liter of urine C.	Gm. per 24 hrs. D.	Index of excretion I.				
			kilos	cc.							per cent	
69	2,316....	Jan. 15	73.0	1,190	0.356	20.7	24.7	108	1.24	0.29	+ 3	
70	2,333 ...	Feb. 7	66.8	3,040	0.356	12.32	37.4	138	1.37	0.26	+13	
71	2,098....	Oct. 26	71.0	1,680	0.358	13.68	22.9	85	1.09	0.33	−10	
72	A.M.C..	May 6	67.5	1,250	0.361	21.2	26.5	125	1.34	0.27	+10	
73	R.A.S...	" 12	70.0	912	0.363	20.7	18.8	84	1.10	0.33	−10	
74	C.C.....	Apr. 18	72.0	2,000	0.363	12.4	24.8	83	1.10	0.33	−10	
75	H.K.A ..	June 4	60.0	1,100	0.363	21.8	23.9	126	1.37	0.29	+ 3	
76	F.C.M ..	Apr. 29	77.2	960	0.363	31.2	29.3	144	1.46	0.25	+16	
77	2,301	Mar. 26	55.0	595	0.366	23.2	13.8	84	1.10	0.33	−10	Post-pneumonia.
78	L.S	May 19	55.0	960	0.385	23.5	22.4	119	1.40	0.27	+10	
79	W.C.M..	" 18	61.0	2,800	0.385	11.9	33.3	112	1.37	0.28	+ 7	
80	R.E.P ...	" 4	66.0	960	0.385	22.6	21.6	94	1.25	0.31	− 3	
81	H.U.....	Apr. 18	45.0	750	0.385	20.1	15.1	91	1.23	0.31	− 3	
82	2,316....	Mar. 22	76.8	1,340	0.386	19.8	26.5	93	1.24	0.31	− 3	
83	C	Jan. 10	70.0	860	0.386	25.6	22.0	96	1.26	0.31	− 3	
84	2,098....	Dec. 11	73.0	1,280	0.388	23.5	30.0	118	1.41	0.28	+ 7	
85	2,333....	Mar. 27	70.0	1,480	0.396	22.8	33.8	132	1.52	0.26	+13	
86	2,333....	Jan. 27	66.4	1,580	0.398	15.1	23.8	80	1.18	0.34	−13	
87	F.C.M ...	May 28	77.2	2,424	0.406	25.2	63.3	224	2.03	0.20	+33	Urea 10 gm.
88	A.T	Apr. 12	70.0	1,520	0.406	22.3	33.8	125	1.51	0.27	+10	
89	2,365....	Mar. 2	59.0	734	0.406	37.5	27.5	153	1.69	0.24	+20	
90	2,365....	Jan. 28	65.8	960	0.416	33.35	32.05	140	1.68	0.25	+16	
91	2,301....	Mar. 22	53.0	1,000	0.416	19.25	19.25	82	1.26	0.33	−10	
92	2,333....	" 22	70.8	1,780	0.427	20.5	36.6	116	1.53	0.28	+ 7	
93	R.A.S...	May 25	70.0	1,872	0.428	23.1	43.0	144	1.72	0.25	+16	
94	A.T	" 21	70.0	1,368	0.428	20.9	20.0	90	1.36	0.31	− 3	
95	L.S	" 23	55.0	1,176	0.428	29.9	23.7	96	1.40	0.31	− 3	
96	2,365....	Mar. 15	57.6	680	0.437	29.6	20.2	90	1.38	0.32	− 7	
97	2,364....	" 31	52.0	1,275	0.437	22.1	28.1	120	1.59	0.27	+10	
98	A.T	May 15	70.0	1,200	0.449	24.1	28.6	89	1.42	0.32	− 7	
99	H.K.A...	" 22	60.0	988	0.470	24.8	24.3	82	1.42	0.33	−10	
100	2,190....	Oct. 16	53.2	780	0.470	27.5	21.5	86	1.45	0.32	− 7	Heart block.
101	2,190....	" 5	53.2	2,480	0.480	13.43	33.3	91	1.51	0.32	− 7	
102	W.C.M..	May 11	61.0	312	0.492	31.0	9.6	33	0.94	0.52	−73	
103	F.C.M...	" 25	77.2	1,536	0.513	35.1	52.6	138	2.01	0.26	+13	Urea 10 gm.
104	A.T	" 28	70.0	3,400	0.513	23.9	81.3	194	2.38	0.22	+27	Urea 10 gm.
105	F.C.M...	" 18	81.6	5,400	0.539	15.7	84.7	122	2.02	0.27	+10	Urea 30 gm.
106	F.C.M...	" 18	81.6	4,320	0.542	16.75	72.5	111	1.90	0.28	+ 7	Urea 30 gm.
107	C.A.L...	" 25	84.0	1,100	0.577	31.6	34.8	63	1.53	0.38	−27	Urea 10 gm.

Examination of the tables shows in general an increase in the value of

$$\sqrt{\frac{D}{Wt}} \times \sqrt{C}$$

parallel with the increase in the concentration of urea in the blood. The result of this parallelism is to keep the formula

$$\frac{Ur}{\sqrt{\frac{D}{Wt}} \times \sqrt{C}}$$

relatively constant. The maximum deviations, tending toward an increase in rate of excretion, nearly all fall in those observations made when the blood urea figure was below 0.300 gram per liter. In this group a number of individuals show an index of 200 or more, corresponding to a deviation of $+33$ per cent from the normal laws. The probable reason for this is discussed below.

Particular interest, both from the physiological and clinical standpoint, attaches to the observations in which the blood urea is between 0.300 and 0.500 grams per liter. It is within these limits that a blood urea figure may be quite normal, or may occur with retention. Within these limits are 71 observations. Of these 71 observations, 52, or 73 per cent, lie within 10 per cent of the standard normal, corresponding to a range in the index from 84 to 125. 69, or 97 per cent, come within a range of 25 per cent of the standard normal. Of the two outside of this limit, one, previously reported, had apparent retention due to dehydration, the other was an experiment following the administration of urea by mouth. Out of the total of 107 observations, only 3 have an index below 80. Of these, Nos. 13 and 102 are directly attributable to the small amount of water ingested and the small amount of urine excreted, and such results are easily avoided in normal individuals by providing a sufficient amount of water before the observation. No. 107 followed ingestion of urea. An index of 80 is to be regarded as the lowest limit of normality, when the index is used as a measure of renal function.

Attention should be called to the conditions following the ingestion of large amounts of urea. In one case a concentration of 0.52 of a gram of urea per liter of blood was attained after 30 grams of urea, and in two observations the index was quite normal, the rate of excretion reaching at one time 84.7 grams of urea per twenty-four hours. We have not observed a concentration of urea in the blood above 0.500 in a normal individual under ordinary circumstances, and this concentration would be very difficult for a normal person to attain without ingestion of urea, on account of the very rapid rate of excretion which takes place when the concentration in the blood approaches this level.

Since we have observed blood urea figures between 0.300 and 0.500 about thirty times in perfectly normal individuals under usual conditions of diet, etc., there can be no question that such figures are within the usual range. During the same time we have found the blood urea below 0.300 only twenty-three times in the same class of individuals. Since the laws of excretion hold very closely for the range from 0.300 and 0.500 and show greater deviations below 0.300, it appears that the lower figures in the blood are due to a rate of excretion, which, while not in any sense abnormal, is higher than that shown by the majority of individuals. These figures occur mainly in young, active adults. It should be noted that the variations in the blood urea figure, as well as in the index, are as great for single individuals as they are for all normals.

We may summarize the findings regarding normal urea excretion as follows: The normal concentration of urea in the blood varies from about 0.200 to 0.500 grams per liter, in the same or different individuals. The rate of excretion is determined by this concentration and by the rate of water output. The laws given above hold closely for concentrations of urea in the blood between 0.300 and 0.500 grams per liter, and somewhat less so for concentrations below 0.300, the tendency in the latter case being toward a higher rate of excretion. Findings simulating those of urea retention are easily avoided by providing for a sufficient output of urine, by administering water before the period of observation.

Normal Chloride Excretion.

Table II summarizes the results obtained from observations on chloride excretion. The class of individuals used and criteria of selection were the same as those in Table I, many of the observations in the two tables being made simultaneously. In this table we have arranged the observations in order according to the calculated value of plasma sodium chloride from the formula given, and have compared in each case the concentration actually found. The difference is expressed as + or —, according to whether the concentration actually found was higher or lower than the concentration calculated from the rate of excretion. We have also calculated the threshold in each case, *i. e.,* the threshold to which

$$\sqrt{\frac{D \sqrt{C}}{4.23 \times Wt}}$$

being added will give the concentration actually found in the plasma. It will be seen that on the whole there is a remarkably close agreement between the actual and calculated plasma sodium chloride, and this agreement is usually within the limit of error in determining the plasma sodium chloride. The threshold is therefore very constant at about 5.62 grams per liter, as originally stated by Ambard and Weill. The actual average of all values for threshold in the table is 5.61 grams +. A certain amount of deviation, as yet unexplained, must be recognized. The maximum variations observed are from 5.24 to 5.84 grams. 65 observations, or 90 per cent, are within a range of 5.52 to 5.72 grams. 50, or 70 per cent, are within range of 5.57 to 5.67 grams, or within range of experimental error of the average threshold of 5.62 grams.

The deviations from the threshold have occurred mainly in young active individuals. It has been a striking feature, not alone in the figures presented, that hospital patients with normal excretion, leading well regulated lives, have an average greater degree of constancy in the chloride threshold and excretion than strictly normal, active individuals.

As previously mentioned, Ambard attributes certain changes in the threshold to the relation of diet, time of meals, etc. Table III

TABLE II.*

The Relation of the Rate of Chloride Excretion (Calculated as Sodium Chloride) to Concentration in Plasma in 72 Observations in Individuals with Normal Excretion, Arranged according to Rate of Excretion, as Modified by Concentration in Urine, and Expressed as Calculated Sodium Chloride in Plasma.

$$\text{Calculated plasma NaCl} = 5.62 + \sqrt{\frac{\text{Gm. per 24 hrs. } \sqrt{\text{Gm. per liter}}}{\text{Wt. in kilos} \times 4.23}}$$

$$\text{Threshold} = \text{actual plasma NaCl} - \sqrt{\frac{\text{Gm. per 24 hrs. } \sqrt{\text{Gm. per liter}}}{\text{Wt. in kilos} \times 4.23}}$$

No.	Subject.	Date.	Weight.	Urine per 24 hrs.	Blood urea per liter.	Index of urea excretion I.	Gm. per liter of urine C.	Gm. per 24 hrs. D.	Sodium chloride. Gm. per liter of plasma.			Thresh-old.	Remarks.
									Calcu-lated.	Actual.	Differ-ence.		
			kilos	cc.	gm.								
1	2.365.....	Mar. 8	59.0	462	0.330	91	2.3	1.18	5.70	5.67	−.03	5.59	Aortic regurgitation.
2	2.365.....	" 2	59.0	734	0.406	153	4.7	3.45	5.78	5.73	−.05	5.57	
3	V.S.....	May 5	72.0	2.300	0.191	170	3.1	7.5	5.82	5.81	−.01	5.61	
4	2.365.....	M. 15	57.6	680	0.437	90	6.2	4.23	5.83	5.83	0	5.62	
5	2.364.....	Apr. 20	46.6	1.285	0.326	160	3.7	4.75	5.84	5.81	−.03	5.59	
6	2.195.....	Nov. 24	53.2	1.220			4.0	4.9	5.85	5.90	+.05	5.67	Post-pneumonia.
7	2.181.....	" 16	45.0	1.100			4.7	5.17	5.86	5.81	−.05	5.57	"
8	2.316.....	Jan. 15	73.0	190	0.356	108	69	8.21	5.88	5.91	+.03	5.65	Aortic regurgitation.
9	F.C.M....	May 4	80.0	3.400	0.254	164	3.7	12.6	5.89	6.05	+.16	5.78	
10	F.C.M....	Jan. 20	80.0	1.073	0.221	191	8.4	9.02	5.90	5.94	+.04	5.66	
11	L.L.V.S...	May 10	58.0	2.360	0.301	75	4.0	0.45	5.90	5.89	−.01	5.61	Auricular fibrillation.
12	2.108.....	Dec. 1	71.0	870	0.336	90	9.5	8.26	5.91	5.99	−.01	5.61	
13	E.S.....	Mar. 10	88.0	730	0.316	110	12.1	8.75	5.91	5.88	−.03	5.59	Exophthalmic goitre.
14	2.364.....	" 31	52.0	1.275	0.437	120	6.0	7.65	5.91	5.87	−.04	5.58	
15	V.S.....	May 5	72.0	5.100	0.224	95	2.9	14.8	5.91	5.87	−.04	5.58	
16	V.S.....	" 18	72.0	600	0.232	70	1.41	7.86	5.93	5.95	+.02	5.64	
17	2.317.....	Jan. 23	39.8	780	0.244	368	7.9	6.16	5.93	5.95	+.02	5.64	Mitral stenosis (digitalis).

* Those with initials italicized were healthy normal adults. The ...rs are hospital ...als.

TABLE II.—*Concluded.*

No.	Subject.	Date.	Weight.	Urine per 24 hrs.	Blood urea per liter.	Index of urea excretion I.	Gm. per liter of urine C.	Gm. per 24 hrs. D.	Sodium chloride. Gm. per liter of plasma. Calculated.	Actual.	Difference.	Threshold.	Remarks.
			lbs.	*cc.*	*gm.*								
18	F.C.M.	May 5	80.0	2,560	0.206	196	5.3	13.6	5.93	6.15	+.22	5.84	
19	2,198	Dec. 11	73.0	1,280	0.388	117	8.7	11.14	5.94	5.95	+.01	5.63	
20	H.T.C.	Mar. 19	65.0	1,282	0.266	190	7.9	10.1	5.94	5.93	−.01	5.61	
21	F.C.M.	Oct. 6	78.0	1,120	0.298	152	9.2	10.3	5.96	5.95	−.01	5.61	
22	A.R.D.	May 5	67.0	1,560	0.206	170	7.5	11.7	5.96	6.08	+.12	5.74	
23	E.S.	" 19	86.0	1,295	0.220	153	11.1	14.4	5.98	6.05	+.07	5.69	
24	2,333	Jan. 27	66.4	1,580	0.398	80	8.2	12.05	5.98	6.00	+.02	5.64	Aortic regurgitation.
25	2,316	Feb. 16	73.2	1,000	0.286	100	12.4	12.4	4.99	5.96	−.03	5.59	
26	F.C.M.	Mar. 19	80.0	2,462	......		7.0	17.25	5.99	6.04	+.05	5.67	
27	V.S.	May 18	72.0	736	0.228	122	15.5	11.4	6.00	5.95	−.05	5.57	Aortic regurgitation.
28	2,236	Jan. 4	71.6	1,180	0.326	100	11.4	13.45	6.00	5.95	−.05	5.57	
29	A.R.D.	May 5	67.0	2,820	0.226	128	5.9	16.7	6.00	6.12	+.12	5.74	" "
30	2,365	" 13	61.0	1,700	0.326	94	7.7	13.1	6.00	5.94	−.06	5.56	Urea 30 gm.
31	2,365	" 26	65.2	4,100	0.324	106	4.5	18.4	6.00	5.94	−.06	5.56	
32	F.C.M.	" 18	81.6	4,320	0.542	111	5.3	22.9	6.01	5.94	−.07	5.55	
33	2,333	Mar. 1	68.4	920	0.276	91	13.3	12.22	6.01	5.99	−.02	5.60	
34	2,316	" 25	77.0	1,880	0.346	102	8.7	16.35	6.01	5.99	−.02	5.60	
35	2,333	Apr. 30	72.2	1,860	0.316	176	8.6	16.0	6.01	6.00	−.01	5.61	
36	A.M.C.	May 19	68.2	4,850	0.280	123	4.5	21.8	6.02	5.86	−.16	5.46	
37	F.C.M.	" 17	81.6	2,880	0.211	220	7.4	20.7	6.03	6.02	−.01	5.61	
38	2,337	Jan. 19	42.0	1,160	0.195	230	8.7	10.1	6.03	6.05	+.02	5.64	
39	2,351	Apr. 26	68.5	3,660	0.281	84	3.9	14.6	6.03	5.99	−.04	5.58	
40	2,175	Nov. 9	49.0	700	......	....	13.5	9.45	6.03	6.02	−.01	5.61	Auricular fibrillation.
41	2,365	Apr. 4	59.8	1,460	0.301	102	9.5	13.9	6.03	6.01	−.02	5.60	Post-pneumonia.
42	E.S.	May 19	86.0	1,125	0.212	228	15.0	16.9	6.04	5.95	−.09	5.53	
43	2,365	" 27	65.2	1,940	0.304	155	8.7	16.9	6.04	5.95	−.09	5.53	
44	2,181	Nov. 24	43.4	1,120	......	....	9.4	10.5	6.04	6.02	−.02	5.60	
45	2,365	Jan. 28	65.8	960	0.416	140	13.7	13.15	6.04	6.04	0	5.62	
46	F.C.M.	Mar. 18	80.0	1,380	0.286	159	12.8	17.65	6.05	6.11	+.06	5.68	

No.	Subject.	Date.	Weight.	Urine per 24 hrs.	Blood urea per liter.	Index of urea excretion I.	Gm. per liter of urine C.	Gm. per 24 hrs. D.	Sodium chloride. Gm. per liter of plasma.			Thresh-old.	Remarks.
									Calcu-lated.	Actual.	Differ-ence.		
			kilos	cc.	gm.								
47	2,333......	Apr. 20	73.6	1,240	0.321	120	12.8	15.9	6.05	5.98	−.07	5.55	
48	F.C.M......	May 5	80.0	1,760	0.206	220	10.9	19.2	6.05	6.20	+.15	5.77	Urea 30 gm.
49	F.C.M......	" 18	81.6	5,400	0.539	122	5.4	29.2	6.06	5.97	−.09	5.53	
50	2,333......	Feb. 7	66.8	3,040	0.356	108	7.1	21.6	6.06	6.04	−.02	5.60	
51	2,316......	Jan. 12	74.0	3,020	0.351	130	8.9	26.9	6.07	6.12	+.05	5.57	
52	2,301......	Mar. 22	53.0	1,000	0.416	82	12.6	12.6	6.07	6.05	−.02	5.60	Post-pneumonia.
53	F.C.M......	Apr. 28	80.0	1,240	0.341	96	14.5	18.0	6.07	6.10	+.03	5.65	
54	A.M.C......	May 6	67.5	1,250	0.301	125	13.0	16.25	6.07	6.16	+.09	5.71	
55	2,337......	Apr. 30	40.6	1,640	0.251	216	8.0	13.1	6.08	6.08	0	5.62	Digitalis.
56	2,203......	Dec. 8	48.4	1,280	.....	.....	10.4	13.3	6.08	6.06	−.02	5.60	Post-pneumonia.
57	P......	May 18	91.0	1,280	0.325	108	16.3	20.8	6.09	5.99	−.10	5.52	
58	G.T.D......	" 29	80.0	1,580	0.305	164	13.2	20.8	6.09	6.11	+.02	5.64	
59	2,337......	Mar. 22	43.0	1,110	0.256	180	11.0	12.2	6.09	6.08	−.01	5.61	
60	2,198......	Nov. 2	71.0	1,580	.....	.....	12.1	19.1	6.09	6.10	+.01	5.63	
61	2,202......	" 16	49.0	1,300	.....	.....	11.1	14.4	6.10	6.04	−.06	5.56	
62	2,202......	" 24	47.2	1,700	.....	.....	9.2	15.6	6.10	6.05	−.05	5.57	
63	2,198......	Oct. 26	71o	1,680	0.358	84	12.35	20.75	6.11	6.11	0	5.62	
64	2,333......	Apr. 16	73o	1,440	0.326	174	14.2	20.4	6.12	6.15	+.03	5.65	
65	2,316......	Mar. 8	75.2	3,280	0.356	84	8.5	27.8	6.12	6.11	−.01	5.61	
66	P......	May 18	91.0	1,300	0.343	80	18.7	24.3	6.14	6.04	−.10	5.52	
67	2,337......	Feb. 26	41.0	1,100	0.246	153	12.4	13.6	6.15	6.23	+.08	5.70	
68	2,333......	Mar. 23	70.8	1,760	0.427	116	13.0	23.2	6.15	6.11	−.04	5.58	
69	2,337......	" 1	41.0	1,300	0.184	240	11.44	1,488	6.16	6.15	−.01	5.61	
70	2,333......	" 8	69.0	1,760	0.336	147	13.2	23.3	6.16	6.19	+.03	5.65	Digitalis.
71	A.M.C.	May 19	68.2	1,125	0.261	175	10.8	28.0	6.16	5.78	−.38	5.24	
72	F.C.M.	Apr. 28	80.0	2,140	0.266	182	17.3	37.0	6.30	6.40	+.10	5.72	Sodium chloride 10 gm.

TABLE III.*

Observations Made on the Same Individuals, on the Morning and Afternoon of the Same Day. All Were Normal, Healthy Adults.

No.	Subject	Time	Date	Weight.	Urine per 24 hrs.	Blood urea per liter.	Index of urea excretion I.	Gm. per liter of urine C.	Gm. per 24 hrs. D.	Sodium chloride, Gm. per liter of plasma.			Threshold.	Remarks.
										Calculated.	Actual.	Difference.		
				kilos	cc.	gm.								
1	F.C.M.	a.m.	Apr. 28	80.0	1,240	0.341	96	14.5	18.0	6.07	6.10	+.03	5.65	Sodium chloride 10 gm.
		p.m.	"	80.0	2,140	0.266	182	17.3	37.0	6.30	6.40	+.10	5.72	
2	F.C.M.	a.m.	May 5	80.0	1,760	0.206	220	10.9	19.2	6.05	6.20	+.15	5.77	
		p.m.	"	80.0	2,560	0.206	196	5.3	13.6	6.93	6.15	+.22	5.84	
3	A.R.D.	a.m.	"	67.0	1,560	0.206	170	7.5	11.7	5.96	6.08	+.12	5.74	
		p.m.	"	67.0	2,820	0.226	128	5.9	16.7	6.00	6.12	+.12	5.74	
4	V.S.	a.m.	"	72.0	5,100	0.224	95	2.9	14.8	5.91	5.87	-.04	5.58	
		p.m.	"	72.0	2,300	0.191	170	3.1	7.5	5.82	5.81	-.01	5.61	
5	F. M.	a.m.	" 18	81.6	5,400	0.539	122	5.4	29.2	6.06	5.97	-.09	5.53	Urea 30 gm.
		p.m.	"	81.6	4,320	0.542	111	5.3	22.9	6.01	5.94	-.07	5.55	" "
6	V.S.	a.m.	"	72.0	600	0.232	70	13.1	7.86	5.93	5.95	+.02	5.64	
		p.m.	"	72.0	736	0.228	122	15.5	11.4	6.00	5.95	-.05	5.57	
7	P.	a.m.	"	91.0	1,300	0.343	80	18.7	24.3	6.14	6.04	-.10	5.52	
		p.m.	"	91.0	1,280	0.325	108	16.3	20.8	6.09	5.99	-.10	5.52	
8	A.M.C.	a.m.	" 19	68.2	4,850	0.280	123	4.5	21.8	6.02	5.86	-.16	5.46	
		p.m.	"	68.2	1,125	0.261	175	10.8	28.0	6.16	5.78	-.38	5.24	
9	E.S.	a.m.	"	86.0	1,125	0.212	228	15.0	16.9	6.04	5.95	-.09	5.53	
		p.m.	"	86.0	1,295	0.220	153	11.1	14.4	5.98	6.05	+.05	5.67	

* The calculations are the same as in Table II.

shows the findings from observations made in the same individuals just preceding and two to three hours after a heavy noon meal. This table includes the maximum variations in the threshold that have been found, in both directions. Out of nine observations, in only two, Nos. 8 and 9, is there evidence of any appreciable change in the threshold between the morning and afternoon periods. In one of these the threshold was already very low in the morning and was still lower in the afternoon. In No. 9 the threshold was apparently raised between the two periods. In none of the cases was any particular attention paid to diuretics, most of the subjects drinking tea with the noon meal.

The individual variations in the chloride threshold and excretion are also of interest. Table IV contains the results of numerous observations made in the same individuals and shows the range of variations that may be encountered under conditions that are approximately identical. Nos. 1, 2, 3, and 4 are normal individuals; 5, 6, and 7 are cardiac cases with full compensation. Nos. 5 and 7 include some observations in subjects under the influence of digitalis; No. 6 includes only the observations made when the patient was not under the influence of any drugs. Subject 1, in whom the greatest number of observations was made, shows the maximum variation. There was a tendency toward a high threshold at most times, and a lowered threshold occurred only under the diuretic effects of large doses of urea.

A comparison of the chloride results with the blood urea figures and the index of urea excretion discloses no interrelationship between the two functions. Either a high or a low threshold for chlorides may exist coincidently with a high or low blood urea or high or low index. No evidence has been found which causes us to believe that the normal variations in chloride and urea functions are parallel. The degree of constancy of chloride excretion is more striking than that of urea, and remains so in the presence of the maximum variations in urea function.

The findings regarding normal chloride excretion may be summarized as follows: The normal and usual range of concentration of chlorides in human plasma is from 5.62 to 6.25 grams of sodium chloride per liter or higher, according to the amount ingested. On

TABLE IV.*

Repeated Observations on the Same Individuals, to Show Normal Variations.

No.	Subject.	Date.	Weight.	Urine per 24 hrs.	Blood urea per liter.	Index of excretion I.	Gm. per liter of urine C.	Gm. per 24 hrs. D.	Sodium chloride. Gm. per liter of plasma. Calculated.	Actual.	Difference.	Threshold.	Remarks.
			kilos	cc.	gm.								
1†	F.C.M......... Normal	1914 Oct. 6	78.0	1,120	0.298	152	9.2	10.3	5.96	5.95	−.01	5.61	
		1915 Jan. 20	80.0	1,073	0.221	191	8.4	9.02	5.90	5.94	+.04	5.66	
		Mar. 18	80.0	1,380	0.286	159	12.8	17.65	6.05	6.11	+.06	5.68	
		" 19	80.0	2,462	......	....	7.0	17.25	5.99	6.04	+.05	5.67	
		Apr. 28, a.m.	80.0	1,240	0.341	96	14.5	18.0	6.07	6.10	+.03	5.65	Sodium chloride 10 gm.
		p.m.	80.0	2,140	0.266	182	17.3	37.0	6.30	6.40	+.10	5.72	
		May 4	80.0	3,400	0.254	164	3.7	12.6	5.89	6.05	+.16	5.78	
		" 5, a.m.	80.0	1,760	0.206	220	10.9	19.2	6.05	6.20	+.15	5.77	Urea 30 gm.
		p.m.	80.0	2,560	0.206	196	5.3	13.6	5.93	6.15	+.22	5.84	" " "
		" 17	81.6	2,800	0.211	220	7.4	20.7	6.03	6.02	−.01	5.61	" " "
		" 18, a.m.	81.6	5,400	0.539	122	5.4	29.2	6.06	5.97	−.09	5.53	
		p.m.	81.6	4,320	0.542	111	5.3	22.9	6.01	5.94	−.07	5.55	
2†	V.S......... Normal	" 5, a.m.	72.0	5,100	0.224	95	2.9	14.8	5.91	5.87	−.04	5.58	
		p.m.	72.0	2,300	0.191	170	3.1	7.5	5.82	5.81	−.01	5.61	
		" 18, a.m.	72.0	600	0.232	70	13.1	7.86	5.93	5.95	+.02	5.64	
		p.m.	72.0	736	0.228	122	15.5	11.4	6.00	5.95	−.05	5.57	
3†	A.M.C......... Normal	" 6	67.5	1,250	0.361	125	13.0	16.25	6.07	6.16	+.09	5.71	
		" 19, a.m.	68.2	4,850	0.280	123	4.5	21.8	6.02	5.86	−.16	5.46	
		p.m.	68.2	1,125	0.261	175	10.8	28.0	6.16	5.78	−.38	5.24	
4†	E.S......... Normal	Mar. 19	88.0	730	0.316	110	12.1	8.75	5.91	5.88	−.03	5.58	
		May 19, a.m.	86.0	1,125	0.212	228	15.0	16.9	6.04	5.95	−.09	5.53	
		p.m.	86.0	1,295	0.220	153	11.1	14.4	5.98	6.05	+.07	5.69	

* The calculations are the same as in Table II.

† Nos. 1, 2, 3, and 4 are healthy normal adults.

No.	Subject.	Date.	Weight.	Urine per 24 hrs.	Blood urea per liter.	Index of urea excretion I.	Gm. per liter of urine C.	Gm. per 24 hrs. D.	Sodium chloride. Gm. per liter of plasma. Calculated.	Actual.	Difference.	Threshold.	Remarks.
		1915	kilos	cc.	gm.								
5‡	2,365 Aortic regurgitation	Jan. 28	65.8	960	0.416	140	13.7	13.15	6.04	6.04	0	5.62	Fever.
		Feb. 24	.....	1,140	0.344	182	9.9	11.3	5.96	5.72	−.24	5.38	Digitalis.
		Mar. 2	.....	734	0.406	153	4.7	3.45	5.78	5.73	−.05	5.57	"
		" 8	59.0	462	0.330	91	2.3	1.18	5.70	5.67	−.03	5.59	
		" 15	57.6	680	0.437	90	6.2	4.23	5.83	5.83	0	5.62	
		May 4	49.8	1,460	0.301	102	9.5	13.9	6.03	6.01	−.02	5.60	
		" 5	59.8	1,900	0.296	116	9.2	17.5	6.08	5.82	−.26	5.36	
		" 13	61.0	1,700	0.326	94	7.7	13.1	6.00	5.94	−.06	5.56	
		" 26	65.2	4,100	0.324	106	4.5	18.4	6.00	5.94	−.06	5.56	
		" 27	65.2	1,940	0.304	155	8.7	16.9	6.04	5.95	−.09	5.53	
6‡	2,333 Aortic regurgitation	Jan. 27	66.4	1,580	0.398	79	8.2	12.95	5.98	6.00	+.02	5.64	
		Feb. 7	66.8	3,040	0.356	108	7.1	21.6	6.06	6.04	−.02	5.60	
		Mar. 1	68.4	920	0.276	91	13.3	12.22	6.01	5.99	−.02	5.60	
		" 8	69.2	1,760	0.336	147	13.2	23.3	6.16	6.19	+.03	5.65	
		" 22	70.8	1,780	0.427	128	13.0	23.2	6.15	6.11	−.04	5.58	
		Apr. 16	73.0	1,140	0.326	174	14.2	20.4	6.12	6.15	+.03	5.65	
		" 20	73.6	1,240	0.321	120	12.8	15.9	6.05	5.98	−.07	5.55	
		" 30	72.2	1,860	0.316	176	8.6	16.0	6.01	6.00	−.01	5.61	
7‡	2,337 Mitral stenosis	Jan. 19	42.9	1,160	0.195	230	8.7	10.1	6.03	6.05	+.02	5.64	Digitalis.
		" 23	39.8	780	0.244	308	7.9	6.16	5.93	5.95	+.02	5.64	
		" 28	39.6	660	0.266	210	6.0	4.0	5.86	5.91	+.05	5.67	
		Feb. 26	41.0	1,100	0.246	153	12.4	13.6	6.15	6.23	+.08	5.70	Digitalis.
		Mar. 1	41.0	1,300	0.184	240	11.44	14.88	6.16	6.15	−.01	5.61	
		" 22	43.0	1,110	0.256	180	11.0	12.2	6.09	6.08	−.01	5.61	Digitalis.
		Apr. 30	40.6	1,640	0.251	216	8.0	13.1	6.08	6.08	0	5.62	Digitalis.

‡ Nos. 5, 6, and 7 are cases with cardiac lesions but with normal excretion.

the excess over a threshold of about 5.62 grams per liter depends the rate of excretion, the laws governing which may be expressed numerically. A concentration below 5.62 grams per liter has not been observed in a normal individual.

DISCUSSION.

In the above laws for both urea and sodium chloride four variable factors are considered; namely, the concentration in blood and urine, the rate of excretion, and the weight of the individual. Indirectly the water content of the blood is considered; inasmuch as it influences the rate of water output and thereby the concentration of urea or sodium chloride in the urine, and the rate of their excretion. The use of body weight in the formula implies that the amount of blood flowing through the kidneys in a given time is relatively constant per kilo of body weight, and also that the amount of functionating kidney tissue is in direct proportion to the body weight. That the factors of blood flow, etc., are subject to variations from time to time, in the same individual, as well as in different individuals, can not be doubted. That the normal variations in the laws are due to such changes can not at present be shown, but it seems that if these factors could be controlled, the degree of constancy might be much higher. Inasmuch as the variations in different individuals of far different weights and proportions are no greater than the variations in the same individual at different times, the relation of body weight to the rate of excretion seems well established. Body temperature has been shown by Ambard to be of some importance, but the normal variations are apparently not sufficient to influence the rate of excretion. That such a number of possible factors influencing a physiological process, such as the rate of excretion of urea or sodium chloride, can be controlled and brought into numerical laws which show a high degree of constancy is of great importance from the standpoint of normal and pathological physiology. It suggests that many of the so called vital processes may work under laws as definite as the better known laws of physics. The rate of flow of liquids under different conditions has long been subject to numerical expression by the laws of hydrostatics. That at least certain functions of body tissues can be as well defined now appears probable.

CONCLUSIONS.

1. The excretion of urea and of chlorides in the normal individual is carried out according to definite laws, capable of numerical expression.

2. The rate of excretion of urea and of sodium chloride is determined by the concentration in the blood, the rate of water output, and indirectly by the weight of the individual.

3. The threshold of sodium chloride excretion is practically constant at about 5.62 grams of sodium chloride per liter of plasma. Slight variations in the threshold occur in normal individuals.

4. The rate of excretion of urea under the conditions found at any time can be measured directly in terms of the normal by an index of urea excretion.

BIBLIOGRAPHY.

1. Ambard, L., *Compt. rend. Soc. de biol.,* 1910, lxix, 411, 506. Ambard, L., and Weill, A., *Jour. de physiol. et de path. gén.,* 1912, xiv, 753.
2. Ambard and Weill, *Semaine méd.,* 1912, xxxii, 217.
3. Ambard, Physiologie normale et pathologique des reins, Paris, 1914.
4. McLean, F. C., and Selling, L., *Jour. Biol. Chem.,* 1914, xix, 31.
5. Marshall, E. K., Jr., *Jour. Biol. Chem.,* 1913, xiv, 283; xv, 487, 495.
6. Van Slyke, D. D., and Cullen, G. E., *Jour. Biol. Chem.,* 1914, xix, 211.
7. McLean, F. C., and Van Slyke, D. D., *Jour. Biol. Chem.,* 1915, xxi, 361.
8. Bayne-Jones, S., *Arch. Int. Med.,* 1913, xii, 90.

EXPLANATION OF PLATES.

Figs. 1 and 2 show a slide-rule adapted to calculation of urea and sodium chloride formulas.

PLATE 27. ,

FIG. 1. I, II, and III show the calculation of the urea index.

Example.

Gm. urea excreted per 24 hrs., $D = 20.0$
Gm. urea per liter of urine, $C = 11.0$
Gm. urea per liter of blood, $Ur = 0.330$
Body weight in kilos, $Wt = 55.0$

I. 55.0 on Wt scale is set opposite 20.0 on D scale (first position).

II. Hair line on runner is moved to 11.0 on C scale (second position).

III. Slide is moved so that 3.30 on Ur scale is at hair line on runner (third position).

Reading is now made at the arrow which points to scale I and is between 99.0 and 100.0. Therefore the index, I, is 100.0.

PLATE 28.

FIG. 2. IV, V, and VI show the calculation of plasma sodium chloride.

Example.

Gm. sodium chloride excreted per 24 hrs., D = 18.0
Gm. sodium chloride per liter of urine,   C = 9.0
Body weight in kilos,   Wt = 40.0

IV. 40.0 on Wt scale is set opposite 18.0 on D scale (first position).

V. Hair line on runner moved to 9.0 on C scale (second position).

VI. Constant at 4.23 on Wt scale moved to hair line on runner.

Reading is made opposite the arrow on E scale, which is at 0.57.

Calculated plasma sodium chloride = 5.62 + E = 5.62 + 0.57 = 6.19.

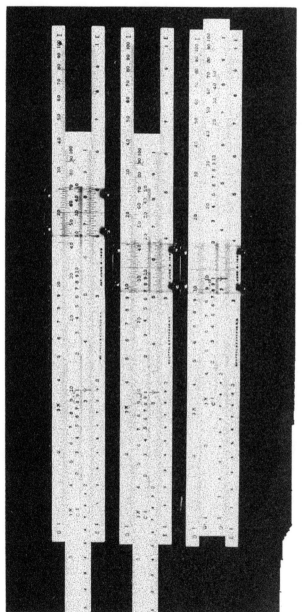

Fig. 1.

(McLean: Laws Governing Rate of Excretion in Man.)

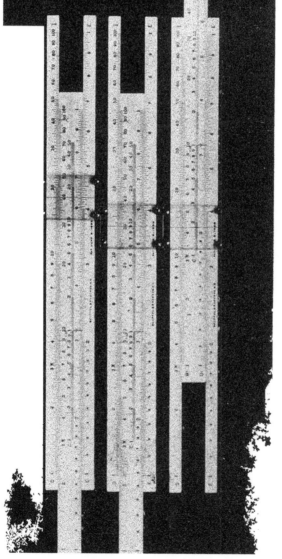

FIG. 2.

(McLean: Laws Governing Rate of Excretion in Man.)

EXTIRPATION OF THE PINEAL BODY.

By WALTER E. DANDY, M.D.

(From the Department of Surgery of Johns Hopkins University, Baltimore.)

PLATES 29 TO 36.

(Received for publication, May 21, 1915.)

The pineal body until recently has been regarded merely as a very curious vestigial inheritance, serving in the chain of evolution as a reminder of a far distant functioning pineal or central eye. There seems to have been nothing in its macro- or microscopic appearance sufficiently noteworthy to stimulate the especial interest of either anatomists or physiologists. Its inaccessible location has, perhaps, deterred experimenters from undertaking investigations of apparently so little promise.

Tumors of the pineal had not infrequently been reported by pathologists, but there was nothing which seemed significantly correlative in the pathological findings and the clinical signs and symptoms. The marvelous story of the thyroid, parathyroids, hypophysis, and adrenals is, no doubt, largely responsible for the recent endeavors to promote the pineal to a position of like importance among the endocrine glands.

In 1898 Heubner[1] presented a case of markedly precocious sexual and less pronouncedly precocious somatic development. The patient was a boy of 4½ years with pubic hair 1 cm. long, penis and testicles as large as the normal at puberty. The mammæ were conspicuous. The body development was equal to that of a boy of 8 or 9. He was abnormally fat. A diagnosis of tumor of the hypophysis was made. The following year this case was reported by Oestreich and Slawyk,[2] who found at autopsy a teratoma of the pineal. Almost

---

[1] Heubner, *Allg. med. Centr.-Ztg.*, 1899, lxviii, 89.

[2] Oestreich, R., and Slawyk, Riesenwuchs und Zirbeldrüsen-Geschwulst, *Virchows Arch. f. path. Anat.*, 1899, clvii, 475.

simultaneously Ogle[3] reported a very similar case, a boy of 6 years
with precocious sexual development.   The precocity was principally
evidenced by an enlarged penis and the presence of pubic hair.
The testicles were about normal in size.   In this instance also a teratoma of the pineal was found post mortem.

Marburg,[4] in 1907, collected from the literature about forty cases
of pineal tumor and added a case of his own.   Marburg's patient
presented no precocity either sexual or somatic; he was merely too
fat.   Marburg endeavored to establish a pineal clinical entity, and
ventured even to pronounce upon the degree, in a given case, of the
activity of this gland.   He classified all pineal glandular symptoms
under the three heads, hypopinealism, hyperpinealism, and apinealism.   Hypertrophy of the genitals and precocity were found in
hypopinealism, adiposity in hyperpinealism, and cachexia in apinealism.

According to this conception, the pineal and pituitary would seem
to have antagonistic activities.   Hypertrophy of the pineal is expected to produce an adiposity indistinguishable from Fröhlich's
dystrophia adiposogenitalis of hypopituitarism.   Atrophy of the
pineal would be accompanied by genital and somatic hypertrophy
and precocity, whereas acromegaly results from hyperplasia of the
pituitary gland.   Cachexia is described with both apinealism and
apituitarism.

It was Marburg's view that the pineal, normally functioning only
during the early years of life, inhibited genital and somatic growth
and sexual characteristics, and that its partial destruction by tumors
with the resultant hypopinealism permitted, uncontrolled, the development of these features.   Consequently, tumors in older youths
and adults would not cause sexual abnormalities, because the pineal
in them had ceased to function.   This view is in accord with the
anatomical evidences of its involution after the early years of life.

[3] Ogle, C., (1.) Sarcoma of the Pineal Body, with Diffused Melanotic Sarcoma of the Surface of Cerebrum.   (2.) Tumor of Pineal Body in a Boy, *Tr.
Path. Soc. London*, 1898-99, 1, 6.

[4] Marburg, O., Zur Kenntnis der normalen und pathologischen Histologie der
Zirbeldrüse; die Adipositas cerebralis, *Arb. a. d. neurol. Inst. a. d. Univ. Wien*,
1908, xvii, 217; Die Adipositas cerebralis. Ein Beitrag zur Pathologie der Zirbeldrüse, *Wien. med. Wchnschr.*, 1908, lviii, 2617.

It must be emphasized that at the time of Marburg's publication there had been absolutely no significant experimental investigations on the pineal and that his classification is based solely on clinicopathological observations; also that of about forty cases of tumor of the pineal gland only in the two above mentioned instances was there any sexual, somatic, or mental precocity. In several cases varying degrees of adiposity had been noted. It is also worthy of note in passing that in both cases of sexual precocity the tumor was a teratoma.

Since Marburg's publication many cases of pineal tumor have been added, the total now being about sixty cases. Von Frankl-Hochwart's[5] case, also a teratoma of the pineal, in a boy of $5\frac{1}{2}$ years, rather large for his age, showed sexual hypertrophy and precocity two months before death. The patient developed a deep voice and the genital hair was equal to that of a boy of 15. There was also mental precocity. Raymond and Claude[6] added a case which presented increasing adiposity, and somatic development but without sexual changes. With the exception of a rather frequent adiposity the other cases have shown but little to indicate glandular influences.

Experimental work has since been added but with contradictory results. In general, attempts have been made to reproduce so called hyperpinealism by feeding pineal extract and apinealism or hypopinealism by extirpation of the pineal. In feeding experiments by McCord[7] on guinea pigs, chickens, and dogs, an increase in weight together with earlier sexual maturity and sexual characteristics resulted. Dana and Berkeleys fed pineal extract of young bullocks and calves to guinea pigs, rabbits, and kittens, and noted a 25 per cent increase in weight over the controls. Later fifty children were injected with pineal extract, but they grew in height and weight

[5] von Frankl-Hochwart, L. Über Diagnose der Zirbeldrüsentumoren. *Deutsch. Ztschr. f. Nervenh.*, 1909, xxxvii, 455.

[6] Raymond, F., and Claude, H., Les tumeurs de la glande pinéale chez l'enfant, *Bull. de l'Acad. de méd. de Paris*, 1910, lxiii, series 3, 265.

[7] McCord, C. P., The Pineal Gland in Relation to Somatic, Sexual and Mental Development, *Jour. Am. Med. Assn.*, 1914, lxiii, 232.

[8] Berkeley, W. N., Dana, C. L., Goddard, H. H., and Cornell, W. S., The Functions of the Pineal Gland, with Report of Feeding Experiments. *Med. Rec.*, 1913, lxxxiii, 835. Dana and Berkeley, The Functions of the Pineal Gland, *Month. Cycl. Pract. Med.*, 1914, xxviii, 78.

less rapidly than the controls, though a distinctly greater mental improvement was observed. These observers think there is no doubt that injections of pineal will clear low grades of mental deficiency. These claims are quite similar to those of McCord, who noted mental precocity in his pineal-fed puppies.

Attempts have been made to remove the pineal by Exner and Boese,[9] Foà,[10] and Sarteschi.[11] The results of these investigations add even more confusion. Exner and Boese found absolutely no changes following complete or partial removal of the pineal. Sarteschi found adiposity, greater somatic and genital development, and sexual precocity in young dogs. Following extirpation of the pineal of chickens, Foà observed a premature development of the primary and secondary sexual characters.

Briefly, therefore, adiposity may result by feeding pineal extract (McCord, Dana, and Berkeley) or by complete or partial removal of the pineal (Sarteschi). Sexual and somatic precocity may result from feeding pineal extracts (McCord, Dana, and Berkeley), or from partial or complete removal of the pineal body (Foà, Sarteschi), or nothing may result from its partial or complete destruction (Exner and Boese). Such is the paradoxical experimental support for Marburg's hypothesis of pineal function.[12]

The foregoing is presented as a purely objective, concise summary of our present knowledge of the pineal. A detailed consideration of the clinical and experimental observations will be given in a subsequent communication. The purpose of this paper is to report briefly the results of pinealectomy in a series of young puppies

[9] Exner, A., and Boese, J., Ueber experimentelle Exstirpation der Glandula pinealis, *Deutsch. Ztschr. f. Chir.*, 1910, cvii, 182; Über experimental Exstirpation der Glandula pinealis, *Neurol. Centralbl.*, 1910, xxix, 754.

[10] Foà, C., Ipertrofia dei testicoli e della cresta dopo l'asportazione della ghiandola pineale nel gallo, *Pathologica*, 1911–12, iv, 445.

[11] Sarteschi, U., La sindrome epifisaria "macrogenitosomia precoce" attenuta sperimentalmente nei mammiferi, *Pathologica*, 1913, v. 707.

[12] It should also be noted that Pellizzi (Pellizzi, G. B., La sindrome "macrogenitosomia precoce," *Neurol., Centralbl.*, 1911, xxx, 870), in 1910, presented two cases of somatic and sexual precocity without substantiation of the diagnosis, as cases of "la sindrome epifisaria macrogenitosomia precoce." In the absence of anatomical proof and the rarity of a correct clinical diagnosis of pineal tumor, this evidence can not be accepted. It was in Pellizzi's laboratory that the experimental work of Sarteschi was conducted.

and to describe the method which has been evolved for extirpation of the pineal body.

Experimental Removal of the Pineal.

In the higher mammals the pineal is so deeply situated, so minute, and so intimately associated with important and easily injured structures that its removal by operation has been regarded as impractical. It is covered by the splenium of the corpus callosum and by the vena Galena magna. It lies between the anterior corpora quadrigemina, and consequently is just above the aqueduct of Sylvius. It is situated almost exactly in the center of the brain. Its removal necessitates opening the third ventricle, the posterior wall of which it forms a part. The greatest of the dangers encountered in the removal of the pineal is hemorrhage, and especially hemorrhage into the ventricles. The greatest difficulty is the definite recognition of the gland.

Foà has successfully excised pineals from chickens, but with the exception of Sarteschi's work, the removal of the pineal in higher mammals has not been successful. Exner and Boese attempted its removal in dogs but soon yielded to the more expedient but objectionable method of cauterization. By these investigators the cautery was inserted blindly through a trephine opening in the skull. This procedure was accompanied by a very high mortality. From a series of 95 animals, death resulted from the operation in 75. The principal cause of death was hemorrhage. This is not surprising, since a successful introduction of the cautery must almost necessarily perforate the vena Galena magna. Sarteschi's operative results were almost equally disastrous. Of 15 dogs operated upon only 3 survived. He ascribes the mortality to hemorrhage, trauma, and anesthesia. He ligates both carotid arteries as a preliminary procedure, divides the superior longitudinal sinus, and arrives at the pineal by separating the cerebral hemispheres and elevating the splenium. The destruction of both carotid arteries and the superior longitudinal sinus seriously complicates the interpretation of results. Destruction of the great vein of Galen may also be an important complication. Into this vein passes practically all the blood from the interior of the brain. The collateral venous supply for the vein is so inadequate that its occlusion results in an internal hydro-cephalus. The amount of cerebral destruction incident to the above mentioned procedure of Sarteschi is also significant.

Evolution of the Operation upon the Pineal Body.

In a series of experiments on internal hydrocephalus by the author[13] three years ago, the vein of Galen was occluded by the application of a silver clip, and in the development of this procedure a field of operation very close to the pineal was made accessible, and thus the possibility of removal of this structure was suggested.

By elaboration of the method of approach, an operation was de-vised which is free from the objections inherent to operations which entail cerebral destruction, vascular injury, or the endangering of the vitality of other cerebral structures. The results which might be noted could be attributed only to the uncomplicated loss of the pineal body. The operation to be described was successfully per-formed on young puppies from ten days to three weeks old under ether anesthesia plus a preliminary, relatively high dose of morphia. Despite the long duration of the operation, which usually required two and one-half to three and sometimes four hours, the animals recovered very quickly, and on the following day were almost as active and playful as the controls (Figs. 5 and 6).

An opening about 2 cm. in length was made in the vault of the skull, extending posteriorly to the inion and mesially to the midsag-ittal line. The dura was opened to and reflected over the superior longitudinal sinus. The occipital lobe was then carefully retracted and following the ligation of a small vein, which bridges the space between the brain and the falx cerebri, the tip of the tentorium cere-belli (osseum) was quickly exposed and the terminus of the vena Galena magna brought into view. Then a very tedious and pains-taking liberation of this vein was begun. By alternately freeing the vein with careful blunt dissection and controlling the hemorrhage with pledgets of cotton, the inferior surface of the vein was liberated

[13] Dandy, W. E., and Blackfan, K. D., Internal Hydrocephalus, an Experi-mental, Clinical and Pathological Study, *Jour. Am. Med. Assn.*, 1913, lxi, 2216; Internal Hydrocephalus, eine experimentelle, klinische und pathologische Unter-suchung, *Beitr. z. klin. Chir.*, 1914, xciii, 392; An Experimental and Clinical Study of Internal Hydrocephalus, *Am. Jour. Dis. Child.*, 1914, viii, 406.

and the corpora quadrigemina exposed. The vein was then carefully elevated in order to work beneath it, and the median groove between the corpora quadrigemina was slowly followed anteriorly until the pineal body was reached. The pineal was then caught in a small biting forceps and removed. From a series of twelve dogs, of varying ages, not one survived the operation longer than one to two hours. Invariably the postmortem examination disclosed the ventricles full of blood, which was presumably the cause of death. The bleeding from the numerous venous and arterial radicles in the enveloping pia always resulted at the time of removal of the gland. Extirpation of the pineal necessitated opening the third ventricle because of the incorporation of this structure in its posterior wall.

Two years later the same plan of attack was again tried, but the pineal was dissected out more thoroughly before removal in order to minimize the hemorrhage when the ventricle was opened. Another addition to the technique of the operation and one which has proved of the greatest importance was to open the third ventricle at a point over the pineal body before attempting its removal. This not only permitted more room by release of the fluid, but, more important still, it collapsed the cerebral ventricles so that if hemorrhage should occur it would not be into the open ventricles and thus cause distention of the brain and possibly rupture of the cortex.

With these modifications this operative procedure became successful. The small amount of bleeding which resulted was sponged away with pledgets of cotton and was always under control.

The operation as performed was, however, very unsatisfactory because of its great difficulty and the length of time required for its accomplishment. The most patient and assiduous care in the control of hemorrhage by wet and dry cotton pressure had to be exercised. It was necessary to hold the head quite motionless during this long and tedious procedure. With practice the pineal could usually be recognized because of its constant and dominant position at the junction of the median quadrigeminal groove and the third ventricle. However, very frequently, despite the greatest caution, the pineal would become covered with blood clot and unrecognizable and, perhaps, occasionally would be sponged away. When this was believed to have occurred an area of tissue from this accurately located position was removed and examined microscopically. In all such cases a second piece of tissue was excised in order to insure complete removal of the gland. This, too, was studied with the microscope. In many instances the pineal area was further treated by an electric cautery needle. At times even with the greatest care, death resulted from bleeding into the ventricles. Successful complete extirpation resulted in only about 25 per cent of the cases operated upon in this manner.

New Method of Pinealectomy.

The foregoing method was very capricious. To be of practical value the pineal body must be more easily reached and removed with greater certainty and less mortality. Consequently a new and simple method of attack has been evolved. Though more delicate and requiring more painstaking care, it can be done almost as easily as a canine hypophysectomy. The new operation can be done in less than one hour. It differs from the preceding operation in that the pineal is reached from in front through the third ventricle rather than from behind. In this way the extensive bleeding consequent to liberation of the vein of Galen is obviated, sidetracked as it were, and the operation can be performed almost bloodlessly. This is accomplished by dividing the splenium of the corpus callosum in the midline for a distance of about 2 cm. from its posterior terminus. This exposes the transparent roof of the third ventricle which is distended by the contained cerebrospinal fluid. A large anemic area is visible in the midline of the roof of the ventricle, between the two small veins of Galen. This is perforated and the opening enlarged backward to the origin of the vena Galena magna by releasing the blades of the forceps. The entire third ventricle is thus brought in full view and the pineal body is readily seen under the origin of this vein, in the median quadrigeminal groove. The pineal body can easily be grasped in the jaws of the cupped biting forceps and completely removed. The accompanying drawings By Mr. Broedel (Figs. 1 to 4) render any description of this operative procedure superfluous.

Practically no bleeding occurs during the exposure of the gland. A little bleeding follows its removal but this can easily be con-

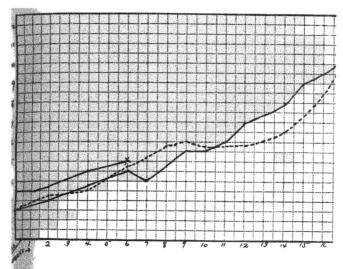

TEXT-FIG. 1. Curve of weights of animals shown in Figs. 6 an
weeks and the abscissa the weight in pounds. The solid lines repres
the that of a pinealectomized dog, all being from the same litter.

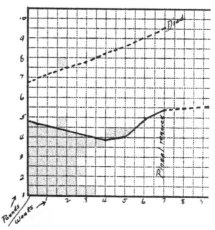

TEXT-FIG. 2. A similar curve of two
ter 6 weeks old. The initial weights sho
The broken line represents the pinealect
solid line the control. At the end of the
was removed from the control dog, whic
normally. The removal of the pineal d
growth.

Pineal Body.

was removed and examined microscopically. In all such
...ond piece of tissue was excised in order to insure complete
...of the gland. This, too, was studied with the microscope.
...instances the pineal area was further treated by an electric
needle. At times even with the greatest care, death resulted
...Successful complete extirpation
...eding into the ventricles.
...in only about 25 per cent of the cases operated upon in this

New Method of Pinealectomy.

...foregoing method was very capricious. To be of practical
...he pineal body must be more easily reached and removed with
...certainty and less mortality. Consequently a new and
...r method of attack has been evolved. Though more delicate
...quiring more painstaking care, it can be done almost as easily
...anine hypophysectomy. The new operation can be done in less
...one hour. It differs from the preceding operation in that the

...ive bleeding consequent
...sidetracked as it were,
...bloodlessly. This is
...rpus callosum in the
...posterior terminus.
...d ventricle which is
...fluid. A large anemic
...oof of the ventricle, between
...en. This is perforated and the
...ard to the origin of the vena Galena magna
...en. The entire third ventricle is
...e blades of the forceps.
...rought in full view and the pineal body is readily seen under
...origin of this vein, in the median quadrigeminal groove. The
...al body can easily be grasped in the jaws of the cupped biting
...ceps and completely removed. The accompanying drawings
...Mr. Broedel (Figs. 1 to 4) render any description of this opera-
...e procedure superfluous.

Practically no bleeding occurs during the exposure of the gland.
little bleeding follows its removal but this can easily be con-

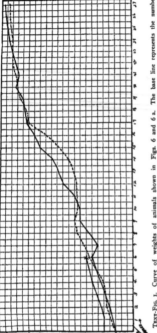

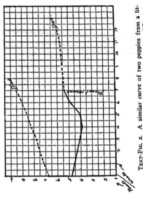

TEXT-FIG. 1. Curve of weights of animals shown in Figs. 6 and 6 a. The base line represents the number of weeks and the abscissa the weight in pounds. The solid lines represent the weights of control dogs, and the broken line that of a pinealectomized dog, all being from the same litter.

TEXT-FIG. 2. A similar curve of two puppies from a litter 6 weeks old. The initial weights show a great difference. The broken line represents the pinealectomized dog, and the solid line the control. At the end of the 6th week the pineal was removed from the control dog, which had not developed normally. The removal of the pineal did not influence the growth.

(Dandy: Pineal Body.)

trolled by a minute tampon of cotton. With collapsed ventricles the bleeding is outward through the wound and is therefore not to be feared. Not infrequently the aqueduct of Sylvius may be filled with blood. This has never caused any mortality because, before closure, the mould of clotted blood may be readily extracted, the aqueduct of Sylvius being in full view. To insure complete excision a second piece of tissue was invariably removed from the pineal region. With this method of operating there has been, as I have said, practically no mortality. It is, however, quite easy to become disoriented, even when following carefully the procedure which I am advocating. If bleeding and laceration of tissue are avoided, as they can be, and the midline is adhered to, there is little danger of losing one's bearings.

In every case a histological examination is made of all the tissue obtained at the operation (Fig. 7 b); and, after death, of the immediate region from which the pineal was removed.

Results Following Pinealectomy.

Our operations for extirpation of the pineal have been mainly upon young puppies, from 10 days to 3 weeks old. Of these one is living 15 months after the operation (Figs. 7 and 7 a); one died of distemper 1 year after operation (Fig. 8); several survived the pinealectomy 3 to 8 months. It is exceedingly difficult to raise puppies in the confined quarters of our laboratory. We were, however, unable to note any difference in the resistance of the operated and the control animals to the usual diseases.

When litters of puppies could be obtained, one or more of the animals were kept as controls. Little importance, however, should be attached to such comparisons because of the great variations found in members of the same family. The pineal was also removed in several adult male and female dogs, and three of these are living longer than four months after the operation (Fig. 9).

Somatic Development, Adiposity, and Mentality.

Careful observations have been made of the growth of the pineal-ectomized animals. Skiagraphs have been taken at various periods.

but there has been no evidence of either superior or inferior somatic development or adiposity (Text-figs. 1 and 2), save perhaps in a single instance. In this animal (Figs. 7 and 7 a) there was a slight increase in weight for a brief period about one year after the operation; this is disappearing, so that now it is scarcely noticeable. It might be attributable to overfeeding. There is nothing in the behavior of the pinealectomized animals to suggest mental precocity.

Sexual Precocity.

We have observed nothing to support the view that the pineal gland inhibits the sexual functions and that its removal is followed by excessive sexual development. Two bitches have lived for one year following the removal of the pineal; both were in heat ten months after the operation, or when about one year old. In neither animal has pregnancy resulted, and in neither was any abnormality observed in the generative organs.

The pinealectomized young male puppies observed for periods of from three to eight months, contrasted with members of the same litter, have given no evidence of sexual precocity or retardation.

Gross and Microscopic Study of the Glands of Internal Secretion.

Examination has been made of the various ductless glands which were obtained at autopsy. In none was a definite macro- or microscopic change observed. The tissues examined include the thymus, parathyroids, thyroid, hypophysis, adrenals, pancreas, liver, spleen, lymph glands, testes, ovaries, and mammary glands.

SUMMARY AND CONCLUSIONS.

1. Following the removal of the pineal I have observed no sexual precocity or indolence, no adiposity or emaciation, no somatic or mental precocity or retardation.

2. Our experiments seem to have yielded nothing to sustain the view that the pineal gland has an active endocrine function of importance either in the very young or adult dogs.

3. The pineal is apparently not essential to life and seems to have no influence upon the animal's well being.

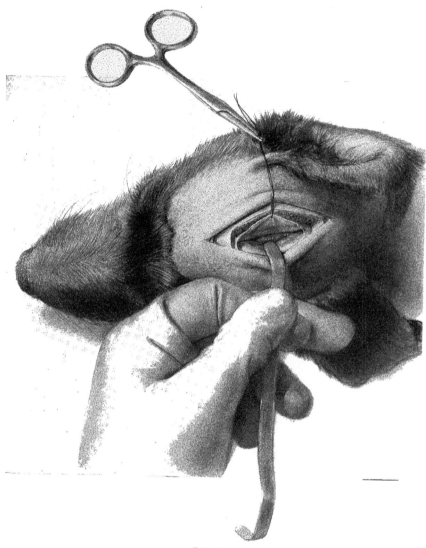

Fig. 1.

(Dandy: Pineal Body.)

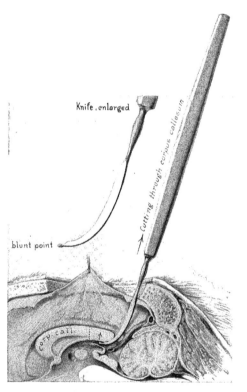

FIG. 2.

(Dandy: Pineal Body.)

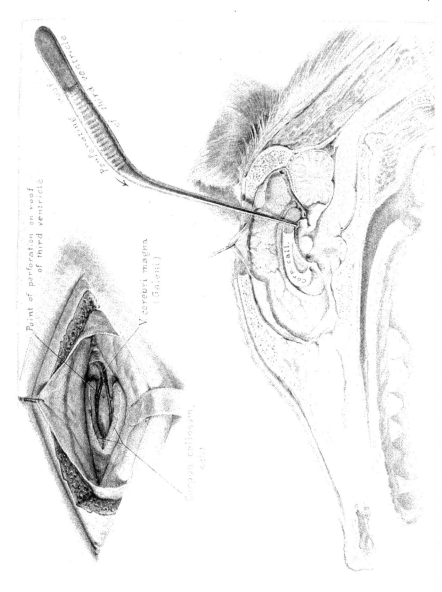

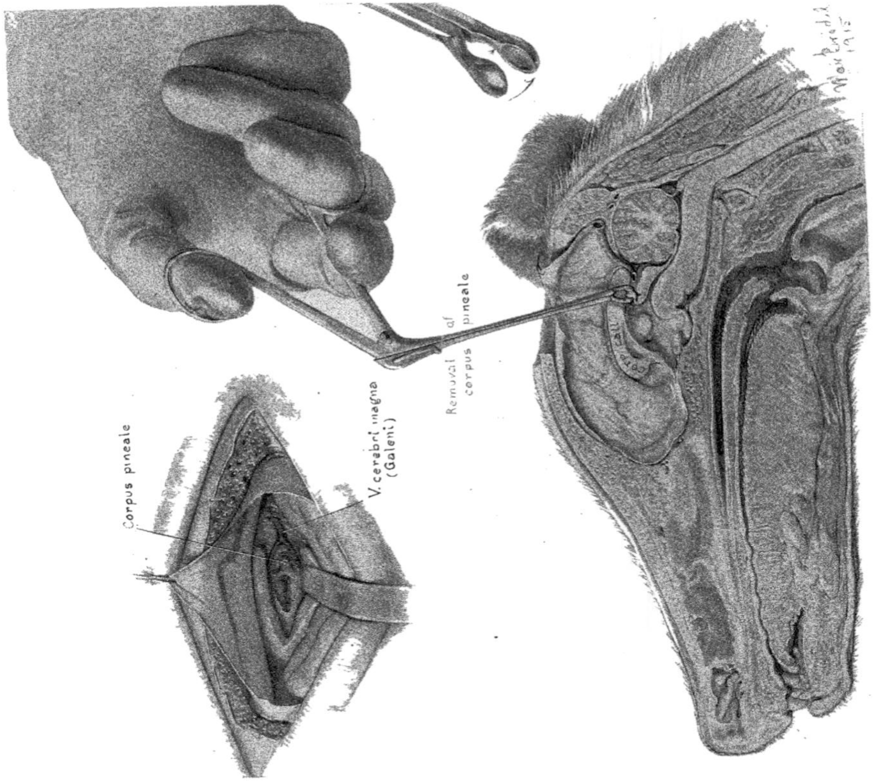

Removal of corpus pineale

V. cerebri magna
(Galeni)

Corpus pineale

Fig. 4.

FIG. 5.

FIG. 6.

Fig. 7.

Fig. 7 a.

(Dandy: Pineal Body.)

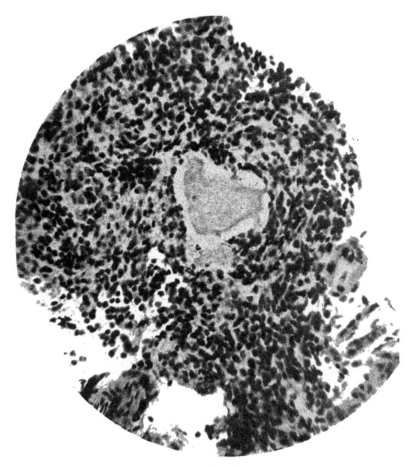

FIG. 7b.

(Dandy: Pir.c.

FIG. 8.

FIG. 9.

(Dandy: Pineal Body.)

It is a pleasure to express my gratitude to Professor Halsted for his interest and suggestions during the course of this investigation.

EXPLANATION OF PLATES.

PLATE 29.

FIG. 1. This demonstrates the location of the bony opening in the skull, the reflection of the dura over the midline, and retraction of the cerebrum, exposing the corpus callosum.

PLATE 30.

FIG. 2. This demonstrates the method of division of the splenium.

PLATE 31.

FIG. 3. The upper plate (enlarged) demonstrates the splenium divided, exposing the roof of the third ventricle with its venæ Galenæ parvæ. The lower plate shows the method of perforation of the roof between these veins.

PLATE 32.

FIG. 4. After opening and widening the defect in the roof of the third ventricle, the pineal is well exposed, as shown in the upper plate, and is easily grasped by the biting forceps, as shown in the lower plate. The jaws of the biting forceps are also shown separately.

PLATE 33.

FIG. 5. Puppies 14 days old. The photograph was taken 24 hours after removal of the pineal. The absence of any operative or glandular effects is well shown.

FIG. 6. Litter of 5 puppies, 3 of which have had the pineal removed and 2 of which are controls. The operated animals are recognized by the shaved heads. Photograph taken 2 weeks after operation.

FIG. 6 a. Pinealectomized animal (black), 4 months after operation, the operation being performed when the dog was 3 weeks old. The white dog is a control. The comparative weights of the animals are shown in Text-fig. 1.

PLATE 34.

FIG. 7. Pinealectomized female dog, 4 months after operation.

FIG. 7 a. The same animal, 14 months after operation. This is the only pinealectomized animal showing any adiposity. This has since disappeared.

PLATE 35.

FIG. 7 b. Photomicrograph of pineal removed from the dog shown in Figs. 7 and 7 a.

PLATE 36.

FIG. 8. Dog in which the pineal had been removed 6 months previously, when the animal was 18 days old.

FIG. 9. This demonstrates the staring expression which occasionally results from injury to the corpora quadrigemina during removal of the pineal.

THE CONCENTRATION OF THE PROTECTIVE BODIES IN ANTIPNEUMOCOCCUS SERUM. SPECIFIC PRECIPITATE EXTRACTS.

By HENRY T. CHICKERING, M.D.

(From the Hospital of The Rockefeller Institute for Medical Research.)

(Received for publication, May 28, 1915.)

It has been noted in a previous communication (1) that the immune substances in antipneumococcus serum may be removed by specific precipitation with extracts of the pneumococcus. It was further shown that the precipitates, when suspended in normal salt solution, and to a less extent when dissolved in a weak solution of sodic hydrate, protect susceptible animals against many times the lethal dose of pneumococcus. The present work is a further study of the action of the precipitates as well as of extracts of precipitates made by various methods.

An attempt was made to render soluble the specific precipitates by the use of weak alkaline salts. As in the preceding experiments (1), it appeared that the amount of sodic hydrate necessary to dissolve the precipitates was so great that in many cases the solution proved either toxic for animals or had suffered diminution in protective properties. Such weak alkalis as sodium phosphate, sodium biphosphate, and sodium carbonate were not strong enough to cause a solution of the precipitate. However, it was noted that when sodium carbonate was added to an emulsion of the whole precipitate in salt solution, a definite flocculation of the suspended particles occurred, and the flocculated particles quickly settled down, leaving an opalescent supernatant fluid which contained protective bodies, agglutinins, and precipitins. This observation suggested the possibility of bringing about a dissociation of the antigen and antibody of the specific precipitate.

It has been shown by Pfeiffer and Friedberger, and Bail and his pupils that the formation of the antigen-antibody complex that takes place in the precipitate

248

which forms when the appropriate bacterial precipitinogen is added to an immune serum, is relatively a loose one, and can be dissociated subsequently by extraction in salt solution. In 1903 Pfeiffer and Friedberger (2) found that when thoroughly washed sensitized cholera vibrios were injected into the peritoneal cavity of a guinea pig, the guinea pig was able to survive a second lethal dose of fresh live cholera vibrios, when injected one or two hours later. This work has been subsequently confirmed and elaborated by Landsteiner and Jagić (3), Hoke (4), Bail and Rotky (5), Bail (6), and Bail and Tsuda (7); and more recently Krauss (8) and Matsui (9) have demonstrated this dissociation of antigen and antibody, both *in vivo* and *in vitro.* They precipitated the immune substances in normal beef serum with live cholera vibrios and with *Bacillus typhosus,* and determined the optimum relation of culture to serum and the effects of varying degrees of temperature on the formation of the precipitate and on the potency of the extracts obtained from such precipitates. In many plate experiments the bactericidal or inhibiting action of these extracts on the growth of cholera vibrios has been demonstrated.

In the present work it will be shown that extracts of the precipitates formed in antipneumococcus serum by the addition of bacterial precipitinogen not only exert an inhibiting influence on the growth of virulent pneumococci *in vitro,* but the extracts contain agglutinins and precipitins and protect susceptible animals, such as the mouse and rabbit, as efficiently as does the original antipneumococcus serum. On the other hand, these extracts contain only a minimal amount of protein as compared with the whole serum. Moreover, considerable experimental evidence suggests that the extracts, and more especially the whole precipitates, produce active as well as passive immunity to pneumococcus infection in mice.

Methods.

As certain modifications have been made in the methods described in the first communication, the preparation of the bacterial precipitinogen and the method of precipitating the serum are reviewed in detail. For instance, acetone instead of alcohol has been used exclusively to kill the bacteria used in the preparation of the extract. Methods have also been employed for the more complete exhaustion of the immune serum.

Preparation of Pneumococcus Extracts (Precipitinogen).

An outline of the method is shown in Text-fig. 1. A twenty-four hour plain broth culture of pneumococcus of Type I or Type

II (10) is centrifuged and the bacterial residue washed twice in normal salt solution. The washed bacterial residue of a liter of broth culture is then emulsified in 5 cc. of normal salt solution and is added slowly to 10 volumes of acetone. There is an immediate flocculation of the bacteria. The mixture is then quickly centrifuged within five minutes, and the supernatant acetone decanted. The bacterial residue is then quickly dried *in vacuo.* The average

PREPARATION OF PNEUMOCOCCUS ANTIGEN

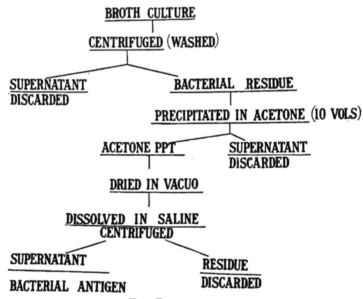

TEXT-FIG. I.

yield of dried bacteria from a liter of broth is about 100 to 150 mg. The acetone-killed bacteria may be stored in dry form until ready for use in precipitating the immune serum.

The dried bacteria are dissolved in normal salt solution, usually 1 to 2 mg. per cc. On shaking thoroughly, the solution of the bacteria is rapid and almost complete. After the bacteria have been dissolved as completely as possible, the solution is centrifuged at high speed for one-half hour. Usually a very small amount of sediment collects at the bottom of the tube. The opalescent supernatant fluid is used as the bacterial extract in precipitating the immune substances from antipneumococcus serum.

Since acetone does not kill the ferments (Van Slyke and Cullen[1] (11)) these bacterial extracts may undergo autolysis. Hence it is advisable to store the bacterial precipitinogen in dry form from which fresh extracts can be readily prepared when needed. However, saline extracts of the bacteria have been used after storage on ice for two months and have shown no appreciable decrease in their ability to precipitate immune sera.

Acetone kills the pneumococcus probably by very rapid dehydration. Cultures of these bacterial extracts are always sterile when the procedure is carried out with aseptic precautions. Smears of the bacterial extracts show a Gram-negative staining amorphous material. Kjeldahl determinations made under Mr. Cullen's direction show that the bacterial extract is practically pure protein, as it contains 16.2 per cent nitrogen.

Method of Specific Precipitation of Antipneumococcus Serum.

An outline of the methods employed, showing the method of obtaining the various fractions, is shown in Text-fig. 2. The presence or absence of immune bodies in the various fractions is also indicated. These facts are shown in detail later.

To an antipneumococcus serum, prepared by actively immunizing a horse to live cultures of pneumococcus of Type I or Type II, a bacterial extract of pneumococcus of the corresponding group is added until the immune serum is exhausted of its antibodies. Testing the absence of agglutinins in the serum from which the precipitate has been removed is a more accurate criterion of complete exhaustion of the serum than is the repeated addition of bacterial precipitinogen until no further precipitate occurs. By this method it is possible to exhaust completely the serum of its protective properties. To exhaust antipneumococcus serum it is necessary to add

---

[1] Buchner was the first to note that acetone did not kill the oxidases. Van Slyke, having prepared an active form of urease by precipitation with acetone, suggested its use in the preparation of pneumococcus extracts.

on the average the bacterial extract of 50 mg. dry weight of bacteria per 100 cc. of serum. It makes little difference, as regards the total amount of bacterial extract necessary to exhaust a given quantity of serum, whether the precipitinogen be added at once or repeatedly in small amounts. The precipitate formed by the addition of the bacterial extract as precipitinogen is apparent at once and is very volu-

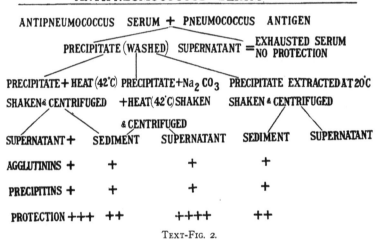

SPECIFIC PRECIPITATION OF ANTIBODIES FROM ANTIPNEUMOCOCCUS SERUM

ANTIPNEUMOCOCCUS SERUM + PNEUMOCOCCUS ANTIGEN

PRECIPITATE (WASHED) SUPERNATANT = EXHAUSTED SERUM = NO PROTECTION

PRECIPITATE + HEAT (42°C) SHAKEN & CENTRIFUGED   PRECIPITATE + Na₂CO₃ + HEAT (42°C) SHAKEN & CENTRIFUGED   PRECIPITATE EXTRACTED AT 20°C SHAKEN & CENTRIFUGED

	SUPERNATANT +	SEDIMENT	SUPERNATANT	SEDIMENT	SUPERNATANT
AGGLUTININS	+	+	+	+	
PRECIPITINS	+	+	+	+	
PROTECTION	+++	++	++++	++	

TEXT-FIG. 2.

minous. The precipitated serum is incubated at 38° C. for two hours to render the reaction complete and is then stored on ice over night, whereupon it is centrifuged. The supernatant serum fluid is pipetted off and is called exhausted serum. The precipitate is washed three times in normal salt solution, in order to free it completely of serum, and is then ready for extraction.

Methods of Extracting the Protective Substances from Specific Precipitates of Antipneumococcus Serum.

Extraction in Normal Salt Solution at 20° C.—The washed specific serum precipitate is emulsified in normal salt solution in one-half to one-fifth the volume of the original serum. It is allowed to stand at room temperature for twenty-four hours, being shaken occasionally. In the following experiments this extract is designated saline extract. This preparation is unsatisfactory, as the extract does not protect animals as highly as do extracts prepared by the methods described below.

Extraction in Normal Salt Solution at 42° C.—The washed serum precipitate is suspended in normal salt solution as described above. It is then heated for one hour at 42° C., being shaken gently at intervals, after which it is centrifuged at low speed. The supernatant fluid from this heated saline emulsion is called heat extract. This extract contains agglutinins and precipitins, and protects mice against many times the lethal dose of pneumococcus.

Extraction in Normal Salt Solution at 42° C. with Sodium Carbonate.—The washed serum precipitate is suspended in salt solution as described above, and about 1 cc. of a 1 per cent sodium carbonate solution is added to the precipitate from 100 cc. of serum, or an amount sufficient to flocculate the suspended particles of precipitate. The flocculated emulsion is then heated for one hour at 42° C. and is gently shaken at intervals, after which it is centrifuged. This supernatant fluid is designated as carbonate extract. With this extract the best results have been obtained, both as regards protection of animals against pneumococcus infection and content of agglutinins and precipitins. In the preparation of the extracts, it is important not to subject the precipitates to prolonged shaking, as this procedure diminishes or destroys their antibody content. Shaklee and Meltzer (12) have previously shown that prolonged shaking destroys the ferments trypsin and pepsin.

EXPERIMENTAL.

Protective Properties of Specific Precipitates and Extracts of Precipitates from Antipneumococcus Serum.

The whole precipitates, and more especially the extracts of the precipitates, protect susceptible animals, rabbits and mice, as well, or almost as well, as the whole antipneumococcus serum.

In Table I, which is a protocol of one of the many protection experiments which have been done upon mice with various specific

TABLE I.*

Protective Experiment with a Fixed Amount of Serum of Type II or Serum Derivatives Formed by Various Methods and Dilutions of a Twenty-Four Hour Bouillon Culture of Pneumococcus of Type II, in Total Volume of 0.5 Cc.   The Mixtures Were Injected Intraperitoneally into Mice.

Mouse No.	Protective fluid.	Culture.	Result.
		cc.	
1	Control........................	0.00001	D., 30 hrs.
2	"	0.000001	" 35 "
3	Original serum, 0.2 cc. in 0.5 cc......	0.2	" 3 days.
4	" " " " " " "	0.1	" 24 hrs.
5	" " " " " " "	0.01	S. 5 days.
6	" " " " " " "	0.001	D., 4½ "
7	Precipitate dissolved in N/10 sodium hydroxide, 0.2 cc. in 0.5 cc........	0.2	" 10 hrs.
8	Precipitate dissolved in N/10 sodium hydroxide, 0.2 cc. in 0.5 cc........	0.1	" " "
9	Precipitate dissolved in N/10 sodium hydroxide, 0.2 cc. in 0.5 cc........	0.01	" 40 "
10	Precipitate dissolved in N/10 sodium hydroxide, 0.2 cc. in 0.5 cc........	0.001	" " "
11	Whole washed precipitate, 0.2 cc. in 0.5 cc..........................	0.2	4 days.
12	Whole washed precipitate, 0.2 cc. in 0.5 cc..........................	0.1	" 56 hrs.
13	Whole washed precipitate, 0.2 cc. in 0.5 cc..........................	0.01	S. 5 days.
14	Whole washed precipitate, 0.2 cc. in 0.5 cc..........................	0.001	" " "
15	Whole washed precipitate, 4 times concentrated, 0.2 cc. in 0.5 cc......	0.2	D., 5
16	Whole washed precipitate, 4 times concentrated, 0.2 cc. in 0.5 cc.....	0.1	" 24 hrs.
17	Whole washed precipitate, 4 times concentrated, 0.2 cc. in 0.5 cc......	0.01	S. 5 days.
18	Whole washed precipitate, 4 times concentrated, 0.2 cc. in 0.5 cc.....	0.001	D., 3 "
19	Carbonate extract, 0.2 cc. in 0.5 cc....	0.2	" 4 '
20	" " " " " " " ...	0.1	S. 5 '
21	" " " " " " " ...	0.01	" " "
22	" " " " " " " ...	0.001	" " "
23	Heat extract, 0.2 cc. in 0.5 cc........	0.2	D., 56 hrs.
24	" " " " " " "	0.1	" 24 "
25	" " " " " " "	0.01	S. 5 days.
26	" " " " " " "	0.001	" " "

* In the tables D. stands for "died," S. for "survived."

precipitates and extracts of precipitates of both Type I and Type II immune serum, it will be noted that the whole precipitate dissolved in sodic hydrate failed to protect. This result with alkaline solutions of the precipitates has been noted several times. As the injection of the dissolved precipitate itself does not harm the mice, it is presumable that the sodic hydrate has a deleterious effect on the immune substances in the precipitate. Of the mice receiving whole

TABLE II.

Protective Experiment with a Fixed Dose of Culture Plus Decreasing Doses of Serum or Serum Derivatives (Same Lot as Experiment I).

Mouse No.	Protective fluid, total volume 0.5 cc.	Culture, pneumococcus Type II.	Result.
		cc.	
1	Control.....................	0.00001	D., 44 hrs.
2	"	0.000001	" " "
3	Original serum 0.2 cc.........	0.01	S. 5 days.
4	" " 0.15 "	0.01	" " "
5	" " 0.1 "	0.01	" " "
6	" " 0.05 "	0.01	" " "
7	" " 0.01 "	0.01	" " "
8	Whole washed precipitate 0.2 cc.	0.01	" " "
9	" " " 0.15 "	0.01	" " "
10	" " " 0.1 "	0.01	" " "
11	" " " 0.05 "	0.01	" " "
12	" " " 0.01 "	0.01	D., 4 "
13	Carbonate extract 0.2 cc......	0.01	S. 5 "
14	" " 0.15 "	0.01	" " "
15	" " 0.1 "	0.01	" " "
16	" " 0.05 "	0.01	" " "
17	" " 0.01 "	0.01	D., 60 hrs.

serum or whole precipitate, those receiving 0.01 cc. of culture survived, while of those receiving the carbonate extract, those receiving as much as 0.1 cc. of culture survived. In other words, the carbonate extract possessed a higher protective power than the original serum, or the whole precipitate. The group of animals receiving heated extract was not as well protected as the group receiving carbonate extract. This is not a sporadic result of the protective action of carbonate extracts, for many experiments with different lots of serum and various samples of extracts of specific precipitates have yielded approximately the same results. Increasing the dose of whole precipitate by concentration does not increase its potency.

This same result has been noted when using whole serum or concentrated globulin fractions of serum as prepared by the method described by Avery (13). Apparently there is a maximum dose of culture against which it is possible to protect a mouse, and increasing the size of the dose of the protective agent does not increase the protection.

On the other hand, if one infects a series of mice with equal doses of culture and attempts to protect them with decreasing doses of serum or serum derivatives, the results have shown that the protective agent is in excess of the needed amount when 0.2 cc. is used as the standard protective dose. Table II illustrates this point. When mice were infected with doses of 0.01 cc. of culture, the mouse that received 0.01 cc. of original serum was protected as well as the one that received 0.2 cc. of serum. In titrating in this manner the protective value of the serum derivatives made up to original volume of serum, it will be seen that there is some loss in potency of the serum derivatives as compared with the original serum. But this can be explained as due to the inevitable slight loss in material during the manipulations incident to the preparation of the specific precipitates and their extracts.

In these experiments the carbonate extract has consistently shown greater protective qualities than the extracts prepared by simple extraction at room temperature or by heating at 42° C.

Further experiments were carried on to determine whether removal of the precipitate from the serum completely exhausted its protective power, and also whether the precipitate still retained protective power after extraction with sodium carbonate.

From the protocol (Table III) it is evident that the exhausted serum and the salt solution used in washing the precipitate free from serum afforded no protection to mice. Apparently the process of washing the precipitate free from the exhausted serum did not diminish appreciably its protective substances. The first carbonate extract protected well; subsequent extractions of the precipitate showed but little potency. While a single extraction removed most of the protective substances that are dissociable from the whole precipitate, some protective power still existed in the residue precipitate which could not be removed by repeated extractions.

TABLE III.

Experiment to Show the Comparative Protective Value of Serum, Exhausted Serum, Wash Water, and Repeated Extraction. The Mixtures Were Made Up in All Cases to a Constant Volume, 0.5 Cc., and Injected Intraperitoneally into Mice.

Mouse No.	Protective fluid.	Culture II 40.	Result.
		cc.	
1	Control...............................	0.00001	D., 42 hrs.
2	"	0.000001	" " "
		Culture I 107	
3	....	0.000001	" 30 "
		Culture II 40	
4	Original serum, 0.2 cc. in 0.5 cc..............	0.1	" 18 "
5	" " " " " " "	0.01	" " "
6	" " " " " " "	0.001	S. 5 days.
7	" " " " " " "	0.0001	" " "
8	Exhausted " " " " " "	0.1	D., 18 hrs.
9	" " " " " " "	0.01	" 20 "
10	" " " " " " "	0.001	" 18 "
11	" " " " " " "	0.0001	" 42 "
12	2d wash water of whole precipitate twice concentrated, 0.2 cc. in 0.5 cc....................	0.1	" 18 "
13	2d wash water of whole precipitate twice concentrated, 0.2 cc. in 0.5 cc....................	0.01	" " "
14	2d wash water of whole precipitate twice concentrated, 0.2 cc. in 0.5 cc....................	0.001	" 29 "
15	2d wash water of whole precipitate twice concentrated, 0.2 cc. in 0.5 cc....................	0.0001	" 42 "
16	Carbonate extract 1st extraction, 0.2 cc. in 0.5 cc..............	0.1	" 18 "
17	" " " " " " "	0.01	S. 5 days.
18	" " " " " " "	0.001	" " "
19	" " " " " " "	0.0001	" " "
20	Carbonate extract 2d extraction, 0.2 cc. in 0.5 cc..............	0.1	D., 18 hrs.
21	" " " " " " "	0.01	" 4 days.
22	" " " " " " "	0.001	" 3 "
23	" " " " " " "	0.0001	S. 5 "
24	Residue of whole precipitate after 2 carbonate extractions, 0.2 cc. in 0.5 cc.......	0.1	D., 18 hrs.
25	Residue of whole precipitate after 2 carbonate extractions, 0.2 cc. in 0.5 cc.......	0.01	" " "
26	Residue of whole precipitate after 2 carbonate extractions, 0.2 cc. in 0.5 cc.......	0.001	S. 5 days.
27	Residue of whole precipitate after 2 carbonate extractions, 0.2 cc. in 0.5 cc.......	0.0001	" " "
		Culture I 107	
28	1st carbonate extract, 0.2 cc. in 0.5 cc..........	0.1	D., 18 hrs.
29	" " " " " " " "	0.01	" " "
30	" " " " " " " "	0.001	" 42 "
31	" " " " " " " "	0.0001	" 72 "

The carbonate extract and heat extracts contained agglutinins and precipitins, but the titer was lower than that of the original serum. A fresh live culture of pneumococcus was agglutinated by the extracts when made up to the original volume of the serum, but not in dilutions of 1 to 5 or 1 to 10. The whole serum itself, on the other hand, agglutinated in dilutions of 1 to 20 or 1 to 30.

The amount of protein in the carbonate extracts is, of course, much less than in the original serum. The whole serum used in this experiment contained 4.98 per cent protein. After removal of the extract the carbonate extract and the residue precipitate each contained 0.08 per cent protein, or about one-sixtieth of the amount in the original serum. It will be further noted that the protection conferred by the use of such extracts is specific, there being no protection against a pneumococcus infection of Group I by an extract of a precipitate from an antipneumococcus serum of Group II, and *vice versa.*

Experiments like the following have been done to ascertain whether the protective substances in antipneumococcus serum can be removed by live washed cultures of pneumococcus. To 25 cc. of antipneumococcus serum, Type I, were added live washed pneumococci of Type I from 150 cc. of a twenty-four hour broth culture. An immediate precipitation occürred. After twenty-four hours the mixture was centrifuged, the precipitate washed in normal saline and emulsified in 12 cc. of normal salt solution. To this emulsion 0.5 cc. of 1 per cent sodium carbonate was added, and flocculation occurred. The mixture was then heated at 42° C. for one hour, being shaken gently at intervals, and then it was heated to 56° C. for one-half hour. After centrifuging, the supernatant fluid was pipetted off and diluted to the original volume of the serum, and is called carbonate extract (Table IV). The sediment was reemulsified and made up to the original volume of the serum with salt solution and is called sediment of precipitate.

This and similar experiments show that antipneumococcus serum can be exhausted of its antibody content by live cultures. To accomplish this, a relatively much larger quantity of live bacteria must be used to exhaust the serum than when a bacterial extract is employed. The washed precipitate of live agglutinated bacteria, after

TABLE IV.

Protective fluid.	Culture I 106. cc.	Result.
Original serum I, 0.2 cc. in 0.5 cc.	0.1	S. 6 days.
" " " " " " " "	0.01	" " "
" " " " " " " "	0.001	" " "
" " " " " " " "	0.0001	
Exhausted " " " " " " "	0.001	D., 18 hrs.
" " " " " " " "	0.0001	" 30 "
" " " " " " " "	0.00001	" 48 "
Carbonate extract, " " " " "	0.1	" 18 "
" " " " " " "	0.01	S. 6 days.
" " " " " " "	0.001	D., 4 "
" " " " " " "	0.0001	S. 6 "
Sediment of precipitate, 0.2 cc. in 0.5 cc.	0.1	D., 18 hrs.
" " " " " " " "	0.01	" " "
" " " " " " " "	0.001	S. 6 days.
" " " " " " " "	0.0001	" " "
Control	0.00001	D., 30 "
"	0.000001	" " '

being killed by heating to 56° C. for one-half hour, as well as carbonate extracts therefrom, protects susceptible animals against many times the lethal dose of pneumococcus. They are not as potent as precipitates and extracts of precipitates formed from antipneumococcus serum by acetone-killed bacterial extracts. It will be seen from Table IV that mice are protected fairly well by both the carbonate extract of the precipitate and the residue of the precipitate after removal of the carbonate extract. In no case did the precipitates and extracts of precipitates of antipneumococcus serum, treated with live cultures, protect as well as the original serum. The carbonate extracts from these precipitates contain agglutinins and precipitins as well as protective substances.

The Development of Active and Passive Immunity to Pneumococcus Infection by the Use of Antipneumococcus Serum Derivatives.

The injection of specific serum precipitates and extracts of such precipitates from antipneumococcus serum into mice produces a certain amount of active, as well as passive immunity to pneumococcus infection. This is to be expected, as the specific precipitate itself contains bacterial antigen, and it is presumable that in the extracts of these precipitates a certain amount of antigen as well as antibody goes into solution.

It is well known that the passive immunity conferred on mice by the injection of antipneumococcus serum is of short duration and

TABLE V.

Protective fluid.	Mouse No.	Culture, pneumococcus Type I.*	Result.	Mouse No.	Culture, pneumococcus Type I.†	Result.
		cc.			*cc.*	
Whole serum, 0.2 cc. in 0.5 cc.	1	0.1	D., 20 hrs.	4	0.1	D., 18 hrs.
" " " " " " "	2	0.01	S.	5	0.01	" 24 "
" " " " " " "	3	0.001	"	6	0.001	" 48 "
Whole precipitate emulsion, 0.2 cc. in 0.5 cc............	7	0.1	D., 20 "	11	0.1	" 18 "
Whole precipitate emulsion, 0.2 cc. in 0.5 cc............	8	0.01	" 24 "	12	0.01	S.
Whole precipitate emulsion, 0.2 cc. in 0.5 cc............	9	0.001	" 44 "	13	0.001	"
Whole precipitate emulsion, 0.2 cc. in 0.5 cc............	10	0.0001	" " "			
Carbonate extract, 0.2 cc. in 0.5 cc...................	14	0.1	" 20 "	18	0.1	D., 18 "
Carbonate extract, 0.2 cc. in 0.5 cc...................	15	0.01	" " "	19	0.01	" 24 "
Carbonate extract, 0.2 cc. in 0.5 cc...................	16	0.001	S.	20	0.001	S.
Carbonate extract, 0.2 cc. in 0.5 cc...................	17	0.0001	"	21	0.0001	"
Exhausted serum, 0.2 cc. in 0.5 cc...................	22	0.1	D., 20 "	25	0.1	D., 18 "
Exhausted serum, 0.2 cc. in 0.5 cc...................	23	0.01	" 26 "	26	0.01	" " "
Exhausted serum, 0.2 cc. in 0.5 cc...................	24	0.001	" 44 "	27	0.001	" 2 days.
Exhausted serum, 0.2 cc. in 0.5 cc...................				28	0.0001	" " "
Control..................	29	0.00001	" 26 "	31	0.00001	" 26 hrs.
"	30	0.000001	" 44 "	32	0.000001	" 48 "

* These cultures were given 5 days later.
† These cultures were given 10 days later.

usually disappears after a period of seven days. On the other hand, the active immunity response of an animal to the injection of a bacterial antigen does not usually appear until the period of passive immunity has passed. To determine the presence or absence of passive and active immunity in mice in response to the injection of antipneumococcus serum and serum derivatives, a series of mice were injected intraperitoneally with 0.2 cc. each of (1) antipneumococcus serum, (2) whole precipitate derived from antipneumococcus serum, (3) a carbonate extract of the whole precipitate, and (4) exhausted serum. After intervals of five and ten days, varying dilutions of live virulent cultures were injected intraperitoneally.

Table V shows that considerable passive immunity was conferred on mice by the intraperitoneal injection of antipneumococcus serum, which immunity persisted for at least five days but had disappeared after ten days. On the other hand, the immunity conferred by the carbonate extract persisted up to ten days at least, when the period of passive immunity had presumably passed. This would seem to indicate that some degree of active immunity had been produced by the carbonate extract. The mice treated with exhausted serum showed, as was to be expected, no evidence of either passive or active immunity.

Of the mice treated with whole specific precipitate, all died when infected with pneumococci five days later. It is possible that the early disappearance of passive immunity in this case might be due to a state of lowered resistance caused by the injection of the large amount of bacterial antigen contained in the whole precipitate. After ten days other mice in this same group showed a considerable resistance which must be interpreted as due to active immunity. Experiments have been carried on to learn whether the early loss of passive immunity after the injection of whole precipitate, as mentioned above, might be prevented by preliminary heating of the bacterial extract used for precipitation. Heating the bacterial extract for one hour at 56° C. does not injure its precipitative power. The precipitates formed with such heated extracts, however, showed no differences from those formed from unheated extracts. The passive immunity produced by the injection of such precipitates lasts no longer than that after the injection of precipitates formed with unheated extracts.

TABLE VI.

Protective Experiment with a Fixed Amount of Polyvalent Serum or Serum Derivatives and Dilutions of a Twenty-Four Hour Bouillon Culture of Pneumococci of Type I and Type II in Total Volume of 0.5 Cc. The Mixtures Were Injected Intraperitoneally in Mice.

Polyvalent serum of Types I. and II + bacterial extract of Type I = Precipitate I.

Polyvalent serum, after removal of Precipitate I, + bacterial extract of Type II = Precipitate II.

Polyvalent serum + bacterial extract of Type II = Precipitate III.

Polyvalent serum, after removal of Precipitate III, + bacterial extract of Type I = Precipitate IV.

Bacterial extract was added in all cases until the serum was exhausted of the corresponding agglutinins.

All precipitates were suspended in saline with sodium carbonate and heated at 42° C. for one hour, shaking gently. Extracts of each were used for protection tests, diluted to the original volume of serum.

Protective fluid.	Culture I 108.	Result.	Culture II 41.	Result.
	cc.		*cc.*	
Original polyvalent serum, 0.2 cc. in 0.5 cc................	0.1	S. 6 days	0.1	S. 6 days.
	0.01	" " "	0.01	D., 18 hrs.
	0.001	" " "	0.001	S. 6 days.
	0.0001	" " "	0.0001	" " "
Extract of Precipitate I, 0.2 cc. in 0.5 cc.................	0.1	D., 18 hrs.	0.1	D., 18 hrs.
	0.01	" 3 days	0.01	" 30 "
	0.001	S. 6 "	0.001	" 18 "
	0.0001	" " "	0.0001	" 30 "
Extract of Precipitate II, 0.2 cc. in 0.5 cc.....:...........	0.1	D., 18 hrs.	0.1	" 18 "
	0.01	" " "	0.01	" 30 "
	0.001	" 30 "	0.001	" 30 "
	0.0001	" 42 "	0.0001	S. 6 days.
Extract of Precipitate III, 0.2 cc. in 0.5 cc.................	0.1	" 18 "	0.1	D., 18 hrs.
	0.01	" " "	0.01	" 56 "
	0.001	" " "	0.001	S. 6 days.
	0.0001	S. 6 days	0.0001	" " "
Extract of Precipitate IV, 0.2 cc. in 0.5 cc.................	0.1	D., 18 hrs.	0.1	D., 18 hrs.
	0.01	" 24 "	0.01	" " "
	0.001	S. 6 days	0.001	" " "
	0.0001	D., 3 "	0.0001	" 24 "
Controls...................	0.00001	" 24 hrs.	0.00001	" 18 "
	0.000001	" " "	0.000001	" " "

TABLE.VII.

The Specificity of Agglutinins Contained in Extracts of Specific Precipitates from Polyvalent Antipneumococcus Serum. The Extracts Are Obtained by Extracting the Specific Precipitates (Experiment VI) in Saline with Weak Sodium Carbonate + Heat.

Polyvalent serum + bacterial extract of Type I = Precipitate I.

Polyvalent serum, after removal of Precipitate I, + bacterial extract of Type II = Precipitate II.

Polyvalent serum + bacterial extract of Type II = Precipitate III.

Polyvalent serum, after removal of Precipitate III, + bacterial extract of Type I = Precipitate IV.

Original polyvalent serum dilution.	+ Culture, pneumococcus Type I.		Original polyvalent serum dilution.	+ Culture, pneumococcus Type II.	
	2 hrs.	24 hrs.		2 hrs.	24 hrs.
1 : 1	+ +	+ +	1 : 1	+ +	+ +
1 : 5	+ +	+ +	1 : 5	+ +	+ +
1 : 10	+	+	1 : 10	+	+
1 : 20	—	—	1 : 20	—	—
Extract of Precipitate I + Culture I			Extract of Precipitate I + Culture II		
5 : 1	+ +	+ +	1 : 1	—	—
1 : 1	—	—	1 : 5	—	—
1 : 2.5	—	—	1 : 10	—	—
1 : 5	—	—	1 : 20	—	—
Extract of Precipitate II + Culture I			Extract of Precipitate II + Culture II		
5 : 1	—	—	5 : 1	+ +	+ +
1 : 1	—	—	1 : 1	—	—
1 : 2.5	—	—	1 : 2.5	—	—
1 : 5	—	—	1 : 5	—	—
Extract of Precipitate III + Culture I			Extract of Precipitate III + Culture II		
5 : 1	—	—	5 : 1	+ +	+ +
1 : 1	—	—	1 : 1	—	—
1 : 2.5	—	—	1 : 2.5	—	—
1 : 5	—	—	1 : 5	—	—
Extract of Precipitate IV + Culture I			Extract of Precipitate IV + Culture II		
5 : 1	+ +	+ +	5 : 1	—	—
1 : 1	—	—	1 : 1	—	—
1 : 2.5	—	—	1 : 2.5	—	—
1 : 5	—	—	1 : 5		

All agglutination tests are done by the macroscopic method. 0.3 cc. of twenty-four hour broth culture + 0.3 cc. of the serum or extract dilution being used. Readings are made after 2 hours in the water bath at 38° C. and after 24 hours on ice. If agglutination occurs it is usually apparent in 5 minutes.

TABLE VIII.

The Specific Adsorption of Agglutinins from a Polyvalent Pneumococcus Immune Serum by Use of Live Washed Cultures of the Pneumococcus and the Dissociation of These Agglutinins from the Specific Precipitates. Live Washed Cultures of Pneumococci of Type I and Type II Were Used in Obtaining the Precipitates and Extracts from a Polyvalent Antipneumococcus Serum.

Polyvalent serum + live culture of pneumococcus Type I = Precipitate I.

Polyvalent serum, after removal of Precipitate I, + live culture of pneumococcus Type II = Precipitate II.

Polyvalent serum + live culture of pneumococcus Type II = Precipitate III.

Polyvalent serum, after removal of Precipitate III, + live culture of pneumococcus Type I = Precipitate IV.

All precipitates were heated for one-half hour at 56° C., and extracted with saline + carbonate.

The extracts were used for agglutination tests.

Dilution of serum.	Culture I.		Dilution of serum.	Culture II.	
	2 hrs.	24 hrs.		2 hrs.	24 hrs.

Polyvalent serum after removal of Precipitate I by live culture of pneumococcus Type I.

Dilution of serum.	Culture I.		Dilution of serum.	Culture II.	
I : I	−	−	I : I	+	+
I : 5	−	−	I : 5	−	−
I : 10	−	. −	I : 10	−	−
I : 20	−	−	I : 20	−	−

Polyvalent serum after removal of Precipitate III by live culture of pneumococcus Type II.

Dilution of serum.	Culture I.		Dilution of serum.	Culture II.	
I : I	+	+	I : I	−	−
I : 5	−	+	I : 5	−	−
I : 10	−	−	I : 10	−	−
I : 20	−	−	I : 20	−	−

Extract I.　　　　　　　　Extract II.

	2 hrs.	24 hrs.		2 hrs.	24 hrs.
Culture I..........	+	+	Culture I..........	−	−
"　II..........	−	−	"　II..........	−	+

Extract III.　　　　　　　Extract IV.

	2 hrs.	24 hrs.		2 hrs.	24 hrs.
Culture I..........	−	. −	Culture I..........	+	+
"　II..........	+	+	"　II..........	−	−

The Specific Adsorption of Immune Substances from Polyvalent Serum by Bacterial Extracts and Live Bacteria.

A polyvalent antipneumococcus serum may be specifically exhausted of its immune bodies for one of the types of pneumococcus by the addition of a bacterial extract of the corresponding type. The immune substances of the other type remain intact and can be removed subsequently by the addition of the appropriate antigen (Tables VI and VII). During the process of fractional precipitation the titer of the immune bodies is somewhat diminished. This diminution may be accounted for partly by the loss of precipitate incident to the manipulation in washing, and partly by the supposition that the antigen-antibody combination is only partially dissociated. Table VIII shows that the specific agglutinins may be removed in like manner from the polyvalent serum by the use of living pneumococci in place of bacterial extracts.

Nature of the Union between Precipitin and Precipitinogen.

In the above detailed experiments it has been shown that the protective substances in antipneumococcus serum can be removed specifically by precipitation with an extract of the homologous type of pneumococcus, and that this precipitate, when suspended in salt solution and injected into susceptible animals, will protect them against many times the lethal dose of pneumococcus.

It has been further shown that the union between the bacterial extract (precipitinogen) and the immune substances in the serum can be to some extent disunited by suitable chemical and physical agents.

The following observations which we have made suggest strongly that in the formation of the precipitate an actual union occurs between the precipitin of the serum and precipitinogen of the bacterial extract. If, before adding the precipitinogen to the serum, the latter be heated at 60° C. for an hour, no precipitation takes place. Nor can the precipitin be reactivated by the addition of unheated normal serum. This fact, that precipitins are inactivated by heat, and then cannot be reactivated by fresh serum, has already been demonstrated by Pick (14) and Kraus and von Pirquet (15). That a union, however, has occurred between the precipitin and precipitinogen, even in the absence of precipitate, is made evident by the fact that if to a mixture of bacterial extract and heated immune serum fresh unheated immune serum be added, no precipitate occurs. The most likely explanation is that the precipitable substance is already saturated with the inactivated precipitins.

By suitable extraction methods as shown above, an apparent dissociation of antigen and antibody occurs. It seems likely, however, that, although a dissociation has occurred, and the agglutinins and protective substances have been set free, a permanent change has occurred in the precipitinogen, for when fresh immune serum is added to the extracts of the specific precipitates, even if the latter be concentrated, no fresh precipitate forms.

DISCUSSION.

These studies suggest that the use of precipitate extracts, prepared as described, may offer certain advantages over the use of whole serum in the treatment of lobar pneumonia, since the extracts possess practically the entire immunizing and protective power of immune serum, and yet contain a very small fraction of the serum proteins. The patient may thus be relieved of the strain incident to the metabolism of the large amounts of protein contained in the large quantities of serum which it is now necessary to employ. It is altogether probable, moreover, that the symptoms of serum sickness, which so frequently follow the use of large amounts of serum, may be lessened, if not prevented entirely, by the use of precipitate extracts. The precipitates and extracts of precipitates are still able to produce the phenomena of anaphylaxis and serum sickness, however, as is shown by the fact that in guinea pigs sensitized to horse serum, acute anaphylactic death may be induced by the injection of precipitates or precipitate extracts. To produce this phenomenon, however, fairly large amounts of these substances are required.

The extracts of precipitates, moreover, may have an additional advantage over serum alone in treatment, in that, in addition to conferring passive immunity, they also induce the production of some degree of active immunity. While the active immunity becomes evident at so late a period that in acute lobar pneumonia it could not be of much therapeutic importance, yet in other, more chronic infections, a similar active reaction induced by suitable extracts might prove of value.

CONCLUSIONS.

1. The protective substances contained in specific precipitates from antipneumococcus serum can be extracted by suitable chemical and physical agents, dilute sodium carbonate at 42° C. being especially advantageous as an extractive agent.

2. The resulting water-clear extracts, when made up to the original volume of the serum used for precipitation, protect animals almost as well as does the whole serum.

3. The bacterial extracts used in precipitating the protective substances from the serum act specifically; that is, a bacterial extract of pneumococcus of Type I removes the protective substances from a Type I immune serum only.

4. In a polyvalent serum of Type I and Type II, the protective substances of each type may be removed independently of each other by the successive addition of the homologous antigens.

5. Extracts of specific serum precipitates contain only one-fiftieth to one-sixtieth of the protein in the original serum, and about one-half the protein of the whole precipitate.

6. Extracts contain not only protective substances but agglutinins and precipitins.

7. Extracts and whole precipitates not only confer passive immunity but stimulate the production of active immunity to pneumococcus infection in rabbits and mice.

BIBLIOGRAPHY.

1. Gay, F. P., and Chickering, H. T., *Jour. Exper. Med.*, 1915, xxi, 389.
2. Pfeiffer, R., and Friedberger, E., *Centralbl. f. Bakteriol., Ite Abt., Orig.*, 1903, xxxiv, 70.
3. Landsteiner, K., and Jagic, N., *München. med. Wchnschr.*, 1903, 1, 764.
4. Hoke, E., *Wien. klin. Wchnschr.*, 1907, xx, 347.
5. Bail, O., and Rotky, K., *Ztschr. f. Immunitätsforsch., Orig.*, 1913, xvii, 378.

6. Bail, O., *Ztschr. f. Immunitätsforsch., Orig.*, 1914, xxi, 202.

7. Bail, O., and Tsuda, K., *Ztschr. f. Immunitätsforsch., Orig.*, 1908–09, i, 546, 772.

8. Krauss, F., *Biochem. Ztschr.*, 1913, lvi, 457.

9. Matsui, I., *Ztschr. f. Immunitätsforsch., Orig.*, 1914, xxiii, 233.

10. Dochez, A. R., and Gillespie, L. J., *Jour. Am. Med. Assn.*, 1913, lxi, 727.

11. Van Slyke, D. D., and Cullen, G. E., *Jour. Biol. Chem.*, 1914, xix, 211.

12. Shaklee, A. O., and Meltzer, S. J., *Am. Jour. Physiol.*, 1909–10, xxv, 81.

13. Avery, O. T., *Jour. Exper. Med.*, 1915, xxi, 133.

14. Pick, E. P., *Beitr. z. chem. Phys. u. Path.*, 1902, i, 393.

15. Kraus, R., and von Pirquet, C. F., *Centralbl. f. Bakteriol., 1te Abt., Orig.*, 1902, xxxii, 60.

THE ACTION OF ETHYLHYDROCUPREIN (OPTOCHIN) ON TYPE STRAINS OF PNEUMOCOCCI IN VITRO AND IN VIVO, AND ON SOME OTHER MICRO-ORGANISMS IN VITRO.

By HENRY F. MOORE, M.B., B.Ch., B.A.O.

(*From the Laboratories of The Rockefeller Institute for Medical Research.*)

(Received for publication, May 25, 1915.)

In 1911 Morgenroth and Levy (1) introduced the drug ethylhydrocuprein (optochin), a derivative of hydroquinine, in the treatment of experimental pneumococcic infection. Subsequent tests have, as a rule, shown the compound to possess a parasiticidal effect on the pneumococci. In the experiments so far reported, different strains of pneumococci, chosen more or less at hazard, were studied; no attempt, however, has been made to investigate the action of the drug on type strains of the various groups of pneumococci. It seemed of interest to investigate the action of the drug from this viewpoint.

Neufeld (2) showed that strains of pneumococci differed among themselves in respect to their immunity reactions (protective bodies). The investigations of Cole (3) and of Dochez and Gillespie (4) have resulted in a serological classification of the pneumococci. It has been shown by these authors that the pneumococci can be divided into at least four groups. The organisms of Groups I and II are specific in their immunity reactions; that is, an immune serum produced against any member of Group I has a specific protective action against, and a specific agglutinative action on, any member of Group I, but has no effect on any members of Groups II, III, or IV. Similarly, a serum immune to any member of Group II will react in the same way with all members of that group, but not with any members of Groups I, III, or IV. In Group III are included all members of the *Pneumococcus mucosus* type. To Group IV belong all pneumococci not belonging to Groups I, II, or III. A serum immune to any member of Group IV exhibits protective and agglutinative reactions with the strain used for its production, but in no instance with any other member of Group IV or with any member of the other three groups. Lister (5) has arrived at a somewhat similar classification from opsonic studies.

EXPERIMENTAL.

The pneumococci used in the present study were representative of these four serological groups and were obtained from the Hospital of The Rockefeller Institute, with the exception of the microorganism designated as South Africa (9), which was obtained from the South African Institute for Medical Research; this organism has been placed in a fifth group by Lister (5) as a result of his opsonic studies. All strains of pneumococci used in the present study were isolated from cases of human pneumonia, except those called I Les and II White, which were obtained from cases of pneumococcal meningitis by lumbar puncture.

The Action of Ethylhydrocuprein in Vitro.

Wright (6) has shown that the action of ethylhydrocuprein hydrochloride on pneumococci *in vitro* is only slightly lessened by the presence of serum, while with antiseptics such as lysol and cresol, under similar conditions, the contrary is the case. Tugendreich and Russo (7) have shown that the hydrochloride of the drug in a concentration of 1 in 16,000 kills the pneumococcus *in vitro* in three hours at room temperature, and that it is much more potent in this respect than the hydrochlorides of the homologous compounds: isopropyl, isobutyl, and isoamylhydrocuprein; this superiority is even more marked over quinine and hydroquinine. In the present study the action of ethylhydrocuprein hydrochloride on type strains of the different groups of pneumococci was worked out and compared with that on other microorganisms under similar conditions.

Technique.—A solution of the drug was made in broth and sterilized by boiling or filtration through a Berkefeld filter. Various dilutions of this stock solution were made; to 2 cc. of each dilution in a test-tube 0.1 cc. of a 24 hour broth culture was added and the tubes were incubated for 18 hours at 37° C. At the end of this period the tubes were examined, and the results, as regards growth or inhibition of growth, noted; a large loopful of the two

TABLE I.

The Action in Vitro of Ethylhydrocuprein Hydrochloride on Some Representatives of the Four Serological Groups of Pneumococci.

Serological group of pneumococcus.	I.			II.			III.			IV.		
Designation of strain of pneumococcus....	106[5]	(B42)1[4]	Lcs	(B21)1[10]	38[3]	White	(A66)14[2]	(E22)4[3]	(B48)1[4]	(M)4[3]	(A67)11[3]	South Africa (9)
Highest dilution causing inhibition of growth......	500,000	1,000,000	500,000	1,000,000	1,000,000	1,000,000	500,000	10,000,000	1,000,000	500,000	1,000,000	1,000,000
Highest dilution causing death........	20,000	1,000,000	100,000	1,000,000	50,000	1,000,000	100,000	10,000,000	100,000	50,000	1,000,000	1,000,000

lowest dilutions showing growth macroscopically, and of all tubes showing no growth, was in each case plated in about 12 cc. of blood agar and incubated. The concentration of the drug was thus enormously diluted. The plates were examined at the end of 24 hours and again at the end of 48 hours. Thus, one could differentiate between an inhibitory effect of the drug on the growth of the microorganism and actual death. Some typical results are given in Table I.

Explanation of Tables.—In the top row the Roman numeral stands for the serological group to which the strain of pneumococcus belongs; a letter and number in brackets, if present, denote the particular strain; the Arabic number represents the number of animal passages of the strain and the exponent the number of cultivations on artificial medium since the last animal passage. In the remaining two rows each number represents the greatest dilution in which the growth of the corresponding microorganism was inhibited or killed, as the case may be. In the case of the pneumococci the range of dilutions of the drug examined was as follows: 1 in 10,000; 1 in 20,000; 1 in 50,000; 1 in 100,000; 1 in 500,000; 1 in 1,000,000; and 1 in 10,000,000.

From the results given in Table I it is seen that ethylhydrocuprein hydrochloride causes in very high dilutions, *in vitro,* an inhibition of growth, and death of the pneumococcus, the latter occurring, generally speaking, in somewhat lower dilutions; and that no constant or considerable differences are seen in these effects on typical representatives of the four groups of pneumococci.

It was thought of interest to compare the action on pneumococci with that on other organisms, more especially streptococci. The results are set forth in Table II.

Description of the Microorganisms Mentioned in Table II.

Streptococcus 1 was a non-hemolytic streptococcus cultivated by Dr. Beattie from a case of rheumatic fever. The culture used was obtained from the Bacteriological Department of the Museum of Natural History, New York.

Streptococci 5 and 7 were two different strains of non-hemolytic streptococci obtained from the Pathological Department of Mt. Sinai Hospital, New York. They were cultivated from the blood of patients suffering from sabacute endocarditis.

TABLE II.

The Action in Vitro of Ethylhydrocuprein Hydrochloride on Microorganisms Other than Pneumococci.

Microorganism.	Micrococcus catarrhalis.	Streptococcus mucosus.	Streptococcus 1.	Streptococcus 5.	Streptococcus 7.	Streptococcus 8.	Streptococcus 59.	Streptococcus R.	Streptococcus C.	B. coli communis.	B. typhosus.	B. paratyphosus B.	Staphylococcus albus.	B. Friedländer.
Highest dilution causing inhibition of growth	10,000	20,000	Not inhibited in 10,000	20,000	10,000	10,000	10,000	20,000	10,000	Not inhibited in 10,000	Not inhibited in 10,000	Not inhibited in 10,000	Not inhibited in 10,000	10,000
Highest dilution causing death	Not killed in 10,000	Not killed in 10,000	Not killed in 10,000	20,000	Not killed in 10,000	Not killed in 10,000	Not killed in 10,000	Not killed in 10,000	Not killed in 10,000	—	—	—	—	Not killed in 10,000

TABLE III.*

The Action in Vitro of Quinine Hydrochloride.

Microorganism.	I105⁴.	II385.	IIIA66.12⁴.	IVM.4³.	Strepto-coccus mucosus.	Micro-coccus catarrhalis.	Strepto-coccus 1.	B. coli communis.	Staphylo-coccus albus.
Highest dilution causing inhibition of growth...	50.000	50.000	50.000	50.000	5.000	5.000	5.000	1.000	1.000
Highest dilution causing death...	5.000	10.000	5.000	10.000	1.000	5.000	2.000	1.000	1.000

* The first four organisms are representatives of the four groups of pneumococci.

Streptococcus 8 was a laboratory strain of hemolytic streptococcus.

Streptococcus 59 was a green-producing, non-hemolytic streptococcus cultivated by Dr. H. F. Swift from the blood of a patient suffering from rheumatic
pericarditis.

None of these strains had a capsule, fermented inulin, or was bile-soluble.

The *Streptococcus mucosus* was a diplococcus growing in short chains; it
showed, with His's stain, a well defined capsule in the body fluids of an infected
animal; it was not bile-soluble; it was pathogenic for mice, causing a sticky
exudate in the peritoneal cavity; it did not ferment inulin; on cultivation for a
few passages on blood agar the growth, which was previously moist on solid
media, became dry in character.

The streptococcus strains called R and C were isolated from normal sputum
and were mucous in type; that is, they gave a mucoid growth on blood agar and
possessed a capsule; these qualities were, however, lost after a few passages on
artificial media. Neither was bile-soluble, and neither fermented inulin.

From Table II it is seen that the action, if any, of ethylhydrocuprein, in the dilutions examined, on the microorganism mentioned
therein was considerably less than that on the pneumococci.

The action *in vitro* of the hydrochloride of quinine (from which
ethylhydrocuprein is derived) on pneumococci and on some other
microorganisms is shown in Table III. It is seen therefrom that,
while the action of quinine hydrochloride is greater on the pneumococci than on the other bacteria, this action is far less in the
former case than that of optochin.

*The Action of Ethylhydrocuprein (Optochin Base) on Represen-
tatives of the Four Groups of Pneumococci in Vivo.*

Owing to the fact that certain biochemical relationships (such as
bile solubility) exist between trypanosomes (8) and spirilla (9),
on the one hand, and the pneumococci (10) alone among the *Co-
caceæ,* on the other, Morgenroth (1) investigated the action on
the pneumococci of that group of compounds which has given results in the therapy of trypanosomal and spirillar infections;
namely, quinine and its derivatives.

Morgenroth and Levy showed that ethylhydrocuprein (a derivative of hydroquinine) exerted a considerable protective action, and a certain degree of curative
action (1) on experimental pneumococcal infection in the mouse. Later Gutmann (11) and Morgenroth and Kaufmann (12) reported experiments in which
this protective action on several strains of pneumococci was shown. Levy (13)
studied the curative effect of the drug on pneumococcal infection of the mouse
by a strain of *Pneumococcus mucosus,* an interval of from two to six hours
having been allowed to elapse between the intraperitoneal infection and the first

administration of the drug. The dosage of the infection does not seem to have been accurately measured in Levy's experiments, nor do the results seem to have been checked by autopsy; an effect of the drug, however, on the infection can be seen; of 16 mice, 10 survived up to the 8th day, when the observation was discontinued. This work is summarized by Rosenthal (14).

So far, the strains studied in this connection were chosen more or less at haphazard, no attempt having been made to study strains of pneumococci in relation to classification. In the following experiments typical representatives of the four groups of pneumococci were used.

Technique.—A 2 per cent solution of the free base ethylhydrocuprein (optochin base) was made in sterile olive oil by rubbing up in a mortar and standing the product so obtained in the incubator at 37° C. over night. Mice of 18 grams' weight and upwards were used. The infection was given in a constant volume of 1 cc. of broth intraperitoneally. The treatment consisted of 0.5 cc. of the 2 per cent optochin base solution in oil (calculated in every case) per 20 grams of mouse given under the skin of the back immediately after the infection; this was followed by an equal dose on the second day of the experiment, and by 0.4 cc. per 20 grams of mouse on the third and fourth days. The mice were observed for a period of from 10 to 14 days. An autopsy was performed on every mouse that succumbed (whether a treated animal or a control), and the heart's blood, and peritoneal fluid if necessary, were examined in a smear preparation, Gram's stain and His's capsule stain being used. If no Gram-positive and capsule-bearing bacteria were found in the smear by these methods, cultures were made on defibrinated rabbit blood agar with abundant inoculation, incubated for 24 hours at 37° C., and then examined.

In tabulating the results the graphic method recommended by Morgenroth is employed. In the text-figures each square denotes one experimental animal; a black square indicates death from pneumococcal septicemia; a white square, survival; an oblique cross denotes that the animal died, but that the examination of the heart's blood was negative, in which instance death was apparently caused by the toxicity of the drug, the infecting pneumococci having been first killed off by its action;[1] and, finally, a square subdivided into

[1] Morgenroth pointed out that the toxic dose for mice is not far removed from the curative dose.

six smaller ones denotes that the animal was sick at the time of observation (Text-fig. 1).

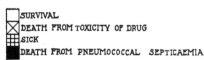

SURVIVAL
DEATH FROM TOXICITY OF DRUG
SICK
DEATH FROM PNEUMOCOCCAL SEPTICAEMIA

TEXT-FIG. 1. Explanation of the method used in the text-figures.

The condition of the animals is first given in the diagram on the second day of the experiment; that is, 24 hours after the infection and first treatment; consequently, the first diagram in each experiment is marked 2d day.

The minimum lethal dose of the infecting microorganism was in each case determined for forty-eight hours unless otherwise stated; in all cases that dose was taken as the m. l. d. which was with certainty fatal within the time limit stated for several mice. In case of doubt reestimations were made. With this statement it is unnecessary to give the estimations in detail.

Throughout the observations the mice were kept at about $75°$ F.[2]

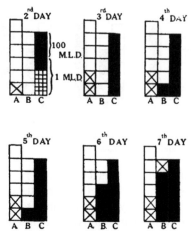

TEXT-FIG. 2. Experiment 1. Strain I 106.[4] The m.l.d. was 0.000,000,001 cc. (48 hrs.). The infecting doses used: 10 m.l.d. (Column A); 1,000 m.l.d. (Column B). Column C, controls (untreated).

[2] The toxicity of the drug is said to be greater when the animals are kept exposed to cold.

Comment on Experiment 1.—(Text-fig. 2.)   Of the 6 treated animals infected with 10 times the m. l. d., 2 died of toxicity, the heart's blood at autopsy being sterile; while the remaining 4 animals remained permanently well; of those infected with 1,000 times the m. l. d. and treated, 1 died of toxicity (showing sterile heart's blood at autopsy), and the remaining 4 of pneumococcal septicemia; the last mentioned 4 animals, however, survived much longer than the untreated controls.

This microorganism was of extraordinarily high virulence for mice, it having had 106 animal passages.   The m. l. d. above stated was regularly fatal and showed 6 to 14 colonies when plated.

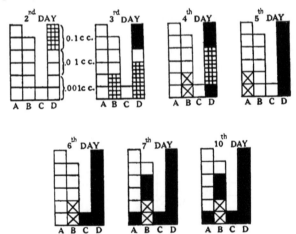

TEXT-FIG. 3. Experiment 2.   Strain I (B42)^{16}.   The m.l.d. was 0.001 cc. The infecting doses used: 10 m.l.d. (Column A); 100 m.l.d. (Column B); 500 m.l.d. (Column C).   Column D, controls (untreated).

Comment on Experiment 2.—(Text-fig. 3.)   This strain was relatively much less virulent than the previous one.   A dose of 0.001 cc. or 0.01 cc. was always fatal within 4 to 5 days, but some-times the animals infected with the larger of these doses survived those infected with the smaller by a day.   In view of the protracted course of the infection in this case the time limit for the m. l. d. was judged as 5 days.   Of those animals injected with 10 times the m. l. d. only 1 died of pneumococcal septicemia, all the others surviving. Of those injected with 100 times the m. l. d. 1 survived; 2 died of toxicity; and 2 of pneumococcal septicemia, but later than the controls.

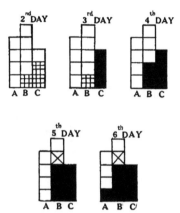

TEXT-FIG. 4. Experiment 3. Strain II 34[19]. The m.l.d. was 0.000.000,001 cc. (48 hrs.). The infecting doses: 10 m.l.d. (Column A); 100 m.l.d. (Column B). Column C, controls (untreated).

Comment on Experiment 3.—(Text-fig. 4.) Of the 4 animals infected with 10 times the m. l. d. and treated, 1 died on the 6th day of pneumococcal septicemia and 3 survived. Of the 5 infected with 100 times the m. l. d. and treated, 3 died of pneumococcal septicemia and 1 survived; the remaining animal died of toxicity. On plating the m. l. d., 6 to 12 colonies resulted. This strain was one of very high virulence, notwithstanding the fact that it had undergone nineteen artificial cultivations since the last animal passage.

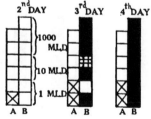

TEXT-FIG. 5. Experiment 4. Strain II (B21) 3[2]. The m.l.d. was 0.0,000,001 cc. (fatal in 4 days). Dose used: 1,000 m.l.d. (Column A). Column B, controls (untreated).

Comment on Experiment 4.—(Text-fig. 5.)   If the m. l. d. is judged to kill in 4 days, 4 mice out of 6 infected with 1,000 times the m. l. d. survived the controls permanently.   Of the 2 given in Text-fig. 5 as dying of toxicity the heart's blood of 1 was sterile, and a Gram-negative bacillus was recovered from that of the other, but no pneumococci.

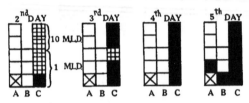

TEXT-FIG. 6.   Experiment 4 a.   Strain II (B21)2^{14}.   The m.l.d. was 0.01 cc. (48 hrs.).   Doses used: 1 m.l.d. (Column A); 50 m.l.d. (Column B).   Column C, controls (untreated).

Comment on Experiment 4 a.—(Text-fig. 6.)   In this experiment 1 animal represented in Column A (1 m. l. d.) and 1 in Column B (50 times the m. l. d.) died on the 8th and 10th days, respectively, the heart's blood showing on examination a Gram-negative bacillus but no pneumococci.   As we were troubled with mouse typhoid at this time, this was probably the cause of death. 5 animals survived, 2 died of pneumococcal septicemia, and 1 of toxicity.

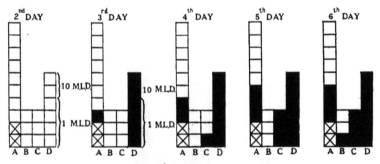

TEXT-FIG. 7.   Experiment 5.   Strain III (A66)12^6.   The m.l.d. was 0.000,000,001 cc. (48 hrs.).   Doses used: 10 m.l.d. (Column A); 100 m.l.d. (Column B); 1,000 m.l.d. (Column C).   Column D, controls (untreated).

Comment on Experiment 5.—(Text-fig. 7.) The Strain III (A 66) 12^6 was a very virulent one. The m. l. d., 0.000,000,001 cc. of a 24 hour broth culture, when plated on blood agar, showed on incubation at least 15 typical colonies. The animal represented by the upper segment of Column C was found almost completely eaten by its fellows on the 6th day; no autopsy could be done, consequently the cause of death is somewhat doubtful. After the 6th day no further deaths occurred in this series. 5 out of 10 animals injected with 10 times the m. l. d. survived; 3 died of pneumococcal infection, and 2 of toxicity (giving a sterile heart's blood). Of 3 injected with 100 times the m. l. d. 2 survived, and the remaining animal succumbed to the infecting pneumococci. Of those injected with 1,000 times the m. l. d. none survived.

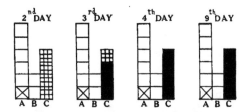

TEXT-FIG. 8. Experiment 6. Strain III (E22) 12^6. The m.l.d. (0 000.001 cc.) was fatal in 48 hrs. 0.00,000,001 cc. of a broth culture was observed to be regularly fatal within four days; the m.l.d. was not, however, at the time of this experiment determined for a longer period than 48 hrs.; consequently the dose of 0.000,001 cc. was probably considerably greater than the m.l.d. for a mouse. Column A, 1 m.l.d. (0.000,001 cc. (48 hrs.)). Column B, 10 m.l.d. (0.00,001 cc. (48 hrs.)). Column C, controls, 1 m.l.d. (untreated).

Comment on Experiment 6.—(Text-fig. 8.) Among the treated animals no death occurred up to the 10th day when observation was discontinued. 1 animal of the series died of toxicity, and none of pneumococcal septicemia.[3]

[3] Further experience with this strain has shown it to be more susceptible to the action of ethylhydrocuprein *in vivo* than any other strain of pneumococcus examined by us. We have been able with the drug to protect against 10.000 times the m.l.d.

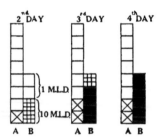

TEXT-FIG. 9. Experiment 7. Strain IV M.4³. The m.l.d. was 0.00,000,001 cc. of a 24 hr. broth culture. Dose used: 10 m.l.d. (Column A). Column B, controls (untreated).

Comment on Experiment 7.—(Text-fig. 9.) No further deaths occurred after the 4th day. Of 8 animals injected with 10 times the m. l. d. and treated, 2 died of toxicity and 6 survived. This was an unusually virulent strain of Group IV. The virulence, however, rapidly fell off on artificial cultivation.

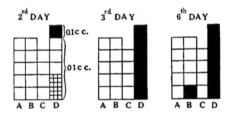

TEXT-FIG. 10. Experiment 8. Strain South Africa (9).5². The m.l.d. was ·0.01 cc. Doses used: 10 m.l.d. (Column A); 20 m.l.d. (Column B); 50 m.l.d. (Column C). Column D, controls (untreated).

Comment on Experiment 8.—(Text-fig. 10.) Of the mice infected with 10 times the m. l. d. and treated all survived; of those infected with 20 times the m. l. d. and treated only 1 died of pneumococcal septicemia, all the others surviving; all those infected with 50 times the m. l. d. and treated survived. The controls infected with 1 m. l. d. all died within 48 hours.

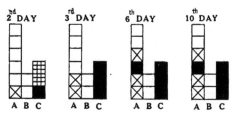

TEXT-FIG. II. Experiment 9. Strain IV (S 10)3[19]. The m.l.d. was 0.01 cc. (48 hrs.). Dose used: 10 m.l.d. (Column A); 50 m.l.d. (Column B). Column C, controls, 1 m.l.d. (untreated).

Comment on Experiment 9.—(Text-fig. II.) Of the 5 animals infected with 1 m. l. d. and treated 2 survived, and 3 died of toxicity. From the heart's blood of one of these latter a Gram-negative bacillus was recovered. One died of pneumococcal septicemia. Both the animals infected with 50 times the m. l. d. and treated recovered.

DISCUSSION.

The results of the present study agree, in the main, with those of the other workers mentioned. We have not had in our experiments such a large percentage of cures as is claimed by Morgenroth, namely 90 to 100 per cent; nor have we seen any definite protective action of optochin *in vivo* with an amount of infection greater than 1,000 times the m. l. d. of a highly virulent strain.[4] Moreover, the greater the virulence of the strain (by passage through mice) the greater was the difficulty in protecting against increasing multiples of the m. l. d. Constant results are difficult to obtain in the mouse, owing to the repeated injections apparently causing the animals a considerable degree of traumatism and consequently rendering them liable to intercurrent troubles. This was true of an epidemic of mouse typhoid which tended to obscure the results. The relative toxicity of the drug for mice is another factor which tends to render results not quite clear. Olive oil, the vehicle in which the drug is given, does not undergo absorption for a week or more, and the disturbance to the tissues caused by the repeated injections of this inert body seems to afford

[4] See page 281, footnote 3.

a suitable nidus for the growth of bacteria which may kill the animal before the observation is concluded. It seems likely that the detrimental factors, toxicity of the drug, the traumatism inflicted by the injections, and the effect of the infection itself, summate in their effects and thus render the probability of the animal surviving less likely, although its body may have been completely sterilized from pneumococcal infection. The effect of ethylhydrocuprein *in vitro* and *in vivo* on the pneumococcus is considerable and specific and is seen on type strains of all four groups of pneumococci.

From the text-figures it is seen that, of 85 mice infected with 100 times the m. l. d. or less and treated, 15 mice, or 17.6 per cent, died of pneumococcal septicemia; 13 mice, or 15.2 per cent, died of toxicity of the drug or some obscure cause, the heart's blood being sterile at autopsy (the corresponding controls invariably died of pneumococcal septicemia) ; 56 mice, or 66.8 per cent, survived. Of these 85 treated mice, 69, or 81 per cent, either recovered or died of causes other than pneumococcal septicemia; such, for example, as the toxicity of the drug.

CONCLUSIONS.

1. Ethylhydrocuprein hydrochloride in very high dilution inhibits the growth of, and in 18 hours kills, representatives of all four groups of pneumococci *in vitro*. The killing effect is generally seen in somewhat lower dilutions than the inhibiting effect. No constant or considerable difference is seen in these actions on representatives of the four groups of the pneumococci. The action of ethylhydrocuprein hydrochloride on the pneumococci *in vitro* is so strongly specific that it may possibly be used as a test for a true pneumococcus.

2. The inhibitory or killing effects of ethylhydrocuprein hydrochloride *in vitro* on bacteria other than pneumococci are slight or absent. The effects are greater on streptococci than on any other organisms examined, but are still much less than on the pneumococci. This action distinguishes between the streptococcus group, including *Streptococcus mucosus* sometimes found in normal

mouths, on the one hand, and the true pneumococcus (including *Pneumococcus mucosus*), on the other.

3. Quinine hydrochloride inhibits the growth of, and kills the pneumococcus *in vitro;* much stronger concentrations, however, are necessary than in the case of ethylhydrocuprein. This effect of quinine hydrochloride is also seen on other organisms, but in a less degree.

4. Ethylhydrocuprein (optochin base) has a well marked protective action against experimental pneumococcal infection in mice in the case of type strains of all four groups of pneumococci; this protective action may be efficient against many multiples of the minimum lethal dose.

BIBLIOGRAPHY.

1. Morgenroth, J., and Levy, R., *Berl. klin. Wchnschr.*, 1911, xlviii, 1560, 1979.
2. Neufeld, F., and Haendel, L., *Ztschr. f. Immunitätsforsch., Orig.*, 1909. iii, 159; *Berl. klin. Wchnschr.*, 1912, xlix, 680; *Arb. a. d. k. Gsndhtsamte*, 1910, xxxiv, 169; in Kolle, W., and von Wassermann, A., Handbuch der pathogenen Microorganismen, 2d edition, Jena, 1913, iv, 513.
3. Cole, R., *Arch. Int. Med.*, 1914, xiv, 56.
4. Dochez, A. R., and Gillespie, L. J., *Jour. Am. Med. Assn.*, 1913. lxi, 727.
5. Lister, F. S., Specific Serological Reactions with Pneumococci from Different Sources, *The South African Institute for Medical Research* [*Publications*], Dec. 22, 1913.
6. Wright, A. E., *Lancet*, 1912, ii, 1633, 1701.
7. Tugendreich, J., and Russo, C., *Ztschr. f. Immunitätsforsch., Orig.*, 1913. xix, 156.
8. Schilling, *Centralbl. f. Bakteriol., 1te Abt., Orig.*, 1902, xxxi, 452.
9. Neufeld, F., and von Prowazek, *Arb. a. d. k. Gsndhtsamte*, 1907. xxv. 494.
10. Neufeld, F., *Ztschr. f. Hyg. u. Infectionskrankh.*, 1900. xxxiv. 454.
11. Gutmann, L., *Ztschr. f. Immunitätsforsch., Orig.*, 1912, xv, 625.
12. Morgenroth, J., and Kaufmann, M., *Centralbl. f. Bakteriol., Ref.*. 1912. liv. Supplement, 69.
13. Levy, R., *Berl. klin. Wchnschr.*, 1912, xlix, 2486.
14. Rosenthal, F., *Ztschr. f. Chemotherap., Ref.*, 1912. i. 1149.

THE EFFECT OF IRRITATION ON THE PERMEABILITY OF THE MENINGES FOR SALVARSAN.

By EDGAR STILLMAN, M.D., AND HOMER F. SWIFT, M.D.

(*From the Hospital of The Rockefeller Institute for Medical Research, New York.*)

(Received for publication, June 9, 1915.)

One of the explanations offered for the beneficial effects of intraspinal injections in the treatment of syphilis of the central nervous system is that there may thus be induced an increased permeability of the meninges following their irritation by the injected substance. The purpose of this investigation was to determine whether the intraspinal, or rather subdural, injection of various substances used in intraspinal treatment would increase the amount of arsenic in the cord and brain of animals which received intravenous injections of salvarsan at the same time.

Sicard and Reilly (1) described a method for increasing the permeability of the brain substance for drugs by trephining the skull and injecting 5 cc. of a 0.5 per cent solution of sodium chloride under the dura with a fine hypodermic needle. They showed that in the cadaver a similar amount of India ink injected subdurally was distributed over an area 8 or 10 cm. in diameter, and suggested from this experimental evidence that two subdural injections of saline, *i. e.*, one in each temporofrontal region, would be sufficient to increase the permeability of a large part of the cerebrum for salvarsan introduced intravenously. They also state that they have injected 0.1 mg. of cyanide of mercury subdurally in the frontal region without harmful effect, and on the following day given salvarsan or neosalvarsan intravenously to the same patient. The cyanide of mercury was used simply to increase the permeability of the tissues. Viton (2) states that after a slight chemical irritation the meninges are rendered more permeable, and he used for this purpose a preparation suggested by Sicard (3) which consists of cyanide of mercury 0.1 mg., and novocain 0.015 gm. in 2 cc. of a 0.5 per cent saline solution. The intraspinal injections were made at intervals of 1 month, mercury or salvarsan being given in the usual way. He states that with this method of treatment the irritative symptoms of tabes are much relieved. Tinel and Leroide (4) injected several drops of a 1 per cent solution of sodium nucleinate into the fourth ventricles of rabbits and immediately afterwards gave the animals 10 cg. of neosalvarsan intravenously. One hour later they again punctured the fourth ventricles and obtained 4 cc. of fibrinous fluid,

286

which contained 0.008 mg. of arsenic. Fluid from the fourth ventricles of control animals contained no arsenic. The fibrinous fluid was evidence of a much more intense irritation than is usually produced by subdural injections in patients, so the increase in permeability which these authors demonstrated cannot be applied to bedside treatment.

The intraspinal injection of normal salt solution, or serum, does produce, however, a temporary irritation of the meninges, as is shown both clinically by pains, and by examination of the cerebrospinal fluid a few hours after such an injection, when several hundred cells per cmm. may be found. It is reasonable to suppose that this irritation might increase the permeability of the tissues, and that salvarsan circulating in the blood would be deposited in larger amounts in the tissues contiguous to the irritated meninges, than when the meninges are in their normal impermeable condition.

Methods.

To determine whether a demonstrable increased deposition of arsenic actually does occur, the following experiments were performed. Cats were selected as the experimental animals, for the following reasons: Intravenous injections can be easily given into the marginal vein of the ear, and there is a fairly large subdural cistern surrounding the cauda equina, so that subdural injections in the cat are similar in nature to those in man in that the injecting needle is not brought into direct contact with the cord. In rabbits the cord extends so low that it is impossible to introduce a needle under the dura and inject a solution without injuring the cord. In the experiments under discussion it was necessary that the nervous tissue be injured in no other way than that possibly resulting from the presence of the substance injected. The intraspinal injections in the cats were made in the following manner: The animals were etherized, and the dura was exposed by laminectomy of the two lower lumbar vertebræ. A fine curved hollow needle attached to a syringe was introduced through the dura, and cerebrospinal fluid was aspirated into the syringe, which was then detached, and another syringe containing the solution to be injected was attached to the needle, and the solution slowly injected. The injections were always preceded by aspiration of cerebrospinal fluid, to be sure that the needle was properly placed. All the animals were injected intravenously with salvarsan in alkaline solution in the proportion of 0.05 of a gram per kilo of body weight. Some of the animals received the intravenous injections one hour before the intraspinal treatment, and in others the order was reversed, so that at the time of intravenous injection the meninges would be already in a state of irritation. On the following day the animals were exsanguinated by opening the jugular veins and carotid arteries, in order to remove as much arsenic-containing blood from the tissues as possible. The cord and brain were immediately removed from the body and the brain was divided into three portions: cerebrum, cerebellum, and midbrain, pons, and medulla; the cord was separated from the medulla, and all the specimens were dried separately. The dried tissue was powdered, thoroughly mixed, and weighed, and duplicates of each specimen were analyzed quantitatively for arsenic.

The analyses were made by a special method devised for the purpose by Vinograd (5). This method depends for its accuracy upon the principle of oxidizing the tissues with small amounts of arsenic-free nitric acid, at 260° C. in a sealed glass bomb. The bomb is then opened, sulphuric acid added, the nitric acid driven off by heating, and the sulphuric acid-arsenic mixture quantitated for arsenic by Sanger and Black's modification of Gutzeit's method. All reagents and utensils were carefully tested for arsenic before using. The standard scales for comparison were made with both arsenious acid and salvarsan. In the results here presented the figures are given in fractions of mg. of salvarsan per gram of dried tissue.

RESULTS.

The operative procedures, substances injected, time relations, and results of the analyses are given in Table I. Four control animals were first treated. Two of them had only intravenous injections of salvarsan. One of these killed one and one half hours later showed no more arsenic in its cerebrospinal axis than one killed after eighteen hours. The comparatively slight neurotropic action of salvarsan demonstrated by Ullmann (6) and by Stühmer (7) probably explains this similarity in arsenic content at such different intervals after treatment. In the other two controls the operative procedure, laminectomy and withdrawal of cerebrospinal fluid, was shown to have no effect. The effect of irritation with the following substances was studied: isotonic salt solution; 50 per cent cat serum diluted with isotonic salt solution; pure cat serum; salvarsanized cat serum obtained by treating cats with 0.05 of a gram of salvarsan per kilo of body weight, bleeding one hour later, separating the serum, and heating to 56° C. This serum contained between 0.015 and 0.025 mg. of salvarsan per cc. of serum. It was injected in 50 per cent dilution and also undiluted. A mixture of cyanide of mercury, novocain, and 0.5 per cent saline, similar to that used by Sicard and by Viton, was also injected.

As a rule, the cords of the ten animals which received intraspinal injections contained no more arsenic than the four controls. The two exceptions are Animals H 104 and H 106, which received intraspinal injections of normal cat serum in a 50 per cent dilution. There are two possible explanations for this variation. First, the 50 per cent serum may have been more irritating than the other substances injected; or, second, the animals were not so completely exsanguinated. We are inclined to attribute the increased amount of arsenic in the cords of these animals to incomplete exsanguination, for in both animals the cerebrum and cerebellum contained considerably more arsenic than the average. Furthermore, if the 50 per cent serum were the important factor, one would expect that the cords of Cats H 110 and H 112, which received intraspinal injections of 50 per cent dilution of salvarsanized serum, would show a similar increase in arsenic, but in these animals both the cords and brains showed the average arsenic content. Upon first thought, one would expect that the cerebrospinal axis of the animals that received subdural injections of salvarsanized serum would contain more arsenic than the animals injected with normal serum or mercury. The rapid diffusion of the small amount of arsenic in the serum through the cerebrospinal fluid, and the rapid excretion of substances from the cerebrospinal fluid into the blood stream, probably explain the fact that the cords and brains of animals which received salvarsanized serum intraspinally contained no more arsenic than the average. Both of these factors have been conclusively demonstrated by Dandy and Blackfan (8). Hall (9) has also shown that the cerebrospinal fluid of patients who received

TABLE I.

Animal No.	Weight.	Operative procedure.	Substance injected intraspinally.	Time relation of intraspinal injection (I.S.) to intravenous injection (I.V.).	Killed after intravenous injection.	Salvarsan per gm. of dried tissue.				Remarks.
						Cerebrum.	Cerebellum.	Midbrain, pons, medulla.	Spinal cord.	
	gm.				hrs.	mg.	mg.	mg.	mg.	
H 115	3,000	None	None		1½	0.025	0.025	0.025	0.017	
H 113	2,560	"	"		18	0.017 (N)*	Traces	0.025	0.025 (N)*	
H 105	3,000	Laminectomy	"		19	0.037	0.037	0.025	0.025	
H 107	1,080	Laminectomy and 1 c. cerebrospinal fluid withdrawn	"	1 hr. before I.V.	18	0.025	0.025	0.017	0.017	
H 116	3,100	Laminectomy and dural inj	1 c 0.9% sodium chloride solution	I.S. 1 hr. after I.V.	17	0.017	Traces	0.025	Traces	
H 104	2,800	"	1 c. 50% normal cat serum	" before "	13	0.037	0.037	0.025	0.075	
H 106	1,90	"	"	" after "	17	0.075	0.075	0.025	0.037 (N)*	
H 109	1,550	"	1 c. 100% normal cat serum	" before "	16	0.025	0.017	0.025	0.025	
H 108	2,050	"	"	" after "	16	0.025 / 0.017	Traces	0.025 / 0.017	0.025	Weakness in both hind legs day after operation.
H 112	2,650	"	1 cc. 50% salvarsanized cat serum	" before "	15	0.017	0.025	0.017	0.025	
H 110	2,250	"	"	" after "	19	0.017	0.017	Traces	0.025	Weakness in right hind leg day after operation.
H 111	1,670	"	1 cc. 100% salvarsanized cat serum	" "	19	0.025	Traces	0.017	0.025	
H 117	2,500	"	Mercury cyanide solution	" before "	19	0.025	0.017 (N)*	0.025	0.025	Weakness in both hind legs day after operation.
H 118	2,850	"	"	" after "	16	0.017	Traces	Traces	0.025	

* (N), duplicate lost. All results are in duplicate unless otherwise noted.

intraspinal injections of neosalvarsan often contains no arsenic twenty-four hours after the injection.

If irritation alone were the important factor in determining the deposition of arsenic in the nervous tissue, the cords, which were more exposed to the irritating effect of the substances injected than the brains, should have shown a higher arsenic concentration. The average arsenic content of the cord and of the cerebrum was practically the same; *i. e.*, 0.029 mg. for the former, and 0.028 mg. for the latter. The cords of the three animals, H 108, H 112, and H 113, which showed clinical evidences of cord or cauda equina injury did not contain more arsenic than the cords of animals which showed no evidence of local lesions. Nor did the more irritating mercury solutions increase the arsenic in the cords or brains of the animals which received it. There was no difference between the animals that received intraspinal injections an hour before the intravenous injections of salvarsan, and those that first received the salvarsan intravenously.

CONCLUSIONS.

The subdural injection of normal salt solution, normal serum, serum salvarsanized *in vivo* or weak solutions of cyanide of mercury does not demonstrably increase the permeability of the spinal cord or brain for salvarsan which is circulating in the blood at the time of the subdural injection.

BIBLIOGRAPHY.

1. Sicard, J. A., and Reilly, *Bull. et mém. Soc. méd. d. hôp. de Paris*, 1913, xxxvi. series 3, 861.
2. Viton, J. J., *Semana med.*, 1913, xx, 1393; abstracted in *Jour. Am. Med. Assn*, 1914, lxii, 582.
3. Sicard, *Bull. et mém. Soc. méd. d. hôp. de Paris*, 1913, xxxvi, series 3, 681.
4. Tinel, J., and Leroide, J., *Compt. rend. Soc. de biol.*, 1913, lxxiv, 1073.
5. Vinograd, M., *Jour. Am. Chem. Soc.*, 1914, xxxvi, 1548.
6. Ullmann, K., *Wien. klin. Wchnschr.*, 1913, xxvi, 216.
7. Stühmer, A., *Arch. f. Dermat. u. Syph.*, 1914, cxx, 589.
8. Dandy, W. E., and Blackfan, K. D. *Am. Jour. Dis. Child.*, 1914, viii, 406.
9. Hall, G. W., *Jour. Am. Med. Assn.*, 1915, lxiv, 1384.

CHANGES IN THE ELECTROCARDIOGRAMS ACCOMPANYING EXPERIMENTAL CHANGES IN RABBITS' HEARTS.

By FRANCIS R. FRASER, M.B.

(*From the Hospital of The Rockefeller Institute for Medical Research.*)

PLATES 37 TO 40.

(Received for publication, June 14, 1915.)

Differences in electrical potential are developed in the structures of the heart during the processes of its action. At any instant of time the waves of the electrocardiogram are the resultant of a number of potential differences developed by the synchronous activities of the cardiac structures. It has been sufficiently demonstrated that the P wave is associated with activity of the auricles, and the Q, R, S, T, and U waves with activity of the ventricles. Clement (1), Erfmann (2), and Lewis (3) have recently shown that the Q, R, and S waves are associated with the spread of the excitation wave from the auriculoventricular junction to the ventricles and throughout the ventricular muscle mass. It is unknown what ventricular structures are associated with the individual waves. It is also unknown whether the direction of the potential difference responsible for a given wave corresponds to any anatomical structure. On theoretical grounds Einthoven (4) and Waller (5) have shown that from the relative sizes of a given wave in the three leads of the human electrocardiogram it is possible to calculate the angle which the direction of potential difference responsible for the wave makes with the anatomical axis of the body. Theoretically then, alterations in the position of the heart in the body that change the direction of the potential difference relatively to the axis of the body will change the size and direction of the waves. Again, if without any change in the position of the heart, alterations occur within the heart so that the distribution of the muscle mass is changed, as for

292

instance by a change in the thickness of the wall of one ventricle relatively to the wall of the other, or by a deviation of the septum to one side or the other, changes in the resultant potential difference would result with corresponding changes in the size and direction of the waves. Other factors are involved in determining the potential differences and character of the waves. These include the direction and rate of travel of the wave of excitation within the conducting structures and the muscle mass.

On account of these considerations it is generally recognized that when changes in the electrocardiogram occur they do so only as the result of definite causes. Instances of such changes are observed in hypertrophy of the human heart.

In patients with hypertrophy of the left ventricle, the R wave is large in Lead I and small or absent in Lead III, while the S wave is large in Lead III and small or absent in Lead I; and in patients with hypertrophy of the right ventricle, the R wave is small or absent in Lead I and large in Lead III, while the S wave is large in Lead I and small or absent in Lead III. These changes correspond to the changes expected theoretically, if in hypertrophy the direction of potential difference responsible for the Q, R, S group is turned to the left or to the right respectively. Other changes have been observed in the course of acute and chronic illnesses; as, for instance, alterations in the size of all or of some of the waves, splitting of waves, and lengthening of the time occupied by a wave or group of waves. These changes are often transitory and so must depend on causes other than increased muscle mass.

The work reported here was undertaken originally in the hope of correlating certain changes in the electrocardiogram with definite changes in the heart.

Method and Material.

For this purpose it was necessary to record changes in the heart during life. For observations on the position, size, and shape of the heart, x-ray plates were made. It was impossible to determine changes in the distribution of the muscle during life. For this and for the condition of the cardiac structures it was necessary to depend upon the examinations made after death. The postmortem examinations were satisfactory for these purposes only in the animals that died or were killed at the time when the electrocardiographic changes were observed. After death the hearts were divided as recommended by Müller (6), and the chambers weighed separately. It was not found advisable to separate the two auricles and auricular septum from each other. Histological examinations were then made.

Rabbits were used in all the experiments. Full growth rabbits were chosen so that changes consequent on growth could be excluded, as it was found necessary to continue the observations in some cases for a period of six months or more.

In all, 33 rabbits were used, of which 4 were kept as controls. The changes were produced by repeated intravenous injection of: (a) spartein sulphate and adrenalin hydrochloride in 7 animals; (b) adrenalin hydrochloride alone in 4; (c) diphtheria toxin in 4; (d) suspensions of living streptococci in 14 animals.[1]

Spartein sulphate and adrenalin hydrochloride were used in the doses recommended by Fleisher and Loeb (7), and the injections were repeated at intervals of three or more weeks. Adrenalin alone was given in increasing doses every two days commencing with 0.1 or 0.2 cc. of 1:1,000 solution for twenty or more doses. Diphtheria toxin was given in varying doses and these were repeated if no definite effect was seen. The streptococci which were injected were isolated from the joints and blood of cases of acute rheumatism, from a tonsillar abscess, a liver abscess, and an abscess about the root of a tooth. Some were hemolytic, others of the *Streptococcus viridans* type. The injections were repeated, as a rule, in increasing doses until effects on the animal or electrocardiographic changes were seen.

The roentgenograms were made with the anticathode of the tube at a distance of 30 inches from the lower end of the sternum. The rabbit was stretched out on its back on a board, in which a gap had been cut so that the plate was in contact with the back of the animal.

Before injection, control electrocardiograms were made with the animal on the right side and on the left side, as well as on its back, so that possible alterations due to changes in the position of the heart

---

[1] The animals injected with streptococci were from a series of experiments undertaken with Dr. Homer F. Swift. The organisms were isolated and the injections made by him.

in the body could be estimated. In only four instances were changes in the size of the waves of more than 1 mm. noted with change of position.

The large Edelmann model of the string galvanometer was employed. On all occasions three leads were recorded. In most cases Lead I (right foreleg and left foreleg) was unsatisfactory, as the waves were very small; and Lead III (left foreleg and left hindleg) was, as a rule, so nearly identical with Lead II (right foreleg and left hindleg) that it has been found sufficient to consider Lead II only.

The resistance of the animal was estimated by the Wheatstone bridge and telephone method, and the string standardized with a constant total resistance in the circuit. The electrocardiograms were made with a resistance added to bring the total resistance in the circuit to a constant value. The deflection times of the different strings employed varied and the times occupied by some of the waves of the rabbits' electrocardiograms are so short that the size of the wave was affected by the variations in deflection time. It has been necessary, therefore, to discard all records in which the deflection time of the string varied from that of the control curves. Records in which the standardization of the string was not satisfactory have also been discarded. More rapid strings than those available might have shown greater changes than have been observed by the methods employed, but such changes as have been recorded would have been shown and shown better by an improvement in the methods.

RESULTS.

The changes in the electrocardiograms to be described are those involving the Q, R, S group only. Changes in the P and T waves also occurred but were so small, except in a very few instances, that accurate comparisons were impossible.

In the course of the observations on the 7 animals injected with spartein and adrenalin the roentgenograms in three animals showed that extensive dilatations to the right or to both sides developed in the course of a few days. Dilatations to the left were accompanied by decrease in the size of the R waves and increase in the size of the S waves; the one instance of marked dilatation to the right by

increased R wave and diminished S wave; while dilatation to both
sides caused little or no change.   These acute changes disappeared
as quickly as they developed.   Of the other 4 animals, 2 died after
the first injection, and 2 after the second, and before significant
changes developed.   After repeated injections and the resultant
acute changes, these 3 animals showed a moderate gradual enlargement to the left associated with decrease in the size of the R waves
and increase in the S waves.

Figs. 1 and 3 represent the changes that occurred in Rabbit 218 E.
Fig. 1 shows the size of the heart at the beginning, and Fig. 3 at the
conclusion of the experiments, and they demonstrated an enlargement to the left side.   Figs. 2 and 4 show the corresponding electrocardiographic changes, and Fig. 5 shows the extreme but transitory
electrocardiographic changes obtained in this animal.   Unfortunately no roentgenogram was obtained on the same day as this
curve.

Figs. 6 and 8 represent the right-sided enlargement that occurred
in Rabbit 211 E, and Figs. 7 and 9 represent the change in the electrocardiograms.   Figs. 10 and 12 represent the extreme change
obtained in Rabbit 278 E and Figs. 11 and 13 give the corresponding
electrocardiograms.   The enlargement in this case was to both sides
while the electrocardiogram was altered but little.   In Table I the
changes in the waves and in the size of the heart throughout the observations in Rabbit 278 E are tabulated.   It will be seen that following the third injection the electrocardiographic changes indicated
an enlargement to the left side, but no roentgenogram was obtained
on that date to correspond to this.   Similar changes occurred after
the last injection.   Daily x-ray observations at this time showed
an enlargement to the left.

Of the 4 animals treated by repeated injection of adrenalin alone,
one, 106 B, showed an acute enlargement to the left accompanied by
increased R wave and increased S wave, but died after the 4th injection, while the others showed slight gradual increase to the left
with similar electrocardiographic changes.

Of the animals injected with adrenalin, or spartein and adrenalin,
those that showed marked distress immediately after injection developed acute changes, while in those that showed little immediate

TABLE I.*

Rabbit 278 E.

Received 6 Injections of Spartein Sulphate 0.012 Gm. per Kilo, and Adrenalin Hydrochloride 0.2 Cc. 1: 1,000 (Parke, Davis and Co.). Total Resistance in Circuit 13,500 Ohms throughout Observations. String Standardized with Resistance in Circuit of 8,700 Ohms.

Date.	Injection.	Rate.	Deflection time.	Q	R	S	Size of heart.	Remarks.
			sec.	mm.	mm.	mm.	mm.	
Dec. 2, 1913		267.3			+3.5	−1.0		On back.
" " "		266.4	0.020		+3.0	−1.0		On right side.
" " "		272.7			+2.5	−1.0		On left side.
" 4, "		.					28 × 23	
" 8, "	1							
" 12, "		246.7	"		+2.5	−1.0		
" 19, "		281.8	"		+2.0	−1.0		
" 30, "		284.3	0.025		+2.5	−3.0		
Jan. 7, 1914	2	267.5	"		+1.5	−1.5		
" " "								
" 12, "		293.7	"		+2.5	−2.0		
" 20, "		216.9	0.015		+2.0	−1.0		
" 22, "							32 × 28	
" 26, "		235.2	0.018		+2.5	−2.0		
" " "	3							
" 29, "		259.2	0.025	− ?	+ ?	−1.5		
Feb. 3, "		261.8	"			−3.5		
" 10, "		274.5	"		+0.5	−2.0		
" 26, "		267.1	0.020		+1.5	−2.0		
Mar. 5, "							38 × 32	
" 17, "		235.4	"		+2.0	−2.0		
" " "	4							
" 23, "		289.1	"		+3.0	−1.0		
" 25, "		273.1	"		+2.0	−0.5		
" 30, "		266.9	"		+2.0	−1.0		
Apr. 6, "							53 × 44	Enlarged to both sides.
" 7, "	5	261.7	"		+2.0	−1.0		
" " "								
" 10, "		255.6	"		+2.0	−1.0		
" 15, "							39 × 33	
" 16, "		230.8	"		+2.0	−2.0		
" 27, "							39 × 31	
" 30, "	6	233.8	"		+2.5	−2.0		
" " "								
May 1, "		254.2	"		+2.0	−1.5	38 × 32	
" 4, "		253.9	"		+1.5	−2.5	37 × 32	
" 5, "		240.0	"		+1.0	−2.0	36 × 33	
" 6, "							35 × 31	
" 7, "		238.4	"		+1.0	−3.0	35 × 30	
" 8, "		244.8	"		+1.0	−3.0	42 × 38	Enlarged to left mainly.
" 11, "		258.2	"		+1.0	−3.5	36 × 29	
" 13, "		268.2	"		+1.0	−2.5	34 × 26	
" 24, "		267.0	"		+1.0	−3.0		
" 25, "							34 × 25	
June 4, "		201.1	"		+2.0	−1.5		
" 5, "		234.8	"		+2.0	−1.5	33 × 25	Enlarged to left mainly.

*The figures in this table and in Table II denoting the size of the heart were obtained by measuring and expressing in mm. the long axis of the heart shadow and the greatest width at right angles to this.

reaction to the injections only slight changes developed in the size of the heart or in the electrocardiograms. In all cases the heart sounds became accentuated and reduplicated and in four cases systolic murmurs developed, but except for restlessness and increased excitability, the animals did not appear to be distressed after recovering from the dyspnea consequent on the injections. This dyspnea never continued for more than a few minutes.

The three animals injected with fatal doses of diphtheria toxin showed slowing of the heart rate before death. Not only did the auricular rate decrease but the auriculoventricular conduction time lengthened, and the time occupied by each phase of the cycle increased. For example, in Rabbit 103 B the heart rate changed from 261.7 to 177.6, the P–R time from 0.07 of a second to 0.10 of a second, the time occupied by the Q, R, S waves from 0.02 to 0.07 of a second, and that occupied by the T wave from 0.12 to 0.16 of a second. Rabbit 105 B, which received three non-fatal doses, showed by x-ray slight enlargement of the heart downwards and to the left with increase in the S wave and decrease in the R wave of the electrocardiogram. Histologically the cases that received fatal doses showed marked degeneration and fatty changes throughout the heart muscle and similar changes in the auriculoventricular bundle. The electocardiographic changes are illustrated in Figs. 14 and 15 from Rabbit 102 B.

No constant changes were seen in the fourteen rabbits that received intravenous injection of streptococci. Lesions resulting from these injections in structures other than the heart occurred but will not be discussed here. Rabbit 115 B was injected with a streptococcus obtained from the blood of a case of acute articular rheumatism. Slowing of the auricular rate occurred and slight lengthening of the auriculoventricular interval and of the Q, R, S time. The changes were similar to those seen in the cases injected with diphtheria toxin. Premature ectopic ventricular contractions also occurred before death. Histologically the heart muscle showed diffuse degenerative changes and numerous necrotic foci. Rabbit 118 B injected with a streptococcus isolated from a tonsillar abscess also developed premature ectopic ventricular contractions, but no focal lesions were found and the muscle showed but slight degenerative changes. Figs. 16 and 17 illustrate the electrocardiographic changes in this animal. Rabbit 119 B injected with the same organism showed lengthening of the Q, R, S period. In the heart there were numerous areas of focal necrosis and degeneration of the muscle. · Rabbit 108 B, injected with a *Streptococcus viridans* obtained at death from the heart's blood of a case of acute rheumatic carditis, showed on an x-ray plate an acute enlargement of the heart to the left; the accompanying electrocardiogram showed a decrease in size of all the waves, but especially of the R wave.

In 9 of the 14 rabbits treated with suspensions of streptococci and in all 4 of those receiving diphtheria toxin, diminution in the size of all the waves was seen. This was never found in the controls and only to a very slight extent in the animals treated with adrenalin or

TABLE II.

Rabbit 109 B.

Received 4 Injections of a Suspension of Streptococcus Isolated after Death from the Heart's Blood of a Patient with Acute Rheumatism. Total Resistance in Circuit for Standardization and for Records, 8,700 Ohms. Deflection Time, 0.02 Seconds throughout Observations.

Date.	Injection.	Rate.	P-R time.	Size of waves.				Q-R-S time.
				P	R	S	T	
			sec.	mm.	mm.	mm.	mm.	sec.
Mar. 20, 1914	I	170.8	0.07	+1.0	+4.0	−2.5	+3.0	0.025
" 23 "		231.6	"	+1.5	+2.5	−3.5	+2.0	0.025
" 24 "		211.4	"	+1.0	+2.0	−2.5	+1.5	0.025
" 25 "		244.8	"	+1.0	+1.0	−2.5	+1.0	0.02
" 28 "		244.5	"	+0.5	+1.0	−1.0	+0.5	0.025
" 30 "		264.5	"	+1.0	+1.5	−2.0	+1.5	0.025
Apr. 2 "		230.7	"	+ ?	+1.0	−1.0	+0.5	0.02
" 6 "		253.3	"	+1.0	+3.5	−2.0	+2.5	0.025
" 8 "		228.8	0.06	+1.0	+3.5	−1.5	+1.5	0.02
" 14 "		256.8	"	+1.5	+3.0	−2.5	+2.0	0.025
" 24 "	2	216.3	0.07	+1.5	+5.0	−2.0	+2.5	0.025
May 3 "		293.0	"	+1.0	+2.0	−3.0	+1.5	0.03
" 4 "		284.8	"	+1.0	+2.0	−1.0	+1.5	0.025
" 7 "		269.1	0.06	+1.0	+3.0	− ?	+2.0	0.025
" 11 "		238.4	0.07	+1.0	+3.0	−1.0	+2.0	0.025
" 18 "	3							
" 22 "		278.2	0.06	+1.0	+2.5	−1.0	+1.5	0.025
" 25 "		241.7	0.07	+0.5	+1.0		+1.0	?
" 27 "		212.3	"	+0.5	+2.0	−1.0	+1.5	0.025
" 28 "	4							
June 8 "		218.6	"	+1.0	+3.0	−1.5	+1.0	0.025

spartein and adrenalin, and in them it was a transient effect.   Table II, giving the observations on 109 B, illustrates this.

DISCUSSION.

Changes in the size and direction of the waves of the·electrocardiogram such as those described may depend, as has been pointed out, on changes in the position of the heart in the body, changes in the relative distribution of the muscle mass, and changes in the condition of the muscle and conducting structures.   It is possible that conditions of the nervous system may exert an influence from outside the heart.   But the influence of the nervous system on the mechanism of the heart beat has been the subject of numerous investigations and changes of this nature have never been demonstrated.   The influence of the position of the heart in the body is definitely excluded, as control curves, taken with the animals in different positions before the injections were commenced, showed that in no instance could the changes that resulted be accounted for in this way.   In some instances also it was ascertained that alterations in the position of the rabbit could not restore the electrocardiogram to the initial form.   The changes in the size and shape of the heart demonstrated by the x-ray examination were always accompanied by changes in the electrocardiogram, and changes in one direction were always accompanied by electrocardiographic changes in a definite direction.   It is probable, therefore, that the alterations seen in the roentgenograms·were the causes of the electrocardiographic changes described.

The observations extended over seven months in some instances. The records were obtained at all stages between the initial normal forms of the electrocardiograms and the extreme qualitative changes usually associated with right- or left-sided enlargement.   These records show that not only the extreme qualitative changes, but also the gradual quantitative decreases and increases in size of the waves of the Q, R, S group seen in the intermediate records are significant of changes in the size and shape of the heart.

Diminution in the size of all the waves was seen in the case of the animals treated with diphtheria toxin.   The hearts of these animals showed extreme degenerative changes after death.   Acute degenerative changes have been described by numerous observers after single injections of adrenalin and of spartein and adrenalin. Following the injections, the electrocardiographic records sometimes showed decrease in size of all the waves before the changes associated with the enlargements of the heart manifested themselves and obscured these more general changes. A toxic condition of the muscle and other structures of the heart and degenerative changes in them must alter their functions, and it is not improbable that their electrical activities may be lowered. All the animals injected with streptococci showed joint lesions or altered heart sounds during life, or degenerative changes in the heart muscles after death, so that a similar reason may be present in the nine animals so treated that showed the decrease in the size of all the waves.

The influence of the nervous system and of pathological lesions of the structures themselves on the function of the pace-maker and the auriculoventricular conducting structures has been repeatedly demonstrated. The slowing of the pace-maker and the lengthening of auriculoventricular conduction time and of the duration of the individual waves, seen in the three rabbits treated with fatal doses of diphtheria toxin and in Rabbits 115 B and 119 B treated with streptococci, may have been caused by changes in the nervous mechanism outside the heart. But the intense degenerative changes seen throughout the conducting structures and the heart muscle afford a more probable explanation. Ectopic ventricular contractions can be caused by mechanical or electrical irritation of the heart muscle. The ectopic contractions seen in Rabbit 115 B may well have been due to the foci of degeneration and necrosis seen. Ectopic ventricular contractions were also obtained in Rabbit 118 B although no such foci were found in the sections examined.

A study of the relative weights of the ventricles of the hearts after death showed no changes sufficiently constant to permit of correlation with the electrocardiographic changes.

CONCLUSIONS.

1. In rabbits transient and also permanent enlargements of the heart occur as the result of the intravenous injection of adrenalin, spartein and adrenalin, diphtheria toxin, and streptococci.

2. Transient and permanent enlargements of the heart to the left are associated with decrease in the size of the upwardly directed waves of the Q, R, S group in Lead II of the electrocardiograms and increase in the downwardly directed. Transient enlargement to the right is associated with changes in the opposite direction.

3. The stages in the process of enlargement have been observed and are associated with gradual changes in the size of these waves.

4. A diminution in the size of all the waves of the Q, R, S group was observed in degenerative conditions of the heart muscle.

5. Extreme degeneration of the structures of the heart was associated with slowing of the heart's action, lengthening of conduction time (P–Q), and lengthening of the time occupied by the individual waves of the electrocardiogram.

6. Ectopic ventricular contractions were seen in rabbits injected intravenously with suspensions of living streptococci.

BIBLIOGRAPHY.

1. Clement, F., *Ztschr. f. Biol.*, 1912, lviii, 110.
2. Erfmann, W., *Ztschr. f. Biol.*, 1913, lxi, 155.
3. Lewis, T., Lectures on the Heart, New York, 1915, 3.
4. Einthoven, W., Fahr, G., and de Waart, A., *Arch. f. d. ges. Physiol.*, 1913, cl, 275.
5. Waller, A. D., *Jour. Physiol.*, 1913, xlvi, p. lix.
6. Müller, W., Die Massenverhältnisse des menschlichen Herzens, Hamburg and Leipzig, 1883.
7. Fleisher, M. S., and Loeb, L., *Arch. Int. Med.*, 1909, iii, 78.

EXPLANATION OF PLATES.

PLATE 37.

FIGS. 1 to 5. Rabbit 218 E. Received injections of 0.012 gm. spartein sulphate per kilo and 0.2 cc. of adrenalin hydrochloride 1 : 1,000 (Parke, Davis and Co.) on Dec. 8, 1913, Jan. 2, Jan. 26, Mar. 17, Apr. 6, and Apr. 30, 1914.

FIG. 1. Roentgenogram of heart, Nov. 28, 1913.
FIG. 2. Electrocardiogram, Lead II, Dec. 4, 1913.
FIG. 3. Roentgenogram of heart, May 5, 1914.
FIG. 4. Electrocardiogram, Lead II, May 5, 1914.
FIG. 5. Electrocardiogram, Lead II, Feb. 3, 1914.

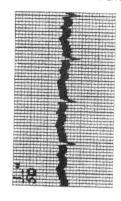

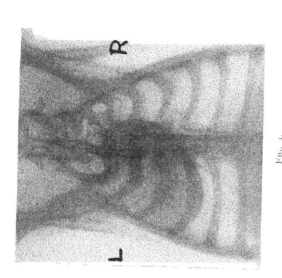

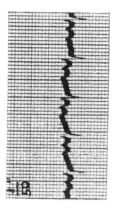

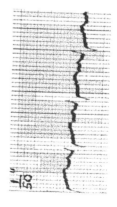

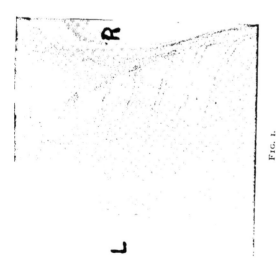

FIG. 1.

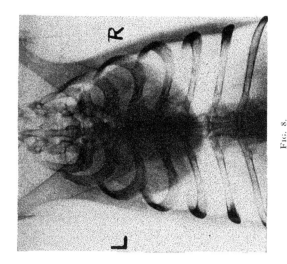

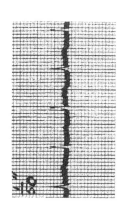

Fig. 8.

Fig. 9.

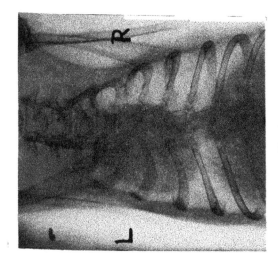

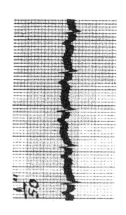

Fig. 6.

Fig. 7.

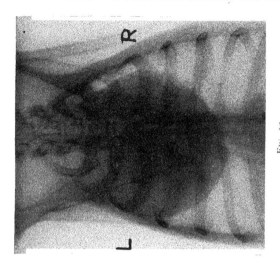

FIG. 12.

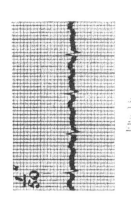

FIG. 13.

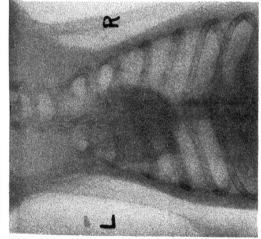

FIG. 10.

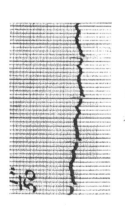

FIG. 11.

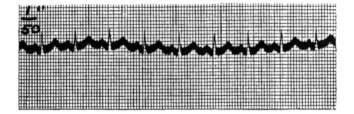

FIG. 14.

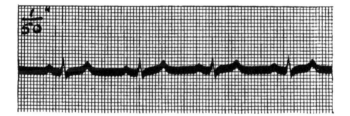

FIG. 15.

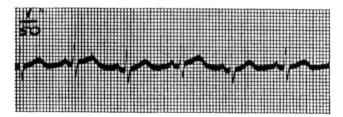

FIG. 16.

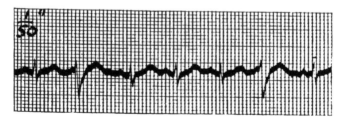

FIG. 17.

(Fraser: Changes in Electrocardiograms.)

PLATE 38.

FIGS. 6 to 9. Rabbit 211 E. Received injections of 0.012 gm. of spartein sulphate per kilo and 0.2 cc. of adrenalin hydrochloride 1 : 1,000 (Parke, Davis and Co.) on Dec. 8, 1913, Jan. 5, Jan. 26, Jan. 30, Mar. 17, Apr. 6, and Apr. 30, 1914.

FIG. 6.   Roentgenogram of heart, Dec. 4, 1913.
FIG. 7.   Electrocardiogram, Lead II, Feb. 3, 1914.
FIG. 8.   Roentgenogram of heart, Apr. 15, 1914.
FIG. 9.   Electrocardiogram, Lead II, Apr. 16, 1914.

PLATE 39.

FIGS. 10 to 13. Rabbit 278 E. Received injections of 0.012 gm. spartein sulphate per kilo and 0.2 cc. adrenalin hydrochloride 1 : 1,000 (Parke, Davis and Co.) on Dec. 8, 1913, Jan. 7, Jan. 26, Mar. 17, Apr. 7, and Apr. 30, 1914.

FIG. 10.   Roentgenogram of heart, Dec. 4, 1913.
FIG. 11.   Electrocardiogram, Lead II, Dec. 2, 1913.
FIG. 12.   Roentgenogram of heart, Apr. 6, 1914.
FIG. 13.   Electrocardiogram, Lead II, Apr. 7, 1914 (before injection).

PLATE 40.

FIGS. 14 and 15. Rabbit 102 B. Received injection of diphtheria toxin, Jan. 29, 1914.

FIG. 14.   Electrocardiogram, Lead II, Jan. 29 (before injection).
FIG. 15.   Electrocardiogram, Lead II, Feb. 1, 1914.

FIGS. 16 and 17. Rabbit 118 B. Received injections of suspensions of streptococcus isolated from tonsillar abscess, Apr. 22, May 2, May 18, and May 28, 1914.

FIG. 16.   Electrocardiogram, Lead II, Apr. 22, 1914 (before injection).
FIG. 17.   Electrocardiogram, Lead II, May 8, 1914.

BACTERIOLOGICAL AND CLINICAL STUDIES OF AN EPIDEMIC OF KOCH–WEEKS BACILLUS CONJUNCTIVITIS ASSOCIATED WITH CELL INCLUSION CONJUNCTIVITIS.[1]

By HIDEYO NOGUCHI, M.D., and MARTIN COHEN, M.D.

(*From the Laboratories of The Rockefeller Institute for Medical Research and the New York Post-Graduate Medical School and Hospital, New York.*)

PLATES 41 TO 43.

(Received for publication, June 21, 1915.)

While engaged in the study of conjunctivitis associated with epithelial cell inclusions, the occurrence of an epidemic[2] of conjunctivitis due to hemoglobinophilic bacilli enabled us to investigate the relationship between this organism and the epithelial cell inclusions.

Since the discovery of the inclusions by von Prowazek and Halberstaedter, who considered them to be the etiological agents of trachoma, an abundant literature has arisen. Opinions are still greatly divided as to the nature of the bodies, although the majority of investigators regard them as an independent organism, as claimed by von Prowazek and Halberstaedter. They do not, however, commit themselves to the statement as to whether the organism represents the causative agent of trachoma or not. We were inclined to the belief that it constitutes an independent organism which produces a true conjunctivitis of a more or less characteristic clinical course. Herzog[3] considered them to be mutation forms of the gonococcus, while Williams[4] views them as the cell inclusions of various organisms, such as the hemoglobinophilic bacillus and allied species.

---

[1] Read before the Ophthalmological Section of the New York Academy of Medicine, May 17, 1915.

[2] Service of Dr. Martin Cohen at the Randall's Island Hospital, New York.

[3] Herzog, H., *Arch. f. Ophth.*, 1910, lxxiv, 520.

[4] Williams, A. W., *Collected Studies from the Bureau of Laboratories, Department of Health, City of New York*, 1912–13, vii, 159–247; *Jour. Infect. Dis.*, 1914, xiv, 261.

Although by means of a special method an organism strikingly similar to the epithelial cell inclusions has been made to grow in pure culture,[5] their pathogenic properties, nevertheless, have not yet been established.

As a result of our previous studies a tentative conclusion had been reached that infection of the conjunctiva with these organisms produces an independent conjunctival disease which may properly be described as cell inclusion conjunctivitis. This organism may be present alone, or it may be associated with other pathogenic organisms.

Outbreak and Course of the Present Epidemic of Koch-Weeks Bacillus Conjunctivitis.

Prior to the outbreak of the Koch-Weeks bacillus epidemic,[6] one of us (Cohen) had under observation for about five months 10 cases of inclusion conjunctivitis and 4 cases at the outdoor department of the Post-Graduate Hospital. In these cases, smears (and in a few cases, cultures) were taken weekly and examined (Noguchi), but no Koch-Weeks bacilli were found. On October 3, 1914, one of the inclusion cases under observation at the Island developed an acute conjunctival inflammation. In the conjunctival smears numerous Koch-Weeks bacilli, as well as epithelial inclusions, could be demonstrated. The acute conjunctival inflammation spread to the remaining nine inclusion cases. The smears from all showed inclusions as well as numerous Koch-Weeks bacilli.

In an adjoining ward there were under treatment 15 cases of follicular conjunctivitis, 3 cases of interstitial keratitis, and 1 case of chronic dacryocystitis. 13 of these contracted the Koch-Weeks bacillus infection. During the entire course, which lasted from two weeks to seven months, no pathogenic organisms but the Koch-Weeks bacilli could be demonstrated in the smears and cultures.

There were 7 patients in the trachoma ward. 2 of these showed simultaneous presence in the conjunctiva of inclusion and of Koch-

[5] Noguchi, H., and Cohen, M., *Arch. of Ophth.*, 1914, xliii, 117; *Jour. Exper. Med.*, 1913, xviii, 572.

[6] In the present article the hemoglobinophilic bacilli found in these cases were designated as Koch-Weeks bacilli, in order to conform with the usage of this term in ophthalmology and bacteriology when speaking of this class of organisms.

Weeks bacilli. The latter organisms were present only in cultures, as there were possibly too few to be recognized in the smears. This condition was present for a short time during this epidemic. The infection with the Koch-Weeks bacillus apparently produced no change in the clinical picture of these trachoma cases. The 4 inclusion cases not treated on the Island showed at no time during the past year the Koch-Weeks bacillus, either in smears or cultures; not even during a relapse or reinfection, of which one of these cases had two, could they be demonstrated. Table I shows the bacteriological findings in the cases on which the present report is based.

TABLE I.

Organisms present.	No. of cases.
1. Epithelial cell inclusions alone	6
2. Epithelial cell inclusions at first, with subsequent Koch-Weeks infection	17
3. Koch-Weeks bacillus alone	13
4. Koch-Weeks bacillus at first, with subsequent appearance of inclusions	2
5. Pneumococcus with inclusions	1
Total	39

In acute catarrhal conjunctivitis it is at times difficult, or impossible, to determine the causative organism from the clinical manifestations alone. This is due to the great variability in virulence of the usual infecting organisms, and to the resisting qualities of the individual patient. Yet in many of the acute infections of the conjunctiva, the etiological diagnosis as made by the clinical appearance could be confirmed subsequently by the bacteriological diagnosis, especially in the subacute state. In mixed infections, such a diagnosis is much more difficult.

In a few instances, even smears and cultures of the conjunctival scrapings fail to clear up the diagnosis, as no pathogenic organisms can be found. That the microscope alone is not sufficient to exclude the presence of an organism is shown by the fact that the scrapings from cases of trachoma which showed neither inclusions nor the Koch-Weeks bacilli (as proved similarly by cultures), when inoculated on the conjunctiva of the higher apes (baboons) produced symptoms similar to those which are seen in inclusion cases in man. They were, however, of a milder type and lasted only about ten days. The conjunctival scrapings from the inoculated animals revealed the presence of epithelial inclusions, but no Koch-Weeks bacilli or allied organisms appeared in the smears or in the cultures.

We shall now discuss the clinical manifestations in the group of cases mentioned in Table I.

The clinical course of inclusion conjunctivitis has already been described by Cohen.[7] In the early stage, the clinical manifestations resemble those seen in the Koch-Weeks bacillus infection, and it is only by the examination of smears or cultures that the diagnosis can be positively made. Briefly, they are as follows: In the beginning there is moderate edema of the lids, with mucopurulent secretion. Somewhat later isolated reddish translucent follicles appear in the lower palpebral conjunctiva and upper folds, these two sites becoming involved simultaneously. At a later stage, the upper palpebral conjunctiva takes on a brick red color and assumes the characteristic granular appearance. These regular and progressive manifestations of the disease retrogress after about two months by absorption of the contents of the follicles and papules. These disappear first from the upper portions of the conjunctiva, when the conditions resemble those of cases of follicular conjunctivitis; and later, in about three months, they begin to disappear from the lower conjunctiva. At the end of four months, the conjunctiva is again in a normal condition.

This is the usual course, but at times there are deviations, due to external or unknown conditions, or to a previous affection of the conjunctiva; as, when grafted on a follicular conjunctivitis, the original follicles become larger and new ones make their appearance. In the stage of retrogression the papules become absorbed and the original follicles, owing to relaxation of surrounding pressure, again become prominent, the course being then one of a follicular conjunctivitis. In 2 cases of inclusion conjunctivitis, there remains solely, at the end of nine months, a fine papillary condition of the upper tarsal conjunctiva; one of these two cases has had two relapses or infections.

It may be mentioned here in passing, that of the 75 cases of inclusion conjunctivitis studied by us during the past five years, 58 cases are still under observation of Dr. Cohen. 2 of these still show a diffuse and linear cicatrization of practically the entire palpebral conjunctiva, but no corneal or other involvement is present. 2 other cases of this group still show a fine papillary hypertrophy of both tarsal conjunctivæ, with no corneal or other complications. The remaining 54 cases have remained normal for the past four and a half years. 17 cases were observed for six months, and were normal when last seen.

4 patients with follicular conjunctivitis became infected with inclusions. The follicles previously present became enlarged and congested, and new papules and follicles appeared on the upper tarsal conjunctiva and the conjunctiva of the lower lid. These cases had the appearance of a severe type of inclusion conjunctivitis.

For several weeks after the beginning of the disease, the smears in all cases showed an abundance of epithelial cell inclusions. Toward the end of the disease, the number of inclusions gradually diminished. Yet after three months' duration, although no inclusions were demonstrable in the smears, the clinical aspect of the disease was usually still evident.

[7] Cohen, M., *Arch. of Ophth.*, 1913, xlii, 29.

On Oct. 23, 1914, one of the inclusion cases on the Island developed a slight conjunctival inflammation, which smears and cultures showed to be due to the Koch-Weeks bacillus. In the succeeding two weeks, as previously mentioned, additional cases of inclusion conjunctivitis contracted the Koch-Weeks bacillus infection. Conjunctival smears and cultures examined weekly for the past several months showed at first the simultaneous presence of inclusion and Koch-Weeks bacilli. In the succeeding two months, the findings varied, at times only one of the two organisms being present, at other times, both. Possibly this was due to relapse or reinfection. In the next two months, although no organism was present in smears and cultures, there was still evidence of inflammation. Ultimately the conjunctiva became normal in all but one case. In this case there is still at the end of a year evidence of an inclusion conjunctivitis in the retrogressive stage.

In an adjoining ward, where mild follicular conjunctivitis cases were being treated, 13 contracted the Koch-Weeks bacillus infection. Previous to the Koch-Weeks infection, all these cases showed over various areas of the conjunctiva follicles varying in number and size, and showing no inflammatory reaction. The first case in this ward developed on Nov. 1, 1914. Only 3 of the 16 cases escaped the Koch-Weeks infection. The Koch-Weeks bacillus was present in great numbers, but no other pathogenic organisms could be demonstrated throughout the entire course of the epidemic, which lasted six months.

The clinical manifestations in these follicular cases were as follows. There were slight edema of the lids and conjunctival congestion with mucopurulent secretion. Both eyes became successively involved. The acute infection lasted six to eight weeks. At the end of this period, the original follicles were still present.

In one of the typical follicular cases it was difficult to determine whether the acute infection was due to the Koch-Weeks bacillus or to the epithelial inclusions. The symptoms were those of a severe inclusion conjunctivitis. But notwithstanding repeated examinations extending over two months, no inclusions could be found. Nevertheless, judging from the clinical course and from the fact that cell inclusions are often absent in the subacute stage, this and similar cases may be considered due to epithelial inclusions.

2 cases of interstitial keratitis on the Island Eye Ward and 1 at the Dispensary had normal conjunctivæ, when they suddenly developed an acute catarrhal conjunctivitis. In the smears and cultures, which were taken weekly for four months, the Koch-Weeks bacillus was present, even at a time when the conjunctiva was free from inflammation. The inflammatory stage lasted from six to eight weeks. Inclusions could at no time be demonstrated in the smears.

The clinical appearance in these cases was moderate edema of the lids, and conjunctival congestion with mucopurulent secretion. In one case there was present a small scleroconjunctival hemorrhage with a small phlyctena. Conjunctival furrows or folds also appeared in those cases when the congestion diminished.

As regards the last group in the table, 2 cases at first showed the Koch-Weeks bacillus in the conjunctival smears. Later, inclusion cells made their appearance, associated with their clinical manifestations. In the first case there was present a mild catarrhal conjunctivitis, papules being present on the lower tarsal conjunctiva alone. In this respect it differed from other inclusion cases where the papules are present also on the upper tarsal conjunctiva. It is possible that the lesions at times can only be detected microscopically. A second examination in this case revealed the epithelial inclusions. Perhaps the inclusions were missed in making the first examination, as it is likely that in some stages of the disease only a few epithelial cells contain inclusions.

The second patient originally had a follicular conjunctivitis. He then contracted an acute catarrhal conjunctivitis, the secretion containing for three weeks the Koch-Weeks bacillus alone. At the end of this time, when the conjunctivitis improved, he again developed an acute conjunctival inflammation, and the smears showed numerous inclusions and a few Koch-Weeks bacilli. This case then followed the usual course of an inclusion conjunctivitis, the conjunctiva ultimately returning to a normal condition.

In a ward assigned to minor eye affections, where there were a few mild Koch-Weeks cases, a routine examination of the conjunctivæ of all the cases was made, in order to determine the presence of the Koch-Weeks bacillus. 2 cases showed the Koch-Weeks bacilli, in spite of the fact that the conjunctivæ were entirely normal throughout the whole period. In another case in the same ward degenerated Koch-Weeks bacilli were found in the conjunctival smears. This case then developed an acute inflammation, showing all the symptoms of an inclusion conjunctivitis with the added findings of epithelial cell inclusions in the smear examination.

In our routine examinations made during the past five years Koch-Weeks bacilli have rarely been found. Recently we encountered eight cases of pneumococcal conjunctivitis. Clinically these cases resembled acute catarrhal conjunctivitis, due to the Koch-Weeks bacillus, and were so diagnosed before the smears and cultures showed the presence of numerous pneumococci without any Koch-Weeks bacillus. The conjunctival secretions in these cases were typically serous, being also associated with diffuse congestion. The duration was from one to two weeks. One of the patients, included already in this paper, had originally an inclusion conjunctivitis. Later, he contracted a Koch-Weeks bacillus infection. After four weeks, when apparently cured, another acute inflammation appeared, which was due to the pneumococcus. This infection was followed by the appearance of inclusion conjunctivitis, the smears showing both pneumococcus and inclusion cells. At present, the clinical symptoms are those of an inclusion conjunctivitis in a state of relapse or reinfection.

The treatment adopted in all these cases was irrigation of the conjunctival sac with a saturated solution of boric acid every two hours, and the application to the conjunctiva, once a day, of a silver nitrate solution, until the acute symptoms had subsided, when irrigation alone was continued. This method of treatment had practically but little effect on the course of the inclusion cases, whereas in the Koch-Weeks infections decided improvement was observable. The disappearance of the organisms from the smears and culture was, however, not at all influenced by the application of the silver nitrate.

Bacteriological and Experimental Studies of the Present Epidemic.[8]

Our technique for obtaining smears and cultures was as follows. The upper lid was everted and then the short end of a sterile slide was gently rubbed over the upper tarsal conjunctiva where there is least likelihood of contamination. By means of a platinum loop, a part of the scrapings removed was used for the purpose of cultivation and animal experiments. The remainder of the materials on the slide was transferred to another slide where a thin spread was made, and examined by means of the Giemsa or Gram stain.

Since our problem deals with the Koch-Weeks infection, the culture media employed were chiefly blood agar and tissue ascitic fluid. The cultural findings are summarized below.

In the present epidemic it was noticed that various strains of the so called Koch-Weeks bacilli or hemoglobinophilic organisms in cultures varied in size to such an extent that they may be divided into a thin (Figs. 4 to 7), a medium (Figs. 8 to 11, 20 to 23), and a coarse type (Figs. 16 to 19), according to their morphology. The coarse variety were bacillary or coccoid (Figs. 12 to 15) and resembled culturally and morphologically some strains of *Bacillus influenzæ* (Figs. 24 and 25) derived from the respiratory organs; while the thinner variety was much narrower and often shorter. After several days' cultivation at 37° C. on blood agar, some of the organisms became somewhat granular and unevenly stainable; when found in large masses they often assumed the appearance of the granular forms of the so called trachoma bodies (Figs. 5 to 7, 9 to 11, 13 to 15, 17 to 19). But these granules did not break down so as to approach the minuteness of the elementary bodies (Figs. 32, 33, 36, 39, 43, 45) which are found in the uncomplicated cases of inclusion conjunctivitis or trachoma. The degenerated or involuted bacilli do not have the sharp contour of the elementary granules. Transplants made from these somewhat degenerated granular masses of the hemoglobinophilic organisms give a good growth of typical bacillary or semicoccoid forms. The character of the colonies on the blood agar is also distinctive, in that they remain minute, sharply elevated with a pointed top, and appear more grayish than the colonies of the

[8] This part of the work was carried out at The Rockefeller Institute.

influenza type varieties, which show a tendency to spread and are of a more dewy aspect. It is quite possible that the minute type constitutes a group by itself, but further study is required to determine its relation to the other groups. At all events several strains of influenza bacillus[9] derived from cases of meningitis and pneumonia appear quite different from the conjunctivitis strains in their morphological features (Figs. 24 to 28). The frequency with which organisms of different type were found in the present epidemic is shown in Table II.

TABLE II.

	Types.		
Organisms present.	Coarse.	Medium.	Thin.
1. Epithelial inclusions first, with subsequent Koch-Weeks infection (17 cases)	1	7	9
2. Koch-Weeks bacillus alone (13 cases)	1	10	2
3. Koch-Weeks bacillus first, with subsequent appearance of inclusions (2 cases)	1 (round type)	1	

As will be seen from Table II, the thin variety occurred very frequently in the cases in which the patients had been previously infected with the inclusions. In the simple Koch-Weeks cases the medium was predominating. It is significant that in the smears taken from the cases infected with the inclusions and Koch-Weeks bacilli it is not difficult to differentiate the initial bodies of the former from the bacilli, because the initial bodies are much larger and oval in shape, and take up a deeper blue stain. Fig. 3 shows a typical inclusion near the nucleus and some Koch-Weeks bacilli in the same cell. The hemoglobinophilic bacilli are seen to be taken up by the polynuclear leucocytes more frequently than by the epithelial cells (Fig. 1), while the inclusion bodies are chiefly epithelial and seldom leucocytic in nature (Figs. 29 to 45). So far as could be ascertained in stained specimens the bacilli remain well preserved within the cells and do not seem to disintegrate into granules (Fig. 2). On the other hand, the initial bodies are often ill defined and do not show any distinct bacillary forms. They are far more

[9] We are indebted to Dr. Martha Wollstein for these strains, for which we here wish to express our thanks.

pleomorphic than the hemoglobinophilic bacilli (Figs. 29, 34, 37, 40, 41, 43, 46, and 47).

The isolation of the hemoglobinophilic organisms, irrespective of the type, was easily accomplished by means of the blood agar; but in many instances the preliminary microscopical search in the smears failed to demonstrate the presence of a few organisms, until they could be found by the cultural procedure. It must be understood that in cases where the inclusions only were found, repeated efforts to isolate the hemoglobinophilic organisms were made, with, however, invariably negative results. In this connection it may be added that a number of cases clinically diagnosed as trachoma was also studied in order to see if any hemoglobinophilic organism could be isolated. The results were uniformly negative. A few cases were also found in the same wards, which may have been infected through contact with the inclusion cases and which, although carefully observed from the earliest stage of the disease, did not at any period show the presence of any hemoglobinophilic bacilli.

Transmission of the Inclusion Bodies to Animals.—In order to find out whether the inclusion virus ever exists in the form of a hemoglobinophilic bacillus, scrapings from the conjunctivæ of these cases were inoculated into the conjunctiva of a baboon. Prior to the inoculation the conjunctiva of the baboon was examined to ensure the absence of such a bacillus. Within seventy-two hours after the inoculation the conjunctiva showed moderate congestion, edema, and a few minute papules; a small quantity of a mucopurulent discharge was present at the inner canthus. The smears and culture were made from the conjunctival scrapings. The result showed that there were numerous cell inclusions, but no hemoglobinophilic organisms[10] (Figs. 46 to 51). The examinations were continued regularly for a period of two weeks and the results were unvarying. As was previously shown by various investigators, the von Prowazek bodies are transmissible to higher apes, but the hemoglobinophilic or Koch-Weeks bacillus is not.

Attempts To Produce Koch-Weeks Bacillus Conjunctivitis in Animals.—Several attempts were made to transfer the hemoglobino-

[10] Cultures made after 24 and 48 hours were also negative in regard to the hemoglobinophilic bacillus.

philic organisms to the conjunctivæ of rabbits (young and adult) and monkeys (baboon and several *Macacus rhesus*), by introducing several loopfuls of the twenty-four hour blood agar cultures of different strains into moderately abraded surfaces of conjunctivæ. The results were completely negative, except in the case of one rabbit, where the organism was still recoverable after twenty-four hours. It seems remarkable that such a large quantity of pure cultures of freshly isolated strains of these organisms from the cases where the inclusions were also present should fail to reproduce the conjunctivitis, in view of the fact that a comparatively small number of the inclusions as contained in the scrapings from a patient can readily reproduce the inclusion conjunctivitis.

From this a conclusion may be warranted that in a conjunctivitis where the inclusion and the hemoglobinophilic bacilli are simultaneously present, two pathogenic factors can be separated by means of transmission of the material into the conjunctiva of a suitable animal (baboon) in which the inclusion virus alone implants itself upon the new host, while the bacilli quickly disappear from the inoculated conjunctiva. For man both organisms are pathogenic, but for the baboon only the inclusion virus is capable of producing infection.

On the other hand, this may not exclude the possibility, as asserted by Williams and her associates, that a conjunctivitis due to the Koch-Weeks bacillus may also show some cell inclusions; since under certain experimental conditions a very suggestive phenomenon, to be related below, has been observed.

Attempts To Produce the Koch-Weeks Bacillus Epithelial Inclusions in Animals.—In order to determine experimentally whether or not the hemoglobinophilic bacilli when taken up by epithelial cells will undergo the morphological changes which lead to the formation of so called inclusion bodies, an intratesticular inoculation of the rabbit with pure cultures of the hemoglobinophilic organisms isolated from the cases already mentioned was resorted to. The local reaction which follows consists of edema of the scrotum and induration of the testicular parenchyma within twenty-four hours. The edema gradually disappears within the next few days, while the testicular induration remains more or less the same for about five days, after which it commences to recede. In some instances the

rabbit succumbed to septicemia and probably to intoxication as a result of the introduction of the cultures. The organs were removed at
intervals of twenty-four hours, three days, and six days, and then
fixed in sublimate alcohol, and stained by Giemsa's acetone method,[11]
and, if overstained, treated with a 10 per cent solution of *Glycerin-
äthermischung* (Grübler) for a few minutes, as advocated by one of
us.[12] The results show that the injection of the bacilli is followed
by an intense leucocytosis, in which the polynuclears invade the
tubules in groups. In the twenty-four hour specimens the organisms
are still well distributed along the interstitial spaces; in the three day
specimens one notices numerous masses of agglutinated bacilli here
and there within the tubular lumina or along the connective tissue.
These masses take on a purplish hue and appear granular in structure and indefinite in outline.   They are on the point of disintegration.   The granules within and about these bacterial masses are
not so minute as to be mistaken for the elementary bodies of the
inclusion.   In the six day specimens some clumped bacteria were
found within the polynuclear leucocytes, but a diligent search failed
to show any typical epithelial cell inclusions.   In these six day preparations the number of the bacterial clumps is smaller than at an
earlier period.   There were no granules small enough to be regarded
as the typical elementary granules.   While a careful comparison of
the inclusions and the bacterial clumps just referred to will reveal
the difference between them, this is not always easy to accomplish
(Figs. 53 to 55).   Fig. 52 shows a mass of granules from a case
of mixed infection of Koch-Weeks bacilli and the von Prowazek
inclusions, and it appears difficult to determine whether they represent the degenerated bacilli or the inclusion granules.   The deep
stain of the mass and the absence of any free bacilli around it seem
to indicate that it belongs to the latter kind.   It is also possible that
some of the clumped granules found in the conjunctival smears
from cases of Koch-Weeks infection might have been interpreted as
the cell inclusions and classified with the von Prowazek inclusions.

*Attempts To Transmit the Inclusion Bodies to a Parenchymatous
Organ in Animals.*—Efforts were also made to transmit the von

[11] Giemsa, G., *Deutsch. med. Wchnschr.*, 1909, xxxv, 1752.
[12] Noguchi and Cohen, *Proc. N. Y. Path. Soc.*, 1910, x, 20.

Prowazek bodies from uncomplicated inclusion cases to the testicles of rabbits, since this organ offers an excellent medium of growth to various highly parasitic organisms which otherwise cannot be easily cultivated.[13] The scrapings of conjunctivæ from 4 different patients were inoculated into the testicles of 8 rabbits; but in spite of the large number of the inclusions contained in the conjunctival scrapings used for this purpose, no success was obtained along this line. The testicles showed within twenty-four hours some induration and edema, but after a few days resumed their normal condition. Tissues removed after twenty-four hours, three days, and six days failed to show any cell inclusions when examined in smears and sections. No hemoglobinophilic organism was found in cultures made from the tissues. This negative finding also tends to strengthen the view that the cell inclusions found in these cases were not the Koch-Weeks bacilli, for if they had been it would have resulted in the production of Koch-Weeks orchitis.

CONCLUSIONS.

1. There are cases in which epithelial cell inclusions may alone be present in the conjunctival smears. In such cases no other pathogenic organisms, such as the Koch-Weeks bacillus or the pneumococcus, can be demonstrated in smears or cultures.

2. The conjunctiva can become simultaneously infected with the inclusion bodies and Koch-Weeks bacilli or other organisms.

3. In cases of acute or subacute conjunctival inflammations due to mixed infections the clinical features of each infection may be present. The course of the inflammation is, however, more prolonged.

4. Within recent years, the Koch-Weeks bacillus has only seldom been found in our routine examinations.

5. The epidemic studied was of a severe type.

6. Clinically it is practically impossible to distinguish pneumococcal conjunctivitis from the Koch-Weeks conjunctivitis. Bacteriological examination of smears and cultures is the only means by which the etiological diagnosis can be definitely established.

7. Conjunctivæ of certain species of monkeys are susceptible to

[13] Noguchi, H., *Jour. Exper. Med.*, 1915, xxi, 539.

the von Prowazek inclusion bodies, but not to the hemoglobino-philic bacilli isolated from cases of epidemic conjunctivitis.

8. The injection of conjunctival scrapings containing the von Prowazek cell inclusions into the testicles of rabbits produces no cell inclusions in the latter, while the injection of a pure culture of the hemoglobinophilic bacilli causes an acute inflammation accompanied by numerous clumps of the organisms, simulating the von Prowazek bodies at certain stages of their evolution.

9. There exists an apparent morphological similarity between the degenerated forms of this variety of the hemoglobinophilic bacilli and the cell inclusions, both in cultures and in experimental orchitis in the rabbit.   But, as a rule, the elementary bodies of the latter are much smaller and more sharply defined than the smallest granules of the former, while the initial bodies are bigger, more intensely stainable, and less definite in their contour than the hemoglobino-philic bacilli found in the infected conjunctivæ.

EXPLANATION OF PLATES.

All the photographs were made from film preparations, except where otherwise stated. They were stained with Giemsa. The enlargement is uniformly 1,000 diameters.

PLATE 41.

FIG. 1.   This shows a mass of the Koch-Weeks bacilli in a film preparation from an uncomplicated case of Koch-Weeks conjunctivitis.   The bacilli belong to the thin type and some scattered examples appear coccoid.   By focusing, the mass is seen to be composed of numerous well defined bacilli which cannot be confused with the inclusion granules.

FIG. 2.   An epithelial cell from a case of uncomplicated Koch-Weeks conjunctivitis.   Around the periphery of the cell and near the nucleus along the lower border numerous bacilli are seen to be attached to, or contained within, the cell body.   There is, however, no difficulty in recognizing the bacilli as such in this instance.

FIG. 3.   A film preparation from a case of mixed infection of the Koch-Weeks and the inclusion organisms.   In the field the epithelial cell is seen to contain a densely stained mass of the initial bodies near the upper right border of the nucleus and numerous Koch-Weeks bacilli to the left, especially where there is a leucocyte.

FIG. 4.   This represents a twenty-four hour old pure culture of a thin type of the Koch-Weeks bacilli on blood agar.

FIGS. 5, 6, and 7. · The appearance of the same organism as in Fig. 4 after 3 days, 5 days, and 8 days, respectively, on the same medium at 37° C.

Figs. 8, 9, 10, and 11. These represent the appearance of a medium type strain of the Koch-Weeks bacilli after 24 hours, 3 days, 5 days, and 8 days, respectively, on blood agar at 37° C.

Figs. 12, 13, 14, and 15. These show the appearance of a strain of round or coccoid type of the Koch-Weeks bacilli in a 24 hour, 3 day, 5 day, and 8 day growth on blood agar, respectively.

Figs. 16, 17, 18, and 19. These show the appearance of a coarse strain of the Koch-Weeks bacilli in a 24 hour, 3 day, 5 day, and 8 day growth on blood agar, respectively. This strain resembles *B. influenzæ* more than the others.

Figs. 20, 21, 22, and 23. These show 4 different strains of the Koch-Weeks bacilli isolated from the mixed infection cases in the present epidemic. They belong to the medium type.

Figs. 24, 25, and 26. These show 24 hour growths of 3 different strains of *B. influenzæ* isolated from cases of pneumonia.

Figs. 27 and 28. These show 24 hour growths of 2 different strains of *B. influenzæ* isolated from cases of meningitis. Both strains were kept on artificial media for several years and tend to form threads more than they originally did (Wollstein).

PLATE 42.

Figs. 29, 30, 31, 32, and 33. These show the inclusions at various stages of evolution in the film preparations made from a case (Tho.) of inclusion conjunctivitis. This patient had been suffering from the inclusion conjunctivitis when he was superinfected with the Koch-Weeks bacilli, which, in due course, disappeared from his conjunctiva leaving the original condition little affected. These inclusions shown here were found in the conjunctiva long after the Koch-Weeks bacilli had disappeared in smears or in cultures. The minute elementary bodies in Figs. 32 and 33 bear no resemblance to the degenerated forms of the Koch-Weeks bacilli.

Figs. 34, 35, and 36. These represent the inclusion bodies in an uncomplicated case (Sh.), that is, without any association with the Koch-Weeks bacilli. The character of the initial bodies in Fig. 34 would definitely dispose of any suggestion as to their being the Koch-Weeks bacilli. Fig. 36 shows a still unburst intracellular aggregation of the elementary bodies.

Figs. 37, 38, and 39. Inclusions from another case (Sat.) of pure inclusion conjunctivitis.

Figs. 40, 41, and 42. Inclusions from a case (Bol.) of trachoma. The initial bodies as shown in Fig. 40 are exceptionally coarse.

Figs. 43, 44, and 45. Inclusions from an uncomplicated case (For.) of inclusion conjunctivitis. In Fig. 43 a thick mass of the initial bodies is seen to embrace the left side of the nucleus of a cell on the left, while on the right numerous elementary granules are scattered around the cell below. Figs. 44 and 45 show the steps of evolution of the inclusion bodies.

PLATE 43.

Figs. 46, 47, 48, 49, 50, and 51. Different stages of the evolution of the inclusion bodies and the conjunctiva of a baboon experimentally infected with the scrapings of the affected conjunctivæ of man. The structure of the initial bodies in Figs. 46, 47, and 48 is quite peculiar and reveals no definite bacillary outlines. They are deeply stained, almost amorphous, and show indistinct coarse granules within the mass. The elementary bodies are even more minute than those in the human cases, as will be seen in the cell occupying the lower half of Fig. 48. In the same figure one sees an irregularly shaped mass of the initial bodies at the upper corner. In Figs. 49, 50, and 51 are shown the enormously distended cells in which the inclusion granules are rapidly multiplying.

Fig. 52. This shows a dense mass of deeply stained granules in a film preparation from a case of mixed infection. This mass appears to be more of the von Prowazek type of body than of the degenerated Koch-Weeks bacilli.

Figs. 53, 54, and 55. Sections of the rabbits' testicles inoculated with pure cultures of Koch-Weeks bacilli isolated from the mixed infection cases during the present epidemic. Fig. 53 shows that the organisms assume a coccobacillary form when examined within twenty-four hours after they were injected into the testicle. In photography they appear somewhat like the initial bodies in the inclusion cases, yet in actual examination of the preparations they are not difficult to distinguish as the Koch-Weeks bacilli. They are freely capable of cultivation when a minute portion of the testicle is transferred to the blood agar, while this is not the case with the initial bodies of the inclusion organism. In Figs. 54 and 55 the sections removed after 3 and 6 days, respectively, are shown in which large masses of agglutinated and somewhat ill defined bacilli are found in clear spaces within the lumen of the testicular tubules. They, too, present an appearance not unlike that of the inclusion organism found in experimental conjunctivitis in monkeys, as well as in an epidemic in man. Here, again, a culture on blood agar is quickly differentiated from the other kind, as the Koch-Weeks bacilli at this stage still thrive very well in culture, while the pure inclusion material does not give any growth with the hemoglobinophilic bacilli.

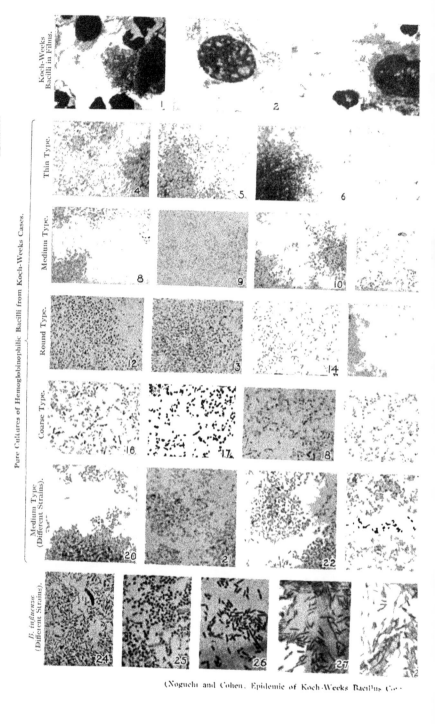

Koch-Weeks Bacilli in Films.

Pure Cultures of Hemoglobinophilic Bacilli from Koch-Weeks Cases.

Thin Type.

Medium Type.

Round Type.

Coarse Type.

Medium Type (Different Strains).

B. influenzae (Different Strains).

(Noguchi and Cohen. Epidemic of Koch-Weeks Bacillus Co..

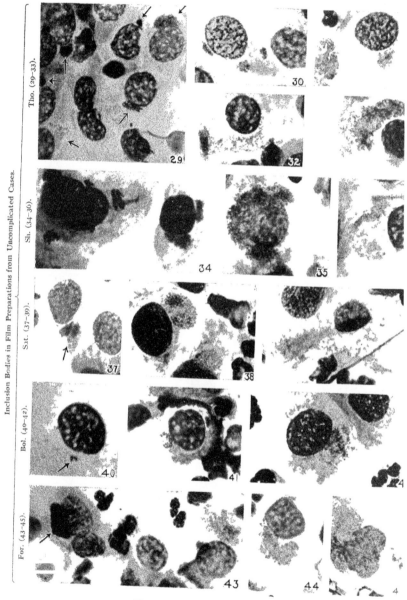

Inclusion Bodies in Film Preparations from Uncomplicated Cases.

Tho. (29–33).

Sh. (34–36).

Sat. (37–39).

Bol. (40–42).

For. (43–45).

THE JOURNAL OF EXPERIMENTAL MEDICINE VOL. XXII.

PLATE 43.

Experimental Inclusion Conjunctivitis in Baboon (46-51).

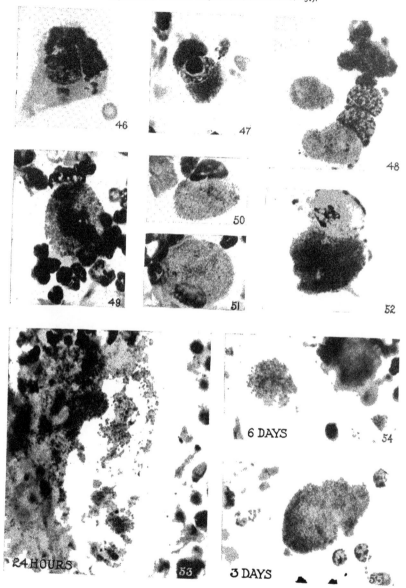

Experimental Koch-Weeks Orchitis in Rabbit (53-55).

(Noguchi and Cohen: Epidemic of Koch-Weeks Bacillus Conjunctivitis.)

STUDIES OF METABOLISM IN THE DOG BEFORE AND AFTER REMOVAL OF THE SPLEEN.

By SAMUEL GOLDSCHMIDT, Ph.D., and RICHARD M. PEARCE, M.D.

(*From the John Herr Musser Department of Research Medicine of the University of Pennsylvania, Philadelphia.*)

(Received for publication, June 4, 1915.)

In contradistinction to the uniformly negative results obtained by those who have studied the influence of the spleen on metabolism are the observations of Richet (1), which seem to indicate that in order to maintain the weight of a splenectomized dog a much larger amount of food is required than is the case with a normal dog. Richet's studies suggested to us a possible explanation of certain contradictory and confusing results which have been obtained in this laboratory during a lengthy study of the influence of diet upon the anemia which follows removal of the spleen. Unfortunately, however, Richet investigated only the weight of the animals in relation to the amount of food given and did not attempt a general study of metabolism. Moreover, a study of the literature of this subject shows that although a very complete metabolism study has been made by Paton (2), who decided that the spleen has no importance in metabolism, no one has made metabolic studies after splenectomy in animals in a state of nitrogen equilibrium. We therefore undertook the present investigation in the hope of throwing some light upon the dietary studies to which we have referred, and also, if possible, to clear up the apparent discrepancy between the results of Richet and of Paton.

Previous Investigations.

Paton's investigation included studies of the nitrogen metabolism and the elimination of salts in a single dog before and after splenectomy. Observations were made during fasting and on (1) meat, (2) oatmeal and milk, and (3) rich nuclein diets. The first postsplenectomy metabolism study was made 26 days and the last 4 months after the operation. Paton's general conclusion is that

319

under the various conditions of his experiments, splenectomy causes no essential difference in the course or nature of the metabolism.

In Richet's first investigation 9 splenectomized dogs were contrasted with 6 normal dogs. No metabolism studies and no examinations of the blood were made. Conclusions were based on records of food taken and the weight of the animals at various intervals. The increased consumption of food by the splenectomized animals is thought by Richet to be due to an increased catabolism in those animals and not to any disturbance of digestion. In a later report (1) he refers to studies of 17 splenectomized dogs, of which 5 were under observation for about 2 years, and confirms the conclusions of his earlier report. In this connection it is a matter of importance that the conclusions are based on the averages of 2 groups of dogs of widely different weights. Richet has not contrasted splenectomized dogs of given weight with normal dogs of the same weight, but if one selects from his tables dogs of the same weight the differences in food consumption are found to be very slight. Only 2 dogs were studied both before and after splenectomy.

Mendel and Jackson (3) who investigated the relation of the spleen to purine metabolism found that in splenectomized dogs and cats no changes occurred.

Verzár (4) has found that extirpation of the spleen in dogs has no appreciable effect upon the respiratory gas exchange. A similar conclusion was reached by Korenchevski (5) as regards both gaseous and nitrogenous metabolism. No other experimental studies of the influence of splenectomy are available except the brief note of Austin and Ringer (6), to the effect that in the dog the absence of the spleen does not in any way modify the course of the glycosuria caused by phlorhizin.

Metabolism studies in man, before and after removal of the spleen, are few in number, and, moreover, are, in all instances, difficult of interpretation because of the existence of pathological conditions, such as Banti's disease, pernicious anemia, or congenital hemolytic jaundice, on account of the presence of which the organ was removed. As the present study has for its object the determination of the influence of the removal of the normal spleen from a normal animal, the literature dealing with changes following its removal in disease (7) need not be considered at this time.

Methods.

The four dogs used in this study were placed upon a constant diet of beef (usually beef heart), lard, and sugar, the amounts of each of which constituents varied according to the caloric needs of each dog. The standard diet contained 0.4 of a gram of nitrogen per kilo and 70 calories per kilo of body weight. A small amount of sodium chloride was given each day, and a sufficient amount of bone ash was added to ensure well formed feces. The water intake for each day was constant. To some animals the beef heart was given raw; in other instances it was boiled. After one or two weeks on the special diet, if the weight of the animal remained constant, a preliminary metabolism study, covering a period of seven days, was made. If the results of this were satisfactory the animal was then splenectomized and the metabolism studies were resumed at various intervals after the operation. In each experiment the diet after operation was always the same as before and was continued without change in the intervals between periods of metabolism study. Analyses were made of all foods for fat and total nitrogen. During the periods of study the animals were kept in the usual metabolism cages. They were catheterized at the end of every twenty-four hours and the feces marked by carmine.

In the analysis of the urine the total nitrogen was determined by the Kjeldahl-Gunning method, ammonia by Folin's method (8), creatine and creatinine by Folin's method (9), and the hydrogen ion concentration according to Henderson's technique (10). In the study of the feces the Kjeldahl-Gunning method was used for total nitrogen, the Folin-Wentworth method for fat (11), and Neumann's method (12) for iron.

The removal of the spleen, an essentially bloodless operation, was done under ether anesthesia. These operations were performed by Dr. Max M. Peet, of the Department of Surgery.

RESULTS.

The details of our studies of nitrogen metabolism are shown in Tables I, II, III, and IV; of fat metabolism in Table V; and of the elimination of iron in Table VI.

Nitrogen Metabolism.—Table I shows the earliest period of metabolic study (three days) after splenectomy. The animal shows no loss of weight, no ill effect of the operation, and the conditions were therefore ideal for the detection of any slight changes in metabolism which might be due to the absence of the spleen. No variations in nitrogen partition were observed, however, and the nitrogen equilibrium was maintained: an average daily balance before operation of 0.45 of a gram and after operation of 0.46 of a gram.

TABLE I.

Dog 57. Nitrogen Metabolism before and Three Days after Splenectomy.

Date.	Weight.	N intake.*	Urine.									Feces total N.	N of urine and feces.	N balance.
			Amount.	Specific gravity.	Reaction to litmus.	H ion concentration.**	Total N.	Ammonia N.	Creatinine.	Creatine.				
	kilos	*gm.*	*cc.*	10				*gm.*	*gm.*	*gm.*	*gm.*	*gm.*	*gm.*	*gm.*
Feb. 28	10.6	4.80	295	29	Acid	6.90	3.45	0.13	0.289	0.478	0.73	4.18	+0.62
Mar. 1	10.8	4.80	175	34	"	6.50	3.76	0.15	0.270	0.357	0.73	4.09	+0.71
" 2	10.8	4.80	225	37	"	6.90	3.44	0.12	0.289	0.326	0.73	4.17	+0.63
" 3	10.8	4.80	225	38	"	6.90	3.57	0.13	0.285	0.321	0.73	4.30	+0.50
" 4	10.8	4.80	190	37	"	6.90	3.65	0.14	0.289	0.335	0.73	4.38	+0.42
" 5	10.9	4.80	200	40	"	6.90	3.86	0.14	0.289	0.335	0.73	4.59	+0.21
" 6	10.9	4.80	220	27	"	6.90	4.02	0.12	0.279	0.377	0.73	4.75	+0.05
Average..	10.8	4.80	219	35		6.84	3.62	0.13	0.284	0.361	0.73	4.35	+0.45
Mar. 7						Splenectomy.								
" 10	10.8	4.80	150	45	Acid	6.15	3.86	0.18	0.311	0.310	0.53	4.39	+0.41
" 11	10.8	4.80	220	35	"	6.50	3.92	0.15	0.311	0.310	0.53	4.45	+0.35
" 12	10.9	4.80	235	37	"	6.50	3.96	0.14	0.332	0.285	0.53	4.49	+0.31
" 13	10.9	4.80	180	40	"	6.50	4.08	0.13	0.302	0.320	0.53	4.61	+0.19
" 14	10.9	4.80	135	45	"	6.15	3.95	0.15	0.289	0.267	0.53	4.48	+0.32
" 15	10.9	4.80	165	37	"	6.15	3.36	0.13	0.289	0.259	0.53	3.89	+0.91
" 16	11.0	4.80	210	42	"	6.90	3.53	0.08	0.311	0.283	0.53	4.06	+0.74
Average..	10.9	4.80	185	40		6.41	3.81	0.16	0.306	0.291	0.53	4.34	+0.46

* Diet: raw beef, 150 gm.; lard, 50 gm.; sugar, 50 gm.
** Expressed as negative logarithms.

Table II shows practically the same results, 13 days and 8 weeks after splenectomy. The animal was in nitrogen equilibrium before splenectomy and maintained that condition after splenectomy. The general metabolism shows entirely normal results. The utilization of nitrogen was in no way interfered with; it was 94 per cent before operation and 95 and 93 per cent in the postsplenectomy periods.

In Table III, which presents observations 2, 6, and 10 weeks after splenectomy, the results in the third and fourth periods (sixth and tenth weeks) are similar to those shown in Tables I and II. In the early period after splenectomy, however, this animal showed a loss of appetite which caused, during the two weeks following operation, a loss in weight of 1.4 kilos. This loss of appetite was not due to infection or other postoperative disturbances, but appeared to be due rather to a dislike of the lard in the diet. When

TABLE II.

· *Dog 48. Nitrogen Metabolism before and Two and Eight Weeks after Splenectomy.*

Date.	Weight.	N intake.*	Amount.	Specific gravity.	Reaction to litmus.	H ion concentration.**	Total N.	Ammonia N.	Creatinine.	Creatine.	Feces total N.	N of urine and feces.	N balance.
	kilos	gm.	cc.				gm.	gm.	gm.	gm.	gm.	gm.	gm.
Nov. 16	13.4	5.60	175	56	Acid		6.70	5.35	0.38	0.300	0.358	0.34	5.69	−0.09
" 17	13.4	5.60	250	47	"		6.90	5.48	0.37	0.338	0.195	0.34	5.82	−0 22
" 18	13.4	5.60	225	50	"		6.70	4.91	0.38	0.352	0.480	0.34	5.25	+0.35
" 19	13.4	5.60	210	43	"		6.80	5.33	0.37	0.368	0.427	0.34	5.67	−0.07
" 20	13.4	5.60	200	44	"		6.90	4.80	0.34	0.361	0.467	0.34	5.14	+0.46
" 21	13.4	5.60	235	46	"		6.90	4.91	0.31	0.368	0.408	0.34	5.25	+0.35
" 22	13.3	5.60	265	37	"		6.90	4.77	0.27	0.385	0.597	0.34	5.11	+0.49
Average.	13.4	5.60	223	46			6.83	5.08	0.35	0.353	0.419	0.34	5.42	+0.18
Nov. 24						Splenectomy.								
Dec. 7	13.2	5.60	195	55	Acid		7.14	5.19	0.31	0.340	0.575	0.30	5.49	+0.11
" 8	13.1	5.60	280	35	"		6.00	5.36	0.34	0.368	0.513	0.30	5.66	−0.06
" 9	13.1	5.60	290	33	"		6.90	4.95	0.29	0.352	0.636	0.30	5.25	+0.35
" 10	13.1	5.60	265	39	"		6.80	4.65	0.29	0.368	0.513	0.30	4.96	+0.64
" 11	13.1	5.60	235	38	"		6.80	4.38	0.29	0.324	0.527	0.30	4.68	+0.92
" 12	13.1	5.60	295	36	"		6.80	4.40	0.27	0.329	0.558	0.30	4.70	+0.90
" 13	13.1	5.60	245	35	"		6.80	4.52	0.23	0.385	0.672	0.30	4.82	+0.78
Average..	13.1	5.60	256	39			6.75	4.78	0.29	0.352	0.570	0.30	5.08	+0.51
Jan. 18	13.6	5.70	275	30	Acid		6.80	4.46	0.22	0.355	0.398	0.40	4.86	+0.84
" 19	13.6	5.70	210	44	"		6.80	4.32	0.24	0.346	0.382	0.40	4 72	+0.98
" 20	13.6	5.70	310	20	"		6.70	4.58	0.22	0.364	0.361	0.40	4.98	+0.72
· " 21	13.5	5.70	300	39	"		6.80	4.92	0.21	0.355	0.474	0.40	5.32	+0.38
" 22	13.5	5.70	250	35	"		6.70	4.49	0.24	0.311	0.362	0.40	4.89	+0.81
" 23	13.4	5.70	300	39	"		6.90	5.12	0.23	0.337	0.463	0.40	5 52	+0.18
" 24	13.4	5.70	300	39	"		6.90	5.51	0.22	0.326	0.405	0 40	5.91	−0.21
Average..	13.5	5.70	278	35			6.80	4.77	0 22	0.342	0 406	0.40	5.47	+0.53

* Diet: raw beef heart, 200 gm.; lard, 60 gm.; sugar, 60 gm.
** Expressed as negative logarithms.

the lard was cut out of the diet the animal ate readily, and later when the lard was again added no trouble was experienced. As may be seen in Table V, this was the only animal which showed a high neutral fat content in the feces, though what relation there may be between this and the dislike of fat is not evident. The practical result of this loss of weight after splenectomy was a moderate retention of nitrogen in the first postsplenectomy metabolism period.

TABLE III.

Dog 52. Nitrogen Metabolism before and Two, Six, and Ten Weeks after Splenectomy.

Date.	Weight.	N intake.*	Urine. Amount.	Specific gravity.	Reaction to litmus.	H ion concentration.**	Total N.	Ammonia N.	Creatinine.	Creatine.	Feces total N.	N of urine and feces.	N balance.
	kilos	gm.	cc.	10			gm.	gm.	gm.	gm.	gm.	gm.	gm.
Nov. 25	10.8	4.77	170	44	Acid	6.50	4.17	0.30	0.368	None	0.38	4.58	+0.22
" 26	10.8	4.77	160	44	"	6.50	4.23	0.26	0.378	"	0.38	4.61	+0.16
" 27	10.8	4.77	225	33	"	6.50	4.49	0.23	0.368	"	0.38	4.87	−0.10
" 28	10.8	4.77	170	58	"	6.50	4.55	0.24	0.368	"	0.38	4.93	−0.16
" 29	10.8	4.77	180	46	"	6.50	4.55	0.25	0.368	"	0.38	4.93	−0.16
" 30	10.8	4.77	215	40	"	6.50	4.68	0.25	0.358	"	0.38	5.06	−0.29
Dec. 1	10.8	4.77	160	43	"	6.50	4.44	0.25	0.351	"	0.38	4.82	−0.05
Average..	10.8	4.77	183	44		6.50	4.44	0.25	0.365		0.38	4.82	−0.05
Dec. 2						Splenectomy.							
" 15	9.4	4.70	225	23	Acid	6.90	2.75	0.16	0.213	None	0.44	3.19	+1.51
" 16	9.4	4.70	180	39	"	6.50	3.73	0.25	0.324	"	0.44	4.17	+0.53
" 17	9.5	4.70	115	43	"	6.80	2.99	0.15	0.281	"	0.44	3.43	+1.27
" 18	9.5	4.70	210	35	"	6.80	2.79	0.15	0.289	"	0.44	3.23	+1.47
" 19	9.6	4.70	85	53	"	6.50	2.48	0.20	0.295	"	0.44	2.92	+1.78
" 20	9.7	4.70	275	20	"	6.70	2.79	0.25	0.291	"	0.44	3.23	+1.47
" 21	9.8	4.70	175	30	"	6.70	2.52	0.23	0.289	"	0.44	2.96	+1.74
" 22	9.8	4.70	225	25	"	6.80	2.88	0.25	0.311	"	0.44	3.32	+1.38
Average..	9.6	4.70	186	34		6.71	2.87	0.21	0.287		0.44	3.31	+1.39
Jan. 12	10.3	4.40	235	28	Acid	6.00	3.33	0.27	0.311	None	0.48	3.81	+0.59
" 13	10.3	4.40	190	35	"	6.30	3.38	0.24	0.311	"	0.48	3.86	+0.54
" 14	10.3	4.40	210	35	"	6.15	3.60	0.26	0.311	"	0.48	4.08	+0.32
" 15	10.4	4.40	180	42	"	6.30	3.45	0.22	0.324	"	0.48	3.93	+0.47
" 16	10.4	4.40	175	39	"	6.30	3.54	0.25	0.324	"	0.48	4.02	+0.38
" 17	10.4	4.40	225	35	"	6.70	3.38	0.25	0.311	"	0.48	3.86	+0.54
" 18	10.4	4.40	205	29	"	6.30	3.57	0.26	0.311	"	0.48	4.05	+0.35
Average..	10.35	4.40	203	35		5.29	3.46	0.25	0.318		0.48	3.94	+0.45
Feb. 10	10.8	4.10	160	35	Acid	6.90	2.96	0.23	0.225	None	0.44	3.40	+0.70
" 11	10.8	4.10	215	25	"	6.80	3.11	0.25	0.228	"	0.44	3.55	+0.55
" 12	10.8	4.10	220	22	"	6.80	3.23	0.20	0.234	"	0.44	3.67	+0.43
" 13	10.8	4.10	200	27	"	6.80	3.20	0.19	0.234	"	0.44	3.64	+0.46
" 14	10.9	4.10	190	32	"	6.80	3.44	0.18	0.231	"	0.44	3.88	+0.22
Average..	10.8	4.10	197	28		6.82	3.19	0.21	0.230		0.44	3.63	+0.47

* Diet: boiled beef heart, 100 gm.; lard, 50 gm.; sugar, 50 gm.
** Expressed as negative logarithms.

However, in the third period, when the animal had returned to exactly the same weight as before operation, nitrogen equilibrium

TABLE IV.

Dog 56. *Nitrogen Metabolism before and Ten Days and Three Months after Splenectomy.*

Date.	Weight.	N intake.*	Urine.								Feces total N.	N of urine and feces.	N balance.
			Amount.	Specific gravity.	Reaction to litmus.	H ion concentration.**	Total N.	Ammonia N.	Creatinine.	Creatine.			
	kilos	gm.	cc.	10			gm.	gm.	gm.	gm.	gm.	gm.	gm.
Feb. 16	8.5	3.43	100	39	Acid	6.30	2.63	0.13	0.221	0.014	0.35	2.98	+0.45
" 17	8.5	3.43	115	44	"	6.00	2.61	0.17	0.270	0.022	0.35	2.96	+0.47
" 18	8.5	3.43	140	38	"	6.15	2.39	0.14	0.279	0.022	0.35	2.74	+0.66
" 19	8.4	3.43	120	42	"	6.30	2.63	0.17	0.319	0.057	0.35	2.98	+0.45
" 20	8.4	3.43	120	47	"	6.30	2.73	0.16	0.261	0.116	0.35	3.08	+0.35
" 21	8.4	3.43	200	39	"	6.30	2.61	0.13	0.265	0.084	0.35	2.96	+0.47
" 22	8.4	3.43	90	50	"	6.15	2.60	0.14	0.253	0.067	0.35	2.95	+0.48
Average..	8.4	3.43	126	43		6.21	2.60	0.15	0.267	0.055	0.35	2.95	+0.48
Feb. 23			Splenectomy.										
Mar. 5	8.0	3.37	165	47	Acid	6.15	3.11	0.13	0.231	0.121	0.41	3.52	-0.15
" 6	8.0	3.37	110	46	"	6.15	3.00	0.14	0.213	0.143	0.41	3.41	-0.04
" 7	8.0	3.37	150	44	"	6.50	3.19	0.14	0.234	0.081	0.41	3.60	-0.04
" 8	8.0	3.37	210	40	"	6.15	3.30	0.16	0.221	0.119	0.41	3.71	-0.34
" 9	7.9	3.37	170	42	"	6.15	3.32	0.16	0.213	0.144	0.41	3.73	-0.36
" 10	7.9	3.37	130	49	"	5.85	3.18	0.16	0.213	0.114	0.41	3.59	-0.22
" 11	7.9	3.37	170	38	"	6.15	3.09	0.12	0.213	0.129	0.41	3.50	-0.13
Average..	8.0	3.37	152	44		6.14	3.17	0.14	0.220	0.122	0.41	3.72	-0.18
May 20	7.9	3.78	130		Acid		2.17		0.213	0.029	0.38	2.55	+1.23
" 21	7.9	3.78	145		"		2.07		0.213	0.014	0.38	2.45	+1.33
" 22	7.9	3.78	130		"		2.24		0.225	0.010	0.38	2 62	+1.16
" 23	7.9	3.78	260		"		2.56		0.238	0.017	0.38	2.94	+0.84
" 24	7.9	3.78	115		"		2.47		0.213	0.014	0.38	2.85	+0.93
Average..	7.9	3.78	156				2.30		0.220	0.017	0.38	2.68	+1.10

* Diet: boiled beef heart, 75 gm.; lard, 40 gm.; sugar, 40 gm.
** Expressed as negative logarithms.

was again maintained. It would seem conclusive, therefore, that the loss of weight and nitrogen retention of the earlier periods were due to an influence other than the absence of the spleen. It is of interest that this dog excreted no creatine.

In the experiments thus far presented there is no evidence that the absence of the spleen influences in any way nitrogen metabolism. In a fourth animal, however, the results were discordant.

This animal (Table IV) had served as a control for the blood

TABLE V.

Fat Determinations before and after Splenectomy.

Dog No.	Period.*	Total intake.	Total output.	Fat utilized.	Total output of fatty acids including soaps.	Fatty acids in total output.	Total output of neutral fats.	Neutral fat in total output.
		gm.	gm.	per cent	gm.	per cent	gm.	per cent
48	I (7 days)	460.6	26.54	94.2	22.45	84.4	4.09	15.6
	Nov. 24	Splenectomy						
	II (7 days)	460.6	20.92	95.5	18.11	86.5	2.81	13.5
	III (7 days)	460.6	15.77	96.6	11.51	73.0	4.26	27.0
52	I (7 days)	374.8	9.19	97.5	6.29	68.4	2.90	31.6
	Dec. 2	Splenectomy						
	II (8 days)	428.32	12.64	97.0	5.93	46.9	6.71	53.1
	III (7 days)	374.8	17.14	95.4	8.44	49.2	8.70	50.8
	IV (5 days)	267.7	7.85	97.1	6.71	85.5	1.14	14.5
56	I (7 days)	298.4	13.59	95.4	9.27	68.3	4.32	31.7
	Feb. 23	Splenectomy						
	II (7 days)	298.4	14.25	95.2	11.11	78.0	3.14	22.0
57	I (7 days)	380.45	23.87	93.7	19.42	81.4	5.04	18.6
	Mar. 7	Splenectomy						
	II (7 days)	380.45	10.56	97.2	7.68	72.7	2.88	27.3

*These periods correspond exactly to those in Tables I, II, III, and IV.

counts of the three animals discussed above, and up to the time of our foreperiod had been for twelve weeks on an adequate constant diet, as was the case in the other animals. Like Dog 52 this animal received boiled meat as a part of the dietary. The effect of splenectomy on the nitrogen metabolism, ten days after the operation, was very slight, but a nitrogen equilibrium of + 0.47 of a gram per day was changed to one of — 0.18 of a gram, figures not beyond the range of normal variations, but which in the light of changes to be discussed later in the paper are suggestive of the influence of anemia. At a later period, three months after splenectomy, the animal had not regained the slight loss (0.5 of a kilo) in weight, but it appeared to be in excellent condition and the anemia which had existed for several months was improving. The plus balance of 1.1 grams of nitrogen per day (upon a slightly higher nitrogen intake) in this period without change in weight suggests the possibility of the utilization of this nitrogen for the repair of the anemia. Utilization of protein was not disturbed, being 90 per cent in Period I, 88 per cent in Period II, and 90 per cent in Period III.

Unlike the other three dogs, we had here in Period II an increase of creatine amounting to 45 per cent. This increase was at the expense of the creatinine, however, for the total creatinine, including preformed creatinine and creatine as creatinine, agrees very closely, amounting to 0.314 of a gram in the foreperiod and 0.325 of a gram in the afterperiod.

During Period III, while the average creatinine output was exactly the same as during Period II, the creatine output fell to a figure lower than either of the preceding periods. The variation in the partition of creatine and creatinine in this animal we are unable to explain.

Fat Utilization.—The utilization of fat (Table V) in all the animals was normal in all periods. The partition of fatty acids (including soaps) and neutral fats shows some variation, especially in Dog 52, but to this we are inclined to ascribe no importance. A thorough search of the literature shows that no studies of fat utilization in animals before and after splenectomy have previously been made.

Iron Elimination.—The elimination of iron in the feces (Table VI) showed no important change after splenectomy in three of the four animals studied. In the fourth (Dog 56) there was an increase of 1.6 mg. per day during the period immediately after splenectomy, amounting to an increase of 21.6 per cent over the foreperiod. In the final period three months after splenectomy the output showed an increase of 148 per cent over the foreperiod and double that of Period II. The intake of iron was not determined, but since the food intake was constant throughout all the periods we have reason to believe that this was a constant factor. The periods of Table VI correspond exactly to those in previous tables. The figures for iron are based on the analyses of samples of feces representing the total feces of the respective periods.

TABLE VI.

Iron Elimination before and after Splenectomy.

Dog No.	Before splenectomy.	After splenectomy.		
	Period I.	Period II.	Period III.	Period IV.
48	17.6*	17.9	16.8	
52	10.5	10.4	9.8	9.9
56	7.4	9.0	18.4	
57	10.9	10.8		

* The figures in this table represent mg. of iron per day in feces.

These studies of iron metabolism are in accord with those of Austin and Pearce (13) reported from this laboratory a year ago, but are opposed to those of Asher and his associates (14). The latter claim that after splenectomy the output of iron in the dog is much higher, often double that of the normal dog. Austin and Pearce, using the Ripper-Schwarzer method, found a slightly increased elimination in three of five dogs within two weeks after splenectomy, but no change in two others of this period or in three dogs studied 1, 9, and 20 months after splenectomy. The results now presented, by Neumann's method, appear to support the conclusion of Austin and Pearce that the spleen does not exert a constant and important influence upon iron metabolism.

DISCUSSION.

These observations show that in three of four animals the removal of the spleen had no effect upon nitrogen metabolism, the utilization of fat, or the elimination of iron, and justify the conclusion that the removal of the normal spleen in a normal animal has no important effect upon general metabolism. It is necessary, however, in order that there may be no question about this conclusion to explain the discordant results in the fourth animal (Dog 56). This animal showed a loss of weight, an increased elimination of iron, and a disturbance of creatine metabolism. The fat metabolism was unaltered. The question arises whether these changes are due to the absence of the spleen, or to the anemia which was present.

In Table VII are presented the blood examinations of each dog at the time of the several metabolism periods. It will be seen that Dogs 48 and 57 showed no appreciable change in the blood picture after splenectomy, but that Dogs 52 and 56 did. The blood changes in Dog 52, however, were relatively slight. The situation in regard

TABLE VII.

*Blood Examinations.**

Dog No.	Period.**	Hemoglobin.	Red cell count
		per cent	
48	I	100	5,900,000
	II	104	5,570,000
	III	100	5,540,000
52	I	105	6,700,000
	II	106	6,840,000
	III	84	5,360,000
	IV	90	5,100,000
56	Initial	105	6,450,000
	I	83	6,020,000
	II	70	5,890,000
	III	72	4,950,000
57	I	95	6,130,000
	II	90	6,130,000

* These examinations were made by Dr. J. H. Austin.
** The periods correspond exactly to those of the previous tables.

to Dog 56 was somewhat different. This animal had been placed on a constant diet, the chief article of which was boiled beef heart, twelve weeks before the first metabolism study. At that time the blood examination showed hemoglobin 105 per cent and red cells 6,450,000. At the time of our presplenectomy period it showed a relatively low hemoglobin content (83 per cent) and 6,020,000 red cells; and after splenectomy the hemoglobin continued to fall, and later examinations, not corresponding to a metabolism period. showed the low level, two and a half months after splenectomy, of hemoglobin 60 per cent, and red cells 4,560,000. It is evident, therefore, that this animal differed from the other three in that it developed an anemia more rapidly and eventually of a more severe grade than was the case in any other animal of this series. As has been shown in previous publications (15) from this laboratory.

anemia of varying severity is a fairly constant result of splenectomy in the dog. The anemia may be slight as in Dog 52, or more severe as in Dog 56, and may, as has been suggested, be prevented by diet. The influence of diet is one of the problems now under investigation in this laboratory, but is not a matter which concerns us at the present time. The essential fact is that Dog 56 developed a severe anemia, already progressive at the time of the first metabolism study, while Dogs 48 and 57 were not anemic, and Dog 52 showed only a slight non-progressive deterioration of the blood. The question naturally arises: Are the increased elimination of iron and the disturbance of the creatine metabolism due to the anemia and not to an influence on metabolism consequent upon the absence of the spleen?

A few words are necessary concerning Richet's statement that the splenectomized dog requires more food to maintain its weight than does the normal dog. In view of our results, Richet's conclusion is not tenable. Dog 57 (Table I) maintained its presplenectomy weight without change in diet and with only a slight change in the nitrogen balance. Dog 48 (Table II) likewise showed only a trifling change during the three weeks after operation and a return to the previous weight after seven to eight weeks. The serious loss of weight in Dog 52 was due to loss of appetite, and that, relatively slight, in Dog 56 was complicated by the coexisting anemia.

Moreover, during the past five years we have frequently noticed a tendency for splenectomized dogs to become obese, and this tendency is mentioned also by several investigators who had studied splenectomized animals for long periods of time. This tendency to put on weight is strikingly shown by two of the dogs (48 and 52) of this series. At the close of the metabolism work presented in Tables II and III these animals were kept for some time on account of the possible necessity of repeating the metabolism studies after longer periods had elapsed. The change from a special diet to the ordinary kennel diet ("table scraps") led to a rapid increase of weight in each instance; in three months the weight of Dog 48 increased from 13.4 to 15.8 kilos, while in two months Dog 52 rose from 10.9 to 12.9 kilos.

Our results are therefore in accord with those of Paton rather than with those of Richet, and demonstrate that in the absence of anemia the removal of the spleen has no influence upon nitrogen or fat metabolism, and in all probability no influence upon iron elimination.

SUMMARY.

Four dogs have been subjected to metabolism studies before splenectomy and at intervals of three days to three months after splenectomy. In three of the four animals the removal of the spleen was not followed by any disturbances of nitrogen metabolism, fat utilization, or iron elimination. Two of these animals showed no anemia, and the third only a slight reduction in hemoglobin and number of red cells.

A fourth animal, studied ten days and three months after splenectomy, developed eventually a definitely progressive anemia of moderate severity. This animal showed a slight loss of weight, a slight disturbance of nitrogen balance, and of creatine-creatinine partition, with a marked increase in the elimination of iron.

We conclude therefore that the spleen has no important influence on metabolism, and that the disturbances occurring in one of our dogs were due to the coexisting anemia and not to the absence of the spleen.

BIBLIOGRAPHY.

1. Richet, C., Des effets de l'ablation de la rate sur la nutrition chez les chiens, *Jour. de physiol. et de path. gén.*, 1912, xiv, 689; 1913, xv, 579.
2. Paton, D. N., Studies of the Metabolism in the Dog before and after Removal of the Spleen, *Jour. Physiol.*, 1899–1900, xxv, 443.
3. Mendel, L. B., and Jackson, H. C., On Uric Acid Formation after Splenectomy, *Am. Jour. Physiol.*, 1900–01, iv, 163.
4. Verzár, F., Die Grösse der Milzarbeit, *Biochem. Ztschr.*, 1913, liii, 69.
5. Korenchevski, V. G., Nitrogenous and Gaseous Metabolism in Spleenless Animals, *Russk. Vrach*, 1910, ix, 1441.
6. Austin, J. H., and Ringer, A. I., The Influence of Phlorhizin on a Splenectomized Dog, *Jour. Biol. Chem.*, 1913, xiv, 139.
7. Goldschmidt, S., Pepper, O. H. P., and Pearce, R. M., Metabolism Studies before and after Splenectomy in Congenital Hemolytic Icterus, *Arch. Int. Med.*, 1915, xvi (in press).
8. Folin, O., Eine neue Methode zur Bestimmung des Ammoniaks im Harne, und anderen thierischen Flüssigkeiten, *Ztschr. f. physiol. Chem.*, 1902–03, xxxvii, 161.

9. Folin, O., Approximately Complete Analyses of Thirty "Normal" Urines, *Am. Jour. Physiol.*, 1905, xiii, 45.

10. Henderson, L. J., and Palmer, W. W., On the Intensity of Urinary Acidity in Normal and Pathological Conditions, *Jour. Biol. Chem.*, 1912–13, xiii, 393.

11. Folin, O., and Wentworth, A. H., A New Method for the Determination of Fat and Fatty Acids in Feces, *Jour. Biol. Chem.*, 1909–10, vii, 421.

12. Neumann, A., Einfache Veraschungsmethode (Säuregemisch-Veraschung) und vereinfachte Bestimmungen von Eisen, Phosphorsäure, Salzsäure, und anderen Aschenbestandtheilen unter Benutzung dieser Säuregemischveraschung, *Ztschr. f. physiol. Chem.*, 1902–03, xxxvii, 115; 1904–05, xliii, 32.

13. Austin, J. H., and Pearce, R. M., The Relation of the Spleen to Blood Destruction and Regeneration and to Hemolytic Jaundice. XI. The Influence of the Spleen on Iron Metabolism, *Jour. Exper. Med.*, 1914, xx, 122.

14. Asher, L., and Grossenbacher, H., Beiträge zur Physiologie der Drüsen. 11te. Mitteilung. Untersuchungen über die Funktion der Milz, *Biochem. Ztschr.*, 1909, xvii, 78. Asher, L., and Zimmermann, R., Beiträge zur Physiologie der Drüsen. 12te. Mitteilung. Fortgesetzte Beiträge zur Funktion der Milz als Organ des Eisenstoffwechsels, *ibid.*, p. 297.

15. Musser, J. H., Jr., An Experimental Study of the Changes in the Blood Following Splenectomy, *Arch. Int. Med.*, 1912, ix, 592. Musser, J. H., Jr., and Krumbhaar, E. B., The Relation of the Spleen to Blood Destruction and Regeneration and to Hemolytic Jaundice. VI. The Blood Picture at Various Periods after Splenectomy, *Jour. Exper. Med.*, 1913, xviii, 487.

STUDIES IN LIVER AND KIDNEY FUNCTION IN EXPERIMENTAL PHOSPHORUS AND CHLOROFORM POISONING.

By E. K. MARSHALL, Jr., Ph.D., and L. G. ROWNTREE, M.D.

(*From the Pharmacological Laboratory of Johns Hopkins University, and the Medical Clinic of Johns Hopkins Hospital, Baltimore.*)

(Received for publication, May 25, 1915.)

INTRODUCTION.

In the course of a somewhat extended study of the value of various tests of liver function in clinical diseases of the liver,[1] it seemed advisable to determine whether a rough approximate standardization of the functional changes could be obtained through comparison of the findings in clinical conditions with graded experimental liver injuries in animals.

Chloroform and phosphorus poisoning have probably been most used in producing experimental liver lesions. The non-specificity of the injury to the liver is at once apparent in the consideration of this method, for it is well known that pathological changes in the kidneys, heart, suprarenals, and voluntary muscles occur as the results of these poisons. The general opinion that the liver lesions are the predominating pathological features and that they are responsible for certain metabolic changes led us to investigate these conditions. However, it may be stated at once that we feel that deductions concerning liver function drawn from such studies are not free from criticism, and may be applied to findings in clinical material only with reserve.

An acceleration of protein metabolism, as evidenced by increased nitrogen output in the urine, has been shown to occur in phosphorus poisoning by Bauer.[2]

[1] Rowntree, L. G., Marshall, E. K., Jr., and Chesney, A. M., *Tr. Assn. Am Phys.*, 1914, xxix, 586. Chesney, Marshall, and Rowntree. *Jour. Am. Med. Assn.* 1914, lxiii, 1533.

[2] Bauer, J., *Ztschr. f. Biol.*, 1871, vii, 63; 1878, xiv, 527.

and in chloroform poisoning by Strassmann.[3] This has been generally confirmed by all subsequent workers.[4] It has been attributed to a decrease in the oxidation processes in the body and to autolysis of tissues.[5] However, Lusk[6] has shown that there is no reduction in the amount of total metabolism in phosphorus poisoning, but there may be rather an increase due to fever and the increased protein metabolism.

In general, the changes occurring in the urine and blood in phosphorus and in chloroform poisoning are similar to those which seem to be characteristic of certain types of liver disease in man.[7] Vidal[8] showed a slight decrease in the urea, and increase in the uric acid, creatinine, and undetermined nitrogen percentages of the total in experimental chloroform poisoning. Paton[9] distinguished between the effects of its inhalation and of its administration by the stomach. In the former he found a normal or increased urea and normal or decreased ammonia percentage; in the latter, a decided decrease in the urea and increase in the ammonia, uric acid, and undetermined nitrogen percentages of the total. He did not think the discrepancy due to size of the dose. Given hypodermically, the same effect, though less pronounced, was noted as when administered by mouth. Howland and Richards,[10] studying delayed chloroform poisoning (anesthesia) in dogs, noted slightly decreased urea and increased ammonia and undetermined nitrogen percentages of the total. Lindsay[11] confirmed these results and found the allantoin, creatine, amino-acids, and polypeptides increased.

The excretion of amino-acids in the urine in phosphorus poisoning has been shown by Abderhalden and Bergell[12] and Wohlgemuth.[13] Fischler and Bardach[14] found in dogs, with and without an Eck fistula, no difference in the nitrogen partition in phosphorus poisoning, low urea and high ammonia percentages, and increase in formol-titratable nitrogen in both. Ishihara[15] has recently studied

[3] Strassmann, F., *Virchows Arch. f. path. Anat.*, 1889, cxv, 1.

[4] Taniguti, K., *Virchows Arch. f. path. Anat.*, 1890, cxx, 121. Savelieff, N., *Virchows Arch. f. path. Anat.*, 1894, cxxxvi, 195. Vidal, E., Influence de l'anesthésie chloroformique sur les phenomènes chimiques de l'organisme, Thèse de Paris, 1897. Paton, D. N., *Proc. Roy. Soc. Edinburgh*, 1908, xxviii, 472. Howland, J., and Richards, A. N., *Jour. Exper. Med.*, 1909, xi, 344. Lindsay, D. E., *Biochem. Jour.*, 1910–11, v, 407.

[5] Jacoby, M., *Ztschr. f. physiol. Chem.*, 1900, xxx, 149. Waldvogel, *Arch. f. klin. Med.*, 1905, lxxxii, 437. Wells, H. G., *Jour. Biol. Chem.*, 1908–09, v, 129.

[6] Lusk, G., *Am. Jour. Physiol.*, 1907, xix, 461. Mandel, A. R., and Lusk, G., *Am. Jour. Physiol.*, 1906, xvi, 129.

[7] For a full discussion of the functional changes occurring in liver diseases, see Rowntree, Marshall, and Chesney, *loc. cit.* Here also will be found a full discussion of the methods employed in our liver functional studies.

[8] Vidal, *loc. cit.*

[9] Paton, *loc. cit.*

[10] Howland and Richards, *loc. cit.*

[11] Lindsay, *loc. cit.*

[12] Abderhalden, E., and Bergell, P., *Ztschr. f. physiol. Chem.*, 1903, xxxix, 464.

[13] Wohlgemuth, J., *Ztschr. f. physiol. Chem.*, 1905, xliv, 74.

[14] Fischler, F., and Bardach, K., *Ztschr. f. physiol. Chem.*, 1912, lxxviii, 435.

[15] Ishihara, H., *Biochem. Ztschr.*, 1912, xli, 315.

subchronic phosphorus poisoning in dogs; and could demonstrate no noticeable changes in the distribution of the nitrogen between the creatine, creatinine, amino-acids, and ammonia.

The most noticeable changes which have been shown in the blood are the marked decrease in fibrinogen and increase in the lipolytic activity. Corin and Ansiaux[16] and Jacoby[17] found the blood incoagulable in phosphorus poisoning, while this decrease in fibrinogen was shown by Doyon and Billet[18] for chloroform poisoning. Whipple[19] and his coworkers have shown the increase in the lipolytic activity of the blood in these conditions, and have further made an extensive study on lipase and fibrinogen in experimental liver injuries (chloroform, phosphorus, hydrazine poisoning, etc.), and in certain clinical diseases of the liver. Ragazzi[20] found that the viscosity and electrical conductivity of the blood were increased in phosphorus poisoning. Frank and Isaac[21] found the blood sugar to be decreased to zero shortly before death from phosphorus poisoning. Opie, Barker, and Dochez[22] showed an increase in the proteolytic enzymes of the blood after chloroform and phosphorus poisonings.

Simultaneous clinical studies by Rowntree, Hurwitz, and Bloomfield,[23] and experimental studies by Whipple, Peightal, and Clark[24] were carried on to determine whether the excretion of phenoltetrachlorphthalein by the bowels in health and disease afforded information relative to liver function. Whipple's studies revealed the facts that in chloroform and phosphorus poisonings marked decrease in the phthalein output resulted, together with the appearance of the drug in the urine, and further that, associated with decrease in the phthalein output, decrease in fibrinogen and increase in blood lipase occurred. In the recent study of lactose tolerance in chloroform and phosphorus poisonings in dogs. Bloomfield and Hurwitz[25] have reviewed the literature on sugar tolerance.

Methods.

Before inducing the chloroform or phosphorus poisoning. each dog was subjected to the series of tests to establish its normals. No

[16] Corin, G., and Ansiaux, G., *Jahresb. f. Thierchem.,* 1894. xxiv. 642.

[17] Jacoby, *loc. cit.*

[18] Doyon. M., and Billet, J., *Compt. rend. Soc. de biol.,* 1905. lvii. 852.

[19] Whipple, G. H., Mason, V. R., and Peightal. T. C., *Bull. Johns Hopkins Hosp.,* 1913, xxiv, 207. Whipple, G. H., and Hurwitz, S. H., *Jour. Exper. Med.,* 1911, xliii, 136. Whipple, G. H., *Arch. Int. Med.,* 1912, ix, 365.

[20] Ragazzi, C., *Jahresb. f. Thierchem.,* 1910, xl. 1276.

[21] Frank, E., and Isaac, S., *Arch. f. exper. Path. u. Pharmakol.,* 1911, lxiv. 274.

[22] Opie, E. L., Barker, B. I., and Dochez. A. R.. *Jour. Exper. Med.,* 1911. xiii, 162.

[23] Rowntree, L. G., Hurwitz. S. H., and Bloomfield. A. L., *Bull. Johns Hopkins Hosp.,* 1913, xxiv, 327.

[24] Whipple, G. H., Peightal. T. C., and Clark. A. H., *Bull. Johns Hopkins Hosp.,* 1913, xxiv, 343.

[25] Bloomfield, A. L., and Hurwitz, S. H., *Bull. Johns Hopkins Hosp.,* 1913. xxiv, 375.

attempt was made to maintain nitrogenous equilibrium except in two animals, although in the others the diet was kept fairly constant throughout the period of observation.[26] The chloroform was given by inhalation, and the phosphorus subcutaneously, in oil. The phenoltetrachlorphthalein test was employed, according to the technique of Rowntree, Hurwitz, and Bloomfield.[27] The fibrinogen determinations were made by the heat coagulation method of Whipple,[28] and the lipolytic activity of the blood plasma by Loevenhart's[29] method. The total non-protein nitrogen, urea nitrogen, and amino-acid nitrogen of the blood serum or whole blood were estimated by the micro-Kjeldahl method of Folin and Denis,[30] the urease method of Marshall,[31] and the nitrous acid method of Van Slyke,[32] respectively. In the urine, total nitrogen was determined by the usual Kjeldahl, the urea by Marshall's,[33] the ammonia by Folin's,[34] and the free amino-acids by Van Slyke's[35] method. The tolerance towards levulose and galactose was determined in the usual way. The hydrogen ion concentration of the blood was determined by the method recently described by Levy, Marriott, and Rowntree.[36]

Liver Function in Phosphorus Poisoning.

In the following protocols of the various animals, the fibrinogen, total non-protein nitrogen, urea nitrogen, and amino nitrogen are expressed in mg. per 100 cc. of blood serum. The urine quantity, unless otherwise stated, is for 24 hours, and the urea nitrogen, ammonia nitrogen, and free amino nitrogen are expressed as percentages of the total nitrogen of the urine. The lipolytic activity of the blood

[26] In the more severe stages of poisoning, the animals refused food for several days.

[27] Rowntree, Hurwitz, and Bloomfield, *loc. cit.*

[28] Whipple, G. H., *Am. Jour. Physiol.*, 1914, xxxiii, 50.

[29] Loevenhart, A. S., *Am. Jour. Physiol.*, 1901–02, vi, 331.

[30] Folin, O., and Denis, W., *Jour. Biol. Chem.*, 1911–12, xi, 527.

[31] Marshall, E. K., Jr., *Jour. Biol. Chem.*, 1913, xv, 487.

[32] Van Slyke, D. D., and Meyer, G. M., *Jour. Biol. Chem.*, 1912, xii, 399.

[33] Marshall, E. K., Jr., *Jour. Biol. Chem.*, 1913, xiv, 283; xv, 495.

[34] Folin, O., *Ztschr. f. physiol. Chem.*, 1902–03, xxxvii, 161.

[35] Van Slyke, D. D., *Jour. Biol. Chem.*, 1913, xvi, 125.

[36] Levy, R. L., Marriott, W. M., and Rowntree, L. G., *Arch. Int. Med.*, 1915, xvi (in press).

plasma is expressed in the number of cc. of N/10 acid produced in 24 hours at 38° C. by a mixture of 1 cc. of plasma, 4 cc. of water, and 0.26 cc. of ethyl butyrate.

Dog I.—Weight 12 kilos. *Normals:* tolerates 0.3 gm. galactose and 1.5 gm. levulose per kilo. Phthalein 32 per cent. *Blood:* fibrinogen, 490 mg.; total non-protein N, 44 mg.; urea N, 20 mg. (46 per cent); amino N, 3.7 mg. *Urine:* 175 cc. containing 4.4 gm. nitrogen; urea N, 83 per cent; ammonia N, 7.4 per cent; amino N, 0.45 per cent.

Mar. 6, 1914. Given 10 mg. phosphorus in oil subcutaneously.

Mar. 8. Given 7.5 mg. phosphorus. 0.2 gm. per kilo of galactose not tolerated. *Urine:* 450 cc. containing 10.3 gm. nitrogen; urea N, 84 per cent; ammonia N, 5.4 per cent; amino N, 0.3 per cent.

Mar. 9. Bled and phthalein injected at noon; died at 3 p. m. *Blood:* lipase, 1.55; fibrinogen, 13 mg.; total non-protein N, 91 mg.; urea N, 33 mg. (36 per cent); amino N, 13.4 mg.

Autopsy.—*Liver:* light brownish yellow in appearance, very friable, and soft. On section the parenchyma has a mottled appearance, areas of depression alternating with those which project. *Lungs:* pale and edematous.

Microscopic Examination.[37]—*Liver:* diffuse peripheral hemorrhagic necrosis. Necrotic cells are found scattered through all parts of the liver lobule. There are no normal liver cells seen. Those remaining alive show extreme parenchymatous and fatty degeneration.

Dog II.—Weight 23 kilos. *Normals:* tolerates 0.25 gm. but not 0.30 gm. galactose and 1.5 gm. levulose per kilo. Phthalein 31 per cent. *Blood:* lipase, 0.25; fibrinogen, 243 mg.; total non-protein N, 51 mg.; urea N, 20 mg. (39 per cent); amino N, 3.8 mg. *Urine:* 390 cc. containing 12 0 gm. nitrogen; urea N, 85 per cent; ammonia N, 5.3 per cent; amino N, 1.5 per cent.

Mar. 21, 1914. Given 13 mg. phosphorus.

Mar. 23. Given 5 mg. phosphorus.

Mar. 24. Phthalein 23 per cent positive in urine.

Mar. 26. Given 5 mg. phosphorus. Tolerates 0.20 gm. galactose per kilo. *Urine:* 600 cc. containing 12.1 gm. nitrogen; urea N, 84 per cent; ammonia N, 5.6 per cent; amino N, 0.99 per cent. Bile in urine, but no jaundice; dog seems sick.

Mar. 27. Given 3 mg. phosphorus.

Mar. 28. Tolerates 1.2 gm. levulose per kilo; phthalein 0 per cent; positive in urine. *Blood:* lipase, 0 60; fibrinogen, 10 mg.; total non-protein N, 110 mg.; urea N, 61 mg. (55 per cent); amino N, 11.3 mg. *Urine:* 1.200 cc. containing 18.2 gm. nitrogen; urea N, 77 per cent; ammonia N, 6.0 per cent; amino N, 4.6 per cent.

Mar. 29. Dog very sick; comatose. Died. *Blood:* (just after death) lipase, 1.30; total non-protein N, 225 mg.; urea N, 179 mg. (79 per cent); amino N, 21 mg. *Bladder urine:* 100 cc. containing 1.6 gm. nitrogen; urea N, 82 per cent; ammonia N, 4.7 per cent; amino N, 2.1 per cent.

Dog III.—Weight 22 kilos. *Normals:* tolerates 0.30 gm. galactose and 1.5

---

[37] For the microscopic reports we are indebted to Dr. G. H. Whipple.

gm. levulose per kilo. Phthalein 30 per cent. *Blood:* lipase, 0.35; fibrinogen, 330 mg.; total non-protein N, 28 mg.; urea N, 8 mg. (29 per cent); amino N, 3.4 mg. *Urine:* 533 cc. containing 10.6 gm. nitrogen; urea N, 81 per cent; ammonia N, 6.7 per cent; amino N, 1.4 per cent.

Mar. 18, 1914. Given 10 mg. phosphorus.

Mar. 19. Tolerates 0.20 gm. galactose per kilo.

Mar. 20. Given 8 mg. phosphorus.

Mar. 21. Tolerates 1.25 gm. levulose per kilo.

Mar. 23. Phthalein 6 per cent; lipase, 0.85; fibrinogen, 200 mg.; total non-protein N, 50 mg.; urea N, 22 mg. (44 per cent); amino N, 3.4 mg.

Mar. 26. *Urine:* 410 cc. containing 7.4 gm. nitrogen; urea N, 83 per cent; ammonia N, 7.4 per cent; amino N, 1.1 per cent.

Mar. 27. Playful. Given 7.5 mg. phosphorus.

Mar. 28. Playful. Does not tolerate 1.2 gm. levulose per kilo. Phthalein 17 per cent positive in urine.

Apr. 2. Does not tolerate 0.20 gm. galactose per kilo.

Apr. 3. Does not tolerate 1.0 gm. levulose per kilo. Given 8 mg. phosphorus.

Apr. 4. Phenolsulphonephthalein 50 per cent for 1¼ hours.

Apr. 5. Does not tolerate 0.18 gm. galactose per kilo. Phthalein 26 per cent. Apparently has been in splendid condition for past week.

Apr. 6. Given in a. m. 11 mg. phosphorus and in p. m. 8 mg. more. *Blood:* lipase, 0.75; fibrinogen, 400 mg.; total non-protein N, 42 mg.; urea N, 18 mg. (43 per cent); amino N, 2.0 mg.

Apr. 7. Vomited. Looks depressed. Wants to lie down. Phenolsulphonephthalein 44 per cent in 1½ hours. *Urine:* 530 cc. containing 8.5 gm. nitrogen; urea N, 82 per cent; ammonia N, 3.2 per cent; amino N, 1.7 per cent. Tolerates 0.8 gm. levulose per kilo.

Apr. 8. Found dead. Weight 17.4 kilos. Has not eaten anything for 2 to 3 days. *Blood:* does not coagulate; total non-protein N, 84 mg.; urea N, 28 mg. (33 per cent); amino N, 11.3 mg.

Autopsy.—Kidney: diffuse epithelial necrosis, involving particularly the convoluted tubules.

The tests on the three dogs suffering from phosphorus poisoning (I, II, and III) showed the following definite agreement in all: The phthalein excretion in the feces was markedly diminished and the dye appeared in the urine; the lipolytic activity of the blood was greatly increased; the fibrinogen sank to a very low level; the total non-protein nitrogen, urea nitrogen, and amino nitrogen of the blood were all more or less increased. The urines showed no conclusive changes in all, while the galactose and levulose tolerance was rather definitely decreased. The urea nitrogen percentage of the total non-protein nitrogen of the blood showed an increase in one case and decrease in the other two. However, this is not at all striking, and no conclusion concerning it can be drawn. The high total non-protein nitrogen and urea nitrogen of the blood in Dog II are worthy of note. The highest blood amino nitrogen (21 mg. to 100 cc.) which we have encountered was seen in this case, in which also the partition of the urinary nitrogen was most markedly disturbed. However, with the possible exception of Dog III (March 23), this was the only time when a severe stage of poisoning (liver injury) as indicated by other tests and the general condition of the animals was coincident with a urine analysis. In Dog II we see the gradual drop of the phthalein to zero, and the liver injury, as shown by the test, was probably in this case the most severe. Dog III apparently had a remarkable resistance to phosphorus. The liver repair indicated on April 5 to 6 is worthy of note.

TABLE I.

Dog IV.

Date (1914).	Phosphorus.	Urine.					Total N.			Blood.			Weight.
		Amount.	Total N.	Urea N.	Ammonia N.	Amino N.	Urea N.	Ammonia N.	Amino N.	Total non-protein N.	Urea N.	Amino N.	
										Per 100 cc.			
	mg.	*cc.*	*gm.*	*gm.*	*gm.*	*gm.*	*per cent*	*per cent*	*per cent*	*mg.*	*mg.*	*mg.*	*kilos*
Nov. 30		112	4.33	3.53	0.18		82	4.2					8.40
Dec. 2 }													
" 3 }	20	205	7.53	6.33	0.41	0.044	84	5.5	0.6	28	12	3.7	8.10
" 4	10	154	3.23	2.52	0.11	0.031	78	3.4	1.0				8.20
" 5		200	5.78	4.91	0.33	0.092	85	5.7	1.6				
" 6	20	218	8.74	6.21	0.26	0.055	72	3.0	0.7				
" 7										88	66	7.7	7.45

Dog V.

Date (1914).	Phosphorus.	Amount.	Total N.	Urea N.	Ammonia N.	Amino N.	Urea N.	Ammonia N.	Amino N.	Total non-protein N.	Urea N.	Amino N.	Weight.
Dec. 11		237	4.19	3.60	0.09		86	2.2					8.40
" 15		78	4.52	3.93		0.05	87		1.1				
" 16	20	110	6.60	5.61	0.76	0.03	85	4.0	0.5	29	12	4.3	8.45
" 17	5	170	4.66	3.90	0.16		83	3.6					
" 19										88	49	7.5	
" 20		115	2.00	1.42	0.16	0.14	71	8.0	7.0	140	90	13.2	7.25

Nitrogen Metabolism of Dogs in Phosphorus Poisoning.

It was considered advisable to obtain data on the nitrogen metabolism in phosphorus poisoning, in view of the cumulative phenomena observed in the blood. Information concerning the relation between the amount of the nitrogenous bodies in the blood and their excretion in the urine is desirable particularly to determine if renal function is involved.

Two dogs (IV and V) were placed on a diet of 200 gm. of ground meat and 150 cc. of water until nitrogenous equilibrium was obtained. Two to three days prior to the administration of phosphorus, the urinary partition (urea N, ammonia N, and amino N) was determined, as well as the total non-protein N, urea N, and amino N of the whole blood. Phosphorus poisoning was induced, and the metabolic studies were continued, the results appearing in Table I.

Dog IV continued to eat until the day of death, whereas Dog V ate nothing for the last two days. Dog IV shows a marked increase in the output of urinary N, with a slight decrease in the urea nitrogen per cent, but no change in the ammonia and amino-acid N per cent, together with an absolute increase in total non-protein N, the urea N and amino N of the blood. Dog V exhibited practically normal urinary N figures until the day of death, when decrease in the total N, an absolute and relative decrease in urea N, and a relative increase in the ammonia and amino N percentages were found. Very marked accumulation of total non-protein N, urea, and amino N occurred in blood.

Renal Function in Phosphorus Poisoning in Dogs.

Since urinary nitrogen changes are utilized in studies of liver function, it is obviously necessary to have some information concerning the effect on renal function of drugs used for the production of these liver injuries. The cumulative phenomena in the blood encountered in our studies on liver function in phosphorus poisoning suggested that the kidney function was affected shortly before death. It was also considered advisable to determine whether an acidosis occurs, as this would affect conclusions drawn from the urinary ammonia findings. The phenolsulphonephthalein output in the urine, and the urea and hydrogen ion concentrations in the blood, were utilized in this connection. The results of studies on four dogs appear in Table II.

In these dogs sufficient phosphorus was given to insure definite functional and anatomical liver injury. Death from the phosphorus resulted in three of them, and jaundice and incoagulable blood were seen in all. These rarely appear except in fatal poisoning.

In Dogs VI and VII, acutely poisoned, no decrease in the phenolsulphonephthalein occurred. The blood urea increased only as a terminal event. Acidosis is indicated in each on the day before death.

Dog VIII, which recovered, showed at one time a phenolsulphonephthalein output considerably decreased,—28 per cent. However, the phthalein rapidly returned to normal. No marked increase in blood urea and no acidosis accompanied this change.

TABLE II.

Dog VI.

Date (1915).	Weight.	Phosphorus.	Sulphone-phthalein, 2 hrs.	Blood urea per 100 cc.	Hydrogen ion concentration of blood.
	kilos	*mg.*	*per cent*	*mg.*	
Mar. 30			78		
" 31	8.0			24	7.70
Apr. 3			68		
" 7	8.4	12.5	70	32	7.45
" 8	8.55		55	35	7.65
" 9	8.40	5.0	80	40	
" 10			60	68	7.25
" 11				176	

Dog VII.

Date (1915).	Weight.	Phosphorus.	Sulphone-phthalein, 2 hrs.	Blood urea per 100 cc.	Hydrogen ion concentration of blood.	Hydrogen ion concentration of serum.
	kilos	*mg.*	*per cent*	*mg.*		
Mar. 31	12.22		60	30	7.55	7.55
Apr. 3			60			
" 7	13.00	20		36	7.50	7.70
" 8	13.50		70	37	7.55	7.65
" 9	12.60	7.5	56	35		
" 10			75	35		
" 11			65	46	7.25	
				127 at death		

Dog VIII.

Date (1915).	Weight.	Phosphorus.	Sulphone-phthalein, 2 hrs.	Blood urea per 100 cc.	Hydrogen ion concentration of blood.
	kilos	*mg.*	*per cent*	*mg.*	
Apr. 13			62	20	7.60
" 14	7.60	5			
" 16		5	67		7.45
" 17			75	40	
" 19				30	7.40
" 20				30	
" 21	6.35		58		
" 22			28	40	
" 23			45		7.45
" 26				36	7.60
" 27			63		
" 28					
" 29	6.30		75	51	7.60
" 30				36	
May 3	5.85		60	22	7.65
" 7			61	17	

TABLE II.—*Concluded.*

Dog IX.

Date (1915).	Weight.	Phosphorus.	Sulphone-phthalein, 2 hrs.	Blood urea per 100 cc.	Hydrogen ion concentration of blood
	kilos	*mg.*	*per cent*	*mg.*	
Apr. 13			64	23	7.60
" 14	5.80	5			
" 16		5	75	30	7.45
' 17			80	27	
' 19				43	7.30
' 20				31	
" 21	5.35	5	73		
" 22			58		
' 23			50		7.35
" 26	5.25	5	65		
" 28		5			
" 29	5.15		53	55	7.55
" 30		5	45	32	7.60
May 3			20	79	7.35
" 4	4.75			112	

Dog IX exhibited a progressively decreasing renal function, as indicated by sulphonephthalein for the week previous to death, a normal increase in the blood urea, and an acidosis early in the poisoning, which reappeared before death.

That considerable injury to renal function does occur occasionally during the course of the intoxication, but that it is usually a terminal event, is evident from the study of these dogs. The renal injury, however, is exceedingly slight, and is late in comparison with the functional injury to the liver. The functional renal injury cannot be absolutely ignored, but it plays only a minor part in the interpretation of the functional changes of the liver resulting from the administration of phosphorus.

Liver Function in Chloroform Poisoning.

In the following protocols the results of liver studies similar to those carried out on the dogs poisoned with phosphorus are reported. No special renal functional studies, however, were made in chloroform poisoning.

Dog X.—Weight 24 kilos. *Normals:* tolerates 0.30 gm. galactose and 1.5 gm. levulose per kilo. Phthalein 30 per cent. *Blood:* lipase, 0.25; total non-protein N, 58 mg.; urea N, 30 mg. (52 per cent); amino N, 3.4 mg. *Urine:* 1,140 cc. containing 15.1 gm. nitrogen; urea N, 91 per cent; ammonia N, 58 per cent; amino N, 1.4 per cent.

Mar. 6, 1914. Chloroform anesthesia for 2 hours.

Mar. 8. Phthalein 22 per cent.

Mar. 10. Does not tolerate 0.20 gm. galactose per kilo.

Mar. 13. Tolerates 1.0 gm. levulose per kilo.

Mar. 14. *Blood:* lipase, 1.20; fibrinogen, 217 mg.; total non-protein N, 56 mg.; urea N, 26 mg. (46 per cent); amino N, 3.0 mg. *Urine:* 250 cc. containing 5.7 gm. nitrogen; urea N, 84 per cent; ammonia N, 6 per cent; amino N, 6.9 per cent.

Mar. 15. Died during chloroform anesthesia.

Dog XI.—Weight 16 kilos. *Normals:* phthalein 35 per cent. *Urine:* 425 cc. containing 7.9 gm. nitrogen; urea N, 82 per cent; ammonia N, 10 per cent; amino N, 1.1 per cent.

Mar. 3, 1914. Chloroform anesthesia for 2 hours.

Mar. 5. Found dead in cage. Body still very warm. *Bladder urine:* 100 cc., containing 2.2 gm. nitrogen; urea N, 40 per cent; ammonia N, 97 per cent; amino N, 3 0 per cent.

Dog XII.—Weight 8 kilos. *Normals:* tolerates 1.5 gm. levulose per kilo. Phthalein 34 per cent, positive in urine. *Blood:* lipase, 0 39; total non-protein N, 33 mg.; urea N, 11 mg. (33 per cent); amino N, 4.3 mg. *Urine:* 198 cc. containing 7.8 gm. nitrogen; urea N, 84 per cent; ammonia N, 6.4 per cent; amino N, 1.0 per cent.

Feb. 14. Chloroform anesthesia for 1½ hours.

Feb. 15. Phthalein 21 per cent, positive in urine. *Blood:* lipase, 0.93; fibrinogen, 395 mg.; total non-protein N, 52 mg.; urea N, 24 mg. (46 per cent); amino N, 4.9 mg.

Feb. 16. *Urine:* 265 cc., containing 13.4 gm. nitrogen; urea N, 84 per cent; ammonia N, 5.3 per cent; amino N, 1.5 per cent.

Died in subsequent chloroform anesthetization.

Dog XIII.—Weight 12 kilos. *Normals:* tolerates 0.22 gm. galactose and 1.5 gm. levulose per kilo. Phthalein 32 per cent. *Blood:* lipase, 0.38; fibrinogen. 485 mg.; total non-protein N, 39 mg.; urea N, 11 mg. (39 per cent); amino N. 4.8 mg. *Urine:* 360 cc. containing 9.1 gm. nitrogen; urea N, 83 per cent; ammonia N, 6.0 per cent; amino N, 1.2 per cent.

Mar. 19, 1914. Chloroform anesthesia for 2 hours. Very light. All subsequent ones light.

Mar. 23. Phthalein 25 per cent.

Mar. 25. Chloroform anesthesia for ¾ hour. Phthalein 27 per cent.

Mar. 26. Tolerates 0.16 gm. galactose per kilo.

Mar. 27. Chloroform anesthesia for 1 hour.

Mar. 28. Does not tolerate 1.2 gm. levulose per kilo.

Mar. 29. Phthalein 30 per cent.

Apr. 1. Chloroform anesthesia for 2 hours.

Apr. 2. Tolerates 0.16 gm. galactose per kilo.

Apr. 3. Does not tolerate 1 gm. levulose per kilo.

Apr. 4. Phthalein 9 per cent, positive in urine. Does not tolerate 0.10 gm. galactose per kilo.

Apr. 6. *Blood:* lipase, 0.85; fibrinogen. 14 mg.; total non-protein N. 25 mg ; urea N, 13 mg. (52 per cent); amino N, 4.0 mg.

Apr. 7. Tolerates 0.8 gm. levulose per kilo. Phthalein 2 per cent, positive in urine.

Apr. 8. *Blood:* lipase, 1.61; fibrinogen, 44 mg.; total non-protein N, 29 mg.; urea N, 10 mg. (34 per cent); amino N, 4.3 mg. *Urine:* 465 cc. containing 6.1 gm. nitrogen; urea N, 70 per cent; ammonia N, 14.4 per cent; amino N, 1.4 per cent.

Killed by bleeding from carotid.

Autopsy.—Liver: normal size, very pale, and friable. Lobules show very plainly. All tissues stained with bile. Kidney looks normal.

Microscopic Examination.—Kidney: cloudy swelling, moderate grade. Cells of convoluted tubules are swollen and granular. The lumina contain pink staining granules. *Liver:* shows central atrophy; extreme fatty degeneration, involving the entire lobule except a thin zone about the portal area; and numerous scattered necrotic liver cells, especially about the central portions of the lobules.

In order to gain some idea of the variations in normal dogs dependent on diet, Dogs XIV and XV were studied with uncontrolled diet (lungs, corn cake, etc.), and then on April 17 placed on a diet of 400 grams of meat per day. The total nitrogen and urea (whole blood) were taken at various times. Table III shows the variations encountered. The urea figures are expressed as mg. of nitrogen per 100 cc. of blood.

TABLE III.

Dog XIV.

Apr., 1914.	4	5	8	15	16	17	18	19	20	21	22	23	24	25	26	27
Total N, urine.........		4.4	9.0		11.2	10.0	5.8	20.0		13.6		15.2		10.1		
Urea N, blood.........	10	14	9	13				21	19				14			14

Dog XV.

Total N, urine.........		5.1	3.7		7.6	4.7	6.9	18.1		17.0		14.4		6.7		
Urea N, blood.........	11	14	9	15				20	25				18			18

Dog XIV.—Weight 11.9 kilos. *Normals:* phthalein 29 per cent. *Blood:* lipase, 0.35; fibrinogen, 470 mg.; total non-protein N, 34 mg.; urea N, 15 mg. (44 per cent); amino N, 3.7 mg.

May 1, 1914. Deep chloroform anesthesia for 2 hours.

May 2. Phenolsulphonephthalein between 60 and 70 per cent for 2 hours. *Blood:* lipase, 2.75; fibrinogen, 90 mg.; total non-protein N, 32 mg.; urea N, 13 mg. (40 per cent); amino N, 4.2 mg. *Urine:* single specimen; urea N, 44 per cent; ammonia N, 8.3 per cent; amino N, 3.1 per cent. Phthalein, only trace in feces.

Dog XV.—Weight 10.2 kilos. *Normals:* tolerates 0.3 gm. galactose and 1.5 gm. levulose per kilo. Phthalein 37 per cent.

May 5, 1914. Deep chloroform anesthesia for 1½ hours.

May 6. *Blood:* lipase, 2.70; fibrinogen, 30 mg.; total non-protein N, 120 mg.; urea N, 80 mg. (67 per cent); amino N, 6.6 mg. *Urine:* 24 hour specimen lost.

May 7. Dead. *Bladder urine:* urea N, 54 per cent; ammonia N, 10.6 per cent; amino N, 3.6 per cent. Phthalein 3 per cent.

In chloroform poisoning in dogs the following changes are found: The tetrachlorphthalein is decreased, and it appears in the urine; lipase is increased; fibrinogen is decreased; the total non-protein nitrogen, urea nitrogen, and amino nitrogen of the blood suffer practically no change. In the urine a more or less pronounced disturbance of the nitrogen partition is seen, while the galactose and levulose tolerance is definitely decreased.

Dog XV was the only one with chloroform poisoning which showed increase in nitrogenous products in the blood. Here, we have a condition somewhat resembling that in phosphorus dogs. The other tests show a very severe type of poisoning in this dog. This brings out the possibility of cumulative phenomena appearing in the course of chloroform poisoning. High amino-acid content of the urine is noted in Dog X (6.9 per cent); Dog XI (3.0 per cent); Dog XIV (3.1 per cent); and Dog XV (3.6 per cent). Dogs XI, XIV, and XV received poisonings which were rapidly fatal, dying within forty-eight hours. In them, a markedly disturbed nitrogen partition occurs in the urine, *i. e.*, a much decreased urea nitrogen, without a corresponding increase in the free amino-acids or ammonia nitrogen.

In Dog XIV, at a time when the liver function was reduced to a minimum, the sulphonephthalein was absolutely normal. It is very unlikely that renal functional changes are greater in chloroform than in phosphorus poisoning.

CONCLUSIONS.

1. We have confirmed the results of previous workers on tetrachlorphthalein, fibrinogen, and lipase.

2. The total non-protein nitrogen, urea, and amino-acids of the blood serum show a definite and sometimes marked increase in phosphorus poisoning. These changes are not so evident in chloroform poisoning, although they sometimes occur. They are usually terminal phenomena.

3. The urinary nitrogen partition between the urea, ammonia, and amino-acids is not always disturbed. The most important changes which occur are an increased amino nitrogen in chloroform and phosphorus poisoning, and a very low urea nitrogen percentage in severe fatal chloroform poisoning.

4. Sugar tolerance towards galactose and levulose is in general markedly decreased in both types of poisoning.

5. In phosphorus poisoning liver functional changes can and do occur without concomitant renal changes. Renal insufficiency usually arises as a terminal event.

6. Increased nitrogenous products in the blood (total non-protein nitrogen, urea, and amino nitrogen) are associated with an increase of these bodies in the urine. Consequently, an increased protein catabolism, as well as renal insufficiency, is necessary to explain this accumulation.

7. A terminal acidosis, as evidenced by increased hydrogen ion concentration in the blood, usually occurs.

THE INTOXICATION OF SPLENECTOMIZED MICE BY FEEDING FRESH SPLEEN AND OTHER ORGANS.

By PAUL A. LEWIS, M.D., and ARTHUR GEORGES MARGOT.

(*From the Henry Phipps Institute of the University of Pennsylvania, Philadelphia.*)

(Received for publication, June 23, 1915.)

As we have previously recorded,[1] splenectomized mice when fed with fresh sheep or mouse spleen frequently show signs of illness. Some animals die within a few hours after eating the organ, and of those that die some show hemorrhages into the peritoneal subserosa or peritoneal cavity. The mice that have been ill are apt to refuse to eat the fresh organ on several succeeding days. When they do eat again they may or may not show signs of illness once more. Splenectomized mice fed with fresh sheep muscle never showed any sign of acute illness. Intact mice fed with spleen also showed no sign of ill effect.

These experiments were made on animals from two weeks to four weeks after the removal of the spleen. The intoxicating effect was readily noticeable in animals infected with the tubercle bacillus. In the non-infected animals the result was definite, but somewhat irregular, so much so that it seemed impracticable to do systematic work which should include a consideration of the negative results.

Passing by the steps which led to a definite and decisive method of work we may say that we found that at four or five days after splenectomy the intoxication is more regular in its occurrence and is apt to be more severe than at later periods. Also, as we continued these observations, we became impressed with the fact that the hemorrhage into the serous membranes and especially into the peritoneum, which we had occasionally noticed, was really an integral part of the process. Following a natural train of thought from this point we were led to test the coagulation time of the blood. We have found that the most constant result, following the feeding of

---

[1] Lewis, P. A., and Margot, A. G., *Jour. Exper. Med.*, 1915, xxi, 84.

spleen to splenectomized mice, is a delayed coagulation time. With these facts as a basis for judgment, we have studied the nature of this intoxication as it relates to the time of its possible development after splenectomy, the rate of its development after feeding, its specificity both as to organ and as to species, the amount of fresh substance necessary to produce the intoxication, and finally the effect of the continuous feeding of less than the fatal amount of substance.

Methods.

The general methods of splenectomy and feeding have been stated in our previous papers. With regard to the feeding we may repeat that the animals are fed with the fresh organs before other food is given on that day. We have found that the change in coagulation time is apt to be somewhat more pronounced after the second day's feeding than on the first day. Therefore, in case the first feeding has given a negative result, we have always fed with the same substance on the next day. The coagulation time has been taken with Bogg's instrument, the blood being drawn from the tail. In the following experiments the first feeding was on the fourth day after splenectomy. The second feeding was on the following day. The coagulation time was taken before feeding and usually from three to four hours after feeding.

The results of these experiments may be summarized as follows:

Splenectomized mice fed on the fourth day and again on the fifth day show no change in the coagulation time of the blood when the following organs are used: lymph nodes, thymus gland, liver, kidney, skeletal muscle, lung, brain, pancreas, pituitary gland, testicle, adrenal, thyroid gland, salivary glands, and the mucosa of the large intestines.

When fed with the following tissues the mice show a delay in the coagulation time varying between two minutes and fifteen minutes: spleen of mouse, rat, guinea pig, rabbit, beef, sheep, cat, and man; gastric mucosa of mouse and cat; and mucosa of the upper small intestines of mouse and cat.

When fed with fresh human bone marrow and with dried sheep blood the mice also show a change in coagulation time. In the experiments so far done the change has been less pronounced than after the feeding of spleen or gastro-intestinal mucosa.

The mice which show the smaller changes in coagulation time seldom show any other evidence of the intoxication. When the delay in coagulation time is marked the animals are apt to be extremely sick, and a considerable number of the animals have died. Death when it occurs usually takes place before the eighth hour. In twenty hours the blood of the surviving animals has usually returned to the normal. Occasionally animals are encountered in which the return to normal is less rapid. In these instances the blood coagulates slowly and the animal is still sick at the end of twenty-four hours. These animals sometimes recover and sometimes die. The animals which die during the course of this intoxication reveal various changes at postmortem examination. In some no gross changes can be made out. In others there are found petechial hemorrhages of the serous membranes, particularly the visceral peritoneum. Some instances of hemorrhage into the intestine have been encountered. The most usual result is a diffuse, rather poorly marked reddening of the peritoneal serosa accompanied by a small amount of blood tinged peritoneal fluid. Occasionally extremely large hemorrhages into the peritoneal cavity have occurred. When this has happened the animal has died quite suddenly. No definite bleeding point has been discovered in any instance. An intense diarrhea accompanies some cases.

We have been unable to distinguish in any way between the intoxication resulting from feeding spleen and that following the feeding of gastro-intestinal mucosa.

The tabulated results of the work on which the foregoing statements are based, together with the necessary control experiments, follow (Experiments I to XXX, Tables I to VIII).

TABLE I.

Experiment I. Rate of Development of the Intoxication after Feeding with Fresh Sheep Spleen.

Mouse No.	Coagulation time.					Remarks.
	Before feeding.	After feeding.				
		30 min.	1 hr.	2 hrs.	3 hrs.	
187	5 min., 10 sec.	5 min., 0 sec.	5 min., 20 sec.	7 min., 10 sec.	9 min., 20 sec.	Died.
188	6 " 10 "	6 " 30 "	6 " 20 "	8 " 0 "	8 " 10 "	Ill.

TABLE II.

Experiment II. Amount of Fresh Spleen (Human) Necessary To Produce Delayed Coagulation.

Mouse No.	Amount eaten.	Coagulation time. Before feeding.	After feeding.	Remarks.
	mg.			
317	100	5 min., 10 sec.	8 min., 10 sec.	Ill.
318	100	4 " 0 "	8 " 20 "	Died.
319	50	6 " 0 "	10 " 10 "	"
320	25	4 " 20 "	6 " 20 "	Ill.
321	10	4 " 40 "	6 " 10 "	"
322	10	5 " 0 "	6 " 20 "	Well.
323	5	6 " 10 "	6 " 0 "	"

TABLE III.

Experiments III to VII. Effect of Feeding Spleen of Other Animals to Splenectomized Mice.

Organ fed.	Mouse No.	Coagulation time. First day feeding. Before.	After	Re-marks.	Second day feeding. Before.	After.	Remarks.
Experiment III.	158	4 min., 30 sec.	4 min., 20 sec.		4 min., 20 sec.	7 min., 10 sec.	Very ill.
Spleen (cat).	159	4 " 20 "	4 " 10 "		4 " 40 "	6 " 20 "	Ill.
	160	5 " 20 "	5 " 30 "		5 " 10 "	5 " 10 "	
	161	5 " 30 "	5 " 40 "		5 " 20 "	8 " 30 "	Died.
Experiment IV.	119	" 10 "	6 " 10 "		6 " 0 "	6 " 10 "	
Spleen (beef).	120	" 30 "	7 " 30 "	Ill	7 " 20 "	11 " 30 "	"
	121	" 20 "	4 " 30 "		4 " 20 "	10 " 10 "	"
	122	6 " 30 "	4 " 20 "		4 " 40 "	6 " 30 "	Ill.
Experiment V.	178	6 " 0 "	6 " 10 "		6 " 30 "	6 " 30 "	
Spleen	179	4 " 10 "	7 " 20 "	"	7 " 0 "	8 " 10 "	Very ill.
(guinea pig).	180	4 " 20 "	4 " 30 "		4 " 20 "	4 " 10 "	
	181	5 " 10 "	5 " 20 "		5 " 0 "	5 " 20 "	
Experiment VI.	174	4 " 10 "	4 " 20 "		4 " 30 "	6 " 10 "	
Spleen	175	4 " 20 "	4 " 20 "		4 " 10 "	7 " 40 "	Ill.
(rabbit).	176	5 " 10 "	5 " 0 "		5 " 30 "	7 " 10 "	"
	177	4 " 30 "	4 " 25 "		5 " 0 "	6 " 30 "	
Experiment VII.	170	" 30 "	7 " 10 "	"	6 " 10 "	7 " 20 "	
Spleen (rat).	171	" 0 "	6 " 20 "		6 " 10 "	6 " 20 "	
	172	" 10 "	6 " 30 "		5 " 40 "	6 " 0 "	
	173	6 " 20 "	8 " 30 "	Very ill	8 " 10 "	8 " 40 "	Did not eat. Died.

TABLE IV.

Experiments VIII to XI. Effect of Feeding Gastro-Intestinal Mucosa to Splenectomized Mice.

Organ fed.	Mouse No.	Coagulation time.						
		First day feeding.			Second day feeding.			
		Before.	After.	Re-marks.	Before.	After.	Remarks.	
Experiment VIII.	34	4 min., 30 sec.	4 min., 20 sec.		4 min., 20 sec.	7 min., 10 sec.		
Stomach	35	5 " 0 "	6 " 10 "		5 " 10 "	7 " 20 "		
(mouse).	36	4 " 10 "	10 " 20 "		4 " 30 "	10 ' 30 "		
	37	6 " 0 "	6 " 10 "		6 " 0 "	9 " 20 "		
Experiment IX.	30	6 " 20 "	7 " 0 "		5 " 20 "	9 " 0 "		
Small intestine	31	4 " 40 "	6 " 5 "		5 " 0 "	8 " 20 "	Died.	
(mouse).	32	6 " 10 "	6 " 20 "		5 " 30 "	6 " 40 "		
	33	5 " 5 "	5 " 20 "		5 " 5 "	7 " 30 "		
Experiment X.	154	5 " 40 "	7 " 10 "	Ill	6 " 0 "	9 " 10 "	Ill.	
Stomach (cat).	155	6 " 20 "	6 " 10 "		6 " 10 "	7 " 20 "		
	156	6 " 10 "	6 " 30 "		6 " 20 "	6 " 20 "		
	157	5 " 10 "	5 " 10 "		5 " 20 "	10 " 40 "	Died.	
Experiment XI.	150	4 " 20 "	6 " 10 "		5 " 0 "	8 " 20 "	Very ill.	
Small intestine	151	6 " 0 "	6 " 10 "		6 " 10 "	6 " 20 "		
(cat).	152	5 " 30 "	5 " 20 "		5 " 20 "	11 " 10 "	Died.	
	153	5 " 20 "	5 " 30 "		5 " 20 "	7 " 30 "	Ill.	

TABLE V.

Experiments XII and XIII (a). Effect of Feeding Dried Whole Blood and Bone Marrow to Splenectomized Mice.

Substance fed.	Mouse No.	Coagulation time.						
		First day feeding.			Second day feeding.			
		Before.	After.	Re-marks.	Before.	After.	Remarks.	
Experiment XII.	182	5 min., 30 sec.	5 min., 20 sec.		5 min., 30 sec.	6 min , 0 sec.		
Dried whole	183	4 " 10 "	7 " 10 "	Ill	7 " 10 "	7 " 40 "	Ill.	
blood (sheep).	184	6 " 0 "	6 " 20 "		6 " 10 "	6 " 10 "		
	185	5 " 20 "	6 " 0 "		5 " 0 "	6 " 40 "	"	
Experiment	327	4 " 10 "	6 " 20 "					
XIII (a).	328	4 " 30 "	5 " 20 "					
Bone marrow	329	5 " 0 "	7 " 0 "	"				
(man).	340	4 " 20 "	6 " 10 "					

TABLE VI.

Experiments XIII (b) to XX.   Effect of Feeding Various Organs to Splenectomized Mice.

Organ fed.	Mouse No.	Coagulation time.			
		First day feeding.		Second day.	
		Before.	After.	Before.	After.
Experiment XIII(b).	83	6 min., 40 sec.	6 min., 20 sec.	6 min., 30 sec.	6 min., 20 sec.
Lymph nodes	84	5 " 30 "	5 " 30 "	5 " 30 "	6 " 0 "
(sheep).	85	4 " 20 "	4 " 30 "	5 " 10 "	5 " 30 "
	86	5 " 10 "	5 " 20 "	5 " 10 "	5 " 10 "
Experiment XIV.	87	6 " 10 "	6 " 0 "	6 " 0 "	6 " 30 "
Thymus (sheep).	88	5 " 30 "	5 " 20 "	5 " 10 "	5 " 20 "
	89	6 " 0 "	6 " 10 "	6 " 20 "	6 " 10 "
.	90	5 " 0 "	5 " 20 "	5 " 10 "	5 " 30 "
Experiment XV.	162	6 " 20 "	6 " 40 "	6 " 30 "	6 " 30 "
Large intestine	163	6 " 10 "	6 " 10 "	6 " 20 "	6 " 10 "
(mouse).	164	5 " 30 "	5 " 20 "	5 " 20 "	5 " 30 "
	165	5 " 40 "	5 " 30 "	5 " 30 "	5 " 30 "
Experiment XVI.	108	5 " 20 "	5 " 30 "	5 " 10 "	5 " 30 "
Liver (sheep).	109	4 " 20 "	5 " 10 "	5 " 10 "	5 " 10 "
	110	5 " 10 "	5 " 20 "	5 " 20 "	5 " 10 "
	111	5 " 30 "	5 " 30 "	6 " 0 "	6 " 20 "
Experiment XVII.	7	5 " 20 "	5 " 10 "	5 " 30 "	5 " 30 "
Pancreas (sheep).	76	· " 10 "	4 " 30 "	5 " 20 "	4 " 0 "
	77	" 20 "	5 " 20 "	6 " 20 "	4 " 10 "
	78	3 " 10 "	4 " 40 "	6 " 10 "	4 " 20 "
Experiment XVIII.	100	6 " 20 "	6 " 10 "	6 " 20 "	6 " 20 "
Pituitary body	101	6 " 10 "	6 " 10 "	6 " 30 "	6 " 10 "
(sheep).	102	5 " 10 "	5 " 0 "	5 " 20 "	5 " 30 "
	103	6 " 15 "	6 " 10 "	6 " 0 "	6 " 0 "
Experiment XIX.	104	6 " 0 "	6 " 10 "	5 " 20 "	5 " 30 "
Brain (sheep).	105	6 " 10 "	6 " 0 "	6 " 0 "	6 " 0 "
	106	5 " 20 "	5 " 10 "	4 " 40 "	5 " 0 "
	107	6 " 20 "	6 " 30 "	6 " 10 "	6 " 30 "
Experiment XX.	112	6 " 0 "	6 " 10 "	6 " 10 "	6 " 30 "
Kidney (mouse).	113	6 " 0 "	6 " 30 "	5 " 30 "	5 " 20 "
	114	5 " 20 "	5 " 10 "	5 " 10 "	6 " 0 "
	115	4 " 30 "	4 " 40 "	4 " 40 "	5 " 0 "

Note slight decrease in coagulation time. Compare with salivary gl (Experiment XXV).

TABLE VII.

Experiments XXI to XXVI. Effect of Feeding Various Organs to Splenectomised Mice.

Organ fed.	Mouse No.	Coagulation time.					
		First day feeding.			Second day feeding.		
		Before.	After.	Remarks.	Before.	After.	Remarks.
Experiment XXI. Lung tissue (mouse).	127	6 min., 10 sec.	6 min., 0 sec.		6 min., 0 sec.	6 min., 0 sec.	
	128	5 " 0 "	5 " 20 "		5 " 10 "	5 " 20 "	
	129	6 " 10 "	6 " 10 "		6 " 10 "	6 " 20 "	
	130	6 " 20 "	6 " 10 "		6 " 30 "	6 " 30 "	
Experiment XXII. Muscle (sheep).	123	5 " 0 "	5 " 10 "		5 " 15 "	5 " 10 "	
	124	4 " 30 "	4 " 20 "		4 " 20 "	4 " 20 "	
	125	5 " 10 "	5 " 30 "		5 " 20 "	5 " 10 "	
	126	6 " 0 "	6 " 10 "		6 " 10 "	6 " 10 "	
Experiment XXIII. Thyroid (sheep).	79	4 " 50 "	4 " 30 "		4 " 30 "	4 " 40 "	
	80	5 " 10 "	5 " 0 "		5 " 20 "	5 " 30 "	
	81	5 " 10 "	5 " 20 "		5 " 10 "	5 " 30 "	
	82	6 " 30 "	6 " 10 "		6 " 10 "	6 " 10 "	
Experiment XXIV. Adrenals (man).	115	6 " 10 "	4 " 10 "	III } Second reading at 6 hrs.	5 " 10 "	5 " 30 "	III } Second reading at 6 hrs.
	116	5 " 30 "	5 " 30 "		5 " 20 "	4 " 10 "	
	117	5 " 30 "	3 " 10 "	" }	5 " 40 "	6 " 10 "	" }
	118	5 " 20 "	4 " 0 "		5 " 0 "	5 " 10 "	
Experiment XXV. Salivary gland (mouse).	313	5 " 10 "	2 " 30 "	Died, 18 hrs.	5 " 30 "	2 " 10 "	Died, 4 hrs. } Second reading at 6 hrs.
	314	6 " 20 "	3 " 10 "	" 10 "	5 " 40 "	3 " 30 "	" 2½ "
	315			Did not eat			
	316			" " "			
Experiment XXVI. Testicle (rabbit).	310	5 " 20 "	5 " 30 "		5 " 30 "	5 " 30 "	Died, 4 hrs. } Second reading at 2 hrs.
	311	5 " 10 "	5 " 0 "		5 " 0 "	5 " 20 "	" 2½ "
	312	6 " 10 "	6 " 30 "		6 " 0 "	5 " 0 "	
	313	4 " 30 "	" 0 "		5 " 0 "	" 0 "	

TABLE VIII.

Experiments XXVII to XXX. Control Experiments Relating to Figures Previously Presented. Intact Mice Fed with Various Organs.

Organ fed.	Mouse No.	Coagulation time. First day feeding. Before.	First day feeding. After.	First day feeding. Remarks.	Second day feeding. Before.	Second day feeding. After.	Second day feeding. Remarks.
Experiment XXVII. Spleen (sheep).	42	5 min., 20 sec.	5 min., 10 sec.	Amount eaten 0.0 gm.	5 min., 5 sec.	5 min., 0 sec.	Amount eaten 0.340 gm.
	43	4 " 10 "	4 " 40 "	" " 0.065 "	4 " 40 "	4 " 10 "	" " 0.185 "
	44	5 " 30 "	5 " 40 "	" " 0.185 "	5 " 10 "	5 " 10 "	" " 0.08 "
	45	5 " 40 "	6 " 10 "	" " 0.08 "	5 " 5 "	6 " 20 "	" " 0.225 "
	46	5 " 30 "	5 " 0 "	" " 0.06 "	6 " 10 "	6 " 20 "	" " 0.068 "
	47	6 " 0 "	6 " 5 "	" " 0.095 "	6 " 20 "	6 " 10 "	" " 0.275 "
Experiment XXVIII. Small intestine (mouse).	45	5 " 5 "	5 " 0 "		5 " 20 "	5 " 10 "	
	46	4 " 20 "	4 " 30 "		5 " 0 "	5 " 10 "	
	47	5 " 0 "	5 " 10 "		5 " 30 "	5 " 40 "	
Experiment XXIX. Stomach (mouse).	42	5 " 30 "	5 " 5 "		5 " 10 "	5 " 30 "	
	43	5 " 30 "	6 " 0 "		5 " 20 "	5 " 30 "	
	44	6 " 30 "	6 " 20 "		6 " 10 "	6 " 0 "	
Experiment XXX. Salivary gland (mouse).	328	6 " 0 "	4 " 30 "	Ill } Readings at	5 " 30 "	5 " 10 "	
	329	5 " 30 "	3 " 0 "	Died } 2 hrs.	5 " 5 "	5 " 30 "	
	330	6 " 30 "	5 " 10 "	"	6 " 10 "	6 " 0 "	

Other experiments were made to determine the effect of feeding the regular diet of oats and bread on the coagulation time of the blood. Neither the intact nor the splenectomized animals showed any change after eating this food.

The intoxication as described bears a certain general resemblance to peptone poisoning. The fact that it is developed from the intestinal tract likewise suggested that the effects might be due to imperfectly detoxicated digestion products. In following up these suggestions we have studied the reaction of mice to Witte's peptone. We have been unable to poison mice by feeding this substance. When intraperitoneal injections of this peptone are given, an intoxication is produced which causes a delay in the coagulation of the blood and which may lead to death. The change in coagulation time develops more slowly. It is not apparent in six hours but is well developed at eighteen hours. No difference could be detected between the reaction of intact and splenectomized mice to Witte's peptone.

If those splenectomized mice, which, having been fed with either spleen or gastro-intestinal mucosa, survive, are fed continuously with the same substance, a certain number of them will die at later feedings. The remainder after ten days or two weeks no longer develop the evidence of illness, even though they may eat large quantities of material. They do not become sick and the coagulation time of the blood remains normal.

Many important questions raised by the demonstration of this tolerance have not as yet been clearly answered in our experiments. We have much reason for considering that this is an acquired tolerance or immunity produced by the repeated feedings in the susceptible animal. Our figures show certain instances in which animals definitely ill with a definitely delayed coagulation time fail to show reactions at subsequent feedings. These instances are, however, small in number and up to the present time few of our observations have been made with this point sufficiently in view. On the other hand, we know that mice immediately after splenectomy differ considerably in their susceptibility to the intoxication. After a number of weeks they become less susceptible spontaneously. In just what degree the natural process operates and in how far we

influence it by repeated feedings we are unable to decide at present.
When we speak of this tolerance as acquired we express our present
belief but recognize it as possible that the repeated feedings may
have served chiefly to select the resistant animals.

We also have certain evidence that this tolerance is in considerable degree specific in the sense that the animals tolerant to spleen
following repeated feedings with this organ are still susceptible to
intoxication with gastro-intestinal mucosa and *vice versa.* We
hope to extend our observations in this direction in the near future.

In most of the experiments tabulated above the result has been
decisive.   Certain instances when this has not been entirely the case
require comment.

1. Thyroid feeding gives no change in the coagulation time of
the blood.   The animals may, however, become ill and some may
die with an acute intoxication.

2. Suprarenal gland when fed to mice gives rise in most instances
to an acute or subacute intoxication with some peculiar characteristics.   The coagulation time of the blood is not slowed.   It may
possibly in some instances be hastened.

3. Administration of pancreas seems to shorten the coagulation
time of the blood appreciably.   In no instance have we been able to
see evidence of illness.

4. The salivary gland of the mouse when fed to mice causes a
violent fatal intoxication.   The mice do not behave exactly as in
the case of the spleen-fed mice and the coagulation time of the blood
is also shortened.

These intoxications with thyroid, suprarenal, and salivary gland
are important in this connection, as they might be confusing to
one who was not aware of the possibility of their occurrence.   We
feel justified in putting them to one side as having no connection
with the subject at hand for the reasons, first, that splenectomized
mice are no more susceptible than normal mice in either instance;
secondly, that the animals which die do not show the hemorrhagic
lesion so frequently found in animals dying after eating spleen or
gastro-intestinal mucosa; and, finally, because the coagulation time
of the blood either is not altered or is diminished.   The intoxications with thyroid and suprarenal glands bear some resemblance to

the picture of such a poisoning which could be drawn on the basis of the well known properties of these organs. That with salivary gland was entirely unforeseen and seems to offer an entirely new subject for consideration.

The experiments presented in which an intoxication of definite character is shown to result from feeding splenectomized mice with either fresh spleen or the mucous membrane of the stomach and upper small intestine are of course difficult if not impossible of interpretation at present.[2] The reaction being so closely limited to these organs, together with the fact that no direct distinction can be made between the intoxication developed in either case, makes it seem reasonable to suppose that there is some close interaction between the spleen and the upper gastro-intestinal mucosa with respect to some function or functions. It is interesting that such an interrelationship has been put forward in the interpretation of experiments of a very different nature.

Luciani[3] reviews the observations of a number of workers in this connection. It is stated to have been shown that an extract of the spleen taken at the height of digestion is capable of activating the zymogens extracted from the pancreas. *In vivo* experiments are said to have rendered it probable that this activation is not only possible but is an important factor in proteolytic digestion. A similar form of interactivity has been supposed to exist between spleen and stomach, but has not been demonstrated experimentally.

Except for the meager suggestion which may be offered by these physiological writings we know of nothing in the literature which is in any way related to the facts as we have observed them. Until more observations are at hand a hypothetical discussion of the subject would be without profit.

SUMMARY.

The intoxication which is developed when splenectomized mice are fed with fresh spleen is more regular in occurrence when the feeding experiment is carried out four or five days after splenectomy than when it is done at later periods. The intoxication is easily recognizable even in its less severe forms by a lengthening of the coagulation time of the blood.

[2] The experiments with dried blood and with bone marrow should in the future receive equal consideration with those in which spleen is used. Up to the present, extraneous circumstances have prevented our giving much attention to the reaction to these substances.

[3] Luciani, L., Physiologie des Menschen, Jena, 1906, ii, 80, 101.

An intoxication of the same character is produced when splenectomized mice are fed with the mucous membrane of the stomach and upper small intestine.   Bone marrow and dried blood probably give the same reaction in a somewhat milder form.   The other organs either give no intoxication at all when fed, or in certain instances the thyroid, adrenal, and salivary gland (mouse) give intoxications of a different character which affect intact mice and splenectomized animals equally.

The spleen or the gastro-intestinal mucosa is equally effective in producing the intoxication, whether it be derived from heterologous or homologous species.

Certain experiments, not reported in detail, indicate that the susceptibility to the intoxication disappears in time and that this time may be shortened by repeated feedings with sublethal amounts of organ substance.

THE FUNCTION OF THE SPLEEN IN THE EXPERIMENTAL INFECTION OF ALBINO MICE WITH BACILLUS TUBERCULOSIS.

THIRD PAPER.

By PAUL A. LEWIS, M.D., and ARTHUR GEORGES MARGOT.

(*From the Henry Phipps Institute of the University of Pennsylvania, Philadelphia.*)

(Received for publication, June 23, 1915.)

In our second paper under this title[1] we were able to state that the increased resistance which splenectomized mice show to infection with the tubercle bacillus is decreased again if the infected animals are fed continuously with fresh spleen. The natural conclusion would follow from this that the feeding of the spleen had served to restore some function of this organ which makes for the normal susceptibility to the infection. The situation was, however, extremely complicated by the fact that splenectomized, uninfected mice were found to suffer from a more or less severe intoxication when fed with fresh spleen. The administration of any poison to mice infected with the tubercle bacillus, even in doses which are considerably below those necessary to cause symptoms in the normal animal, is apt to shorten life. We were at that time of writing, therefore, unable to reach a satisfactory conclusion in regard to the issues involved.

Since then we have made a more extensive study of the intoxication to which splenectomized mice are subject when fed fresh spleen.[2] As a result of this study we attained command of a method which seemed to make it possible to determine whether or not feeding of fresh spleen would restore a function to the body which affects its susceptibility to infection in definite degree.

[1] Lewis, P. A., and Margot, A. G._ Jour. Exper. Med.,_ 1915. xxi. 84.

[2] Lewis and Margot, _ibid.,_ 1915. xxii. 347.

The facts on which we rely for the purpose of the present paper are the following. When recently splenectomized mice are fed with fresh spleen or with the fresh mucous membrane of the stomach and upper small intestine an acute intoxication is produced. In its immediate manifestations the intoxication is the same, whichever of these organs is administered. A number of weeks after splenectomy the mice are less susceptible to this intoxication and in time they can no longer be affected. This return to the normal condition of tolerance can be hastened by the continuous feeding of the fresh organ in sublethal doses, beginning a few days after splenectomy. In this way we have been able to produce groups of mice which between two and three weeks after splenectomy can be fed fresh spleen and fresh gastro-intestinal mucosa with entire impunity.

It is unimportant for the purposes of the present discussion to decide upon the nature of this tolerance: whether, for example, it is an acquired immunity or a selection of naturally resistant specimens. It is, on the other hand, absolutely essential to a proper interpretation of the results, that the tolerance exist at the time stated, two or three weeks after splenectomy, for the following reason.

We have previously stated that the resistance given against infection with the tubercle bacillus gradually disappears, being entirely gone in six months. We present figures in this paper (Experiment II. Group k) to show that in as short a time as ten weeks the specific effects of splenectomy have disappeared, or at least have been reduced to an insignificant residuum. It is obvious that experiments which depend for their interpretation on the presence or ·absence of this resistance must be carried out in such a way that the animals are infected from two to three weeks after splenectomy, at a time when the resistance is at a maximum and when, as we have repeatedly shown, it is almost uniformly present. On the basis of these considerations we have carried out the following experiment.

A number of mice were splenectomized and separated into three main groups. The first group (Table I, Experiment I, b) was retained on the regular diet. The second group was fed fresh spleen continuously in addition to the regular diet, beginning on the fourth day after splenectomy and throughout the course of the experiment. A considerable number of this group died, following the first few feedings, of the acute intoxication we have considered. Those surviving in the third week appear in Table I (Experiment I) as Groups c, d, e, and f, according to the source of the spleen with which they were fed. The third group was fed in exactly the same way with stomach or small intestine, and the survivors appear in Table I (Experiment I) as Groups g and h. The animals of all the groups were now, in the third week, inoculated intraperitoneally with 1 mg. of a culture of the tubercle bacillus, Bovine C. A group of intact, normal mice was also inoculated in the same way, as controls. The feedings were continued during the course of the infection. The results of the experiment are shown in Table I (Experiment I).

TABLE I.

Experiment I. Intraperitoneal Infection with 1 Mg. of Culture Bovine C.

Group.	Mice.	Treatment.	Days lived.
a	Intact	None	19, 21, 29, 29, 32, 34, 39, 42.
b	Splenectomized	"	3, 9, 42, 44, 61, 87, 89.
c	"	Fed beef spleen	29, 38, 52.
d	"	" sheep "	26, 28, 30, 36, 37, 61.
e	"	" rabbit "	24, 28, 40, 41.
f	"	" human "	29, 30, 34, 37, 39.
g	"	" stomach (mouse)	9, 57, 71, 94.
h	"	" small intestine (mouse)	2, 78, 79, 90, 126, one still living.

The results of this experiment seem clear. The normal mice (Group a) were all dead by the forty-second day. Disregarding. as we are accustomed to do in this work, the animals dying in less than ten days, the splenectomized mice which received only the normal diet (Group b) lived much longer than this (forty-two to eighty-nine days). The results with these groups repeat our fundamental experiment showing the increased resistance following splenectomy, and furnish the standards for comparison with the following groups.

The mice fed with fresh spleen continuously. with two exceptions were dead in the same time as the normal controls. The resistance given by splenectomy is therefore abolished by feeding spleen. The two exceptions cannot be held of serious account. The experiment contains so many possibilities for failure that it is rather surprising that a larger number of exceptional cases do not occur. The most obvious possibility in this connection is that these exceptional mice may not have eaten sufficiently or with sufficient regularity to produce the result attained with the larger number.

The animals fed with the gastro-intestinal mucosa lived as long as or longer than the unfed, splenectomized mice. This result seems to dispose entirely of the objection which might be offered on the basis of the spleen feeding alone that the shortening of life could be the result of an added low grade intoxication.

The experiment as a whole is convincing evidence that as a specific result of feeding fresh spleen the resistance to tuberculous infection is lowered to the normal level again.

Because of its ability to produce in splenectomized mice an intoxication with the same general character and intensity as that produced by the administration of spleen, the feeding of the gastro-intestinal mucosa forms the most convincing evidence that the effects produced are in fact the restoration of a true splenic function as related to the specific infection. This being clear, another experiment which we had carried out at an earlier date may be re-

TABLE II.

Experiment II. Intraperitoneal Infection with 1 Mg. of Culture Bovine C.

Group.	Mice.	Treatment.	Days lived.
a	Intact	None	10, 15, 17, 20, 21, 22, 22, 24.
b	Splenectomized	"	40, 48, 48, 50, 58, 89.
c	"	Fed sheep spleen	14, 19, 22, 26, 26, 29.
d	"	" watery extract of sheep spleen	13, 29, 32, 36, 36.
e	"	" residue after water extraction	28, 30, 32, 34, 36, 87.
f	"	" sheep liver	1, 1, 58, 58, 68, 80.
g	"	" " thymus gland	1, 3, 48, 58, 73, 84.
h	"	" " thyroid "	5, 13,* 17,* 29,* 47, 50.
i	"	" " lymph nodes	1, 3, 51, 54, 58, 59.
j	"	" " pancreas	5, 40, 48, 64, 82, 90.
k	"	None	10, 19, 21, 24, 24, 25, 28, 30
	10 weeks before		

* Died immediately after feeding.

ported, as it is now susceptible of interpretation. In this case, reported here as Experiment II (Table II), other important organs were fed, in comparison with the spleen and certain spleen products.

This experiment differs from the first one in that no effort was made to create a tolerance to spleen before the infection was established. The result shows that while the tolerance is a necessary feature in eliminating an important objection to our interpretation of this sort of experiment, it is not at all essential to the success of the experiment itself.

Table II shows that the loss of resistance as a consequence of spleen feeding is due to the restoration of a function probably peculiar to that organ. The resistance is not affected by feeding liver, thymus gland, lymph nodes, pancreas, or thyroid gland. On the same point our first experiment (Table I) shows that it is not affected by feeding gastro-intestinal mucosa. In our earlier papers it was shown that when fresh muscle was fed the resistance was likewise unaffected. It would be desirable from this point of view to feed certain other organs or tissues, notably the bone marrow.

The results of Experiment I and those reported in our second paper[3] show that the source of the spleen is a matter of indifference; whether it is derived from mouse, sheep, rabbit, beef, or human being, the result is essentially the same.

DISCUSSION.

We have reached a point in our consideration of the relation of the spleen to the tuberculous infection in the mouse where it seems possible and advantageous to discuss the subject in more abstract terms. Until it is shown by experiment to be otherwise, we shall in the future attribute the specific properties of the spleen in its relation to tuberculosis to the activity of a single substance. For convenience we may call this substance tuberculosplenatin, a name suggesting merely its origin and its apparent relationship to tuberculosis.

Tuberculosplenatin we consider to be a substance peculiar to the spleen in the same way that adrenalin is peculiar to the suprarenal gland. It is found in the spleen of several different mammals. Its action can be demonstrated by the use of mice as we have described, either by removing the organ and following the course of the infec-

[3] Lewis and Margot, *ibid.*, 1915. xxi. 84

tion, or by following the course of the infection in splenectomized mice to which the substance is restored by feeding. While tuberculosplenatin exists in other mammals than the mouse we have not so far been able to demonstrate its activity when these other mammals are infected, presumably because of the presence of other factors which obscure its action.

Since the feeding experiments bring results, we must conclude that tuberculosplenatin is able to resist the activity of the gastric juices to a certain extent at least. We have made a beginning in the study of its physical and chemical properties. The results reported in Experiment II above with Groups d and e indicate that the substance is soluble in cold water but that it is difficult to extract completely in this way.

Tuberculosplenatin acts by producing an increased grade of susceptibility or by diminishing the resistance of the animal to infection. There is very little if any evidence as to the mechanism of the action. Murphy and Ellis[4] believe that the increase in resistance following splenectomy is a consequence of an increase in the circulating lymphocytes that follows the operation. If this is true, the substance we are considering might possibly act in restraining the freedom of action of these cells. We have made some blood counts to control this point in connection with Experiment I. We find, as did Murphy and Ellis, that there is a definite lymphocytosis following splenectomy. This has persisted, however, up to the time of infection in spite of the feeding of spleen, and while our observations are not extensive enough to be final we cannot at present attribute the loss of resistance in our experiments to a depression of the number of circulating lymphocytes. The enlargement of the lymph nodes as a result of splenectomy has not been sufficiently marked in our experience to be significant. It has certainly been no less marked in those animals receiving spleen in the food than in the others. For the present, in view of the almost unlimited possibilities for experimental observation in connection with this line of work, we are disposed to regard attempts to explain the reactions we have encountered as of secondary interest.

[4] Murphy, Jas. B., and Ellis, A. W. M., *Jour. Exper. Med.*, 1914, xx, 397.

SUMMARY.

Experiments are reported which show that in all probability the increased resistance to tuberculous infection which is imparted to mice by the removal of the spleen is a consequence of the loss of a function of the organ. This function can be restored by the feeding of fresh spleen. For the present we attribute these changes to the removal and restoration, as the case may be, of a particular substance for which the designation tuberculosplenatin is suggested. This substance is assumed to be related to the spleen as adrenalin is related to the adrenal gland. It is peculiar to the organ but not to the species. It is not found in other organs of the body so far as our observations have extended. The absence of the substance from the lymphatic glands seems of especial importance in this connection.

THE NUMERICAL LAWS GOVERNING THE RATE OF EXCRETION OF UREA AND CHLORIDES IN MAN.

II. The Influence of Pathological Conditions and of Drugs on Excretion.

By FRANKLIN C. McLEAN, M.D.

(From the Hospital of The Rockefeller Institute for Medical Research, New York.)

(Received for publication, April 17, 1915.)

INTRODUCTION.

In a previous communication (1) we have considered the laws governing the rate of excretion of urea and chlorides and their variations in individuals with normal excretion. The present paper deals with the application of these laws to the study of pathological conditions, and to the study of the action of certain drugs on excretion. The aim is for the present not primarily to draw conclusions regarding the diseased conditions themselves, but to offer a means of obtaining more accurate information regarding these conditions than has hitherto been possible. Certain cases which illustrate the nature of the observations made possible by the method have been selected for presentation.

The laws formulated by Ambard for the excretion of urea are already in clinical use in France (2). In the clinic the method has been used almost solely as a measure of renal function, the variations in the values obtained with Ambard's coefficient serving as an indication of disturbed function. Widal, Ambard, and Weill (3) have studied the excretion of chlorides in edematous subjects, and have made some observations on the effect of diuretics. Bauer and Habetin (4) and more recently Bauer and von Nyiri (5), working in Ortner's clinic in Vienna, have published observations, obtained by means of Ambard's laws, on urea and chloride excretion in a considerable number of cases. Inasmuch as they have used an altogether unreliable method for the determination

366

of urea in the blood, and publish figures as high as 0.779 gm. of urea per liter of serum in individuals with normal excretion, a figure which a person with normal excretion could attain only after ingestion of enormous quantities of urea, our figures can in no way be comparable with theirs.

With the more accurate and rapid methods for urea and chloride determinations now available, requiring only a small amount of blood, it is possible to make frequent and repeated simultaneous observations on urea and chloride excretion in relation to the concentration in the blood in the same individual, and at a minimum of discomfort to him. A wider field is thus opened for investigation. Quantitative studies of human bodily functions could heretofore be carried out only to a limited extent. With the methods here presented, we have a direct measurement of one of the most important excretory functions, elimination of urea through the kidneys, and also a standard for judging the concentration of chlorides in the plasma in its relation to normality. That these methods are more delicate than other methods used for study of disturbed renal function is apparent from the data presented.

Methods of Study.

Simultaneous observations on urea and chloride excretion are made in patients in a manner identical with that described in the preceding paper (1). One-half hour after the patient drinks 150 to 200 cc. of fluid, the bladder is emptied and the subject takes no further fluid or food until a carefully timed period, usually of seventy-two minutes, is ended. The urine excreted during this period is collected, and at the middle of the period about 10 cc. of blood are withdrawn from an arm vein, clotting being prevented by a small amount of powdered potassium oxalate. The choice of a period of seventy-two minutes is merely for the sake of convenience, seventy-two minutes being one-twentieth of twenty-four hours. A one or two hour period may, of course, be used. all calculations in any case being made on a basis of twenty-four hours. In case an error of a few minutes is made in the time of collection of the second specimen, the calculation should be made on the basis of the time actually elapsed between the voiding of the first and second specimens. The amount of urea in the whole blood, and total chlorides, estimated as sodium chloride, in the oxalated plasma after centrifugalization. are determined. Both urea and chlorides are determined in the urine. By substituting the values obtained in the proper formulas the relationship of the rate of excretion of these substances to their concentration in the blood is determined. In the case of urea the rate of excretion under the conditions found is directly measured in terms of the normal. by the use of the following formula:

$$\text{Index of urea excretion} = \frac{8.96 \times \text{Gm. urea per 24 hrs.}}{\text{Wt. in kilos} \times (\text{Gm. urea per liter of blood})^2} \times \text{Gm. urea per liter of urine}$$

In the normal individual this formula gives a value of about 100 for the index. The theoretical concentration of sodium chloride in the plasma, under the conditions found, is calculated from the following formula:

$$\text{Calculated plasma NaCl} = 5.62 + \sqrt{\frac{\overline{\text{Gm. NaCl per 24 hrs. } \sqrt{\text{Gm. NaCl per liter of urine}}}}{4.23 \times \text{Wt. in kilos}}}$$

For the derivation and significance of these formulas the reader is referred to the foregoing paper. Analyses are carried out by the methods there indicated, and calculations are made with the slide-rule there described.

Observations are made as frequently as desirable. Since they involve no discomfort to the patient, other than the slight inconvenience incident to taking blood, daily observations are usually made when changes in the state of the patient are taking place. The results may be obtained in full within an hour after collection of the specimens, so that information is furnished more rapidly than in other ways. As the observations involve neither additions to the diet nor introduction of foreign substances, they interfere in no way with the dietetic régime or with the progress of the case. The calculations are independent of nitrogen and chloride intake. Gastro-intestinal disturbances, such as vomiting and diarrhea, therefore, do not affect the results, and it is unnecessary to keep the patient on an analyzed diet. Observations can be made at any desired time in almost every case.

Catheterization has been performed only when necessary or advisable in order to avoid danger of losing the specimen, as, for instance, during delirium. It should be practiced in cases in which there is residual urine. Water is given before the seventy-two minute period in order to avoid conditions simulating retention due to dehydration. In the great majority of cases a rate of flow of urine for the period corresponding to 1,000 cc. or more per twenty-four hours is obtained. In practice we make it a rule not to accept results on urea excretion calculated from a urine flow at a rate of less than 500 cc. in twenty-four hours, since results obtained under the condition of a very low rate of urine excretion are apt to be misleading. The chloride formula appears to hold even for very low rates of urinary output.

Pathological Conditions Affecting Excretion.

Conditions Associated with Retention.—Retention of nitrogen and salt have long been recognized as occurring in certain forms of cardiac and renal disease. By the use of the present methods new light is thrown on the mechanism of this retention. The salt retention of pneumonia, for example, has been shown to be entirely different in mechanism from the salt retention of chronic nephritis. For our purpose we propose to consider retention only in the sense of its relation to excretion. That is to say, our conception of retention, in the sense just defined, is a relatively high concentration of substances in the blood in order to induce sufficient excretion. That retention in the tissues themselves occurs is often lost sight of in studies of renal function by methods which require a comparison of intake and output of certain substances. It is manifestly wrong to consider the kidneys responsible for the failure to excrete a certain amount of salt given by mouth, if the salt is taken up by the tissues and the concentration in the plasma remains low. But if the concentration of urea or chlorides in the blood or plasma remains proportionately high, and the rate of excretion proportionately low, it is correct to speak of retention in the sense of a failure to excrete properly. This is the condition in certain types of cardiorenal, or renal disease.

Retention, thus defined, is not an accumulation, but a higher level of substances in the blood, to compensate for increased difficulty in passing through the outlet. It does not mean that the kidney is not able to excrete the normal amount of urea, for the daily excretion, as has repeatedly been shown, may be quite normal. As much may, in fact, be excreted as is absorbed, and the individual be in perfect nitrogen balance (6), although the blood urea or blood nitrogen is much increased. But it does mean that an increased amount in the blood is required in order that increased pressure may be provided to cause the same rate of excretion through diseased kidneys that would be carried on through normal kidneys with a smaller amount in the blood. Actual accumulation may and does occur, under certain conditions. It differs from retention in the above sense, because there is a failure to excrete substances as rapidly as they are absorbed or formed. The picture is different from the purely compensatory phenomenon just described.

The abnormal individual who has retention comes to an equilibrium at a level which is different from that of the normal individual. By comparing his rate of excretion with the standard normal, under the conditions of concentration found in the blood and urine, we obtain a measure of the degree of compensation required to secure the necessary rate of excretion. The necessity for increased pressure may be due to diminished outlet, because a certain amount of the kidney tissue is absent or is not functionating. Thus a person with only one normal kidney, the other being removed, should have an index of urea excretion of approximately 50, 100 representing normality.   On the other hand, the kidney cells present may be functionally deficient, though still able to carry out their functions if the pressure under which they work is increased.   In either case the index measures the relative functional capacity for the excretion of urea.   An index of 100, or 100 per cent, is the standard normal, based on a considerable number of observations on normal individuals.   An index of 25 indicates that the individual under observation is excreting urea at 25 per cent of the normal rate under the conditions of concentration of urea found in the blood and urine.   An index of 80 is considered as the lower limit of normality, and no normal individual should give a lower figure, providing the fluid intake is sufficient to prevent dehydration.

Observations.

· Case I (Text-fig. 1) provided opportunity for making repeated observations during the course of acute nephritis, occurring as a sequel of pneumonia and empyema.   Besides fluid intake and output, weight, concentration of albumin in the urine, systolic and diastolic blood pressure, P-R time measured electrocardiographically, pulse rate, and temperature, Text-fig. 1 illustrates the findings as regards urea and chloride excretion, with occasional observations on the rate of excretion of phenolsulphonephthalein.   The concentration of urea in the blood and the rate of excretion are shown, together with the index of urea excretion, calculated as above indicated.   It will be seen that during the first two months of the disease, though the level of the blood urea fluctuated considerably, due to changes in the nitrogen intake, the index remained remarkably constant at 6 to 8.   The first fall in blood urea followed an initial restriction of diet, and was accompanied by a proportionate decrease in the rate of excretion.   As the nitrogen intake increased, the blood urea rose steadily, reaching a level of 1.3 grams per liter and maintaining this level for some time.   The rate of excretion at the same time increased proportionately, so that the index remained constant.   When functional improvement did occur it was accompanied by a rapid

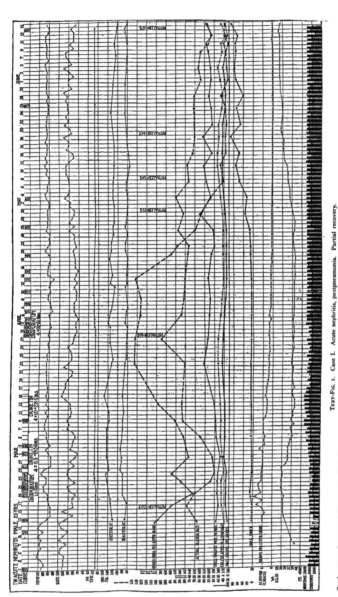

Text-Fig. 1. Case I. Acute nephritis, postpneumonia. Partial recovery.

fall in the blood urea, and an increase in the index toward normal. The rate of excretion during and following this fall remained practically the same as when the high concentration of urea in the blood was found. The maintenance of a constant relationship of rate of excretion to the concentration in the blood, at 6 to 8 per cent of normal, under fluctuating conditions of nitrogen intake, indicates that the process of excretion was being carried out under laws just as definite as those governing the normal excretion. During this period the actual rate of excretion maintained was as high as later in the disease when the conditions were more nearly normal. During recovery the only changes were the fall in concentration of urea in the blood and the rise in the value of the index. One must conclude, therefore, that the high concentration of urea in the blood was maintained in order to compensate for the difficulty in passing through a damaged outlet in the form of kidneys which were functionating at a rate only 6 to 8 per cent of normal. When functional conditions improved, it was no longer necessary to maintain the concentration of urea in the blood at so high a level, and it promptly fell to within normal limits. The relation of the rate of phthalein elimination to the urea index is shown in this case. Both increased at the same time, the increase being parallel. The failure of the index to reach the normal figure for some time, in spite of the fact that the blood urea was within normal limits, shows that the blood urea figure alone was not sufficient in this case to indicate the functional disturbance which still existed.

Text-fig. 1 also shows the rate of chloride excretion, maintained at a low level by a salt-poor diet; the theoretical concentration of sodium chloride in the plasma, calculated from the rate of excretion from the formula given above; and the concentration of sodium chloride found in the plasma by analysis. A marked discrepancy is seen between the actual and calculated concentrations of sodium chloride in the plasma, the former being always higher, and reaching at one time a level of over 7 grams per liter. A return of the plasma chlorides toward the theoretical concentration was much slower than the fall in blood urea, and confirms previous experience that chloride function may be disturbed for some time after apparent recovery. The case also shows the independence of the urea and chloride functions, since the urea function improved so rapidly some time before any change in the chloride relations was visible.

Drug experiments in this case were entirely negative. Edema was present at the start, but disappeared rapidly under free catharsis and sweat baths. The course of the edema is indicated by the weight chart. The later increase in weight was a healthy increase and was not accompanied by any edema.

Case II (Table I) is one of acute nephritis following pneumonia, similar to, but much milder than, the previous one. The urea index remained fairly constant at about 75 for a time, and then rose to a point considerably higher than normal. The chloride excretion followed a similar course, the concentration of sodium chloride found in the plasma being higher than the theoretical concentration during the first stage, and lower during the stage in which the urea index was high. This represents a stage of vascular irritability, which has been described qualitatively by Schlayer (7) and others, but this method offers a means of measuring such changes quantitatively. Recovery in this case is accompanied by a return of both urea and chloride functions toward normal.

TABLE I.

Case II.

Date (1915).	Weight.	Urine per 24 hrs.	Urea.				Sodium chloride.					Urine findings.		
			Gm. per liter of blood Ur.	Gm. per liter of urine C.	Gm. per 24 hrs. D.	In-dex ofex-cre-tion I.	Gm. per liter of urine C.	Gm. per 24 hrs. D.	Gm. per liter of plasma.			Blood.	Albumin.	Casts.
									Cal-cu-lated.	Ac-tual.	Dif-fer-ence.			
	kilos	*cc.*												
Mar. 9	51.0	1,100	0.562	26.2	28.8	83	0.9	0.99	5.69	5.60	−.09	++	H.T.*	+++
" 10	51.0	2,070	0.577	20.9	43.3	104	0.7	1.5	5.70	5.65	−.05			
" 11	51.0	1,195	0.627	27.0	32.3	75	0.6	0.7	5.67	5.81	+.14	o	F.T.	+
" 12	51.0	1,200	0.557	23.3	28.0	77	2.0	2.4	5.75	5.97	+.22		V.F.T.	
" 13	51.0	1,600	0.506	16.4	26.3	71	5.2	8.3	5.92	6.15	+.23	o	V.F.T.	+
" 18	51.0	2,034	0.547	15.35	31.2	72	9.3	18.9	6.14	6.17	+.03		V.F.T.	Few.
" 22	53.0	1,500	0.557	17.1	26.6	58	8.3	12.5	6.02	6.08	+.06		V.F.T.	"
" 26	53.8	1,440	0.457	16.3	23.4	75	10.9	15.7	6.10	6.06	−.04		V.F.T.	++
Apr. 2	56.0	6,000	0.321	9.0	54.0	250	7.3	43.8	6.33	6.03	−.30		V.F.T.	++
" 9	56.2	6,300	0.308	10.0	63.0	330	9.8	61.8	6.52	6.21	−.31		V.F.T.	+
May 30	57.0	1,500	0.250	15.9	23.8	240	13.5	20.2	6.17	6.19	+.02		V.F.T.	Few.

* In Table I, H.T. signifies heavy trace; F.T., faint trace; and V.F.T., very faint trace.

Case II.—(Table I.) D. C., male, aged 16. Diagnosis, lobar pneumonia, acute nephritis. Admitted Mar. 8, 1915, on 6th day of pneumonia. Temperature 105.5° F. Consolidation posteriorly in right lower lobe. Blood pressure 112 systolic, 60 diastolic. Mar. 10. Crisis, followed by rapid disappearance of physical signs of consolidation. No edema at any time; no further abnormal physical signs, except for slight pallor. Mar. 26. Discharged from hospital, feeling well except for slight weakness. Blood pressure 102 systolic, 60 diastolic. May 30. Patient has been working for a month; feels perfectly well. Blood pressure 108 systolic, 70 diastolic.

In Case III (Table II) enormous pressure was required in order to induce the necessary rate of excretion of urea and sodium chloride. The close agreement of the urea index with the rate of phthalein excretion is also seen in this case.

Attention has been called by Widal and Javal (8), and more recently by Folin, Denis, and Seymour (9), and by Frothingham and Smillie (10) to the reduction in blood urea or blood nitrogen brought about by giving a low protein diet. They state that in cases with high blood nitrogen, such as in this case (Case III, Table II), no marked reduction occurs on a nitrogen-poor diet. That any considerable reduction is difficult in a case such as this may be shown mathematically. Were this patient to attain a blood urea figure of 0.42 grams per liter, that is to say, one within normal limits, the total urea excretion per twenty-four hours in 1 liter of urine would be 1 gram, the index remaining at 1.2. The formula would then read $\text{Index} = \dfrac{8.96 \times 1 \sqrt{1}}{42.1 \times (0.420)^2} = 1.2$. As the concentration of urea is practically the same in all the tissues as in the blood, before the concentration of over 2 grams per liter could be reduced to 0.42 grams per liter this patient would have to excrete nearly 80 grams of urea already in the body. As the rate of excretion would diminish in proportion to the square of any diminution in the concentration in the blood, it would require an infinite time for the blood urea figure to reach the low level, even if the diet could be kept so that only 1 gram of urea were formed daily. The practical impossibility of reaching such a result is obvious. The comparative ease with which a patient with an index of urea excretion more nearly normal would reach a blood urea figure within normal limits when on a low nitrogen diet may easily be seen.

Case III.—(Table II.) J. O. S., male, aged 13. Diagnosis, chronic nephritis, mitral stenosis. Admitted Jan. 19, 1915, with generalized edema, dyspnea, frequent nausea, and vomiting. Blood pressure 185 systolic, 135 diastolic. Urine clear, straw colored, specific gravity 1008, acid, no sugar. Albumin, 6.5 gm. per liter (Esbach). Many large granular casts, hyalin and epithelial casts, epithelial cells, and leucocytes. No red cells. Eye grounds normal. Jan. 20. Phenolsulphonephthalein, 2 per cent excreted in 2 hours. Jan. 26. Left hospital unimproved. Died one month later.

TABLE II.

Case III.

Date (1915).	Weight.	Urine per 24 hrs.	Urea.				Sodium chloride.				
			Gm. per liter of blood Ur.	Gm. per liter of urine C.	Gm. per 24 hrs. D.	Index of excretion I.	Gm. per liter of urine C.	Gm. per 24 hrs. D.	Gm. per liter of plasma.		
									Calculated.	Actual.	Difference.
	kilos	cc.									
Jan. 20...	42.1	889	2.147	8.8	7.82	1.0	1.5	1.33	5.71	6.55	+.84
" 22...	42.1	960	2.203	9.37	9.0	1.2	2.0	1.92	5.75	6.59	+.84

Case IV (Text-fig. 2) illustrates an actual accumulation of urea and chlorides, due to restriction of fluids while on a full diet. That the power of excreting substances in high concentration in the urine is diminished in certain forms of nephritis is well recognized. It is quite apparent that a fluid output insufficient to carry off the waste products formed will lead to actual accumulation of these products in the body. This condition is manifested, in the case presented, by the increasing urea and chloride content of the blood and the diminishing index of urea excretion. When the fluid intake is increased, urea, chlorides, and index return rapidly to their former state. This case also illustrates the necessity of careful control of the fluid intake in any experiments on nitrogen or chloride balance. That a washing out of urea and chlorides occurs in both normal and pathological individuals with a high fluid output is recognized and expressed in the formulas. Any experiment aiming to determine the balance between nitrogen or chloride intake and output, which does not take the fluid output into account, is apt to be misleading.

Conditions similar to those occasioned by disturbance of function due to renal disease may also be found in heart failure, with passive

congestion of the kidneys. In case the circulation alone is at fault
the conditions return to normal when the congestion is relieved by
digitalis or otherwise.

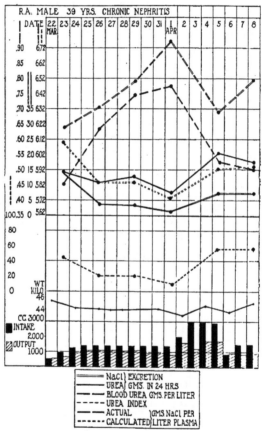

TEXT-FIG. 2. Case IV. Chronic nephritis. Blood pressure 185 systolic, 130
diastolic. Urine clear, amber, specific gravity 1023, acid, no sugar, albumin 2
gm. per liter (Esbach). Numerous hyalin, few granular casts. Red blood cells
and white blood cells. Mar. 23 to Apr. 2. Fluids restricted, food and salt *ad
libitum.* After Apr. 2 fluids increased and food and salt restricted.

Case V (Table III) illustrates the findings in a typical case of
heart failure with passive congestion of the kidneys, which was

restored to a functionally normal state by digitalis. During the condition of passive congestion, as evidenced by cyanosis, edema, and albuminuria, the blood urea was high, the urea index low, and the plasma chlorides high in relation to the rate of excretion. Under the action of digitalis the blood urea fell, the index rose to normal, and the plasma chlorides fell until the concentration agreed with the theoretical concentration as calculated from the rate of excretion. The urine became albumin-free and no casts were to be found, indicating complete relief from the congestion which had caused disturbed function.

Case V.—(Table III.) M. H., male, aged 53. Diagnosis, aortic and mitral insufficiency, cardiac failure, passive congestion of kidneys. Admitted Jan. 2, 1915, with dyspnea, cyanosis, edema of extremities. Blood pressure 143 systolic, 110 diastolic. Urine dark amber, cloudy, specific gravity 1027, acid, albumin + + +, numerous hyalin and granular casts. Jan. 8. Condition unchanged by rest in bed; digitalis therapy started (digipuratum 0.1 gm., 4 times daily). Jan. 12. Patient much more comfortable, edema disappearing, fluid output in excess of intake. Jan. 17. Digitalis discontinued; total 2.9 gm. Jan. 19. No edema, or dyspnea. Urine albumin-free, no casts. Blood pressure 120 systolic, 75 diastolic. Patient continued in good condition up to time of discharge on Apr. 11.

TABLE III.

Case V.

Date (1915).	Weight.	Urine per 24 hrs.	Urea.				Sodium chloride.						Urine albumin, gm. per liter (Esbach).	Medication.
			Gm. per liter of blood Ur.	Gm. per liter of urine C.	Gm. per 24 hrs. D.	Index of excretion I.	Gm. per liter of urine C.	Gm. per 24 hrs. D.	Gm. per liter of plasma.				
									Calculated.	Actual.	Difference.		
	kilos	*cc.*											
Jan. 4..	83.1	500	0.567	27.3	13.65	24	3.3	1.65	5.71	5.84	+.13	1.5	
" 8..	77.0	420	0.597	30.8	12.94	25	3.35	1.41	5.70	6.15	+.45	0.5	
" 12..	74.0	3.020	0.351	11.84	35.7	130	8.9	26.9	6.07	6.12	+.05	0.25	Digitalis.
" 15..	73.0	1,190	0.356	20.7	24.7	100	6.9	8.21	5.88	5.91	+.03	0	"
" 27..	71.0	1,620	0.328	15.7	25.5	110	7.7	12.5	5.96	5.85	−.11	0	
Feb. 16..	73.2	1,000	0.286	16.42	16.42	100	12.4	12.4	5.99	5.96	−.03	0	

Case VI (Table IV) is a case of heart failure in which actual accumulation of urea and chlorides in the blood occurred. This is evidenced by the very rapid rise in the concentration of both urea and chlorides in the blood, without any increased rate of excretion. This picture, with its rapidly diminishing urea index, is quite different from that presented by Cases I and III, in which a persistently high blood urea was accompanied by a constant value for the index.

Case VI.—(Table IV.) A. W., female, aged 13. Diagnosis, mitral insufficiency, auricular fibrillation, cardiac failure, death. First admitted Apr. 17, 1913. Last admission, Dec. 3, 1914, with edema, dyspnea, and cyanosis. Urine amber, cloudy, specific gravity 1030, acid, albumin + + +,. few granular casts. Blood pressure 120. Dec. 12. Phenolsulphonephthalein, 68 per cent excreted in 2 hours. Patient's condition failed to improve materially under digitalis, and about Dec. 28 she became decidedly worse, gradually stuporous, and more cyanotic. Edema rapidly disappeared during last few days, and blood contained a very high proportion of cells and a low percentage of plasma. Jan. 10. Died, after having been unconscious two days. Autopsy revealed enormously enlarged and dilated heart, with lesion of mitral valve. The kidneys were those of passive congestion.

TABLE IV.

Case VI.

Date (1914).	Weight.	Urine per 24 hrs.	Urea.				Sodium chloride.					Urine albumin, gm. per liter (Esbach).
			Gm. per liter of blood Ur.	Gm. per liter of urine C.	Gm. per 24 hrs. D.	Index of excretion I.	Gm. per liter of urine C.	Gm. per 24 hrs. D.	Gm. per liter of plasma.			
									Calculated.	Actual.	Difference.	
	kilos	cc.										
Dec. 12 ...	37.9	680	0.460	23.75	16.15	88	7.0	4.75	5.90	6.15	+ .25	
" 21 ...	38.8	540	0.432	22.05	11.9	70	3.3	1.8	5.76	5.95	+ .19	3.0
(1915)												
Jan. 5 ...	37.0	1,850	0.837	11.13	20.6	24	2.4	4.44	5.83	7.24	+1.41	1.5
" 7 ...	32.0	1,000	1.38	12.25	12.25	6	3.7	3.7	5.85	7.85	+2.00	2.5
" 8 ...	30.0	500	2.18	17.85	8.92	2	1.1	0.55	5.69	8.44	+2.75	0.5
" 9 ...			2.70							8.24		
" 10 ...			3.36									0.1

We may summarize the findings in cases with retention as follows. In individuals with defective elimination of urea or chlorides, an increased concentration of these substances in the blood is required to compensate for the faulty excretion. Under the conditions of increased concentration in the blood, the actual rate of excretion and the balance between intake and output may be quite as good as in a perfectly normal individual. The increased concentration in the blood is not to be regarded as an accumulation unless the concentration in the blood is increasing and is associated with a diminished elimination. In any given case the index of urea excretion measures, in its relation to normality, the rate of excretion under the conditions found. Increased chloride concentration in the plasma is shown by comparing the amount actually found with the theoretical concentration, calculated from the rate of excretion. The concentration in the blood and the rate of excretion of urea and chlorides depend not only on the nitrogen and chloride intake, but on the fluid intake and output. Deficient elimination due to passive congestion may be restored to normal if the circulatory failure is relieved.

The Chlorides in Pneumonia.

That the chloride retention in pneumonia is associated with a diminished chloride content in the blood has been shown by Peabody (11) and others. Peabody has also shown that the concentration of chlorides in the blood increases at the time excretion begins. This is directly opposed to the condition in nephritis, where chloride retention is usually associated with a high chloride content in the plasma, and suggests at once that the chloride retention occurring in pneumonia and other fevers is not due to a failure on the part of the kidneys to excrete chlorides, as some writers still hold (12).

We have examined the chloride content of the plasma and its relation to the rate of chloride excretion in 60 observations in 13 cases of lobar pneumonia, and have found that the failure to excrete chloride during the acute stage of the disease is almost always associated with a concentration of sodium chloride in the plasma below 5.62 grams per liter (the normal threshold).[1] One must believe that this is the cause of failure to excrete chlorides, since excretion begins with a rise in the chloride content of the plasma.

Case VII (Text-fig. 3) illustrates the findings regarding chloride excretion in a typical case of uncomplicated pneumonia. During the acute stage of the disease, when only a very small amount of chlorides appeared in the urine, the plasma content of sodium chloride was considerably below 5.62 grams per liter. At the crisis the plasma chlorides rose abruptly, and excretion began and continued

---

[1] Similar findings have been reported by J. Snapper (*Deutsch. Arch. f. klin. Med.*, 1913, cxi, 429) in 5 observations during the chloride retention of pneumonia.

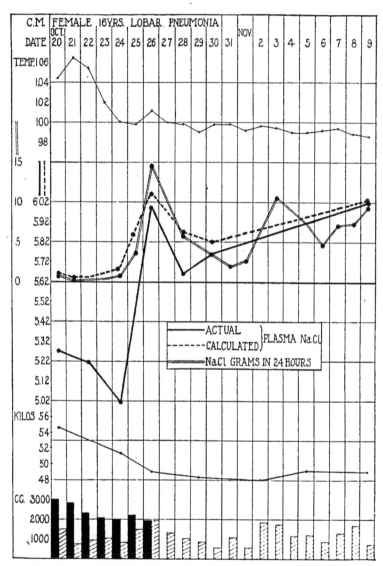

TEXT-FIG. 3. Case VII. Uncomplicated lobar pneumonia, with recovery.

in proportion to the concentration in the plasma. That the low rate of excretion of chlorides in pneumonia is not due to chloride starvation has been repeatedly shown (11). That the low chloride content of the plasma is not due to low intake is shown by the rapid rise in Case VII at the time of crisis, without any increase in the chloride intake. While the finding of a low concentration in the plasma shows that the kidneys are not responsible for salt retention in pneumonia, it does not explain the mechanism of the retention, which requires further investigation.

The Chloride Threshold in Fever.

We have calculated the threshold in cases with fever from the formula used in normal individuals (1):

$$\text{Threshold} = \text{Plasma NaCl} - \sqrt{\frac{\text{Gm. NaCl per 24 hrs.} \sqrt{\text{Gm. NaCl per liter of urine}}}{4.23 \times \text{Wt. in kilos}}}$$

For the most part the threshold, as calculated by this formula, is actually lowered during the chloride retention of pneumonia, the average threshold in our cases being 5.42 as opposed to the normal threshold of 5.62 grams of sodium chloride per liter of plasma. That a similar condition occurs in other fevers is shown by Case VIII (Table V).

Case VIII.—(Table V.) B. G., female, aged 20. Diagnosis, acute rheumatic fever. Admitted Nov. 20, 1914. Nov. 27. Pain and swelling in joints. Temperature 103° F. Urine amber, cloudy, specific gravity 1019, alkaline, faint trace of albumin, no casts, few leucocytes.

TABLE V.
Case VIII.

Date (1914).	Weight.	Urine per 24 hrs.	Gm. per liter of urine C.	Gm. per 24 hrs. D.	Sodium chloride. Gm. per liter of plasma. Calcu-lated.	Actual.	Differ-ence.	Thresh-old.
	kilos	cc.						
Nov. 27........	60.0	1,285	2.8	5.1	5.80	5.54	−.26	5.36

In some instances the threshold was never materially lowered, but in no case was it appreciably raised during the acute stage of the disease. In one case of chronic interstitial nephritis, dying of pneumonia, with high blood urea, the chloride threshold was considerably lowered. In one case it was raised temporarily during convalescence. As a rule, a lowered threshold returns to normal at the time excretion begins. One case showed a threshold still lowered at the time of discharge from the hospital, after complete recovery.

The findings regarding chlorides in pneumonias and other fevers may be summarized as follows: Chloride retention in pneumonia is associated with a lowered concentration of chlorides in the plasma, and failure of excretion is apparently due to this cause. The threshold is often considerably lowered during fever, and usually returns to normal after the temperature becomes normal. A raised threshold has not been observed during acute fever.

The Chlorides in Diabetes (13).

In 78 observations in 28 cases of diabetes mellitus, mainly severe cases, we have found in a majority a lowering of the chloride threshold. Such a condition is shown in Case IX (Text-fig. 4). In this case, under varying conditions of diet, excretion of sugar, and concentration of sugar in the blood, the plasma chlorides remained proportionately very low, corresponding to a threshold of about 5.12 instead of 5.62 grams of sodium chloride per liter of plasma. Only when the patient developed an edema, such as that to which diabetics are subject, did the plasma chlorides show a relative increase, returning to their previous relationship to the rate of excretion with the disappearance of the edema.

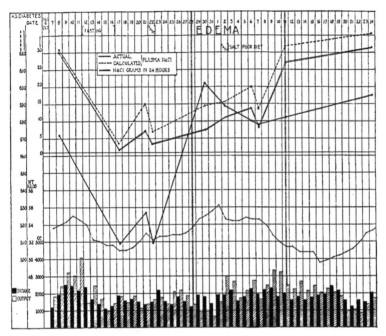

TEXT-FIG. 4. Case IX. Oct. 7. Urine clear, amber; sugar, 75 gm. in twenty-four hours; albumin 0; casts 0. Ferric chloride reaction heavy. Fasted Oct. 12 to 18, with whiskey. No alkalis. Oct. 18. Sugar 0; albumin 0. Ferric chloride reaction very faint. Urine remained sugar-free, with the exception of occasioual traces when the increase in diet was not tolerated. Nov. 1. Well marked edema of both feet and ankles, which remained until Nov. 11. Urine, albumin 0, no casts. Had similar edema at subsequent times.

Case X (Table VI) illustrates the findings in a typical case with constantly lowered threshold. In this case the actual concentration of sodium chloride in the plasma is almost invariably below 5.62 grams per liter, the normal threshold.

The average of the threshold calculated by the above formula in 78 observations in diabetics is 5.34, in contradistinction to the average of 5.61 + grams in all normal individuals. Up to the present time we have not succeeded in associating this change in chloride threshold and excretion with any other peculiarities of individual cases of diabetes. It seems to bear no relation to the amount of sugar excreted nor to acidosis. It seems to occur most commonly in the more severe cases, but has also been seen in milder cases.

Case X.—(Table VI.) J. U., male, aged 44. Diagnosis, diabetes mellitus.

TABLE VI.

Case X.

Date (1914).	Weight.	Urine per 24 hrs.	Sodium chloride.						Sugar in 24 hrs.	Ferric chloride reaction.	Diet per 24 hrs.	
			Gm. per liter of urine U.	Gm. per 24 hrs. U.	Gm. per liter of plasma.			Threshold.			Carbohydrate.	Total calories.
					Calculated.	Actual.	Difference.					
	kilos	*cc.*							*gm.*		*gm.*	
Dec. 3..	45.0	860	1.9	1.63	5.73	5.00	−.73	4.89	2.73	+++	Fasting with alcohol 3d day.	
" 7..	45.0	655	2.4	1.6	5.73	5.08	−.65	4.97	0	+	0	637
(1915)												
Jan. 6..	41.3	1,850	11.13	20.6	6.23	5.35	−.88	4.74	0	+	0	466
May 4..	42.6	950	10.1	9.6	6.03	5.87	−.16	5.46	0	+	40.1	343.5
" 7..	44.0	800	9.8	7.85	5.98	5.22	−.76	4.86	0	+	70	600.5
" 12..	44.0	1,320	9.2	12.1	6.06	5.48	−.58	5.04	Trace	+	10	553
" 13..	43.8	1,590	9.4	14.9	6.12	5.60	−.52	5.10	0	+	20	929
" 18..	44.0	690	10.0	6.9	5.96	5.40	−.50	5.12	Trace	+	15.1	1,526.9
" 22..	44.0	700	14.3	10.0	6.07	5.70	−.37	5.25	0	+	0	1,427
June 2..	43.0	800	9.5	7.6	5.98	5.42	−.56	5.06	0	+	0	1,962

The Influence of Drugs on Excretion of Urea and Chlorides.

Study of the effects of drugs has so far been confined mainly to the action of digitalis. We have found, in the majority of cases examined, a lowering of the chloride threshold as the result of giving digitalis, both in individuals with normal excretion and in cases with passive congestion. The threshold may be brought merely to normal, as in Case V (Table III), or it may be brought temporarily far below normal, as in Case XII (Table VII).

Case XII.—(Table VII.) G. M., male, aged 50. Diagnosis, cardiac hypertrophy, auricular fibrillation. Patient has been under observation since May 10, 1914. Rate remains slow and patient feels well when taking digitalis, but rate becomes rapid and dyspnea and edema appear when digitalis is omitted. Oct. 5, 1914. Patient has had no digitalis since Sept. 27. Pulse rate gradually increasing. Urine clear, amber, specific gravity 1025, neutral, albumin 0.25 gm. per liter (Esbach), no casts found. Oct. 16. Marked dyspnea, face puffy, heart rate rapid. Output of fluid has diminished and weight increased. Digitalis started. Oct. 23. Fluid output increased and patient much better. No edema.

Nov. 2. Patient feels well; rate remains slow. Two similar series of observations have since been made on the same patient.

TABLE VII.

Case XII.

Date (1914).	Weight.	Urine per 24 hrs.	Urea.				Sodium chloride.			Gm. per liter of plasma.			Phthalein excreted.		
			Gm. per liter of blood Ur.	Gm. per liter of urine C.	Gm. per 24 hrs. D.	Index of excretion I.	Gm. per liter of urine C.	Gm. per 24 hrs. D.		Calculated.	Actual.	Difference.	1st hr.	2d hr.	Total.
	kilos	*cc.*											*per cent*	*per cent*	*per cent*
Oct. 5..	71.0	760	0.400	18.2	13.85	53	13.0	9.8		5.96	6.11	+.15	38	30	68
" 16..	73.0	685	0.500	26.82	18.37	47	5.1	3.5		5.78	5.97	+.19	Medication.		
" 23..	72.6	1,600	0.352	13.15	21.0	76	10.0	16.0		6.03	5.51	−.52	Digitalis.		
" 26..	72.0	1,680	0.358	13.68	22.9	82	12.3	20.7		6.11	6.11	0	"		
Nov. 2..	72.4	1,580	0.396	14.02	22.3	66	12.1	19.1		6.09	6.10	+.01			

Urea excretion is increased after passive congestion is relieved by digitalis, but digitalis seems to have no specific direct effect on urea excretion. Case XI (Text-fig. 5) shows the effect of both digitalis and theocin, as separate observations on the same individual. Both the chloride and urea findings are illustrated, as in Text-fig. 1. Except for the temporary fall in the urea index at first, caused by insufficient fluid, the index remains practically unchanged, and shows no response to drug therapy. The constantly low index is apparently due to a disturbance in function due to chronic nephritis, with hypertension. When admitted, the patient was under the influence of large doses of infusion of digitalis, and the plasma chlorides were very low, the concentration being below the normal threshold. As the influence of digitalis disappeared, the plasma chlorides rose rapidly, edema appearing and increasing with the increase in plasma chlorides. Diuresis with theocin increased the chloride excretion and brought down the plasma chlorides temporarily to the theoretical concentration. Conditions rapidly returned to the previous state, until digitalis was again administered. At this time the plasma chlorides did not fall to the previous low level, though they responded to digitalis therapy with a decrease in concentration, approaching but not reaching the theoretical concentration.

Fig. 5. Case XI. Chronic nephritis, hypertension, and cardiac failure. Urine clear, ecific gravity 1007, acid, albumin heavy trace, numerous hyalin and granular casts. under influence of infusion of digitalis. Later treated with theocin and with digi-

Edema, however, disappeared and the patient gave other evidences of a favorable reaction to digitalis. Since this time the patient has been out of the hospital, and has returned with edema. At this later time, not shown in Text-fig. 5, digitalis caused the excretion of 11,000 cc. of urine in forty-eight hours, and the plasma sodium chloride fell to 5.37 grams per liter and remained low for some time. The index of urea excretion was exactly the same during this enormous diuresis as on the preceding day, when the fluid output was only 900 cc. The length of time for which the chloride threshold remains low after digitalis varies in different cases, but it may remain low for ten days or longer after digitalis is discontinued.

Digitalis and diuretin were both without effect in acute nephritis (Case I), so far as could be told by our observations. They were also without effect in Case IV in later observations, not shown in Text-fig. 2. We cannot at present make any general statements regarding the use of diuretics in nephritis, but the method offers opportunity for exact study of the action of various diuretic drugs when applied to diseased conditions.

Relation of Chloride Excretion to Edema.

We are not yet prepared to take up in detail the subject of the relation of chloride excretion to edema, but certain observations here presented refer also to this subject. Cases I, IV, and XI with nephritis; V, VI, and XII with heart failure; and XI with diabetes all showed edema, and all showed a relatively increased concentration of chlorides in the plasma. Case XIII (Text-fig. 6) had a pronounced generalized edema, occurring during serum sickness after pneumonia, after treatment with serum. Though his plasma chlorides had been low during pneumonia, they rose to a high point during the edema, and returned to the theoretical concentration as the edema disappeared. This case is of interest in that it illustrates two types of salt retention occurring in the same patient within a few days. During pneumonia, excretion was low, with a low concentration in the plasma. During serum sickness, excretion was again low, with a high concentration in the plasma. The normal period following the edema serves as a control for the other periods.

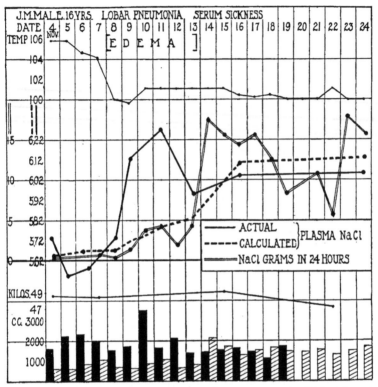

TEXT-FIG. 6. Case XIII. Lobar pneumonia; serum-treated. Crisis followed by serum sickness with generalized edema. Nov. 11. Urine, no albumin. Complete recovery occurred.

In Cases V, XII, IX, XI, and XIII disappearance of edema was accompanied by the return of the plasma chlorides to practically the theoretical figure. In Case I edema disappeared long before chloride excretion became normal, though the patient was constantly on a salt-poor diet. Case IV has never been free from edema and has always a relatively high concentration of chlorides in the plasma. Cases with persistent edema and low plasma chlorides have not been seen. Whether the relationship is one of cause and effect, or merely one in which the change in chloride excretion is due to the same cause as the edema, we cannot at present state.

SUMMARY.

A method of measuring the rate of excretion of urea, in terms of the normal, as presented in a preceding paper, has been applied to the study of diseased conditions and of the action of drugs. Simultaneous observations on chloride excretion have been made by comparing the concentration of chlorides actually found in the plasma with the theoretical amount, calculated from the rate of excretion. Observations made in this way have been found applicable to the study of various diseased conditions.

CONCLUSIONS.

1. The rate of excretion of urea in diseased conditions may be measured in terms of the normal by the index of urea excretion, and direct evidence as to the state of one of the more important excretory functions may be thus obtained.

2. The rate of excretion of chlorides gives a basis for calculating the theoretical concentration of chlorides in the plasma. By comparing the concentration actually found with the theoretical concentration, changes in the function of chloride excretion may be studied.

3. Increased concentration of urea in the blood is, as a rule, a compensatory phenomenon, in order to provide sufficient pressure to cause its excretion through a damaged outlet. Under certain conditions actual accumulation occurs.

4. Relatively increased concentration of chlorides in the plasma occurs in certain conditions, especially in certain forms of cardiac and renal disease.

5. Under certain conditions, notably in fevers or in diabetes, or as the action of diuretics (digitalis), the chloride threshold may be temporarily or permanently lowered. This may result in a decrease in the concentration of chlorides in the plasma to a point lower than the lowest point which is seen in normals.

6. Failure to excrete chlorides in pneumonia is associated with a lowered concentration of chlorides in the plasma. Excretion begins at the time this concentration increases.

7. Edema is usually accompanied by a relatively increased con-

centration of chlorides in the plasma. The relations ordinarily return to the normal state when edema disappears.

8. Chloride and urea functions may be quite independent of one another.

BIBLIOGRAPHY.

1. McLean, F. C., *Jour. Exper. Med.,* 1915, xxii, 212.
2. For a review of French work, see Ambard, L., Physiologie normale et pathologique des reins, Paris, 1914.
3. Widal, F., Ambard, L., and Weill, A., *Semaine méd.,* 1912, xxxii, 361.
4. Bauer, R., and Habetin, P., *Ztschr. f. Urol.,* 1914, viii, 353.
5. Bauer, R., and von Nyiri, W., *Ztschr. f. Urol.,* 1915, ix, 81.
6. Volhard, F., and Fahr, T., Die Brightsche Nierenkrankheit, Berlin, 1914, 168.
7. Schlayer, *Med. Klin.,* 1912, viii, Supplement, 211.
8. Widal, F., and Javal, A., *Compt. rend. Soc. de biol.,* 1904, lvi, 301.
9. Folin, O., Denis, W., and Seymour, M., *Arch. Int. Med.,* 1914, xiii, 224.
10. Frothingham, C., Jr., and Smillie, W. G., *Arch. Int. Med.,* 1915, xv, 204.
11. Peabody, F. W., *Jour. Exper. Med.,* 1913, xvii, 71.
12. von Fürth, O., Probleme der physiologischen und pathologischen Chemie, Leipzig, 1913, ii, 609.
13. A preliminary report of this portion of the work was published in the *Am. Jour. Physiol.,* 1915, xxxvi, 357.

THE CHEMOSEROTHERAPY OF EXPERIMENTAL
PNEUMOCOCCAL INFECTION.

By HENRY F. MOORE, M.B., B.Ch., B.A.O.

(*From the Hospital of The Rockefeller Institute for Medical Research.*)

(Received for publication, July 8, 1915.)

In a former communication (1) we reported that the action of ethylhydrocuprein (a derivative of hydroquinine introduced by Morgenroth and Levy (2)) was well marked *in vitro* and *in vivo* on type strains of the four groups of pneumococci. Since we possess in certain types of antipneumococcus serum the means of conferring on experimental animals considerable specific protection against very many multiples of the minimal lethal dose of a virulent pneumococcus culture, it seemed to us of interest to study the combined effect of chemo- and serotherapy *in vivo,* and to endeavor to get some numerical values representing the results.

Neufeld (3) showed that strains of pneumococci differed among themselves in respect to their immunity reactions (protective antibodies).

The work of Cole (4) and of Dochez and Gillespie (5) has resulted in a serological classification of the pneumococci. The pneumococci can be divided into at least four groups; an immune serum produced against any member of Group I has a specific agglutinative action upon, and a specific protective action against, any member of Group I, but has no such effects on any member of the other three groups. In like manner, an immune serum against any member of Group II behaves similarly with respect to any member of Group II, but has no effect on any member of Groups I, III, or IV. To Group III belong all the micro-organisms of the *Pneumococcus mucosus* type. To Group IV belong all those strains of pneumococci which do not fall into the other three groups; an immune serum produced against any member of this group has a specific agglutinative action upon, and a specific protective action against, the strain used for its production, but has no such effects on any other member of this group, or of the other groups.

Neufeld and Engwer (6) and Engwer (7) have studied the effect of combining the serum and chemotherapy in guinea pigs infected with pneumococcal pleural exudates; these observers saw an increased effect by the combination. Their protocols do not show, however, to what group either the pneumococcus or antiserum belonged; it is not clear whether or not the antiserum used was produced against a member of the group to which the infecting pneumococcus belonged. Boehncke (8), too, saw a markedly increased effect in mice by the combination of serum and drug therapy over and above the effect of either of these two separately; in the protocols of his experiments, also, we have little information as to the group to which the pneumococcus or antiserum belonged.

We deemed it of importance to study the results of combined chemotherapy and serum therapy in the light of the serological classification of the pneumococci mentioned above; and, since the results of serum therapy in man have not been at all as satisfactory in the case of lobar pneumonia due to pneumococci belonging to Group II as when the disease is caused by a strain of Group I, we have given the greater part of our attention, in this respect, to infection with pneumococci belonging to Group II. Further, the threshold value of our immune horse serum to Group I is more than ten times greater than the corresponding serum to Group II. Hence a satisfactory result with an antiserum to Group II should mean an even better result with an antiserum to Group I, in the case of infection with a strain of the homologous group.

Since toxic symptoms have been several times noticed in human patients and in experimental animals treated with ethylhydrocuprein, we have endeavored to reduce to a minimum the amount of ethylhydrocuprein given to the animals in the present study. The dose of ethylhydrocuprein (optochin base) recommended by Morgenroth (9) in experimental pneumococcal infection in the mouse is 0.5 cc. of a 2 per cent solution in olive oil given under the skin of the back immediately following the infection, and repeated on the next day; this is followed on the third day, and on the fourth day if desirable, by a similar injection of 0.4 cc. of the same solution. The dosage used in the present study is considerably below this. The experimental animals used were mice.

EXPERIMENTAL.

The strain of pneumococcus of Group II used in the present study was a stock strain and was maintained in a condition of maximal virulence throughout the experiments; that is to say, 0.000,001 cc. of a twenty hour broth culture regularly killed mice within forty-eight hours. Several controls of virulence were done in the case of each experiment. A twenty hour broth culture of the pneumococcus was used in every case. In the designation of the culture, the Roman numeral represents the group to which the pneumococcus belongs, the Arabic figure the number of animal passages which the strain had had, and the exponent the number of cultivations on artificial media since the last animal passage.

Mice of 18 grams' weight and upwards were used; the dosage of the drug was calculated in every case according to the weight of the animal. An autopsy was performed on every mouse that succumbed, unless otherwise stated in the protocols (owing to the body having been eaten by the survivors).

A constant amount of immune horse serum, 0.2 cc., was used in the experiments; it was mixed with the infecting dose of culture, and the mixture, making a total volume of 1 cc., was given intraperitoneally.

The ethylhydrocuprein (optochin base) was administered in the form of a 2 per cent solution in olive oil, given under the skin of the back. This treatment was given, in the first instance, immediately

Experiment 1. Titer of Antipneumococcus Serum II.—(Table I.) This serum protected regularly against 0.001 cc. of a 20 hour broth culture of any strain of pneumococcus belonging to Group II.

TABLE I.

Culture II 42⁵; Virulence Maximal.

Amount of culture.	Result.	Amount of culture.	Result.	Amount of culture.	Result.
cc.		*cc.*		*cc.*	
0.001	S	0.02	■ 48 hrs.	0.08	■ 24 hrs.
"	"	"	■ 60 "	"	■ " "
"	"	0.04	S	"	■ " "
"	"	"	"	"	■ " "
"	"	"	"	"	■ " "
"	"	"	■ 48 "	"	■ " "
"	"	"	■ " "	"	■ " "
0.01	"	0.06	■ 24 "	"	■ 48 "
"	"	"	■ 48 "	"	S
"	"	"	■ " "	0.1	■ 24 "
"	"	"	■ " "	"	■ " "
"	■ 72 hrs.	"	■ " "	"	■ 48 "
"	■ " "	"	■ 72 "	"	■ "
0.02	S	"	■ " "		
"	"	"	■ " "		
"	■ 30 "	"	■ " "		

after the infection; if a second dose was given it was given after an interval of 24 hours.

In the protocols the letter S stands for survival of the corresponding animal; a black square (■) means death of the animal from pneumococcal septicemia, capsulated diplococci having been found in its heart's blood by His's capsule stain, or, if this examination proved negative, Gram-positive diplococci having been recovered by culture (abundant inoculation on defibrinated rabbit blood agar) from the heart's blood at autopsy. An oblique cross (X) means death of the animal, the bacteriological examination of the heart's blood (and peritoneal cavity, if necessary) by smear and culture being negative; in this case the death of the animal may have been due to fortuitous influences, the toxicity of the drug, traumatism inflicted in the experiment, etc. When an autopsy was not possible on an animal, its death is indicated by the sign +.

Comment on Experiment 1.—From the protocol it is seen that, while the intraperitoneal injection of 0.2 cc. of antipneumococcus serum II gives to mice a sure protection against 0.001 cc. of a virulent culture (Strain II 42⁵), this amount of serum only gives a very doubtful protection against 0.01, 0.02, and 0.04 cc. of culture, and none at all against 0.06 cc. or more.

Experiment 2.—Effect of a single dose of 0.45 cc. and of a single dose of 0.5 cc., respectively, of a 2 per cent solution of ethylhydrocuprein (optochin base) in olive oil per 20 gm. of mouse on pneumococcal infection. (Table II, Strain II 42⁶.)

TABLE II.

Mouse No.	Amount of culture.	Normal horse serum.	Ethylhydrocuprein 2 per cent solution.	Result. 2d day.	Result. 3d day.
	cc.	cc.	cc.		
1	0.08	—	0.45	Sick	■
2	''	—	''	''	■
3	''	—	''	''	■
4	''	—	''		■
5	''	0.2	''	Sick	■
6	''	''	''	''	■
7	''	''	''	''	■
8	''	''	''	''	■
9	0.001	—	0.5		■
10	''	—	''	''	■
11	''	—	''	Well	■
12	0.0001	—	''	Sick	■
13	''	—	''	■	■
14	''	—	''	Well	■

Comment on Experiment 2.—A single treatment with either 0.45 cc. or 0.5 cc. of the 2 per cent solution of the drug is without effect on the lethal issue due to pneumococcal septicemia, not alone when the dose of infection approaches that used in the following experiments (0.08 cc. to 0.5 cc.), but also when it is 800 to 5,000 times smaller (0.0001 cc.).

Experiment 3. Combined Therapy.—(Table III.) Strain II 42⁰. Treatment: 0.45 cc. of a 2 per cent solution of ethylhydrocuprein in oil, per 20 gm. of mouse, and 0.2 cc. of antipneumococcus serum II.

TABLE III.

Mouse No.	Dose of culture.	2d day.	3d day.	4th day.	5th day.	6th day.	7th day.	8th day.
	cc.							
1	0.08	Well	Well	Well	Well	Well	Well	Well.
2	"	"	"	"	"	"	"	"
3	"	"	"	"	"	"	"	"
4	"	"	"	"	"	"	"	"
5	"	"	"	"	"	"	"	"
6	"	"	"	"	"	"	"	"
7	"	"	"	"	"	"	"	"
8	"	"	"	"	"	"	"	"
9	"	"	X					
10	"	"	Well	"	■			
11	0.1	"	"	"	Well	"	"	"
12	"	"	"	"	"	"	"	"
13	"	"	"	"	"	"	"	"
14	"	"	"	"	"	"	"	"
15	0.2	"	"	"	"	"	"	"
16	"	"	"	"	"	"	"	"
17	"	"	"	"	"	"	"	■
18	"	"	"	■				

Experiment 3 a. Controls to Experiment 3.—Table IV shows (1) controls infected, and treated with 0.2 cc. of antipneumococcus serum alone; (2) controls of virulence.[1]

TABLE IV.

(1) Controls treated with 0.2 cc. of antipneumococcus serum alone.			(2) Controls of virulence.		
Mouse No.	Dose of culture.	Result.	Mouse No.	Dose of culture.	Result.
	cc.			*cc.*	
1	0.06	■ 24 hrs.	1	0.00001	■ 18 hrs.
2	"	■ 48 "	2	0.000001	■ " "
3	0.08	■ " "	3	"	■ " "
4	"	■ 24 "	4	"	■ "
5	"	■ " "	5	"	■ "
6	0.1	■ " "			

[1] For controls treated with ethylhydrocuprein see Experiment 2.

Comment on Experiments 3 and 3 a.—8 of 10 animals infected with 0.08 cc. of culture and treated with 0.45 cc. of the 2 per cent solution of the drug and 0.2 cc. of serum (each of which of itself is without effect with this dose of culture) survived. One animal died on the third day and pneumococci were not recovered from the heart's blood. The remaining animal died of pneumococcal septicemia on the fifth day (delayed death). All the mice infected with 0.1 cc. of culture and similarly treated survived; 50 per cent of the series infected with 0.2 cc. of culture and also similarly treated died of pneumococcal septicemia; the others survived. The control animals all died of pneumococcal septicemia within forty-eight hours.

Experiment 4. Combined Therapy. —(Table V.) Strain II 43³. Treatment: 0.45 cc. of a 2 per cent solution of ethylhydrocuprein in oil per 20 gm. of mouse, and 0.2 cc. of antipneumococcus serum II.

TABLE V.

Mouse No.	Dose of culture.	2d day.	3d day.	4th day.	5th day.	6th day.
	cc.					
1	0.3	Well	Well	■		
2	"	"	"	■		
3	"	"	"	■	.	
4	"	"	"	■		
5	"	"	"	■		
6	"	Sick	+ (eaten)			
7	0.5	Well	Well	Well	Well	Well.
8	"	"	"	"	"	"
9	"	"	+ (eaten)			
10	"	■				
11	0.7	Well	Well	"	"	"
12	"	"	■			
13	0.8	"	Well	Sick	■	
14	"	"	"	■		

Comment on Experiment 4.—In this experiment in which a larger dose of infection was given than in Experiment 3 the results did not come out uniformly or clear. The majority of the animals died, although in most cases death was delayed. We thought that a small dose of optochin given on the second day might serve to tide the animals over the third and fourth days and so help ultimate recovery; the result of this modification in dosage of the drug is shown in the next experiment. Virulence controls were done, as in Experiment 3 a, but are not shown in the table.

Experiment 5. Combined Therapy.—(Table VI.) Strain II 43[6]. Treatment: 0.45 cc. of 2 per cent ethylhydrocuprein in oil immediately after infection and 0.4 cc. after 24 hours' interval per 20 gm. of mouse; 0.2 cc. of antipneumococcus serum II at the time of infection.

TABLE VI.

Mouse No.	Dose of culture.	2d day.	3d day.	4th day.	5th day.	6th day.
	cc.					
1	0.3	Well	Well	Well	Well	Well.
2	"	"	"	"	"	"
3	"	"	"	"	"	"
4	"	"	"	"	"	"
5	"	"	"	"	"	"
6	"	"	"	"	■	
7	"	"	"	"	X	
8	"	"	"	■		
9	"	Sick	+ (eaten)			
10	0.5	Well	Well	Well	Well	"
11	"	"	"	"	"	"
12	"	"	"	"	■	
13	"	"	"	"	■	
14	"	"	"	■		
15	"	"	"	■		
16	"	"	Sick	■		
17	"	Sick	+ (eaten)			

Comment on Experiment 5.—50 per cent of the animals infected with 0.3 cc. of culture, and treated as stated, survived; of those that died, pneumococci were not recovered from the heart's blood of one; one was eaten by its fellows and could not, therefore, be autopsied; two gave positive blood cultures. In the case of infection with 0.5 cc. of culture, only two animals recovered; pneumococci were recovered from the heart's blood of all those that died except one, the body of which was so far eaten that it could not be bacteriologically examined. In this experiment the altered dosage of the drug ensured a better result than in the former case, but it is evident that we are at the limit of protective power for the conditions obtaining therein. Consequently we proceeded to study the effect of a single larger dose of drug combined with the serum therapy; namely, 0.5 cc. of the 2 per cent oily solution per 20 grams of mouse. Virulence controls were done, as in Experiment 3 a, but are not shown in the table.

Experiment 6. Combined Therapy.—(Table VII.) Strain II 44⁴. Treatment: 0.5 cc. of 2 per cent ethylhydrocuprein base in oil per 20 gm. of mouse, and 0.2 cc. of antipneumococcus serum II.

TABLE VII.

Mouse No.	Dose of culture.	2d day.	3d day.	4th day.	5th day.	6th day.	7th day.	8th day.
	cc.							
1	0.2	Well	Well	Well	Well	Well	Well	Well.
2	"	"	"	"	"	"	"	"
3	"	"	"	"	"	"	"	"
4	"	"	"	"	"	"	"	"
5	"	"	"	"	"	"	"	"
6	"	"	"	"	"			
7	"	"	"	"	+ (eaten)			
8	"	+ (eaten)						
9	0.3	Well	"	"	Well	"	"	"
10	"	"	"	"	"	"	"	"
11	"	"	"	"	"	"	"	"
12	"	"	"	"	"	"	"	"
13	"	"	"	"	"	"	"	"
14	"	"	"	"				
15	"	"	"	+ (eaten)				
16	"	"	"	" "				
17	0.4	"	"	Well	"	"	"	"
18	"	"	"	"	"	"	"	"
19	"	"	"	"	"	"	"	"
20	"	"	"	"	"	"	"	"
21	"	"	"	"	"	"	"	"
22	"	"	"	"	"	"	"	
23	"	"	"	"	"	"	+ (eaten)	
24	"	"	"	X				
25	"	■						

Two mice infected with 0.000001 cc. and one with 0.00001 cc. of Strain II 44⁴ (controls of virulence) died of pneumococcal septicemia in thirty-six hours. Thirteen controls infected with 0.08 cc. of culture and treated with 0.5 cc. of the ethylhydrocuprein solution alone immediately after the infection all died on the third day (forty-eight hours after infection) from pneumococcal septicemia.

Comment on Experiment 6.—In this case we increased the single dose of ethylhydrocuprein to 0.5 cc. of the 2 per cent solution of the base in oil per 20 grams of mouse. Six of eight animals infected with 0.2 cc. of culture, and six of eight animals infected with 0.3 cc. of culture, and treated as stated, survived. The four mice which died were eaten by their fellows and therefore could not be autopsied; in three of these cases death was delayed (fourth and fifth days). Of the nine animals infected with 0.4 cc. of culture, six survived; one died on the second day with a sterile heart's blood (abundant inoculation) but pneumococci were recovered from the peritoneal cavity; one died on the fourth day and showed a sterile heart's blood on cultivation; and one was eaten by its fellows on the seventh day. The thirteen controls infected with 0.08 cc. of culture and treated with the dose of drug used in the actual experiment all died in forty-eight hours, pneumococci being present in the heart's blood of all. These controls are not shown in the protocol.

Experiment 7. Combined Therapy.—(Table VIII.) Strain II 45³. Treatment: 0.5 cc. of 2 per cent solution of ethylhydrocuprein base in oil per 20 gm. of mouse, and 0.2 cc. of antipneumococcus serum II.

TABLE VIII.

Mouse No.	Dose of culture.	2d day.	3d day.	4th day.	5th day.	6th day.	7th day.	8th day.
	cc.							
1	0.4	Well	Well	Well	Well	Well	Well	Well.
2	"	"	"	"	"	"	"	"
3	"	"	"	"	"	"	"	"
4	"	"	"	"	"	"	"	"
5	0.5	"	"	"	"	"	"	"
6	"	"	"	"	"	"	"	"
7	"	"	"	"	"	"	"	"
8	"	"	"	"	"	"	"	"
9	"	"	"	"	"	"	"	"
10	"	Sick	"	"	"	"	"	"
11	"	"	"	"	"	"	"	"
12	"	"	"	■				
13	0.6	"	"	Well	"	"	"	"
14	"	"	"	"	"	"	"	"
15	"	"	"	"	"	"	"	"
16	"	Sick	Sick	Sick	■			
17	"	Well	■					
18	"	Sick	■					

Three virulence controls (0.000001 cc.) all died in forty-eight hours of pneumococcal septicemia.

Comment on Experiment 7.—All the animals infected with 0.4 cc. of culture and treated, and all but one infected with 0.5 cc. of culture and treated, survived; from the heart's blood of the one which died (fourth day) pneumococci were recovered. 50 per cent of those infected with 0.6 cc. of culture and treated survived; the remainder died of pneumococcal septicemia (third tc fifth day). We have here, evidently, reached the limit of protection to be gotten by the simultaneous administration of antiserum and drug in the manner and quantities indicated above.

In addition to the experiment mentioned above, we have carried out a series of experiments in which the mice were treated with ethylhydrocuprein and a non-homologous serum. For example, we have used an antiserum to pneumococcus of Type I in the case of infection with from 0.0001 cc. to 0.1 cc. of a broth culture of a virulent pneumococcus of Type II (Strain II 34[20]), the minimal lethal dose of which was 0.000001 cc.; and we have given the ethylhydrocuprein solution both as one dose (0.5 cc.) at the time of infection, and as three doses on the three first days of the experiment (0.25 cc., 0.25 cc., and 0.2 cc.), the administration of the first dose following the infection immediately; by thus combining the drug treatment with non-homologous serum therapy, we have seen no effect such as we have described above when the serum was homologous. Again, in the case of treatment of infection with a virulent strain of Group IV (Strain IV A 67.11[3]), the minimal lethal dose of which was 0.000001 cc., we have had no such success on giving the drug and a potent antiserum to pneumococci of either Group I or Group II. The protocols of these experiments need not be exhibited.

DISCUSSION OF RESULTS.

The results of this study of the treatment of experimental pneumococcal infection in the mouse by simultaneous administration of ethylhydrocuprein and antipneumococcus serum are clear. The protocols show that 0.5 cc. of a 2 per cent oily solution of ethylhydrocuprein (optochin base), or 0.2 cc. of the type homologous antiserum, are each, by themselves, powerless to protect against 0.06 cc. of a highly virulent culture of a pneumococcus of Group II; but that if both these bodies, which are chemically far removed from each other, be exhibited simultaneously, protection is given to the animal against as large a dose of infection as 0.5 cc. of culture. The potency of the antiserum of Type II is such that 0.2 cc. confers a certain protection on mice against simultaneous infection with an amount of a virulent culture lying somewhere between 0.001 cc. to

0.01 cc.; with a higher dosage of culture this protection is quite uncertain, while it is altogether absent in the case of infection with 0.06 cc. of culture. The dose of ethylhydrocuprein generally used in the present study (0.5 cc. of the 2 per cent solution) is unable to protect against a dose of infection as small as 0.0001 cc. of a virulent culture. Hence, the exhibition of such a small dose of this drug as 0.5 cc. of the 2 per cent solution (0.01 of a gram) per 20 grams of mouse is capable of raising the threshold value of the serum more than fifty times; that is to say, mice are protected against 0.5 cc. of a virulent culture by the combined therapy, while 0.2 cc. of serum alone will not invariably protect against 0.01 cc. of culture, and the dose of ethylhydrocuprein used does not even protect against five thousand times less than 0.5 cc. of culture. Thus the protective value of simultaneous administration of the drug and serum, even though they be given by different routes, is many times greater than,

TABLE IX.

Combined Ethylhydrocuprein and Serum Therapy.

Type II Antiserum. Pneumococcus, Stock Strain of Group II.

Culture.	Ethylhydro-cuprein 2 per cent solution.	Antiserum II.	Normal horse serum.	Survivals.	Deaths due to pneumo-coccal septi-cemia.	Deaths due to doubtful cause (no autopsy).
cc.	*cc.*	*cc.*	*cc.*	*per cent*	*per cent*	*per cent*
0.001	—	0.2	—	100	0	0
0.01	—	"	—	66.6	33.3	0
0.02	—	"	—	40	60	0
0.04	—	"	—	60	40	0
0.06	—	"	—	0	100	
0.08	—	"	—	0	100	
0.08	0.45	"	—	0	100	
"	"	—	0.2	0	100	
"	0.5	—	—	0	100	
"	"	—	0.2	0	100	
0.0001	"	—	—	0	100	
0.08	0.45	0.2	—	80	10	10
0.1	"	"	—	100	0	0
0.2	"	"	—	50	50	
0.3	"	"	—	0	100	
0.2	0.5	0.2	—	75		25
0.3	"	"	—	75	0	25
0.4	"	"	—	76.9	7.6	15.3
0.5	"	"	—	87.5	12.5	0
0.6	"	"	—	50	50	

and out of all proportion to, the protective value of either alone. The preceding epitome of the results (Table IX) will make this clear.

It is important to note that, where a non-homologous antiserum is used in this combined method, *e. g.*, a Type I antiserum with a pneumococcus of Group II or a Type I or II antiserum with a pneumococcus of Group IV, no such effect as that just stated is seen. Indeed, no more effect is obtained than if normal, and not immune, serum be used. It seems that, if the dose of ethylhydrocuprein given in the first instance be below a certain quantity in relation to the amount of the infection, subsequent readministration of the drug on the next day is powerless to prevent a fatal result from pneumococcal septicemia.

CONCLUSIONS.

1. A single small dose of ethylhydrocuprein (optochin base), which by itself has practically no protective effect against experimental pneumococcal infection in mice, is capable of increasing the threshold value of the type homologous antipneumococcus serum at least fifty times.

2. This effect is proportionately many times greater than a simple summation of the protective effects of these two bodies.

3. No such effect is obtained when the antiserum used is one produced against a strain of pneumococcus from a group other than that to which the infecting pneumococcus belongs.

BIBLIOGRAPHY.

1. Moore, H. F., *Jour. Exper. Med.*, 1915, xxii, 269.
2. Morgenroth, J., and Levy, R., *Berl. klin. Wchnschr.*, 1911, xlviii, 1560, 1979.
3. Neufeld, F., and Haendel, *Ztschr. f. Immunitätsforsch., Orig.*, 1909, iii, 159; *Berl. klin. Wchnschr.*, 1912, xlix, 680; *Arb. a. d. k. Gsndhtsamte*, 1910, xxxiv, 169; in Kolle, W., and von Wassermann, A., Handbuch der pathogenen Microorganismen, 2d edition, Jena, 1913, iv, 513.
4. Cole, R., *Arch. Int. Med.*, 1914, xiv, 56.
5. Dochez, A. R., and Gillespie, L. J., *Jour. Am. Med. Assn.*, 1913, lxi, 727.
6. Neufeld and Engwer, *Berl. klin. Wchnschr.*, 1912, xlix, 2381.
7. Engwer, T., *Ztschr. f. Hyg. u. Infectionskrankh.*, 1913, lxxiii, 194.
8. Boehncke, K. E., *München. med. Wchnschr.*, 1913, lx, 398.
9. Morgenroth, J., and Kaufman, M., *Ztschr. f. Immunitätsforsch., Orig.*, 1913, xviii, 145.

THE MECHANISM OF ANAPHYLACTIC SHOCK.

Studies on Ferment Action. XXIII.

By JAMES W. JOBLING, M.D., WILLIAM PETERSEN, M.D., and
A. A. EGGSTEIN, M.D.

(*From the Department of Pathology, Medical Department, Vanderbilt University,
Nashville.*)

(Received for publication, May 5, 1915.)

Despite the intensive study which the phenomena of anaphylaxis
have received, especially by immunologists, with considerable the-
orizing about the probability of protein splitting, and the source and
the mechanism of production of the intoxicating agent, there have
been published few experiments dealing with the actual ferment
changes and the amounts of protein split products in the serum
during shock.

Abderhalden and Pincussohn and their various coworkers (1) noted by means
of the polariscope an increase in the power of the serum of sensitized animals
to split the specific antigen. Pfeiffer and Mita (2) worked along similar lines.
A summary of their work, together with detailed experiments carried out by
means of the dialysis method, is found in a paper of Pfeiffer and Jarisch (3).

Auer and Van Slyke (4) have reported experiments in which they determined
the amino-acid content of lungs of guinea pigs before and during shock. They
found no noteworthy change.

In a recent paper Zunz and György (5) have confirmed the findings of Abderhalden and of Pfeiffer and Mita. Following the primary injection they found
a gradual increase in a protease for beef protein, reaching a maximum in about
15 days; this ferment, according to their observations, disappears following
shock, only to reappear on a further injection if made from 4 to 10 days after
recovery from shock. They definitely determined an increase in amino-acids
during acute shock in dogs, this being quite contrary to the supposition of
de Waele (6).

In previous papers we have shown that the homologous serum may be toxic
for animals under conditions which remove the antiferment and permit serum
protease to act (7) ; that anaphylatoxin formation depends upon the adsorption
of the serum antiferment (8) ; while an increase in antiferment tends to diminish
the susceptibility to shock (9). During trypsin shock, which closely resembles
anaphylactic shock, we have noted a marked increase in the serum protease and
lipase (10).

The present experiments deal with the serum ferments and the protein split products during anaphylactic shock in dogs. The changes that occur are striking and have not, so far as we are aware, been recorded by other workers. We believe that the facts developed confirm the idea of a protein cleavage, in contradistinction to the supposition of a purely physical change, as advanced, for instance, by von Behring (11) and previous workers. But this cleavage concerns the proteins of the animal itself, rather than the antigen.

While these observations have been made on the serum, the deduction that the primary site of the reaction is to be found in the serum is not therefore warranted, for the serum changes may simply express the end results of cellular changes.

Technique.

Dogs have been used exclusively in our experiments, not only because of the larger amounts of serum which are available and which are needed in work of this kind, but because the greater susceptibility of the bronchial musculature of the guinea pig is lacking to a large degree, so that death from asphyxia does not complicate the picture before serum changes become established. The dog, too, probably more closely parallels the human condition. The methods used to determine the ferments and the split products have been fully described in a previous paper (12). The proteoses present in the serum have been determined as follows: 5 cc. of serum are diluted with about 10 cc. of normal saline and 2 cc. of a mixture of 10 per cent acetic acid and 20 per cent salt solution are added. The serum is then boiled for 10 minutes, filtered through hard paper, and the filtrate saturated while boiling with sodium sulphate. The precipitate is collected and after washing with water saturated with sodium sulphate it is dissolved from the filter paper by means of a small amount of water made slightly alkaline with sodium carbonate. The dissolved proteoses are collected directly in the large Jena glass tubes used for oxidizing, and nitrogen determinations are made. The results charted therefore represent the amount of nitrogen present as proteoses in 5 cc. of serum. The majority of the determinations made to determine the amount of non-coagulable nitrogen in the serum, together with the amount of protease activity, have been made in duplicate.

Purified horse serum albumin has been used as an antigen. This preparation, containing antiferment, shows considerable resistance to tryptic digestion. Injections have been intravenous; blood samples have been obtained from ear veins; during shock recourse to a suction pump has been found necessary to obtain blood in sufficient quantity.

The Effect of Primary Injection and Sensitization.

The first experiments illustrated the changes that occur in the serum following the first or sensitizing dose.

Dog 25.—Weight 8 kilos. Sensitized Feb. 27, 1915, and Mar. 2, 1915, with subcutaneous injections of horse serum albumin. On Mar. 3, 1915, at 9.30 a. m. 0.5 gm. of horse serum albumin was injected intravenously. There resulted an immediate rise in temperature to 106° F. at 11.10 a. m., together with an immediate leucopenia. The antiferment index was practically unaltered, as was also the amount of non-coagulable nitrogen in the serum. The serum protease showed no change until the afternoon, when a moderate increase was noted. The serum lipase increased during the same time. Blood drawn the following morning showed a slight increase in antiferment, a decrease in protease and lipase, and a slight increase in non-coagulable nitrogen.

Dog 21.—Weight 4 kilos. Sensitized in the same manner as Dog 25. 0.5 gm. horse serum albumin injected at 9.50 a. m., Mar. 5, 1915. The results are similar to those of the first animal, except that the rise in protease was noted a little sooner than in the previous dog. The non-coagulable nitrogen showed a considerable increase after 24 hours.

Inasmuch as the interest in anaphylaxis centers chiefly about the protease and the amount of split products, we have collected the results of a series of experiments dealing with the injections of the antigen at various intervals after sensitization in Text-fig. 1. From this it will be noted that there is a progressive decrease in the time interval of the mobilization of the protease in the serum. Thus in Dog 15, a single injection in a non-sensitized dog has resulted in a decrease in protease, and practically no change in the non-coagulable nitrogen. The dogs which received injections of the antigen from four to ten days after sensitization (Nos. 25, 31, 21, 33, and 63) showed an increase in protease at an interval after the injection which becomes progressively less, while a rise in the non-co-agulable nitrogen in the serum becomes evident in those animals which are sensitized longest. When the animal finally becomes fully sensitized the ferment mobilization is immediate, as is also the increase in non-coagulable nitrogen.

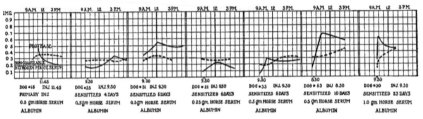

Text-Fig. 1. Mobilization of serum protease during the course of sensitization.

It should be emphasized that the ferment which we demonstrate is not specific, for when the serum is emulsified with chloroform and incubated, the ferment digests the serum proteins. The only evidence of specificity lies in the fact that the ferment is mobilized by the injection of the specific antigen.

The Effect of Acute Anaphylactic Shock.

The following experiments serve to illustrate the effect of the reinjection of the antigen after complete sensitization.

Dog 15.—Weight 4.5 kilos. Sensitized Feb. 9, 1915, with 1.0 gm. of horse serum albumin intravenously. The primary injection resulted in the usual rise in temperature and fall in leucocytes. The antiferment, which had a relatively high titer, fell slightly, then made a gradual recovery. The non-coagulable nitrogen showed practically no change, while there was a decided fall in the protease content. The lipase showed a gradual increase from 1.5 to 5 cc. N/100 sodium hydrate after 24 hours. When we now contrast this picture with that following the injection after sensitization, there has been noted a much greater fall in the antiferment titer, an immediate rise in non-coagulable nitrogen (from 0.27 to 0.45 mg.), together with a large increase in serum protease (from 0.47 to 0.87 mg.). The lipase increased from 0.8 to 6.8 cc. The animal died 4 hours after the injection.

Dog 33.—Weight 9 kilos. Sensitized, beginning Mar. 8, 1915, with three sub-cutaneous injections of horse serum albumin. The effect of the primary intravenous injection before complete sensitization is shown in Text-fig. 2. The injection (0.5 gm. of horse serum albumin) was made at 9 a. m., Mar. 15, 1915 (7 days after the first sensitizing dose). As a result there was noted a distinct rise in the protease, a slight increase in the lipase in the afternoon; a slight fall,

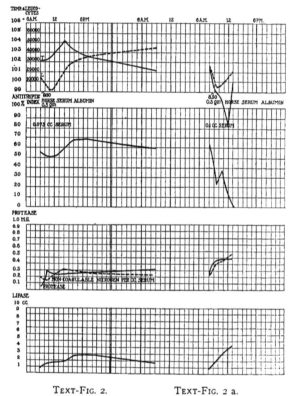

TEXT-FIG. 2.   TEXT-FIG. 2 a.

TEXT-FIGS. 2 AND 2 a. Serum changes following antigen injection before and after complete sensitization.

followed by a rise in the antiferment titer; the non-coagulable nitrogen first showed a slight fall followed by a rise in the afternoon. The intoxicating injection was made Mar. 29, 1915, at 8.50 a. m. (0 5 gm.) (Text-Fig. 2 a). As a result there was a typical fall in the antiferment titer, and an immediate increase in the protease, lipase, and non-coagulable nitrogen. The animal died at 1.30 p. m. Of the total amount of the non-coagulable nitrogen in the blood serum at this time 33 per cent represented nitrogen present in the form of proteoses. It is possible that some of these proteoses may be toxic.

Dog 30.—Weight 7.7 kilos. Sensitized Mar. 7, 1915. Reinjected at 9.30 a. m.,

Mar. 30, 1915, with 1.0 gm. horse serum albumin. Died at 1.30 p. m. (Text-fig. 3). In this experiment a larger amount of antigen was injected for the intoxicating dose than in the two previous animals. The ferment picture is typical

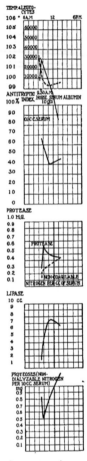

TEXT-FIG. 3. Serum changes during acute anaphylactic shock.

and similar to the other experiments. In this animal the amount of proteoses present in the serum was determined by dialyzing the filtrate from 10 cc. of serum after the coagulable nitrogen had been removed by acid and heat. Nitrogen determinations were then made on the undialyzable portion. This represents the higher split products (proteoses). As will be observed from Text-fig. 3, there resulted an almost immediate drop in the amount present in the serum with an increase following later.

Protracted Shock.

The following experiments illustrate the conditions when the animal does not die within a few hours, but lives over a period of about twenty-four hours following the second injection.

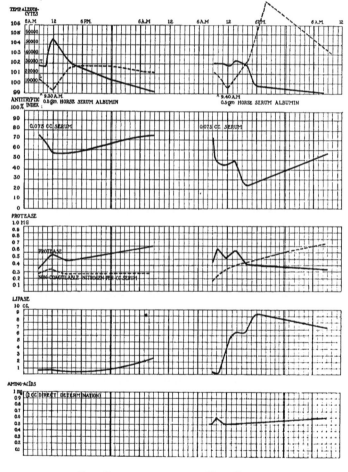

TEXT-FIG. 4.   TEXT-FIG. 4 a.

TEXT-FIGS. 4 AND 4 a. Serum changes during protracted anaphylactic shock.

Dog 31.—Weight 11 kilos. Sensitized Mar. 7, 1915. First intravenous injection at 9.30 a. m., Mar. 12, 1915 (Text-fig. 4). On Mar. 30, 1915, at 9.40 the second intravenous injection was made. The dog showed the usual malaise and prostration, but apparently recovered somewhat in the afternoon. The animal died at 8.30 the following morning. As will be observed from Text-fig. 4 a there was noted a marked leucocytosis (97,000) in the evening. The antiferment showed the usual fall with a recovery the next morning. The non-coagulable nitrogen increased from 0.17 mg. per cc. to 0.75 mg. the following morning. The protease showed an immediate increase with some fluctuations following, and a gradual decline. The lipase increased to a marked extent. The amino-acids (Van Slyke method) increased immediately after the injection, then fell to the original figure, and increased the following morning. Whatever may have been the factor which delayed exitus in this dog, it seems to have had some influence on the ferments and the antiferments, and was manifested possibly first in the fall in protease at noon. It will be observed that all the curves in the experiment are broken and in so far differ from the normal course of events during shock.

The Effect of Reinjection During the Refractory Stage.

In the following experiment the effect of a single shock in the sensitized animal, followed by a second injection after two days, has been studied.

Dog 65.—Weight 6.2 kilos. Sensitized by an intravenous injection, Apr. 2, 1915. On Apr. 20, 1915, at 8.37 a. m., 0.5 gm. of horse serum albumin was injected, which dose was sufficient to cause a profound intoxication, but not death. The animal was bled at 8.50 and 10.00 a. m., 1, and 3 p. m., and the following morning. The resulting serum changes are quite typical (Text-fig. 5). The antiferment showed the usual immediate drop, with a following recovery and second drop. The non-coagulable nitrogen showed only a slight increase and remained low, thus differing from the other animals so far studied. The rise in protease was instantaneous and very marked, fell rather rapidly, and then remained low. The lipase showed the usual rise, but not to such a marked extent. The serum proteoses showed the immediate diminution which is characteristic, while the amino-acids increased quite markedly.

On reinjection two days later (9 a. m., Apr. 22, 1915) it will be noted (Text-fig. 5 a that the change in antiferment was absent. The protease showed some decrease and remained low, as did the lipase. The proteoses showed a slight decrease after 15 minutes with a distinct increase later. The amino-acid increased gradually and returned to normal the following morning.

Shock in the Pregnant Animal.

During pregnancy animals are known to show a relative immunity to anaphylactic shock; the mechanism of the resistance is not understood. It is possible that the increase in antiferment which usually accompanies pregnancy might in a measure account for this

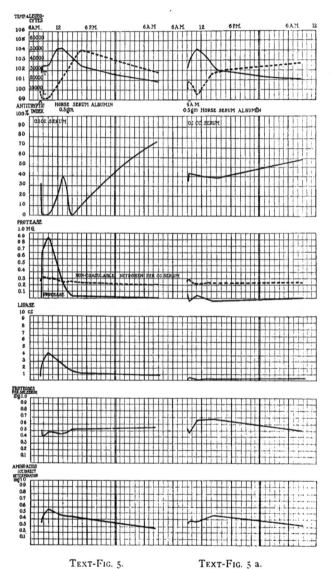

TEXT-FIG. 5.   TEXT-FIG. 5 a.

TEXT-FIGS. 5 AND 5 a. Serum changes during acute anaphylactic shock and following antigen injection during the refractory period.

resistance, since a certain amount of resistance can be imparted to an animal by increasing its antiferment, as we have shown previously (9).

The following experiment was made on a pregnant animal.

Dog 56.—Weight 6 kilos. Sensitized by an intravenous injection of horse serum albumin Apr. 2, 1915. A second injection, made three days later, resulted in no marked alterations with the exception of a rather sharp drop in the afternoon in the amount of proteoses contained in the serum. The intoxicating dose was made at 8.30 a. m., Apr. 22, blood being collected at 8.45 and 9.30 a. m., 12, and 3 p. m. The animal was somewhat nauseated and vomited several times, but remained in fair condition until the afternoon; it died suddenly at 4.30 p. m. In the first place it was observed that the antiferment instead of being high was quite low (0.15 cc. of serum inhibiting only about 10 per cent of the digestion). The protease, however, rose sharply, as in the other experiments, although there was no increase in non-coagulable nitrogen until towards noon. The lipase curve was very irregular, but continued to rise. The proteoses were practically unaltered for 15 minutes, later increased, decreased toward noon, and increased markedly toward the terminal stage. The amino-acids at first declined progressively, but increased during the afternoon.

The Relation of Serum Proteoses and Amino-Acids to Shock.

When the antigen is injected into the circulation of the sensitized animal proteolysis takes place. This fact has been definitely established through the work of Zunz and György in demonstrating an increase in amino-acids by the Van Slyke method during shock. The various experiments which we have detailed have also illustrated this phase, inasmuch as an increase in the non-coagulable nitrogen is an almost invariable accompaniment of the acute shock. The question arises whether Zunz and György and the other workers who have advanced the idea of specific protease action are justified in assuming that the increase in amino-acids proves that the specific antigen has been split.

Before the antigen reaches the amino-acid stage it must pass through the various cleavage phases,—proteoses, peptones, etc.,—and we should expect that there would be a preceding or concomitant increase in these split products in the serum. The contrary is true. This is illustrated in Text-fig. 6, in which the amino-acid and protease curves are illustrated immediately after the injection of the antigen at 8.37 a. m. (Dog 65). The immediate rise in amino nitrogen is apparent, but is accompanied by an actual decrease of the amount of proteoses in the serum. We have repeatedly observed this relation and have never noted an increase in proteoses during the first fifteen minutes following the antigen injection,

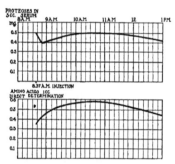

TEXT-FIG. 6. Relation of serum proteoses and amino nitrogen during acute anaphylactic shock.

when the anaphylactic symptoms are, of course, already well advanced. The decrease of the proteoses at the instant of the shock, together with an increase in amino-acids, seems to warrant the assumption that the primary cleavage that occurs has its substrate in the proteoses already present in the serum.

Non-Specificity of Serum Protease.

In the preceding paragraph attention has been called to the fact that when the increase in amino-acids occurs immediately after the injection of the antigen into the sensitized animal, there is a simultaneous decrease in the higher split products of the serum, a condition which is contrary to the view that it is the injected antigen that is being split. For if that were true we should expect an increase in the higher split products, produced during the process of cleavage of the antigen to the amino-acid stage. The idea that it is the antigen that is split has seemed in accord with the data at hand up to the present time. Thus Abderhalden and Pincussohn and Pfeiffer and Mita report positive results when the serum of sensitized animals is permitted to act upon the specific antigen *in vitro*.

And the experiments of Zunz and György confirm these findings by means of the increase in amino-acids demonstrable with the Van Slyke apparatus *in vivo* and when the specific serum is permitted to "digest" the antigen *in vitro*. So, too, they also report that immediately following shock this property disappears, affording an attractive explanation for the state of anti-anaphylaxis. This entire hypothesis concerning the mechanism of anaphylactic shock is based, therefore, upon the specificity of the serum proteases. In a previous paper we have shown that there is no element of specificity in so far as the proteases are concerned in the digestion occurring in the Abderhalden reaction (13). The digest is obtained from the serum proteins and is brought about by changes in the colloidal state of dispersion of the serum induced by the antigen introduced as a substrate, resulting in what may be termed local areas of antiferment deficiency, due either to an actual adsorption of antiferment by a formed substance, such as a precipitate, or to a change in the degree of dispersion of the unsaturated lipoids upon which the antiferment property depends, as we have previously demonstrated (14). If a protease is present in such sera a positive reaction can occur; if not, the digestion does not take place.

Since it is well known from various immunological experiments that minimal amounts of the antigen can call forth a specific reaction, as, for instance, in the precipitin test or in complement deviation, we can, by means of dilution of the specific substrate, show that the amount of digestion brought about by the addition of such a specific substrate may be greater by many times than the total amount of nitrogen introduced, thus positively proving that the digestive products are not derived from the specific substrate and can only be formed from the serum itself. Such an experiment is shown as follows: Fresh serum is used from a fully sensitized dog. A 1 per cent horse serum albumin solution, to which the dog has been sensitized, is used as a substrate and varying dilutions to 0.001 per cent are made. If we now add a constant amount of the ferment (the specific serum) to decreasing dilutions of the substrate, we should expect that with minimal dilutions there would not be enough substrate present for the ferment to act upon, and as a result the amount of digestion products should become smaller. In the following experiment the serum was allowed to " digest " the horse serum albumin solution for twenty hours under toluol (Table I).

TABLE I.

Tube No.	Serum 56/5.	Horse serum albumin solution.		Nitrogen available as substrate.	Amino-acids determined in 1 cc. of mixture (direct).	Increase in amino-acids.
		cc.	per cent	mg.	mg.	mg.
1	1		1	0.74	0.279 ⎱	
2	1 (inactivated)		1	0.74	0.245 ⎰	0.034
3	1		0.1	0.074	0.245 ⎱	
4	1 "		0.1	0.074	0.190 ⎰	0.055
5	1	1	0.01	0.0074	0.256 ⎱	
6	1 "		0.01	0.0074	0.178 ⎰	0.078
7	1		0.001	0.00074	0.268 ⎱	
8	1 "		0.001	0.00074	0.190 ⎰	0.078

As will be observed, with an available total nitrogen from the specific substrate of about 0.00074 mg. (Tubes 7 and 8) we obtain a digest of 0.078 mg. of amino-acids per cc. of the mixture. This experiment, which we have repeatedly made, affords absolute proof that the theory of specific protease action is without warrant. The various methods of demonstrating specific protease action all have a common error in that the digestion of the serum proteins and of the higher split products present in the serum (proteoses) by the non-specific serum proteases is ignored. The serum protease should be regarded as a simple ferment, polyvalent in character, and in so far resembling the ordinary tryptic ferment. This is precisely the view that Fermi (15), working with a variety of methods, has reached. Serum ereptase should, however, be strictly distinguished from the protease.

The Serum Esterase.

An invariable accompaniment of the shock has been a marked mobilization of the serum esterase. While the first injection of horse serum albumin may cause some increase in the ferment, it never causes so sudden nor so extensive a rise in the esterase curve. This increase is not specific for anaphylactic shock, but occurs following trypsin and peptone shock, as well as following intoxication by bacterial proteins of varied derivation. This increase probably represents a mobilization from tissue cells in general rather than the secretion of any one organ.

In a previous paper (8) we stated:

" The arguments made against the protein-intoxication conception, such as the time element and the minute amount of substance necessary, are not convincing, for we know that ferment action may be very rapid; and as far as the quantity of substrate is concerned the argument fails if we place the matrix of the poison in the serum proteins themselves. There is some reason to believe, however, that if ferment action is the basis of anaphylactic shock, these ferments may have a much wider range of action than merely on the introduced protein."

The facts brought out during the course of this work, have, we believe, established this view as essentially correct. In brief, these results can be stated as follows: The injection of the antigen in the non-sensitized animal is practically without influence on the serum ferments or the split products present in the serum.

During the period following the sensitizing injection and preceding the development of complete sensitization the organism responds to the injected antigen by a progressively increasing rapidity of mobilization of ferments, the amount of protease mobilized becoming greater as the maximum of sensitization is reached. This protease is not specific.[1]

The acute shock is accompanied by an immediate and marked increase in serum protease; by a fall in antiferment; by a rise in the amino-acid content and non-coagulable nitrogen of the serum, together with a primary decrease in the serum proteoses. The esterase rise is constant and progressive. When death occurs after a few hours the serum may contain a large amount of proteoses (33 per cent of the total non-coagulable nitrogen).

Before discussing further the mechanism of the various phenomena we must emphasize the fact that whatever change takes place during the course of the reaction is not due to specific pro-

[1] In a paper which has appeared after the completion of our work Pfeiffer (Pfeiffer, H., *Ztschr. f. Immunitätsforsch., Orig.*, 1915, xxiii, 515) has demonstrated an increase in the peptolytic power of the serum during anaphylactic shock. Because of the extent of the paper it cannot be adequately reviewed at this time.

teases (*Abwehrfermente*). The antigen which we used (purified horse serum albumin) is exceedingly resistant to the ordinary tryptic ferment, a ferment which is not only polyvalent but also very active; and it seems unreasonable that this same protein should be readily attacked by a specific protease when we know that the antiferment, to which the horse serum albumin owes its resistance, inhibits not only the tryptic ferment but also the so called specific ferment. The actual digestion is due to an entirely different mechanism,—to a non-specific ferment acting upon serum proteins. This digestion is brought about through colloidal changes which take place when the antigen is brought into contact with the serum. Either a precipitate is formed which may act as an adsorbing substance for the antiferment, or else the dispersion of the antiferment itself is also altered, so that local areas of antiferment deficiency are formed with a resulting immediate cleavage either from the serum proteins (especially the globulins) or from the proteoses already present in the serum. Bearing these conditions in mind, the mechanism of the production of the toxic substances which produce the acute shock becomes relatively simple, but involves not only the serum, but the cellular elements of the organism as well, and the latter probably primarily.

The mobilization of the protease must be a cellular phenomenon and during the course of sensitization this response is increased not only in intensity, but also in rapidity. This need not result in an intoxication, for the isolated cell, as the various investigators who have studied cellular phenomena in anaphylaxis have shown (Dale, Schultz, Weil, etc.), may be capable of repeated response to specific stimuli; so we must concede that the antigen is not in itself toxic for the cell either before or after sensitization. Associated, however, with the cellular response we find serum changes that do result in an intoxication. There is an immediate fall in the antiferment titer. This may in part be due to a saturation of the antiferment by the mobilized ferment. It is more probably due to a colloidal change with a resulting lessening of the dispersion of the unsaturated lipoids upon which the antiferment property depends. This lessening of the antiferment titer will, of course, facilitate proteolysis. Furthermore, definite changes in the antibodies of the serum (especially the precipitins) occur at the time of shock. Whether or not an actual change to the final picture which is demonstrable *in vitro* with the precipitin reaction occurs is immaterial. The precipitin content which before the shock is present in the serum disappears instantly when shock is induced (Joachimoglu (16)). There is evident at least a colloidal rearrangement which must result in local areas of antiferment adsorption or diminution in which proteolysis can take place.

This proteolysis is made evident by the immediate increase in amino-acids. The only question relates to the substrate from which the amino-acids are derived. Contained in the serum of normal guinea pigs and dogs one finds a relatively large amount of higher split products (5 cc. of serum containing from 0.3 to 0.5 mg. of nitrogen in this form). Immediately after shock these split products are diminished in amount, but later increase progressively until death takes place. It is quite evident that if we were dealing with a shock primarily due to protein split products derived from the introduced antigen we should expect to find an immediate increase in the higher split products derived from that antigen.

Briefly, then, the mechanism of acute intoxication may be stated as follows: An immediate mobilization of a non-specific protease in large amounts; the formation of a state of antiferment deficiency through colloidal changes; a simultaneous cleavage of serum proteins (proteoses) through the peptone stage to amino-acids; an intoxication by these peptones with a resulting cellular injury, made evident by an increase in serum lipase, the fall in temperature, and other manifestations of acute shock. The elements of specificity lie in the mobilization of the non-specific cellular ferment and the colloidal changes in the serum, not in the production of specific ferments.

In a later paper we shall discuss more fully the mechanism of anti-anaphylaxis and other manifestations of protein intoxication.

CONCLUSIONS.

1. The serum ferments are practically unaltered by a primary injection of foreign protein.

2. During the course of sensitization the injection of the antigen is followed by the mobilization of a non-specific protease which increases in rapidity and intensity as the maximum period of sensitization is reached.

3. Acute shock is accompanied by:

(a) The instantaneous mobilization of a large amount of non-specific protease; (b) a decrease in antiferment; (c) an increase in non-coagulable nitrogen of the serum; (d) an increase in amino-acids; (e) a primary decrease in serum proteoses.

4. Later there is a progressive increase in the non-coagulable nitrogen, in proteoses, and in serum lipase.

5. The acute intoxication is brought about by the cleavage of serum proteins (and proteoses) through the peptone stage by a non-specific protease.

6. The specific elements lie in the rapid mobilization of this ferment and the colloidal serum changes which bring about the change in antiferment titer.

BIBLIOGRAPHY.

1. Abderhalden, E., Cammerer, G., and Pincussohn, L., *Ztschr. f. physiol. Chem.*, 1909, lix, 293. Abderhalden and Pincussohn, *ibid.*, 1909, lxi, 200. Abderhalden, E., Medigreceanu, F., and Pincussohn, L., *ibid.*, 1909, lxi, 205. Abderhalden, E., London, E. S., and Pincussohn, L., *ibid.*, 1909, lxii, 139. Abderhalden and Pincussohn, *ibid.*, 1909, lxii, 243; 1910, lxiv, 100, 433; 1910, lxvi, 88, 277. Abderhalden, E., Pincussohn, L., and Walther, A. R., *ibid.*, 1910, lxviii, 471. Abderhalden and Pincussohn, *ibid.*, 1911, lxxi, 110. Abderhalden, *München. med. Wchnschr.*, 1912, lix, 1305.
2. Pfeiffer, H., and Mita, S., *Ztschr. f. Immunitätsforsch.*, Orig., 1910, vi, 18.
3. Pfeiffer, H., and Jarisch, A., *Ztschr. f. Immunitätsforsch.*, Orig., 1912–13, xvi, 38.
4. Auer, J., and Van Slyke, D. D., *Jour. Exper. Med.*, 1913, xviii, 210.
5. Zunz, E., and György, P., *Ztschr. f. Immunitätsforsch.*, Orig., 1915, xxiii, 402.
6. de Waele, H., *Ztschr. f. Immunitätsforsch.*, Orig., 1912, xiii, 605.
7. Jobling, J. W., and Petersen, W., *Jour. Exper. Med.*, 1914, xix, 480.
8. Jobling and Petersen, *ibid.*, 1914, xx, 38.
9. Jobling and Petersen, *ibid.*, 1914, xx, 468.
10. Jobling, J. W., Petersen, W., and Eggstein, A. A., *Jour. Exper. Med.*, 1915, xxii, 141.
11. von Behring, E., *Deutsch. med. Wchnschr.*, 1914, xl, 1857.
12. Jobling, Petersen, and Eggstein, *Jour. Exper. Med.*, 1915, xxii, 129.
13. Jobling, Eggstein, and Petersen, *ibid.*, 1915, xxi, 239.
14. Jobling and Petersen, *Jour. Exper. Med.*, 1914, xix, 459.
15. Fermi, C., *Centralbl. f. Bakteriol., 1te Abt., Orig.*, 1914, lxxii, 401.
16. Joachimoglu, G., *Ztschr. f. Immunitätsforsch., Orig.*, 1911, viii, 453.

A CONTRIBUTION TO THE BIOLOGY OF PERIPHERAL NERVES IN TRANSPLANTATION.[1]

By RAGNVALD INGEBRIGTSEN, M.D.

(*From the Pathological Institute of the University Clinic, Christiania.*)

PLATES 44 TO 46.

(Received for publication, May 11, 1915.)

The large amount of experimental and clinical work in transplantation that has been done during the last few years has elucidated many essential points, and the transplantation of bone, skin, connective tissue, blood vessels, and glandular organs has assumed a practical value.   This is not the case with peripheral nerves, possibly because their transplantation in human beings has been mainly heteroplastic.

The processes leading to the degeneration of a divided peripheral nerve are well known, and the work of Nageotte has recently added minute and important histological details to our knowledge of the cells of Schwann and the part played by them in the degeneration and regeneration of a peripheral nerve.   I shall not comment upon attempts at transplantation of nerves in human beings.   Of the experimental work in this field I shall mention only the results of those investigators who have made histological examinations of the transplanted pieces, especially for the Wallerian degeneration, and who have attempted, in the case of a negative result, to determine the variations in these processes; that is, in the Wallerian degeneration, produced by transplantation.

Huber[2] has examined histologically mainly heteroplastic transplanted segments of nerves, and concluded from his experiments that the degenerative process, which occurred in these pieces was very much like the true Wallerian degeneration, the various stages of which, however, succeeded each other more rapidly in the graft than in the peripheral part of a divided nerve.

[1] Aided by a grant from The Rockefeller Institute for Medical Research.
[2] Huber, G. C., *Jour. Morphol.*, 1895, xi, 629.

Ballance and Stewart[3] in their experiments on transplantation are reserved in their conclusions, expressing themselves as follows: The degeneration appears in the graft exactly as in the peripheral part of a divided nerve. The graft itself is a piece of dead tissue and is gradually absorbed and replaced like a clot by living tissue. The regeneration occurs later, but is not a result of the work of the cells of the graft itself. This conclusion seems to me to express two conflicting views, of which one, if true, necessarily excludes the other. If the degeneration of the graft appears exactly as in the peripheral segment of a divided nerve, then the segment is not dead. For the peripheral segment of a divided nerve is not dead. The axis cylinder dies and degenerates, but the cells of Schwann, on the other hand, after the division give evidence of their life by their proliferation.

Neither Ballance and Stewart nor Huber discriminated between the auto-, homo-, and heteroplastic transplantation. This was first done by Merzbacher.[4] In 1905 Merzbacher observed that the graft in auto- and homoplastic transplantations only degenerated in a typical way, whereas in heteroplastic transplantations there occurred in the fibers various regressive processes that resulted in the necrosis of the piece. In auto- and homotransplantations the grafts survive and therefore are capable of a true Wallerian degeneration, which can take place only in nerves in a condition of survival. In heterotransplantations the grafts die and cannot degenerate, but become necrotic; in these cases there is no, or only an insignificant formation of myelin ovoids. The conclusions of Merzbacher were confirmed by the experiments of Segale.[5]

Verga[6] in 1910 performed a series of homo- and heterotransplantations, bridging the central and peripheral part of a divided nerve by the graft. Verga found that the segments healed and always degenerated, and that from an anatomical standpoint there was no difference between the homo- and the heteroplastic graft.[7]

In 1911 Maccabruni[8] in Golgi's laboratory made a number of homo- and heteroplastic transplantations. His results are in accord with those of Huber and Verga, recording a typical Wallerian degeneration in homoplastic as well as in heteroplastic grafts. The center of both kinds of graft he found necrotic, possibly due to the lacking supply of nourishment. Thin grafts degenerated completely without necrosis. From the 8th to the 14th day he found proliferation of the cells inside the nerve fibers by means of karyokinetic division. The process has a slower course than in a degenerating nerve. Regarding the origin of the cellular elements inside the fibers, Maccabruni expresses some reservation. His preparations do not show whether these cells represent the syncytium of Schwann, or whether they are connective tissue cells.

[3] Ballance, C. A., and Stewart, P., *Rev. neurol.,* 1902, x. 860.

[4] Merzbacher, *Neurol. Centralbl.,* 1905, xxiv, 150.

[5] Segale, L., cited by Maccabruni, F., *Folia neuro-biol.,* 1911, v, 598.

[6] Verga, *Jahresber. f. Chir.,* 1910, xvi, 481.

[7] This publication is not available. In the summary of the article in the *Jahresber. f. Chir.* the cells of Schwann are not mentioned.

[8] Maccabruni, *loc. cit.*

Besides the work just mentioned, experimental transplantation of nerves has been performed by Gluck,[9] Kilvington,[10] and Duroux.[11]   The results of these investigators are encouraging as far as the function is concerned.   Neither Kilvington nor Duroux, however, made histological examinations of the transplanted pieces, and the interpretation and the conclusions drawn from his material by Gluck concerning the processes of regeneration are not convincing.   The lack of proof, in the work of Gluck, that the graft is different from dead material has caused Kölliker to remark that the nerve bridge must be supposed to prevent the regeneration of the peripheral part instead of facilitating it, and Kolliker advocates the bridging of the defect by strands of catgut or tubes as superior to the application of a graft of nervous tissue.

In the problem of transplantation of nerves the question of the fate and survival and multiplication of the cells of Schwann is of importance.   The solution of this point, which is the only reliable sign of the survival of the transplanted piece, gives the key to the problem and will influence the procedure of surgeons in cases of nerve defects.   If the grafts die and become necrotic they are no more suitable for bridges than strands of catgut.   If it is true, on the other hand, that the grafts do survive, the statement of Kölliker lacks support, and in bridging nerve defects grafts of peripheral nerves must be preferred to dead material.

I have experimented on the sciatic nerve of rabbits, from which pieces 2 to 3 cm. long are taken out and then either reimplanted into the same animal, united to the cut ends of the nerve by means of a single silk suture (autoplastic), or implanted into the sciatic nerve of another rabbit (homoplastic), or into guinea pigs (heteroplastic). I have made three series of experiments.   In each series I have operated on several animals and the transplanted pieces were removed for histological examination at different intervals (4, 8, 12, 16, or 20 days) after the transplantation.

. Then the pieces were treated in the way indicated by Nageotte, which has given excellent results in his study of the Wallerian degeneration.   The grafts were hardened in Dominici's solution, next they were dissociated by means of needles as far as possible, and stained by hematoxylin before passing through alcohol and mounted in cedar oil.   From some cases sections were prepared and treated according to the method of Marchi.

[9] Gluck, *Jahresber. f. Chir.*, 1895, i, 282.
[10] Kilvington, B., *Brit. Med. Jour.*, 1908, i, 1414.
[11] Duroux, E., *Lyon Chir.*, 1912, viii, 562.

I wish to emphasize the necessity of making dissociation preparations, if one wishes to make indisputable observations on the cells of Schwann. In such preparations only we are sure that a certain cell belongs to a certain fiber, and in this case only can we count the accurate number of cells in each individual fiber. I shall give a summary of my results from each of the three series, beginning with the autoplastic transplantations.

Autoplastic Transplantations.

In this series the examination of the graft four and six days after transplantation (Fig. 4) reveals a process which is not very different from the ordinary Wallerian degeneration (Figs. 1, 2, and 3). The nuclei of the cells of Schwann have fallen in towards the center of the fibers between two myelin ovoids. The nuclei are richly provided with chromatin and are embedded in protoplasm. In a few fibers from the sixth day two or three nuclei are observed close to each other, indicating that multiplication of these cells has already begun. Between and in the myelin ovoids there are immigrated mononuclear cells of a type quite different from the cells of Schwann. The nuclei of these cells are smaller and richer in chromatin than the nuclei of the cells of Schwann. They present phagocytic properties filling their cell bodies with fragments of myelin, and Nageotte, who found them in the Wallerian degeneration, called them "*corps granuleux.*" Probably they are lymphocytes.

The only difference between the autoplastic graft from the fourth and sixth days and the peripheral part of a divided nerve is that in the latter the formation of myelin ovoids is more advanced than in the graft. In the examination of the grafts in later stages, we find this feature again and again. After eight days we find a degenerative process resembling a somewhat delayed Wallerian degeneration. A graft from the eighth day is in as degenerative a stage as a peripheral nerve from the fifth to sixth days. But in some fibers we observe no degeneration at all; there is no formation of myelin ovoids and they look perfectly normal. These fibers belong to the central parts of the graft, and judging from the appearance of Marchi preparations, these fibers later become necrotic. This is true about the central fibers of homoplastic grafts as well as autoplastic. On the twelfth day we find as pronounced a Wallerian degeneration as on the eighth or ninth day. A large number of myelin ovoids has been formed, and there are many "*corps granuleux*" and numerous nuclei of the cells of Schwann.

After the sixteenth and twentieth days (Figs. 5 and 6) the nuclei of the cells of Schwann are arranged in long rows inside the sheaths of Schwann with continuous protoplasmic bridges about and between them; the only difference from the Wallerian degeneration of the same stage is that in the latter the absorption of myelin fragments is a good deal more advanced than in the graft.

Homoplastic Transplantations.

I shall next describe the results of my homoplastic transplantations.

In almost every respect the preparations from the fourth and fifth days (Figs. 7 and 8) resemble the picture of a nerve on the fourth and fifth days of Wallerian degeneration. The nuclei of the cells of Schwann have fallen in towards the center of the fibers, and are richly provided with chromatin. A formation of myelin ovoids has started and only a few immigrated cells are seen. In one of the preparations from the fourth day we find a cell of particular interest. In one of the fibers (Fig. 8) we observe three oval nuclei of Schwann close to each other, and also a large darkly stained cell including a nucleus in mitotic division. This cell gives proof that the cells of Schwann multiply in a homoplastic transplanted graft. The increased number and the long rows of nuclei of Schwann in individual fibers already give strong evidence for the probability of such a conclusion, but the mitotic figure presents a picture from a stage of the process itself. We do not hesitate in the identification of this cell. The well outlined protoplasmic body of the cell, which is never observed in the immigrated cells during their mitotic division, determines that the cell is really a cell of Schwann.

I have never observed in the grafts in autoplastic transplantation such a mitotic division of the cells of Schwann. But mitotic divisions are not easily found and are not frequently seen in the cells of Schwann during their proliferation in the ordinary Wallerian degeneration. Accordingly there is no reason to doubt that mitotic divisions may also be observed in autoplastic transplantations, since we know that grafts in such conditions are best fit for survival.

During the first ten to eleven days the degenerative process in the homoplastic transplanted grafts appears mainly as in autoplastic transplantations; that is, like a Wallerian degeneration, only a little more slowly. Myelin ovoids are formed, the cells of Schwann multiply, and immigrated phagocytic cells loaded with fatty granules are seen in the fibers (Fig. 9). From the eleventh to twelfth days, these cells are present in a number considerably exceeding those in the autoplastic grafts. They are steadily increasing in number, and from the sixteenth to eighteenth days (Fig. 10) they form a marked feature in the whole picture.

It is possible that the presence of these numerous immigrated cells in homoplastic grafts—cells provided with phagocytic properties and probably of lymphocytic origin—are playing some part in the mechanism of immunity against homoplastic transplantation of tissue in general. For in transplantation of organs and tissue there is a marked difference between the final result of homoplastic and autoplastic transplantation (the kidneys, for instance). But I do not wish to enter upon further discussion of the importance of the phagocytic cells. I wish only to add that in homoplastic nerve grafts the nuclei of Schwann from the eighteenth to twentieth days are pale and faintly stained and are evidently in a necrobiotic condition, which is possibly dependent upon the presence of the lymphocytes.

Heteroplastic Transplantations.

In heteroplastic transplanted nerves an abundant formation of myelin ovoids occurs during the first four to five days. But later these grafts do not resemble either the autoplastic or homoplastic transplanted pieces or nerves in Wallerian degeneration. There is no proliferation of the cells of Schwann. These cells, on the contrary, are faintly stained or have completely disappeared. From the eighth and tenth days the contents of the fibers consist mostly of irregularly broken up pieces and fragments of myelin and proto-plasm and the whole fiber looks necrotic (Fig. 11).

In later stages we find between and in the fibers numerous immi-grated cells, and from the sixteenth to eighteenth days the graft on gross examination is yellowish, soft, and necrotic, or it is encap-sulated by young connective tissue.

SUMMARY.

In autoplastic transplanted nerves a degenerative process occurs which resembles the ordinary Wallerian degeneration, but appears a little more slowly than the latter. The cells of Schwann are in a condition of survival and are capable of multiplication after the transplantation.

In homoplastic transplanted nerves I have found a degenerative process resembling a Wallerian degeneration, somewhat delayed. The cells of Schwann multiply, and for some time at least are in a condition of survival. After twelve to fourteen days an abun-dant and increasing immigration of lymphocytes is observed, and from the eighteenth day the cells of Schwann develop a necrobiotic appearance.

In heteroplastic transplanted nerves numerous myelin ovoids are formed during the first four to five days, but there is no prolifera-tion of the cells of Schwann, and no Wallerian degeneration is seen. The graft becomes necrotic within about two weeks.

The formation of ovoids that occurs during the first four to five days after the performance of the heteroplastic transplantation does not reveal the condition of the life of the graft. This for-mation of myelin ovoids is found in the nerve fibers when they have been kept in an incubator for twenty-four hours in Ringer solution (Nageotte[12]) or in homologous or heterologous serum, but it is not found in the fibers after their incubation in isotonic salt solution (Nageotte), the presence of calcium being necessary for the occur-rence of ovoid formation.

[12] Nageotte, J., *Compt. rend. Soc. de biol.,* 1910, lxix, 556.

CONCLUSIONS.

Heteroplastic transplanted nerves become necrotic. They are unsuitable for bridges in cases of nerve defects, and my results explain the failure of the attempts at heteroplastic transplantation of nerves in human beings.

If we wish to bridge a nerve defect by implantation we must use autoplastic or homoplastic grafts. The occurrence of a Wallerian degeneration in these grafts during the first two to three weeks after the transplantation should make bridging a promising operation; for in this period the grafts resemble the peripheral part of a divided nerve and must be assumed to be capable of regeneration, and thus are very different from dead material.

I have studied the process of regeneration, and shall communicate in a future article my results of bridging defects, which are encouraging as far as the function is concerned.

My results with homoplastic transplantation of nerves have a bearing on the homoplastic transplantation of limbs, which has been successfully performed in dogs by Carrel. None of his dogs lived long enough to show any function of the transplanted leg. The practical value of this operation is dependent, of course, upon the return of function, and especially on the regeneration of the nerves in the transplanted leg. The results with homoplastic transplantation of nerves seem to indicate the possibility of a regeneration of the nerves in a homoplastic transplanted leg.

EXPLANATION OF PLATES.

PLATE 44.

Figs. 1, 2, and 3 are nerve fibers from the peripheral part of divided nerves. Wallerian degeneration in different stages.

FIG. 1. Wallerian degeneration, 4th day.

FIG. 2. Wallerian degeneration, 7th day.

FIG. 3. Wallerian degeneration, 14th day. Multiplication of the nuclei of Schwann, numerous immigrated cells ("*corps granuleux*").

FIG. 4. Nerve fibers from a graft, 6 days after transplantation (autoplastic). Multiplication of the nuclei of Schwann.

FIG. 5. Nerve fiber from a graft, 16 days after transplantation (autoplastic). Long rows of nuclei of Schwann. Some immigrated cells ("*corps granuleux*").

PLATE 45.

FIG. 6. Nerve fiber from a graft, 18 days after transplantation (autoplastic). Multiplication of the nuclei of Schwann. Reduction of the myelin ovoids. Numerous immigrated cells.

FIG. 7. Nerve fibers from a graft, 4 days after transplantation (homoplastic). The nuclei of Schwann have fallen in towards the center of the fibers, embedded in protoplasm.

FIG. 8. Nerve fibers from a graft, 4 days after transplantation (homoplastic). In one of the fibers a mitotic figure is seen in a cell of Schwann.

FIG. 9. Nerve fibers from a graft, 10 days after transplantation (homoplastic). Multiplication of the nuclei of Schwann. Immigrated small darkly stained cells (*" corps granuleux "*).

PLATE 46.

FIG. 10. Nerve fiber from a graft, 18 days after transplantation (homoplastic). Multiplication of the nuclei of Schwann. These nuclei are pale and faintly stained. A large number of immigrated cells is seen.

FIG. 11. Nerve fibers from a graft, 10 days after transplantation (heteroplastic). No cell is seen. The fibers appear necrotic.

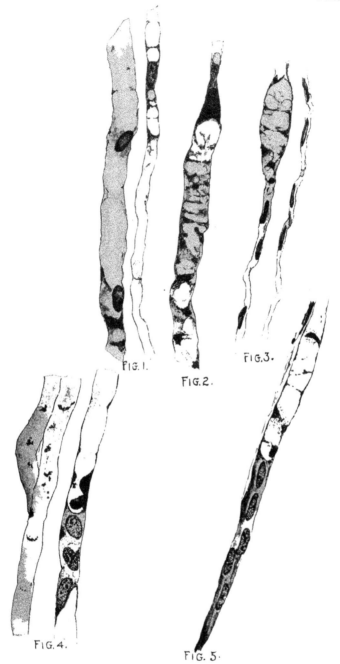

FIG. 1.

FIG. 2.

FIG. 3.

FIG. 4.

FIG. 5.

Ingebrigtsen: Biology of Peripheral Nerves in Transplantation.

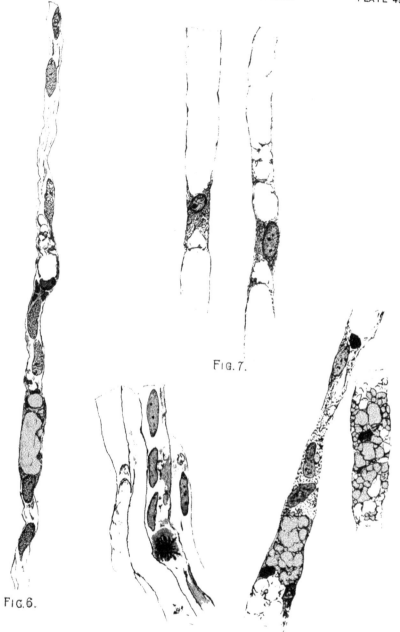

FIG.7.

FIG.6.

FIG.8.

FIG.9.

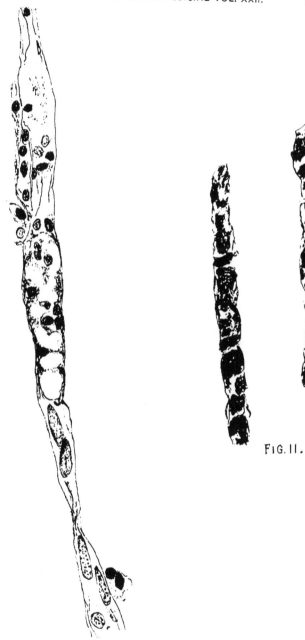

FIG. 11.

FIG. 10.

THE DIAGNOSTIC VALUE OF THE PLACENTAL BLOOD FILM IN ÆSTIVO-AUTUMNAL MALARIA.

By H. C. CLARK, M.D.

(*From the Board of Health Laboratories, Ancon.*)

PLATE 47.

(Received for publication, May 22, 1915.)

Any additional factor that can be placed in the hands of the diagnostician to aid in the differentiation of diseases seems worthy of introduction; therefore, the examination of the placental blood film is recommended to indicate better the presence of æstivo-autumnal malaria when it complicates labor and in differentiating it in the early days of the puerperium from puerperal sepsis.

The cause of a fever in the puerperium is not very difficult to find when the case is under competent observation in a non-malarial region, but the diagnostic significance of such a fever in the tropics may require further analysis to eliminate latent or active protozoal disease.

Malaria, as a cause of death in the Panama Canal Zone, ranks well below several other factors, yet it has been a very important one in morbidity statistics, and all divisions of the hospital service at Ancon have been obliged to keep it in mind as the single or a possible associated factor in seeking to establish a diagnosis.

Marchiafava and Bignami (1) feel that the diagnosis of malaria can be made without a puncture of the spleen if examinations of the peripheral blood are made often enough and with sufficient care. This statement has been largely confirmed by James (2) in his experience with malaria on the Canal Zone, in so far as it relates to individuals who are incapacitated by this disease and in many cases where it seemed to be merely a coincident factor; but there are examples of a latent type, or individuals who may be called malarial carriers, that will not reveal the presence of the parasite in a single examination of the peripheral blood and sometimes even over a long

427

period of time unless some factor of an intercurrent nature lowers that individual's resistance and encourages a malarial outburst.

It occasionally happens that the physician suddenly assumes the responsibility of a case in labor associated with a severe attack of malaria which is at the moment between the periods of sporulation; in such instances the examination of the placental blood film will expedite the diagnosis and permit the immediate institution of treatment.

Puerperal fever in regions where æstivo-autumnal malaria is endemic always offers some reason for anxiety to the conservative practitioner who wishes at the earliest moment to differentiate malaria and puerperal sepsis; while to the other type malaria affords one of the safe retreats in diagnosis to cover all fevers which appear in the puerperium.

The evil relation of malaria to the course of pregnancy and the puerperium has long been given consideration by various writers on tropical medicine, but the primary intention of this paper is to advocate the routine examination of the stained fresh placental blood film as an immediate means of identifying the presence of malaria in nearly every case when it is associated with labor or the early days of the puerperium.

It has long been known that it is possible to find an abundance of parasites in a blood film made from the maternal surface of the placenta in a woman whose labor was complicated by a serious attack of malaria, but as far as I can learn no practical application of this fact has ever been made, it being simply considered a curious feature sometimes encountered.

I believe that the Ancon clinical and pathological data soon to be mentioned, together with the results of 400 observations on placental blood films, will show that it has a practical application and is a worthy test for routine practice in tropical obstetrics and in other regions where this type of malaria is found. This is the only instance in which the body voluntarily offers for examination an intricate capillary or blood sinus system that closely approximates such vascular systems as the spleen, bone marrow, and liver, and the fact is well established that the adult forms of the æstivo-autumnal parasite remain, as a rule, stationary in such systems.

Unfortunately the value of this feature did not impress itself at a time in the course of the construction of the Panama Canal when it could have been applied to the largest number of individuals who were the most susceptible to malaria, and also before the sanitary system was operating so successfully against this disease.

However, the study of this short series of 400 cases, made under the present adverse circumstances, may yet reveal sufficient evidence to impress the importance of such an examination, and it will at the same time offer other points of interest.

Pathological Observations.

Some time ago a collection was made, for general study and analysis of all fatal cases of æstivo-autumnal malaria that had come into my hands through the routine pathological service of the Board of Health Laboratories during a period of about four years. There were included in this series of 107 cases 31 females of whom 7 were children. Among the remaining 24 adult women, 12 were found to have been either in the late stages of pregnancy or not far advanced in the puerperium. This fact seemed so significant that the general anatomical records covering the same period of time were again searched to see what further relation could be found between other conditions and the pregnant state.

192 autopsies were found that had been performed on adult females. 26 of this number were either pregnant or had been recently delivered. The following relationship was found:

```
Malaria, æstivo-autumnal ............................. 12 cases.
Malarial hemoglobinuric fever ........................  3    "
Puerperal septicemia .................................  4
Eclampsia ............................................  3
Typhoid fever ........................................  2
Chronic nephritis ....................................  1
Extra-uterine pregnancy ..............................  1
```

The race of the cases showing an association of the pregnant state with æstivo-autumnal malaria and malarial hemoglobinuric fever is shown in Table I.

TABLE I.

Race.	No. of cases of malaria.	No. of cases of malarial hemoglobinuric fever.
Negroes, West Indian	8	1
White, Spain	2	2
Mestizo, Panama-Colombia	1	0
White, United States; never under medical care	1	⌐

Clinical Observations.

Through the kindness of Dr. A. B. Herrick, Chief of the Surgical and Obstetrical Clinics, and of Dr. W. E. Deeks, Chief of the Medical Clinic, their Ancon Hospital Records covering the same period of time as the pathological records just mentioned were next appealed to along the same lines, and after excluding those cases contained in the anatomical report the following additional data were assembled (Table II).

TABLE II.

Race.	No. of cases of miscarriages and abortions.	No. of cases of normal pregnancy.	No. of other accidents of pregnancy and labor.	Total.	No. of positive malarial cases.
Americans.....................	85	340	12	437	1
West Indian negroes...........	54	295	36	385	31
Europeans....................	9	66	1	76	2
Latin-Americans..............	14	102	6	122	3
Chinese......................	0	2	0	2	1
Syrians......................	0	1	0	1	0
Total.....................	162	806	55	1,023	38
No. of associated malarial attacks.	14	16	8		38

Among the 38 cases there were 5 deaths not included in the list of autopsies.

It is not possible to confirm entirely by these statistics the more or less general opinion that in the tropics where malaria is endemic the great increase in abortions and miscarriages is chiefly due to malaria. There are too many other factors which, no doubt, play a part in this and have not been ruled out by the authors of such an opinion. However, there is present in the data a very striking relationship between many of these accidents and associated malaria as compared to other diseases such as pneumonia, typhoid, tuberculosis, and even syphilis.

Tuberculosis leads all other factors as a cause of death in the

Canal Zone during this period of time, in so far as a record of autopsies can be taken for a guide. Pneumonia also stands well ahead of malaria as a cause of death. Syphilis is a common disease on the Zone in these days, but armed with a clinical test it can be well identified; yet the combined anatomical statistics on the known evil association of other diseases with pregnancy and the puerperium do not exceed the results which malaria alone has accomplished, and this fact is well supported by the clinical data.

The various opinions expressed by Scheube (3), Deaderick (4), Williams (5), Craig (6), and others led to the taking of further observations before expressing any view based on the local data just mentioned.

Conditions have rapidly changed on the Canal Zone since the clinical and pathological reviews were made, there being a marked reduction of people in the Canal Zone and also of the malarial rate. Many villages have been abandoned from which such cases were received. In addition to these factors the new observations have been conducted chiefly in the dry season. Nevertheless, the examination of 400 cases of labor has been made on individuals from various parts of the Canal Zone, the period of time beginning August 21, 1914, and ending March 30, 1915.

Methods Employed in Making the Observations.

At the end of labor some blood films were made from the mother's peripheral blood by the physician in charge of the case, and these films and a history of the case together with the carefully protected placenta and umbilical cord were sent to the Board of Health Laboratories at Ancon, where an immediate opportunity was given me to prepare the placental and fetal blood films.

Great care must be taken in preparing the fetal blood film from the blood in the umbilical cord to prevent contamination of this specimen with the placental blood. False impressions of the infected fetal blood can easily be obtained unless precautions are taken.

The maternal surface of the placenta when it was received at the Laboratory was usually covered over with the membranes, and the folds of these membranes often contained a large amount of maternal blood. It has, therefore, been the custom to prepare first the film from the umbilical cord. The cord is thoroughly cleaned and then cut near its middle, and the oozing blood caught and spread in the usual manner. This film is marked and laid aside before disturbing the placenta and membranes.

The placental blood film is made from the maternal surface of the placenta after all free blood and clots have been removed from its surface and from the folds in the membranes. Blood was made to ooze from the clean surface by grasping a portion of it with a hemostat. The specimen was then removed by pressing down and dragging the end of a glass slide across the bleeding surface, and from this material the films were made. The three types of blood films were then dried and stained with a polychrome stain (Hastings) in the usual way that the ordinary blood film is treated. Then followed a routine examination for the presence of the malaria plasmodium.

The Wassermann test was applied to many of the cases and invariably to all suspicious cases whenever the state of the blood contained in the fetal membranes was such that it could be used for the test. If a dead fetus accompanied the placenta and cord and its condition permitted the application of the test, it was performed. Suspicious lesions in the placenta or in a dead fetus were prepared by the Levaditi method to permit a search for *Treponema pallidum.* This organism was also kept in mind when the blood films were searched. The series was closed after the receipt of 400 specimens that were in a suitable condition for examination.

The results are presented in Tables III and IV.

TABLE III.

400 cases examined.	No. of positive cases of malaria.
Blood films from maternal surface of placenta	19
Blood films from mother's peripheral blood	8
Blood films from umbilical cord	1

400 Routine Cases of Labor.

Race of the women examined.	No. examined of each race.	No. of positive identifications of malaria.	Per cent of positive cases.
North Americans (white)	118	0	0.0
Latin-Americans (mestizo)	92	3	3.26+
Europeans (white)	17	1	5.88+
West Indian negroes	173	15	8.67+
Total	400	19	4.75

TABLE III.—*Concluded.*

Local Residence of the Positive Cases.

Residence.   No.

Unknown suburban division of Panama City 4 ⎫
Calidonia .... 4 ⎪
Guachapali .... 3 ⎬ Panama City ....15 ⎫
San Miguel .... 2 ⎪ ⎪
Maranon .... 1 ⎪ ⎪
Chorrillo .... 1 ⎭ ⎬ Colon
  ⎪
Mount Hope .... 1 ⎫ ⎪
Colon .... 2 ⎬ Colon .... 3 ⎪
  ⎭ ⎪
Culebra .... 1 ⎫ C. Z. .... 1 ⎭

Total 19

Seasonal Incidence of the Positive Cases.

August 0	December 3
September 2	January 2
October 4	February 1
November 3	March 4

Labor Accidents That Were Known To Have Occurred in the Series.

Abortion 9
Premature labor (fetus dead in 13 of these cases) 23
Full term still-births 11
Full term children dying inside of 48 hrs. 1

A total of 44 accidents in the series of 400 cases.

There were 19 positive cases of æstivo-autumnal malaria in the 400 cases, and 7 of these 19 cases (a percentage of 36.84 +) were found associated with some of the accidents named above.

Besides the 19 cases positively identified as malaria, there were 5 other cases with a reliable history of having been treated for malaria in recent months, and in 2 of these, full term still-births were recorded; while in a 3d case a grave anemia and premature labor were noted. Thus, if 24 be accepted as the actual number of cases with a malarial influence present, and 10 the number of accidents, a percentage of 41.66 is revealed. Other factors found in connection with some of these accidents where no malaria influence was established are given in Table IV.

TABLE IV.

A calculus in the left ureter and left pyonephrosis 1
Grave anemia 1
Acute endometritis, suppurative 3
Eclampsia 1
Typhoid 1

27 of the accidents thus remain in which no plausible factor at the time of labor could be named as the cause.

The racial classification of the women in which the total number of accidents occurred is as follows: Americans, 14; Europeans, 6; Latin-Americans, 6; West Indian negroes, 18.

Discussion of Tabulated Features.

The placental films revealed the presence of malaria in eleven cases whose peripheral blood appeared negative. On the other hand, the peripheral blood was never found positive without the placental film showing a much heavier infection. This is an important difference and one which might be expected when the general circumstances, surrounding the origin of the two films, are given consideration. The adult parasites, especially of the æstivo-autumnal type, are prone to remain stationary in the more complex capillary systems of the deep viscera, and after the third (7) or fourth month of its development the placenta belongs to that class of viscera. It rapidly increases in importance as a favorable site for the development of adult parasites as its maturity is reached. In most instances the first field examined in the placental films will be sufficient to establish the diagnosis of malaria in untreated cases, even though a tedious search may have been required to locate parasites in the peripheral blood of the same case.

The malarial carriers identified in this series by examination of the placental film and not found positive in their peripheral blood required not more than one to five minutes in locating the first parasites. Seldom will it be necessary to make a protracted search of this film if malaria is present during labor and quinine has not been previously administered.

In some of the cases identified in this group the infections were so heavy that a fatal issue seemed certain for both the mother and child, yet in all instances the mother lived and in nearly all cases the child lived. The vitality of some of these jaundiced, premature children is remarkable. A few of the patients among the positive cases had been admitted just before or at the time of labor and were apparently between the periods of sporulation, since there was no temperature, yet heavy infections were noted in the placental films. Even with this knowledge of these cases in mind, a tedious search of the peripheral blood was required to find the ring forms. When under short observation these cases can occasion unpleasant surprises.

The average numerical relationship between the parasites found during a clinical attack of malaria in these three types of films can be more easily appreciated by inspection of Figs. 1 and 2 which have been drawn from the actual preparations made in such a case. Fig. 2 representing the placental film contains parasites drawn from three cases in order to present some common features of interest that will be explained in the key to the illustrations. The numerical relationship there represented is, however, accurate for the conditions named.

The commonest types encountered in the placenta are the segmenting and presegmenting forms. It also frequently happens that the majority present are younger adult forms. Crescents were seldom found, yet they are common forms in the films made from bone marrow and the spleen. Sometimes the large forms seem entirely free from red blood cells, but expert variation in the method of staining might have shown blood cell remnants.

The phagocytosis of pigment in the placenta is not so extensive as it is in the spleen, liver, and bone marrow, but it is quite marked and often even the polymorphonuclear leucocytes take part in this activity.

It is a much easier task to prepare and examine a placental film at the time of labor than to spend perhaps long periods of time on each of two or three consecutive days following labor in an effort to catch the products of a sporulation in the peripheral blood of the patient; because, in the placenta, the parasites are quite large and very abundant and there is much evidence of phagocytosis, while in the peripheral blood a few small ring forms are all that can usually be expected.

The study of the fetal blood from the umbilical cord presents a feature of interest in regard to the evidence it affords in support of the view that the mother's blood and the fetal blood never intermingle under normal conditions.

A single instance occurred in this series where an undoubted moderate infection of the fetal blood existed at the time of birth. The mother's blood and the placenta bore evidence of a very heavy infection. There was the history and the evidence of an associated accident of pregnancy in relation to the placenta so that it seems wise to attribute this infection to the placental accident. The mother and child were discharged in such a short time from the institution in which the case occurred that no further observations

on the child's blood could be taken.   The final result is not known,
but the mother was a black woman and the probabilities are strong
that both mother and child are well.   This is the only case in the
last five years that has come under my observation that can certainly be established as an intra-uterine transmission of malaria to
the fetus.   Our few cases of malaria in infants are all surrounded
with so many circumstances favorable to postnatal infection that
they cannot be safely considered cases of congenital malaria.   As a
matter of fact it seems reasonable to assume that the observations
made in this series alone offer good support to the long standing
belief that the blood streams of the mother and the fetus never actually intermingle unless some rare accident to the placenta occurs.

It seems necessary to impress again the fact that three important
features existed during the collection of these cases which have
greatly lessened the number of infections otherwise to be expected
in such a number of individuals.   The sanitary system continues
with ever increasing success to exert a strong influence against this
disease.   The reduction in the population of the Canal Zone has
been great.   The observations have also been carried on partly in
the dry season, at which time this disease is supposed to be at a
low ebb.

White women coming from the temperate zone of the United
States and Europe, who live on the Canal Zone, are believed to be
the most susceptible women to malaria, but nearly all European
women have gone and the Americans avail themselves of all the opportunities offered by the system of sanitation as well as professional care throughout their course of pregnancy.

This leaves only the negro women, whose race seems to enjoy a
relative immunity to malaria, and the native women, who possess
even a higher grade of immunity than the West Indian negro.
Under the circumstances, however, it does not seem unusual for the
highest rate of infection to be found in this series in the negro.   The
negress is the domestic servant and she lives under limited circumstances chiefly in the suburban divisions of Panama City where the
rent is lowest and the malarial rate highest.

These conditions no doubt explain the high rate in the relatively
immune negress and the absence of the disease in the non-immune

white women. The same rate and character of infections in the white woman that have been found in the black woman in this instance would certainly have been accompanied with more serious consequences.

The negress is very indifferent to her pregnant state and seldom can tell within a month or two the correct time for her course of pregnancy to end. It is therefore quite possible, judging from the appearance of some of the children, that several more premature labors associated with the malarial cases in this series should be recorded; but the histories offered in these instances have been adhered to and they were not counted among the accidents.

It seems incredible that a fetus can live where nutrition is derived from such heavily infected placental blood, yet in a remarkable number of cases (negroes at least) they do survive. In some cases this may be due in part to the fact that latent malaria many times does not show a terrific outburst until very late in pregnancy when the patient's vitality is low and the placenta a better harbor to assist in the development of parasites. At this time the child may be sufficiently developed to fight for an independent existence forced upon it by miscarriage. The probabilities are that its chance for life is better under even these conditions than to continue its intra-uterine existence during the climax of a severe malarial infection.

The placenta is one malarial battle-field over which the body holds the ruling influence, for by expulsion of the uterine contents the field and its contained enemies are quickly eliminated.

The later influences of such a birth as this on the infant can only be a subject for conjecture. It would certainly appear to offer a predisposing cause to the diseases of early infancy. When a fetus is developed throughout its uterine life in association with latent malaria in the mother and at least in contact with one acute attack in the mother it would seem that there should be some degree of immunity gained to this disease by the child in after life even though the parasites were never admitted to its circulation from that of the mother. This thought recalls an opportunity afforded me a few years ago to study microscopically the tissues of a still-birth associated with fatal malaria in the mother. The tissues happened in this instance to be in good condition. Its blood films and those made from its spleen were negative for parasites, yet when the histological study was made, a condition comparable to the cloudy swelling associated with deaths due to febrile diseases and abundant microscopic evidence of pigment was found that I am not able to tell from the melanin of malaria.   It is difficult to say whether this pathology was due to toxins in the mother's blood stream or simply to interference with the nutrition of the fetus.   The deposition of the pigment in its tissues is also open to several lines of thought.

GENERAL DISCUSSION.

Scheube (3) states that disorders of menstruation and a decrease in fertility are more frequent in European women, and miscarriages also occur more often with them in the tropics than when they live in Europe, and he adds the further statement that climate is less to blame for this than malaria.   This opinion is given more or less general support and there is no doubt that malaria does bear a very important relation to these conditions, yet there are many other factors in the tropics where malaria is endemic that help to make these features more common.

Williams (5) offers a counter-statement: Despite the somewhat wide-spread opinions to the contrary, it would appear that ordinary forms of malaria have but little influence upon the course of pregnancy; it is probable, however, that the pernicious forms of malaria may have a much more deleterious effect.   He admits the tendency to malarial outbreaks in pregnancy and the puerperium.

A compilation of the various records in regard to the malarial influence in pregnancy and an additional valuable opinion is found in Deaderick's (4) writings.

The experience at Ancon indicates that æstivo-autumnal malaria does exert frequently an evil influence in the late stages of pregnancy and in the puerperium, sometimes being fatal to both mother and child, but there yet remains a very large number of miscarriages, abortions, instances of sterility, and menstrual disturbances that cannot be explained, and least of all can they be ascribed to malaria on any other than circumstantial data of a high rate of malarial morbidity in the tropics.

The anatomical structure of the placenta in the late stages of pregnancy and the opportunity it offers for the development of the malarial parasite makes it a point of strategic importance for the parasite in interrupting the late course of pregnancy, and it is surprising that it does not always do so. What harm may come to the menstrual process and to fertility in untreated malarial carriers who are apt to suffer from numerous small outbursts of malaria is less easy to answer. Such a circumstance would probably play an important part.

The results and statements of the physicians in charge of obstetrical work at Ancon Hospital offer confirmation to Williams' (5) statement that quinine can be given with impunity in such conditions. The accidents number about the same with or without its administration, and certainly the mother's prognosis is better when it is given.

It is hard to dismiss the idea that children born under such conditions, if they survive, do not gain at least a temporary immunity from the disease. It is extremely rare to receive at the morgue in Ancon a child under one or two years of age dying of malaria. Observations have been made, however, by competent people on large amounts of clinical material who claim that all malarial immunity is acquired.

These children, born of mothers with a positive malarial infection, will be followed, as far as possible, to determine their future relation to the disease and to see whether their birth under such conditions plays a predisposing part in making them more easily the victims of other diseases of infancy.

The observations of Bastianelli and Bignami (8) on the punctiform hemorrhages, which sometimes occur about the vessels in the brain in some of the severe cases of æstivo-autumnal malaria, support the theory of a viscid property possessed by this parasite, and Bass (9) has offered further proof of this fact. These small punctate hemorrhages were said to consist of uninfected red blood cells, while the cells, visible in the capillaries about which the hemorrhages were found, were all infected cells. An explanation offered is that the infected cell anchors itself to the endothelium of the vessel, while flexible non-infected cells under pressure escape by diapedesis. I have personally observed instances of these hemorrhages and can confirm their statement in regard to their composition and the nature of the cells in the capillaries near them.

A somewhat parallel circumstance occurred in another instance in which a fatal termination of a severe case of æstivo-autumnal malaria was brought about by the spontaneous rupture of the spleen and hemorrhage into the peritoneal cavity. The free blood in the peritoneal cavity unexpectedly revealed the discoid and ring forms commonly found in the peripheral blood, while films from the splenic pulp revealed an abundance of the larger adult forms. This circumstance, I believe, also supports the theory that these larger parasites or the infected cells possess a greater degree of viscidity and can maintain their position in the favorable locations mentioned. The higher degree of body temperature probably assists in the development.

Labor, associated with malaria, spontaneously offers the morphologist an opportunity to study at the same moment the parasites of the peripheral and deep circulation. Even at the end of twenty-four hours a heavily infected placenta can often be subjected to an examination with good results, but this necessitates a knowledge of the degenerated forms of the parasite and a more intense method of staining.

SUMMARY.

1. The placental blood film examination is worthy of routine application wherever æstivo-autumnal malaria is endemic. This type of malaria when associated with labor and the early days of the puerperium can be·more easily and certainly diagnosed by the use of this film and a polychrome stain than by employing the usual films made from the mother's peripheral blood at the time of labor.

The placental film in such an infection offers an abundance of adult parasites and far more evidence of the presence of pigment, while the peripheral blood film frequently offers but a scant number of the small ring or discoid forms of a parasite. The examination of the present series revealed positive placental films in nineteen cases, while but eight of these same cases were positive in the peripheral blood film examination. On the other hand, no peripheral blood films were found positive in which the associated placental films did not reveal a far more abundant evidence of the infection.

2. The early days of the puerperium can by this method be protected many times from a malarial outburst, and, as a rule, puerperal sepsis can be differentiated.

3. The intricate vascular architecture of the mature placenta rivals that of the spleen, liver, and bone marrow as a harbor for adult malarial parasites of this type and as a storage for pigment.

4. The localization of parasites in the placenta is unique. Here is the one vascular system which particularly favors the development of the parasites but which at the same time is so situated that it may be spontaneously discarded by the body at the climax of the attack. By this simple act late in pregnancy the prognosis for both mother and child may be improved.

5. The pregnant state encourages attacks of malaria by lowering bodily resistance and by furnishing an additional harbor for the development of parasites. A tenable theory in regard to most attacks of this nature, occurring in cases under professional care, would appear to be the development of latent malaria (malarial carriers) into acute attacks toward the close of the pregnant state. The women who expose themselves (as the negroes in this series) offer favorable conditions to the introduction of a primary infection.

Malaria frequently interrupts the late stages of pregnancy and sometimes causes the death of the mother and the fetus, more often the latter. The records at Ancon indicate that it more frequently exerts a harmful influence than other types of infectious diseases in this locality.

6. Most of the children in this series that were delivered while malaria was present in the mother, were of a race that seems to possess a relative immunity to the ravages of malaria. This may account for the fact that the negro fetus more nearly approximates the full term of development when associated with this disease and is comparatively subjected to a less number of the accidents of pregnancy. Many of them revealed evidence of prematurity and were jaundiced, but, as a rule, they developed rapidly.

The commonest mishap is miscarriage late in pregnancy. Occasioual still-births occur and sometimes there is a fatal issue to both the mother and child.

7. Cases diagnosed as congenital malaria probably indicate that

some accident occurred to the placenta, because it practically never happens that fetal blood is positive at the time of birth, regardless of the degree of infection in the mother.

Many of the cases now reported in the literature as congenital malaria suggest immediate postnatal infection as their history, as our pathological and clinical records testify.

8. The size of the intervillous spaces of the placenta and their adaptability in the localization of parasites seem to disprove to a certain extent the old idea that the localization depends on the smallness of the capillary caliber.   If this were the case the brain should be more often the seat of an extensive localization than the spleen, bone marrow, and placenta, yet our anatomical records will not support that theory.   A sluggish blood sinus with a large endothelial surface, a higher internal body temperature, and red blood cells burdened with parasites of a certain age beyond the ring form seem to be important factors in the localization and development of the æstivo-autumnal parasite.

9. The racial disparity of malarial infections shown in this series is believed to be due to local conditions and a wrong impression is apt to be given by our statistics in regard to the relative immunity of the negro race.

The white women on the Canal Zone avail themselves of all the opportunities the sanitary system affords; they live well and place the entire course of their pregnant state under competent professional care, while the negro woman is indifferent to her pregnant state, works as a domestic servant, and lives in the cheapest unprotected quarters that can be rented in the suburban divisions of Panama City where the malarial rate is highest and the sanitary control is difficult.

It should be noted that these negro women can carry an infection with little manifestation of its presence that would produce serious results in the white women brought from the temperate zone regions of Europe and the United States.

Thanks are due Dr. Samuel T. Darling for help rendered in preparing the paper, and to various members of the sanitary staff who have made it possible to assemble the material for study.

BIBLIOGRAPHY.

1. Marchiafava, E., and Bignami, A., in Twentieth Century Practice of Medicine, New York, 1900, xix, 24.
2. James, W. M., The Practical Value of the Ross "Thick Film" Method in the Diagnosis of Malaria, *Proc. Med. Assn. Isthmian Canal Zone,* 1911, iv, pt. 1, 49.
3. Scheube, B., The Diseases of Warm Countries, 2d edition, London, 1903, 152.
4. Deaderick, W. H., A Practical Study of Malaria, Philadelphia, 1909, 55, 242–243, 353–354.
5. Williams, J. W., Obstetrics, 3d edition, New York, 1912, 485.
6. Craig, C. F., The Malarial Fevers, Haemoglobinuric Fever and the Blood Protozoa of Man, New York, 1909, 280.
7. Piersol, G. A., Human Anatomy, Philadelphia, 1907, 53.
8. Marchiafava, E., and Bignami, A., *loc. cit.,* pp. 231–232.
9. Bass, C. C., Cultivation of Malarial Plasmodia *in Vitro, Am. Jour. Trop. Dis.,* 1914, i, 546.

EXPLANATION OF PLATE 47.

The comparative appearance of the types of films used in the study of the 400 cases are here represented by Mr. George Newbold. The primary intention of the drawings is to show the numerical relation of the parasites in each type of the films. The case selected for the drawings was one in which the individual on admission to the hospital gave no evidence of the infection, but soon after labor showed a moderate rise in temperature. Fig. 1 was drawn true to the actual conditions found in the films, but Fig. 2 was made into a composite drawing of parasites found in three separate cases, in order to represent some common phases which occur in this type of film. However, the figures are true to the average numerical relationship found in a simple attack of æstivo-autumnal malaria complicating labor.

Fig. 1. A film from the mother's peripheral blood stream stained with Hastings' polychrome modification. This film revealed an average of about 3 parasites to each microscopic field examined (obj. 1/12, oc. No. 8).

A, B, and C represent three forms of the young parasite.

D represents a large lymphocyte with a clump of phagocyted pigment.

E is a polymorphonuclear leucocyte.

Fig. 2. A composite drawing made from the placental blood films of 3 cases. Two of the cases show phases of degeneration of the parasites due to the interval of time which passed before films could be prepared from the specimens received.

A is a large lymphocyte showing phagocytic activity. These cells are more commonly observed here than in the peripheral blood.

B and C represent the commonest types of parasites encountered in the fresh placental film, segmenting and presegmenting forms. Many times the field will be crowded with these forms, while in lighter infections 1 to 5 parasites to the field are found, and in these instances it seldom happens that any parasites can be found in a peripheral blood film.

D represents the appearance of such parasites as C when 24 hours have elapsed following labor, before the film could be prepared. By the use of expert methods of counterstaining the Hastings' preparation with Giemsa stain, these parasites can be made to resemble presegmenting types.

E is the smallest parasite usually found in this film.

The remaining parasites, which show poor definition and stain deeply and irregularly, represent films made from a case 10 or 12 hours after the end of labor. Magnification the same as in Fig. 1.

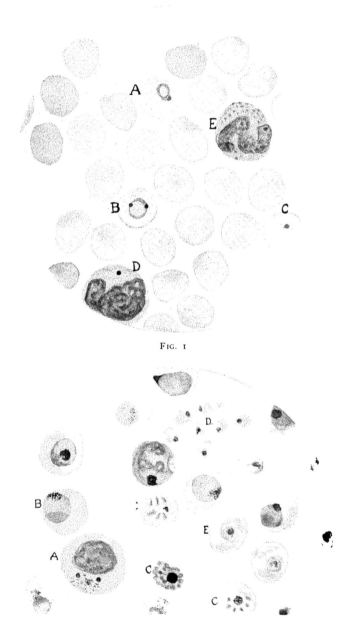

FIG. 1

AN IMMUNOLOGICAL STUDY OF BACILLUS INFLUENZÆ.

By MARTHA WOLLSTEIN, M.D.

(From the Laboratories of The Rockefeller Institute for Medical Research.)

PLATE 48.

(Received for publication, June 15, 1915.)

Bacillus influenzæ is a frequent invader of the human body, where it either causes or complicates important pathological processes. Among the pathological conditions which it produces are bronchopneumonia, empyema, and leptomeningitis; and among those in which it accompanies streptoccoci and pneumococci are the common laryngeal, tracheal, and bronchial affections. Moreover, it is sometimes associated with these cocci in lobular and lobar pneumonia.

According to the location and severity of the infections the influenza bacilli remain confined to the local lesions, or at the same time invade the blood stream. When they are confined to the local lesions in the respiratory organs, the bacilli tend to be of low virulence for animals. When they invade the blood, as they do in leptomeningitis and in some instances of pneumonia, they tend to be of higher virulence. Along with the blood invasion, influenzal suppurative arthritis may occur. These facts suggested the problem as to whether the strains of influenza bacilli isolated from various sources are identical, or whether they may be differentiated into groups on sound biological or serological grounds.

Sources of the Cultures.—The strains of B. *influenzæ* employed in this study came from two main sources; first, the respiratory mucous membrane or the lungs; second, the cerebrospinal fluid. The respiratory strains were usually slightly virulent, the other strains more virulent for laboratory animals. The bacilli isolated from the blood both before and after death were always compared with the respiratory and meningeal cultures obtained during life from the same case. This applies also to the strains isolated from suppurating joints occurring as a complication of influenzal bacteriemia with meningitis.

Criteria of Virulence for Animals.

The subject of virulence or pathogenicity of *B. influenzæ* for animals has been dealt with in previous papers.[1] It will suffice to recapitulate the main facts in this place. The white mouse succumbs to intraperitoneal injections of cultures irrespective of their origin. A peritoneal exudate arises which contains large numbers of the influenza bacilli, as does the heart's blood. Guinea pigs of 200 grams' weight, on the other hand, succumbed to the intraperitoneal injection of one blood agar culture of all the meningeal, and to about one-half of the respiratory, strains tested.

It is by means of the rabbit that the distinction of the virulent or pathogenic strains from those lacking this quality is accomplished. Rabbits of about 1,000 grams' weight are employed. Using one blood agar culture injected intravenously as the test dose, nineteen of the twenty meningeal strains tested caused death within eighteen to thirty hours. The heart's blood in fatal cases contained large numbers of the bacilli. The respiratory strains are for the most part non-pathogenic for rabbits. Among several score of such strains I encountered only six which were virulent for these animals. Four of them came from the lungs and heart's blood and two from the lungs alone (the heart's blood being free) of infants at the Babies' Hospital. In none of the six cases were the leptomeninges involved. In this connection it may be mentioned that Batten[2] described a strain of *B. influenzæ* obtained from the meninges which was wholly devoid of pathogenic properties for mice, guinea pigs, and rabbits.

Morphology.

B. influenzæ is subject to marked variations in morphology. Two main forms are met with, the short and the long. The respiratory strains usually belong to the first class. In films from the bronchial secretion influenza bacilli are short and rather thick, but regular in

[1] Wollstein, M., *Am. Jour. Dis. Child.*, 1911, i, 42; *Jour. Exper. Med.*, 1911, xiv, 73.

[2] Batten, F. E., *Lancet*, 1910, i, 1677.

size. On blood agar plates dew-drop colonies are developed which can be readily differentiated from colonies of all other bacteria by their translucent, colorless appearance. The edges are quite regular, they always remain small, and they never cause hemolysis in the surrounding medium. The bacilli in such colonies are very short and regular, often coccoid in size, with deeply staining poles. No threads are formed (Fig. 4). On blood agar slants the growth is very profuse, but the small translucent colonies do not coalesce. In the condensation water of the tubes threads are formed, but they are never long. The meningeal strains belong to the second class. The bacilli in the cerebrospinal fluid are sometimes remarkably long and thick, and so little do they resemble the usual forms of *B. influenzæ* that the diagnosis of influenzal meningitis from films alone becomes difficult (Fig. 1). In other cases short, almost round forms appear in the films so that the presence of a coccus may be suspected (Fig. 2). Grown on moist blood agar plates dew-drop colonies are produced, in which the individuals may be swollen and atypical (Fig. 3). When transplanted to blood agar in which there is less free fluid, the forms become typical. Similarly, an excess of bronchial secretion tends to cause swelling of the bacilli in respiratory strains, but the typical form appears in subcultures. The matter of moisture present affects the morphology of both types. Even so the meningeal strains tend toward larger, more definitely bacillary forms, and incline to the formation of larger threads. When the meningeal strains are recovered from the peritoneal cavity of inoculated animals they are always small and more regular. But the next subculture again shows the larger bacillary forms.

The form tends to be constant for each strain. Thus, a coccoid strain is apt to remain small and regular and to produce only short threads. The bacillary forms, on the other hand, tend to long thread formation. However, after long periods of artificial cultivation (two years or more) both the coccoid and bacillary forms, as they existed, on the one hand, in respiratory and, on the other, in meningeal lesions, grow larger and acquire the power to give rise to long threads (Figs. 5 and 7). If the later generations are examined within the first twenty-four hours of growth some minute bacilli will be detected, indicating the original type to which the strain belongs.

A medium ill suited to growth leads to greater irregularity of form.   Thus plain agar tubes upon which a small amount of human blood was placed gave rise upon inoculation of various strains to typical bacilli on the one hand, and, on the other, to long interlacing threads, recalling the leptothrix threads described by Ritchie[3] (Fig. 8).   The latter were produced by one meningeal and several respiratory strains.   Subcultures in the usual medium from these specimens yielded the typical bacilli.   While meningeal and respiratory strains of *B. influenzæ* differ morphologically they present no points of difference in their method of growth on blood agar plates and slants.

Serological Reactions.

The main purpose of this investigation was, as stated, a minute study of the immunological or serum reactions of a considerable number of strains of *B. influenzæ,* with the object of determining whether the strains compose one or more groups, irrespective of virulence.   For this purpose immune sera were employed, of which the decisive ones were those prepared in the rabbit with selected strains of the cultures.

Monovalent sera were obtained by immunizing rabbits to virulent and non-virulent strains of *B. influenzæ.*   The virulent strains isolated from cases of influenzal meningitis were well borne by the animals in increasing doses over a period of three to five months. The respiratory strains, on the other hand, were badly borne, the rabbits becoming emaciated and dying after a dose which animals inoculated with virulent cultures were well able to bear.   From these results it is fair to argue that the non-virulent respiratory strains do not produce immune bodies in sufficient quantities to protect rabbits against repeated and increasing doses of the bacilli.

Opsonins.—The opsonic content of the monovalent sera was fairly high, phagocytosis of the organisms being present in dilutions of 1 to 1,000.   No specific reaction was obtained, however, the heterologous strains being taken up by the leucocytes in dilutions as high as those in which homologous strains were phagocyted.

Agglutinins.—The monovalent rabbit sera did not agglutinate

[3] Ritchie, J., *Jour. Path. and Bacteriol.,* 1910, xiv, 615.

homologous strains *in vitro* in higher dilutions than they agglutinated heterologous strains, and no reaction was obtained above a
dilution of 1 to 100.

Bull's[4] experiments have shown that agglutination *in vitro* may
give different results from agglutination of the same bacteria *in
vivo*. Therefore, his method was used to compare the strains of
influenza bacilli. Two young rabbits were inoculated intravenously
with a non-virulent respiratory and a virulent meningeal culture,
respectively. Blood was taken from the heart at intervals of thirty
seconds to eighteen minutes, and slides were prepared. The respiratory culture showed marked clumping within one minute after inoculation, and in five minutes no bacilli could be demonstrated in the
blood. The meningeal culture was not agglutinated at all by the
blood of the inoculated rabbit, and at the end of eighteen minutes
the bacilli were as numerous as they had been half a minute after the
injection. The difference in the two results was very striking and
clean cut. On killing the animals the bacilli of the respiratory
strain were found within leucocytes in the liver, spleen, and lungs,
while the virulent (meningeal) bacilli were found only in small
numbers in the organs and were extracellular. A third rabbit was
inoculated with a meningeal culture which had been isolated two
years before, and which had lost its virulence for rabbits. The
result was identical with that obtained with the non-virulent respiratory strain, proving again that the difference between the two
strains is not absolute but only relative.

Complement Deviation.—Antigens were prepared from seven
non-virulent respiratory strains and from seven virulent meningitis
strains, according to the method used in the research laboratories
of the New York Board of Health. I am indebted to Miss Olm-
stead for a description of this method, which is similar to the one
described by Schwartz and McNeil[5] for gonococcus antigens. except
that the suspensions of influenza bacilli are kept at 55° C. over night.
since they undergo autolysis very slowly. At the end of the period
of autolysis, which lasted from fourteen to eighteen hours. a differ-

[4] Bull. C. G., *Jour. Exper. Med.,* 1915, xxii. 484.
[5] Schwartz, H. J., and McNeil. A., *Am. Jour. Med. Sc.,* 1912, cxliv. 815.

ential point between respiratory and meningeal strains of influenza bacilli was noted in the appearance of the fluid. The suspensions of the meningeal cultures were invariably turbid, with a comparatively small precipitate in the tube. The suspension of the respiratory strains, on the other hand, showed a perfectly clear fluid above a large amount of precipitate. In other words, the bodies of the non-pathogenic bacilli underwent less perfect dissolution than did the bodies of the pathogenic strains. The appearances of the precipitates as revealed by stained films under the microscope were quite similar. The bacilli no longer stained deeply and were more or less disintegrated.

All the antigens were tested against two monovalent rabbit sera immune to the meningeal strains, and against two rabbit sera immune to the respiratory strains. The results are given in Tables I to IV.

TABLE I.

Complement Deviation.

Immune Serum of Respiratory Strains + Antigens of Respiratory Strains.

Complement. Guinea pig serum 1:40 dilution.	Antigen.		Immune serum (Robinson).	Anti-sheep rabbit serum 1:1,000 dilution.	Sheep corpuscles 1:20 dilution.	Result.
cc.		cc.	cc.	cc.	cc.	
0.1	1. R.	0.3	0.1	0.25	0.25	No hemolysis.
0.1		0.2	0.1	0.25	0.25	Some hemolysis.
0.1		0.1	0.1	0.25	0.25	Hemolysis.
0.1		0.2	0.05	0.25	0.25	"
0.1	2. S.	0.3	0.1	0.25	0.25	No hemolysis.
0.1		0.2	0.1	0.25	0.25	" "
0.1		0.1	0.1	0.25	0.25	Hemolysis.
0.1		0.2	0.05	0.25	0.25	"
0.1	3. C.	0.3	0.1	0.25	0.25	No hemolysis.
0.1		0.2	0.1	0.25	0.25	" "
0.1		0.1	0.1	0.25	0.25	" "
0.1		0.05	0.1	0.25	0.25	Hemolysis.
0.1		0.2	0.05	0.25	0.25	"
0.1	4. F.	0.3	0.1	0.25	0.25	No hemolysis.
0.1		0.2	0.1	0.25	0.25	Hemolysis.
0.1	5. M.	0.3	0.1	0.25	0.25	No hemolysis.
0.1		0.2	0.1	0.25	0.25	Hemolysis.
0.1	6. M.O.	0.3	0.1	0.25	0.25	No hemolysis.
0.1		0.2	0.1	0.25	0.25	Hemolysis.
0.1	7. L.	0.3	0.1	0.25	0.25	"

TABLE II.

Complement Deviation.

Immune Sera of Respiratory Strains + Antigens of Meningeal Strains.

Complement. Guinea pig serum 1:40 dilution.	Antigen.		Immune serum (Robinson).	Anti-sheep rabbit serum 1.1,000 dilution	Sheep corpuscles 1:20 dilution.	Result.
cc.		*cc.*	*cc.*	*cc.*	*cc.*	
0.1	1. D.	0.3	0.1	0.25	0.25	No hemolysis.
0.1		0.2	0.1	—	—	Hemolysis.
0.1	2. B.H.	0.3	0.1	0.25	0.25	No hemolysis.
0.1		0.2	0.1	0.25	0.25	" "
0.1		0.1	0.1	0.25	0.25	Hemolysis.
0.1		0.2	0.05	0.25	0.25	"
0.1	3. N.Y.H.	0.3	0.1	0.25	0 25	No hemolysis.
0.1		0.2	0.1	0.25	0.25	Some hemolysis.
0.1		0.1	0.1	0.25	0.25	Hemolysis.
0.1		0.2	0.05	0.25	0.25	"
0.1	4. Ch.	0.3	0.1	0.25	0.25	No hemolysis.
0.1		0.2	0.1	0.25	0.25	Some hemolysis.
0.1		0.1	0.1	0.25	0.25	Hemolysis.
0.1		0.2	0.05	0.25	0.25	"
0.1	5. P.	0.3	0.1	0.25	0.25	No hemolysis.
0.1		0.2	0.1	0.25	0.25	Hemolysis.
0.1	6. F.A.	0.3	0.2	0.25	0.25	"
0.1		0.3	0.1	0.25	0.25	"
0.1	7. L.F.	0.3	0.2	0.25	0.25	"
0.1		0.3	0.1	0.25	0.25	"

Thus it follows that all the sera made by immunizing rabbits with meningeal or with respiratory strains of influenza bacilli contained immune bodies capable of binding complement in the presence of antigens made from both virulent and non-virulent strains. But the sera obtained with the virulent organisms gave binding in higher dilutions than did the sera made from non-virulent bacilli. In other words, the sera obtained by immunizing rabbits with virulent influenza bacilli contained immune bodies capable of uniting with their homologous antigens in comparatively high dilutions, and with heterologous antigens in lower dilutions: while the sera resulting from the inoculation of rabbits with non-virulent influenza bacilli contained less complement binding body for all antigens.

An antigen made from a particular respiratory strain reacted with

TABLE III.

Complement Deviation.

Immune Sera of Meningeal Strains + Antigens of Meningeal Strains.

Complement. Guinea pig serum 1:40 dilution.	Antigen.		Immune serum.		Anti-sheep rabbit serum 1:1,000 dilution.	Sheep corpuscles 1:20 dilution.	Result.
cc.		*cc.*		*cc.*	*cc.*	*cc.*	
0.1	1. B.H.	0.1	1. B.H.	0.1	0.25	0.25	No hemolysis.
0.1		0.1		0.05	0.25	0.25	" "
0.1		0.1		0.03	0.25	0.25	" "
0.1		0.1		0.01	0.25	0.25	Hemolysis.
0.1	2. Ch.	0.1		0.1	0.25	0.25	No hemolysis.
0.1		0.1		0.05	0.25	0.25	" "
0.1		0.1		0.03	0.25	0.25	Some hemolysis.
0.1		0.1		0.01	0.25	0.25	Hemolysis.
0.1	3. N.Y.H.	0.1		0.1	0.25	0.25	No hemolysis.
0.1		0.1		0.05	0.25	0.25	" "
0.1		0.1		0.03	0.25	0.25	" "
0.1		0.1		0.01	0.25	0.25	Hemolysis.
0.1	4. F.A.	0.1		0.1	0.25	0.25	No hemolysis.
0.1		0.1		0.05	0.25	0.25	" "
0·1		0.1		0.03	0.25 .	0.25	Hemolysis.
0.1	5. D.	0.1		0.1	0.25	0.25	No hemolysis.
0.1		0.1		0.05	0.25	0.25	Hemolysis.
0.1	6. P.	0.1		0.1	0.25	0 25	No hemolysis.
0.1		0.1		0.05	0.25	0.25	" "
0.1		0.1		0.03	0.25	0.25	Hemolysis.
0.1	1. B.H.	0.1	2. N.Y.H.	0.1	0.25	0.25	No hemolysis.
0.1		0.1		0.05	0.25	0.25	" "
0.1		0.1		0.03	0.25	0.25	Some hemolysis.
0.1		0.1		0.01	0.25	0.25	Hemolysis.
0.1	2. N.Y.H.	0.1		0.1	0.25	0.25	No hemolysis.
0.1		0.1		0.05	0.25	0.25	" "
0.1		0.1		0.03	0.25	0.25	" "
0.1		0.1		0.01	0.25	0.25	Hemolysis.
0.1	3. D.	0.1		0.1	0.25	0.25	No hemolysis.
0.1		0.1		0.05	0.25	0.25	" "
0.1		0.1		0.03	0.25	0.25	Hemolysis.

a heterologous serum (*i. e.*, from a meningeal strain) in lower dilution than with its own serum. However, still another respiratory strain failed to bind at all with any immune serum. Since these results were obtained in repeated tests with antigens made from the two strains and at different times, they are probably not to be regarded as accidental, but as indicating that the respiratory strains differ among themselves in strength of antigenic power. With this conclusion the protection experiments also agree.

TABLE IV.

Complement Deviation.

Immune Sera of Meningeal Strains + Antigens of Respiratory Strains.

Complement. Guinea pig serum 1:40 dilution.	Antigen.		Immune serum.	Anti-sheep rabbit serum 1:1,000 dilution.	Sheep corpuscles 1.20 dilution.	Result.
cc.		cc.	cc.	cc.	cc.	
0.1	1. R.	0.2	1. N.Y.H. 0.1	0.25	0.25	No hemolysis.
0.1		0.2	0.05	0.25	0.25	" "
0.1		0.2	0.03	0.25	0.25	" "
0.1		0.2	0.01	0.25	0.25	Hemolysis.
0.1	2. S.	0.2	0.1	0.25	0.25	No hemolysis.
0.1		0.1	0.1	0.25	0.25	Hemolysis.
0.1		0.2	0.05	0.25	0.25	"
0.1		0.3	0.05	0.25	0.25	"
0.1	3. C.	0.3	0.1	0.25	0.25	No hemolysis.
0.1		0.2	0.1	0.25	0.25	" "
0.1		0.1	0.1	0.25	0.25	Hemolysis.
0.1		0.05	0.1	0.25	0.25	"
0.1	4. F.	0.3	0.1	0.25	0.25	Some hemolysis.
0.1		0.2	0.1	0.25	0.25	Hemolysis.
0.1	1. M.	0.3	2. B.H. 0.1	0.25	0.25	No hemolysis.
0.1		0.2	0.1	0.25	0.25	Hemolysis.
0.1	2. Ma.	0.3	0.1	0.25	0.25	No hemolysis.
0.1		0.2	0.1	0.25	0.25	Hemolysis.
0.1	3. S.	0.2	0.1	0.25	0.25	No hemolysis.
0.1		0.1	0.1	0.25	0.25	Some hemolysis.
0.1		0.2	0.05	0.25	0.25	Hemolysis.
0.1		0.3	0.05	0.25	0.25	"
0.1	4. C.	0.2	0.1	0.25	0.25	No hemolysis.
0.1		0.1	0.1	0.25	0.25	" "
0.1		0.05	0.1	0.25	0.25	Hemolysis.
0.1	5. L.	0.3	0.1	0.25	0.25	"

While the test of complement deviation brings out a difference between the strains of influenza bacilli isolated from the respiratory tract and those obtained from the meninges, the difference is simply one of degree and not of kind. The sera and antigens made from respiratory strains were both much weaker than were those made from the meningeal strains. This fact is probably explained by the imperfect autolysis of the cultures derived from the respiratory tract and by their inability to produce more than a small amount of protective immune bodies in inoculated rabbits. The antigenic properties of the respiratory cultures are far weaker than are those of the strains isolated from the meninges.

Protection.—In order to determine whether the non-virulent respiratory strains which did not kill rabbits elicit the production of immune bodies, the surviving rabbits were reinoculated at various intervals with virulent meningeal strains.

Ten strains of *B. influenzæ*, isolated from the respiratory tract, were found to be totally unable to afford any protection to young rabbits against a subsequent inoculation with virulent strains. On the other hand, two respiratory cultures which did not kill rabbits in the ordinary lethal dose (one blood agar slant) of the standard virulent strains did protect the animals from a lethal dose of a virulent culture injected after two to four weeks. As was to be expected, sublethal doses of virulent cultures, whether of meningeal or of respiratory origin, protected rabbits against full doses given twelve days to three weeks later.

It is apparent that the respiratory strains are not identical, but that they differ among themselves in the amount of protective immune bodies they are able to develop in rabbits, just as they differ in the amount of complement binding body they produce.

The following protocols illustrate these results.

Protocols.

Experiment I.—Jan. 15. Rabbit, weight 962 gm. Inoculated intravenously with 1 blood·agar slant of respiratory strain R of 24 hours' growth suspended in 1 cc. of salt solution.

Jan. 16. Rabbit alive, apparently well.

Jan. 18. Rabbit quite well.

Jan. 30. Rabbit reinoculated intravenously with 1 blood agar slant of meningeal strain N of 24 hours' growth, suspended in 1 cc. of salt solution.

Jan. 31. A. m. Rabbit ill.

P. m., 29 hours after inoculation, rabbit died.

Experiment II a.—Jan. 16. Rabbit, weight 975 gm. Inoculated intravenously with 1 blood agar slant of meningeal strain H of 24 hours' growth, in 1 cc. of salt solution.

Jan. 17. Rabbit dead. Profuse growth of *B. influenzæ* from heart's blood.

Experiment II b.—Jan. 19. Rabbit, weight 975 gm. Inoculated intravenously with ½ blood agar slant of meningeal strain H of 24 hours' growth, suspended in 1 cc. of salt solution.

Jan. 20. Rabbit quite ill.

Jan. 22. Rabbit well.

Feb. 3. Inoculated with one culture of meningeal strain N.

Feb. 5. Rabbit well.

Experiment III.—Jan. 19. Rabbit, weight 970 gm. Inoculated intravenously with 1 blood agar slant of respiratory strain C of 24 hours' growth, suspended in 1 cc. of salt solution.

Jan. 20. Rabbit well.

Feb. 10. Inoculated intravenously with 1 culture of meningeal strain N in 1 cc. of salt solution.

Feb. 15. Rabbit well.

SUMMARY.

Influenza bacilli isolated from various pathological processes in man differ widely in pathogenic power for animals, especially rabbits. While the cultures derived from the leptomeninges and blood, and rarely from the pneumonic lung are pathogenic, those generally derived from the respiratory tract exhibit little or no virulence for rabbits.

The two types of cultures as indicated by virulence for animals do not differ in kind, but only in degree, in relation to the serological tests of agglutination, complement deviation, and opsonification.

The two types of cultures do, however, differ with respect to their ability to undergo autolysis. While the virulent cultures autolyze almost completely, yielding a turbid supernatant fluid and little sediment, the non-virulent cultures give rise to an abundant sediment and a clear supernatant fluid.

The non-virulent cultures incite far less antibody production in rabbits. Hence, rabbits inoculated with non-virulent strains yield sera possessing low antibody content. Conversely, rabbits inoculated with virulent strains yield sera possessing a higher content of antibody.

In keeping with and possibly because of the low antibody content of the sera of rabbits inoculated with the non-pathogenic strains, the rabbits so treated are not, as a rule, protected against subsequent inoculation with virulent strains.

Influenza bacilli therefore vary in pathogenic effect both for man and animals, but they are not distinguishable by means of serological reactions into different types. Apparently all influenza bacilli belong to one class or race irrespective of origin or virulence.

EXPLANATION OF PLATE 48.

FIG. 1. Large forms of *B. influenzæ* in a film from cerebrospinal fluid in a case of seropurulent leptomeningitis.

FIG. 2. Small forms of *B. influenzæ* in a film from cerebrospinal fluid in a case of seropurulent leptomeningitis. Some phagocytosis.

FIG. 3. Impression from a colony of *B. influenzæ* on a moist blood agar plate. Note long and swollen forms, also polar staining. Meningeal strain.

FIG. 4. Characteristic small type culture of *B. influenzæ;* 20 hours' growth. Respiratory strain.

FIG. 5. Large type culture of *B. influenzæ* after 2 years of artificial cultivation.

FIG. 6. Same strain as Fig. 5 from peritoneal cavity of guinea pig 24 hours after inoculation.

FIG. 7. Small type of *B. influenzæ* after 2 years of artificial cultivation.

FIG. 8. Culture of *B. influenzæ* on blood smeared agar, showing interlacing threads.

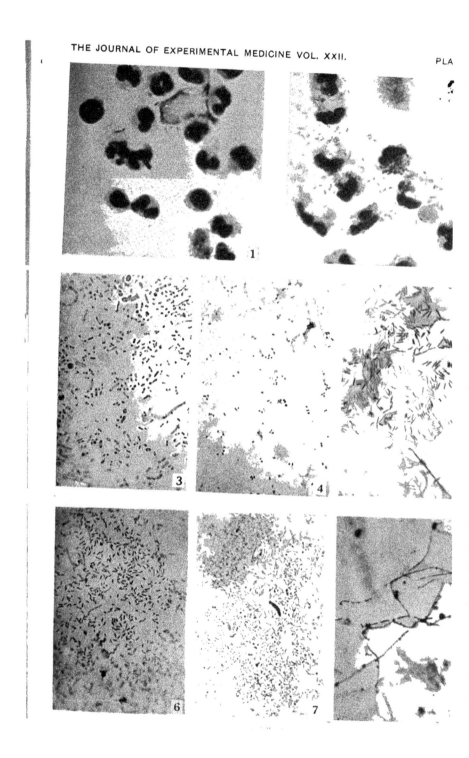

THE MECHANISM OF THE CURATIVE ACTION OF ANTIPNEUMOCOCCUS SERUM.

By CARROLL G. BULL, M.D.

(From the Laboratories of The Rockefeller Institute for Medical Research.)

PLATES 49 TO 52.

(Received for publication, June 1, 1915.)

In a previous paper[1] I reported on the course of the septicemia in the rabbit which follows an intravenous injection of pneumococci. In brief it is as follows: By making blood cultures at frequent intervals it was found that the course was determined largely by the number and virulence of the bacteria injected. The injection of a relatively small number of pneumococci of low virulence is followed in a few hours by a sterile condition of the blood, and the bacteria do not reappear in the circulation. When large numbers of similar pneumococci are injected or when a small number of a virulent strain is given, the initial decrease in the number of bacteria in the blood takes place less rapidly and completely and, after an interval of five or six hours, an increase occurs. When, however, large numbers of virulent pneumococci are injected, a slight initial fall occurs and death results in a few hours from severe bacteremia.

These results, it was noted at the time, could be completely changed if a small quantity of active antipneumococcus serum was injected. Under these conditions the bacteria even when highly virulent were abruptly swept from the blood. No explanation of the phenomenon was attempted at the time. It is well known that the immune serum is devoid of bactericidal properties *in vitro*. On the other hand, bacteriotropic substances are demonstrable both *in vitro* and *in vivo* and Neufeld and others believe that a relation exists between bacteriotropic value and protective action.[2] The

---

[1] Bull, C. G., *Jour. Exper. Med.*, 1914, xx, 237.

[2] For previous work and literature on these points the following references are given: Neufeld, F., and Rimpau, W., *Ztschr. f. Hyg. u. Infectionskrankh.*, 1905, li, 283. Boehncke, K. E., and Mouriz-Riesgo, J., *Ztschr. f. Hyg. u. Infectionskrankh.*, 1915, lxxix, 355. Cole, R., *Arch. Int. Med.*, 1914, xiv, 56.

question therefore arises as to what the abrupt disappearance of the pneumococci from the circulation is due and whether it is the result of the action of one or more of the antibodies contained in the blood and the immune serum, and which of them plays the essential part.

EXPERIMENTAL.

The investigation of the pneumococcus was rendered far easier than it otherwise would have been by the fact that the Hospital of The Rockefeller Institute kindly supplied several types[3] of the pneumococcus with their corresponding immune horse serum. Although the majority of the experiments were made with a virulent pneumococcus of Type I, Types II and IV were also employed to supplement and extend the study. Since the investigation of the blood of the rabbits, employed for the tests, during life proved an essential part of the experiments, a brief description of the procedures will be introduced here. They will not be repeated in connection with the illustrative experiments given.

Blood Cultures.—The number of bacteria in the blood stream was determined by taking a small quantity of blood from the heart with a graduated pipette to which a needle was attached by means of a thick-walled rubber tube. A measured quantity of the blood was immediately diluted with a large quantity of sterile physiological salt solution and portions of the dilutions were added to Petri dishes containing sterile defibrinated rabbit blood. The colonies were counted after twenty-four hours' incubation. The plating was done in this manner throughout the work.

Microscopic Examination of the Blood.—For microscopic examination the blood was obtained as described above and heavy films were made on slides. The films were allowed to dry in the air, fixed in methyl alcohol, and stained with a dilute solution of Manson's stain. In this way every bacterium is rendered visible, independently of the thickness of the smear.

Microscopic Examination of the Tissues.—Tissues were prepared for microscopical study by crushing and sectioning. For the crushed preparations, the fresh tissues were finely crushed in tissue crushers, portions of the pulp spread on slides, fixed in a flame,

[3] Dochez, A. R., and Gillespie, L. J., *Jour. Am. Med. Assn.*, 1913, lxi, 727.

and stained by Gram's method. Sections were stained by Gram's method, eosin and methylene blue, and hematoxylin and eosin.

Immune Serum in Pneumococcus Septicemia in Rabbits.—It has been stated that the immediate effect of the injection of an active immune serum into the blood of rabbits in which many pneumococci are present is to bring about an abrupt disappearance of the bacteria from the circulation, although they may later return. This effect is illustrated by Experiment I in which a bacteremia is produced by the injection of a culture of virulent pneumococcus.

Experiment I.—4 rabbits received each in the ear vein an intravenous injection of 0.4 cc. per kilo of an 18 hour bouillon culture of a Type I pneumococcus. 2 of the 4 animals received an intravenous injection 1 minute later of the homologous immune serum in the opposite ear vein, while the remaining 2 were untreated. Blood cultures, by the method described, were made from all the animals. The result is summarized in Table I.

TABLE I.

Rabbit No.	Weight.	Immune serum per kilo.	Colonies of bacteria 1 minute after serum injection.	Colonies of bacteria 15 minutes after serum injection.	Remarks.
	gm.	*cc.*			
I	1,550	0.2	1,320,000	0	Bacteremia after 12 hrs.; death in 36 hrs
II	1,650	0.5	1,250,000	0	Bacteremia; death in 48 hrs.
III	1,850	0	1,260,000	1,100,000	Bacteria did not leave blood; death within 24 hrs.
IV	1,575	0	1,330,000	1,275,000	Same as Rabbit III.

As Experiment I shows, the immune serum brings about a rapid disappearance of the bacteria from the blood. Other experiments directed to the answer of the question indicated that the time required for the blood to be rendered sterile depends, first, on the number of the bacteria injected, and, second and especially, on the amount of serum given. The phenomenon of abrupt removal seems to be independent of the virulence of the pneumococci, although the end result is determined by the virulence. It can be predicted that had the pneumococci been of lower virulence Rabbits I and II would have recovered and the bacteria would not have reappeared in the blood at all.

Mechanism of the Removal Process.

The immediate effect, therefore, of the immune serum is to cause the pneumococci to be rapidly removed from the blood stream. This action is exercised independently of the end result; namely, whether the bacteria are to be permanently suppressed, or whether they reappear later and cause the death of the animals. The question which presents itself for solution is the manner in which the removal is accomplished.

It is known that an immune serum even in the fresh state is devoid of bactericidal action on pneumococci *in vitro*. It was desirable to ascertain the effect of the whole blood under the same conditions. Tests were made by adding to fresh rabbit's blood, either hirudinized or defibrinated, varying quantities of immune serum and a convenient number of pneumococci and then plating the mixture at stated periods during twenty-four hours. It sufficed to add two drops of a bouillon culture, twenty-four hours old, to each cc. of the blood and immune serum mixture.

The effect of the immune serum was to cause great reduction in the number of colonies developing on the plates made from the second to the twelfth hour. The later plates showed increasing numbers of colonies and those made at the expiration of twenty-four hours corresponded with the control plates. The smallest number of colonies on a plate was ten to twenty; sterility was never achieved.

In following with the microscope the changes taking place in the test-tubes it was found that what appeared to be destruction of the pneumococci was actually an agglutination of the bacteria through which the number of colonies was reduced. Definite clumps were discovered in tubes in which the proportion of immune serum was 1 to 500. This fact is noteworthy since the macroscopic agglutination titer of the serum was 1 to 50.

Agglutination and Phagocytosis in Vivo.

It now became desirable to investigate the same points under the conditions of the living body. Experiments 2 and 3 were performed for that purpose.

Experiment 2.—A rabbit having been inoculated by intravenous injection with a small quantity of a bouillon culture of a virulent pneumococcus was permitted to develop a severe bacteremia which required about 10 hours. At that time 1 cc. of immune serum was injected into the ear vein, after which specimens of blood were taken from the heart at intervals of from 30 seconds to 5 minutes. Film preparations were prepared with each specimen, fixed in methyl alcohol, and stained with Manson's stain.

The specimen taken before the serum was injected showed numerous diplococci distributed evenly throughout the blood. The specimen taken 1 minute after the serum was given revealed large clumps of diplococci (Figs. 1 and 2). The next specimen removed showed progressively fewer clumps until the 5 minute specimen was reached, when no diplococci whatever were found.

At the expiration of 30 minutes after the injection of the immune serum the rabbit was killed. The lungs, liver, spleen, and kidney were put through the crusher, and film preparations were made from each. Clumps of diplococci were found in the lungs, liver, and spleen (Figs. 3, 4, and 5), while very few clumps were present in the kidney. The masses of diplococci were sometimes free, but usually they were contained within phagocytic cells and chiefly within polymorphonuclear leucocytes, especially in the lung, liver, and spleen (Figs. 6, 7, 8, and 9). Sections of the organs were also prepared, but they were less informing than the crushed tissues. The sections exhibited, however, groups of leucocytes in the capillaries and other small vessels of the lungs, the sinusoids of the liver, and the blood spaces of the spleen, of which a part enclosed clumps of diplococci (Figs. 12, 13, 14, 15).

Experiment 3.—In this animal the equivalent of a developed bacteremia was induced by injecting into an ear vein the virulent pneumococci derived by centrifugalization from 50 cc. of a twenty-four hour old bouillon culture. After 2 minutes a film of the blood taken from the heart was made and then 1 cc. of immune serum injected. Subsequently film preparations of the heart's blood were prepared at periods of from 30 seconds to 5 minutes.

The results were identical with those described in Experiment 2. The blood film taken before the serum was injected showed many diplococci but no clumps; 30 seconds after the serum injection large clumps were present; the crushed organs showed clumped diplococci within phagocytes.

The two experiments were repeated many times and with unvarying appearances. They were also modified in such a manner as to permit the treated rabbits to survive 1, 2, 3, and 4 hours after the injection of the immune serum. A marked difference was observed in the number and condition of the clumps of bacteria according to the period of survival. The greatest number of free clumps of diplococci were present in the animals killed at the 30 minute period, and the greatest number of phagocytes carrying the clumps were found in the animals killed at one and at two hours (Fig. 10). The preparations of animals killed at two and one-half hours show-

ing the phagocytes containing diplococci are fewer and the form of
the bacteria is irregular and the staining faint and poor. Later
preparations show very few phagocytes enclosing disintegrated bacteria (Fig. 11). The preparations at four hours contained a small
number of free, single diplococci which were the last bacteria to
disappear.

So far as the blood phenomena are concerned—the abrupt disappearance of the cocci from the circulating blood, agglutination, and
phagocytosis—active and passive immunity are similar. The actively immune rabbits probably possess a tissue immunity which
passive immunity does not confer. This point is under investigation.

DISCUSSION.

It is now possible to define clearly the effects exerted by an antipneumococcus serum when it brings about an amelioration of the
infective pneumococcus process in rabbits, at least. It may be
well to state at the outset that the effects were, in our experiments,
limited to the interaction of the specific type or group of pneumococcus and homologous serum. No crossed or heterologous action
was ever observed. Although a few typical or illustrative experiments in which Type I pneumococcus and serum were employed are
given, similar experiments were made and identical results obtained
with pneumococci and serum of Types II and IV.

Setting out with the condition of pneumococcus bacteremia of
rabbits, developing on the one hand after a small inoculation of
virulent pneumococci and directly produced on the other by an injection of massive quantities of cultures, the first effect of an injection of immune serum is to cause an almost instantaneous agglutination of the diplococci in the blood and the immediate removal of
the clump formed by the spleen, liver, lungs, and, to a smaller extent, the kidneys. Probably still other organs participate in this
process.

This agglutination and removal by the organs of the pneumococci
is quickly followed by a process of active phagocytosis, chiefly
through the medium of the polymorphonuclear leucocytes, in the
course of which the bacteria are taken up by the cells in enormous

numbers. A single leucocyte may contain from fifty to one hundred diplococci. The act of phagocytosis is directed, apparently, exclusively or almost so against the clumps, while the few single diplococci remain free within the vessels. The act of phagocytosis follows quickly upon the agglutination and removal and, it would appear, never takes place in the blood stream itself; for in the study of hundreds of blood films made from the heart only one leucocyte containing diplococci was encountered. The process extends, however, over several hours and consists of two distinct phases: one the mere englobing of the masses of diplococci, and the other their digestion or disintegration. For the unagglutinated diplococci which remain free retain form and staining powers very much longer than those contained within phagocytes. The diplococci within the phagocytes show at first normal morphology and staining (Figs. 6, 7, 8, and 9), while, later, bizarre forms and imperfect staining are met with (Fig. 11). After a few hours only very few diplococci and those disintegrating are still visible in the phagocytes. Cells other than polymorphonuclear leucocytes play a minor part in the phagocytic process.

The intimate interaction of agglutination and protection which the experiments indicate is emphasized in other ways. I have found that the smallest quantity of immune serum which influences a pneumococcic infection in the rabbit by diminishing the number of diplococci in the circulating blood or by prolonging the life of the animal brings about clump formation *in vivo;* and Avery has noted that in the fractionated and purified antipneumococcic serum the protective bodies are always accompanied by agglutinating substances.

But in order that the agglutinated diplococci may be rendered harmless it is not enough that they shall be removed from the blood by the organs; they must be further withdrawn, apparently, into the phagocytes where they are quickly disintegrated. As the experiments *in vitro* with the whole blood and immune serum showed, the clumped diplococci are capable of further multiplication. Hence, along with the removal of the clumps of diplococci by the organs it is necessary that an accumulation of polymorphonuclear leucocytes in the same organs be brought about simultaneously.

The fact is well known that the intravenous injection of foreign protein causes a leucopenia. I found that the intravenous injection in rabbits of one cc. of antipneumococcus horse serum led, within five minutes, to a reduction of the number of circulating leucocytes by one-half. The low number persisted from three to four hours, after which it rose usually somewhat above the normal. This reduction and rise in number was not participated in by all the white cells but chiefly by the polymorphonuclear leucocytes. Normal horse serum produced similar effects. The injection of bouillon cultures of pneumococci causes a greater fall still in the number of circulating leucocytes and, according to the final result in death or recovery, there was continued depression or later rise.

The manner in which the leucocytes react to the foreign protein as contained in the horse serum provides a mechanism for bringing large numbers of polymorphonuclear leucocytes into close proximity with the agglutinated diplococci within the organs. That the leucocytes are not destroyed but merely accumulate in the organ in certain forms of leucopenia has been shown by Goldschieder and Jacob. On the other hand, an homologous immune serum,—for example antipneumococcus serum prepared in the rabbit,—would also bring about agglutination of the diplococci but whether it would equally produce leucopenia has still to be determined. Some other means of attracting leucocytes from the blood and the bone marrow to the capillaries of the organs may be necessary and this means may well have to do with the presence of the clumps of diplococci themselves. But this constitutes a point of further study.

SUMMARY.

An active antipneumococcus serum causes agglutination of pneumococci *in vitro* and *in vivo*.

The antiserum acts in far greater dilution in causing agglutination as determined by the microscopic than by the macroscopic test.

The antiserum when injected into the circulation of rabbits suffering from pneumococcus bacteremia causes a rapid disappearance of the diplococci from the blood. This disappearance is brought about by instantaneous clumping of the diplococci *in vivo* and the removal of the clumps by the liver, spleen, lungs, and possibly other organs.

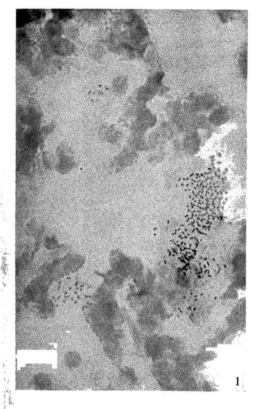

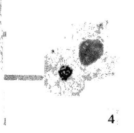

(Bull: Curative Action of Antipneumo

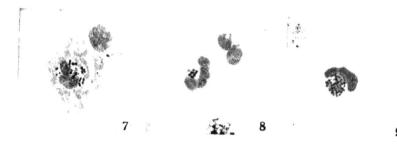

7　　　　8　　　　9

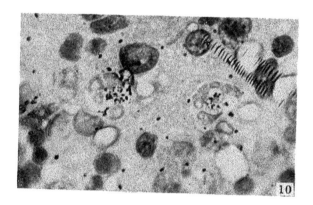

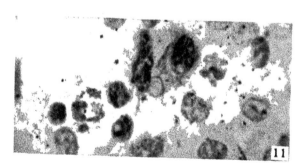

(Bull: Curative Action of Antipneumococcus Serum.)

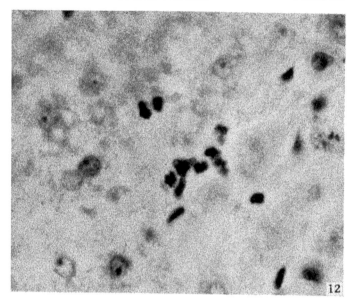

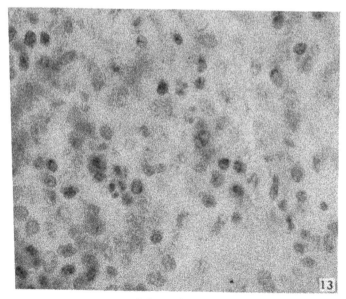

(Bull: Curative Action of Antipneumococcus Serum.)

14

15

(Bull: Curative Action of Antipneumococcus Serum.)

The same means which cause the clumping and removal from the blood of the diplococci and their accumulation in the organs cause also leucopenia with accumulation of polymorphonuclear leucocytes in the same organs. The leucocytes act as phagocytes and ingest the clumped diplococci which undergo rapid dissolution within the cells.

The diplococci which fail of agglutination tend not to be phagocyted and persist longer in viable form than the ingested clumps.

The protection afforded by the immune serum is specific for the type organism and is determined by the presence of agglutinins which prepare the pneumococci for removal from the blood and phagocytosis on a large scale.

Normal rabbit blood is devoid of agglutinating and of protective effect against virulent pneumococci.

Agglutination is not merely an incidental phenomenon, but constitutes an essential process in association with phagocytosis in the protection of the rabbit against pneumococcus infection.

EXPLANATION OF PLATES.
PLATES 49 AND 50.

FIG. 1. Clump of pneumococci in heart's blood of a rabbit having a pneumococcic septicemia; specimen was taken 1 minute after 2 cc. of immune serum were given. Gram's stain.

FIG. 2. Clump of pneumococci in heart's blood of a rabbit 1 minute after 0.1 cc. of serum was given.

FIGS. 3, 4, and 5. Clumps of pneumococci in the liver, lungs, and spleen, respectively, of a rabbit 30 minutes after serum had been administered. Gram's stain.

FIGS. 6, 7, 8, and 9. Polymorphonuclear leucocytes containing clumps of pneumococci from the liver, spleen, lungs, and kidney, respectively, of a rabbit having a pneumococcic septicemia. The tissues were removed 1 hour after an injection of immune serum. Gram's stain.

FIG. 10. Crushed preparation from the spleen of a rabbit having a pneumococcic septicemia. 2 leucocytes are present which contain clumps of pneumococci. The tissue was removed 1½ hours after an injection of immune serum had been given. Gram's stain.

FIG. 11. Same as Fig. 10. Tissue removed 2 hours after the serum was given. 1 leucocyte contains a clump of pneumococci. The bacteria stain poorly.

PLATES 51 AND 52.

FIGS. 12, 13, 14, and 15. Sections of tissue from liver, spleen, and lungs, respectively, of a rabbit 2 hours after an intravenous injection of 1 cc. of a bouillon culture of pneumococci. All sections show accumulated polymorphonuclear leucocytes. Stained with hematoxylin and eosin.

A METHOD OF SERUM TREATMENT OF PNEUMO-COCCIC SEPTICEMIA IN RABBITS.

By CARROLL G. BULL, M.D.

(From the Laboratories of The Rockefeller Institute for Medical Research.)

PLATE 53.

(Received for publication, June 1, 1915.)

In a separate paper[1] I have reported experiments which indicate that the removal of pneumococci from the circulation in rabbits and their destruction are determined by three main factors: (1) the agglutination of the bacteria in the blood stream; (2) the assembling of the clumps in the internal organs; and (3) the inclusion of the masses by and digestion within polymorphonuclear phagocytes. In order that these several processes may occur it has been found necessary, in the case of virulent pneumococci, to employ a suitable immune serum.

In this paper I propose to describe still other experiments which bear upon the above conception of the manner of interaction of immune serum and the animal body. It has hitherto been ascertained that the activity of an immune antipneumococcic serum is quite strictly limited. Beyond a given infecting dose of the pneumococci any practicable quantity of immune serum ceases to be protective. Indeed, as the dose of serum is increased, especially when it comes from a foreign species of animal, a point is reached at which the serum becomes deleterious rather than helpful. The precise conditions upon which this change of action depends are not known; it is surmised that a heterologous serum, in small experimental animals especially, may exert a toxic action which becomes noticeable in effect when the dose is relatively large. In any case a striking disparity exists between the unlimited neutralizing action of the antitoxic sera, diphtheria and tetanus sera, for instance, and the strictly limited anti-infectious action of the antibacterial sera, among which the antipneumococcus serum is to be classed.

[1] Bull, C. G., *Jour. Exper. Med.*, 1915, xxii, 457.

466

Although the literature on the anti-infectious power of antipneumococcic serum is voluminous, the chief recent important contributions to the subject are those of Neufeld and Haendel[2] and of Cole.[3] The former investigators established the fact of the existence of specific types of pneumococci which are subject to influence only by the corresponding immune sera; and the latter, together with his coworkers, besides extending the knowledge of the types of pneumococci and their specific antisera, observed also that up to a certain degree of infection the protective dose of serum is parallel to the quantity of culture inoculated, after which the quantity of serum required to save life is proportionally large, and finally, a degree of infection may be reached against which no amount of serum will protect. For example, o.2 cc. of serum of Type I will regularly protect a mouse against o.1 cc. of a Type I culture, no matter how high its virulence. That is the largest dose of culture, however, against which the immune serum will protect, no matter how much serum is employed. 1 or even 2 cc. are no more effective than o.2 cc. (Cole).

In view of these data, it has been long believed that the animal body supplied a necessary substance which cooperated with the immune serum in overcoming bacterial infections. The nature of this substance has been conjectured merely; there exists no actual knowledge of its nature. The experiments reported in the previous paper referred to[4] seem to indicate quite definitely that the part which the rabbit's body itself plays in overcoming pneumococcus and some other bacterial infections is supplied by the phagocytes. The experiments to be related here bear on this conception of the protective mechanism.

EXPERIMENTAL.

As has just been stated, experiments seemed to indicate that the phagocytic cells function as the chief defensive agents against infection. If this is true, then the limit of the effective action of an immune serum may possibly be extended by giving the phagocytes assistance and providing them sufficient time in which to do their work. We observed that when a rabbit suffering from septicemia is given a

[2] Neufeld, F., and Haendel, *Ztschr. f. Immunitätsforsch., Orig.,* 1900, iii, 150.
[3] Cole, R. I., *N. Y. Med. Jour.,* 1915, ci, 1, 59.
[4] Bull, *loc. cit.*

large injection of immune serum intravenously, the bacteria are agglutinated into massive clumps (Figs. 1 and 2) most of which accumulate in the lungs where they come imperfectly under the influence of the phagocytes and may even, through extensive capillary obstruction, interfere with the circulation. The experiments to follow were performed on rabbits with a strain of pneumococcus of Type I maintained at high virulence by continuous passage through these animals. The corresponding immune serum was prepared in the horse and kindly supplied by the Hospital of The Rockefeller Institute.

The virulence of the culture was not accurately titrated. It was such, however, that one drop of the blood of a rabbit succumbing to infection sufficed to kill in twelve hours a rabbit weighing two kilos. The pneumococci were grown in beef infusion broth and were eighteen to twenty-four hours old when used. When large quantities were to be injected the bacteria were thrown out of the culture in the centrifuge, the greater part of the fluid was poured off, and the remainder suspended for injection. The bacteria were not washed. The injections both of bacteria and serum were made intravenously.

Experiment 1. Effect of Large Serum Injection in Rabbits Suffering from Severe Pneumococcus Septicemia.

Rabbit A.—Weight 1,800 gm. 0.1 cc. of culture given. 8 hours later the blood contained large numbers of pneumococci. 4 cc. of immune serum injected. The animal died within 1 minute of respiratory failure.

Autopsy.—Large clumps of pneumococci in the heart's blood; the lungs were distended, and many clumps of bacteria were in the vessels; clumps of bacteria also in the vessels of the choroid plexus (Figs. 1 and 2).

Rabbit B.—Weight 1,600 gm. 0.1 cc. of heart's blood of a rabbit just dead of pneumococcus septicemia given. 6 hours later the blood revealed the existence of a severe septicemia with the diplococci uniformly scattered throughout. 2 cc. of immune serum were injected and films from the heart's blood were made every minute for 3 minutes.

The immediate effect of the serum was to produce collapse of the animal attended by labored respiration and urination. The 1 minute film showed many large clumps of bacteria, the 2 minute fewer clumps, and the 3 minute few small clumps. The rabbit survived, temporarily recovered, but died 6 hours later. The blood and organs contained many pneumococci.

Rabbit C.—Weight 2,200 gm. The sediment from 50 cc. of a bouillon culture injected. 2 minutes later 10 cc. of immune serum given in opposite ear vein. The blood was quickly cleared of the pneumococci. Death occurred after 12 hours.

Autopsy.—Lungs edematous, emphysematous, and hemorrhagic. Small hemorrhages in surface of the kidneys and peritoneal surface of the intestines. The blood, lungs, spleen, and liver contained very great numbers of pneumococci.

Rabbit D.—Weight 2,300 gm. The sediment from 50 cc. of bouillon culture injected. No further treatment. Death in 4½ hours. The blood and organs contained very many pneumococci.

The experiments indicate that when the blood of the rabbit contains a very large number of pneumococci the intravenous injection of a large amount of immune serum causes certain definite effects. In the first place, the effect may be to cause almost immediate death. This accident results from the rapid agglutination in large clumps of the pneumococci in the circulation and the massing of the clumps in the lungs and brain where, acting as emboli, they produce respiratory failure and death. In the second place, the blood may be temporarily cleared of the bacteria and life be prolonged. Finally, however, the life of the animal is not spared, as the bacteria reinvade the blood and cause death.

From the foregoing it appears that the formation of large clumps is a disadvantage to the animal, since they tend to be held back chiefly in the lungs, the circulation of which they obstruct; and, besides, the large bacterial masses are not readily phagocyted.

These disadvantages can be avoided by the injection of small quantities of immune serum which produce clumps containing twenty to thirty pneumococci and to their quite regular distribution in the lungs, liver, and spleen, where they come under the influence of phagocytes under favorable conditions. Once removed from the circulation and lodged in the organs, the effect of a larger quantity of serum was studied.

Experiment 2. Effect of First Small and Later Larger Doses of Immune Serum in Pneumococcus Septicemia.

Rabbit A.—Weight 1,800 gm. The sediment from 75 cc. of a bouillon culture given. Immediately 0.5 cc. of immune serum injected into opposite ear vein. At the examination 25 minutes later there were no bacteria in the blood. 5 cc. of immune serum injected. Animal showed no symptoms for 2 hours, after which symptoms appeared, and death occurred at end of 5 hours.

Autopsy.—The lungs were edematous and hemorrhagic. Hemorrhages in surface of kidneys and peritoneal coat of intestine. The blood was free of diplococci and very few were found in the spleen, lungs, and liver. Almost all had been destroyed.

Rabbit B.—Sediment from 75 cc. bouillon culture administered. No treatment. Died in 6 hours. Blood and organs teeming with bacteria.

The experiment recorded under Rabbit A was repeated several times but without changing the end result. In spite of the practical destruction of the pneumococci, death quickly resulted. The appearance of the organs, and especially the hemorrhages, were taken to indicate a severe grade of intoxication,[5] and this effect was provisionally related to the large second dose of serum leading to too rapid disintegration of the diplococci. Hence it was decided to bring about a more gradual disintegration of the bacteria if possible.

Experiment 3. Effect of Repeated Small Doses of Immune Serum in Pneumococcic Septicemia.

Rabbit C.—Weight 2,000 gm. The sediment from 150 cc. of a bouillon culture given. Two minutes later 0.5 cc. of immune serum injected. At the end of 30 minutes the bacteria had left the blood and 1 cc. of the serum was administered. From now on 1 cc. of the serum was administered every 2d hour. Although symptoms appeared at the end of the first 2 hours they disappeared and the rabbit was in good condition until the 22d hour, when restlessness and rapid respiration followed by rigidity of hind leg developed. Later opisthotonos supervened. A lumbar puncture yielded cerebrospinal fluid containing many diplococci. Death occurred in the 28th hour.

Autopsy.—Blood, spleen, liver, and lungs were free of bacteria. The meninges were inflamed and contained very large numbers of pneumococci. Death obviously resulted from meningitis.

Rabbit D.—The sediment from 100 cc. of bouillon culture was given. No treatment. Died in 4 hours. The blood and organs contained very large numbers of diplococci.

The experiment described in Rabbit C was made several times but always with the same final result. The success of the experiment was frustrated by the intervention of the meningitis. The possibility is not excluded that this complication may itself be prevented by bringing a proper dose of the immune serum into the meninges which are not reached from the circulating blood.[6] The effect of reducing the infecting dose of bacteria was next tried. The sediments from 75 cc. and from 50 cc. of bouillon culture only delayed the onset of meningitis. The sediments from 35 cc. of culture gave partially successful results.

[5] Sprunt, T. P., and Luetscher, J. A., *Jour. Exper. Med.*, 1912, xvi, 443.

[6] Flexner, S., *Jour. Am. Med. Assn.*, 1913, lxi, 447.

Experiment 4.

4 rabbits weighing from 1,900 to 2,200 gm. were given the sediment from 35 cc. of bouillon cultures. 2 minutes later 0.5 cc. of immune serum were injected into the opposite ear vein. 30 minutes later and every 2d hour thereafter 1 cc. of the serum was given. This was continued up to the 56th hour. The final results are shown in Table I.

TABLE I.

Rabbit 1.		Rabbit 2.		Rabbit 3.		Rabbit 4.	
Time.	Observation.	Time.	Observation.	Time.	Observation.	Time.	Observation.
hrs.		*hrs.*		*wks.*		*hrs.*	
67	Right hind leg paralyzed	6	Diarrhea		In a light stupor for 1st day. No other symptoms	50	Both forelegs paralyzed.
83	Both hind legs completely paralyzed; severe opisthotonos; very restless and excitable	24	"			53	Severe opisthotonos, restless, excitable, breathing labored
		48	Diarrhea improved. Rabbit showed no further symptoms	3	Agglutinating titer of serum, 1:125		
96	Comatose; lying on side; breathing labored	3 wks.	Serum agglutinates pneumococci in a dilution of 1:150	6	Living and in perfect condition	60	Comatose; lying on side.
122	Died					80	Died.
	Autopsy.—Brain and cord intensely injected. Smears from surface of brain and lateral ventricles showed very many pneumococci. No pleurisy, no pericarditis. Blood, lungs, liver, and spleen free of bacteria	6 wks.	Living and in perfect condition				*Autopsy.*—Same as Rabbit 1.

Experiment 4 shows that it is possible by employing properly graduated doses of an immune serum to cure rabbits of a massive pneumococcus infection. The experiment indicates also that it is only the chance diplococci excluded from the influence of the circulating immune serum which escape destruction and multiply. It is of interest to know that the probability of the pneumococci escaping the destructive action of the serum is reduced by one-half by diminishing the infecting dose to a point which is, however, still massive according to the grade of virulence of the strain of pneumococcus employed.

Experiment 5. A Comparison of the Three Methods of Treatment Employed.

2 rabbits were used for each method. The infecting dose was the sediment from 20 cc. of bouillon cultures. 2 rabbits were treated as in Experiment 4.

with frequent small doses of serum; both recovered. 2 were treated first with a small dose of serum and then with a' large dose. They died. 2 were given one large dose of serum immediately after the infection; both died of septicemia.

DISCUSSION.

The experimental data presented in the foregoing pages should be considered from several points of view: from their bearing on experimental pneumococcus infection in the rabbit, and from their bearing on the theory of the anti-infectious action of immune serum in general.

The most striking fact regarding the experimental pneumococcus infection is the one that a single large dose of the immune serum given at the beginning of the infection is far less effective in overthrowing the infection than small repeated doses of which the maximal amount may not equal the single large quantity unsuccessfully employed.

The reason for this disparity has also been made clear in large part. The serum does not protect by a process of neutralization of a true toxin as antidiphtheritic serum does; or if any neutralization occurs it is a minor, and not the decisive process. The anti-infectious serum performs two things: it brings about an agglutination of the bacteria, and it prepares them for phagocytosis in the organs. Hence, whatever favors this process will be beneficial, and whatever hinders it will be detrimental. A certain concentration of the serum likewise promotes the assembling of the leucocytes in the organs.[7] On the other hand, higher concentration causes the formation of such large clumps as to escape phagocytosis. The free bacteria then quickly multiply, escape into the blood, and cause fatal infection. An excess of serum acts disadvantageously in another way not yet explained. Even when, through a small dose of serum, the small clumps have been formed in the blood and removed by the organs, a large following dose of the serum brings about a fatal issue. In this instance. the bacteria do not begin to multiply and invade the organs. The blood and the organs may be quite or nearly sterile. Death appears to result from intoxication. But just how the serum acts in producing the intoxication has not been determined.

When the pneumococci have all been destroyed in the organs

[7] Bull, *loc. cit.*

through the operation of the small doses of the serum repeated at intervals, only part of the rabbits are saved. The effect of the serum thus administered is clearly to provoke the destruction of the bacteria under conditions which avoid intoxication. But only those pneumococci are destroyed and rendered harmless that come under the direct influence of the serum which can reach all essential parts of the body except one; namely, the subdural space. When, under any circumstances, the pneumococci reach the subdural space they are not restrained by the treatment but develop rapidly and in great numbers and thus cause a fatal meningitis. The conditions leading to the meningitis are two: (a) large dosage and (b) survival for a sufficiently long period of time. The dose must be larger than the small doses of serum capable of destroying in a certain time period, and the animal must survive even the large doses about 20 hours in order that the meninges may become infected.

But the quantity of virulent pneumococci which rabbits can be made to support under the influence of the method of repeated serum injections is still so very large that the question may be raised whether the limits of activity of this anti-infectious serum have not been greatly underestimated. It remains, of course, to be determined whether still other anti-infectious sera are capable of having their powers enhanced by a similar method of administration. In any case, the subject is one that calls for restudy and perhaps revision.

On the other hand, the experiments affirm nothing as to the efficacy of the method in the serum treatment of lobar pneumonia in man. It may be supposed that so far as the pneumococci in the circulating blood in lobar pneumonia are concerned. small doses of the serum would suffice to bring about their removal. What effect the small doses might have upon the pneumococci in the lungs cannot be predicted. But since the method is one that is readily carried out in man it will doubtless receive attention in due time.

SUMMARY.

The treatment of pneumococcic septicemia in the rabbits by large doses of immune serum is detrimental. since the serum causes the formation of large clumps of bacteria in the blood which are taken out chiefly by the vessels of the lungs in which they accumulate and impede the circulation.

The large doses of serum are also detrimental when they follow upon small ones through which the small clumps formed are deposited in the spleen, liver, and other organs. In this instance, the large amount of serum leads to the destruction of the pneumococci under conditions which promote an intoxication. The precise mechanism of this action is not known.

The treatment of pneumococcic septicemia in rabbits by small repeated doses of immune serum can be successfully carried out. The number of pneumococci capable of being brought to destruction through phagocytosis in the organs in this way is very great.

Not all the rabbits treated with small repeated doses of the serum survive. Those that succumb do so not to a general infection but to a pneumococcus meningitis. The explanation of this phenomenon is simple. When the number of pneumococci originally inoculated is very great a small number penetrate into the subdural space. Those in this space do not come under the influence of the serum, hence they are not agglutinated and prepared for phagocytosis, whence they multiply and set up a fatal meningitis.

The activity of the immune serum administered in this way against virulent pneumococci is so great that a revision of our notions in the limit of powers of the anti-infectious sera seems necessary. It is patent that the problem is not simply a relation between quantity of immune bodies and number of bacteria. It is more complex than that conception indicates. The factor of the leucocytes and the degree of their possible activities under the conditions of the experiment come into play. Hereafter, in defining the mode and power of action of anti-infectious sera the condition of cooperation of the body-forces will have to be more strictly considered.

EXPLANATION OF PLATE 53.

Fig. 1. A large mass of pneumococci in a blood vessel. Tissue was taken from the region of the choroid plexus of a rabbit dying 1 minute after receiving 3 cc. of immune serum. The rabbit had a severe pneumococcus septicemia when the serum was given.

Fig. 2. A large clump of pneumococci. Preparation from the lung of the rabbit described in Fig. 1.

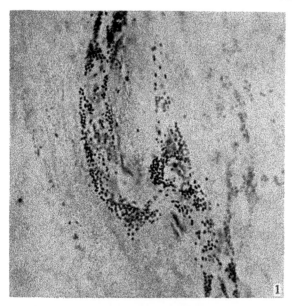

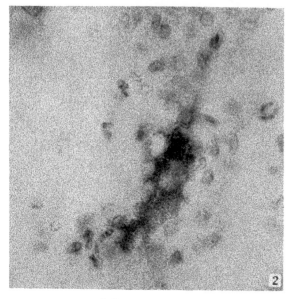

THE FATE OF TYPHOID BACILLI WHEN INJECTED INTRAVENOUSLY INTO NORMAL RABBITS.

By CARROLL G. BULL, M.D.

(From the Laboratories of The Rockefeller Institute for Medical Research.)

PLATE 54.

(Received for publication, June 1, 1915.)

The course of typhoid bacillus infection in the rabbit has been followed by many investigators. Certain facts are established: The typhoid bacilli injected into the blood do not long remain in the circulation; the gall bladder frequently becomes invaded quickly after the injection; and the bacilli survive and multiply there for a long period of time.[1] Ultimately they may disappear from all parts of the body except the gall bladder.

The present study was undertaken to determine more accurately than had hitherto been done the manner in which typhoid bacilli are removed from the blood of normal rabbits. This study was part of a more general study of the mechanism of bacterial immunity in the rabbit,—whether native or acquired. The rabbit may be regarded as possessing a high degree of natural immunity for the typhoid bacillus, and the fate of the bacilli injected was traced from the blood through the various organs in arriving at an explanation of their disappearance.

Source of the Cultures.

The strain of typhoid bacilli used in the mass of the work was obtained from a capsule of Besredka's sensitized vaccine kindly supplied by Captain H. J. Nichols of the Medical Corps. U. S. Army. Three other typical strains from widely different sources were used to check the results obtained with the Nichols strain. All four strains had been under artificial cultivation for some time and should

[1] Nichols, H. J., *Jour. Exper. Med.*, 1914, xx. 573.

be considered as essentially non-virulent, although death of the animal due to intoxication sometimes ensued. The bacilli used for inoculation were cultivated on plain agar and used when twenty-four hours old. For injections they were washed from the medium with 0.85 per cent sodium chloride solution.

Disappearance of Bacilli from the Blood.

From one-thirtieth to one-fiftieth of the bacilli from an agar slant were injected into the ear vein and Petri plates were made from the heart's blood at various intervals, beginning as early as thirty seconds after the injections. It was found that the first specimen taken always contained the largest number of bacilli and that they left the blood with remarkable rapidity. Cultures made ten minutes after the injections contained only a few colonies. Specimens taken fifteen to twenty minutes after injecting the bacteria were often sterile. In some cases a few colonies developed in cultures made several hours after the inoculations or at the time of the death of the animals. Even when the bacilli from an entire agar slant were given and death resulted in two hours the blood was frequently sterile. The following instance will serve as a typical experiment.

Experiment 1.—A rabbit weighing 2,000 gm. was given ¼₀ of a 24 hour agar slant of typhoid bacilli in the ear vein. Blood cultures made from the heart at stated intervals gave the results indicated in Table I.

TABLE I.

Time after injection. min.	No. of colonies per cc.
1	10,000,000
2	2,500,000
5	100,000
15	40
20	1

In several experiments of this kind the same general results were obtained. In some instances the bacilli left the blood somewhat more slowly than in others, but the variation was a matter of a few minutes only.

The abrupt disappearance of typhoid bacilli from the blood stream was investigated. Why should the bacilli leave the blood so much more rapidly than other bacteria, namely, streptococci, pneumococci, or dysentery bacilli of the Shiga type, which were studied? If the process was merely a matter of filtration by the capillary systems of the various organs as Wyssokowitsch[2] concluded, this difference should not exist. The notion prevailing has been that the typhoid bacilli are destroyed (dissolved) by the rabbit's blood within a short time. Indeed, it is known that normal rabbit's whole blood or serum *in vitro* kill the bacilli in relatively brief periods of time. Our next inquiry was directed to the organs to determine whether the bacilli taken out of the blood accumulated in them.

Relation of Bacilli to the Tissues.

From one-thirty-fifth to one-fiftieth of the growth from an agar slant was injected into the ear vein of normal rabbits and specimens of blood were taken for culture from the heart one minute after the inoculation to determine the number of bacilli injected, and again at ten minutes to make sure that the bacilli had left the circulating blood. The rabbits were then killed by a stroke on the neck at ten, twelve, or twenty minutes, and the various organs removed and crushed finely in tissue crushers, after which definite quantities of the pulps were thoroughly shaken in sterile salt solution, and plated. The results of this test are given in illustrative Experiment 2. It may be stated here, however, that in no instance were as many bacilli found in any of the organs per unit of measure as had been present in the blood, and moreover that most of the tissues contained very few bacilli at all.

Experiment 2.—A rabbit was given 200,000,000 typhoid bacilli, as determined by test plating, into the ear vein, and the following data were determined: blood from the heart 1 minute after the injection gave 3,000,000 colonies per cc.; 6 minutes after, 80; and 16 and 20 minutes after, no colonies per cc. The tissues removed at 21 minutes gave colonies per cc. of crushed tissue pulp as follows: spleen, 2,000,000; liver, 1,600,000; lung, 100,000; mesenteric lymph node. 1,000; skeletal muscle, brain, and other tissues, comparatively very few. The weights of the organs were determined and the total number of bacteria recovered, with the finding that it was far below the actual number injected. In none of the organs was the number of colonies per unit of measure as great as the number originally found in the blood.

[2] Wyssokowitsch, W., *Ztschr. f. Hyg. u. Infectionskrankh.,* 1886. i. 3.

From this and similarly conducted experiments it was concluded that the bacilli did not accumulate in the tissues, and it was therefore considered desirable to follow the course of events in an organ for several hours after the bacilli had been injected into the blood. The liver was selected for this study since it is possible to remove portions of the organ at various intervals over a period of time. A typical result is given in Experiment 3.

Experiment 3.—A rabbit was given $\frac{1}{35}$ of an agar slant of a culture of typhoid bacilli into the ear vein and estimates were made as follows: Blood 1 minute after the injection gave 15,000,000, and 10 minutes after, 10 colonies per cc.; liver 3 minutes after the injection gave 12,000,000, 14 minutes 6,000,000, 1 hour 700,000, 2 hours 80,000, and 3 hours 1,000 colonies per cc. of crushed pulp.

The data supplied by Experiment 3 support, on first examination, the prevailing notion that certain bacteria and, in this instance, typhoid bacilli fall a ready victim to the destroying and presumably the dissolving effect of the blood, whether in the general circulation or in the capillaries of the organs. This view is based in large part on the classical studies of Wyssokowitsch, and is upheld by the well known fact that fresh serum and the whole blood of rabbits are highly destructive *in vitro* to the typhoid bacilli. Without, however, adopting the usual explanation of the disappearance of the typhoid bacilli from the blood and organs, as illustrated by the experiments on the liver, it was deemed desirable to repeat certain of the experiments on the bactericidal action of normal rabbits' blood.

Action of the Serum and Whole Blood of the Rabbit in Vitro on Typhoid Bacilli.

Agglutinins.—The serum of normal rabbits was tested for agglutinating value. The bacilli were grown upon agar slants and washed off with 5 cc. of normal saline solution. One drop of the suspension of bacilli was added to one cc. of the serum dilution. The agglutinating value of normal serum varied from zero in full serum to positive in 1 to 100 dilution. About one-fifth of the sera examined were devoid of demonstrable agglutinins. The agglutinins are thermolabile and are destroyed at 56° C.

Bactericidal Power.—The fresh serum obtained from the heart's blood after coagulation was diluted with saline solution and from 2,000 to 4,000 typhoid bacilli (estimated by test plating) were added to one cc. of the dilution. The mixtures in test-tubes were incubated at 37° C. for two hours and the entire contents of the tubes were plated. In order to eliminate the clumping of the bacilli and adhesion to the sides of the tubes, melted agar was poured into them after the plating and they were incubated and examined.

The results can be stated as follows: The sera of different rabbits vary considerably in destructive effect on typhoid bacilli. Full serum and serum diluted 1 to 10 usually destroyed all the bacilli. Serum diluted 1 to 50 usually caused diminution but never complete destruction of the bacilli. The defibrinated whole blood and hirudinized blood act in a manner similar to the serum.

There can, therefore, be no doubt that fresh rabbit serum and the whole rabbit blood destroy considerable numbers of typhoid bacilli *in vitro*. Whether the same means operate in the destruction which takes place *in vivo* remains to be determined.

Fate of Typhoid Bacilli in the Blood and Organs of Inoculated Rabbits.

The experiments performed with the pneumococcus[3] suggested that a similar study be made of typhoid bacilli injected into the circulation of rabbits, especially in view of the fact that while rabbits readily succumb to typhoid intoxication they are highly resistant to typhoid infection. Experiment 4 will serve as an illustration of this class of tests.

Experiment 4.—An agar slant of typhoid bacilli was suspended in about 5 cc. of salt solution and injected into the ear vein of a normal rabbit. Specimens of the heart's blood were taken and films prepared at 30 seconds, 1, 2, 3, 5, and 7 minutes. The films were stained by Manson's method. Clumps of bacilli were found even in the first film (Fig. 1). The second film showed a larger number of clumps, while the number diminished in the next specimen, and none were found in the last, or 7 minute, specimen. At the end of 7 minutes the rabbit was killed by a stroke on the neck and the organs were immediately removed and finely crushed in tissue crushers. Films were prepared from the pulp and stained by Manson's method. Microscopical examination of the slides showed that clumps of bacilli had accumulated in the capillaries, sinusoids, and blood spaces

---

[3] Bull, C. G., *Jour. Exper. Med.*, 1915, xxii, 457.

of the various organs, especially the liver, lungs, and spleen, and a large proportion of the clumps had already been phagocyted by the polymorphonuclear leucocytes (Figs. 2, 3, and 4) which had accumulated in the organs following the injection of the bacteria. Free and unclumped bacilli were also found.

This experiment was repeated in its essential aspects several
times with wholly concordant results. The experiment was varied
in such a manner as to permit the inoculated rabbits to survive for
different periods of time, after which the blood and organs were
examined. It was found that the largest number of leucocytes
enclosing clumps of bacilli were present in the organs of animals
killed from thirty to ninety minutes after inoculation. The organs
of animals permitted to survive three or four hours still contain
many leucocytes, but very few leucocytes containing distinct bacteria
are met with. Leucocytes containing granules of disintegrated
bacilli or bacilli which have lost power of staining may be found in
small number. On the other hand, the three hour specimens still
show a certain number of unagglutinated bacilli, outside cells with
staining properties unimpaired. (Fig. 5).

Still another variation of the experiment was to remove portions
of the liver from ten minutes to two hours after injecting the
bacilli. The findings just described were corroborated by this procedure. Specimens taken from ten to ninety minutes contained
many phagocyting cells; while later specimens showed fewer phagocyting cells and finally cells which enclosed only disintegrating
bacteria.

DISCUSSION.

If we review the findings described in this paper we shall arrive
at somewhat conflicting results as to the manner in which typhoid
bacilli are disposed of, respectively, by the body and the blood of
the rabbit.

Directing attention first to the phenomena observed outside the
body it may be affirmed, in keeping with usual knowledge, that the
fresh blood serum as well as the fresh whole blood of the rabbit is
capable of destroying, apparently by a process of solution, considerable numbers of typhoid bacilli. There is no reason, moreover,
to doubt that the process of destruction in this instance is the

common one of bacteriolysis in which amboceptor and complement play the decisive part.

But the essential question at issue is not the extracorporeal but the intracorporeal method of destruction of typhoid bacilli. It is upon that point that light is especially needed. To apply directly the results of test-tube experiments to the explanation of what takes place in the body itself has not proven wholly illuminating. We already know that typhoid bacilli may appear and survive in the blood of human typhoid fever patients at a time when the shed blood is highly bacteriolytic for the bacilli.

The observations which this paper records indicate a wide disparity between the processes involved in the destructions of the bacilli in test-tubes and in the living body. In the latter, the bacilli introduced into the blood are quickly agglutinated, after which they are removed by the organs. In the interstices of the organs they come into close relation with polymorphonuclear leucocytes (themselves assembled in the organs as result of the bacterial injection) which englobe and destroy them. There is no evidence at hand which connects ordinary bacteriolysis with *intra vitam* destruction of typhoid bacilli. The unagglutinated and unphagocyted bacilli in the organs resist longest and stain best; and complement has yet to be detected in the circulating plasma.

This view of the process of *intra vitam* destruction of typhoid bacilli may serve to explain the fact that typhoid bacilli sometimes circulate in the blood of typhoid patients. It is known that bacilli cultivated from the circulating blood are often inagglutinable. We have found no indications that phagocytosis of the bacteria studied by us takes place in the blood or on a grand scale in the unagglutinated state. Hence, as the bacilli cannot be agglutinated and removed by the organs and also cannot be phagocyted in the blood stream, they continue to circulate, under some conditions. until they are removed and destroyed by the phagocytes.

The indications therefore are that in the body the destruction of typhoid bacilli by means of bacteriolysis does not take place. The question arises, however, whether in the tests of the survival of the bacilli in the pulp the conditions produced do not render the operation of the bacteriolytic processes impossible. To test this point, crushed liver, spleen, and kidney were added to fresh normal rabbit sera just previous to introducing the typhoid bacilli. The test-tubes were incubated for two hours, after which plates were made. It developed that the action of the spleen and kidney pulps was negligible, while the liver pulp caused complete inhibition of bactericidal effect. Further tests indicated that it is the biliary constituent of the liver that is responsible for this action, since bile in quantities themselves non-hemolytic inhibits the activity of a hemolytic system apparently through its anticomplementary effect.

SUMMARY.

Typhoid bacilli are agglutinated promptly in the circulating blood of normal rabbits and quickly removed from the blood stream.

The clumped bacilli accumulate in the organs and are taken up by assembled polymorphonuclear leucocytes in the liver, spleen, and possibly other organs.

The phagocyted clumps of bacilli are digested and destroyed by the phagocytes.

Hence, destruction of typhoid bacilli *intra vitam* is brought about by an entirely different process than is the destruction by serum and whole blood *in vitro*. While the latter is caused by bacteriolysis, the former results from agglutination and intraphagocytic digestion.

Lysis by fresh blood serum is not appreciably affected by spleen or kidney pulp, but it is inhibited by liver pulp. The action of the liver is referable to its biliary constituents, which exert anticomplementary action.

Probably in certain examples of typhoid fever in man the typhoid bacilli in the circulating blood being inagglutinable cannot be removed by the organs and hence are not phagocyted and destroyed.

The observed disparity between the ready destruction of typhoid bacilli by serum and shed blood and the resistance sometimes offered by the bacilli in the infected body is explained by the essential differences in the destructive processes in operation within and without the body.

1

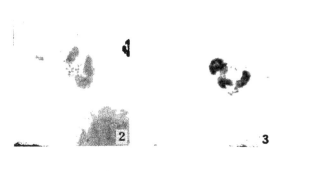

2   3   4

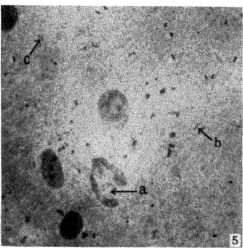

(Bull: Fate of Typhoid Bacilli Injected into Normal Rabbits.)

EXPLANATION OF PLATE 54.

FIG. 1. Heart's blood from a rabbit containing 2 clumps of typhoid bacilli. The specimen was taken 30 seconds after the bacteria were injected into the ear vein. Stained by Manson's method.

FIGS. 2, 3, and 4. Polymorphonuclear leucocytes from lung, liver, and spleen, respectively, containing clumps of typhoid bacilli. The tissues were removed from the animal 30 minutes after the bacteria were injected. Manson's stain.

FIG. 5. A smear of crushed liver tissue. The specimen was removed from a rabbit 2½ hours after an intravenous injection of typhoid bacilli. (a), leucocyte containing disintegrated bacilli; (b) and (c), free bacilli staining deeply. Manson's stain.

THE AGGLUTINATION OF BACTERIA IN VIVO.

By CARROLL G. BULL, M.D.

(*From the Laboratories of The Rockefeller Institute for Medical Research.*)

(Received for publication, June 1, 1915.)

Among the factors to which bacterial immunity is ascribed the phenomenon of agglutination occupies the most subordinate place. Indeed, the general view held by immunologists is to the effect that however valuable agglutination may be *in vitro* in identifying bacteria or discovering specific forms of infection, yet the process plays no essential part in the protection of animals as such. The appearance of agglutinins in the sera or body fluids of infected individuals or inoculated animals is considered a sign of the existence of a greater or less degree of immunity; the agglutinins themselves being merely incidental accompaniments of the true immunity factors,— lysins, bacteriotropins, and antitoxins. We have, however, made certain observations within the last few months which indicate that agglutinins play quite an important part in at least certain instances of active and passive immunity.

In reporting his classical work on the specificity and practical application of the phenomenon of agglutination, Gruber[1] expressed the opinion that agglutinins were quite essential properties of an immune serum. He believed that the phenomenon of agglutination was due to the fact that the agglutinins (Glabrificine) increased the viscidity of the bacterial bodies and caused them to adhere to one another. Gruber advanced the idea also that this increased viscidity aided in the englobement of the bacteria by the phagocytes, and would probably explain the accumulation of bacteria in the organs when they were injected into the circulation. Gruber's hypothesis has never been further developed or proved by himself or others.

Metchnikoff[2] contends that agglutinins play no rôle, however small, in active or passive immunity processes. "The phenomenon of agglutination is of no great importance from the point of view of natural immunity."[3] "We have already given the arguments which render it impossible for us to attribute to the agglutinative property of the fluids of the body any rôle, however unimportant,

[1] Gruber, M., *Wien. klin. Wchnschr.*, 1896, ix, 183, 204.

[2] Metchnikoff, E., Immunity in Infective Diseases, Cambridge, 1905.

[3] Metchnikoff, *loc. cit.*

in natural immunity against micro-organisms."[4] "The part played by agglutination in this immunity (acquired) is merely accidental and subordinate."[5]

After extensive investigation, Salimbeni[6] concluded that bacterial agglutination never takes place within the animal organism. This author worked with animals immunized to the cholera vibrio. The sera of the animals were strongly agglutinative *in vitro*, but he was unable to find any evidence indicative of agglutination *in vivo*.

Sawtschenko and Melkich[7] believed that agglutinins were present and active in the plasma of cases of recurrent fever. They found small clumps of spirochætes in blood immediately after its removal from the patients. It was suggested by them that the agglutination in the plasma was incomplete because of the rapid movement of the blood.

A glance at the literature given above readily convinces one of the slight consideration agglutinins as protective agents have received. Gruber's suggestions seem to have gone by as such and Sawtschenko's observations have not been considered sufficient proof of the functioning of agglutinins *in vivo*. Salimbeni has done the most direct and extensive work on this subject and his results led him to conclude that agglutination is strictly a phenomenon of the test-tube. And Metchnikoff is quite firmly convinced of the unimportance of agglutinins in both natural and acquired immunity. Our observations made in a more or less incidental manner seem to justify a different conception of agglutination as an immunity factor, for which reason they have been put together in this paper.

Pneumococci and Agglutination in Vivo.

In the course of the study of the manner of the rapid disappearance of pneumococci from the blood stream of rabbits following the injection of antipneumococcus serum the observation was made that an almost instantaneous agglutination of the pneumococci was produced in the blood by the serum introduced and, furthermore, that the clumped or massed cocci quickly accumulated in the internal organs.[8]

These facts in themselves are not only new, as is evident from the literature previously quoted, but they are important as bearing on our conception of the biological processes which come into play in the course of bacterial immunity. Aside from this consideration, however, the facts are of interest in connection with the following points: (1) the time required for the reaction to take place; (2) the

---

[4] Metchnikoff, *loc. cit.,* p. 258.

[5] Metchnikoff, *loc. cit.,* p. 263.

[6] Salimbeni, A. T., *Ann. de l'Inst. Pasteur,* 1897, xi, 277.

[7] Sawtschenko and Melkich, *Ann. de l'Inst. Pasteur,* 1901, xv, 497.

[8] Bull, C. G., *Jour. Exper. Med.,* 1915, xxii, 457.

degree of its specificity; (3) the concentration of serum necessary to effect agglutination; and (4) the relation of the quantity of serum employed to the course of a bacterial septicemia.

EXPERIMENTAL.

When the agglutination tests are made in test-tubes it is customary to incubate the tubes for two hours at 37° C. and then to set them aside at a low temperature for another period of several hours before the results are read.

In contradistinction to the slowness with which agglutination occurs and becomes evident *in vitro* is its instantaneous occurrence *in vivo*. To demonstrate this point we have proceeded as follows: The bacteria (pneumococci) from 50 cc. of bouillon are injected into the ear vein of a rabbit, after which a specimen of blood is taken from the heart to ascertain if sufficient bacteria are present to be easily found with the microscope and to determine the absence of clumps. The antiserum is then injected and a specimen of blood taken from the heart not later than thirty seconds afterwards. Other specimens are taken at one, two, three, five, and ten minute intervals. The sample of blood removed at twenty seconds usually contains the largest and greatest number of clumps. The second sample may also show many clumps of fair size, but the later samples show only a few small clumps. As a rule, no clumps or free bacteria whatever can be found at the expiration of five or ten minutes, particularly if from 1 to 2 cc. of serum have been injected.

Why bacteria agglutinate so much more rapidly *in vivo* than *in vitro* is not evident. The temperature conditions do not offer a satisfactory explanation, because the reaction does not occur in the test-tube immediately after body temperature is reached. The effect of the agitation caused by the circulation of the blood is not known, but would not seem to suffice to produce the great differences noted. It is quite possible that some constituent of the blood of the host aids the specific serum in producing those changes in the bacterial bodies which precede agglutination, but of this we are wholly ignorant.

The specificity of antipneumococcic agglutinins is as strict *in vivo* as *in vitro*. A heterologous serum causes no agglutination in a concentration fifty times as great as that in which a homologous serum gives a positive reaction.

In following the reaction quantitatively *in vivo* we found that quantities of serum as small as 0.05 cc. per kilo of body weight of the rabbit caused the formation of small clumps, and that that quantity was the least amount that exercised any appreciable influence on the course of the septicemia or sufficed to prolong the life of the inoculated animals. Since the agglutinating titer of the antiserum employed was 1 to 50 by the macroscopic method, this result was wholly unexpected. But the result was rendered more comprehensive when it was ascertained that by merely modifying the method somewhat the titer could be increased to 1 to 500. Thus, when from 0.08 to 0.1 cc. of a bouillon culture of the pneumococci is added to 1 cc. of the diluted serum, clumps are formed which may be large enough to be seen by the naked eye, provided the fluids are clear. Stained preparations viewed with the microscope also reveal the clumps. But even so, agglutination within the body seems to take place with dilutions of the serum which are wholly ineffective in the test-tube.

Typhoid Bacilli and Agglutination in Vivo.

When typhoid bacilli are injected into the circulation of normal rabbits, they quickly leave the blood. The precise manner of the disappearance has not been investigated. Because the typhoid bacilli are subject to lysis it has been supposed that they are in fact disintegrated.

When, however, the blood was examined as described in connection with the pneumococci it was learned that the bacilli agglutinate within a few seconds after entering the blood stream. The rapidity with which agglutination takes place is affected by the power of the serum to cause agglutination in the test-tube. With rabbits whose blood does not possess this power a single bleeding could be performed before clumps were formed; in all others, agglutination had taken place within the brief period of thirty seconds.

The heating of typhoid bacilli to 80° C. for half an hour rendered them more readily agglutinable *in vivo* than unheated bacilli. While bacilli heated to from 65° to 80° C. proved inagglutinable in test-tubes. our experience here confirmed the observation of others.[9] In this respect, therefore, agglutination *in vivo* differs from that *in vitro*.

Dysentery Bacilli and Agglutination in Vivo.

As a rule, normal rabbit serum does not agglutinate *in vitro* in dilutions of 1 to 10 any of the varieties of dysentery bacilli. The sera of all the rabbits we tested were inactive in this dilution. We employed the following strains: Shiga, Strong, Hiss, and Flexner.

The strains of the Shiga type do not agglutinate and remained in the blood for twenty minutes distributed uniformly; while the other strains undergo immediate agglutination, clumps being found twenty seconds after the bacteria were injected. By means of this reaction it was possible to determine within two minutes whether the culture tested belonged to the Shiga or Flexner groups of dysentery bacilli.

The fate of the dysentery bacilli could also be followed. When the rabbits are killed as early as seven minutes after receiving an inoculation of one of the Flexner group of bacilli, large numbers of leucocytes carrying clumps of one hundred or more bacilli were already present in the lungs, liver, and spleen.

When an immune serum for the Shiga type of bacillus is injected into the circulation the Shiga bacilli, which otherwise do not agglutinate, become immediately clumped. The distinction, therefore, in the behavior of the Shiga and the Flexner groups depends on the presence in the normal rabbits of agglutinins for the one and not for the other group.

Bacillus influenzæ and Agglutination in Vivo.

A non-virulent strain of *Bacillus influenzæ* isolated from the respiratory tract agglutinated in the circulation of normal rabbits one minute after injection. A virulent strain isolated from a case of influenzal meningitis was not agglutinated and was still in the circulating blood twenty minutes after injection. The distinction of virulent and non-virulent strains is determined by the fact that the former causes a fatal septicemia in young rabbits, while the latter does not.[10]

[9] Porges, O., *Ztschr. f. exper. Path. u. Therap.*, 1905, i, 621.

[10] Wollstein, M., *Jour. Exper. Med.*, 1915, xxii, 445.

The results described in the preceding pages emphasize the occurrence as well as the significance of agglutination of bacteria in the blood once they have gained access to the circulation.

In the first place, the power of the blood in normal animals to cause agglutination determines, apparently, in large measure whether the bacteria are to be promptly removed and septicemia avoided or to remain and to bring about without restraint that serious condition. The importance of this power as far as the normal rabbit is concerned is well illustrated by the examples afforded, on the one hand, by pneumococci, dysentery bacilli of the Shiga type, and virulent influenza bacilli, and, on the other hand, by typhoid bacilli, dysentery bacilli of the Flexner group, and non-virulent influenza bacilli.

The normal blood of the rabbit fails to agglutinate the Shiga dysentery bacilli outside (*in vitro*) or even inside the circulation, with the result that they are not converted into clumps in the blood stream and hence are not promptly removed from it, and while the normal blood does not agglutinate the Flexner group of dysentery bacilli in the test-tube in a 1 to 10 dilution even after two hours' incubation at 37° C., it does agglutinate them immediately in the circulation, whence they are quickly removed. Similarly, the circulating blood of the rabbit does not agglutinate pneumococci or the virulent form of influenza bacilli, with the result that they remain and multiply there, leading to a septicemia to which the animal succumbs; while typhoid bacilli and the non-virulent form of influenza bacilli are instantly agglutinated and promptly removed from the circulation.

In view of these facts we can discern in the phenomenon of agglutination in the blood an essential mechanism through which bacteria are so changed that they accumulate in the capillaries and sinuses of the viscera; and perceive in this mechanism the decisive act which determines whether protection is to be afforded or fatal septicemia supervene.

Indeed, the facts at hand illuminate the mechanism of protection from infection even further. It would appear that in the instances given the bacteria neither accumulate in the organs to any degree when unagglutinated nor are they taken up by the leucocytes within the circulation when they remain there. Moreover, those bacteria which enter the organs before agglutination takes place remain in the single form and escape phagocytosis. When, however, agglutination has taken place the bacteria are quickly removed from the circulation and are englobed by leucocytes in the lungs, liver, spleen, and other organs, and, as may be inferred from the rapidity with which they are disintegrated in the phagocytes as compared with their persistence when free, are quickly destroyed, and possibly finally detoxicated. In other words, in order that the bacteria may be promptly removed from the blood stream it is requisite that they be first agglutinated, which condition is also required in order that they be destroyed *en masse* within the organs, a process achieved, apparently, chiefly through phagocytosis.

This phenomenon of protection in the normal animal is paralleled by what happens when an effective antiserum is employed to prevent or combat a bacterial infection, as is illustrated by the examples given of serum protection in pneumococcus and Shiga bacillus infection in the rabbit. It may be inferred that similar processes occur in other infections and in other animals, including man, but many more particular cases will need to be investigated before general deductions are made.

SUMMARY.

1. Small quantities of antiserum bring about instantaneous agglutination of pneumococci in the circulation of the rabbit; the reaction is specific and occurs in every case in which sufficient serum is given to influence the course of the septicemia or to prolong the life of the animal.

2. The agglutinating titer of antipneumococcus serum can be made considerably higher by adding only a small quantity of culture to the tests, thus making the test a finer differential.

3. Typhoid bacilli agglutinate spontaneously in the circulation of the normal rabbit; the reaction is positive *in vivo* even in cases in which undiluted serum gives a negative result *in vitro;* heating the bacilli to 80° C. for thirty minutes renders them more agglutinable *in vivo.*

4. Dysentery bacilli of the Shiga type do not agglutinate in the blood stream of the normal rabbit, but a small quantity of antiserum injected into the circulation causes immediate agglutination; while all strains of the Flexner group undergo spontaneous agglutination.

5. Non-virulent influenza bacilli agglutinate spontaneously in the circulation of the normal rabbit; virulent strains remain in the blood unclumped.

6. In all instances so far investigated of both passive and natural immunity, agglutination of the bacteria within the blood of the infected animal was followed by a rapid removal of the bacteria from the circulation, and by phagocytosis and destruction of the agglutinated bacteria in the capillary systems of the viscera; while those bacteria which are not agglutinated remain in the circulation and produce a progressive septicemia.

7. Hence the agglutinins seem to play the decisive part in at least certain instances of bacterial infections.

FATTY DEGENERATION OF THE CEREBRAL CORTEX IN THE PSYCHOSES, WITH SPECIAL REFERENCE TO DEMENTIA PRÆCOX.

By HENRY A. COTTON, M.D.

(From the Pathological Laboratory of the Royal Psychiatric Clinic, Munich, and the Laboratory of the New Jersey State Hospital for the Insane, Trenton.)

PLATES 55 AND 56.

(Received for publication, May 12, 1915.)

Object and Scope of the Investigation.

Although there exist in the literature many references to the normal and abnormal accumulation of fatty pigment or lipoid material in the nerve cells and neuroglia, it cannot be said that the subject has been treated exhaustively, or that our knowledge of the source and character and occurrence of this phenomenon is adequate. The significance of lipoid degeneration of nervous tissue and its relation, if any, to definite mental diseases has not been thoroughly investigated, and therefore it was thought important enough to make a thorough investigation of the subject and to ascertain if any possible relation existed between fatty degeneration of the nervous elements of the cortex and the various psychoses.

In order to determine this question, many difficulties arose, as the fatty content of the nervous elements in normal individuals had to be thoroughly studied before one could decide whether the accumulations of fatty pigment were pathognomonic or not. We know that these accumulations become more pronounced as the individual becomes older, so that it is important to consider the age of the individual before deciding upon the pathologic significance of this phenomenon. It is also possible that various poisons, especially alcohol, may have a potent effect upon the occurrence of fatty degeneration of the nervous elements, even in the adolescent and adult years of life. Therefore in order to establish any pathologic significance to this phenomenon only a pronounced increase in the lipoid and fatty pigment out of proportion to the age of the patient can be regarded as patently pathologic. And this abnormal condition can be brought into definite relation to the morbid process only when it occurs regularly in many cases of the same disease. The object of this investigation then was to determine the relation of fatty deposits in the elements of the central nervous system as found in both normal and abnormal brains of human species. In order to determine whether it was a necessary constituent of nervous tissue, the brains of various wild and tame animals were also studied.

Methods.

The Herxheimer method[1] of staining the fatty pigment by means of Scharlach R was the principal method used, as this method is far superior to any other for demonstrating the fatty pigment, and it also enables us to make a comparative estimate of the character and amount of the pigment present.

For this purpose formalin-hardened material was used, and sections were cut with a freezing microtome. No alcohol material can be used, as the alcohol extracts a large amount of the fatty pigment, and the picture is therefore inaccurate.

Old formalin material can be used, although freshly hardened tissue is preferable.

[1] *Herxheimer's Stain for Fat by Scharlach R.*

1. Formalin fixation.
2. Frozen sections.
3*. Stain in Scharlach R solution 2 to 5 minutes, with or without heating. The stain is perhaps a trifle more intense with heat, but precipitation is more likely to occur.
4. Wash in water.
5. Stain in dilute Ehrlich's hematoxylin (diluted 1 to 4) for 1 to 3 minutes; usually a minute or a few seconds are sufficient if the solution is well ripened.
6. Wash well in tap water, 3 to 5 minutes or more.
7. Transfer to a slide by dipping the slide in the water beneath the section.
8. Dry off the water about the margins of the section with filter paper, but do not blot.
9. Mount in glycerine and seal the edges of the cover-glass with wax.

* *Scharlach Staining Solution.*

Make up just previous to use.

Saturate the following solution with Scharlach R in a test-tube by agitation and heat; filter carefully.

Absolute alcohol .... 70 cc.
10 per cent sodium hydrate solution .... 20 "
Distilled water .... 10 "

The Nissl stain was also used to determine other pathologic changes in the nervous elements. But frequently the picture obtained by the Nissl method gave no indication of the presence of the fatty pigment. Only in cases of far advanced degeneration such as that found in senile dementia was the fatty pigment demonstrable by this method.

The neurofibrils were studied by means of the Bielschowsky and Cajal methods, but the results obtained were not considered trustworthy on account of their difficulty and often unreliability. It was therefore impossible to form any opinion of the relations of the neurofibrils to the fatty degeneration, with the exception of a few types, notably the pathologic changes found in central neuritis and occasionally in senile dementia.

There can be no doubt that changes in the neurofibrils associated with fatty degeneration of the nerve cells exist, but with our present unreliable silver impregnation methods it is impossible to demonstrate this relation.

The Occurrence of Fatty Pigment in Brains of Normal Individuals.

This investigation was begun in the laboratory of the clinic at Munich in 1905–06. The difficulty of obtaining normal brain material in a community where the excessive use of alcohol was the rule soon became apparent. A number of normal brains were examined, but changes demonstrated by the Nissl as well as by the Scharlach method forced us to discard such material. In the interval (1906–11) normal material was collected principally from accident cases which came under the supervision of the county physician and from electrocuted cases at the State Prison. In this way we were able to examine the brains of about thirty cases of supposedly normal individuals of various ages. Many were not available because of excessive alcoholism, but we were able to form some opinion as to the occurrence of fatty pigment in various decades from the new-born to old age.

In a new-born no lipoid pigment was found in the glia, the ganglion cells, or in the cerebral cortex. In the case of a young man 16 years old, killed by an electric shock, lipoid granules were found in many ganglion cells of the cerebral cortex, but always in very small quantity only; they lay in the pyramid cells mostly at the base of the cells by the side of the nucleus, and produced the impression that they did not permeate the entire cell plasma but included only a comparatively small part of the cells. In the glia cells as well as in the vessels were found no single isolated fatty granules, and none were found in the superficial glia.

A considerable increase of the lipoid pigment was found in the case of an electrocuted convict 30 years old; here appeared likewise only very few fatty granules in the ganglion cells; usually only 3 to 10 granules lay at the base of the cells near the border. As in the former case, so here in the glia and even in the glia cells of the most superficial strata no deposit of lipoid pigment was found. Neither were lipoid granules found in the ganglion cells of a 27 year old negro, electrocuted. In the case of a 36 year old suicide (because of unhappy family relations) somewhat more fatty deposit was found, still the amount was not very considerable. In a 70 year old man who had shown no mental defect the mass of the lipoid pigment in the ganglion cells was noticeably augmented. In many ganglion cells the entire base of the cells seemed to be filled with lipoid materials; the nucleus was also somewhat displaced toward the apical dendrite; and in the glia cells a large quantity of fat, likewise in the vessels, was demonstrated. Other brains of individuals, appearing mentally normal, were not available because during life a considerable abuse of alcoholic beverages had occurred, and a multiplication of the lipoid pigment in certain of these cases might probably have been connected with chronic alcoholism.

The result of these investigations may be summarized as follows: that lipoid materials are not to be found in the glia cells, the ganglion cells, and the vascular walls of the cerebral cortex of the new-born; that in the second decade of life lipoid materials already occur normally in the ganglion cells of the cerebral cortex, yet still in very small amount and only at particular points of the ganglion cells; that in the third period or decade they do not yet seem to be materially increased, but thereafter in later life the quantity assumes a considerable augmentation. While during the middle period of life only little fat is found in the glia cells and in the cells of the vascular walls together with a considerable increase in the ganglion cells, in advanced age the fat is increased in the glia and in the vessels; even in the most superficial strata of the glia there was found in middle age no accumulation of the lipoid materials. In the large Betz cells of the motor area, however, there is usually present a large amount of pigment even in normal individuals of various ages. Also in the cells of the anterior horn of the spinal cord fatty pigment is usually found in fairly large amount.

Hence we must conclude that in the large motor cells a considerable amount of fatty pigment is normal. It does not occur in the axis cylinder or even in the neighborhood of the base, but is usually found to one side of the nucleus and apparently does not interfere with the functioning of the cell.

The Occurrence of the Lipoid Pigment in Mental Disorders.

We have investigated the following mental disorders:

Amentia .... 1 case.

Intoxications $\begin{cases} \text{veronal poisoning} \\ \text{pbosphorus poisoning} \end{cases}$ .... 2 cases.

Chronic alcoholism

Manic depressive insanity 10

Dementia præcox 20

General paralysis 12

Senile dementia .... 11 "

Anxiety psychosis .... 1 case.

Central neuritis .... 1 "

Circulatory psychosis .... 1 "

Idiocy .... 4 cases.

Epilepsy .... 19 "

Alcoholic insanity .... 10 "

In a case of infectious psychosis no augmentation of the lipoid substance in the cerebral cortex could be demonstrated out of proportion to the amount corresponding to the age of the patient. The same was true of the lipoid materials in the case of one dead of acute veronal and acute phosphorus poisoning. In individuals, victims of chronic alcoholism, an increase of the lipoid materials in all the elements of the cerebral cortex correspondent to the age of the patient appeared quite regularly. In the psychotic condition of chronic alcoholism, especially in the condition of dementia, the increase appears peculiarly rich, still none appears for the individual disorders of characteristic peculiarity.

In manic depressive insanity of the young there appeared no particular diminution of the amount nor change in the position of the lipoid materials compared with the normal. On the other hand, in the cases of later years a considerable amount of the lipoid materials was without doubt demonstrable, correspondent to the age of the person. Thus some cases which had repeatedly exhibited acute manic depressive conditions showed already at fifty years of age as rich a quantity of lipoid materials in the cerebral cortex as otherwise are usually to be seen only after sixty years of life.

Dementia Præcox.

Twenty cases of dementia præcox were examined. In this disease the estimation of the fatty degeneration was very difficult, because most of the cases ended fatally through exhaustion, especially from tuberculosis. Yet still there occur some cases among these in which death occurred very early or resulted from diseases which could hardly have exerted any essential influence on the cortical cells. We may therefore regard the pathological findings as a characteristic disease-process, in addition to the fact that these cases show remarkable similarity in the amount and distribution of the pigment. Because of the significance of the fatty pigment found in this group several cases are given in detail, with short abstracts. Cases dying of various intercurrent diseases are given to show the similarity of the cortical findings.

Case 1.—M. C. Female, 30 years old; married. Psychosis appeared after pregnancy, lasted four years. Paranoid type. Moderate dementia. Cause of death tuberculosis of the lungs. Section 15½ hours after death. No macroscopic alterations to be seen.

With the Nissl preparation there is seen a remarkable alteration of the cerebral cortex. Many cells are in a sclerotic state; in the superficial strata the glia are proliferated, and change in the architecture of the cerebral cortex is demonstrable. Several glia cells have penetrated many of the ganglion cells; the cellular body is slightly stained—almost invisible—and no more Nissl bodies are to be seen. With Scharlach preparations one sees a considerable increase of the lipoid substance. In the central lobe the small and medium pyramid cells are not so noticeably altered, but the Betz cells are strongly affected. The fatty pigment is dispersed in the entire cell body and is in great accumulation. Many cells are completely degenerated, and the cell is indicated only by the accumulation of the fatty pigment. In the superficial strata and also in the inferior strata the glia is largely fat. In the frontal lobes one sees almost all the small pyramid cells nearly wholly fat, and affected by this peculiar degeneration that is described later. The fatty pigment is deposited in the entire cellular body, especially at the base, and in many cells has penetrated the axis cylinder (Fig. 1). Frequently the nucleus is laterally displaced. The vessels contain much fat in the adventitia cells and in many small vessels the glia cells are found clumped about the vessel in which much fat is contained (Fig. 7).

Case 2.—A. D. Female, 24 years old; single. Duration of the disease, 1½ years. Teacher. Katatonic type. Refusal of food. Sense deceptions, delusions. Poison in food, etc. Moderate dementia. Died after 8 months in institution. Wasted. Cause of death tuberculosis of the lungs. Section 14¾ hours after death. No significant results are macroscopically demonstrable. Brain negative. On microscopic examination, with Nissl preparations one sees in the superficial as well as in the other layers that the glia cells are full of fatty pig-

ment. Mostly one sees a considerable increase in the glia cells, large proto-
plasmatic body, and fatty pigments. Usually they lie on the vessels, in many
places six or eight consecutively.

The ganglion cells show marked fatty degeneration; this alteration corresponds to the sclerotic type (darkly stained nucleus and narrow cell body).
The protoplasmic prolongations can be followed out very distinctly and most
of the cells in the third stratum are plainly sclerotic, very slightly stained, and
only shadows. These walls are attacked by large glia cells; usually one sees
three to six glia cells in the cellular body of the ganglion cells. With the Nissl
method the vessels showed no very considerable alteration. With the Scharlach
method one sees at once in all cortical elements pathological alterations similar
to those described in the foregoing case.

Case 3.—A. B. T.   Female, 35 years old; married.   Duration of the disease
4 years.   Katatonic type.   Two years before death cellulitis of the arms and
legs.   Profound dementia.   Died suddenly, after some days, of volvulus and
chronic nephritis.   Section 5½ hours after death.   Slight pachymeningitis, moderate chronic septic meningitis, moderate brain atrophy, pronounced ependymal
granulation of fourth ventricle.   The pathological findings in this case were, of
course, not so well defined as in the two cases described above; still they are
in accord in that there is present a pronounced fatty degeneration of all the
nervous elements, which normally does not correspond to this age.

Case 4.—J.   Died very suddenly of peritonitis, at 34 years of age.   This case
shows a far-reaching accord with the first case.   Irregular arrangement of the
fat in isolated little heaps over the entire cell, or string-like arrangement along
the border of the ganglion cell.   In most glia nuclei are found only a few fatty
granules.   The fatty granules of the glia satellite cells are often so dispersed
over the cell that the fatty granules of the ganglion cell itself cannot be differentiated.   Here and there one meets with individual glia cells richly endowed
with fatty granules.   In the vessels the fatty degeneration is not so marked.

From the study of our cases we conclude that in dementia præcox
a profound degree of fatty degeneration in all cells of the cerebral
cortex is found.   In the ganglion cells the augmentation is often as
rich as in senile dementia.   Nevertheless it is indisputable that in
most preparations much less fat is found than in senile dementia.
It probably results therefrom that the fat is not found lying together
in such thick clots as in senile dementia.   And in general in the Nissl
preparation they appear far paler and less yellowish than in the
others.   The profoundly fatty degenerated ganglion cells in dementia præcox exhibit a honeycomb structure, and the comb walls which
enclose the structure slightly stained yet still recognizable through
the powerful refraction of the light are much thicker than in senile
dementia, where they are formed only of rows of fine granules, and
finally may wholly vanish.   Also in fibrillar pictures which have

been stained by the Cajal or Bielschowsky method we see relations different from senile dementia. We were not able to represent in such cells any finding or any deviations whatever of the fibrils, as is so easily done in senile dementia. But we notice the absence of fibrillar alterations, either because they really are not present, or because on account of the imperfection of the method they cannot be represented.

As a characteristic of all these cases must be noted that in all the cells of the cerebral cortex the lipoid materials were considerably augmented. The increase affected the entire cerebral cortex, but appeared most distinct in the second and third layers. Particularly striking pictures are obtained from the third stratum of the frontal lobe, where are found many ganglion cells whose entire cellular body is filled with lipoid substance. A frequent occurrence was the lipoid pigment continued on into the axis cylinder prolongation at its point of origin to the point of medullation. Such considerably extensive pictures of a similar sort can elsewhere be obtained only in senile dementia. With comparative preparations which were fixed with alcohol and stained with basic aniline, it is shown that all these cells present an advanced stage of sclerotic alterations, so that here also the fatty degeneration was connected with the sclerosis, as very frequently happens in chronic morbid conditions. In these preparations it is shown that the prolongations of the cells are stained remarkably far, often narrowed, often serpentine (especially in dendrites), the cellular body somewhat shriveled, the exterior forms remarkably angular. The nucleus is stained dark, often drawn out to an extraordinary length, often triangular or pyramid-formed, fitted to the shriveled cellular body. Often, too, in the exterior pyramid strata may be noticed a certain arrangement of the cortical elements in that the cells exhibit their apical dendrites not vertical to the superior surface, but placed in various directions, and are advanced noticeably far out of the cells. Not always did the accumulation of the fatty materials in the ganglion cells correspond to the accumulations in the glia. Sometimes they were found in regions in which the cells were profoundly fatty with only regressive alterations of the glia nuclei without lipoid granules. In other places, however, the glia showed rich fatty pigmentation and the same condition shows itself also in the cells of the vascular wall. In the inferior strata where fewer acutely altered ganglion cells were found, more lipoid materials could, for the most part, be determined in the glia cells. Also the toluidin blue preparation showed that the glia nuclei in the second and third cortical strata could be recognized as affected by very considerable retrogressive alterations. They were in part extraordinarily pycnotic. Now and then such nuclei were found in the middle of a sclerotic ganglion cell. The glia cells of the superficial layer showed mostly a considerable fatty deposit, and in vessels of the inferior strata was found a large quantity of fat.

From this investigation there can indeed be no doubt that in cases of dementia præcox a considerable degenerative alteration in the cerebral cortex occurs, which alterations appear with special distinctness in the second and third cortical strata, and which lead to an extensive fatty pigmented sclerosis of the cortical elements. It can indeed be asserted that in many cases the greater part of the second and third strata is in such a condition that no further functioning can be expected. We have not been able to institute thorough examinations of the entire cerebral cortex, but the preparations from our cases give us the impression that the alteration in the frontal lobe is essentially more far-reaching than in the central convolutions. In this respect it is to be mentioned that in Case I, where the ganglion cells in the frontal lobes were greatly loaded with fat, almost no accumulation of fatty pigment was found in the ganglion cells of the second and third strata of the central lobe. Also very little fat is found in the glia in either the superficial or deeper strata. The Betz cells are irregularly altered. In many cells there is shown a considerable accumulation of the fatty pigment which is not particularly distinguished from the normal condition, but in other cells the pigment is deposited around the nucleus. This irregular distribution of the fatty pigmented degeneration of the cortical elements of the cerebral cortex in dementia præcox possibly differentiates this process from senile dementia, since in the latter the whole cerebral cortex is affected equally.

Other Psychoses.

In a case of so called "anxiety psychosis," a disease of the climacteric period leading quickly to death, no pronounced increase of the lipoid pigment was exhibited, only it was striking that the fatty granules were deposited in the ganglion cells not in a collected accumulation, but distributed over the entire cellular body.

Very remarkable and unusual was the appearance of lipoid material in the nerve cells in a case of central neuritis, a peculiar terminal disease which was first described by Adolf Meyer[2] and later by others.[3] We will summarize the case briefly.

> The symptoms of central neuritis gradually developed upon an apparent depressed phase of manic depressive insanity. They were general physical weakness, diarrhea, inclination to fall on the floor, later involuntary twitchings of all muscles of the extremities, followed by atrophy, stiffness in muscles, and pain upon movement. The peculiar confusion delirium occurred in connection with the twitchings and became even more profound with the progress of the other symptoms; later, unable to speak or to swallow, with profound hallucinations of a terrible or fearful type, and occupation delirium. Autopsy was made three hours after death and showed nothing specific macroscopically. In the cerebral cortex of the central lobe one sees at once the peculiar alterations of the Betz ganglion cells, the so called "axonal reaction" of Nissl, and first described by Adolf Meyer as characteristic of this disease. These alterations are accompanied by a pronounced degeneration of the white substance, as is illustrated by the Marchi stain.

On viewing the preparations with low magnifying power we are at once struck by the immense accumulation of lipoid materials in the cerebral cortex. Especially the exterior and the interior strata of the large pyramid cells are presented in many sections as a red stripe. In other sections the fatty degeneration seemed to be not entirely of so high a degree. When we examine the individual elements it is shown for the most part that the whole basic part of the ganglion cells even up to the processes is filled full with fatty granules, which appear, however, in many cases as no longer sharply isolated, but seem clustered together. The nucleus is then frequently displaced toward the apical process, and frequently one

[2] Meyer, A., On Parenchymatous Systemic Degeneration Mainly in the Central Nervous System. *Brain*, 1901, xxiv, 47.

[3] Cotton, H. A., and Southard, E. E., A Case of Central Neuritis with Autopsy. *Am. Jour. Insan.*, 1909, lxv, 633.

notices that on the border of the cell body large fatty spheres seem to project. It is not possible to determine here with certainty, however, whether we are confronted with an extrusion of the lipoid materials out of the ganglion cells, or whether these fatty globules belong to the satellite cells of the ganglion cells. In individual cells it is even possible definitely to assert that these large fatty globules, which in part reach to the compass of small glia grains, are encamped on glia nuclei. In isolated cases similar large fatty globules are found also in glia cells which are not satellite cells.

In the Betz pyramids we see no such masses of fatty deposits, but on the other hand we receive the impression that in their exterior part they contain many fine fatty granules which like a cloak veil over the fat-free interior of the cellular body (Fig. 6).

In the Nissl picture these Betz pyramids show changes which remind us of that cellular alteration which is observed in a transverse section of the peripheral nerves, in their ganglion cells (axonal reaction).

The nucleus has been shoved out to the border of the ganglion cells, and the Nissl bodies have undergone a dissolution. Where the ganglion cells are undergoing so advanced a stage of the disintegration progress we find also rich masses of fatty globules in the adventitial cells of the vessels. Besides this particular alteration of the Betz pyramid cells the small cells are pronouncedly fatty, but the type of fatty degeneration differs materially from the process described in dementia præcox. This mixed character of the fatty pigment of central neuritis is shown for comparison with dementia præcox (Fig. 6).

A case of circulatory psychosis,[4] or, as termed by the author, cardiogenetic psychosis,[5] was also thoroughly studied, and because it presented certain pathological features we report it here.

The patient was a negro woman, aged 55 years, who showed no symptoms until 8 days before admission to the State Hospital at Trenton. The psychosis was characterized by profound depression, motor restlessness, and extreme apprehensiveness, especially at night; marked variability of moods was present;

[4] Jakob, A., Zur Symptomatologie, Pathogenese und pathologischen Anatomie der " Kreislaufspsychosen," *Jour. f. Psychol. u. Neurol.,* 1909, xiv, 209.

[5] Cotton, H. A., and Hammond, F. S., Cardio-Genetic Psychoses, Report of a Case with Autopsy, *Am. Jour. Insan.,* 1911, lxvii, 467.

and frequently she was profoundly delirious, and at times seemed in a condition of hypomania. She died three months after admission, the cause of death being myocardial degeneration and edema of the lungs, and subacute adherent pericarditis. Autopsy three hours after death showed no macroscopic alteration of the brain.

In a general review it is at first to be determined that no focal or macroscopic organic disorders were present, and no inflammatory processes in the cerebral cortex. Only the nervous elements were attacked by the process,—the large elements of the inferior cerebral cortex, especially the Betz pyramids, and the small and medium cells. The most important alterations are in the Betz cells, and the type of the axonal reaction, as has already been described in the above case of central neuritis.

The findings here accord completely with the above cases. An extraordinary deposit of fat was found in the cerebral cortex, in conditions which suggest central neuritis. Only in this case the phenomena were not so prominent, the individual fatty granules had not coalesced into great masses, but the fatty granules had remained isolated as we have observed in senile dementia and the usual forms of fatty degeneration.

Senile Dementia.

We have examined eleven cases of this psychosis. The lipoid materials are multiplied in most of the cells where normally they lie in the ganglion cells. We see, however, in very many cells the fatty materials finally filling out the whole cell body. It is frequently noticed thereby that the cells are widened out by reason of the fatty deposit at the base, and that the cell nucleus is displaced toward the apical process. Also occasionally fat is seen in the cellular prolongations, and not very seldom the axis cylinder prolongations are indicated through a series of fatty granules. The fatty granules, even in senile dementia, are comparatively large; the entire cell body may have been filled full of fatty granules and the prolongations become invisible. In the Nissl preparation the spindles and Nissl bodies are lacking in the region of the fatty granules; on the other hand one often sees in their arrangement dark granules forming meshes between which the fatty granules are deposited.

The fibril picture at the point of the fatty degeneration of the ganglion cells is altered in a wholly characteristic manner. One may see that at the place of the fatty degeneration the fibrils are lacking; the fibrils do not run their otherwise customary course. but they lie much rather at the place of the fatty degeneration and run their course in the border much thicker than usual or normal. We see single fibrils enter in the fatty locality and there pass over into meshwork. This meshwork may indeed not answer to a meshwork of fibrils, but a scaffolding of a web-like ground substance in whose meshes the lipoid materials are deposited. The fatty degeneration seems to affect in like manner all ganglion cells of all strata of the cerebral cortex. Particularly in many small cells one can recognize only the ganglion cellular body and the surrounding fatty granules. In many senile cortices there is found no ganglion cell which does not contain a large amount of fatty granules. The glia cells show in like manner a rich accumulation of fatty granules. A peculiarly rich fatty accumulation is found in many glia satellite cells. The fatty granules here often lie so close and massive on the ganglion cells that it is difficult even to separate them from the fatty granules of the ganglion cell.

Frequently those belonging to the glia cell are larger than those belonging to the ganglion cell, and thus the differentiation is facilitated. Often we find that the fatty granules belonging to the glia nucleus are situated somewhat on the side of the nucleus; so that occasionally also we meet with fatty granules which apparently seem to be lying free. On closer study of the section it becomes probable that also these little heaps belong to the glia nuclei (Fig. 5).

In senile dementia we seldom find glia cells without fatty deposit; both the large light staining cells and the small dark cells contain fatty granules. In the former the fat often stretches far away in a long tail of granules, or lies arranged in various gradations; though mostly so that the nucleus itself for the most part lies free. In the small dark nuclei we find often only single fatty granules. Usually the glia cells which lie in the neighborhood of the vessels are particularly rich in fat. In the vessels themselves there is found an extraordinarily large amount of fatty granules. They lie in large part in the adventitial cells; also in single intimal cells the whole cell is filled with fatty granules. In the adventitial space are found fatty granules which do not lie in the cellular body.

Progressive Paralysis.

Also in paralysis there is found a large amount of fat in the cerebral cortex, yet not so much as in senile dementia. Here and there we find cells in which the fat fills up the entire cell as in senile dementia. In general, however, a different arrangement is to be seen. The fatty granules seem to be much more strewn over the entire cell, while at no point do we reach such dense accumulations as are the rule in senile dementia (Fig. 1 a).

We have not seen fat in the prolongations of the cell, especially in the axis cylinder prolongations. Also a complete degeneration of the entire cell, as is frequent with senile dementia, seldom occurs. In our cases we have to deal with instances of paralysis predominantly running their course rapidly yet not wholly demented.

Also in glia cells fat is frequently found, but not in such an arrangement as is the rule in senile dementia. Occasional "*Stäbchen*" cells contain fat in the polar prolongations. In many cases the blood vessels contain large accumulations of fat, especially in the adventitial cells.

Epilepsy.

We have examined nineteen cases, among which were found various types of degeneration. We owe a part of our examined material to the Institution for Epileptics at Stettin, Würtemberg. For the most part the patients were epileptics in youth, and died at an early age. For the examination cases were selected in which no microscopic disease foci were found. Our material also included cases of status epilepticus from the Psychiatric Clinic at Munich.

Special interest is due to Case P., sixteen years old, who died in status epilepticus after he had suffered in earliest youth from teething convulsions. On the day before his death he had his first attack of status epilepticus, but previously had shown no epileptic symptoms. In the majority of the ganglion cells there is found a large quantity of fat in large grains which mostly lie in small groups at various points of the ganglion cells. In general the large cells are considerably more fatty than the small ones.

The glia cells of the first cortical stratum are richly endowed with fat. In the rest of the cortex the fat in the glia cells is very sparse.

Mostly one sees only single grains, especially in the satellite cellular body, while, on the other hand, the glia nuclei in the marrow contain a rich amount of fat. We see single cells which are surrounded with a large number of fine fatty granules. Here and there are found single adventitial cells surrounded with fat; most of the vessels exhibit less fatty deposit. In the lymph spaces of the large vessels of the marrow are found cells richly endowed with fat, but specifically fatty granular cells are not to be seen. Very rich is the fatty content of a second case of epilepsy which died in status epilepticus. We find here in all ganglion cells fatty granules remarkable by reason of their exceptional size. As a rule they lie singly, but dispersed over the entire cell, even far up to the polar prolongations. This was not observed in the other case.

In place of the small granules one sees small and large accumulations of fat, until a greater part of the cell is apparently filled with granules. The profound fatty degeneration of the ganglion cells corresponds also to considerable fatty deposit in the glia cells. A peculiarly beautiful appearance is made by the glia cells of the superficial layer which shows prolongations filled with very numerous fatty granules, stretching out from the cellular body in various directions. The abundance of glia cells in the first cortical stratum is noteworthy.

In the ganglion cells of the deeper layers and also in the glia cells we find fatty granules for the most part lying isolated. Their relation to the glia nucleus corresponds to the shriveled protoplasmic body. Very numerous are the fatty granules in the glia cells of the white substance where they often form rather coarse little heaps around the nucleus or are arranged in the form of a comet's tail springing away only in one direction. Finally the fatty content of the vessels is extraordinarily rich. All vessels are extremely prominent by reason of the large number of fatty granules deposited in them. Individual vessels are imbedded in the fatty granules. The lymph sheaths contain numerous cells with fine grains of deposited fat and also the adventitial as well as the intima cells seem to be completely filled with fat.

The fatty degeneration in the case of another young epileptic who presented an advanced grade of epileptic dementia and who died in status epilepticus, while not in every way similar, is of a somewhat similar character. In individual ganglion cells the fatty degeneration reached a very high degree. Here too the fatty granules reach out far in all dendrites. There were found cells in which the dendrites seemed distended far from the nucleus through the deposit of fatty granules. Also here again our attention is struck by numerous cells in the first cortical stratum which send out dendrites in all directions richly endowed with fatty granules.

The glia cells of the whole cortex likewise show prolongations for some distance from the cell and filled with fatty granules. It often proved to be very difficult to differentiate the fatty granules belonging to the ganglion cells from those belonging to the glia cells. The glia cells of the white substance exhibit the same conditions as those in the preceding case. The vessels in general contain but little fat.

In still another case, an epileptic who died when eight years old of status epilepticus, the fatty deposits were of a peculiar nature. The ganglion cells here, through their form and the dark, often triangular nucleus, produce the impression of far advanced sclerosis. Many show very rich deposits stained red which, however, for the most part have not assumed the form of round granules. In general they are irregular, sometimes handle-formed, and often one end lies in the cell while the other projects out of it. In individual cells there occur also peculiar sausage-shaped formations. In many cases these reached a considerable length; they lie usually at the base of the cell and at the side of the nucleus. In between the usual fatty granules are often found. In general the fatty degeneration reaches a very advanced degree. Here again we see in many cells single fatty granules in the polar prolongations far from the cell nucleus.

In the glia cells of the deeper layers of the cortex are found very often fine fatty granules. Also in the glia cells of the white substance the fatty granules show an arrangement similar to that in the cases of epilepsy already described.

In the adventitial cells of the cortex and the marrow there are for the most part extremely coarse fatty granules in the vessels. In the adventitial lymph sheaths of the vessels there are rich and numerous granular cells.

Cerebral Syphilis.

In the two cases of cerebral syphilis that we have studied, one was a case of meningitis with peculiar participation of the spinal cord and the base, while over the hemispheres was to be demonstrated only a high grade thickening of the pia, with but very few infiltration cells. We were able to demonstrate also in the adventitial sheaths of the vessels of the superficial cortical stratum sparse lymphocytes lying wholly isolated. Before his death the patient presented the picture of a profound dementia. In almost all ganglion cells of the cerebral cortex there is a rich amount of fat. In most of the cells it lies at the base, so that the nucleus appears somewhat displaced upward, frequently mounts upward by the side of the nucleus, and still further occupies the region above the nucleus. The arrangement is very similar to that found in senile dementia. The fatty granules are large and give one the impression that the fatty accumulation pervades the protoplasm of the cell in its total depth. Around the glia nuclei lies abundant fat in very irregular form, and then again it is found deposited in single tails. Also in the glia of the white substance fat is found in like manner. The vessels are in part without fatty deposit. In other vessels fat is found in the adventitial cells, though not in excessive quantity.

The second case of cerebral syphilis is that of a woman in whom in the basal ganglia small and large focal softenings are found. In the hemispheres are shown no infiltration whatever and no alterations of the vessels. A long time before her death the patient presented the picture of profound dementia. The cortex here is shown to be extraordinarily rich in fat. The ganglion cells are for the most part full of coarse fatty granules, so that only a little of the remaining cellular body is to be seen.

The fatty granules appear to penetrate the entire cell and the whole cellular body. The large and small ganglion cells show the same degree of fatty degeneration. The picture of the glia is here peculiar. Around the small dark glia nuclei lie frequently little heaps of coarse fatty granules equally distributed in all directions. As a rule, elsewhere one does not see such an arrangement of the fat, since the fatty granules are mostly situated at the side of the nucleus, or lie arranged in some prolongations. In contrast with the extraordinarily rich fatty deposits in the nerve and glia cells there is found but little fat in the vessels.

Alcoholism and Alcoholic Psychoses.

In M., a case of chronic alcoholism, the fatty degeneration of the ganglion cells is rather irregular. We find in all regions cells which contain but very little fat, and the small ganglion cells contain quite a large amount. In the large pyramids the fat lies at the base or at the side of the nucleus. Now and then we meet with a cell which contains a great quantity of fat. On the other hand, there is found an exceptionally high grade of fatty deposit in the glia. In all strata of the cerebral cortex are seen glia cells which are surrounded by coarse fatty granules. Sometimes the fat lies round about the nucleus; then again it is arranged in grape-cluster form at one side of the nucleus; and again it lies together in one or several tails running far on.

Not infrequently in several of them these fat-besieged cells lie side by side, so that large clusters of fatty granules lie together. Through the extraordinary accumulation of fat in the glia of the cerebral cortex the picture receives a wholly peculiar imprint. In the vessels there is quite a large amount of fat in the adventitial cells. "*Körnchen*" cells are nowhere to be seen.

In G. G., a case of alcoholic hallucinosis, aged 35 years, who died of dysentery, no fat was found in the ganglion cells of the frontal lobe or in the glia. In the temporal lobes, however, a moderate amount of fat was found in the ganglion cells, but none in the glia or adventitial cells of the blood vessels. It is important to note that no fat was found in the glia of the superficial layers of the cortex in any region.

In the other cases of chronic alcoholism and alcoholic dementia, varying in age from forty-eight to sixty-seven years, usually there was found a considerable amount of fatty pigment in ganglion cells, glia, and blood vessels, also in the superficial glia which resembled the fatty degeneration found in senile dementia, except that it was not of such a high grade. In no case was the peculiar fatty degeneration of the axis cylinders of the ganglion cell which we have described in dementia præcox to be found.   In the cases below sixty years of age the fatty deposits were greater than those found in normal brains of that age.   So we must conclude that alcohol has a marked influence in the amount of fatty deposits in the cortex, when its use extends over a long period in the individual's life.

The Occurrence of the Lipoid Substance in the Cerebral Cortex of Animals.

We have had the opportunity to examine many examples of the cerebral cortex of various animals.   In rabbits fat was not present in the cerebral cortex in any noteworthy amount.   Just as little was to be found in an ape, although it had died of a general anemia on account of long confinement.   On the other hand, a slight amount was found in the ganglion cells and the glia of a calf.   A somewhat greater amount was found in the ganglion cells and in the glia of a full grown cow.   But fat was found in the cerebral cortex of a deer which we had used for purposes of comparison in order to examine both domestic animals and those living in the open.   The age of the deer could not be determined.

It is very unusual to find a large amount of fat in the ganglion cells and in the glia cells, and even when small amounts are found in ganglion cells and glia no fat is found in the adventitial cells. The results with a swine and a sheep were similar to those with the brain of the calf.   In both of these cases the animals were young.   In all these animals the deposit of fatty granules was collected at the base of the cells and was of small extent.   When the fatty granules lay in the glia, as a rule only one or two were found in the cells.   It could not be determined which cortical layers were rich or poor in fat.   A much greater amount of fat was found in the brain of a horse which had reached the age of twenty-six years. Especially in the vessels there was a large amount of fatty pigment; it lay in the adventitial cells, and around the vessels were many glia cells containing a large amount of fat.

DISCUSSION.

The most important results of this study seem to be the demonstration of definite changes in the cortex in dementia præcox, which have hitherto escaped notice and have not been previously described.

Kraepelin[6] gives a reproduction of nerve cells from dementia præcox, with a high grade of fatty degeneration, but does not give credit to anyone in particular for these findings, although he evidently considers them of importance. The sclerotic changes in the nerve cells, as seen by the Nissl method, are also evidence of a profound pathologic change which cannot be ascribed to the age of the patient, for these cases are all from the early years, the third to the fourth decade.

We have been able to demonstrate these cell changes in practically all cases of dementia præcox, especially in those dying of an intercurrent disease; notably, tuberculosis. But in the case of patients affected with other psychoses, dying of tuberculosis, notably general paralysis, these changes are not to be found. It would therefore seem that tuberculosis alone could not account for these changes. What then is the significance of these changes? And the answer to that question cannot, as yet, be fully determined, although many interesting hypotheses could be discussed.

We cannot assume that the fatty degeneration or fatty deposits in the nerve cells, glia, and blood vessels are pathognomonic for dementia præcox, as we find similar changes in other psychoses. such as alcoholic insanity and senile dementia. But in no other psychosis do we find these changes at such an early age in the patient and never so pronounced. The type shown (Figs. 1 and 2) with the axis cylinder at its point of origin filled with fat appears to be found more commonly in dementia præcox. Occasionally one finds an indication of such a change in senile dementia (Fig. 8), but it is not so pronounced. Whole cells completely degenerated are found in both conditions. The fatty deposits in the superficial glia are about the same in both conditions, and are also found in alcoholic cases in advanced years of life. But the degeneration of

---

[6]Kraepelin, E., Psychiatrie. Ein Lehrbuch für Studierende und Ärzte. 8th edition, Leipzig, 1913. iii. pt. 2, 903, figure 191.

the axis cylinder and the fatty deposits in the superficial glia have not been observed in normal brains up to the sixth decade of life.

It is fair then to assume that these changes while not pathognomonic of dementia præcox are at least characteristic changes for that disease. The next question that arises is naturally: What is the cause of this fatty deposit in the cortex? And here the biochemical metabolism seems to play a very important part. With a better understanding of the function of the neuroglia and its relation to the nerve cells, this question does not seem so difficult. The classical work of Alzheimer upon the function of the glia gives us a better idea of just how the glia is concerned in the nourishment of the nerve cells, as well as its function in the removal of the products of metabolism in nervous tissue. It is fair to assume that the fatty deposits occur as a product of faulty metabolism, and also that it may be the end product of the activity of the cell, which under normal conditions is removed or neutralized by certain biochemical agents.

When this mechanism is at fault these end products are not disposed of, and hence accumulate in the cell bodies, glia, or walls of the blood vessels. The fact that the glia cells forming the superficial layer are loaded with fatty deposits would seem to substantiate this view. For the glia here usually show a marked reaction to any changes in the cortex, and it is possible to suppose that in the attempt to dispose of this product of metabolism they become the bearers of the fat, carrying it to the surface where an outlet can be obtained.

Without more experimental data one cannot hazard an opinion as to the nature and source of this product.

The findings of these changes, especially in dementia præcox, it is to be hoped will at least furnish the basis for further work in the field of metabolism, particularly that concerned with the glands of internal secretion. ·

In establishing these histopathological changes in dementia præcox, we believe that a foundation has been laid for considering dementia præcox an organic brain disease, and it is to be hoped that this work will stimulate investigation into the origin and nature of the lipoid pigment found in the cortex in this disease. The relation of the glands of internal secretion to the pathological process of dementia præcox offers the most fruitful field for investigation at the present time.

In conclusion I wish to express my sincere thanks and deep appreciation to Professor Alois Alzheimer for his valuable assistance to me in the laboratory of the Psychiatric Clinic in Munich. It was at his suggestion that this work was undertaken, and the greater part of it was done under his helpful direction. Without his cooperation and assistance this paper would not have been completed. I am also indebted to Dr. F. G. Scammell, County Physician at Trenton, for furnishing me with a greater part of the normal material, through accident and criminal cases coming under his jurisdiction; and to Dr. F. S. Hammond, Pathologist of the New Jersey State Hospital at Trenton, for his valuable assistance in the preparation of material, and in preparing the colored lumière plates from which the illustrations are made. The drawings were made by Dr. Peter F. Mallon from the colored photomicrograph.

CONCLUSIONS.

From the data provided by our investigations we may conclude that in all pathologic processes of the cortex which end in dementia and death, the fatty degeneration of the elements of the cortex plays a not unimportant part. The characteristic change for most of the psychoses is found in a great increase in amount of fatty deposits when compared to normal individuals of the same age. In some processes such as senile dementia and dementia præcox the fatty substance appears to fill completely the cell body, and these cells have apparently lost all their functioning power. It is not common to find the fatty deposits in the processes of the ganglion cells except in dementia præcox, and to a limited extent in senile dementia.

In other cases the pathological variety of the fatty deposits in the ganglion cells is seen to be diffused over the whole ganglion cell. We were able especially to observe this in infectious psychoses, in general paralysis, and in epilepsy. The so called central neuritis assumes a peculiar attitude in that it plainly leads swiftly to an acute fatty degeneration of the ganglion cells, in which there exists an inclination of the fatty granules to flow together into large masses. Frequently the fatty degeneration of the ganglion cells appears to be connected with the sclerosis of the cells, especially when it is a matter of slowly progressing alterations of degeneration. The behavior of the glia is not wholly uniform in the various disease processes. In chronic disease processes we often find that the extent of the fatty accumulations in the ganglion cells does not correspond to an equal increase in the glia cells, while the otherwise acutely degenerative alteration in the nuclei of the glia is noticeable. In acute processes we see regularly an equal accumulation of the fat in both species of cell. The conditions of the cells in the vascular wall are wholly similar to those of the glia cells. We must therefore assume that in chronic diseases the fatty substance has been carried out of the glia and the vascular walls while it has been retained longer in the ganglion cells. Among all the disease processes amaurotic idiocy assumes a peculiar position. We have observed that in addition to the fatty materials of the scarlet fat stain, still other fatty materials, lipoid in character, have made their appearance. While the study of the fatty deposit in the cerebral cortex offers some points for a differential diagnosis, yet it is not adopted in all cases, since the distinction in individual disease processes is not always characteristic. From the preceding examination, however, in many cases there result important findings which briefly we summarize as follows:

1. In all degenerative alterations in the cerebral cortex the mass of the lipoid materials in the ganglion cells in comparison with that in healthy individuals of equal age is found to be considerably augmented. In the alteration of the lipoid materials in the ganglion cells two types in general may be distinguished: (a) An augmentation of the lipoid materials in the ganglion cells, in places where normally a small amount of fat is found. (b) An augmentation of the lipoid materials over the entire cell.

2. The first type we find also characteristic in senile dementia. The second type occurs in acute infectious psychoses, general paralysis, and well advanced epilepsy.

3. While the advanced lipoid degeneration of the ganglion cells in senile dementia has already been described in many ways, it has

appeared from our investigations that also in the young chronic cases of dementia præcox far-reaching fatty degeneration of the ganglion cells, especially in the second and third cortical strata, likewise occurs. These findings should constitute an important contribution to the pathological anatomy of dementia præcox.

4. The so called central neuritis represents a peculiar disease process according to the appearance of the fatty degeneration, since this fatty degeneration reaches a very advanced degree, and also in so far as it deviates from other disease processes in that here there comes out very distinctly in the picture an inclination of the fatty granules to flow together.

5. Amaurotic idiocy also represents a particular disease process in respect to the lipoid degeneration, since here in addition to otherwise distributed scarlet stain lipoid materials, still other specific lipoid materials make their appearance.

EXPLANATION OF PLATES.

The drawings were made with an Eddinger drawing apparatus from colored photomicrographs of the original sections. The fatty material is stained red by Scharlach R and Ehrlich's hematoxylin is used as a counterstain for the cell body and nucleus.

> ax. axis cylinder of cell.
> nl. nucleolus.
> gl. neuroglia.
> g. ganglion cell.
> adv. adventitial cells.

PLATE 55.

FIG. 1. Three ganglion cells from the 3d layer of the cerebral cortex of the frontal lobe, from a case of dementia præcox 27 years of age(all three of these cells were found in one fluid ½₂ oil immersion). a. Entire cell body filled with fat, only nucleus and nucleolus free. b. and c. Characteristic fatty deposit in base of cell and in axis cylinder which is much swollen and affected to the point of myelinization.

FIG. 2. A medium sized ganglion cell from third layer of cortex in another case of dementia præcox, 30 years of age, with fatty axis cylinder and base of cell filled with fat. In normal brains of this age practically no fat is found in cells of this type, shown in Figs. 1 and 2.

FIG. 3. Giant Betz cell from the motor cortex in the case of senile dementia. aged 69. A considerable amount of fat in the base of the cell. In normal brains these large motor cells contain some fat, but never as much as shown in this cell.

Fig. 4. Section from the superficial layer of the cortex, showing neuroglia cells filled with fat, from a case of dementia præcox, aged 30. Normal brains at this age show absolutely no fat in the superficial neuroglia. The amount of fat is only found in senile dementia.

PLATE 56.

Fig. 5. A group of ganglion (g.) and neuroglia (gl.) cells from a case of senile dementia, showing the usual form of fatty degeneration in this disease. The cell body of the ganglion cell is filled with fat which surrounds the nucleus, but the axis cylinder is rarely affected. The glia cells contain more fat than is the case in dementia præcox (found in one field $\frac{1}{12}$ oil immersion). The free fat outside of the cell probably belongs to the neuroglia cells.

Fig. 6. Giant Betz cells from motor cortex from central neuritis. The nucleolus is eccentric and stains heavily; the center of the cell has lost its staining properties, surrounded by fatty material. Typical "axonal reaction."

Fig. 7. Blood vessel from the white substance of the brain from a case of dementia præcox, with considerable amount of fat in adventitial (adv.) cells. The glia cells lie close to the wall of the capillary and are filled with fat. These are not "*Körnchen*" cells, but ordinary glia cells, apparently transferring the fat to the blood vessels. This is a common finding in dementia præcox and is found only to great extent in senile dementia.

Fig. 8. Section of cortex from case of senile dementia, showing fatty degeneration of ganglion cells and glia. In one ganglion cell the axis cylinder (ax.) has some fat, but not to the extent as seen in dementia præcox. (Compare Fig. 1.)

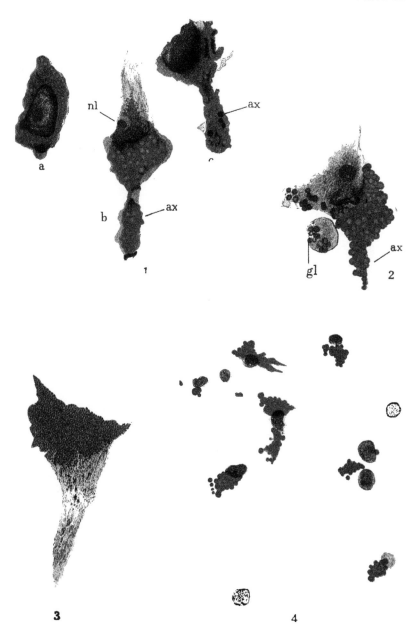

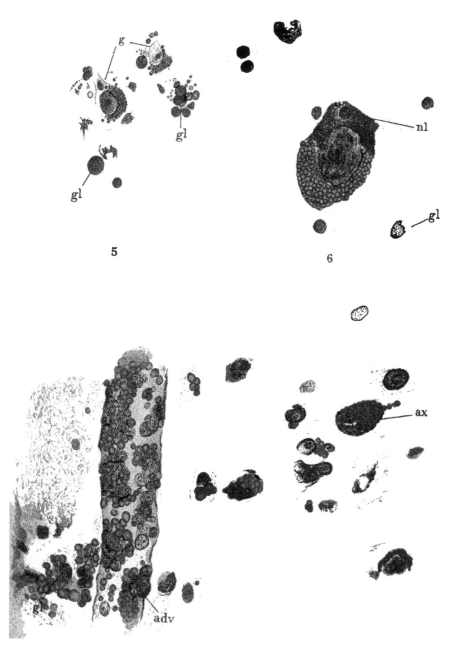

CHEMOPATHOLOGICAL STUDIES WITH COMPOUNDS OF ARSENIC.

I. Types of the Arsenic Kidney.

By LOUISE PEARCE, M.D., and WADE H. BROWN, M.D.

(From the Laboratories of The Rockefeller Institute for Medical Research.)

Plates 57 to 60.

(Received for publication, May 15, 1915.)

During the past year, while investigating the pathological action of a large number of organic preparations of arsenic,[1] it was noted that the gross appearance of many of the kidneys resulting from the administration of these compounds did not conform to the type usually described as the arsenic kidney. The wide variations in the type of kidney observed, and at the same time the constancy of the type produced by a given compound in a given animal species, were so marked as to suggest that the classical picture was not a constant one for all compounds of arsenic. To obtain a more comprehensive idea of these variations and some conception of their cause, a series of experiments was carried out with a number of well known arsenicals, and the gross pathological results obtained serve as the basis for this report. The microscopical studies are reported in a separate paper.

EXPERIMENTAL.

Dogs were used for these experiments, because the dog kidney is large enough to show the gross pathological picture quite well. This series of experiments was supplemented, however, by similar experiments carried out with guinea pigs, intended principally as a

[1] For several years, The Rockefeller Institute for Medical Research has been interested in the chemotherapy of infections. The work now in progress is being carried out jointly by a chemical division and a biological division. The chemical work is being done by Dr. W. A. Jacobs and Dr. Michael Heidelberger. The series of studies contemplated under the title "Chemopathological Studies with Compounds of Arsenic" are an outgrowth of the work on chemotherapy.

control upon the histological changes produced, since the dogs obtainable for this work so frequently showed spontaneous kidney lesions.[2] It may be stated here that the gross appearances presented by the guinea pig kidneys were essentially the same as those of the dogs.

TABLE I.

Dog.	Drug.	Dose per kilo.	Strength of solution	Route of administration.	Remarks.					
					1 day.	2 days.	3 days.	4 days.	5 days.	6 days.
A	Arsenious acid	*mg.* 7	1:100	Subcutaneous	Dead					
B	Arsenious acid	5	1:333	Intraperitoneal	"					
C	Atoxyl	40	1:50	Subcutaneous	Slightly ill	Well	Well	Weak	Better	Dead
D	"	20	1:10	Subcutaneous	Well	"	40 mg. Ill	Killed		
E	"	50	1:50	Intraperitoneal	Ill; killed					
F	"	20	1:75	Intraperitoneal	"	"				
G	Arsacetin	300	1:10	Intraperitoneal	Ill	Better	Better	Well 600 mg.	Dead	
H	"	400	1:25	Intravenous	Ill; killed					
I	"	600	1:10	Intravenous	"	"				
J	"	400	1:40	Intravenous	Ill	Dead				
K	Arsenophenylglycine	225	1:10	Intraperitoneal	"	Better	3 days Well 225 mg.	19 days Well 250 mg.	20 days Well; killed.	
L	Arsenophenylglycine	500	1:10	Intravenous	Dead					
M	Arsenophenylglycine	300	1:20	Intravenous	"					
N	Arsenophenylglycine	200	1:20	Intravenous	Well; killed					
O	Arsenophenylglycine	500	1:15	Intravenous	Dead					
P	Salvarsan	100	1:400	Intraperitoneal	Ill	Ill; killed				
Q	"	100	1:333	Intravenous	Dead					
R	Neosalvarsan	150	1:30	Intravenous	"					
S	Galyl	100	1:100	Intravenous	"					

[2] Dayton, H., *Jour. Med. Research*, 1914, xxxi, 177.

TABLE II.

Structural Formulæ of Arsenicals.

*Preparation No. 606 in Ehrlich's series designated as Salvarsan is the dihydrochloride of this compound. The formula here given, however, is that most frequently encountered in the literature and is the substance derived from the dihydrochloride by the addition of alkali.

Solutions of arsenious acid, atoxyl, arsacetin, arsenophenylglycine, salvarsan, neosalvarsan, and galyl were injected without an anesthetic into young adult dogs. The injections were made intravenously, intraperitoneally, or subcutaneously; no difference was

noted in the results obtained by using these different routes.   Doses
of various size were used, as shown in Tables I and II, but in the
majority of instances amounts approximating the lethal dose were
injected, in order to accentuate the lesion produced by the particular compound.   In a few instances much smaller amounts were
used, in order to determine the point of greatest susceptibility and
the characteristics of the essential lesions.   The animals were usually killed at the end of twenty-four hours, if they were not already
dead, so that we might obtain lesions comparable in time (length
of survival) as well as dose.   In each instance the dog was lightly
anesthetized with chloroform and bled to death from the carotid
artery.

RESULTS.

Arsenious Acid.

Injection of arsenious acid into dogs produces a uniformly red
kidney (Fig. 1).   The organ is enlarged, is purplish red in color,
and on section drips blood.   The cut surface is reddened from the
outermost edge to the pelvis, through the cortex, boundary zone, and
medulla.   All the zones or territories of the kidney are practically
uniformly involved.   The kidneys of dogs that survive large
doses of arsenious acid for a sufficient length of time show opaque,
yellowish white striations in the inner portion of the cortex resulting from tubular necrosis which, however, is not an essential feature of the change.   The outspoken picture is that of a diffusely
reddened kidney indicative of uniform vascular injury and is one
of the two principal types produced by arsenical compounds.

Salvarsan, Neosalvarsan, and Galyl.

The type of kidney produced by injections of these drugs falls
into the general class of red kidneys, although certain differences
relate them to the other main type, the pale kidney.   These differences consist in a relatively greater degree of tubular necrosis and
in the character of the vascular injury.   This group of kidneys, as
shown by the salvarsan and neosalvarsan kidneys (Figs. 3 and 8,
respectively), shows a considerable amount of necrosis of the cortex
which is most conspicuous in the inner half.   On the other hand,

although the kidney is reddened throughout, the congestion and hemorrhage are most prominent in the outer half of the cortex and the inner portion of the boundary zone.

Arsacetin.

Injection of arsacetin into dogs produces an exceedingly pale kidney (Fig. 2), that differs strikingly from the red type just described. The organ is enlarged, soft, and very friable. It is grayish or yellow in color; in some instances it may be pink, with slight congestion of the surface vessels. It cuts very easily and on section the cut edge is everted. The cortex bulges, is widened, and the striations are yellowish, opaque, and raised as a result of wide-spread tubular necrosis; the glomeruli may or may not be prominent. The boundary zone practically always shows some congestion and the same may be said of the medulla, but to a much less degree. The dominant feature of this type of kidney, therefore, is its paleness due to tubular necrosis, as contrasted with the redness of the kidney of arsenious acid resulting from vascular injury. Kidneys of dogs that survive large doses of arsacetin for a sufficient length of time show a greater degree of congestion or even some hemorrhage in the boundary zone. The hemorrhage may extend somewhat into the medulla and to a lesser degree into the cortex.

Arsenophenylglycine and Atoxyl.

Both of these drugs produce pale types of kidneys, but the amount and extent of vascular injury are somewhat greater than in the arsacetin kidney. The cortical necrosis of the arsenophenylglycine kidney (Fig. 4) is extreme and is quite comparable to that produced by arsacetin. On the other hand, the congestion and hemorrhage of the medulla are relatively prominent and especially so in the inner portion of the boundary zone. We may, therefore, consider the kidneys produced by these drugs as related to both the pale and the red types of kidneys.

DISCUSSION.

In studying the action of this group of arsenicals upon the kidney, the object in view has necessitated the subordination of detail to the broader conception of types of renal injury. Further, we can not now attempt a complete description and grouping of all the types of arsenic kidneys which we have observed. Our object, at present, is to lay emphasis upon the difference in the character of the renal injury produced by different compounds of arsenic as expressed in the gross appearance of the organ. For this purpose we have chosen two extreme types.

The classical red kidney resulting from injections of arsenious acid into dogs contrasts sharply with the pale kidney produced by arsacetin. The red kidney is one in which uniform vascular injury predominates; all zones of the kidney seem to be affected equally, cortex as well as medulla. The pale kidney, on the other hand, presents striking evidence of the dominance of tubular necrosis, and the reddening of the boundary zone is a relatively inconspicuous feature. With the other arsenicals this difference in the gross appearance of the kidneys is not so marked, and we may perhaps look upon the kidneys produced by salvarsan, neosalvarsan, and galyl, on the one hand, and by atoxyl and arsenophenylglycine on the other, as constituting transitions or subgroups between the two principal types. The salvarsan group of kidneys is preeminently red, but there is also a relatively great degree of tubular necrosis; the arsenophenylglycine group of kidneys is pale, yet it shows relatively marked vascular injury, especially in the boundary zone and contiguous territory.

The type of kidney produced by a given compound can not be regarded as fixed in all its details, but the size of the dose and the rate of action as measured by the length of survival of the animal are factors that exercise a distinct influence upon the final result, as may be illustrated by atoxyl. The lethal effects of atoxyl are developed slowly in the dog and it would probably require many times the minimal lethal dose to kill within twenty-four hours. The tolerable dose of this drug for dogs as given by Mesnil and Brimont[3] is 10 mg. per kilo of body weight. Dog F, given double this dose (20 mg. per kilo), showed slight symptoms of intoxication at the end of twenty-four hours and when killed the kidneys were quite pale and slightly enlarged. The cortex showed opaque striations,

[3] Mesnil, F., and Brimont, E., *Ann. de l'Inst. Pasteur*, 1908, xxii, 856.

but the striking feature was a pencil line of hemorrhage in the inner boundary zone (Fig. 6). This animal illustrates well the character and location of the essential changes produced by atoxyl. Dog E, given 50 mg. of atoxyl per kilo, was more toxic at the end of twenty-four hours and was killed. The kidneys (Fig. 7) were markedly enlarged and the opaque yellow gray surface was thickly stippled with red. On section the cortex showed alternate striations of red and opaque yellowish gray; the glomeruli were prominent. The line of red in the medulla of Dog F had given place to an intense reddening of the entire medulla, which was practically a mass of clotted blood. The massive dose of atoxyl used in this dog exaggerates the effects produced in Dog F and at the same time serves to indicate the character of changes referable to dosage. Dog C received 40 mg. of atoxyl per kilo and although this animal passed large amounts of blood in the urine, other evidence of intoxication was relatively slight. It finally died on the sixth day. The kidneys of this dog were practically identical with those of Dog D (Fig. 5) which received a dose of 20 mg. per kilo, three days later a second dose of 40 mg. per kilo, and was then killed after two days. With these two dogs the influence of time as well as dose is introduced. This group of atoxyl kidneys illustrates well the elasticity of a type but at the same time emphasizes the sharp differentiation of even a transitional group from the classical arsenious acid kidney.

As regards the factors that endow compounds of arsenic with the power of producing constant and characteristic types of renal injury, one can hardly escape the conclusion that such activities are intimately related to the chemical constitution of the compounds. However, both organic and inorganic compounds of arsenic in which the arsenic is trivalent or pentavalent are known to produce red kidneys of the arsenious acid type. On the other hand, while both trivalent and pentavalent compounds of arsenic produce pale kidneys, we know of no inorganic arsenical that acts in this way. Further, substances as closely related as atoxyl and arsacetin, or as salvarsan and arsenophenylglycine, may act quite differently. Again, compounds as different as arsenious acid and salvarsan, or as arsacetin and arsenophenylglycine, may simulate each other quite closely in their effects.

For the present, therefore, we can go no further than the conclusion that the character of the renal injury produced by compounds of arsenic is determined either directly or indirectly by the chemical constitution of the compound.

CONCLUSIONS.

1. All arsenical compounds do not produce the same type of renal injury.

2. In general, there are two broad groups of kidneys produced, the red and the pale, with a variety of subdivisions of each group depending upon modifications in the chemical constitution of the compound, dosage, and length of survival of the animal.

EXPLANATION OF PLATES.

PLATE 57.

FIG. 1. Kidney of arsenious acid (natural size). Dog A. Red type of kidney. Uniform congestion and hemorrhage throughout cortex and medulla. Tubular necrosis in inner half of cortex.

FIG. 2. Kidney of arsacetin (natural size). Dog I. Pale type of kidney. Necrosis of cortex. Congestion of boundary zone.

PLATE 58.

FIG. 3. Kidney of salvarsan (natural size). Dog Q. Reddening of kidney throughout, especially of outer cortex and inner portion of boundary zone. Tubular necrosis of inner half of cortex.

FIG. 4. Kidney of arsenophenylglycine (natural size). Dog L. Tubular necrosis of entire cortex. Diffuse congestion of medulla with a band of hemorrhage.

PLATE 59.

FIG. 5. Kidney of atoxyl (natural size). Dog D. Necrosis of cortex. Diffuse congestion of medulla with hemorrhage of boundary zone and inner half of cortex.

PLATE 60.

FIG. 6. Kidney of atoxyl (reduced). Dog F. Pencil line of congestion and hemorrhage along inner boundary zone.

FIG. 7. Kidney of atoxyl (reduced). Dog E. Necrosis of cortex. Hemorrhage of medulla.

FIG. 8. Kidney of neosalvarsan (reduced). Dog R. Necrosis of cortex. Congestion and hemorrhage of boundary zone.

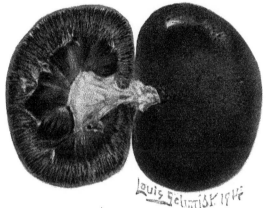

FIG. 1.

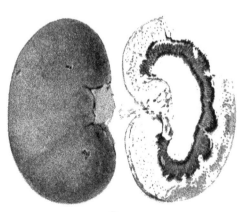

FIG. 2.

(Pearce and Brown: Types of Arsenic Kidney

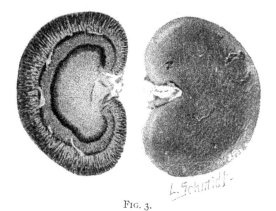

FIG. 3.

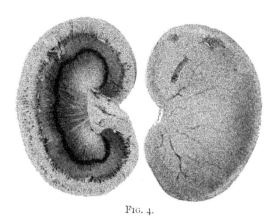

FIG. 4.

(Pearce and Brown: Types of Arsenic Kidney.)

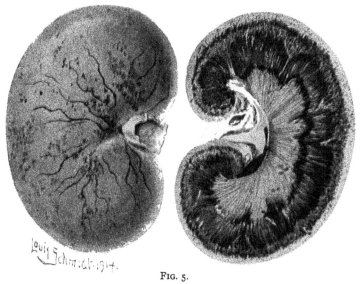

FIG. 5.

(Pearce and Brown: Types of Arsenic Kidney.)

PLATE 60.

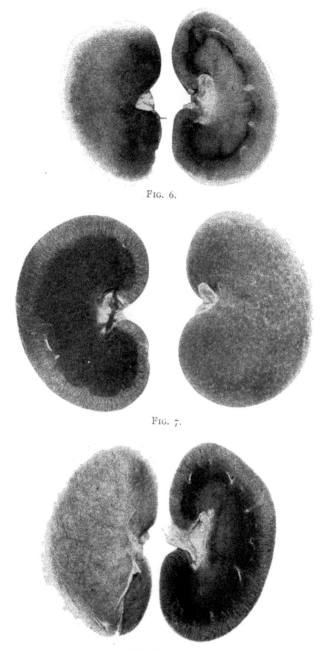

FIG. 6.

FIG. 7.

FIG. 8.

(Pearce and Brown: Types of Arsenic Kidney.)

CHEMOPATHOLOGICAL STUDIES WITH COMPOUNDS OF ARSENIC.

II. Histological Changes in Arsenic Kidneys.

By LOUISE PEARCE, M.D., and WADE H. BROWN, M.D.

(From the Laboratories of The Rockefeller Institute for Medical Research.)

Plates 61 to 67.

(Received for publication, June 24, 1915.)

The idea that arsenicals as a class produce a vascular type of injury in the kidney has gained general acceptance, and workers in experimental nephritis have substantiated this idea for such compounds as have been employed to produce so called arsenical nephritis. However, as we have shown in the preceding paper,[1] all compounds of arsenic do not produce the same type of injury, and from the gross appearance of the kidneys of dogs poisoned with various well known arsenicals, we are able to recognize two widely separated types of kidney, the red and the pale, with many transitional varieties or subgroups, each being more or less characteristic for a given compound. In like manner, we have found that the histological changes in these given types are equally different and characteristic.

Experimental.

Tissues from the kidneys described and illustrated in the preceding paper[2] were fixed in Zenker's fluid, sectioned in paraffin, and stained with hematoxylin and eosin. The histological changes, therefore, may be related directly to the descriptions and illustrations in this paper.

Red Kidneys.

Arsenious Acid.—The most characteristic acute lesions produced by arsenious acid consist in a uniform dilatation and congestion of

[1] Pearce, L., and Brown, W. H., *Jour. Exper. Med.*, 1915, xxii, 517.
[2] Pearce and Brown, *loc. cit.*

the blood vessels of the kidney with an escape of blood into the interstitial tissues throughout the cortex and medulla (Fig. 1). The glomeruli are swollen, and the tuft completely fills the capsular space. Occasionally there is desquamation of swollen, disintegrating cells of the capsular epithelium and a slight albuminous precipitate in the capsular space. The tuft vessels are widely dilated and filled with blood; some contain hyaline plugs. The changes in the tubular epithelium are less marked, the cells are greatly swollen and vacuolated, often occluding the lumen of the tubule and are frequently stripped up from the basement membrane. Parenchymatous and fatty degeneration are marked, but necrosis of tubular epithelium is relatively slight. Some of the tubules contain an albuminous precipitate; others, especially in the boundary zone, contain blood.

Salvarsan.—While vascular injuries dominate the histological changes produced by poisoning with salvarsan, the lesions differ in essential respects from those produced by arsenious acid. Throughout the cortex and medulla the vessels are dilated and congested, but patches of hemorrhage are more numerous in the outer cortex and are especially prominent in the boundary zone. The glomeruli are all large, with the tuft practically filling the capsular space; the tuft capillaries are widely dilated and filled with blood; and there is a slight accumulation of polymorphonuclear leucocytes in the capillaries. The cells of the outer portion of many of the tufts are markedly swollen, granular, and pale-staining, and many of these swollen cells appear to be desquamating into the capsular space (Fig. 4). Hyaline thrombi are numerous (Fig. 3). Occasionally there is a slight amount of albuminous precipitate in the capsular space. The changes in the tubules are relatively more pronounced than with arsenious acid. The epithelium of the tubules shows an extreme degree of parenchymatous and fatty degeneration with definite necrosis. The cells are very granular and ragged with frayed edges, and many are actually disintegrating (Figs. 3 and 4). In many areas there is a protrusion of swollen tubular epithelium into the glomerular space. Practically all the tubules contain albuminous precipitate and some of the cortical and many of the medullary tubules contain blood (Fig. 3). Of especial interest is a fairly extensive and regular edema of the labyrinth, particularly of the lower two-thirds of the cortex extending but little into the medulla (Fig. 4).

Neosalvarsan.—The histological changes produced in the kidney by neosalvarsan are still further removed from those produced by arsenious acid, both in type and degree, although they conform in general to the changes observed in other red kidneys. Here again vascular dilatation and congestion are present, and hemorrhage, as with salvarsan, is most marked in the outer cortex and boundary zone.

The glomeruli are large, but the tuft itself is small and compressed, occupying approximately only one-third to one-half of the capsular space. The remaining space is filled with an extensive albuminous precipitate (Fig. 5). The tuft vessels are irregular, many being enormously dilated and filled with blood and numerous hyaline plugs; in consequence, other vessels are completely collapsed. There is an increase in the number of polymorphonuclear leucocytes in the capillaries of the tuft. The injury to the tubular epithelium is distinctly more prominent than with either arsenious acid or salvarsan. The epithelium of the convoluted tubules. particularly of the outer half of the cortex, is almost completely necrotic, the epithelium of many of the tubules being converted into homogeneous. pink-staining masses (Fig. 5). There is less tubular necrosis in the inner cortex, but there is marked parenchymatous and fatty degeneration with an albuminous precipitate in the lumen of the tubules. A considerable number of the ascending limbs of the loops of Henle shows an extreme degree of degeneration occasionally going on to necrosis. Many of the tubules in the boundary zone and medulla contain blood. In the lower portion of the cortex and in the boundary zone there are a few irregular patches of interstitial edema.

Galyl.—In general. sections from kidneys after injection of galyl resemble those of salvarsan and neosalvarsan. but there are certain relative differences. The tuft capillaries are uniformly dilated and contain blood and irregularly distributed hyaline thrombi. There is a considerable amount of swollen cells desquamated from the capsule into the capsular space (Fig. 2). The epithelium of the convoluted tubules. especially in the outer half of the cortex. shows disintegration, desquamation. and necrosis. but there is very little necrosis *en masse*. The epithelium of the loops of Henle shows degeneration, although comparatively little necrosis. The cortical tubules contain much albuminous precipitate, and the majority of the tubules of the medulla contain blood.

Pale Kidneys.

Arsacetin.—In the group of pale kidneys, of which the arsacetin kidney is taken as the type, the relative degree of vascular and tubular injury is the reverse of that observed in the red kidney. Following the injection of arsacetin, the vessels and capillaries of the kidney are not usually dilated or congested except to a slight degree in the boundary zone, and there is no hemorrhage. The glomeruli are practically normal in appearance; the tuft fills most of the capsular space but is not swollen; the capillaries of the tuft are moderately dilated and contain some blood. There is only a slight amount of albuminous exudate. On the other hand, practically all the epithelium of the convoluted tubules is necrotic, the tubules appearing as large, swollen, granular, pink-staining masses (Fig. 6). The cells which are not actually necrotic are extremely degenerated with pyknotic and fragmenting nuclei. The tubules are choked with masses of disintegrating cells or contain an albuminous precipitate. The epithelium of the loops of Henle is markedly degenerated, the cells are swollen, ragged, and vacuolated. In some instances, when a very large dose of arsacetin is given (Dog I),[3] the wide-spread necrosis of tubular epithelium includes that of the ascending limb of the loop of Henle. The collecting tubules show a moderate amount of parenchymatous and fatty degeneration.

Arsenophenylglycine.—The injury produced in the kidney by arsenophenylglycine combines both vascular and tubular changes. The vessels and capillaries throughout the kidney are moderately dilated and congested. There is slight escape of blood, if any, from the vessels in the cortex, but in the boundary zone there is a considerable amount of hemorrhage which extends in streaks into the medulla. The glomeruli are uniformly large, the tuft filling approximately one-third to one-half the capsular space. The capillaries of the tuft are irregular in appearance; some are moderately

[3] The designations of the dogs correspond to those given in the preceding paper (Pearce and Brown, *loc. cit.*).

dilated and filled with blood, but many are completely collapsed. The epithelium of the capsule and of the adjoining portion of the connecting tubule is enormously swollen, homogeneous, and pale pink. Many cells of the capsule and connecting tubule have desquamated into the capsular space (Fig. 7), and in certain instances the mass of degenerated and disintegrating cells in the space apparently comes almost entirely from the tubule. Practically all the convoluted tubules are necrotic and appear as solid, homogeneous, pink-staining cylinders, or the tubule is filled with disintegrating hydropic cells. The epithelium of the loops of Henle in the outer cortex is similarly necrotic but in the inner portion there is less actual necrosis. Here the cells are markedly swollen and granular with many large vacuoles and pyknotic nuclei. The tubules contain an albuminous precipitate. The epithelium of the collecting tubules is also degenerated. The lumen of almost all the tubules is completely occluded by necrotic cell masses or by enormously swollen, degenerated, and desquamated cells. In the medulla hyaline casts are quite abundant and many tubules contain blood. There are a few irregular patches of interstitial edema in the inner cortex.

Atoxyl.—Following a small dose of atoxyl (Dog F), the only distinctive pathological changes consist in parenchymatous and fatty degeneration of tubular epithelium and slight congestion and hemorrhage in the boundary zone. The epithelium of the convoluted tubules stains palely, the cells are ragged and granular with irregular vacuolization, as shown in Fig. 8. There is some fragmentation of nuclei and many tubules contain albuminous precipitate. With a larger dose of atoxyl, however, (Dog E) a greater variety of changes is encountered. The vessels throughout are dilated and congested and there is some escape of blood into the interstitial tissue and tubules of the cortex. There is an extensive hemorrhage into the medulla, obscuring much of its structure. The glomeruli are swollen, but the glomerular tufts are irregular in size, some filling the capsular space while others occupy only a portion of it. Most of the tuft capillaries are dilated and filled with blood, and there is an albuminous precipitate in the capsular spaces. The tubular epithelium shows a peculiar series of changes. In general, it stains poorly; the cells of the convoluted tubules are swollen, granular.

ragged, and very hydropic; many have desquamated, choking the lumen of the tubule (Fig. 9). There is great irregularity in the preservation of the nuclei; many are fragmented or pyknotic, others are entirely gone. An occasional mitotic figure is seen. Practically all the tubules contain albuminous precipitate and there are a few hyaline casts. In the inner portion of the cortex there is an extreme degree of irregularity in the tubules, so that it is difficult to identify with certainty the various types. In this region the loops of Henle show a striking change; they are markedly enlarged, the cells are widely separated from one another, and are swollen and hyaline with pyknotic nuclei as shown in Fig. 9. Some of these cells are necrotic and desquamated. Another prominent feature of the atoxyl kidney is a profuse exudation of polymorphonuclear leucocytes which is most marked in the inner half of the cortex and the boundary zone.

An increase in the length of survival of the animal (Dogs C and D) gives an opportunity for the development of still further pathological changes in the kidney. In such cases, the hemorrhage invades the inner cortex as well as the medulla (Fig. 11). The tubular epithelium shows more marked disintegration of cells, many of which are hyaline while others are hydropic and swollen. There is an increase in the exudate in the tubules which consists of albuminous precipitate, colloid droplets, red blood cells, leucocytes, and casts, causing, in many instances, a marked compression of the tubular epithelium (Fig. 10). The number of casts is distinctly greater in these dogs of longer survival. In the medulla the tubular epithelium is almost completely desquamated, and the cells are intensely hyaline with pyknotic nuclei. There is a marked exudate between the tubules consisting of serum, fibrin, red blood cells, and a few leucocytes; no interstitial structures can be distinguished. Numerous mitotic figures are present in the epithelium of both cortex and medulla. Moreover, there is a very marked increase in the interstitial leucocytic exudate in these dogs of longer survival (Fig. 11).

DISCUSSION.

As far as we are able to determine from a pathological study of various arsenic kidneys, the idea that arsenical compounds as a class produce vascular, in contradistinction to tubular, injury must be modified.

It is quite true that with a certain group of these compounds vascular lesions predominate, causing red kidneys, of which arsenious acid is the type. Even within this group of arsenicals, however, we are able to recognize certain differences in the character of the action of the different compounds. For instance, with salvarsan and neosalvarsan, hemorrhage tends to be restricted to the outer cortex or the boundary zone; the formation of hyaline thrombi in the glomerular capillaries is more pronounced; and there is a profuse albuminous exudate in the capsular space with desquamation of epithelial cells. Interstitial edema and degeneration and necrosis of the tubular epithelium are distinctly more pronounced than with arsenious acid.

As we study kidney lesions produced by other arsenicals, however, we find, as with arsacetin, that the predominating change is one of degeneration and necrosis of tubular epithelium. These pale kidneys offer a striking contrast to the red kidneys of vascular injury. Moreover, other pale kidneys, such as those produced by such substances as arsenophenylglycine and atoxyl, in which the tubular injury is the predominant feature, may show well marked vascular injury. In these cases, the hemorrhage may be zonal in character, involving much of the medulla or even the lower portion of the cortex. The tuft capillaries may be dilated and congested with more or less albuminous precipitate in the capsular space, and there may be an exudation of serum, fibrin, red blood cells, and leucocytes into the interstitial tissues.

The histological changes in the kidneys produced by a particular arsenical compound, while elastic, are quite characteristic of the action of the compound and accord with the gross appearance of the organ. Therefore, in surveying this series of arsenic kidneys from a microscopic as well as from a gross pathological point of view, we are able to differentiate two extreme types: *i. c.*, the red and the pale. The red kidney is essentially one of vascular injury, the pale kidney is predominantly one of tubular necrosis. In addition, various transitional or subgroups exist, in which the kidney, although belonging to the red type, shows relatively a great degree of tubular injury and *vice versa.* In regard to the tubular necrosis in these kidneys, the prompt and active regeneration of the tubular epithelium following the injection of such compounds as arsacetin, arsenophenylglycine, and atoxyl, seems to preclude the possibility that the wide-spread tubular necrosis produced by these substances can be regarded as a purely anemic phenomenon. Hence, we cannot ascribe to arsenical compounds, as a class, the property of producing a purely vascular nephritis; but we must recognize the fact that arsenical compounds produce characteristic renal lesions which may be either predominantly vascular or tubular in type and that the mode of action and the character of the lesions produced are bound up with the chemical constitution of the compound.

SUMMARY.

1. We have shown that the type of renal lesion produced by compounds of arsenic varies widely: while some arsenicals produce changes in which vascular injury predominates, others.produce an equally dominant tubular injury.

2. In either of these groups the character and degree of the vascular or tubular injury produced by different compounds shows further variation, such that the lesions of different arsenicals of the same group are not identical. Each compound of arsenic that we have tested, therefore, produces a lesion-complex in the kidney that is relatively characteristic for that compound.

3. The mode and character of the action of arsenicals are dependent upon the chemical constitution of the compound.

EXPLANATION OF PLATES.

The illustrations are all from untouched photomicrographs. Magnification, × 208.

PLATE 61.

Fig. 1. Arsenious acid. Dog B. Section from the outer cortex. There is marked dilatation and congestion of vessels, including the glomerular capillaries, with hemorrhage into the interstitial tissues. The tubular epithelium is swollen and degenerated.

Fig. 2. Galyl. Dog S. Section from the outer cortex. The vessels are congested and there is a slight interstitial hemorrhage. The tuft capillaries are uniformly dilated and contain blood and there is a slight albuminous precipitate in the capsular space. Tubular degeneration and disintegration are marked.

PLATE 62.

FIG. 3. Salvarsan. Dog Q. Section from the outer cortex. Dilatation and congestion of vessels with slight hemorrhage. The glomeruli are swollen and the tuft capillaries contain blood and numerous hyaline thrombi. There is an albuminous precipitate in the capsular space. Tubular epithelium shows degeneration and slight necrosis.

FIG. 4. Salvarsan. Dog Q. Section from the inner cortex. Marked interstitial edema. Partial disintegration of glomerular tuft and accumulation of epithelial cells and cell detritus in the capsular space. Degeneration and necrosis of tubular epithelium.

PLATE 63.

FIG. 5. Neosalvarsan. Dog R. Section from the outer cortex. The glomeruli are swollen; the tuft is compressed; the glomerular capillaries are congested and contain numerous hyaline thrombi. There is an abundant albuminous precipitate in the capsular space. Many tubules show a massive necrosis, others show degeneration and disintegration of epithelial cells with pyknotic nuclei.

PLATE 64.

FIG. 6. Arsacetin. Dog I. Section from the midcortex. Extensive necrosis *en masse* of the epithelium of the convoluted tubules with marked degeneration and slight necrosis of the loops of Henle. Vessels and glomeruli normal.

PLATE 65.

FIG. 7. Arsenophenylglycine. Dog L. Section from the outer cortex. The glomeruli are large, the tufts are compressed, and the glomerular vessels are partially collapsed. The capsular epithelium is swollen and desquamated and the capsular space is filled with necrotic cellular debris. Degeneration and necrosis of tubular epithelium are marked.

PLATE 66.

FIG. 8. Atoxyl. Dog F. Section from the outer cortex. The vessels and glomeruli are normal, except for a slight albuminous precipitate in the capsular space. The epithelium of the convoluted tubules is extremely ragged and degenerated with slight necrosis.

FIG. 9. Atoxyl. Dog E. Section from the midcortex. The glomeruli are slightly enlarged, the capillaries of the tuft are slightly congested, and the capsular space contains an albuminous precipitate. The epithelium of the convoluted tubules is swollen, granular, and hydropic, and in places shows necrosis. The loops of Henle are markedly dilated; the cells are hyaline with pyknotic nuclei and are widely separated from one another. Many cells are desquamated. There is a diffuse hemorrhage and polymorphonuclear exudate into the interstitial tissues and occasionally into the tubules.

PLATE 67.

FIG. 10. Atoxyl. Dog D. Section from the outer cortex. The glomeruli are enlarged. The tuft is slightly compressed and the capsular space contains an abundant albuminous precipitate. The epithelium of the convoluted tubules is extremely ragged and hydropic and in some areas the cells are necrotic and desquamated. Other tubules are filled with a granular precipitate, hyaline droplets, and an occasional cast, and the epithelium of these tubules is compressed.

FIG. 11. Atoxyl. Dog D. Section from the inner cortex. The glomeruli are as in Fig. 10. The tubules throughout are necrotic, and filled with cellular detritus or granular casts. There is a marked interstitial hemorrhage with an exudate of polymorphonuclear leucocytes.

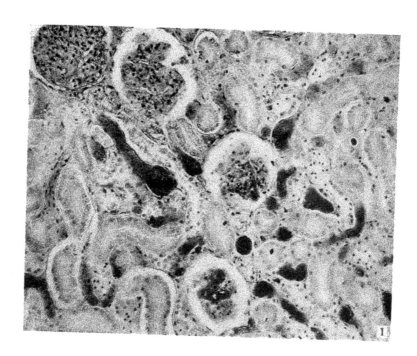

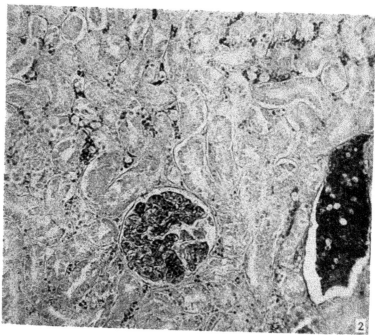

(Pearce and Brown: Changes in Arsenic Kidneys.)

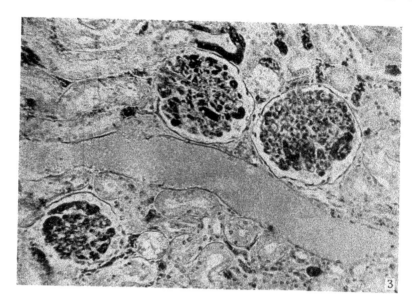

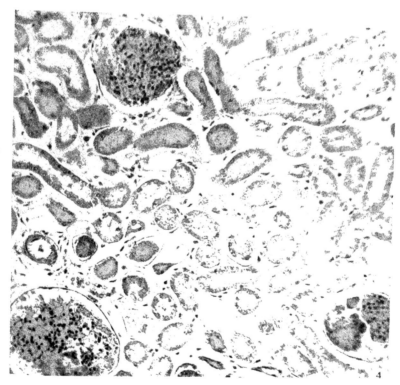

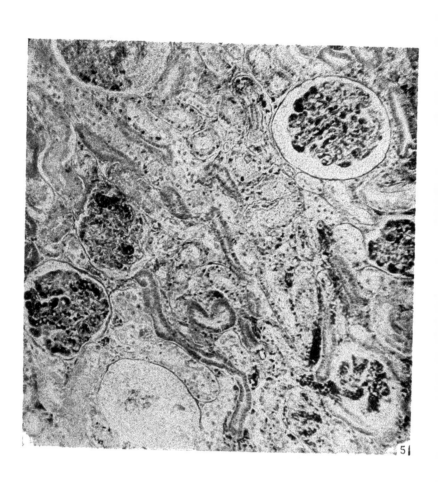

5

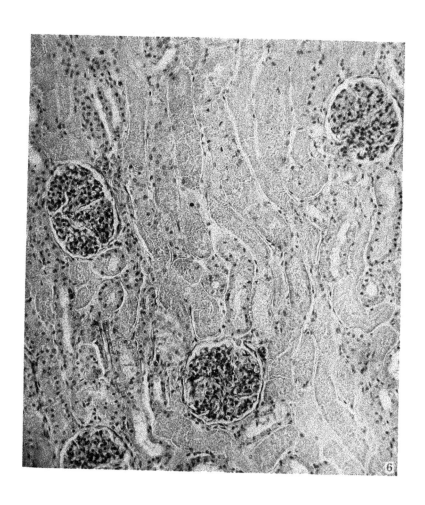

(Pearce and Brown: Changes   Arsenic Kidneys.)

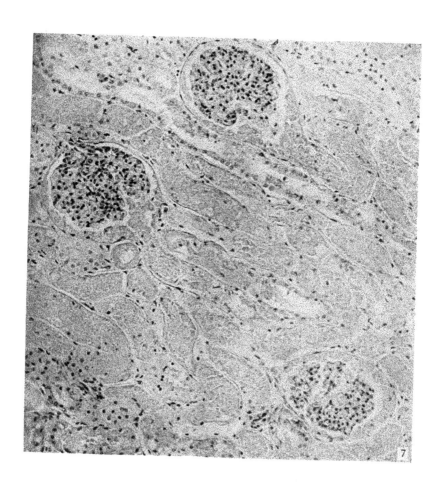

7

(Pearce and Brown: Changes in Arsenic Kidneys

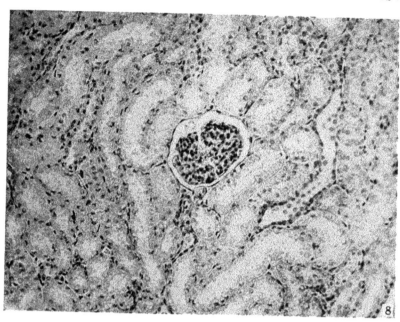

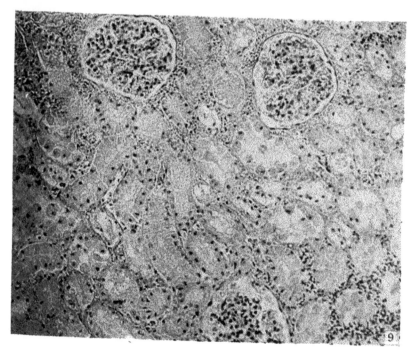

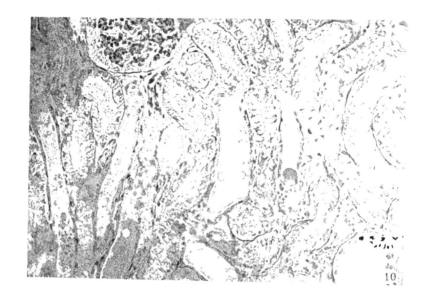

10

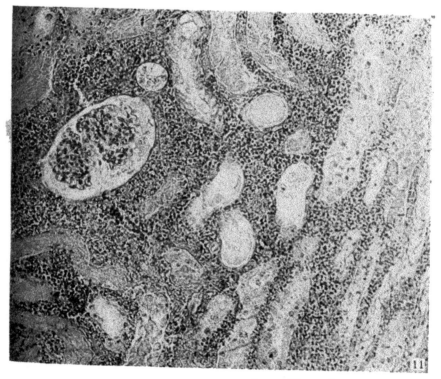

11

CHEMOPATHOLOGICAL STUDIES WITH COMPOUNDS OF ARSENIC.

III. On the Pathological Action of Arsenicals on the Adrenals.

By WADE H. BROWN, M.D., and LOUISE PEARCE, M.D.

(From the Laboratories of The Rockefeller Institute for Medical Research.)

PLATES 68 TO 73.

(Received for publication, June 24, 1915.)

In spite of the unusual interest that has been manifested in both the ductless glands and experimental arsenic therapy during recent years, the action of arsenicals upon the adrenals, which is one of the most constant and perhaps important features of arsenical intoxication, has remained practically unnoted.

This phase of arsenical action was first observed by us in the course of our chemotherapeutic studies. The adrenals were so constantly involved in the toxic action of our series of arsenicals that we were led to suspect that the tendency to adrenal injury was not peculiar to our compounds but was shared, to a greater or less degree, by other arsenicals. We instituted a series of experiments, therefore, to determine whether such substances as arsenious and arsenic acids, sodium cacodylate, atoxyl, arsacetin, arsenophenyl-glycine, salvarsan, and neosalvarsan possessed a similar tendency to adrenal injury. These substances were chosen as representative and accessible types of organic and inorganic compounds of trivalent and pentavalent arsenic.

EXPERIMENTAL.

Materials.

The routine tests of the above named drugs were made upon male guinea pigs weighing 400 to 500 grams, though supplementary evidence was obtained from both rabbits and dogs which were used in order to facilitate intravenous administration of the drugs.

Technique.

Sterile solutions of the substances were injected intraperitoneally. The arsenious and arsenic acids were dissolved in the theoretical amount of sodium hydroxide and given in a 1 per cent solution. The solutions of salvarsan and neosalvarsan were prepared as for clinical use. All the other compounds were given in as dilute solution as possible, but the concentration necessarily varied somewhat with the amount of the drug given.

Dose.

On account of a lack of accurate information as to the toxic action of some of the arsenicals, we were forced to employ a wide range of doses before obtaining results that could be regarded as comparable. The following figures (Table I) were found to represent roughly the doses in grams per kilo of body weight causing death in guinea pigs in one to three days, except in the instance of arsacetin and salvarsan, where the largest doses used are stated, though they did not kill.

TABLE I.

Drug.	Dose per kilo of body weight. gm.
Arsenious acid	0.010 to 0.012
Arsenic acid	0.015 " 0.020
Sodium cacodylate (Merck)	0.700 " 1.000
Atoxyl	0.080 " 0.100
Arsacetin	0.300
Arsenophenylglycine	0.250 (or less)
Salvarsan	0.150
" rabbit (intravenous)	0.100 to 0.150
Neosalvarsan	0.300 " 0.350

The effects of three types of intoxication were studied with most of the compounds; *i. e.,* the effect of an acutely fatal dose, the effect of single large, sublethal doses, and the effect of repeated large doses.

Gross evidences of injury were noted, and with each animal the right adrenal was preserved in Zenker's fluid; the anterior half of the left adrenal in Müller's fluid, to preserve the chromaffin; and the posterior half in 10 per cent formalin. Sections were always taken from corresponding levels. The Zenker and Müller material was sectioned in paraffin and stained with hematoxylin and eosin. Frozen sections were made from the formalin-fixed tissue and stained for lipoids with Herxheimer's Scharlach R counterstained with aqueous alum hematoxylin.

As the prime object of the experiments was to determine a single fact,—the presence or absence of adrenal injury as a result of arsenical intoxication—the scope of the experiments does not justify a detailed description of the action of each compound. The description given is a composite one based on the results obtained with the various arsenicals upon the adrenals of the guinea pig. Since all the compounds do not affect the adrenals in precisely the same manner, however, such peculiarities in their action as we have noted may be briefly indicated.

The Normal Adrenal of the Guinea Pig.

To facilitate description and as a basis for comparison, we wish to refer briefly to certain features of the normal adrenal of the guinea pig. The adrenals of adult male guinea pigs of the weights used show certain differences that accord in general with differences in the color of the animals. The cortex is grossly divisible into two zones: an outer waxy zone and an inner pigmented zone. In black guinea pigs the adrenals are relatively large with a narrow rim of waxy cortex sharply demarcated from a broad, intensely pigmented, inner zone. In white guinea pigs the other extreme exists: namely, small adrenals with a relatively broad rim of waxy cortex which gradually merges with a relatively narrow and slightly pigmented inner zone. These markings are doubly important since they furnish landmarks and standards for gross pathological changes in the adrenal cortex and correspond quite accurately with the normal distribution of lipoids (Figs. 1 and 2).

It so happens that the microscopic as well as the gross changes produced by arsenicals conform so closely to these divisions of the cortex that they can be most conveniently described upon such a basis of division. Without raising the question, therefore, of the relation of these zones to the generally recognized microscopic divisions, the glomerulosa, the fasciculata, and the reticularis, we have used the designations, Zones 1, 2, and 3. Zone 1 corresponds with the zona glomerulosa; Zone 2 with the waxy or lipoid-containing fasciculata; and Zone 3 with the pigmented portion of the cortex which is also easily recognizable with the microscope on account of the difference in the character of the cells of Zones 2 and 3.

Gross Pathological Changes.

Large doses of all the arsenicals tested cause an acute swelling of the adrenals, usually accompanied by congestion of the surface vessels with scattered foci and streaks of hemorrhage. On section the organs are very soft; the waxy cortex appears slightly gray and translucent and may be streaked with red. Frequently a distinct red line separates the waxy and pigmented zones. The latter area is usually very soft and depressed, but rarely shows any early change in its extent or the intensity of its pigmentation. The medulla is apt to be poorly defined, soft, and somewhat congested.

Later (after forty-eight to seventy-two hours) the adrenals, though still swollen, are quite pale. On section the organ is firmer, and the waxy zone of the cortex is usually distinctly widened, gray, and translucent. The pigmented cortex is narrowed and the intensity of pigmentation is decreased. The original line of contact between these two zones is indicated by a conspicuous narrow band of opaque yellow, gray, or pink tissue. The medulla is usually normal in appearance or somewhat more conspicuous than normally.

Microscopic Changes.

Lipoids.—We have observed two types of alteration in the lipoid content of the adrenal: change in the size of the droplets and change in the amount and distribution of the lipoids. Very early, large droplets of lipoid material appear among the fine granules normally present in the cells of Zone 2 of the cortex. These large droplets are most numerous at the outer and inner edges of the zone and in the latter location, the junction of Zones 2 and 3, there appears to be an actual increase in the amount of lipoid present (Fig. 3).

Subsequently lipoid material increases in Zone 1 and begins to accumulate in considerable amounts in Zone 3 (Fig. 3), while at the same time the amount of lipoid diminishes in Zone 2; this decrease is first apparent in the middle and inner half of the zone (Figs. 3, 4, and 5). Finally, a general depletion sets in and the lipoids disappear completely from the inner half of Zone 2, then from Zones 1 and 3, leaving only scattered droplets in the outer half of Zone 2 and possibly fine granules at the line of separation of Zones 2 and 3 (Figs. 4, 5, and 6).

Very rarely have we been able to demonstrate lipoid material in true medullary cells. In a few instances, fine granules have been observed in these cells.

Vascular Changes.—In the early stages of arsenical poisoning, the vessels of both the cortex and medulla of the adrenal are dilated and filled with blood. Hemorrhage is marked with some compounds (arsenious and arsenic acids and sodium cacodylate) especially in Zone 2 of the cortex, at the junction of Zones 2 and 3, and between the cortex and medulla (Fig. 7). Hyaline or leucocytic thrombi are occasionally seen in the vascular spaces and a slight leucocytic exudate and interstitial edema are frequently present.

Cellular and Structural Alterations.—The cortical cells of the adrenal show a variety of degenerative changes. The cells of Zone 1 are usually swollen; their cytoplasm is granular and slightly vacuolated. The cells of Zone 2 suffer most: the sharp outlines and spongy cytoplasm of these cells are soon lost; they become irregular, ragged, granular, and vacuolated; some are enormously swollen with pale-staining cytoplasm while others are shrunken. Hyaline cells are seen here and there and necrotic cells are numerous in extreme cases. All these cellular changes are again most pronounced at the junction of Zones 2 and 3 (Fig. 8). In Zone 3, cells that are usually finely granular show some small vacuoles and a coarsely granular or hyaline cytoplasm. Necrotic cells in this zone are most numerous at its inner edge.

The architecture of the cortex is frequently disturbed. The cell columns may be broken up and the cells appear completely isolated or in irregular clumps and columns.

The cells of the medulla also show evidence of injury. They are frequently ragged with scant, pale-staining, granular cytoplasm and shrunken, pyknotic nuclei. The most striking feature of the change, with some compounds, is the presence of large numbers of colloid droplets within and among the cells of the medulla (Fig. 9). Necrotic cells may be numerous.

Chromaffin.—All the arsenicals that we have tested exercise some influence upon the chromaffin content of the adrenal. Some compounds (arsenious and arsenic acids) seem to cause but a slight reduction in the chromaffin content as judged by the color of the medullary cells of tissue fixed in Müller's fluid. Judged by the same standards, other arsenicals produce an extreme depletion of the chromaffin (sodium cacodylate, salvarsan, and neosalvarsan) as shown by comparing Figs. 10, 11, and 12.

Recovery.—The injury phase of the action of arsenicals upon the adrenals develops with a varying rapidity and persists for a variable length of time; with some compounds the changes develop rapidly and recovery is equally rapid, while with others the reverse is true. Regeneration of cortical cells is usually rapid and mitotic figures may be seen as early as twenty-four hours after the administration of the drug but are not numerous until forty-eight to seventy-two hours (Fig. 13). Here activity is most marked in the outer half of Zone 2. Six to twelve mitotic figures frequently occur in a single high power field of the microscope. Mitotic division occurs to a limited degree in the cells of Zones 1 and 3. The cells of the medulla also regenerate by mitotic division (Fig. 14). We observed mitotic figures in these cells in several instances after sodium cacodylate, atoxyl, and arsenophenylglycine, but they were never numerous.

Perhaps as a sequel to the injury produced in the medulla, round-celled or polyblastic infiltration is of frequent occurrence and is usually accompanied by some fibroblasts and an increase in endothelial cells (Figs. 15 and 16). To a less degree changes similar to those in Fig. 16 have been seen in control animals, which leaves some doubt as to the significance of this particular type of infiltration.

DISCUSSION.

Thus far, all the compounds of arsenic that we have tested have exhibited a definite action upon the adrenals of the guinea pig, but it by no means follows that the character and degree of such action are identical for all arsenicals. On the contrary, while certain features of the action appear to be common to a number of compounds, other features of the action may be quite distinctive. Thus, arsenious acid and sodium cacodylate show a strong tendency to produce congestion and hemorrhages in the adrenals, while such compounds as arsenophenylglycine and arsacetin produce most marked disturbances in the lipoid content, with but slight tendency to congestion or hemorrhage. In like manner, arsenious acid seems to cause relatively slight alteration in the chromaffin content, while sodium cacodylate acts strongly on the medullary cells.

We must also recognize the fact that the relative importance of adrenal injury in the lesion-complex is another variable. With some arsenicals the effect upon the adrenals may be distinctly overshadowed by injury to other organs, while with other compounds the injury inflicted upon the adrenals plays a prominent part in the toxic manifestations.

We may conclude, therefore, that some factor other than the mere presence of arsenic must exercise a distinct influence upon the activity of these compounds; namely, their chemical constitution.

Finally we must differentiate between the action of toxic doses and the action of therapeutic doses of arsenicals upon the adrenals. Since the action of toxic doses indicates a strong selective affinity of the adrenals for compounds of arsenic, we may legitimately infer that with therapeutic doses injury might give place to stimulation, a conception which clinical experience with some arsenicals seems to justify.

However, generalizations as to the action of arsenicals on the adrenals should be made with caution. From these experiments concerning the action of compounds of arsenic upon the adrenals, we believe that a wide field of investigation has been opened up and that future work will justify our belief in the importance of the action of arsenicals upon the adrenals.

SUMMARY.

1. Toxic doses of all arsenicals of which we have any knowledge produce definite pathological changes in the adrenals of guinea pigs. These changes include congestion, hemorrhage, disturbances in the

lipoid content, cellular degenerations and necroses, and reduction in the chromaffin content.

2. The character and severity of the injury produced by different arsenicals varies with the chemical constitution of the compounds.

3. From these facts, we believe that adrenal injury is an important factor in arsenical intoxication and suggest that therapeutic doses of some arsenicals may produce adrenal stimulation.

EXPLANATION OF PLATES.

The illustrations are all from untouched photomicrographs. Those showing lipoids are from frozen sections stained with Herxheimer's Scharlach R counterstained with aqueous alum hematoxylin. Figures showing chromaffin are from tissue fixed in Müller's fluid and stained with hematoxylin and eosin. All doses of drugs are expressed in gm. per kilo of body weight.

Changes in the Lipoids of the Adrenal Cortex.

PLATE 68.

FIG. 1. Lipoids of normal adrenal cortex. Black male guinea pig. The lipoid, which appears as black granules, is sharply confined to the relatively narrow outer rim of the cortex. $\times$ 80.

FIG. 2. Lipoid of normal adrenal cortex. White male guinea pig. The lipoid extends over more than one-half of the cortex and is irregularly demarcated at the inner edge of the zone. $\times$ 80.

FIG. 3. Increase in the amount of demonstrable lipoid and an abnormal distribution produced by arsenophenylglycine 0.250 gm. Animal killed after 48 hours. The lipoid is distributed over the entire cortex and is especially abundant at the junction of Zones 2 and 3. The photographic intensity of the lipoid in this figure had to be considerably suppressed in the original photomicrograph in order to preserve some detail. $\times$ 80.

PLATE 69.

FIG. 4. A later stage of the same type of change as that in Fig. 3. A distinct decrease of the lipoids is here apparent in the inner half of Zone 2. 2 doses of arsenophenylglycine 0.100 gm., followed on the 5th day by arsenophenylglycine 0.050 gm. Animal died in 24 hours. No postmortem decomposition. $\times$ 80.

FIG. 5. Still more pronounced change in the amount, distribution, and character of the lipoid droplets. Neosalvarsan 0.300 gm. Guinea pig killed after 48 hours. $\times$ 80.

FIG. 6. Extreme depletion of lipoids produced by 4 doses of arsacetin (0.100, 0.200, 0.200, 0.200 gm.) within 16 days. Guinea pig killed 3 days after the last dose. The lipoid is mostly in the form of large droplets with some very fine granules in the cells through the middle of the cortex. $\times$ 80.

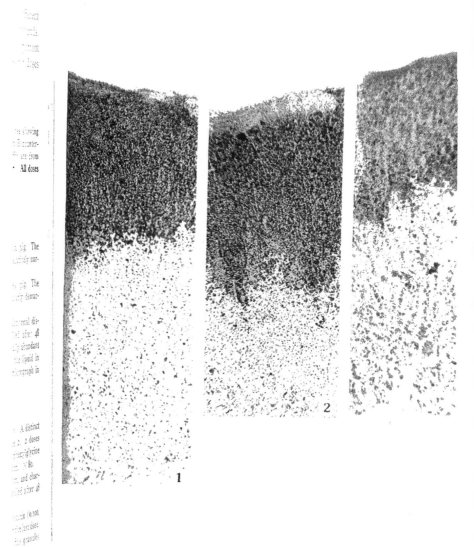

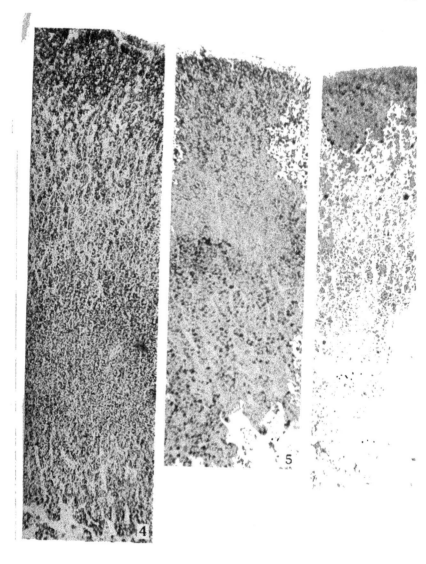

PLATE 7

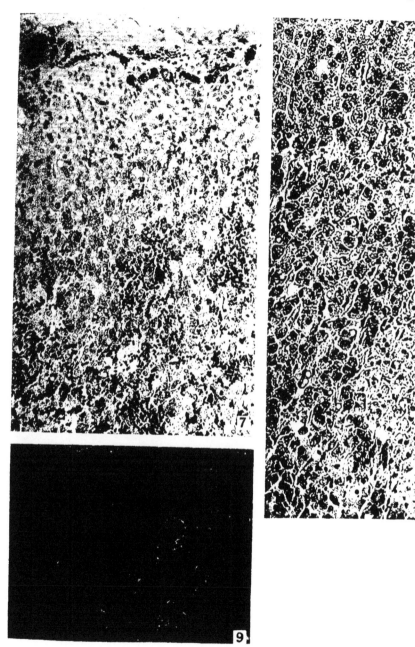

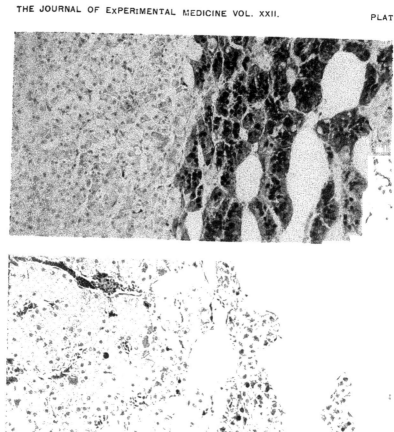

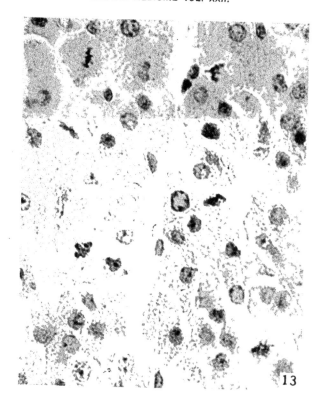

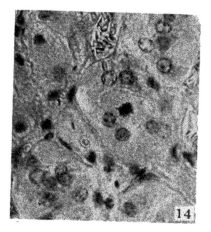

(Brown and Pearce: Action of Arsenicals on Adrenals

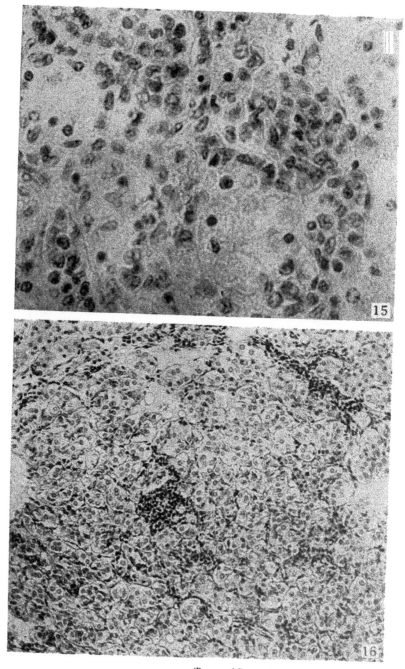

(Brown and Pearce: Action of Arsenicals on Adrenals.)

Congestion, Hemorrhage, and Degeneration.

PLATE 70.

FIG. 7. Congestion and hemorrhage in the cortex of the adrenal produced by 2 doses of sodium cacodylate 0 700 gm. in 4 days. Guinea pig died on the 5th day. Hemorrhage is most marked in the inner half of Zone 2. The cortical cells also show degeneration and vacuolization. × 200.

FIG. 8. Degeneration of the cells of the midzone of the adrenal cortex with slight leucocytic infiltration resulting from arsenophenylglycine. Same adrenal as in Fig. 4. × 200.

FIG. 9. Colloid degeneration of medulla of the adrenal. Same animal as in Figs. 4 and 8. × 775.

Changes in the Chromaffin.

PLATE 71.

FIG. 10. Chromaffin of adrenal of normal guinea pig. × 200.

FIG. 11. Reduction in the chromaffin content of the medulla with shrinkage and degeneration of medullary cells. Sodium cacodylate 0 700 gm. Guinea pig killed after 48 hours. × 200.

FIG. 12. Reduction of chromaffin after 2 doses of neosalvarsan (0 190 gm. and 0.500 gm.) on the 5th day. Died within 18 hours. No postmortem decomposition. × 200.

Regeneration of Adrenal Cells.

PLATE 72.

FIG. 13. Cortical regeneration. Neosalvarsan 0.300 gm. Killed after 48 hours. × 700.

FIG. 14. Mitosis in cell of medulla. Atoxyl, 3 doses of 0.050 gm. in 5 days. Guinea pig killed 3 days after last dose. × 600.

Infiltration in the Medulla.

PLATE 73.

FIG. 15. Infiltration of polymorphonuclear leucocytes and polyblasts into a degenerated area of the medulla. Arsenophenylglycine. Same adrenal as in Figs. 4, 8, and 9. ×700.

FIG. 16. Infiltration of lymphocytes and fibroblasts in the medulla of the adrenal. Arsenious acid 0.005 gm. Killed after 3 days.

A NOTE ON THE IMMEDIATE EFFECTS OF REDUCTION OF KIDNEY SUBSTANCE.[1]

By H. T. KARSNER, M.D., H. A. BUNKER, Jr., A.B., and
G. P. GRABFIELD, M.D.

(*From the Department of Pathology (Phillips Fund) of the Harvard Medical
School, Boston.*)

(Received for publication, June 11, 1915.)

Several investigators have studied the nitrogenous metabolism of animals following reduction of kidney substance; notably, Tuffier,[2] Bradford,[3] Bainbridge and Beddard,[4] Pearce,[5] and Pilcher.[6] Studies of nitrogen balance and partition have been made and somewhat contradictory results obtained. de Paoli[7] observed that large quantities of kidney substance could be removed without endangering life and that the minimum necessary for life is one-half of one kidney, or one-quarter of the total kidney substance. Tuffier, on the other hand, stated that life is possible with only 1.5 gm. of kidney substance per kilo of body weight (the total is about 7 gm. per kilo in the dog). Bradford found the danger limit reached at 2 gm. per kilo; this was confirmed by Pearce and by Pilcher.

As regards function Tuffier, as quoted by Bradford, states: "It is possible to remove quantities of kidney substance equal in weight and in volume to those of the sum of the two kidneys in such a way as yet to leave behind a considerable amount of kidney substance, as shown by postmortem examination; that this diminution in the amount of kidney substance is *not* accompanied by any functional disturbance of the urine, and that the urine and urea, after oscillations due to the operation, return to their normal amount."

Bradford found that excision of less than two-thirds of the kidney volume

[1] Aided by a grant from The Rockefeller Institute for Medical Research.

[2] Tuffier, T., Études expérimentales sur la chirurgie du rein, Paris, 1889; cited by Bradford, J. R., The Results Following Partial Nephrectomy and the Influence of the Kidney on Metabolism, *Jour. Physiol.*, 1898–99, xxiii, 415.

[3] Bradford, *J. R., loc. cit.*

[4] Bainbridge, F. A., and Beddard, A. P., The Relation of the Kidneys to Metabolism, *Proc. Roy. Soc., London, Series B*, 1907, lxxix, 75.

[5] Pearce, R. M., The Influence of the Reduction of Kidney Substance upon Nitrogenous Metabolism, *Jour. Exper. Med.*, 1908, x, 632.

[6] Pilcher, J. D., On the Excretion of Nitrogen Subsequent to Ligation of Successive Branches of the Renal Arteries, *Jour. Biol. Chem.*, 1913, xiv, 389.

[7] de Paoli, E., *Centralbl. f. Chir.*, 1892, xix, 78; cited by Bradford, *loc. cit.*

resulted in temporary increase of the watery part of the urine, which became lasting when two-thirds were removed; removal of "approximately three-quarters of the total kidney weight is followed by a very great increase in the amount of urinary water, and also by an increase in the amount of urea excreted." He found also a considerable increase in the nitrogenous extractives of the blood and tissues, particularly muscles, most marked after excision of three-quarters, but quite apparent after excision of two-thirds of the kidney substance. This excess was distributed just as after double nephrectomy and after the intravenous injection of urea; he states that the quantity in the muscles is too great to be accounted for by the mere products of normal metabolism.

Bainbridge and Beddard working on cats found that removal of approximately three-fourths of the total kidney substance is not constantly followed by an increased output of nitrogen, and that it takes place only in cats which have lost 22 per cent or more of their initial body weight at the time of its onset. They concluded further that the kidneys have no direct influence upon nitrogenous metabolism and that the increased output of nitrogen is simply the result of inanition. They found the cats still able to pass a concentrated urine and that the amount of urine is not necessarily increased beyond the normal.

Pearce, using dogs, confirmed the general results of Bainbridge and Beddard and concluded further that the metabolism condition of starvation is apparently the result of the gastro-intestinal disturbance constantly associated with extensive kidney reduction and not of a disturbance of general nitrogenous metabolism. He found that the gastro-intestinal disturbance was not due to diminished absorption and found no evidence to support a theory of internal secretion on the part of the kidney.

Pilcher, using both cats and dogs, reduced the kidney substance by ligation of branches of the renal arteries. He states that with but one-fourth of the kidney substance functioning the quantity of urine was practically normal; there is marked temporary prostration, anorexia, loss of weight, and increased nitrogen output with finally a return to normal, with a slight tendency to nitrogen retention.

REPORTS OF EXPERIMENTS.

Our experiments have been conducted because of the availability of the Folin methods[8] for the determination of the non-protein nitrogen of the blood.

Dogs were kept on a constant diet and determinations were made of the total non-protein nitrogen of the blood, obtained by aseptic heart puncture. The urine was collected in the ordinary metabolism cages and the total non-protein nitrogen and urea nitrogen were determined. All analyses were made in duplicate or triplicate.

[8] Folin, O., and Denis, W. Protein Metabolism from the Standpoint of Blood and Tissue Analysis, *Jour. Biol. Chem.*, 1912, xi. 161; New Methods for the Determination of Total Non-Protein Nitrogen, Urea and Ammonia in Blood. *ibid.*, p. 527.

All the blood results are reported in mg. per 100 cc. of blood. All the urine results are reported in mg. per twenty-four hours.

Five dogs were used, but the results can be summarized in the protocols of three (Tables I, II, and III).

TABLE I.

Dog 1.

Female, Weight 6,500 Gm. Urine Negative for Albumin. Diet 200 Gm. of Dog Biscuit.

Date.	Blood. Total non-protein nitrogen	Urine.		
		Total non-protein nitrogen.	Urea nitrogen.	Amount.
				cc.
Apr. 29	36	1,840	—	100
" 30	44	2,520	1,861	63
May 1	45	2,010	1,903	201
" 2	Left kidney (weight 17 gm.) removed under ether anesthesia			
" 3	44	1,192	859	159
	Slight diarrhea			
4	43	1,451	1,357	41
	Diarrhea very severe			
7	51	—	—	—
9	44	—	—	—
	Diarrhea finally cleared up. Experiment resumed			
" 24	38	—	—	—

At 12.30 p. m. the right kidney (weight 22 gm.) was removed and the animal bled at more frequent intervals, as follows:

Total non-protein nitrogen in blood.

May 24........12.30 p. m....................Operation
     7.30 p. m... 44
' 25........ 1.00 a. m... 56
     1.00 p. m... 85
     7.00 p. m... 111
     11.30 p. m... 133
' 26........ 2.00 p. m... 196
     11.00 p. m... 227
" 27........10.00 a. m... 285
     2.00 p. m.............................Animal found dead.

Autopsy shows slight superficial infection of the abdominal wound; the abdominal cavity is clean; a few cc. of gray, slightly turbid fluid are found in the left pleural cavity.

TABLE II.

Dog 2.

Female, Weight 6,000 Gm. Urine Negative for Albumin. Diet 2 1/2 Dog Biscuits.

Date.	Blood. Total non-protein nitrogen.	Urine.		
		Total non-protein nitrogen.	Urea nitrogen.	Amount.
				cc.
Aug. 7.........	24	—	—	—
" 8.........	18	4,005	3,337	89
" 10.........	—	2,278	1,960	245
' 11.........	Left kidney (weight 35 gm.) was removed under ether anesthesia			
" 12.........	43	—	—	0
" 13.........	27	2,156	1,705	55
" 14.........	26	3,149	2,350	94
" 15.........	21	3,393	3,237	390
" 16.........	—	2,470	2,041	130
" 18.........	Right kidney was removed under ether anesthesia			

Total non-protein nitrogen in blood.

Aug. 18........11.45 a. m.....................Operation		
11.45 a. m..	31
11.45 p. m..	72
Aug. 19........11.45 a. m..	95
11.45 p. m..	114
" 20........11.45 a. m..	143
6.00 p. m..	154
12.00 midnight.................................	165
" 21........12.00 noon..	208
10.00 p. m..	225
' 22........ 9.00 a. m..	282
5.00 p. m.................................Found dead.	

Autopsy showed no sepsis or other abnormalities.

TABLE III.

Dog 3.

Female, Weight 4,000 Gm. Urine Negative for Albumin. Diet 120 Gm. Chopped Beef and 2 Dog Biscuits.

Date.	Blood. Total non-protein nitrogen.	Urine.		
		Total non-protein nitrogen.	Urea nitrogen.	Amount.
				cc.
Aug. 24.........	23	1,823	1,568	49
" 25.........	—	3,725	3,053	192
" 28.........	Upper pole of right kidney (weight 6 gm.) was removed under ether anesthesia through lumbar incision by Dr. W. C. Quinby			
" 29.........	26	4,480	3,360	224
" 30.........	20	4,480	4,032	224
" 31.........	22	3,800	2,800	200
Sept. 1.........	Entire left kidney (weight 22 gm.) was removed under ether anesthesia			
" 2.........	28	—	—	150
3.........	41	4,620	3,135	330
4.........	35	3,995	2,820	235
" 8.........	41 Diarrhea			
" 11.........	33 Diarrhea			
" 12.........	26 Diarrhea			
" 14.........	33 Diarrhea			
" 18.........	Remainder of right kidney (weight 17 gm.) was removed under ether anesthesia			

		Total non-protein nitrogen in blood.
Sept. 18........	3.00 p. m.....................Operation	
	3.00 p. m.....................................	35
19........	12.00 noon.....................................	68
" 20........	11.30 a. m.....................................	152
	5.00 p. m.....................................	172
21........	9.30 a. m.....................................	227

2.00 p. m. Death occurred as the blood was being withdrawn. The last operation wound was infected; otherwise the autopsy showed nothing abnormal.

DISCUSSION OF RESULTS.

From these data it may be said that removal of approximately one-sixth of the kidney substance (Dog III) results in no marked alteration of the non-protein nitrogen of the blood or of the urine, although there is, as Pilcher expresses it, a slight tendency to accumulation[9] in the blood and increased output in the urine for a

[9] The term accumulation is used instead of retention to avoid confusion with the use of the latter term as expressing the holding of nitrogen in the body for metabolic and constructive purposes.

period of twenty-four (blood) or forty-eight hours (urine), probably to be explained by the operation. Excision of one-half the total kidney substance in the case of Dog II was followed by a distinct accumulation in the blood during the first twenty-four hours, associated, however, with anuria; such accumulation, however, did not appear in the corresponding period in Dog I, which was not anuric. As seen in both dogs, however, the output for the first forty-eight hours appears to be diminished, thus again showing a slight tendency toward accumulation for at the most forty-eight hours. The removal of approximately two-thirds of the kidney substance (Dog III) was followed by a very slight increase in output, and after twenty-four hours a slight but distinct increase in the non-protein nitrogen of the blood which lasted for three days.

An examination of the amount of urine and its concentration shows that reduction of the kidney substance down to approximately one-third does not prevent that fragment from excreting urine in normal amounts and concentrations.

Diarrhea was found to occur frequently and followed the excision of only one-half the kidney substance. None of the dogs lost markedly in weight, and this reduction in no case reached the critical 22 per cent of Bainbridge and Beddard.

The results following the complete removal of kidney substance are shown graphically in Text-fig. 1. In Dogs I and III the rate of increase was considerably less in the first twenty-four hours than in the succeeding periods. On the second and third days the rate of increase was practically the same and from rough calculations based on the output for twenty-four hours was about what it should be per 100 cc. of blood for a twenty-four hour period. It is easily possible that operative shock might account for the relatively slight increase in the first twenty-four hours.

In Dog II the results are more constant, although in the last twenty-four hours there was a slight exacerbation of rate. In other words, for this animal the rate of increase per 100 cc. of blood should be roughly 57 mg. per day per 100 cc., so that the amounts would be for each succeeding day 88, 145, 203, and 260, whereas they actually were 95, 143, 208, and 282 (allowing twenty-one and one-half hours for the last day). It is regretted that the

actual intake of nitrogen following the total nephrectomy was not
estimated and the results here presented are regarded as suggestive

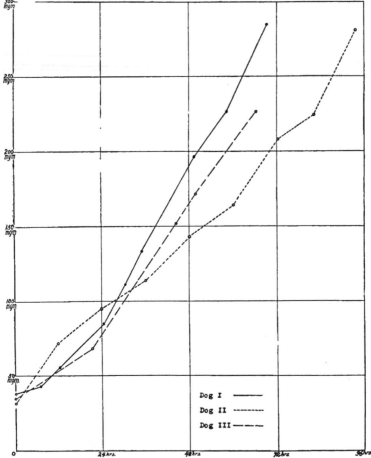

TEXT-FIG. I. Chart showing total non-protein nitrogen in blood following
complete nephrectomy.

rather than conclusive. Further studies are being carried out in an
attempt to elaborate on and elucidate this increase in total non-
protein nitrogen in relation to intake, distribution in the body, and
other factors which are known to influence protein metabolism.

A FURTHER STUDY OF THE BACTERICIDAL ACTION OF ETHYLHYDROCUPREIN ON PNEUMOCOCCI.

By HENRY F. MOORE, M.B., B.Ch., B.A.O.

(*From the Hospital of The Rockefeller Institute for Medical Research.*)

(Received for publication, September 1, 1915.)

In a previous communication[1] we stated that ethylhydrocuprein inhibits the growth of, and kills, pneumococci *in vitro* in very considerable dilutions of the drug, and that it exerts a considerable protective action in experimental pneumococcal infections in mice. The present study was undertaken with the object of gaining some information as to the rate of absorption of the drug into the circulation in the animal body, as to how long the resulting bactericidal effect, if any, of the serum on pneumococci lasted, and as to the mode of action of the drug on these microorganisms.

Ethylhydrocuprein, a derivative of hydroquinine, was introduced by Morgenroth[2] in 1911 in the treatment of experimental pneumococcal infection in mice. It has since been subjected to study by many observers, who, generally speaking, agree that it has a bactericidal action on pneumococci, *in vitro* and *in vivo*. Wright[3] showed that the blood serum of mice previously treated with the drug killed pneumococci in the test-tube.

In the present study we considered it advisable to make use of rabbits as our experimental animals, and, having determined the tolerated and toxic doses of the hydrochloride of the drug (optochin hydrochloride) and of the free base (optochin base) for these animals, we proceeded to study the action of each of these preparations when given by different routes as set forth below. The hydrochloride was given to the rabbits subcutaneously, dissolved in 5 cc. of

[1] Moore, H. F., *Jour. Exper. Med.*, 1915, xxii, 269.

[2] Morgenroth, J., and Levy, R., *Berl. klin. Wchnschr.*, 1911, xlviii, 1560, 1979.

[3] Wright, A. E., Morgan, W. P., Colebrook, L., and Dodgson, R. W., *Lancet*, 1912, ii, 1633, 1701.

distilled water; intravenously, dissolved in 10 cc. of physiological salt solution; and by mouth, dissolved in 25 cc. of distilled water. The free base was given subcutaneously and intramuscularly dissolved in from 5 to 6 cc. of sterile olive oil.

EXPERIMENTAL.

Our experiments show that normal rabbits of approximately 2,000 grams tolerate a single dose of 0.1 gram of the hydrochloride given subcutaneously and 0.125 gram of the base in oil given in the same manner, per kilo of body weight. It seems, however, that the tolerance of normal rabbits of greater weight—3,000 grams and upwards—is less than this. The tolerance of normal rabbits of about 2,000 grams' weight for a single dose of the base in oil when this solution is given intramuscularly (into the erector spinal mass) is still lower—0.075 gram per kilo of body weight. Finally, the animals are able to bear, without showing signs of toxicity, only a small dose of the hydrochloride dissolved in normal salt solution given slowly intravenously; namely, a dose lying between 0.02 and 0.05 of a gram per kilo of body weight. In this case the drug must be given well diluted and very slowly; for, if given in any considerable concentration, a reaction between it and the blood plasma takes place with a resulting heavy precipitate which causes speedy death.

We have observed that the tolerance of rabbits previously infected with pneumococci seems to be somewhat less than that of normal animals.

The drug in a toxic dosage gives rise to certain characteristic symptoms in rabbits, the variety and intensity of which depend in part on the dose, in part on the route by which it is given, and in part on the location of the injection. In order of severity, in the case of subcutaneous or intramuscular injection, the symptoms are as follows: quietness of the animal and disinclination to eat; halting movement of the legs nearest the side of injection; spastic and incoordinated movements of the same; and complete paralysis of the extremities nearest the injection site. Finally, the paralysis may spread to the other limbs and the animal may lie on the floor collapsed, and die. The respirations are at first hurried and, with a larger dose, later diminish in rate, and symptoms of dyspnea may appear. When a toxic dose is given intravenously, the animal shows convulsive movements, more or less severe according to the size of the dose, and may finally die with convulsions and exophthalmos. The bactericidal action of the drug described below is seen in the serum of animals showing severe toxic signs as well as in animals which received a dose well below the toxic limit.

Technique.—Normal rabbits weighing about 2,000 grams were used. Each rabbit was bled from the marginal ear vein, the ear having previously been thoroughly cleansed with bichloride of mercury and alcohol and wiped dry with a sterile sponge. The drug was then administered in the manner and amount mentioned in each protocol below. The animals were bled either as before from the marginal ear vein, or directly from the heart—which latter may be done with safety many times—at stated intervals after the administration of the drug. The blood in each case received in centrifuge tubes was, after the final bleeding, placed in the ice chest until the following morning, when the clot was loosened and centrifugalized. 3 cc. of serum from each centrifuge tube were pipetted off the clot into a test-tube, a separate pipette being used in each case, and inactivated for one-half hour at 56° C. in a water bath. (In a few special cases the active serum was used with the object of comparing the bactericidal power of such serum with the same serum inactivated, as described above.) The 3 cc. of serum in each test-tube were inoculated with 0.5 cc. of a dilution of an 18 hour broth culture of a pneumococcus of from 1 in 100,000 to 1 in 1,000,000. Stock strains of pneumococci of Groups I and II were used, generally the latter. The pneumococci having been thoroughly mixed with the serum, 0.5 cc. of the mixture was plated in about 10 cc. of 1 per cent glucose agar, which had been previously melted and cooled to 40° C. By this means the drug, even if present in the serum in sufficient concentration to prevent the growth, or cause death, of the pneumococci, was so diluted in the plate as to leave the pneumococci free to grow unhindered. The Petri dishes used were 10 cm. in diameter. The plates were incubated for from 20 to 24 hours, at 37° C., and at the end of this period the number of colonies in each was counted. The tubes containing the inoculated serum were also incubated at 37° C. for definite periods, as stated in the tables, at the end of each of which periods 0.5 cc. of the contents of each tube was plated as before, and these plates were incubated in the same way. In this way we were enabled to gain information on the bactericidal action on pneumococci of the serum of animals treated with ethylhydrocuprein in relation to the points already stated.

Explanation of the Protocols.—The numerals in the protocols represent the number of colonies resulting from plating 0.5 cc. of the inoculated serum from the corresponding test-tube either immediately after the tubes were inoculated, or after a definite period of incubation; the figures in the vertical columns correspond to the intervals between the administration of the drug and the various bleedings of the animal; the figures in each horizontal row represent the number of colonies per 0.5 cc. of serum from each particular bleeding after a definite period of incubation.

Experiments Illustrating the Effects on Pneumococci of the Serum of Animals Treated with Ethylhydrocuprein Hydrochloride (Optochin Hydrochloride) Dissolved in Water, Given Subcutaneously (under the Skin of the Back).

Experiment 1.—(Table I.) Rabbit 98 E; weight 1,600 gm. Received 0.1 gm. of ethylhydrocuprein hydrochloride per kilo of body weight, in 5 cc. of distilled water, subcutaneously. No toxic symptoms. Pneumococcus: Stock strain of Group II.

TABLE I.

Serum obtained.	No. of colonies per 0.5 cc. when plated immediately after inoculation.	No. of colonies when plated after 7 hrs.' incubation.	No. of colonies when plated after 22½ hrs.' incubation.
Before giving drug......	119	Almost infinity	Infinity.
1½ hrs. after............	105	36	0
4 " "	126	Almost infinity	Infinity.
5 " "	116	" "	"

Experiment 2.—(Table II.) Rabbit 115 D; weight 1,900 gm. Received 0.1 gm. of ethylhydrocuprein hydrochloride per kilo of body weight, in 5 cc. of distilled water, subcutaneously. No toxic symptoms. Pneumococcus: Stock strain of Group II.

TABLE II.

Serum obtained.	No. of colonies per 0.5 cc. when plated immediately after inoculation.	No. of colonies when plated after 7½ hrs.' incubation.	No. of colonies when plated after 20 hrs.' incubation.
Before giving drug......	410	Infinity	Infinity.
1⅙ hrs. after............	445	29	3
2½ " "	492	176	Infinity.

Experiment 3.—(Table III.) Rabbit 95 E; weight 2,200 gm. Received 0.1 gm. of ethylhydrocuprein hydrochloride per kilo of body weight, in 5 cc. of distilled water, subcutaneously. No toxic symptoms. Pneumococcus: stock strain of Group II.

TABLE III.

Serum obtained.	No. of colonies per 0.5 cc. when plated immediately after inoculation.	No. of colonies when plated after 6½ hrs.' incubation.	No. of colonies when plated after 21 hrs.' incubation.
Before giving drug......	329	Almost infinity	Infinity.
1 hr. after.............	268	46	18
2 " "	273	135	146
4⅝ " "	316	Several thousand	Infinity.

Experiment 4.—(Table IV.) Rabbit 141 D; weight 2,130 gm. Received 0.1 gm. of ethylhydrocuprein hydrochloride per kilo of body weight, in 5 cc. of distilled water, subcutaneously. No toxic symptoms. Pneumococcus: stock strain of Group I rendered highly virulent for rabbits by passage.

TABLE IV.

Serum obtained.	No. of colonies per 0.5 cc. when plated immediately after inoculation.	No. of colonies when plated after 6 hrs.' incubation.	No. of colonies when plated after 22 hrs.' incubation.
Before giving drug......	281	Almost infinity	Infinity.
1⅝ hrs. after...........	351	62	0
2¼ " "	285	134	Several thousand.
3⅓ " "	282	188	Infinity.
4⅜ " "	297	506	"
5⅝ " "	306	Almost infinity	——

Experiment 5.—(Table V.) Rabbit 2; weight 2,300 gm. Received 0.2 gm. of ethylhydrocuprein hydrochloride per kilo of body weight, in 5 cc. of distilled water, subcutaneously. Severe toxic appearances; died immediately after last bleeding. Pneumococcus: stock strain of Group II.

TABLE V.

Serum obtained.	No. of colonies per 0.5 cc. when plated immediately after inoculation.	No. of colonies after 2½ hrs.' incubation.	No. of colonies after 4 hrs.' incubation.	No. of colonies after 6½ hrs.' incubation.	No. of colonies after 10½ hrs.' incubation.
Before giving drug.............	456	478	Almost infinity	Infinity	—
2½ hrs. after.................	460	382	245	35	1
7½ " "	478	456	390	370	127

Experiment 6.—(Table VI.) Rabbit 3; weight 1,700 gm. Received 0.15 gm. of ethylhydrocuprein hydrochloride per kilo of body weight, in 5 cc. of distilled water, subcutaneously. Severe toxic symptoms; died immediately after last bleeding. Pneumococcus: stock strain of Group II.

TABLE VI.

Serum obtained.	No. of colonies per 0.5 cc. when plated immediately after inoculation.	No. of colonies after 2 hrs.' incubation.	No. of colonies after 5 hrs.' incubation.	No. of colonies after 11 hrs.' incubation.	No. of colonies after 24 hrs.' incubation.
Before giving drug.........	187	188	Almost infinity	Infinity	Infinity.
1½ hrs. after.....	194	127	64	0	0
2½ " "	187	175	123	56	46
4¾ " "	192	181	212	Several thousand	Infinity.

Comment on Experiments 1 to 6, Inclusive.—A study of the fig-ures shown in the protocols of these experiments reveals certain facts. The serum of animals given a suitable dose (*e. g.,* 0.1 gram per kilo of body weight) of ethylhydrocuprein hydrochloride subcu-taneously has a strongly bactericidal action *in vitro* on the pneumo-cocci, these microorganisms growing freely in normal rabbit serum. This property is possessed by serum obtained as early as one hour after the administration of the drug—in fact, it is at its maximum about this time, or a little later; having attained a maximum potency the bacteriolytic effect gradually falls off, and, in the case of a dose of 0.1 gram per kilo of body weight (a dose well tolerated), it passes into an inhibitory effect on the growth of the pneumococci within from 2½ to 3 hours after the animal received the drug; the inhibitory effect seems to restrain the free growth of the microorgan-isms for some hours, after which this action seems to be overcome; this inhibitory effect on growth, in its turn, disappears about 5 hours after the drug has been given. Further, the killing off of the pneu-mococci in the serum of animals treated with the drug is a gradual process, lasting several hours, according to the dosage of the drug and the amount of inoculation, etc. The same effects are to be seen in the serum of animals which have received a toxic dose of the drug, except that the duration of these actions is longer, and, per-haps, more powerful.

Experiments Illustrating the Effects on Pneumococci of the Serum of Animals Treated with the Free Base Ethylhydrocuprein (Optochin Base) Dissolved in Sterile Olive Oil, Given Subcutaneously (under the Skin of the Back).

Experiment 7.—(Table VII.) Rabbit 142 D; weight 2,200 gm. Received 0.075 gm. of ethylhydrocuprein base per kilo of body weight, in 6 cc. of olive oil, subcutaneously. No toxic symptoms. Pneumococcus: stock strain of Group I rendered highly virulent for rabbits by passage.

TABLE VII.

Serum obtained.	No. of colonies per 0.5 cc. when plated immediately after inoculation.	No. of colonies after 4 hrs.' incubation.	No. of colonies after 22 hrs.' incubation.
Before giving drug......	301	Almost infinity	Infinity.
1½ hrs. after...........	305	116	0
2½ " "	290	227	Almost infinity.
3½ " "	297	189	" "
4½ " "	351	408	Infinity.
5 " "	358	Several thousand	—–-

Experiment 8.—(Table VIII.) Rabbit 91 E; weight 2,200 gm. Received 0.1 gm. of ethylhydrocuprein base per kilo of body weight, in 6 cc. of olive oil, subcutaneously. No toxic symptoms. Pneumococcus: stock strain of Group II.

TABLE VIII.

Serum obtained.	No. of colonies per 0.5 cc. when plated immediately after inoculation.		No. of colonies after 4 hrs.' incubation.		No. of colonies after 22 hrs.' incubation.	
	Serum inactivated.	Serum not inactivated.	Serum inactivated.	Serum not inactivated.	Serum inactivated.	Serum not inactivated.
Before giving drug.....	321	—	Almost infinity	—	Infinity	—
1 hr. after...........	342	292	94	74	1	0
2½ " "	382	350	158	162	98	37
4¾ " "	305	323	207	257	Several thousand	Almost infinity.

Experiment 9.—(Table IX.) Rabbit 105 D; weight 1,750 gm. Received 0.1 gm. of ethylhydrocuprein base per kilo of body weight, in 6 cc. of olive oil, subcutaneously. No toxic symptoms. Pneumococcus: stock strain of Group I.

TABLE IX.

Serum obtained.	No of colonies per 0.5 cc. when plated immediately after inoculation.	No. of colonies after 3 hrs.' incubation.	No. of colonies after 7½ hrs.' incubation.	No. of colonies after 20 hrs.' incubation.
Before giving drug...	573	Several thousand	Infinity	—
2 hrs. after........	349	200	41	0
3 " "	402	224	76	3
4⅙ " "	400	270	123	17
5⅙ " "	410	238	138	8

Experiment 10.—(Table X.) Rabbit 106 D; weight 1,700 gm. Received 0.1 gm. of ethylhydrocuprein base per kilo of body weight, in 6 cc. of olive oil, subcutaneously. No toxic symptoms. Pneumococcus: stock strain of Group I.

TABLE X.

Serum obtained.	No. of colonies per 0.5 cc. when plated immediately after inoculation.	No. of colonies after 3 hrs.' incubation.	No. of colonies after 7½ hrs.' incubation.	No. of colonies after 20 hrs.' incubation.
Before giving drug...	397	Several thousand	Infinity	—
2 hrs. after........	330	197	59	0
3 " "	338	229	125	14
4⅙ " "	358	298	190	552*
5⅙ " "	374	460	—	—

* Macroscopically there appeared to be no growth in the corresponding tube.

Experiment 11.—(Table XI.) Rabbit 104 D; weight 1,650 gm. Received 0.125 gm. of ethylhydrocuprein base per kilo of body weight, in 6 cc. of olive oil, subcutaneously. Animal showed a tendency to lie quiet for some hours after the injection. Pneumococcus: stock strain of Group I.

TABLE XI.

Serum obtained.	No. of colonies per 0.5 cc. when plated immediately after inoculation.	No. of colonies after 3 hrs.' incubation.	No. of colonies after 7½ hrs.' incubation.	No. of colonies after 20 hrs.' incubation.
Before giving drug...	594	Several thousand	Infinity	Infinity.
2 hrs. after........	520	149	51	0
3½ " "	419	159	107	0
4½ " "	368	256	1,200	Infinity.

Experiment 12.—(Table XII.) Rabbit 101 D; weight 1,720 gm. Received 0.15 gm. of ethylhydrocuprein base per kilo of body weight, in 6 cc. of olive oil, subcutaneously. Animal apparently well up to 6 hours after injection; died during the last bleeding. Pneumococcus: stock strain of Group II.

TABLE XII.

Serum obtained.	No. of colonies per 9.5 c. when plated immediately after inoculation.	No. of colonies after 5½ hrs.' incubation.	No. of colon es after 8½ hrs.' incubation.	No. of colonies after 25 hrs.' incubation.
Before giving drug...	976	Almost infinity	Infinity	Infinity.
2 hrs. after........	1,100	180	57	0
3½ " "	1,328	386	292	0
5 " "	1,152	403	283	Almost infinity.
6 " "	1,284	Several thousand	Infinity	Infinity.

Experiment 13.—(Table XIII.) Rabbit 102 D; weight 1,770 gm. Received 0.15 gm. of ethylhydrocuprein base per kilo of body weight, in 6 cc. of olive oil, subcutaneously. No toxic symptoms. Pneumococcus: Stock strain of Group II.

TABLE XIII.

Serum obtained.	No. of colonies per 0.5 cc. when plated immediately after inoculation.	No. of colonies after 5½ hrs.' incubation.	No. of colonies after 8¼ hrs.' incubation.	No. of colonies after 25 hrs.' incubation.
Before giving drug...	1,176	Infinity	Infinity	Infinity.
2 hrs. after........	1,232	452	224	10
3½ " "	958	339	169	Infinity.
5 " "	1,152	500	441	Almost infinity.

Experiment 14.—(Table XIV.) Rabbit 121 F; weight 2,050 gm. Received 0.15 gm. of ethylhydrocuprein base per kilo of body weight, in 6 cc. of olive oil, subcutaneously. No toxic symptoms. Pneumococcus: Stock strain of Group II.

TABLE XIV.

Serum obtained.	No. of colonies per 0.5 cc. when plated immediately after inoculation.	No. of colonies after 5¼ hrs.' incubation.	No. of colonies after 23 hrs.' incubation.
Before giving drug......	68	Almost infinity	Infinity.
2 hrs. after............	58	0	0
3 " "	64	0	0
4 " "	59	7	0
5 " "	43	18	0

Comment on Experiments 7 to 14, Inclusive.—These protocols, like those of Experiments 1 to 6, show that the serum of animals given subcutaneously a suitable dose of ethylhydrocuprein base (optochin base) dissolved in olive oil has a strongly bactericidal action *in vitro* on the pneumococci, that this property is possessed by serum obtained as early as 1 hour after the administration of the drug, and that having attained a maximum potency the bacteriologic effect gradually falls off. In the case of a dose of 0.1 gram per kilo of body weight (a dose well tolerated), the bactericidal action passes into an inhibitory effect on the growth of the pneumococci in about four hours after the animal received the drug; here again, the inhibitory effect seems to restrain the free growth of the microorganism for some hours, after which this action seems to be overcome. The inhibitory effect, in its turn, ultimately disappears. As before, the protocols show that the killing off of the pneumococci in the serum of the treated animals is a gradual process. The effects are more lasting than when the hydrochloride in water is given by the same route.

Experiments Showing the Effects on Pneumococci of the Serum of Animals Treated with Ethylhydrocuprein Hydrochloride Dissolved in 10 Cc. of Normal Saline Solution and Given Slowly Intravenously.

Experiment 15.—(Table XV.) Rabbit 9; weight 2,000 gm. Received 0.02 gm. of ethylhydrocuprein hydrochloride per kilo of body weight, in 10 cc. of normal saline, slowly, intravenously.

Slight convulsions toward end of injection; immediately after injection the animal lay stretched out on floor and apparently could not rise. Complete recovery in a few minutes. Pneumococcus: stock strain of Group II.

TABLE XV.

Serum obtained.	No. of colonies per 0.5 cc. when plated immediately after inoculation.	No. of colonies after 7 hrs.' incubation.	No. of colonies after 18 hrs.' incubation.
Before giving drug......	405	Infinity	Infinity.
5 min. after..........	426	533	About 2,000.
1½ hrs. "	360	Infinity	Infinity.

Experiment 16.—(Table XVI.) Rabbit 10; weight 2,700 gm. Received 0.078 gm. of ethylhydrocuprein hydrochloride per kilo of body weight, in 10 cc. of normal saline, slowly, intravenously.

Convulsions and exophthalmos toward the end of the injection; immediately after injection the animal lay on floor and seemed collapsed; twitching of legs. Recovered in a few minutes. Pneumococcus: stock strain of Group II.

TABLE XVI.

Serum obtained.	No. of colonies per 0.5 cc. when plated immediately after inoculation.	No. of colonies after 3½ hrs.' incubation.	No. of colonies after 7 hrs.' incubation.	No. of colonies after 20 hrs.' incubation.
Before giving drug...	304	249	Infinity	Infinity.
5 min. after.......	329	235	87	10
1½ hrs. "	340	315	Infinity	—
4½ " "	350	292	"	—

The serum of rabbits which had been given intravenously doses of the hydrochloride smaller than those given to Rabbits 9 and 10 showed no bactericidal or inhibitory effect on the pneumococci. An animal given 0.05 gm. per kilo of body weight intravenously died immediately after the injection with convulsions.

Comment on Experiments 15 and 16.—The serum of rabbits given a non-fatal dose of hydrochloride of the drug intravenously does not show such a strong or prolonged effect on pneumococci as when the drug is given subcutaneously. When given intravenously, toxic signs are more easily obtained than when other routes are used.

Experiments Showing the Effects on Pneumococci of the Serum of Animals Treated with Ethylhydrocuprein Base (Optochin Base), Dissolved in Olive Oil, and Given Intramuscularly.

Experiment 17.—(Table XVII.) Rabbit 120 D; weight 1,620 gm. Received 0.075 gm. of ethylhydrocuprein base per kilo of body weight, in 6 cc. of olive oil, intramuscularly. No toxic symptoms. Pneumococcus: stock strain of Group II.

TABLE XVII.

Serum obtained.	No. of colonies per 0.5 cc. when plated immediately after inoculation.	No. of colonies after 10 hrs.' incubation.	No. of colonies after 22 hrs. incubation.
Before giving drug......	61	Infinity	Infinity.
1½ hrs. after..........	60	142	"
3⅙ " "	71	Almost infinity	"
4⅙ " "	77	Infinity	
4⅚ " "	73	"	

Experiment 18.—(Table XVIII.) Rabbit 121 D; weight 1,850 gm. Received 0.075 gm. of ethylhydrocuprein base per kilo of body weight, in 6 cc. of olive oil, intramuscularly. No toxic symptoms. Pneumococcus: stock strain of Group II.

TABLE XVIII.

Serum obtained.	No. of colonies per 0.5 cc. when plated immediately after inoculation.	No. of colonies after 10 hrs.' incubation.	No. of colonies after 22 hrs.' incubation.
Before giving drug......	74	Infinity	Infinity.
1½ hrs. after............	64	15	2
3½ " "	91	Several thousand	Infinity.
4¼ " "	91	Almost infinity	"
5 " "	89	Infinity	"

Experiment 19.—(Table XIX.)   Rabbit 169 D; weight 2,000 gm.   Received 0.1 gm. of ethylhydrocuprein base per kilo of body weight, in 6 cc. of olive oil, intramuscularly.   Distinct toxic symptoms within 2 hours after administration of drug.   Recovered next day.   Pneumococcus: stock strain of Group II.

TABLE XIX.

Serum obtained.	No. of colonies per 0.5 cc. when plated immediately after inoculation.	No. of colonies after 4½ hrs.' incubation.	No. of colonies after 21 hrs.' incubation.
Before giving drug......	1,012	Almost infinity	Infinity.
1½ hrs. after............	984	124	1
2½ " "	952	413	Infinity.
4½ " "	1,010	Several thousand	"

Comment on Experiments 17 to 19, Inclusive.—These experiments, illustrating the results of giving the base of the drug dissolved in oil intramuscularly, show that the drug is more toxic when given by this route than when given subcutaneously in the same vehicle and that the bactericidal effect on the pneumococci is not so prolonged, and probably not so intense, as when the subcutaneous route is used.

In addition to the experiments mentioned above, we have studied the bactericidal action of the serum of rabbits into the stomachs of which the hydrochloride of the drug, dissolved in 25 cc. of distilled water, was introduced by means of a stomach tube; the tolerance is greatest by this route, but the bactericidal effects are slight or absent; even by giving doses as large as 0.3 to 0.4 gram per kilo of body weight to rabbits of 2,000 grams' weight, and otherwise using the same technique as described above, we have not been able to demonstrate any bactericidal action of the serum of these animals on pneumococci within a period of from 2½ to 6 hours after administration of the drug, and only a slight inhibitory effect on their growth.

In order to gain some idea of the bactericidal action on pneumo-cocci of the serum of man treated with the drug the experiments immediately to be described were carried out. The technique was the same as that described above.

Experiment 20.—(Table XX.) Normal man, M.; weight 50.8 kilos. Received, *per os*, in capsules, 0.5 gm. of ethylhydrocuprein hydrochloride. No toxic signs or symptoms. Pneumococcus: stock strain of Group II.

TABLE XX.

Serum obtained.	No. of colonies per 0.5 cc. when plated immediately after inoculation.	No. of colonies after 2½ hrs.' incubation.	No. of colonies after 8 hrs.' incubation.	No. of colonies after 12 hrs.' incubation.	No. of colonies after 21 hrs.' incubation.
Before giving drug..	395	3,000 (approximately)	Infinity	Infinity	—
1 hr. after.........	458	382	391	1,500	Several thousand.
2 " "	560	311	337	3,000 (approximately)	Several thousand.
5 "	374	400	Several thousand	Almost infinity	Almost infinity.

Experiment 21.—(Table XXI.) Normal man, A; weight 68.96 kilos. Received *per os*, in capsules, 0.5 gm. of ethylhydrocuprein hydrochloride. No toxic signs or symptoms. Pneumococcus: stock strain of Group II.

TABLE XXI.

Serum obtained.	No. of colonies per 0.5 cc. when plated immediately after Inoculation.	No. of colonies after 4¾ hrs.' incubation.	No. of colonies after 9⅘ hrs.' incubation.	No. of colonies after 21 hrs.' incubation.
Before giving drug...	517	Almost infinity	Infinity	Infinity.
1 1/3 hrs. after.......	560	540	Almost infinity	"
2½ " "	521	230	225	Almost infinity.
5¾ " "	518	483	Several thousand	Infinity.

Experiment 22.—(Table XXII.) Man, B.; weight 52 kilos. Received 0.5 gm. of ethylhydrocuprein hydrochloride dissolved in 5.0 cc. of sterile redistilled water, subcutaneously, in left flank. Next day complained of pains in legs, had a slight rise of temperature, and the area around the injection site was hyperemic and showed a slight superficial edema. Pneumococcus: stock strain of Group II.

TABLE XXII.

Serum obtained.	No. of colonies per 0.5 cc. when plated immediately after inoculation.		No. of colonies after 3 hrs.' incubation.		No. of colonies after 8¾ hrs.' incubation.		No. of colonies after 21 hrs.' incubation.	
	Serum inactivated.	Serum not inactivated.	Serum inactivated.	Serum not inactivated.	Serum inactivated.	Serum not inactivated.	Serum inactivated.	Serum not inactivated.
Before giving drug....	76	112	1,300	2,000	Almost infinity	Infinity	Infinity	Infinity.
1 hr. after..........	68	54	25	40	34	19	Almost infinity	Infinity.
" " ....:	65	52	29	27	22	140	Infinity	Infinity.
3	61	68	38	37	153	467	"	Infinity.
4½ " "	56	59	24	31	713	431	"	Infinity.

Experiment 23.—(Table XXIII.) Woman, S.; weight 51 kilos. Received 0.5 gm. of ethylhydrocuprein hydrochloride dissolved in 5 cc. of sterile redistilled water, subcutaneously, in abdominal wall. Next day the patient complained of general malaise and headache, had a slight rise of temperature, and had a painful hyperemic area of infiltration and edema about 3 inches in diameter around the injection site which disappeared in a few days. Pneumococcus: stock strain of Group II.

TABLE XXIII.

Serum obtained.	No. of colonies per 0.5 cc. when plated immediately after inoculation.	No. of colonies after 3 hrs.' incubation.	No. of colonies after 8¾ hrs.' incubation.	No. of colonies after 21 hrs.' incubation.
Before giving drug......	60	51	Almost infinity	Infinity.
1 hr. after............	56	35	1,320	"
2⅙ " "	74	34	439	"

Comment on Experiments 20 to 23, Inclusive.—The figures show that the serum of a man to whom had been given a single dose of 0.5 gram of the hydrochloride of the drug either by the mouth or subcutaneously, has a decided inhibitory effect on the growth of the pneumococcus lasting for several hours, after which this action seems to be overcome, and has a bactericidal action which is much less strong than that shown in the protocols of the animal experiments described above. Similar effects are seen whether the drug be administered subcutaneously or by mouth. The employment of the former method, however, subjects the patient to considerable discomfort.

The serum of a normal man to whom 1.8 grams of the hydrochloride of the drug were given by the mouth in 24 hours, that is, in four doses of 0.45 gram each given at regular intervals, showed a cumulative effect; namely, the bactericidal action was much stronger 2¾ hours after the last, than at a similar period after the first, dose.

DISCUSSION OF RESULTS.

In serum, the pneumococci show a tendency to form short chains composed of from two to four diploid forms. The figures in the foregoing protocols consequently must be regarded, not as giving the total number of pneumococci per 0.5 cc. of the inoculated serum, but as an index to that number. The figures, in other words, are relative, not absolute, but the conclusions drawn with regard to the bactericidal action of ethylhydrocuprein are perfectly valid, for the growth in normal serum serves as a control. Moreover, we have cumulative evidence of this bactericidal action of the drug in the fact that the progressive diminution in the number of pneumococci is seen in a considerable number of different experiments. Again, this progressive diminution frequently goes to the point of complete disappearance of living pneumococci in a few hours.

The protocols, therefore, set forth above show that the serum of rabbits treated with ethylhydrocuprein exerts a bactericidal action on the pneumococci in the test-tube, and later inhibits their growth. The intensity and the time of appearance and disappearance of these actions depend on the dosage, upon the route by which the drug is given, and on the method of administration. With the dosage used in the experiments above described, these effects last longest when the base is given in oil subcutaneously; they are not quite so lasting when the hydrochloride is given in water, subcutaneously; and are still less lasting when the base is given in oil, intramuscularly. To obtain bactericidal effects in rabbits by giving the hydrochloride dissolved in physiological saline intravenously, one must give toxic doses dangerous to the life of the animal. The giving of the hydrochloride directly into the stomach of rabbits does not seem to be efficient from the point of view of a resulting bactericidal action of the serum on the pneumococci.

It was thought that a possible explanation of the progressive decrease in the number of colonies in those tubes in which this was demonstrated might be an agglutinative effect of the drug on the pneumococci. We have frequently examined the contents of the tubes microscopically, macroscopically, and by cultural methods for sterility (the latter in the cases in which the plates showed no colonies), and we have never found any evidences of agglutination.

It will be noticed in the protocols of several experiments that, in the case of serum obtained when, apparently, the optochin concentration in the blood is diminishing, the number of colonies in the incubated tubes at first shows a progressive decrease (the number in the control normal serum, obtained before giving the drug, showing, at the same time, a progressive increase) ; but, that, later, after a few hours' incubation, the pneumococci progressively increase, until they become too numerous to count in the plates. This phenomenon is more apparent in tubes lightly inoculated than in tubes more heavily inoculated with pneumococci. Evidently, if the optochin concentration in the serum falls below a certain point in relation to the number of pneumococci present, the pneumococci which survive the bactericidal action may, after a few hours in contact with the drug in the incubator, acquire the property of overcoming this action of the drug and, therefore, grow freely. It would be interesting to investigate this property, so rapidly acquired, from the point of view of fastness. It is not due to a destruction of the bactericidal action of the serum caused by simple incubation, because serum incubated for 24 hours, and then inoculated and studied as above described, shows the same intensity of bactericidal action on pneumococci as samples of the same serum not previously subjected to incubation before inoculation. This overcoming of the action of the drug seems to be an entirely new, and, relatively, quickly acquired property of the pneumococci themselves.

It does not seem that there is any considerable difference in the bacteriolytic or inhibitory effects of serum obtained after administration of the drug whether such serum be used after having been inactivated for ½ hour at 56° C., or active.

CONCLUSIONS.

1. The serum of rabbits which have been previously treated with a single dose of ethylhydrocuprein (optochin) exerts a bactericidal action on, and, later, inhibits the growth of pneumococci in the test-tube.

2. These actions are most evident in the serum of rabbits when the base (optochin base) is given in oil subcutaneously; somewhat less when the hydrochloride of the drug is given in water subcutaneously; slight when the base is given in oil intramuscularly: and least evident, or absent, when the hydrochloride in water is introduced directly into the stomach. To get these effects by the intravenous route, toxic doses must be given, and, even with toxic non-fatal doses, the effects do not last long.

3. In the case of the base given in oil subcutaneously to rabbits in a dosage of 0.1 gram per kilo of body weight, the bactericidal action of the serum is at its maximum about one hour after administration. and it passes into an inhibitory effect about four hours after the drug has been given.

4. In man the same inhibitory and bactericidal actions of the serum are present when a single dose of 0.5 gram of the hydrochloride of the drug is given by the mouth or subcutaneously. but the bactericidal action is not so marked as in rabbits.

5. When the optochin concentration in the serum has. apparently. diminished to a certain point in relation to the number of pneumococci present, the pneumococci which have survived the bactericidal action for a few hours acquire the power of growing freely.

THE SERUM FERMENTS AND ANTIFERMENT DURING PNEUMONIA.

Studies on Ferment Action. XXIV.

By JAMES W. JOBLING, M.D., WILLIAM PETERSEN, M.D., and
A. A. EGGSTEIN, M.D.

(From the Department of Pathology, Medical Department, Vanderbilt University, Nashville.)

(Received for publication, May 15, 1915.)

Our knowledge concerning the factors that induce the crisis in pneumonia is still unsatisfactory, despite the innumerable attempts which have been made in the endeavor to advance our understanding of the processes by various immunological methods of investigation. There have been demonstrated only slight changes in the serum complement and serum antibodies before or after crisis; the opsonic and leucocyte activity, while increased, shows no striking difference; the infecting organism may show no change in its virulence or its resistance (Rosenow); and yet in a period of a few hours a phenomenon occurs, which in its sharp demarcation more nearly resembles a chemical reaction *in vitro* than a biological process *in vivo*. In so far as we may consider the involved tissue as being isolated from the general circulation, as Kline and Winternitz (1) have recently pointed out, the process must of necessity be largely local in its origin and effect. Recovery is coincident with the destruction not only of bacteria, but also with the removal of the great mass of fibrinous and cellular detritus. It is, therefore, in all of its essentials an autolytic process, and our hope of therapeutic results must be based not only on the idea of overcoming the infecting organism but also of favorably influencing the autolytic changes. The isolation of the lung tissue from the general circulation favors autolysis, for if the tissue were freely supplied with blood serum, with its great concentration of antiferment, no autolysis could take place.

Inception of Autolysis.

In a general way we can consider the factors that influence the inception of autolysis as depending on a balance between the amount of ferment, on the one hand, and the inhibiting elements, on the other, both of which are variables and subject to changes either through chemical or physical alteration. With the disintegration of the accumulated polymorphonuclear leucocytes, a large amount of a powerful polyvalent ferment is liberated, capable of dissolving the fibrin and cellular debris. This ferment is not active in the presence of blood serum because of the excess of antiferment. This inhibiting factor can be influenced either by a change in dispersion, whereby the unsaturated lipoids are rendered less disperse, *i. e.,* when the reaction becomes acid (Opie (2)); by an increased oxidation, as, for example, by iodine (3), whereby the unsaturated bonds of the antiferment are reduced and the antitryptic activity is lessened; by an increased metabolic demand on the lipoids, as, for instance, during starvation, when a marked fall in antiferment occurs (4); and by a saturation with an excess of ferment. During the course of the pneumonia the leucocytes are undergoing disintegration with a resulting liberation of ferment. The blood serum, on the other hand, is subject to various of the above mentioned factors which tend to lower the antiferment power (after the original toxic rise has occurred). We should therefore expect that at some moment the balance would be destroyed and autolysis actively commence. Almagià (5) has recently studied this relation and has called attention to the coincidence of evidences of autolysis with the crisis. He considers the very production of an autolyzing fluid as the factor which inhibits the further multiplication of the pneumococci, for he found that the growth of pneumococci was completely inhibited by autolyzing fibrin. We can thus readily see that the pneumonic process, apart from complications, is actually a self-limited one, for even if the organism is so virulent that it will readily kill the leucocytes this very property will set free a large amount of ferment in a short space of time and thus will shorten the course of the disease.

We must, however, take cognizance of another possible factor: *i. e.,* the question whether or not the blood serum itself may contain an increased amount of ferment derived from the tissue cells of the body, which, mobilized, would be brought to bear on the diseased organ. That during the course of a pneumonia serum ferments may be found which digest lung tissue and placenta has repeatedly been noted.[1] These ferments are not specific; experiments which might be so interpreted are based on the fact that ordinary immunity reactions occur which by means of adsorption phenomena permit proteolysis of the serum proteins. Guggenheimer (6) has studied the problem in another way, in that he has determined the inhibitory or accelerating effect of the pneumonic serum on liver autolysis. He noted that pneumonic sera occasionally acted in an auxiliary way and regards this as evidence of a mobilization of cellular ferments. Other sera, however, showed a marked inhibiting effect, as did the normal serum.

Early during the course of antiferment investigations Ascoli and Bezzola (7) studied the conditions obtaining during pneumonia. With the Gross-Fuld method they noted an early rise in antiferment titer, with a decline after the crisis.

The serum investigations so far reported have lacked any quantitative basis, because the methods used to determine the amount of ferment and antiferment are largely qualitative. In order to obtain a more definite insight into the ferment-antiferment balance of the serum we have undertaken a study of a number of patients during the course of pneumonia.

The serum protease has been estimated by the amount of non-coagulable nitrogen formed per cc. of serum when emulsified with chloroform (8); the lipase by the usual ethyl butyrate method; and the antiferment titer by determining the amount of casein digestion by means of the Folin method, instead of the unreliable estimation by inspection of the precipitate. In the majority of the cases the type of organism has been determined by agglutination by means of specific sera.[2]

[1] Falls (Falls, F. H., *Jour. Infect. Dis.*, 1915, xvi, 466) has recently demonstrated such ferments in a series of cases. He concludes that the source of the ferments is not to be found in the leucocytes.

[2] These specific sera were kindly supplied by Dr. Cole of The Rockefeller Institute. We wish furthermore to express our appreciation to Drs. McCabe, Manier, and Weaver, and the members of their respective staffs of the City, Vanderbilt, and Industrial School Hospitals, who have placed their patients at our disposal for study and have facilitated our work in every way.

James W. Jobling, William Petersen, and A. A. Eggstein.   571

Case Records.

The following case records and their accompanying text-figures are typical of the relations.

Case 1 (19).—(Text-fig. 1.) J. H., male, colored, aged 28. Became sick Apr. 6, 1915, when he had a chill lasting a half hour, together with severe pain in both sides of chest. Entered hospital Apr. 7, 1915.

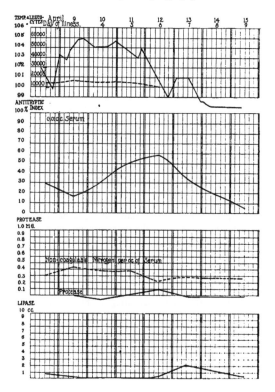

TEXT-FIG. 1. Serum changes during pneumonia (Case 1).

Physical Examination.—Well nourished colored man. Mucous membrane of mouth red, tongue coated, pharynx engorged. Pulse rapid. Lungs: diminished expansion on the left side, distinct tubular breathing, also vocal fremitus and dullness on percussion. Apr. 8, 1915. Urine negative except for a few granular casts.

The case progressed normally. On the 6th day the temperature began to decline and after a slight increase the following day reached normal the same evening.

The following facts will be noted from Text-fig. 1. The antiferment curve reached its maximum on the sixth day of illness, following which it showed a progressive decline. The protease reached a maximum on the sixth day; *i. e.,* when the temperature began to decline. At this time also the non-coagulable nitrogen showed a considerable drop. Following the fall in temperature the protease remained at a low level. The serum lipase declined progressively until immediately after the crisis when it increased almost to the strength of normal serum, with a following decline. The proteoses contained in the serum gave the following results (per 5 cc. of serum):

Date.	mg.
Apr. 8, 1915	0.25
" 10, "	0.21
" 13, "	0.13

The organisms isolated from the sputum agglutinated with serum of Type I.

Case 2 (16).—(Text-fig. 2.) J. W., male, colored, aged 18. Present illness began about 2 p. m., Mar. 16, 1915, with a chill and headache, aching of limbs and body, pain in left side and chest below the heart, accompanied with shortness of breath and expectoration of a dark brown sputum, and sometimes fresh blood. Admitted to the hospital Mar. 20, 1915.
Physical Examination.—Strong, well developed young negro. Head examination negative. Heart rapid; pulse rather weak and running in character. Blood pressure, systolic 98. Lungs: increased voice and breath sounds in both lower and middle lobes of right lung; grunting at the end of expiration; some bubbling râles over right lung anteriorly. Abdomen soft and slightly tender. Liver and spleen negative.
Pneumococcus isolated from the sputum agglutinated with serum of Type I.
On Mar. 23 the patient became delirious and remained irrational for the following 6 days, during which time it was necessary to restrain him in bed.

Text-fig. 2 shows that the greatest rise in antiferment occurred after the 6th day of illness and remained high until the 13th day. The non-coagulable nitrogen reached the maximum on the 12th day; after that it declined slowly. The protease change is quite striking. There was a lack of protease action during the entire time before the crisis. On the day before the crisis the total non-coagulable nitrogen of the serum incubated under chloroform decreased from 0.27 to 0.17 mg. per cc. of serum. A control made by means of the Van Slyke method gave the following result, thus checking very well with the Folin determination:

$$mg.$$

Amino nitrogen, 2 cc., direct 0.872
"   "   after incubating, 2 cc., direct 0.692
Loss ...... 0.180 mg. per 2 cc.

ar	21	22	23	24	25	26	27	28	29	30	31	Apr1	2	3	4
Illness	6	7	·8	9	10	·11	12	13	·14	15	·16	1ᵗ	18	19

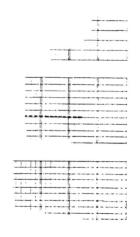

TEXT-FIG. 2. Serum changes during pneumonia (Case 2).

On the following day, however, we noted a marked protease action, after which the ferment was again practically negative. The lipase reached the lowest titer on the 7th day, after which there was a gradual increase until the 14th day.

Case 3 (12).[3]—O. B., male, white, aged 18. Mar. 23, 1915, patient had a chill followed by a headache; pain in chest, especially on the right side; coughed a good deal, expectorated colored sputum. Admitted to the hospital Mar. 25, 1915.

Physical Examination.—Strong, robust boy. Head, heart, arteries, and pulse negative. Lungs: increased fremitus and voice sounds, marked increase over the right lobe, breath sounds also increased, percussion very flat. Abdomen slightly tender. Liver and spleen not palpable. Blood pressure, systolic 128. Urine contained albumin and granular casts. Clinical course was uneventful. The temperature fell by lysis, reaching normal on the 8th day.

The antiferment titer, which was relatively low, reached a maximum on the 6th day after which it declined progressively. The maximum of non-coagulable nitrogen was reached on the 6th day, declining slowly after that time. The protease was strongly positive on the 6th and 8th days of illness, corresponding to the period of lysis. The lipase reached the maximum on the 6th day, after which it increased almost to normal.

Case 4 (17).—G. T., male, white, aged 18. Patient was well and at work when without seeming cause he had a hard chill about 3 p. m. on Apr. 3, 1915. The chill continued through the night and until about 9 o'clock the next morning, together with pain in axillary region. After the chill he states that he perspired; had dyspnea; developed a cough at the time of the chill which became worse after the chill. Admitted Apr. 5, 1915.

Physical Examination.—White boy, fairly well nourished. Head negative; throat somewhat engorged. Heart normal; P. M. I. in the fifth interspace, nipple line; pulmonic second accentuated; arteries in good condition. Inspection shows decreased expansion on the left side and lower lobe both in front and behind; increased tactile fremitus; vocal resonance and tubular breathing on auscultation. Vocal fremitus is at first absent over the lower lobe on the left side, but is noted on change of position and after coughing. Abdomen normal. Urine negative. The crisis occurred on the 7th day.

In this case the antiferment did not reach a maximum until the day following the crisis, after which it declined. The non-coagulable nitrogen was never very high and showed little change. The protease before the crisis was negative; at the time of the crisis it became normal. The lipase showed the usual change. Proteoses per 5 cc. of serum:

Date.	*mg.*
Apr. 6, 1915	0.14
" 9, "	0.0

[3] This case was from the City Hospital.

Case 5 (20).[4]—R. M., male, colored, aged 20. Present illness commenced the last week in March with pain in the right side. He had no chill but vomited the next day. Since he became ill he has been in bed at a railroad camp; is somewhat constipated; has coughed a good deal, expectorating much sputum, sometimes blood tinged; has pain, now worse than at the onset on each inspiration and when he coughs, more especially over abdomen and lower chest on right side; never had a chill.

Physical Examination.—Negro, poorly nourished. Head negative. P. M. I. inside of nipple line in fifth interspace; sounds are regular, no murmurs; second pulmonic accentuated; pulse regular and full, and of good volume. Blood pressure, systolic 103, diastolic 60. Lungs: respiration rapid, decreased expiration on the right side; tactile fremitus increased over right lower lobe together with decreased voice sounds, dullness, and tubular breathing. There is a friction sound behind and over the right lower lobe. The middle lobe of the right lung is quite flat with some tubular breathing, increased voice sounds, and fremitus. The left lung is normal. The abdomen is rigid and tender and the lymph glands especially are enlarged. Apr. 9, 1915. Urine examination: albumin negative; numerous casts. Apr. 14, 1915. Some hyaline and granular casts. Apr. 8, 1915. Blood examination: red cells 4,800,000; hemoglobin 70 per cent. Apr. 9, 1915. Condition worse; temperature has risen to 103° F., and pulse to 140. Examination shows no change over right middle lobe; right lower lobe still shows impairment with bronchovesicular breathing and increased voice sounds, except the lower part in posterior axillary line, where there is flatness with diminished breath and voice sounds. The upper right lobe shows slight impairment with a tendency to bronchovesicular breathing; posterior shows nothing as yet definite. Left lung clear throughout. Heart sounds rapid and of only fair quality, though the second pulmonic is still accentuated. Needle 'was inserted next to right inner area but no fluid obtained. Blood pressure, systolic 116. Apr. 10, 1915. Patient much improved, temperature having fallen this morning to 99.4° F. and pulse to 110. Patient expectorated yesterday considerable brownish material. Examination to-day shows much less impairment over middle lobe except in region of anterior axillary line where it is still flat with tubular breathing and exaggerated voice sounds; many moist râles over whole lobe. Lower lobe is clearing up except at base where there are still diminished breath sounds and voice sounds. Blood pressure, systolic 105, diastolic 60. Apr. 11, 1915. General condition excellent. Temperature is normal and pulse 100. Examination shows but little impairment over middle lobe, with only slight voice sounds and breath changes. Apr. 12, 1915. Blood pressure, systolic 95.

As will be noted, the patient entered the hospital practically at the end of his illness, the antiferment and non-coagulable nitrogen declining very rapidly, while the protease tended to rise. 5 cc. of serum contained 0.2 mg. of proteoses on April 11, 1915.

[4] This case was from the Vanderbilt Hospital.

Case 6 (21).[5]—H. H., male, white, aged 18. In 1913 patient had rheumatism in both arms and legs; was ill in bed about a month. Present illness began Apr. 6, 1915, with pain on left side in nipple region. Apr. 7, 1915. Had chill and high fever, expectorated, cough was dry and hacking. Since then cough has been deep seated and he has expectorated more than at first. Entered hospital Apr. 9, 1915.

Physical Examination.—Fairly well nourished. Head negative. P. M. I. normal, arteries soft and compressible, accentuated pulmonic second sound. Lungs: deficient expansion on left side, decreased fremitus over left upper lobe in front, dullness and feeble breath sounds over left upper lobe, tubular breathing in the left axilla. Blood pressure, systolic 95. Crisis on the 8th day. General condition excellent throughout.

The antiferment titer reached a maximum on the 6th day, after which it declined until the day following the crisis. The non-coagulable nitrogen reached its maximum on the 5th day, at which time the protease was lowest. The protease reached + 0.02 mg. per 1 cc. on the day following the crisis. The lipase showed the usual curve. Proteoses per 5 cc.:

Date.	*mg.*
Apr. 9, 1915	0.19
" 13, "	0.12
" 14, "	0.13

Pneumococcus of Type I was isolated from the sputum.

Case 7 (18).—T. A., male, colored, aged 30. Present illness began with headache about a week ago. On Apr. 5, 1915, 3 or 4 days following headache, he had chill about 10 a. m., which lasted for about 3 hours; no vomiting, but fever following chill; pain in the right side. Admitted Apr. 6, 1915.

Physical Examination.—Well nourished colored man. Head negative; some pyorrhea; pulse 120, but full in volume. P. M. I. fifth interspace mammary line; no murmurs. Increased fremitus and voice sounds on the right side in front, no dullness on percussion. Abdomen rigid, otherwise negative.

The patient showed a rather irregular course. On the 7th day the temperature dropped to normal, but he complained of a severe sore throat the same evening and.a marked tonsilitis was noticed. This continued for 3 days. After this time the temperature again reached normal. Apr. 17. Pulse 118, respiration 30. Tonsils engorged, with some exudate. Lungs: fremitus increased over upper and middle right lobes; very dull; occasional moist râle. Breath sounds bronchial. This case is rather interesting because of the period on the 7th day when it appeared clinically that the patient was having a crisis.

At this time the non-coagulable nitrogen had decreased and the protease reached + 0.05 mg. per 1 cc. The antiferment also

[5] This case was from the City Hospital.

showed a slight fall from the height reached on the previous day. Then, however, the tonsilitis set in and we found the antiferment continuing at a high titer, while the non-coagulable nitrogen increased again and the protease fell. The resolution seems to have been delayed, for the clinical signs began to clear only after April 17, when the protease again reached + 0.05 mg. per 1 cc. of serum. The proteoses determined per 5 cc. of serum gave the following figures:

Date.		*mg.*
Apr. 6, 1915	....	0.125
" 14, "	....	0.26
" 16, "	....	0.13
" 18, "	....	0.00

Case 8 (14).[6]—W. A., male, colored, aged 35. Present illness began Mar. 31, 1915, with pain in the left side, coughing, and expectoration of a bloody sputum : he lost his appetite but has not been nauseated; there was diarrhea; no chill. Admitted to hospital Apr. 1, 1915.

Physical Examination.—Well nourished young man. Teeth bad. Heart normal; second pulmonic accentuated. Lungs: tactile fremitus normal on both sides, except in front on left side where it is increased downward and outward from the left nipple line. Percussion is normal except for the above mentioned area, which is very dull. On auscultation, vocal fremitus is normal except over the same area where it is increased; there is bronchophony and tubular breathing on inspiration and expiration, together with coarse râles. Abdomen is negative. Blood pressure, systolic 90 and diastolic 50. Urine shows few hyaline and granular casts. Apr. 1, 1915. Red blood cells 5,000,300. Apr. 4, 1915. Condition good. Temperature dropped suddenly to normal at 9 a. m., pulse and respiration dropping with it. Signs over original area practically cleared up, beyond moist râles. No impairment and no tubular breathing: no signs over rest of lower lobe, beyond a few râles. Over left upper lobe there is flatness, with exaggerated tactile and voice frictions and bronchial breathing; moist râles are heard over this area. Blood pressure, systolic 80, diastolic 50. Apr. 5, 1915. Temperature went up again last evening to 101° F. Examination this morning shows complete consolidation of the upper lobe with exaggerated tactile fremitus, vocal resonance, and bronchial breathing: original area shows no change in signs. Abdomen quite distended. Apr. 6, 1915. Temperature down around 99° F., pulse slow and of better volume Patient is resting better. Pain in side not severe; signs still present on left side. Blood pressure, systolic 95, diastolic 60. Apr. 7, 1915. Blood pressure, systolic 80. Apr. 8, 1915. General condition good. Temperature has been practically normal for last 2 days. Signs over lower left lobe have practically cleared up: there is some slight impairment over left upper lobe with bronchovesicular breathing and a few râles. Apr. 11, 1915. Condition good. Temperature has been normal

[6] This case was from the Vanderbilt Hospital.

now for almost 4 days. Examination of lung to-day shows no impairment or evidence of previous trouble except occasional râles at end of inspiration.

The antiferment curve reached its maximum at the time the temperature first fell to normal. It is quite evident, however, that at this time the patient had no crisis, for the leucocytes continued to rise until the following day. The non-coagulable nitrogen, however, was much lower after the fifth day of illness. The protease was positive on the sixth and ninth days of illness; after that time the patient made an uneventful recovery. The lipase was rather high throughout.

Case 9 (29).[7]—H. G., male, colored, aged 35. Illness began Apr. 6, 1915, when he began to cough a great deal, which caused pain in left side; expectorated much thick brownish sputum; had no chill. Entered the hospital Apr. 11, 1915.

Physical Examination.—Icteric; mucous membrane of mouth red; tongue coated; heart and arteries normal. Lungs: left lung shows deficient expansion, tactile fremitus increased, and large râles and spoken voice sounds heard in both upper and lower lobes, together with dullness on percussion. Abdomen negative. Reflexes normal.

This patient is the only one in the series that we have studied who showed a constant amount of protease in the serum during the disease and showed no change during the crisis, unless it was a slight fall. The antiferment remained high even after the crisis. The lipase remained low throughout. Proteoses per 5 cc. of serum:

Date.	*mg.*
Apr. 16, 1915	0.21
" 17, "	0.12

Case 10.[8]—(Text-fig. 3.) J. M. J., male, white, aged 40. Became ill Jan. 31, 1915, when he had a chill. Entered hospital Feb. 5, 1915.

Physical Examination.—Typical right lower lobe involvement. Urine contained numerous casts. White blood count 10,200. Sputum contained numerous pneumococci (type not determined). The patient showed evidence of profound intoxication, was stuporous, with rapid pulse and respiration. On the 10th day the temperature dropped by crisis, but the physical condition, especially the pulse rate, continued unfavorable. From the 9th to the 12th day the patient was delirious. After the 12th day the recovery was rapid.

[7] This case was from the City Hospital.
[8] This case was from the City Hospital.

The course of the disease was marked by a rather low leucocytosis and a large amount of non-coagulable nitrogen in the serum. The temperature fell as the amount of protease increased, but there was considerable intoxication for the three days following, as evidenced

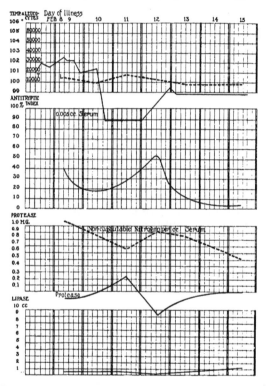

Text-Fig. 3. Serum changes during pneumonia (Case 10).

by the increase in antiferment, the drop in protease, and the rise in non-coagulable nitrogen. The lipase began to return toward normal after this second period of intoxication.

Case 11 (22).[9]—E. H., male, colored, aged 30. Patient came to hospital because of pain in lower right chest, which increased on deep inspiration; this was first noted Apr. 4, 1915, when coughing accompanied pain. Apr. 7, 1915.

[9] This case was from the City Hospital.

Had a chill and headache since that time. Admitted to the hospital, Apr. 9, 1915.

Physical Examination.—Head negative. Heart: apex slightly down and out from the normal point; no murmurs; second pulmonic increased; pulse full, rather hard but regular. Lungs: right, slight dullness of lower lobe in front; no increase in voice sounds and tactile fremitus; left lung normal. Blood pressure, systolic 110, diastolic 65.

There was a progressive rise in the serum protease, reaching a maximum on the seventh day of illness, at which time the non-coagulable nitrogen had largely decreased. The antiferment decreased after the seventh day. Proteoses per 5 cc. of serum:

Date.	*mg.*
Apr. 12, 1915	0.13
" 13, "	0.30
" 14, "	0.29
" 15, "	0.15

Pneumococci from sputum agglutinated with serum of Type I.

Case 12 (23).[10]—G. F., male, colored, aged 20. Apr. 4, 1915. Pain in back and left side on breathing; no chill; has had some fever and expectorates blood tinged sputum; some coughing; shortness of breath. Entered hospital Apr. 11, 1915.

Physical Examination.—Eyes and head negative. Arteries soft, heart normal. P. M. I. fifth interspace inside of nipple space. Lungs: dullness, tubular breathing, increased voice sounds, tactile fremitus over left lower lobe, friction sound heard over left lower chest. Abdomen rigid. Liver and spleen not palpable, otherwise normal.

The patient became irrational and delirious the second day after admission to the hospital. On the 11th day of illness the pneumonic process involved the upper lobe on the same side, evidenced by area of tubular breathing and dullness. The patient died on the 15th day. No autopsy was obtained. A pneumococcus agglutinating with serum of Type II was obtained from the sputum.

There was a rather constant increase in antiferment, which reached a maximum on the eleventh day and remained high after that time. The protease was low during the course of the disease but showed a gradual rise. The non-coagulable nitrogen remained relatively low throughout. Proteoses per 5 cc. of serum:

Date.	*mg.*
Apr. 12, 1915	0.1
" 16, "	0.0

[10] This case was from the City Hospital.

Case 13.[11]—(Text-fig. 4.) J. H. D., male, colored, aged 14. Jan. 28, 1915, 3 a. m. Became suddenly ill with a severe chill and pain on the right side of chest. Has coughed for the past few days. Admitted to hospital 9 a. m., Jan. 28, 1915.

Physical Examination.—Well nourished negro boy. Physical findings negative. No evidence of consolidation. White blood count 40,000. Jan. 29. Dullness and bronchial breathing were noted over entire right lung. The patient became very restless and coughed considerably. Jan. 30. All evidences of a complete consolidation of the entire right lung were present. There was con-

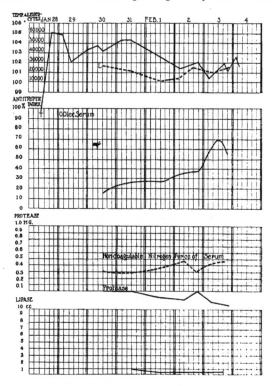

TEXT-FIG. 4. Serum changes during pneumonia with fatal termination (Case 13).

siderable pain and restlessness. Feb. 3. In the morning the patient showed definite signs of improvement, was quiet, and numerous large râles were noted over the involved area. During the following night, however, he became much worse and died at 5.30 a. m.

[11] This case was from the Vanderbilt Hospital.

582   *Serum Ferments and Antiferment during Pneumonia.*

Data from the Postmortem Protocol.—Body is that of a negro boy 14 years old, 172 cm. in length, well developed, and well nourished.

Lungs.—The right lung weighs 1,080 gm. The surface is smooth with the exception of the posterior part of upper lobe. There are many soft, friable adhesions. There is a small amount of amber-colored fluid in the pleural cavity. The lung when removed from the chest maintains its form. There are definite areas of consolidation in the lower part of the upper lobe. The middle lobe is free. On section through these consolidated areas the surface yields very little fluid and the lung tissue is definitely consolidated, of a grayish appearance, and around the margins of each there is a red zone. The left lung weighs 740 gm. The pleura is smooth. The upper lobe is normal. The lower lobe shows areas of consolidation similar to those described in the right.

Anatomical Diagnosis.—1. Lobar pneumonia of both lower lobes and bronchopneumonia of the right upper lobe. 2. Fibrinous pleuritis. 3. Acute splenitis. 4. Fatty changes and cloudy swelling of the liver, kidneys, and myocardium.

It will be seen from Text-fig. 4 that there was a progressive increase in the non-coagulable nitrogen until the evening of February 2, when a fall occurred. At this time there was also a distinct change in the ferment, so that on incubation a loss of non-coagulable nitrogen no longer occurred, although the serum did not actually become proteolytic. Instead, the non-coagulable nitrogen again increased and the ferment fell while the antiferment increased markedly. It will be observed that clinically the patient was in good condition during the night and morning after the temporary fall in non-coagulable nitrogen and the rise in ferment. The lipase showed simply a continued decrease.

Case 14.[12]—H. M., male, colored, aged 30. Became ill with chill and thoracic pain on Dec. 11, 1914. Entered hospital Dec. 15, 1914.

Physical Examination.—Breath sounds diminished on left side, with vocal fremitus increased over right upper lobe posteriorly. Pulse rate 108 to 120. Heart normal. Patient continued to run a typical course and the temperature dropped to normal on the 9th day, the respiration dropping from 38 to 24, and the pulse rate from 132 to 108 from the afternoon of the 19th to the morning of the 20th day. The temperature, however, again rose and continued to be irregular. Dec. 28, 1914. Patient developed meningeal symptoms. Lumbar puncture at this time yielded a small amount of turbid fluid. The pressure was not increased, nor were organisms obtained on examination. Dec. 31, 1914. Patient died.   •

Autopsy.—Dec. 31, 1914.

Heart.—Weight 450 gm. The tricuspid valve measures 14 cm. The pulmonary valve measures 7 cm. The cusps of the mitral valve contain vegetations measuring from 2 cm. in diameter to others which are smaller. These vegeta-

[12] This case was from the City Hospital.

tions are firmly attached to the cusps. The distal surface is roughened and corrugated; otherwise they present a smooth surface. The cusps of the aortic valve are smooth and shining. The myocardium is brownish red in color and presents several grayish white patches, which are evidently fibrous in character. The left ventricle measures 17 mm., the right ventricle 7 mm. in thickness.

Lungs.—The left lung weighs 840 gm. The lung is entirely covered with a fibrous and fibrinous exudate, which binds it to the chest wall. It is bluish black in color. The consistency is increased and the organ contains no air except in the apex. The bronchi are bright red in color and contain a mucopurulent exudate. On section the surface is dark gray in color. In the lower lobe the process is diffuse. In the upper lobe it is nodular in character. The centers of the nodules present a grayish color and the peripheries are red. The cut surface is rather smooth. The right lung weighs 500 gm., and is bluish black in color. The edges are rounded. The surface is smooth. The upper and lower lobes are bound together by old fibrous adhesions. The lung is markedly emphysematous. On section the surface is pinkish gray in color and dry.

Brain.—On removing the dura the cortex is seen to be covered with a greenish, cloudy, fibrinous exudate. This involves the frontal and parietal lobes of the cortex. Occiput shows more congestion than the other two named. The fluid is increased along the sulcus and especially along the Rolandic fissure. In the fluid are flakes of a purulent material. The intima is firmly adherent to the convolutions of the brain and the exudate extends between the convolutions. The base of the brain shows the same condition as the cortex but apparently more marked in the infundibulum and the medulla oblongata. The lateral ventricle contains a thick, cloudy fluid. The choroid plexus is covered with an inflammatory exudate. The third ventricle is filled with the above described fluid. Upon section the blood vessels in the cerebellum are engorged and in many places there are many small petechial hemorrhages. The blood vessels medially to the external capsule are very much engorged. The purulent exudate is also found in the fourth ventricle.

Anatomical Diagnosis.—1. Lobar pneumonia of the left upper and lower lobes. 2. Pneumococcus meningitis. 3. Vegetative endocarditis. 4. Acute parenchymatous nephritis. 5. Cloudy swelling of the myocardium. liver, and spleen. 6. Fatty changes of the liver. 7. Fibrous and fibrinous pleuritis. 8. Emphysema of the right lung. 9. Fibrous myocarditis.

Microscopical Diagnosis.—Heart shows an increase in fibrous tissue. the muscle fibers are pale, enlarged, and some of them, especially beneath the pericardium, show fatty degeneration. One lung showed congestion, anthracosis. and compensatory emphysema; in the other lung the pleura was thickened. and many of the alveoli of the lung were collapsed, some of them showing fibroblasts and fibrous tissue. There was intense congestion and the bronchi were filled with an acute inflammatory exudate. Liver showed congestion, a few areas of necrosis, and some cloudy swelling. The capsule of the spleen is thickened; there is much congestion; and an increase in parenchymatous tissue. The capsule of the kidney is thick, glomeruli are swollen, the epithelium of the tubules is swollen, granular, and contains fat. Pancreas is negative. Stomach shows no definite change. The cerebrum is covered with an inflammatory exudate. rich in polymorphonuclear leucocytes. The pia mater is markedly thickened.

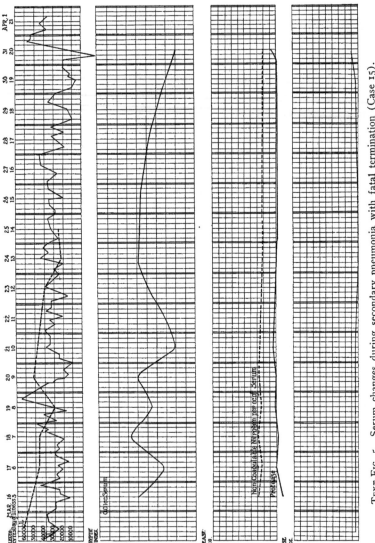

TEXT-FIG. 5.   Serum changes during secondary pneumonia with fatal termination (Case 15).

the cortex is edematous, and there is marked perivascular infiltration of polymorphonuclear leucocytes.

This case showed the usual progressive rise in antiferment to the seventh day of illness, and the temperature reached normal on the ninth day. The protease was only determined three times, showing a progressive rise, however, about the time of improvement of the condition clinically. No signs of resolution were noted, however; the antiferment again rose, as did also the non-coagulable nitrogen. The lipase, except for an early rise, remained low. The high antiferment index may possibly be associated with the lack of autolysis in the lung described above.

Case 15.[13]—(Text-fig. 5.) *J. T.*, female, colored, aged 14. Mar. 6. 1915. Abortion of 3 months' fetus, followed by fever 3 days later, together with dyspnea, cough, and general malaise. Entered hospital Mar. 12, 1915.

Physical Examination.—Young mulatto girl, well nourished. Heart action vigorous and rapid; second pulmonic sound accentuated; slight systolic thrill; no murmurs. Lungs: diminished expansion on left side; increased tactile fremitus over left upper lobe; dullness over lower portion of left upper lobe posteriorly and in axilla; increased voice sounds, bronchial breathing; and a few râles over lower portion of upper left lobe. Mar. 16. The entire left lobe became involved, with more numerous râles over the upper lobe. Mar. 17. There was distinct evidence of involvement of the lower right lobe. and the pulmonic second sound was found less accentuated. Mar. 18. Some signs of resolution were noted in the left lower lobe. Mar. 19. Hemorrhage. Mar. 20. Heart sounds very weak; patient in poor condition. The following days showed considerable resolution, but the cough continued, and the sputum became rather offensive. Apr. 1. Patient in a very critical condition. Patches of bronchial breathing and bronchophony over posterior part of both upper lobes. and in front on left side, with a patch in the right axilla. Great amount of offensive sputum. Died in the evening after a rather profuse hemorrhage.

In this case of septic pneumonia the fluctuating involvement of the lungs finds its reflection in the constantly changing antiferment curve, and the non-coagulable nitrogen also shows an almost daily variation. The protease remained low throughout. despite the clinical evidence of resolution in part of the involved areas. An autopsy was not obtained.

Analysis of Cases.

On examination of Text-figs. 1 to 5 there will be noted a very constant increase in the serum protease about the time of crisis. together

[13] This case was from the Vanderbilt Hospital.

with a decline in the antiferment titer. These two factors would indicate that at this time conditions for autolysis would be most favorable, and we must remember that the sera are after all only an index of the condition that obtains in the local areas involved.

It will be noted that the non-coagulable nitrogen decreases, as a rule, after the crisis, as Rzetkowski (9) has already pointed out, and the proteoses also diminish markedly.

When we average the various changes that have occurred in the eight cases that ended by crisis we obtain the figures given in Table I.

TABLE I.

Average Antiferment Titer.

Day before crisis. *per cent*	Day of crisis. *per cent*	Day following crisis *per cent*
57	50	46

Average Total Non-Coagulable Nitrogen per Cc.

mg.	*mg.*	*mg.*
0.50	38	40

Average Protease Action per Cc. (Autolysis under Chloroform).

mg.	*mg.*	*mg.*
— 0.05	+ 0.05	+ 0.02

Average Lipase per Cc.

0.45 cc. $\frac{N}{100}$ NaOH	0.7 cc. $\frac{N}{100}$ NaOH	0.9 cc. $\frac{N}{100}$ NaOH

The protease activity which is demonstrated is probably not derived from the leucocytes undergoing autolysis in the pneumonic lung. If that were true we would expect that the ferment would continue to be present after the autolysis had once begun. That this is not the case is best illustrated in Text-fig. 5, when clinically all signs of resolution were present while the protease curve remained negative. The ferments are probably mobilized from the tissue cells in general.

DISCUSSION.

In a paper discussing the crisis in pneumonia, Hektoen (10) has recently reviewed the present status of our knowledge and concludes that the crisis is the effect of the prompt destruction of the infecting organism at a time when the antipneumococcal reaction reaches a certain height. But the mere concentration of antibodies can hardly explain the whole picture, for in pneumonia, as in typhoid, such immune reactions may be quite marked, but the organism nevertheless may remain viable. He remarks that the crises cannot be wholly dependent on resolution, for, while they usually concur, the crises may precede resolution. But autolysis may take place in a lung without affording immediate evidence by physical finding, just as readily as the early infection may be well established in the lung before signs of consolidation can be elicited.

Dick (11) and Rosenow (12) have studied specific proteases both in the immune animal and in clinical material, Dick determining an increased proteolytic power for pneumococcus protein at the time of the crisis, and associating it with an increase in the complement strength. Even if such proteolysis were specific, rather than based on colloidal changes following in the wake of the usual immunity reactions, it would have no influence on living pneumococci, for proteolytic ferments, as is well known, are without action on intact organisms (13).

That autolysis plays a large share in the pneumonic picture has been shown by Müller (14). Flexner (15) has discussed more especially the conditions present in unresolved pneumonia.

Before entering into a discussion of the relation of autolysis to the crisis, we believe that our ideas concerning the factors that produce the intoxication must be amplified. Quite naturally the entire emphasis has so far been placed on the pneumococcus. Kruse (16) and Cole (17) have discussed the work so far accomplished in the study of pneumococcus toxicity. Vaughan (18) has shown that pneumococcus proteins are relatively less toxic than those of other organisms. The toxicity is manifested on simple solution, to which the pneumococcus is rather liable, e. g., with bile salts (Neufeld), and soaps (Lamar); by freezing (Cole); lipoid solvents (Jobling and Strouse); and also during autolysis, as Rosenow has demonstrated (19). Rosenow determined the rate of autolysis by means of the Sorensen method.

Using the Folin method and coagulating by means of heat and acid, we have not been able to convince ourselves that the pneumococcus undergoes autolysis more readily than other organisms. In one experiment the following rate of autolysis to non-coagulable forms of nitrogen was noted:

	hrs.	per cent
Pneumococcus	24	6
"	48	11
Typhoid	24	4
"	48	25
Staphylococci	24	30
"	48	45

It is possible that other strains might show some difference. The organisms used contained from 3.6 to 4.7 per cent of total lipoids: the degree of unsaturation was not determined because the amount of lipoids obtained was too small. If the pneumococcus at the time of crisis was subject to extensive lysis and was the sole factor in the causation of the toxemia, we should expect an increase rather than a lessening in intoxication, for during lysis toxic material is liberated from the organism.

The fact which we wish to emphasize is the great mass of fibrin and leucocytic debris which dominates the pneumonic picture. This to all intents and purposes represents foreign proteins as far as the lung is concerned and while undergoing solution must give rise to toxic split products. While the inhibitory factors are in the ascendency, this material is undergoing a very slow autolysis with the splitting proceeding only to the higher and toxic products which are absorbed as such. The balance being once destroyed, and the autolytic process allowed full sway, the splitting proceeds much farther to non-toxic products. This period is characterized clinically by the greatest excretion of nitrogen following immediately upon the crisis. We believe that it is this factor (to which Cole (20) has already referred) that must be taken into consideration when we seek to define the various potentially pyrogenic sources.

We regard the amount of proteoses present in the serum as an indication that such a condition actually obtains. It will be noted from the various determinations made that these reach two or three times the amount present in the early serum samples and fall off quite markedly after crisis. These must be derived either from the resolving lung tissue, or the tissue cells in general, for their amount is too large for the pneumococcus protein to be considered their source. It is recognized clinically that fibrin and blood clots or aseptic tissue autolysis in general give rise to a febrile condition as readily as an accumulation of leucocytes among which the infecting organism may have completely disappeared. The source of the toxin here is quite palpably the autolyzing homologous proteins. It seems only reasonable, therefore, that the same condition, even in an exaggerated form, should, during the pneumonic process, give rise to an intoxication. Under such circumstances the pneumococcus intoxication proper would be only an accessory factor, and would account for the fact that the rapid destruction of the organisms which must occur during and after the crisis is accompanied by no further evidence of intoxication.

CONCLUSIONS.

1. The crisis in pneumonia is usually accompanied by (a) decrease in the serum antiferment; (b) the mobilization of a nonspecific protease in the serum; (c) an increase in serum lipase; (d) a decrease in the non-coagulable nitrogen, and of the proteoses in the serum.

2. The crisis is associated with the beginning of an active autolysis, the latter depending on an altered relation between the ferment-antiferment balance.

3. The fibrin and leucocytic debris must be considered as one of the potential sources of toxic substances. With rapid autolysis proceeding, only non-toxic materials are absorbed.

BIBLIOGRAPHY.

1. Kline, B. S., and Winternitz, M. C., *Jour. Exper. Med.*, 1915, xxi. 311.
2. Opie, E. L., *Jour. Exper. Med.*, 1905, vii, 316.
3. Jobling, J. W., and Petersen, W., *Arch. Int. Med.*, 1915. xv. 286.
4. Jobling and Petersen, *Ztschr. f. Immunitätsforsch , Orig.*, 1915. xxiv (in press).
5. Almagià, M., abstracted in *Centralbl. f. Biochem. u. Biophys.*, 1913–14. xvi. 283.
6. Guggenheimer, H., *Deutsch. Arch. f. klin. Med.*, 1913, cxii. 248.
7. Ascoli, M., and Bezzola, C., *Berl. klin. Wchnschr.*, 1903. xl. 391.
8. Jobling, J. W., Eggstein, A. A., and Petersen, W., *Jour. Exper. Med.*, 1915. xxi, 239.
9. Rzetkowski, K., *Gaz. lek.*, 1912. xxxii, series 2, 419.
10. Hektoen, L., *Jour. Am. Med. Assn.*, 1914. lxii. 254.
11. Dick, G. F., *Jour. Infect. Dis.*, 1912, x, 383.
12. Rosenow, E. C., *Jour. Infect. Dis.*, 1912, xi, 286.
13. Jobling and Petersen, *Jour. Exper. Med.*, 1914. xx, 452.
14. Müller, F., *Verhandl. d. Kong. f. inn. Med.*, 1902, xx, 192.
15. Flexner, S., *Univ. Penn. Med. Bull.*, 1903–04. xvi. 185.
16. Kruse, W., *Allgemeine Mikrobiologie*, Leipzig, 1910. 958.
17. Cole, R., *Jour. Exper. Med.*, 1914. xx, 346.
18. Vaughan, V. C., Vaughan, V. C., Jr., and Vaughan, J. W., *Protein Split Products in Relation to Immunity and Disease*, Philadelphia and New York, 1913.
19. Rosenow, *Jour. Infect. Dis.*, 1911, ix, 190; 1912, x, 113.
20. Cole, *Arch. Int. Med.*, 1914. xiv, 56.

SERUM CHANGES FOLLOWING KAOLIN INJECTIONS.

Studies on Ferment Action. XXV.

By JAMES W. JOBLING, M.D., WILLIAM PETERSEN, M.D., and A. A. EGGSTEIN, M.D.

(*From the Department of Pathology, Medical Department, Vanderbilt University, Nashville.*)

(Received for publication, May 19, 1915.)

The observation of Keysser and Wassermann (1), that the serum of certain animals becomes toxic when incubated with an inert substance such as kaolin, has stimulated considerable interest in the mechanism of the reactions that occur. Keysser and Wassermann considered the phenomenon due to an adsorption of amboceptor. Gengou had noted that quite indifferent substances (he used barium sulphate and calcium fluoride) were hemolytic and that this effect was held in abeyance by serum.

Friedberger and Kumagai (2) made similar observations with kaolin and determined that the serum albumin and globulin protected against the hemolytic effect, while the split products of the proteins protected in a degree corresponding to their complexity. They decided that the inhibiting substance of the serum was not lipoidal, for they found that if they washed the serum ten times with ether the serum still protected. They do not explain the loss of the inhibitory effect of the serum, for if this is due solely to protein bodies we can hardly see why single or even repeated treatments with the kaolin should exhaust the serum.

Friedberger at first considered the effects of intravenous injection of inert substances, such as kaolin, as purely mechanical, *i. e.*, plugging of the capillaries of the brain and lung, in this way leading to an acute death. Later, however, together with Tsuneoka (3) he admitted that the toxic effect could not be so interpreted, for after treatment with serum the kaolin was non-toxic.

In previous papers (4) we have developed the idea that it is the serum antiferment which is most readily adsorbed from the serum by inert substances such as kaolin and barium sulphate. By such adsorption local areas of antiferment deficiency are formed where proteolysis can take place, with a resulting production of toxic split products. Inasmuch as this is the mechanism in anaphylatoxin formation we might expect that the intravital injection might give evidences of a similar condition. The following experiments have been carried out with this point in view.

James W. Jobling, William Petersen, and A. A. Eggstein. 591

Dog 11.—Weight 5.5 kilos. 0.05 gm. of kaolin suspension was injected intravenously at 9.15 a. m. The animal showed evidence of malaise; was nauseated; temperature increased slightly after 1 hour but fell in the afternoon.

From Text-fig. 1 it will be observed that there was at first a sharp rise in the antiferment, followed by a gradual decline. The serum lipase showed a slight decrease, as did also the protease. A second

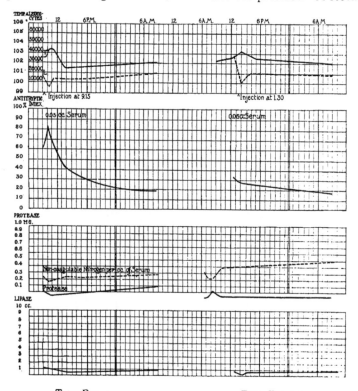

TEXT-FIG. 1.   TEXT-FIG. 1 a.

TEXT-FIGS. 1 AND 1 a. Serum changes following intravenous kaolin injections.

injection was made 4 days later (Text-fig. 1 a). double the amount of the first injection being used (0.1 gm.). The temperature and the leucocytic reaction were similar, although the antiferment

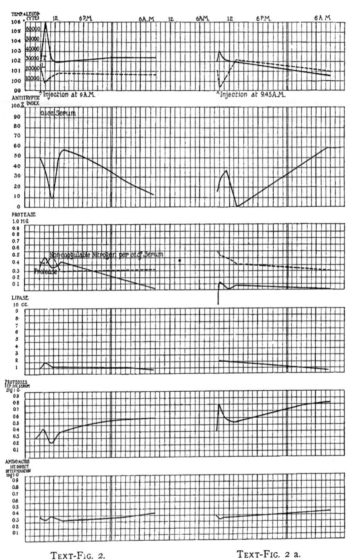

TEXT-FIG. 2.   TEXT-FIG. 2 a.

TEXT-FIGS. 2 AND 2 a. Serum changes following intravenous kaolin injections.

showed less alteration. In the afternoon there was a considerable increase in the non-coagulable nitrogen in the serum. The protease was only temporarily increased.

Dog 66.—Weight 6 kilos. 0.5 gm. of kaolin was injected at 9.45 a. m. (Text-fig. 2 a). This amount caused complete prostration with nausea, vomiting, and diarrhea. The blood taken 15 minutes after the injection remained fluid.

As will be noted from Text-fig. 2 a there was an immediate rise in antiferment, followed by a secondary fall with recovery the following morning. The non-coagulable nitrogen decreased while there was some increase in the protease, which remained low throughout. The lipase showed a gradual decrease. There was an immediate increase in the proteoses in the serum, while at the same time a slight decrease in amino-acids was noted.

Dog 63.—Weight 6.3 kilos. Sensitized Apr. 2, 1915, with horse serum albumin. 0.5 gm. of kaolin was injected at 9 a. m., Apr. 21, 1915.

This animal differed in its response in the greater rise in temperature and an immediate fall in antiferment (Text-fig. 2); indeed the antiferment curve is exactly the reverse of that of Dog 66. There was noted an immediate rise in proteoses, while the amino-acids decreased slightly as in Dog 66.

The following dogs (Table I) received doses which were almost immediately followed by evidence of a profound shock and an early death. In each case the symptoms were identical.

TABLE I.

Dog No.	Weight. kilos	Injection.				Result.
67	5.0	1	gm. kaolin in suspension			Died, 10 min.
65	6.2	0.8	"	"	"	"	"	20 "
68	6.2	0.6	"	"	"	"	"	5 "
69	6.0	0.4	"	"	"	"	15

The blood examinations immediately before the injection and at the time of death are given in Table II.

It will be observed that there is a distinct increase in the amount of protease in the serum, as also of the higher split products, while the amino-acids show practically no change. The antiferment titer is markedly increased.

TABLE II.

Dog No.	Time.	Total non-coagulable nitrogen per cc.	Pro-tease action per cc.	Proteoses per 5 cc. of serum.	Amino-acids per cc. (direct determination).	Lipase per cc.	Antiferment inhibition.
		mg.	*mg.*	*mg.*	*mg.*	*cc. N/100NaOH*	
67	Before injection	0.28	0.22	0.5	0. 2	0.5	17% per 0.1 cc.
	After "	0.32	0.30	0.74	0.30	0.2	50% " " "
65	Before "	0.28	0.04	0.45	0.41	2.2	33% " 0.2 "
	After "	0.28	0.37	0.41	0.42	2.2	75% " " "
69	Before "	0.31	0.37	0.55			
	After "	0.32	0.68	0.83			

DISCUSSION.

The experiments which have been detailed above give evidence that the toxicity of an inert substance such as kaolin depends on adsorption phenomena which result in protein splitting. Friedberger and Tsuneoka consider that an adsorption of certain substances from cells essential to life takes place. If that were the case we should expect that the serum would protect against the toxic effect, or, if this protection were not sufficient, that the red cells would be the first to be injured with a resulting hemolysis. This, however, never takes place *in vivo,* even during the acute shock. On the other hand, the shock bears much resemblance to the intoxication brought about by protein split products in its effect on coagulation, gastro-intestinal intoxication, the immediate leucopenia, etc. If this is the case we should expect to find an increase in protein split products in the serum and an increase in the serum protease.

That this is actually the case will be observed from the text-figures and Table II. The only difference to be noted is the absence of any rise in the serum lipase which follows shock in anaphylaxis and peptone poisoning. If the adsorption that occurs when the kaolin reacts with the serum *in vitro* or *in vivo* were one of proteins, as Friedberger suggests, there would be no occasion for protein splitting. If, on the other hand, an adsorption of antiferment occurs (the antiferment being lipoidal is adsorbed most readily from the serum), we can easily understand that wherever such adsorption is in progress a localized area of antiferment deficiency must for the moment exist in which splitting can occur whenever ferments are present.

Just as in anaphylatoxin formation the matrix of the split products is to be found in the serum proteins, the kaolin acts merely as an agent which for an instant alters the normal ferment-antiferment balance and in this way brings about a protein intoxication.

Considering the fact that the first effect of acute shock is an increase of proteoses with practically no effect on the amino-acids, it would seem probable that the splitting takes place only to the higher stages. The splitting evidently differs from that occurring during anaphylactic shock (5), when we find first a decrease in proteoses and an increase in amino-acids, indicating a splitting through the peptone stage.

Whether other formed bodies, such as bacteria, ever bring about a toxic effect in this manner is of interest rather from a theoretical than a practical point of view, for the number of bacteria present in the serum would have to be very great to bring about marked changes. Bacteria do, however, readily adsorb the serum lipoids (6), and it seems probable that many of the phenomena which we classify under *"Serumfestigkeit"* are due to the changes so brought about.

CONCLUSIONS.

1. The intoxication produced by the intravenous injection of inert substances such as kaolin is due to protein split products derived from the serum proteins.

2. The kaolin acts as an adsorbing medium for the serum antiferment, bringing about an alteration in the ferment-antiferment balance.

3. The intoxication is accompanied by an increase in serum protease, and of proteoses.

4. The serum lipase, the amino-acids, and the total non-coagulable nitrogen show relatively little change.

5. The antiferment shows an initial increase, followed by a loss.

BIBLIOGRAPHY.

1. Keysser, F., and Wassermann, M., *Ztschr. f. Hyg. u. Infectionskrankh.*, 1911, lxviii, 535.
2. Friedberger, E., and Kumagai, T., *Ztschr. f. Immunitätsforsch.*, *Orig.*, 1912, xiii, 127.

3. Friedberger, E., and Tsuneoka, R., *Ztschr. f. Immunitätsforsch., Orig.*, 1913–14, xx, 405.
4. Jobling, J. W., and Petersen, W., *Jour. Exper. Med.*, 1914, xix, 459, 480; xx, 37.
5. Jobling, J. W., Petersen, W., and Eggstein, A. A., *Jour. Exper. Med.*, 1915, xxii, 401.
6. Jobling and Petersen, *Jour. Exper. Med.*, 1914, xx, 452.

THE EFFECT OF PROTEIN SPLIT PRODUCTS ON THE SERUM FERMENTS AND ANTIFERMENT.

STUDIES ON FERMENT ACTION. XXVI.

BY JAMES W. JOBLING, M.D., WILLIAM PETERSEN, M.D., AND A. A. EGGSTEIN, M.D.

(*From the Department of Pathology, Medical Department, Vanderbilt University, Nashville.*)

(Received for publication, June 1, 1915.)

In previous papers we have discussed the ferment changes that occur in dogs during trypsin, anaphylactic, and kaolin shock, noting more particularly the relation of the serum protease to the antiferment and to the split products contained in the serum, together with changes in the lipase content. In the present paper we shall present the resulting ferment changes following the injection of various protein derivatives.

The fundamental ideas of Vaughan[1] relating to the toxicity of protein split products have been fully developed during the course of the last few years; the only element of uncertainty lies in the interpretation of certain phases of specificity of ferments which are supposed to split the native protein to toxic fragments. According to Vaughan's hypothesis, specific protease is produced capable of such function, a supposition which has received much support from the work of the Abderhalden school. There is, however, certain experimental evidence to the contrary, indicating that the specific element is not due to specific ferment action, but rather that the splitting that occurs is due to a non-specific ferment: that the split products are largely derived from the serum proteins and not from the injected antigen; while the element of specificity lies in the colloidal changes which bring about a lowering of the antiferment titer and in the rapid mobilization of the protease.

[1] Vaughan, V. C., Vaughan, V. C., Jr., and Vaughan, J. W., Protein Split Products in Relation to Immunity and Disease, Philadelphia and New York, 1913.

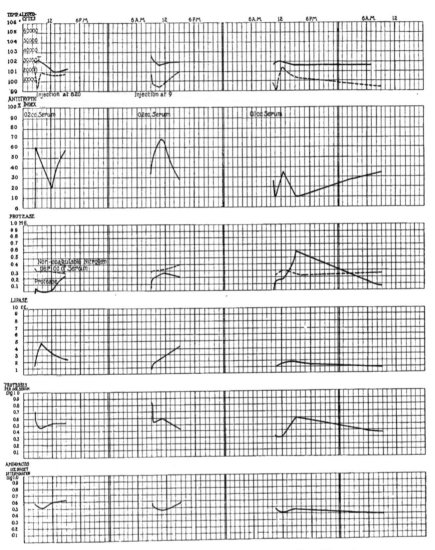

TEXT-FIG. I.   TEXT-FIG. I a.   TEXT-FIG. I b.

TEXT-FIGS. I, I a, AND I b. Serum changes following intravenous injections of (Text-fig. I) proto-albumoses, (Text-fig. I a) primary proteoses, (Text-fig. I b) secondary proteoses.

While it is true that every protein when hydrolyzed will yield toxic substances, there is considerable variation in the degree of toxicity of the derivatives from different proteins. According to Baehr[2] this difference has some relation to the chemical configuration of certain components of the fragments. It has been found that of the various split products the primary proteoses are in general most toxic (Zunz, Jobling and Strouse, Zunz and György), the secondary proteoses being much less toxic. While shock produced by the split products is usually called peptone shock, the peptones, as a rule, are less toxic than the primary proteoses. The peptone preparation which we have used was, however, an exception to this statement, displaying an instantaneous and marked toxicity. It was prepared from dog muscle by the usual method.

The technique used in these experiments has been fully described previously, so that a repetition will be superfluous.

EXPERIMENTAL.

Effect of Primary Proteoses.

Dog 28.—Weight 7 kilos. 0.2 gm. of proto-albumoses (prepared by peptic digestion of dog serum) was injected intravenously at 9.45 a. m. The dog showed practically no evidence of malaise, although there was a slight primary fall in the leucocytes with a rise during the afternoon.

There was practically no change in the ferments and the antiferments; the non-coagulable nitrogen decreased slightly, although the amino nitrogen and the proteoses were unaltered.

Dog 70.—Weight 8.1 kilos. 0.5 gm. of proto-albumoses (from Witte's peptone) was injected intravenously at 8.20 a. m. The dog became nauseated and vomited almost immediately; it showed extreme prostration during the afternoon and died during the night (Text-fig. 1).

Dog 71.—Weight 6.6 kilos. 0.5 gm. of primary proteoses (from Witte's peptone) was injected intravenously at 9 a. m. Manifestations of intoxication were the same as in Dog 70. Death occurred during the night (Text-fig. 1 a).

In both these experiments there will be noted an immediate rise in the antiferment with a subsequent fall; a mobilization of protease more marked in Dog 71; a rather sharp rise in the serum lipase; and a distinct decrease in the serum proteoses and amino nitrogen. It is

[2] Bachr, G., *Proc. N. Y. Path. Soc.*, 1913, xiii, 151.

rather curious that the injection of 0.5 of a gram of proteoses should be followed by the decrease in the amount of the proteoses in the serum.

Effect of Secondary Proteoses.

Dog 73.—Weight 6 kilos. 0.5 gm. of secondary proteoses (from Witte's peptone) was injected at 10.50 a. m. The dog became somewhat nauseated but did not show the extreme prostration that characterized the two previous dogs; it showed no ill effects the following morning.

It will be observed (Text-fig. 1 b) that there occurred a well marked rise in serum protease, while the lipase was not affected. Proteoses were increased during the afternoon, but returned to normal the following morning. The amino nitrogen showed only a slight primary decrease.

Effect of Peptone.

Dog 74.—Weight 15 kilos. 0.25 gm. of muscle peptone injected at 9 a. m. (Text-fig. 2). The animal was immediately prostrated and was nauseated, after which it rapidly recovered and showed no further ill effects.

As will be observed from Text-fig. 2 there was almost an immediate rise in protease with a following decline. The antiferment showed rather a marked fluctuation, while the lipase showed only a slight rise with a fall in titer later. The proteoses were at first decreased but increased during the afternoon. The amino nitrogen showed only a slight increase.

Dog 43.—Weight 7 kilos. 0.17 gm. of muscle peptone was injected intravenously at 11.30 a. m. (Text-fig. 2 a). The animal became ill in a manner similar to Dog 74, and later developed a paresis of the hind legs. Died at 5 p. m.

In this animal the intoxication was more evident, both in its effect on the temperature and leucocyte curve, and in its effect on the serum ferments.

Dog 47.—Weight 4.5 kilos. This animal was given a very small dose (0.05 gm.) of muscle peptone intravenously at 9 a. m., without ill effects of any marked degree except an initial nausea and prostration (Text-fig. 2 b).

In contrast to the two previous dogs, however, there was a constant fall in the antiferment titer with a recovery the following day,

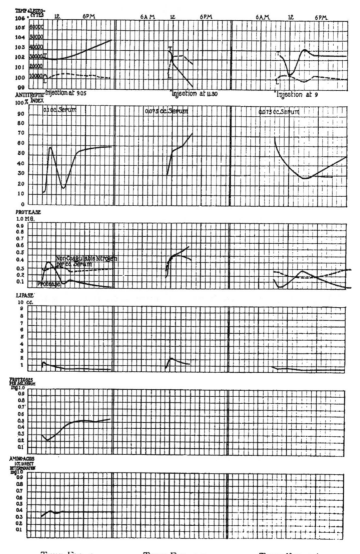

TEXT-FIG. 2.   TEXT-FIG. 2 a.   TEXT-FIG. 2 b.

TEXT-FIGS. 2, 2 a, AND 2 b. Serum changes following the intravenous injection of dog muscle peptone.

while the protease showed at first a decline and an increase during the afternoon. The lipase was not increased.

When introduced into the stomach or rectum (0.5 of a gram) the peptone was without toxic effect, and produced no alteration in the serum. When injected directly into the lumen of the small intestines, however, death resulted within five to ten minutes, the animal presenting the picture of profound shock.

CONCLUSIONS.

1. In dogs the toxic effect of primary proteoses is usually associated with the following serum changes: (a) an increase in serum antiferment, with a following fall in titer; (b) some increase in serum protease; (c) an increase in serum lipase; (d) a decrease in serum proteoses and amino nitrogen.

2. Secondary proteoses produce (a) less marked changes in the antiferment titer; (b) a marked increase in serum protease; (c) an increase in serum proteoses; (d) only a slight change in serum lipase; (e) a primary decrease in amino nitrogen.

3. The peptone which we have used (prepared from dog muscle) caused (a) a change in antiferment titer similar to that produced by the primary proteoses; (b) a marked increase in serum protease; (c) only a slight increase in serum lipase; (d) a primary decrease in proteoses, followed by an increase later; (e) an increase in aminoacids.

4. A very small dose of peptone resulted in a decrease in antiferment titer, together with a primary decrease in serum protease.

5. The peptone preparation was non-toxic when introduced into the stomach or rectum, while the intestinal injection was followed by an immediate intoxication.

We desire to express our appreciation to Messrs. T. B. and Paul Christian, who have materially assisted in various phases of this work.

THE EFFECT OF KILLED BACTERIA ON THE SERUM FERMENTS AND ANTIFERMENT.

STUDIES ON FERMENT ACTION. XXVII.

BY JAMES W. JOBLING, M.D., WILLIAM PETERSEN, M.D., AND A. A. EGGSTEIN, M.D.

(*From the Department of Pathology, Medical Department, Vanderbilt University, Nashville.*)

(Received for publication, June 1, 1915.)

In a study of the serum reactions brought about by the injection of foreign substances, interest naturally centers on changes produced by the injection of various bacteria. That, apart from the well known immunity reactions, certain changes occur in the ferments of the serum when bacteria cause a reaction on the part of the host has become evident during the past few years, largely through a study of anaphylaxis and its correlation with immunological phenomena. The work of Vaughan, of Kraus, Pfeiffer, Friedberger, de Waele, Zunz, and many others, has emphasized the importance of protein intoxication, and, as a corollary, that of the ferments which bring about a lysis of the bacterial protein. In this way the study of the phenomena of infection and immunity has gradually centered about the idea that the hydrolyzed protein of the bacterium was responsible for the intoxication, while the unaltered protein was the cause of the immunity reaction.

More recently the subject has been somewhat confused rather than aided by the endeavor to demonstrate specific protease action and in this way possibly to connect the simple antibody reactions with the hydrolysis which results in an intoxication. In a recent paper (1) we have sought to demonstrate that the methods used and the results obtained in the course of such studies were seriously to be questioned; that the serum protease was not specific, but polyvalent; that in many instances the antigen was not the substrate

603

which yielded the split products, but that they were derived from the serum supposed to contain the specific ferment.

The older idea of Pfeiffer that the bacterium contained a preformed toxic substance which became evident on lysis has been gradually superseded, and the emphasis placed on the theory that the non-toxic proteins of the bacteria cell are hydrolyzed to toxic substances. There are, however, certain facts which have not been taken into consideration in the development of this hypothesis. They are as follows: (1) bacteriolysis has been confused with proteolysis, despite the fact that intact organisms are quite resistant to proteolytic ferments; (2) bacteria adsorb antiferment from the serum and in this way become even more resistant to proteolytic influence; (3) the serum contains an antiferment which prevents the splitting of native proteins, but does not prevent the splitting of the higher split products to amino-acids; (4) simple lysis of cells, as, for instance, by grinding and freezing, as Cole (2) has recently shown for pneumococci, may free substances which are toxic in a degree and manner similar to the protein split products derived in other ways; (5) the amount of preformed split products in the cell, which are set free by simple lysis, has been ignored.

In the following experiments we have used dried, killed organisms so that we might work with constant amounts and the factor of intravital multiplication be obviated.

The technique used in the determinations of the serum ferments, antiferment, and split products has been previously described (3).

EXPERIMENTAL.

Typhoid Bacteria.

Dog 20.—Weight 7 kilos. 20 mg. of dried typhoid organisms were suspended in saline solution and injected intravenously at 9.30 a. m. (Text-fig. 1). The animal became ill shortly after the injection, was prostrated gradually, and died during the course of the night.

As will be observed from Text-fig. 1, the temperature, except for a slight initial rise, decreased during the afternoon, while the leucocyte curve at first declined and later increased. The antiferment showed little alteration. The most striking changes are noted in the serum protease and lipase which increased to more than ten times their original titer. The non-coagulable nitrogen also increased after a slight initial decrease.

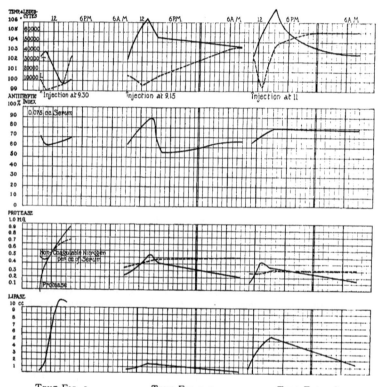

TEXT-FIG. I.   TEXT-FIG. I a.   TEXT-FIG. I b.

TEXT-FIGS. I, I a, AND I b. Serum changes following intravenous injection of typhoid bacilli; (Text-fig. I), 20 mg., (Text-fig. I a), 10 mg., (Text-fig. I b), 10 mg., second injection.

The next dog was given a similar dose but in two injections.

Dog 38.—Weight 4 kilos. 10 mg. of dried typhoid bacteria were injected at 9 a. m.; a second injection of 10 mg. was made at 11 a. m. Died at 5 p. m.

This experiment showed a similar increase in serum lipase and protease, although following the second injection there was no

further increase in protease, and a drop occurred.   The antiferment
showed no marked changes.

In the following dog a small dose was given after an interval of
several days.

Dog 22.—Weight 6.5 kilos.   10 mg. of dried typhoid bacteria were injected at
9.15 a. m., Feb. 23, 1915 (Text-fig. 1 a).   A second injection was made after 8
days, on Mar. 3, at 11 a. m. (Text-fig. 1 b).

The animal in both instances showed the usual evidences of illness.
From Text-figs. 1 a and 1 b it will be observed, however, that, following the second injection. the mobilization of the protease was more
rapid than after the first injection, and the rise in lipase was more
marked.   This is the condition usually occurring during sensitization (4).   The antiferment showed less change the second time
than the first, while the temperature and the leucocyte count showed
evidence of even a more marked intoxication.

Various Organisms.

Dog 35.—Weight 4 kilos.   10 mg. of dried subtilis bacilli were injected at
9 a. m.

In this instance. while there was some change in the leucocyte
count and a slight rise in temperature the only marked change occurred in the antiferment, which fell following the injection but
increased later.   The lipase increased slightly.

Dog 21.—Weight 4 kilos.   50 mg. of dried diphtheria bacilli were injected at
10.30 a. m. (Text-fig. 2).

The animal responded by a distinct rise in temperature and leucopenia.   There was an immediate rise in protease, while the lipase
increased progressively until the following morning.   The antiferment was unaltered.

These two organisms differ widely in their resistance to tryptic
digestion.   *Bacillus subtilis* contains only a small amount of lipoids,
autolyzes readily. and contains preformed a considerable amount of
non-coagulable nitrogen.   Diphtheria bacilli, on the other hand, contain considerable antiferment. are very resistant to tryptic digestion,
and contain only a very small amount of non-coagulable nitrogen.

Typhoid organisms occupy rather an intermediate position (5). It is interesting that the rise in lipase following the diphtheria injection occurs at a much later period than after typhoid injections.

Dog 52.—Weight 4.5 kilos. 20 mg. of dried pneumococci (Neufeld strain) were injected at 11 a. m.

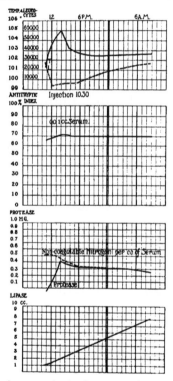

TEXT-FIG. 2.   Serum changes following intravenous injection of diphtheria bacilli.

There resulted a rise in temperature to 105° F., and the leucocytes showed a primary rise. The serum, however, showed no marked changes; it is quite evident from this experiment that the temperature curve bears no direct relation to the ferment changes.

In the following experiments the serum proteoses and amino nitrogen have been determined in addition to the ferment changes.

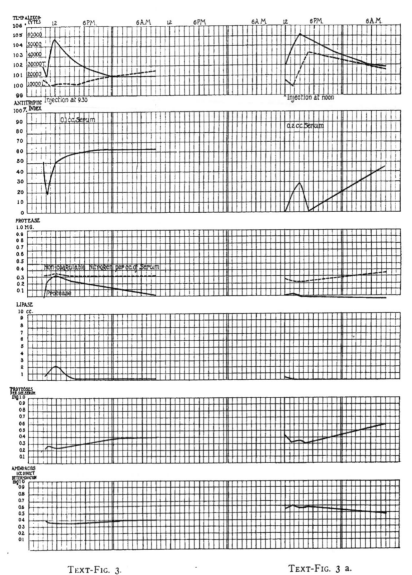

TEXT-FIG. 3.   TEXT-FIG. 3 a.

TEXT-FIG. 3.   TEXT-FIG. 3 a.

TEXT-FIGS. 3 AND 3 a.   Serum changes following intravenous injection of (Text-fig. 3) staphylococci and (Text-fig. 3 a) tubercle bacilli.

Dog 76.—Weight 5.4 kilos. 10 mg. of typhoid bacteria were injected at 9 a. m. Death at 4 p. m.

The ferment changes were typical, the serum protease increasing from 0.05 to 0.7, the lipase from 1.2 cc. to 7 cc. at the time of death. There was a slight increase in serum proteoses while the amino nitrogen showed a primary decrease.

Dog 78.—Weight 6 kilos. 50 mg. of dried *Staphylococcus aureus* were injected at 9.30 a. m. The animal remained well and showed no evidence of the intestinal irritation always observed after injections of typhoid bacteria.

There will be observed, however, a sharp rise in temperature (Text-fig. 3), together with a marked fall in the antiferment titer. Protease and lipase increased, but returned to normal the following day. Serum proteoses increased, while the amino nitrogen showed a slight decrease.

Dog 79.—Weight 5.5 kilos. 100 mg. of dried tubercle bacilli (human) were injected at noon (Text-fig. 3 a). While the temperature and leucocyte curve rose considerably, the animal remained quite well and was free from nausea or gastro-intestinal symptoms.

The serum ferment remained unaltered, while the antiferment titer showed some changes. The proteoses at first decreased, but returned to slightly above normal the following day. The amino nitrogen showed a slight decrease after twenty-four hours.

DISCUSSION.

Incidental to the study of anaphylaxis (4), we have shown that the primary injection of protein, while it may be associated with malaise, a rise in temperature, and a leucopenia, is not associated with a change in protease, whereas the split products of proteins, as illustrated in the preceding paper,[1] may give rise to such a mobilization. It becomes quite evident, therefore, when we examine the effects produced, for instance, by typhoid bacteria, that the toxicity cannot be due to the effect of the native protein unless that protein is immediately hydrolyzed to its toxic components. We must assume either that such splitting takes place immediately on injection, or that

[1] Jobling, J. W., Petersen, W., and Eggstein. A. A., *Jour. Exper. Med.,* 1915. xxii, 597.

a simple lysis due to other causes sets free preformed toxic substances. Inasmuch as specific ferments are quite out of question at the time of the primary injection, we must acknowledge that whatever splitting may occur is due to a non-specific protease. Pfeiffer (6) has recently made observations following intoxication by burns and photodynamic agents, in which he determined the proteolytic strength of the serum with glycyltryptophane. He noted an increase following the intoxication, and discusses the possible relation of such ferments to the so called "*Abwehrfermente.*"

It is quite evident from the text-figures that if the development of toxicity depends on soluble substances derived from the killed organisms, these substances are at least in part preformed.

The relative toxicity bears some relation to the degree of resistance of the organism to the proteolytic ferment. This is noted with the tubercle bacillus (Text-fig. 3 a), which in large doses has caused a rise in temperature but no change in the ferments. The diphtheria organism (Text-fig. 2), which is also quite resistant, shows a maximum rise in lipase only after twenty-four hours, while the staphylococcus (Text-fig. 3), less resistant, shows a prompter effect. When we compare the toxicity of subtilis and typhoid bacilli, however, resistance to digestion cannot be the most important factor, for the subtilis bacillus is rather easily digested, at least as easily as the typhoid bacillus, whereas the toxic manifestations are quite out of proportion in the case of typhoid injection.

Under these circumstances we are rather of the opinion that proteolysis, in so far as it relates to the production of toxic substances from bacteria *in vivo,* is not to be emphasized as the sole agent in the mechanism, but rather that certain organisms contain preformed toxic split products which may be liberated during bacteriolysis, a reaction in which proteolysis has no part as yet demonstrable. On the contrary, it would seem warrantable to suppose that in the serum the protease might act rather as a detoxicating agent, in that it is able to split the toxic fragments to the non-toxic amino-acids. In this reaction the antiferment plays no part, for it seems to act only in inhibiting the splitting of the native proteins to toxic fragments. The mobilization of the protease, as of the lipase, occurs as a result of cellular injury, but is not necessarily associated with the febrile reaction.

CONCLUSIONS.

1. The intravenous injection of killed organisms is followed by the mobilization of a non-specific protease and lipase; the rapidity and extent of this reaction depend upon the toxicity of the organism and on the resistance of the organism to proteolysis.

2. The temperature and leucocytic curve bear no relation to the ferment changes.

3. The serum antiferment is usually increased after the injection.

4. Of the organisms studied, the typhoid bacilli produced the most marked ferment changes, and the tubercle bacilli the least.

5. The toxicity of the dried organisms cannot depend wholly upon proteolysis *in vivo,* but must depend in part on the preformed toxic substances liberated on lysis.

6. Serum protease should not be considered as the sole exciter of intoxication through the production of protein split products; it seems possible that its function may in part be one of detoxication.

BIBLIOGRAPHY.

1. Jobling, J. W., Eggstein, A. A., and Petersen, W., *Jour. Exper. Med.,* 1915, xxi, 239.
2. Cole, R., *Jour. Exper. Med.,* 1914, xx, 346.
3. Jobling, Petersen, and Eggstein, *Jour. Exper. Med.,* 1915, xxii, 129.
4. Jobling, Petersen, and Eggstein, *ibid.,* p. 401.
5. Jobling and Petersen, *Jour. Exper. Med.,* 1914, xx, 452.
6. Pfeiffer, H., *Ztschr. f. Immunitätsforsch., Orig.,* 1915, xxiii, 473.

A STUDY OF THE CULTIVATION OF THE TUBERCLE BACILLUS DIRECTLY FROM THE SPUTUM BY THE METHOD OF PETROFF.

By ROBERT A. KEILTY, M.D.

(*From the McManes Laboratory of Pathology of the University of Pennsylvania, Philadelphia.*)

(Received for publication, June 14, 1915.)

The purpose of this paper is to confirm the work of Petroff[1] in a recent paper on the cultivation of the tubercle bacillus and to add a little from my own experience with the method.

Briefly, as outlined by Petroff, the sputum is mixed with a 3 per cent sodium hydroxide solution and incubated for a period of thirty minutes at 37° C. · This is neutralized to sterile litmus paper with normal hydrochloric acid, centrifugalized, and the sediment inoculated on a veal-egg medium to which gentian violet in the dilution of 1 to 10,000 has been added.

From the standpoint of the general pathologist, since I had had variable results from antiformin, it seemed to me that if this method were available many avenues were opened up for the study of this organism in pure culture. It was therefore tried out in a series of twenty-five known positive sputa. At the same time other media, concerning which I hope to report later, were used as controls.

Many tubes were planted from each case, some with a platinum loop, and others by means of a sterile pipette. The latter is very satisfactory for inoculating a series of tubes rapidly with considerable amounts of material.

My results in the use of this method at first were discouraging, but, as often, the fault or faults were mine and not the method's.

---

[1] Petroff, S. A., A New and Rapid Method for the Isolation and Cultivation of Tubercle Bacilli Directly from the Sputum and Feces, *Jour. Exper. Med.,* 1915, xxi, 38.

Out of the twenty-five cases, from the last series of seven planted on twenty-one tubes I was able to secure ten tubes of gentian violet-egg-veal media with a pure culture of tubercle bacillus. Since my purpose was to test the method not so much for its rapidity as for its availability in isolating tubercle bacilli from generally contaminated material I think it may be stated that the method offers unlimited possibilities.

My failures may be grouped under three headings: drying, no growth, and contamination. Drying of this medium in the incubator and even at room temperature is very marked, so that it may be necessary to add a few cubic centimeters of sterile water to unused media kept only for a short time. Petroff also advises leaving the freshly inoculated tubes in the incubator for a few days until the moisture carried from the centrifuge tubes is absorbed and then paraffining the tubes. This has eliminated the question of drying. Secondly, no growth took place where known positive organisms existed. From my experience the reason for this probably was the neutralization of the sodium hydrate solution. As the neutral point is reached the liquid assumes a milky appearance. It would seem that it is better to stop just here, leaving the solution possibly a little alkaline in reaction. Thirdly, contamination is somewhat harder to explain. Contamination usually resulted in liquefaction of the medium. In my experience it occurred mostly in tubes without gentian violet, although some of the latter also suffered. Here (aside from the question of technique, for some of my contaminations, *e. g.*, with subtilis and spores, were due to errors in technique) there still remain some organisms which this method will not eliminate. These contaminations in some instances were due, in spite of care, to the frequent daily examinations of many of the tubes. The temperature of the incubator is an important point in successful cultivation; the maximum of range should be between 37° and 38° C.

As to the rapidity of growth much cannot be stated as yet. At the end of three weeks' incubation, however, well marked colonies appear which will continue to grow at room temperature. They start as small pin points, becoming multiple small, raised, grayish yellow points about 1 mm. in diameter at the end of two months.

Where they become confluent they preserve their original colonies, growing upon one another.

Microscopically these cultures bear out the statement, which is often made, that the tubercle bacillus has no morphology. It is acid-fast, resisting 30 per cent hydrochloric acid for at least thirty seconds; no pleochromatic forms are encountered. It occurs as rods, straight, curved, long, short, broad, thick, and thin; as small, large, and mostly single cocci; occasionally it resembles a strepto-coccus; forms similar to the diphtheria bacillus are seen, but larger, dumb-bell-shaped, clubbed, barred, and granular. The beaded appearance of the original organism from the sputum is con-spicuous by its absence. Branching forms are encountered and, aside from the question of overlying, a Y-shape is the most fre-quent. The predominating type is a slightly curved, short, moder-ately thin, solid rod with denser granules at one or both ends.

CONCLUSIONS.

1. Probably in the majority of cases it is possible, with scrupulous technique, to isolate the tubercle bacillus from contaminated material such as the sputum, by the method outlined by Petroff.

2. It is necessary to use several tubes.

3. Neutralization, drying, temperature of the incubator, and contamination after the growth has started are important points to be noted in the process.

4. The colonies start as pin points, becoming larger and confluent, but still preserving individual groups of a dry appearance which sim-ulate the medium closely in color.

5. The bacillus of tuberculosis in young culture is tinctorially acid-fast and of a polymorphous morphology.

EXPERIMENTAL ARTHRITIS IN THE RABBIT. A CONTRIBUTION TO THE PATHOGENY OF ARTHRITIS IN RHEUMATIC FEVER.

By HAROLD KNIEST FABER, M.D.

(Received for publication, July 16, 1915.)

The significance of experimental arthritis in the rabbit has for many years been a matter of dispute. This may be attributed to the lack of data bearing upon the mechanism involved in the production of the lesions. It may be of interest to scrutinize the literature for facts which may relate to the conditions under which arthritis is, or fails to be, produced.

The first to produce arthritis in animals with pure cultures of streptococci was Loeffler (1), who published his results in 1884. Making cultural examinations of the exudate in the throats of patients with scarlet fever and diphtheria he isolated, besides the bacillus which now bears his name, several strains of streptococci, or micrococci, as he called them. Two of these strains, one from a fatal case of scarlet fever, the other from a fatal case of diphtheria, produced suppurative arthritis upon intravenous inoculation in 9 out of 12 rabbits. With Fehleisen's original strain of *Streptococcus erysipelatis* and by the same method he then produced a similar arthritis in 2 out of 3 rabbits. Smears from the pus in all cases showed abundant organisms and cultures made from 13 of them showed growth in 4, or 30 per cent—a high figure, considering the relatively crude methods employed. Most of the rabbits died.

Attention is called to the following points in Loeffler's work. There was between the time of injection and the development of arthritis an incubation period of 4 to 8 days, during which the animals were apparently normal. Arthritis followed only those injections which were given intravenously. Subcutaneous injections were followed by local infection usually ending in death without the evolution of joint lesions. The cultures, as shown by the mortality, were highly virulent. The infecting organism could be easily demonstrated in smears from the joint exudate and in a considerable percentage the cultures were also positive.

The streptococcus isolated by Wassermann (2) gave similar results. though it was probably not so virulent. The results are not reported in detail.

The diplococcus of Poynton and Paine (3) produced arthritis in 4 out of 9 rabbits injected (original report). The incubation period was 2 to 4 days. Two of the affected rabbits died on the 10th and 20th days. respectively. and two were killed. Only one recovered. Smears and cultures from the joint exudate were constantly positive. Here again the strain was evidently highly virulent.

Meyer (4) cultivated streptococci from tonsillar crypts of patients with rheumatic fever and produced arthritis in rabbits with these strains. The incubation period was 6 to 8 days and it was usually possible to demonstrate the organisms in numbers by smear and culture. Protocols and details are lacking.

Cole (5) in 1904 made a more careful study of experimental streptococcal arthritis. Using several strains isolated from widely varying conditions he produced with all a definite arthritis in rabbits,—also by intravenous inoculation. The organisms could usually be easily demonstrated in the joint exudate by smear and culture.

The most careful study of the one injection arthritis produced with virulent streptococci was made by Jackson (6). Using a highly virulent streptococcus isolated from the milk epidemic in Chicago she studied the joint reaction histologically at different periods after the injection. This report shows that 2 hours after injection streptococci could be found in the vessels of the periarticular tissues, that at 10 hours intravascular collections of leucocytes were present, while at 24 hours exudation and migration of leucocytes into the joint cavity had occurred. There is then a refractory period following the arrival of organisms at the joint during which no evident inflammatory reaction occurs, and this we may call the incubation period. It must be inferred that this reaction depends for its promptness and severity upon the virulence of the infecting organism. This is in accordance with the experiments of Dreyer (7) who found that the degree of reaction after injection of organisms into the joint of the rabbit is a good measure of their virulence. The point of most interest to us in Cole's paper is, however, the fact that two of his rabbits, inoculated with a less virulent strain, developed arthritis only after a second injection. In one of these rabbits smears and cultures were negative from all the affected joints except one and in this both smear and culture were positive. In several of the other animals spontaneous recovery occurred and in these Cole was able to provoke a relapse by another intravenous injection of the same organism. One attempt to cause a relapse with another strain of streptococcus failed.

Shaw (8) in 1904 and Schloss and Foster (9) in 1913 found that in monkeys arthritis occurred only after the second intravenous injection. Rothschild and Thalhimer (10) in 1913 note in their protocols that several of their rabbits with arthritis following intravenous inoculation of *Streptococcus mitis,* a slightly virulent organism, received five and six injections. In about one-third of these, cultures from the joint were positive.

It appears, then, that according to the virulence of the streptococcus used and possibly also according to the variety, two types of reaction may occur after intravenous inoculation. In one type, produced by streptococci of high virulence, the joint is attacked after the first injection. It is to be noted, however, that even in these instances a period of incubation is always observed. In this type of reaction cultures and smears from the affected joint are usually positive and the organisms are present in considerable numbers. In the second type of reaction for which less virulent streptococci are used, no demonstrable joint lesion follows the first injection but a very definite arthritis does follow the second or, as will be shown, a still later injection. In other words the development of arthritis in these cases is conditioned probably upon some anterior process set up in the affected joint. In this type, cultures and smears from the joint frequently show no bacteria and, if bacteria are found, they are in much smaller numbers than in instances of the first type.

The number of definitely controlled and carefully reported experiments bearing upon the second type of reaction is quite small. A larger series is reported below.

In 1913 a Belgian investigator, Herry (11), reported a series of experiments which indicated the existence of a third type of reaction. In a remarkably large number of cases of rheumatic fever he was able to isolate a streptococcus from the blood or articular exudate. From this organism he obtained by a process of extraction with normal saline, desiccation, trituration, reextraction. and centrifugalization, an endotoxin soluble in water. By injection of this into the joint followed by an intravenous injection of the living organism 8 to 15 days later he was able to produce constantly in rabbits a definite arthritis in the joint originally treated. Cultures and smears from the joint were positive for about a week after the intravenous injection. Those experiments throw a clearer light upon the process of joint localization, but unfortunately they suffer from incompleteness.[1]

EXPERIMENTAL.

Experiments were therefore instituted with the object of treating joints with streptococci so that they might be more reactive to later intravenously injected streptococci and with the hope that arthritis might be made to localize in the joint so treated. It was thought that if this could be done the mechanism of joint localization might be partly explained.

[1] I have attempted without success to repeat Herry's experiments with extracts of the *S. mitis* used for the other experiments in this paper. Herry's method calls for the use of the supernatant fluid of a bacterial suspension after centrifugalization and it is probable that a certain number of organisms remained in this material. However, the author states that the same results may be obtained with endotoxin passed through a Chamberland filter. It should be stated that the subject of endotoxins in streptococci has been thoroughly studied and that present day opinion denies their existence. Herry's experiments are. however, illuminating.

A strain of *Streptococcus mitis seu viridans* was used for most of the experiments which was kindly lent by Dr. E. Libman of Mt. Sinai Hospital, New York. It was isolated by him from the blood of a patient with subacute endocarditis. It does not hemolyze blood, and grows on blood agar in small, discrete, dry, grayish colonies. It clouds glucose-ascitic agar. In broth it clumps and sinks to the bottom of the tube, leaving the medium clear. It is insoluble in rabbit bile and has no capsule. Since coming into our hands its pathogenicity for rabbits has been slight, a single intravenous inoculation of the contents of two 12 ounce Blake bottles of agar[2] failing to cause more than the slightest and most transient symptoms. This strain is designated No. 7. No. 4 is a similar organism from a similar source. No. 59 is a *Streptococcus viridans* obtained one year ago by Dr. Homer F. Swift from a case of pericarditis. It does not fall readily into any of the common groups of Andrewes and Horder. It is only slightly pathogenic for rabbits. Pn. I is a Type I pneumococcus isolated at the Hospital of The Rockefeller Institute. Pn. S. Afr. is a pneumococcus of Type IV isolated in South Africa from the Kaffir epidemic. The latter is not very pathogenic for rabbits. The *Bacillus typhosus* is a laboratory strain.

TABLE I.

Rabbits Receiving One Intravenous Injection of Streptococcus 7.

Rabbit No.	Amount injected.	Period of observation after injection.	Arthritis.	Remarks.
		days		
1	3 agar slants	49	0	
3	1 " slant	49	0	Intra-arterial.
5	½ Blake bottle	31	0	
6	¼ " "	39	0	
7	" " "	13	0	
9	" " "	39	0	
19	1 agar slant	14	0	
38	" " "	37	0	Slight limp on 16th day. No swelling or other evidence of arthritis.

Total .... 8 rabbits.
Arthritis .... 0.

[2] The surface of the medium in the average Blake bottle is equal to that of 10 agar slant tubes.

Rabbits of medium size (of 1,200 to 1,800 grams' weight), usually females, were used. All injections were given in the ear vein except one or two into the femoral artery. The latter method was found to have no localizing effect in the corresponding leg.

Table I shows the effect of one injection, Table II of two injections, and Table III of three or more injections.

TABLE II.

Rabbits Receiving Two Intravenous Injections of Streptococcus 7.

Rabbit No.	Amount of 1st injection.	Interval between 1st and 2d injections.	Amount of 2d injection.	Arthritis.	Period of observation after 2d injection.
		days			*days*
8	¼ Blake bottle	9	⅓ Blake bottle	0	45
36	½ agar slant	6	1 " "	0	20
K	⅔ " "	15	⅓ " "	0	1
L	" " "	15	" " "	0	13

Total .... 4 rabbits.
Arthritis .... 0.

It appears that with the streptococci used two sensitizing doses were needed before any of the rabbits developed arthritis. Smears from the joint exudate in a few cases showed a few streptococci and in three cases the cultures were positive. Cultures were made in most of the cases by planting the fluid in tall tubes of glucose-ascitic agar and in the others in ascitic bouillon, both methods appearing to be equally efficacious.

The fluid in different rabbits varied from a thick, practically purulent exudate to one showing only a moderate opalescence. All the exudates examined, with the exception of that from Rabbit 17, were viscid. All contained numerous polymorphonuclear leucocytes, and some large mononuclear lymphocytes and large endothelial cells. The last named and occasionally the other two types of cell frequently contained inclusions which were interpreted as phagocyted cocci. They were Gram-negative and somewhat larger than the cocci injected. Unaltered cocci were rarely seen within the cells. It is interesting to note that similar inclusions were described by Bosc and Carrieu (12) and have been seen by the writer in cells in exudate from human rheumatic fever.

TABLE III.

Rabbits Receiving Three or More Intravenous Injections of Streptococci.

Organism.	Rabbit No.	Amount of 1st injection.	Interval between 1st and 2d injections.	Amount of 2d injection.	Interval between 2d and 3d injections.	Amount of 3d injection.	Interval between 3d and 4th injections.	Amount of 4th injection.	No. of injections.	Arthritis.	Joint affected.	Remarks.
Streptococcus 7	14	½ Blake bottle	9 days	½ Blake bottle	63 days	¼ Blake bottle	— days	—	3	+	Both knees	Positive culture from joint fluid.
"	15	"	9	"	63	"	—	—	3	+	"	Positive culture from joint fluid.
"	16	⅓	1	⅓	1	⅓	33	⅓ Blake bottle	4	o	—	Positive culture from joint fluid.
"	17	"	1	"	1	"	33	⅓ Blake bottle	4	+	Left wrist	Joint attacked 23 days after 3d injection. After recovery, a 4th injection caused a relapse.
"	24	"	3	"	1	"	27	⅓ Blake bottle	4	o	—	
"	25	"	3	"	1	"	27	⅓ Blake bottle	4	o	—	
"	J	1/10 Blake bottle each day for 8 days. Arthritis on 8th day.							8	+	Both knees	—
59	48	3 cc. broth culture	2	3 cc. broth culture	1	3 cc. broth culture	—	—	3	o	—	
"	77	1 agar slant	1	1 agar slant	1	1 agar slant	—	—	3	+	Right shoulder	Positive culture from joint fluid.
4	49	4 cc. broth culture	3	"	3	"	1	—	3	o	—	

Total rabbits injected with Streptococcus 7.......... 7.　　Arthritis.......... 4.

" 　　 " 　　 " 　　 59.......... 2.　　 " 　　 1.

" 　　 " 　　 " 　　 4.......... 1.　　 " 　　 o.

The next series of experiments was made in an attempt to sensitize a joint with streptococci so that arthritis in that joint would follow a later intravenous injection of the same organism.

The left knee was used in all cases. The injections were made by the following technique:

A suspension of the organism was made by scraping the surface of an agar growth into sterile salt solution. In a few cases a broth culture was employed. The suspension was drawn into a syringe and the needle inserted through the patellar ligament just below the patella. It was carefully pushed proximally, avoiding the bone surfaces as much as possible until it was felt to slip forward easily. This indicated that the point of the needle was in the synovial pocket under the quadriceps. The suspension was slowly injected and a little of it made to flow back into the syringe by pressure over the lower part of the quadriceps (piston test). This was returned to the joint and the needle quickly withdrawn.

After a few trials it was possible to inject the joint without infecting any of the periarticular tissues, thus giving a sharply localized reaction. This procedure caused an arthritis which usually subsided in 2 to 4 weeks, the joint and its contents returning to their normal state as far as could be determined microscopically. Dead bacteria were rapidly phagocyted, while the living resisted destruction and removal for a longer time. In all but a few cases dead bacteria were used for sensitization.

At a varying period after the inflammation and its products had been shown by examination of the synovial fluid to have disappeared an intravenous injection of the same organism was given. The results are given in Table IV.

The exudate in all cases except Rabbit 84 failed to show growth and recognizable streptococci were seen in the smears only once. The picture seen in the smears had the same characteristics as in the cases of successive intravenous inoculation without direct sensitization. In Rabbit 58 a second intravenous injection after the reaction following the first had subsided was also followed by a transient articular reaction.

The effect of similar procedure with other bacteria was then investigated with the results shown in Table V.

The condition of susceptibility in a joint to intravenous injections of streptococci has been referred to as one of sensitization, but it

TABLE IV.

Rabbits Receiving an Intravenous Injection of Streptococcus 7 after Treatment of the Left Knee with the Same Organism.

Rabbit No.	Amount injected into knee.	Inter-val.	Amount injected intravenously.	Arthritis (gross signs).	Joint fluid.	
					Culture.	Smear.
		days				
26	½ Blake bottle, living	53	2 cc. broth culture	+	o	
34	1 cc.,* killed	28	1.5 cc. broth culture	+	o	P.n.l. +++ Inclusions +
35	" " living	28	1.5 cc. broth culture	+	o	P.n.l. +++ Inclusions +
45	0.2 " "	15	1 agar slant	+	o	P.n.l. ++ A few cocci.
53	0.5 " killed	60	2 cc.	+	o	P.n.l. ++ No bacteria.
58	0.2 " "	35	½ agar slant	+	o	P.n.l. +++
58†		65	1 " "	+	o	P.n.l. ++ No bacteria.
61	" " "	19	" " "	+	o	P.n.l. ++ No bacteria.
65	10 mg. dried cocci, precipitated with alcohol	17	" " "	+	o	P.n.l. +
66	10 mg. dried cocci, precipitated with alcohol	31	2 cc.	+	o	P.n.l. ++
84	0.2 cc., killed	36	½ Blake bottle	o§	+	No p.n.l. No inclusions.
111	" " "	14	" " "	+	o	P.n.l. ++++ Inclusions +
112	" " "	14	" " "	±	o	P.n.l. ++ Inclusions +
113	" " "	14	" " "	++	o	P.n.l. +++ Inclusions +
114	" " "	14	" " "	+	o	P.n.l. +++ Inclusions +
115	" " "	14	" " "	++	o	P.n.l. ++ Inclusions +

Total rabbits treated with Streptococcus 7.......... 15. Arthritis.......... 14.

*Unless otherwise specified, figures given indicate amounts of a bacterial suspension consisting of the contents of a Blake bottle agar growth at 24 hrs. taken up in 10 cc. of normal saline. The total is equivalent to the growth on 10 agar slants.

†65 days after the arthritis following the 1st intravenous injection had subsided a 2d intravenous injection was given and was likewise followed by marked inflammation of the joint.

§4 days later slight reddening and swelling of the joint appeared and polymorphonuclear leucocytes were found in the fluid, but the culture showed no growth.

TABLE V.

Rabbits Receiving Intravenous Injections of Streptococcus 59, Bacillus typhosus, or Pn. S. Afr. after Treatment of the Left Knee with the Homologous Organism.

Rabbit No.	Amount injected into knee.	Interval.	Amount injected intravenously.	Arthritis (gross signs).	Joint fluid. Culture.	Smear.	Organism.
		days					
69	½ agar slant, killed	16	2 loops	o	—		B typhosus.
91	0.5 cc., killed	29	½ agar slant	o	+	Some blood No excess cells	" "
92	" " "	31	⅔ " "	o	o	Normal, except that one cell shows inclusions	" "
93	" " "	31	" " "	o	o	Normal, except for one small clump of bacilli	" "
95	" " "	29	½ " "	o	o	Normal	" "
72	1 mg. bacteria sensitized by Gay's method	21	¼ " "	±	o	Cells increased, 80% p.n.l. No bacteria	S. 59
73	1 mg. bacteria sensitized by Gay's method	29	1 " "	+	+	P.n.l. ++ Inclusions +	
83	0.2 cc., killed	35	½ Blake bottle	o	o	A few p.n.l. Inclusions +	
85	0.2 cc.	35	" " "	o	o	A few p.n.l. Inclusions +	
96	0.5 "	31	1 " "	o	o	Normal	Pn. S. Afr.
97	" "	31	1 broth tube	o	o	A few p.n.l. No inclusions	" " "

Total rabbits treated with B. typhosus .... 5.   Arthritis .... 0.
"   "   "   " Streptococcus 59 .... 4.   "   .... 2.
"   "   "   " Pn. S. Afr. .... 2.   "   .... 0.

may be fairly asked whether any inflammation of the joint may not predispose to later joint affections. In other words, is the predisposing factor simply a non-specific inflammation or is it a specific sensitization? In order to answer this question a series of control experiments were made by injecting one sort of bacteria into the knee and following this with an intravenous injection of another sort (Table VI). By using two closely related strains of streptococci two doubtful (certainly very slight) reactions were obtained. By crossing with streptococcus and pneumococcus no reactions were obtained. It seems fair to conclude from these experiments that the reaction is due to a specific sensitization showing some evidence of a

TABLE VI.

Attempts at Cross-Sensitization.

Rabbit No.	Inoculation into knee — Amount	Organism	Interval (days)	Intravenous inoculation — Amount	Organism	Arthritis	Joint fluid — Culture	Smear	Remarks
52	0.5 cc. susp., killed by heat	Streptococcus 7	69	⅔ agar slant, living	Streptococcus 59	?	0	Moderate number cells, 90% p.n.l. No bacteria	Slight or absent palpable swelling of knee.
67	10 mg. bacteria precipitated with alcohol	"	24	¼ "	Pn. II	0	0	Normal	No swelling.
76	0.2 cc. susp., killed by heat	" 59	21	⅔ "	Streptococcus 7	?	0	Moderate number cells, 9% p.n.l. No bacteria	Very doubtful palpable swelling of knee.
83	0.2 cc. susp., killed by heat	" 7	35	½ Blake bottle, living	" 59	0	0	Few cells. Rare p.n.l. No inclusions	No swelling or other evidence of arthritis.
84	0.2 cc. susp., killed by heat	" 59	35	"	" 7	0	0	Normal. No p.n.l. No bacteria	No swelling or gross evidence of arthritis.
106	0.3 cc. susp., killed by heat	Pn. S. Afr.	14	"	"	"	0	Rare cells. 3 p.n.l. No bacteria seen.	No swelling or other gross evidence of arthritis.
107	0.3 cc. usp., killed by heat	" "	14	"	"	"	0	Rare cells. 1 p.n.l. No bacteria seen.	No swelling or other gross evidence of arthritis.
108	0.3 cc. susp., killed by heat	" "	14	"	"	"	0	No ? cess cells No bacteria	No swelling or other gross evidence of arthritis.
109	0.3 cc. susp., killed by heat	" "	14	"	"	"	0	Very few & No p.n.l. No bacteria	No swelling or other gross evidence of arthritis.
110	0.3 cc. susp., killed by heat	" "	14	"	"	"	0	A few cells. 10% p.n.l. No bacteria	No swelling or other gross evidence of arthritis.

Total rabbits tested for heterologous sensitization 10

    "   showing joint reaction .... 2 (doubtful).

group specificity comparable with the group agglutinations of certain bacteria.

Pathological Anatomy.

The arthritis following intravenous inoculation alone and that following intravenous inoculation after local sensitization presented the same gross and microscopic picture. An exception must be made in the case of Rabbits 14 and 15, in which a chronic arthritis with the continued presence of demonstrable living streptococci in the exudate followed the last injection. Here erosions in the articular cartilages were found. In the other rabbits the joint at autopsy showed the following changes: The synovial surface was moderately congested, the villi usually more so than the other parts. An excess of stringy fluid was usually present. The capsule was somewhat swollen. In cases which had reacted severely there was occasionally found after inflammation a small amount of inspissated exudate in the upper angle of the joint cavity. Sections showed a marked leucocytic infiltration of the villi and to a slighter degree of the subendothelial layer of the capsule. Examination of the adjacent cartilage and bone failed to show any noticeable change. Examination of the articular and periarticular tissues for bacteria by the Gram stain did not reveal the presence of organisms after a careful search. The vessels of the capsule and of the villi were moderately distended but no thrombi were seen. The reaction as seen under the microscope appeared to be mainly in the villi with the synovial membrane also playing a part.

DISCUSSION.

It is believed that the above experiments may throw some light upon the mechanism of acute arthritis. It seems to be established that the first attack of a highly virulent streptococcus can cause an arthritis and that the arthritis-so produced is usually of a severe type with an exudate containing large numbers of viable organisms.

On the other hand, streptococci of lower virulence are frequently not able to produce arthritis at the first attack but at this time prepare the way for such an effect in a later attack.

It seems to be clearly proved that this preparatory or sensitizing process is, within narrow limits, a strictly specific one; *i. e.,* the organism used for the exciting, intravenous injection must be the same as that used for the sensitizing, intra-articular injection, else the reaction fails to occur.

It further appears that this preparation may be made by the introduction into the joint of the organism, living or dead, and it seems fair to conclude that the arthritis produced by successive intravenous injections results from a similar process; *i. e.,* a preliminary deposit of organisms in the joint leading to sensitization but not to gross inflammatory lesions, the latter resulting from a subsequent deposition of organisms. It is also shown that this reaction in a sensitized joint may be repeated several times in the same animal.

A close analogy between this reaction and the relapses in rheumatic fever can readily be drawn. Relapses are so common in this disease as to be one of its distinguishing characteristics. Thus in the figures given by Mosler and Valentin (13) 60 out of the 142 cases studied were suffering from their second or a later attack at the time of admission to the hospital. Cole produced 8 successive attacks, each followed by recovery, in a single rabbit.

Without assuming the specificity of any one organism the conception of relapsing arthritis as the effect of a virus upon an homologously sensitized joint may be fairly applied to the relapsing cases of human rheumatic fever.

The reaction of a sensitized joint in rabbits inoculated intravenously may properly be designated as an induced relapse and it will be seen from the protocol of Rabbit 17 that a second relapse can be induced by a second intravenous injection. From the ease and constancy with which lesions can be induced in sensitized joints by the method above outlined it is strongly suggested that the relapses, if not the primary arthritis, in rheumatic fever result from a virus reinvading the blood stream and provoking a reaction in a previously sensitized locus.[3]

[3] It may not be too wide a step to pass from the consideration of the effects induced in the joints to the phenomena observed in the endocardium and possibly even the myocardium in rheumatic fever. That they too are of relapsing nature is admitted; and that they also result from sensitization may be suggested.

In respect to the primary attack an analogy may be drawn from the fact above noted, that in rabbits and monkeys there is needed either a period of incubation, or, in most cases, a series of injections of the exciting organism before the joint is attacked. This fact suggests the hypothesis that some degree of sensitization, or of heightened activity on the part of the fixed cells is necessary before a definite and marked tissue reaction occurs.

It is not desired to state any opinion as to the identity of the organism causing rheumatic fever. In view of the fact that several different organisms can cause arthritis in the rabbit, that several different organisms can cause arthritis in man, and that the clinical manifestations of rheumatic fever vary widely it may well be that no one organism is constantly at fault. Nevertheless the streptococcus, and in most cases a member of the *viridans* group, has been the one most often found when cultures are positive. Further, this organism shows greater and more constant arthrotropic properties than any other now known. The preponderance of the evidence now available, therefore, is with the streptococcus. At this point the case must rest until further proof is offered.

SUMMARY.

By a process of sensitization described it was found possible to cause arthritis in rabbits constantly after one intravenous injection of the streptococcus.

This reaction is specific.

By intravenous inoculation, without previous sensitization, of the streptococcus used in these experiments it was possible to cause arthritis in rabbits only after three or more injections.

An analogy is suggested between the arthritis induced by sensitization and the relapses in human rheumatic fever.

A further analogy is suggested between the development in rabbits of arthritis after repeated intravenous injections and the development of the primary lesion in human rheumatic fever.

BIBLIOGRAPHY.

1. Loeffler, F., Untersuchungen über die Bedeutung der Mikroorganismen fur die Entstehung der Diphtherie beim Menschen, bei der Taube und beim Kalbe, *Mitt. a. d. k. Gsndhtsamte.*, 1884, ii, 421.

2. Westphal, Wassermann, and Malkoff, Ueber den infektiösen Charakter und den Zusammenhang von akutem Gelenkrheumatismus und Chorea, *Berl. klin. Wchnschr.*, 1899, xxxvi, 638.

3. Poynton, F. J., and Paine, A., The Etiology of Rheumatic Fever, *Lancet*, 1900, ii, 861.

4. Meyer, F., Zur Bakteriologie des acuten Gelenkrheumatismus, *Ztschr. f. klin. Med.*, 1902, xlvi, 311.

5. Cole, R. I., Experimental Streptococcus Arthritis in Relation to the Etiology of Acute Articular Rheumatism, *Jour. Infect. Dis.*, 1904, i, 714.

6. Jackson, L., Experimental Streptococcal Arthritis in Rabbits. A Second Study Dealing with Streptococci from the Milk Epidemic of Sore Throat in Chicago, 1911–12, *Jour. Infect. Dis.*, 1913, xii, 364.

7. Dreyer, L., Ueber Virulenzprüfung mittels intraartikularer Impfung, *Centralbl. f. Bakteriol., 1te Abt., Orig.*, 1912–13, lxvii, 106.

8. Shaw, W. V., Acute Rheumatic Fever and Its Etiology, *Jour. Path. and Bactcriol.*, 1904, ix, 158.

9. Schloss, O. M., and Foster, N. B., Experimental Streptococcic Arthritis in Monkeys, *Jour. Med. Research*, 1913–14, xxix, 9.

10. Rothschild, M. A., and Thalhimer, W., Experimental Arthritis in the Rabbit, Produced with *Streptococcus mitis, Jour. Exper. Med.*, 1914, xix, 444.

11. Herry, Contribution à l'étude du rhumatisme articulaire aigu; essai de pathogénie et de sérothérapie; étude clinique, anatomique, et expérimentale, *Bull. de l'Acad. roy. de méd. de Belgique*, 1914, xxviii, 76.

12. Bosc, F. J., and Carrieu, M., Inclusions intracellulaires dans le liquide articulaire du rhumatisme articulaire aigu, *Compt. rend. Soc. de biol.*, 1913, lxxiv, 1262.

13. Mosler, E., and Valentin, B., Zur Pathologie des akuten Gelenkrheumatismus, *Berl. klin. Wchnschr.*, 1910, xlvii, 1778.

THE ACTION OF THE LETHAL DOSE OF STROPHAN-THIN IN NORMAL ANIMALS AND IN ANIMALS INFECTED WITH PNEUMONIA.

By ROSS A. JAMIESON, M.B.

(*From the Hospital of The Rockefeller Institute for Medical Research.*)

PLATES 74 TO 76.

(Received for publication, July 15, 1915.)

The following experiments were undertaken for the purpose of ascertaining whether the action of a digitalis body when administered to animals suffering from pneumonia differed from its action when no infection or fever was present.   That there is a difference of opinion regarding the action of the drug when given in acute infections is shown in the practice of clinicians.

For example, Mackenzie (1) doubts its value and says that digitalis " is often used when patients are dying of some grave affection, as pneumonia, as a last resort, probably more for the purpose of doing something than with expectation of great benefit. I have never seen much good follow the administration of digitalis in acute febrile states." Gibson (2) in his article on acute pneumonia also says, regarding the use of digitalis, that in the worst cases of pneumococcal poisoning the heart absolutely refuses to respond. On the other hand, in Krehl's clinic (3) digitalis is used in all cases of pneumonia with the belief that it has a beneficial effect.

On account of this difference of opinion, Gunn (4) recently performed experiments with the view of determining what influence the height of the temperature has on the action of digitalis. In his experiments he perfused excised rabbits' hearts with solutions of strophanthin at different temperatures. The perfusion was carried to the point of arrest of the heart's action. By this procedure Gunn demonstrated that with variations in temperature ranging from 28° to 41° C. strophanthin acts more quickly on the isolated heart as the temperature is raised. The conditions in Gunn's experiments are not analogous to those in pneumonia, since in the presence of this infection toxemia accompanies the increased temperature.

In the experiments now reported the attempt was made to duplicate the usual human conditions as nearly as possible.   A digitalis body was administered to animals suffering from an acute pneumonic infection and the effect compared with that obtained when the drug was given to healthy animals of the same species. A comparison of the results in the two groups we believe should yield information regarding the influence of infection on the action of the drug.

Methods.

The animals employed were cats and dogs. In each species one group was infected and another group was used as a control. The animals were infected by the method of intrabronchial insufflation described by Lamar and Meltzer (5). The organism used was an eighteen to twenty-four hour old broth culture of virulent pneumococci[1] belonging either to Type I or II (6), and was obtained from the laboratory of the Hospital. Of these cultures the cats received from 10 to 20 cc., and the dogs from 20 to 30 cc. The temperature of the animals was taken daily for several days before infection and also during the course of the disease.

The digitalis body employed was Thoms crystalline g-strophanthin (Merck). An alcoholic stock solution of the drug was made and from this an aqueous solution was prepared for immediate use. In the experiments on cats the aqueous solution was of such a strength that 10 cc. of solution contained 0.1 mg. of strophanthin. For the dogs a solution double this strength was employed. In all experiments the strophanthin was given intravenously. The animal was lightly anesthetized with ether, the femoral vein exposed, and a small glass cannula was inserted. This was connected with a burette containing the solution. The anesthetic was not continued beyond the time necessary for the insertion of the cannula. In injecting the strophanthin, an endeavor was made to maintain a uniform rate of flow. A lethal dose was given to each animal in approximately one hour, unless the animal died before the time required for administering the lethal dose had elapsed. The amount of strophanthin injected differed slightly in cats and dogs. In a large series of experiments on cats, Hatcher and Brody (7) and, later, Eggleston (8) found the lethal dose of strophanthin to be 0.1 mg. per kilo of

[1] The virulence was such that 0.000001 cc. was fatal in white mice.

body weight. In the present experiments on cats this amount was injected in both the control and the infected animals. In the dogs it was necessary first to estimate the minimal lethal dose (m.l.d.) in normal animals; it was found to be 0.12 mg. per kilo of body weight. Autopsies were performed on all the infected cats and on all cats in the control group which succumbed to the lethal dose of strophanthin. All the dogs, whether normal or infected, were autopsied. In a number of cats and dogs electrocardiograms were taken before infection and also just before injecting strophanthin. A series was also made during the period of injection.

TABLE I.

Normal Cats Injected with Strophanthin.

Date (1914).	Cat.	Sex.	Weight at time of injection.	Stro-phanthin injected.	Amount of m. l. d. + or −.	Result.	Duration of experiment.
			kilos	*mg.*	*mg.*		*min.*
Nov. 3	A	F.	2,200	0.22	L*	D†	52
" 3	B	M.	2,400	0.24	L	D	60
" 3	C	F.	2,100	0.21	L	D	110
" 7	E	M.	3,150	0.315	L	S	—
" 7	F	M.	3,000	0.3	L	D	50
" 7	G	F.	2,500	0.25	L	D	59
" 7	H	F.	2,850	0.285	L	D	70
" 10	I	F.	2,000	0.2	L	S	—
" 10	J	M.	3,650	0.365	L	D	65
" 10	K	F.	2,300	0.23	L	S	—
" 14	L	F.	2,925	0.253	L−0.0395	D	48
" 14	M	F.	4,050	0.405	L	D	63
" 14	N	M.	1,700	0.17	L	S	—
" 20	P	M.	3,300	0.33	L	D	60
" 20	Q	F.	1,975	0.187	L−0.01	D	27
" 20	R	F.	1,700	0.17	L	S	—
" 20	S	M.	1,800	0.18	L	S	—
" 25	T	F.	3,050	0.305	L	D	78
" 25	U	M.	2,800	0.28	L	S	—
" 25	V	M.	2,425	0.2425	L	S	
" 25	W	M.	2,250	0.225	L	S	

Total = 21.
Died = 12 = 57.1 per cent.
Survived = 9 = 42.9 per cent.

*In the tables L represents the estimated lethal dose. L + or − indicates the amounts more or less than the estimated lethal dose.
† D = died; S = survived.

In the experiments on cats, three groups were studied: a control group and two groups of infected animals. The control group consisted of normal, uninfected animals, each of which received 0.1 mg. of strophanthin per kilo of body weight. This is the amount reported by Hatcher and Brody as the minimal lethal dose. The second group consisted of infected animals. They were otherwise untreated, and served the purpose of observing the course of the disease. The animals in the third group were also infected. Each of these received an amount of strophanthin equal to that given to the first group.

Group I (Table I) formed the control series of experiments. This group consisted of twenty-one cats, all of which were apparently normal and healthy. All received the estimated lethal dose of strophanthin (0.1 mg. per kilo of body weight) in the manner described. Of the twenty-one cats, twelve, or 57.1 per cent, died, while nine, or 42.9 per cent, survived. In the cats that died, the time that elapsed from the commencement of the injection until death occurred ranged from 27 to 110 minutes, the average time being 62 minutes. All the cats received the estimated lethal dose with the exception of animals L and Q. These two died when slightly less than the estimated amount had been injected. The low mortality in this group is a subject which will be discussed later.

TABLE II.

Cats Infected with Pneumococci, but Receiving No Strophanthin.

Date of infection (1914).	No. of cat.	Sex.	Type of organism.	Highest temperature before and after infection.	Result.
				° C.	
Nov. 23	34	F.	II	38.5–38.8	D Nov. 26.
" 23	35	M.	II	38.4–39.6	S
" 23	36	M.	II	37.7–39.6	D " 27.
" 23	37	M.	II	38.2–38.6	S
" 24	38*	F.	I	38.2–38.8	D Dec. 14.
" 24	39	F.	I	38.6–39.3	S
" 24	40	M.	I	39.0–39.0	S
" 24	41	F.	I	38.4–39.2	S
" 27	42	M.	II	38.2–39.4	S
" 27	43	M.	II	39.0–39.6	S
" 27	44	M.	II	39.0–39.4	S
" 27	45*	F.	II	37.8–39.0	D " 14.

Died = 16.6 per cent.

* The two cats, Nos. 38 and 45, which died three weeks after infection, evidently not from pneumonia, but from some intercurrent infection, are not considered in estimating the mortality for the series.

Group II (Table II) comprised twelve cats, each of which was infected by an intrabronchial insufflation of pneumococcus culture. Six of the animals developed very severe symptoms. The other six were only moderately ill. Of the former, one died on the third and one on the fourth day. The symptoms of the disease appeared eight to twelve hours after infection. The animals became listless and prostrated. Almost all showed a rise in temperature ranging from a fraction of a degree to $3.6°$ C. The symptoms of the disease apparently reached their height on about the second or third day after infection. At this point convalescence set in. Practically all the animals lost weight.

Group III (Table III) consisted of forty-three cats, all of which were infected. In seven, the culture used was obtained at autopsy from the heart's blood of infected cats; the rest were infected with organisms obtained in the manner already described. The symptoms of disease corresponded to those described under Group II. Ten cats failed to show a rise in temperature, but the temperature was not taken sooner than twenty-four hours after infection, and it is possible that, had readings been made at an earlier period, a rise might have been observed. A few animals which showed only a very slight reaction to the infecting organisms after the first insufflation, were reinfected. Strophanthin was given as nearly as possible at the height of infection; that is to say, on the second or third day, according to the observations made in Group II. Following the injection, twenty-three cats, or 53.5 per cent, died, and twenty, or 46.5 per cent, survived. The cats that died during or following the injection did so in from 25 to 108 minutes, the average time being 62 minutes. Nine cats succumbed before the injection was completed. They received slightly less than the estimated lethal dose. Three cats, Nos. 17, 48, and 53, through an error in calculation, received slightly more than the estimated lethal dose, but as these cats all survived the error does not affect the results.

A comparison of Groups I and III, the animals in both of which received strophanthin, shows that the mortality in each series, following the injection of the adopted lethal dose of the drug, was practically the same; *viz.*, 57.1 and 53.5 per cent. The course of the disease in the infected animals, however, was short; in many in-

TABLE III.

Cats Infected with Pneumococci, and Later Injected with Strophanthin.

Date (1914).	No. of cat.	Sex	Type of organism.	Weight on day of injection.	Stro-phanthin injected.	Amount of m. l. d. + or −.	Re-sult.	Dura-tion of experi-ment.	Highest tempera-ture before and after infection.
				gm.	*mg.*	*mg.*		*min.*	*° C.*
Oct. 26	1	F.	I	2,250	0.203	L −0.022	D	47	37.8–39.2
" 26	2	F.	I	2,000	0.20	L	S	—	36.2–39.8
" 28	3	M.	II	1,750	0.175	L	D	57	——
" 28	4	F.	II	2,550	0.245	L −0.01	D	62	——
" 29	6	F.	I	2,000	0.15	L −0.051	D	45	36.7–39.0
" 29	7	F.	I	1,300	0.13	L	S	—	36.0–38.0
" 30	9	—	II	2,700	0.27	L	S	—	37.0–38.1
" 30	10	F.	I	2,250	0.225	L	S	—	37.2–37.8
Nov. 2	11	F.	II C*	1,500	0.15	L	D	57	37.8–37.8
" 11	12	M.	II C I	3,100	0.31	L	D	66	38.1–38.6
" 11	13	F.	II C I	1,875	0.1875	L	S	—	38.6–40.0
" 11	14	M.	I	2,575	0.2575	L	D	76	38.8–39.0
" 12	16	F.	II C	3,000	0.30	L	S	—	39.0–39.4
" 12	17	M.	II C	2,200	0.225	L +0.005	S	—	38.8–38.8
" 12	18	M.	II C	2,250	0.225	L	D	88	39.2–40.0
" 13	19	M.	II	3,600	0.3595	L −	S	—	39.0–39.2
" 13	22	M.	II	3,325	0.254	L −0.0785	D	42	39.2–39.0
" 18	23	F.	I	2,425	0.2425	L	D	80	39.1–38.6
" 18	24	M.	I	3,500	0.319	L −0.031	D	62	39.0–38.2
" 18	26	M.	I	2,475	0.2475	L	D	60	38.4–39.4
" 19	15	F.	II C Nov. 10 I " 17	2,900	0.183	L −0.107	D	25	38.6–39.6
" 19	27	M.	I	1,700	0.17	L	D	92	39.0–39.8
" 19	28	F.	I	2,225	0.2225	L	D	73	38.4–37.8
" 19	29	F.	I	2,350	0.235	L	D	108	38.2–39.0
" 23	30	F.	II	3,050	0.275	L −0.03	D	48	38.1–38.2
" 23	31	M.	II	3,750	0.375	L	S	—	38.0–40.6
" 23	32	F.	II	2,500	0.25	L	S	—	38.2–39.0
" 23	33	F.	II	2,575	0.2575	L	S	—	38.8–39.6
" 30	48	F.	II	2,700	0.271	L +0.001	S	—	38.8–40.6
Dec. 2	50	F.	II	3,000	0.30	L	D	60	38.5–39.0
" 2	51	F.	II	2,500	0.25	L	S	—	39.0–39.8
" 2	52	F.	II	2,900	0.29	L	D	60	37.5–38.3
" 2	53	M.	II	3,325	0.354	L +0.0215	S	—	38.2–39.3
" 3	55	M.	II	3,850	0.385	L	S	—	39.2–39.0
" 3	56	M.	II	2,950	0.295	L	S	—	38.2–41.2
" 4	57	F.	II	3,100	0.31	L	D	65	38.2–38.8
" 7	62	F.	II	3,000	0.3	L	S	—	38.2–39.4
" 8	63	M.	II	3,400	0.34	L	S	—	38.0–40.0
" 9	64	M.	I	2,500	0.25	L	S	—	37.6–39.2
" 10	65	F.	I	1,550	0.155	L	S	—	38.2–36.6
" 11	66	F.	I	3,050	0.265	L −0.04	D	63	38.0–38.0
" 15	60	F.	II	2,000	0.13	L −0.07	D	37	38.6–38.4
" 18	59	F.	II	1,600	0.16	L	D	46	38.4–39.4

Total   = 43.
Died   = 23 = 53.5 per cent.
Survived = 20 = 46.5 per cent.

* C indicates that the culture used was obtained from the heart's blood of a cat previously infected.

stances the animals were not seriously ill, and a number failed to show any increase in temperature, or showed an increase which was comparatively slight. It was considered inadvisable, therefore, to draw conclusions from these experiments. The attempt was accordingly made to enlarge our experience by performing experiments in dogs.

The Action of Strophanthin in Normal and Infected Dogs.

The experiments on dogs were undertaken with the expectation that they would show a greater degree of reaction to infection than cats. No standardization of strophanthin in dogs has been made (in so far as could be ascertained), so that it was necessary first to determine the lethal dose of the drug. Experiments were, therefore, carried out on ten dogs (Table IV) to determine the minimum lethal

TABLE IV.

Dogs Injected with Strophanthin, To Estimate the Lethal Dose.

Date (1915).	Dog.	Sex.	Weight at time of injection.	Total strophanthin injected.	Strophanthin injected per kilo of body weight	Duration of experiment.	Deviation from average lethal dose.
			gm.	*mg.*	*mg.*	*min.*	*per cent*
Jan. 20	A	F.	8,000	0.98	0.122	47	2
" 21	B	F.	5,100	0.57	0.111	42	10
" 22	C	M.	7,350	1.07	0.145	110	16
" 25	D	F.	9,000	0.9	0.1	65	17
" 28	E	M.	5,700	0.704	0.123	53	1
Feb. 10	F	F.	5,750	0.75	0.130	95	5
" 10	G	M.	5,000	0.59	0.118	57	5
" 15	H	F.	5,300	0.606	0.114	68	8
" 15	I	F.	4,150	0.55	0.132	90	6
" 16	J	F.	4,750	0.71	0.149	93	20

Average dose per kilo = 0.124 mg.

dose. Before injection the animals were observed for a time under uniform conditions and all were apparently normal and healthy. The injections were made during the months of January and February. The females used were neither pregnant nor lactating. The animals were not fed for a period of eighteen hours preceding the experiment. This short fast was enforced so that the presence of food in the stomach should not influence the weight of the animal, and so that vomiting during the preliminary anesthesia might be prevented.

A fresh alcoholic solution of strophanthin was prepared and the aqueous solution made from this was of a strength such that 5 cc. represented 0.1 mg. of strophanthin. The solution was tested on three cats and was found to have the same effect as had the first stock solution. The lethal dose for these three animals was 0.108, 0.127, and 0.107 mg. per kilo of body weight, respectively.

The injection of strophanthin in dogs was made in a manner identical with that described for cats. The lethal dose for the ten dogs used was found to average 0.124 mg. per kilo of body weight. Death occurred in approximately 60 minutes. The dose ranged from 0.1 to 0.149 mg. per kilo of body weight. Of the ten dogs, seven showed a deviation of 10 per cent or less from the average m. l. d. The remaining three dogs, C, D, and J, gave respectively a deviation of 16 per cent above, 17 per cent below, and 20 per cent above the average. If the three dogs showing the greatest variation be deducted from the series, the result is not materially altered and the m. l. d. still remains 0.12 mg.

TABLE V.

Dogs Infected with Pneumococci and Later Injected with Strophanthin.

Date (1915).	No. of dog	Sex.	Type of organism.	Weight before infection.	Weight at time of injection.	Total strophanthin injected.	Strophanthin injected per kilo of body weight.	Dura tion of experiment.	Deviation from average lethal dose.	Highest temperature before and after infection.
				gm.	gm.	mg.	mg.	min.	per cent	°C.
Jan. 15	1	F.	II	9,000	7,900	1.237	0.137	108	10	38.8–39.2
" 18	3	F.	II	8,150	7,300	1.105	0.151	85	20	39.0–40.9
" 26	4	M.	II	9,400	8,800	1.146	0.130	78	4	39.8–39.5
Feb. 1	5	M.	II	8,500	7,800	0.82	0.105	55	16	39.8–39.5
" 3	6	M.	II	9,000	7,900	0.88	0.111	73	11	39.6–39.2
" 4	8	F.	II	8,300	8,050	1.01	0.125	95	0	39.8–39.6
" 17	10	F.	II	5,800	5,750	0.68	0.118	73	6	38.2–40.4
" 19	11	M.	II	6,350	6,150	0.614	0.0998	55	20	38.8–39.8
" 24	12	M.	I	8,600	7,500	1.08	0.144	115	15	39.0–39.8
" 26	14	M.	I	8,500	8,250	1.1	0.133	90	6	38.5–41.5

Average lethal dose per kilo = 0.125 mg.

A series of ten experiments was next carried out to ascertain the lethal dose in pneumonic dogs (Table *V*). These animals were infected by the method described. In the reaction to the infection, the dogs displayed symptoms similar to those described for the cats. They became listless and usually refused to eat. The respirations appeared to be somewhat labored and were increased in rate. In all the dogs there was a loss in weight, amounting in two to more than 1,000 grams. The temperature following infection usually rose. In seven dogs the increase ranged from a fraction of a degree to 3° C. Three dogs, however, failed to show an increase, but in these the temperature was not taken until twenty-four hours after infection with pneumococci. The injection of strophanthin was made during the second twenty-four hours following infection because this was the time when the animals were most severely ill. Lamar and Meltzer (5) and, later, Wollstein and Meltzer (9) also give this time as the period of the maximum reaction. The average amount of strophanthin per kilo of body weight injected was 0.125 mg., which was practically the same as the lethal dose required in the normal dogs. The range of the dosage for the infected dogs varied from 0.099 to 0.151 mg. per kilo of body weight. In six of these dogs the amount of strophanthin injected fell within 11 per cent of the average. The remaining four dogs, Nos. 3, 5, 11, and 12, showed respectively a variation of 20 per cent above, 16 per cent below, 20 per cent below, and 15 per cent above. If the four animals showing the greatest variation are omitted from the series it is found that the average m. l. d. of the remaining six dogs is still 0.12 mg. The time which elapsed between the commencement of the injection and the death of the animal varied from 55 to 115 minutes, the average time being 82 minutes. A comparison, then, of the normal and infected dogs shows the lethal dose of strophanthin in each to be the same.

Autopsy Findings.—Autopsies were performed on all the cats and dogs in both the normal and infected series, except on the cats in Group I that survived the injection of strophanthin and on the cats in Group II that survived the infection. Autopsies performed on cats of the control series (Group I) killed by strophanthin showed the lungs to be normal. Two cats of Group II, Nos. 34 and 36, died during the height of the infection. No. 34 showed a massive consolidation of the entire left lung, and also of the middle and posterior lobes of the right. No. 36 showed scattered areas of consolidation throughout all the lobes of both lungs, the left lung being almost entirely consolidated. A well marked fibrinous exudate covered the pleura and a moderate amount of fluid was found in both pleural cavities. Two other cats belonging to this group died at a period three weeks after infection, but their death was evidently due to an intercurrent disease. Autopsies on these animals showed small, scattered, abnormal areas in both lungs suggestive of an old pneumonic process.

Autopsies were performed on all the animals of Group III. The lungs in each instance showed definite signs of consolidation. The involvement varied in extent from a portion of one lobe to an entire lung. The gross examination showed the area involved to be deeply congested and the overlying pleura to be somewhat dull and lacking in luster. In a number of the animals the pleura showed the presence of a small amount of serofibrinous exudate. Two animals, Nos. 33 and 55, developed a fibrinopurulent pleurisy, and one, No. 33, also showed a well marked pericarditis. The portion of the lung involved was usually firm, inelastic, and friable. On section the lungs were firm; the cut surface was uniformly dark red with a tendency to be dry. On compression, a small quantity of bloody fluid exuded from the surface. Small portions of tissue cut from the consolidated areas sank in water. The gross appearance of the lungs in dogs was similar to that in the cats. Blood cultures were made at the time of autopsy from the heart's blood in fourteen cats; of these, twelve contained pneumococci.

Microscopic examination of the lung tissue showed the alveoli to be filled with an exudate consisting chiefly of leucocytes. A few large lymphocytes, red blood cells, and desquamated epithelial cells were also present. Only small quantities of fibrin were contained in the alveolar exudate. The amount of fibrin found in the pneumonias in cats was slightly more than that found in the dogs. The walls of the alveoli were only moderately congested. The bronchi contained small amounts of exudate, consisting chiefly of red and white blood cells with particles of desquamated epithelium and small amounts of fibrin.

Electrocardiograms.—Electrocardiograms were made of ten infected cats, of seven normal dogs, and of all the infected dogs. Non-polarizable electrodes were placed on the right fore and the left hind leg. Control curves were obtained before infection of the animals, before injection of strophanthin, and also at ten minute intervals after the injection was begun. The principal changes found to follow the injection were increase in the conduction time, alterations in the size and form of the T wave, and the production of extrasystolic irregularities. Changes in the P-R time occurring during the course of the strophanthin injection were seen in four of the ten cats. In these there was an increase of 0.02 of a second or more. In the remaining six cats there was no alteration in the conduction time. Changes in the P-R time were also seen in five of the normal and in six of the infected dogs. In all of these there was an increase of 0.02 of a second or more (Figs. 1 B and 3 C). The remaining dogs of both groups showed no important change in the P-R time. Blocking of the auricular impulse occurred in two of the normal and in three of the infected dogs (Figs. 2 B and 3 B).

Alterations in the T wave have recently been shown to occur in man during the administration of digitalis (10). The curves of cats and dogs displayed similar changes (Figs. 4, 5, and 6). Of the ten cats, a negative T wave was present in the control curves of two, and became deeper after the injection (Fig. 5 B). In two others a diphasic T wave became negative, and in a fifth an upwardly directed T wave became markedly diminished in height (Fig. 6). In the remaining five cats, the strophanthin produced no demonstrable change in this wave. In two of the seven normal dogs, the T wave in the control electrocardiogram was positive. After injection it became diphasic in one, and negative (Fig. 1 B) in the other. The remaining five dogs all showed negative T waves in the control curve. On the form of these strophanthin appeared to have no influence. In the ten infected dogs changes occurred more often. In six animals the T wave was positive in the control curve and became negative under the influence of strophanthin (Figs. 3 B and 3 C). In the seventh a negative T wave became diphasic. In the remaining three dogs negative T waves in the control curve became deeper under the influence of the drug (Fig. 4 B). Changes in the T wave were observed, therefore, five times in ten cats, and in twelve of seventeen dogs.

Extrasystolic (11) irregularities occurred in all the animals following the injection (Figs. 3 D and 6 B). Several of the cats which received the average lethal dose of strophanthin and survived showed an irregularity of this type to a marked degree. When electrocardiograms were made of these animals, twenty-four hours, later, the heart rhythm was again normal (Fig. 6). In one cat that survived the injection of strophanthin, persistence of a change of shape in the T wave continued for a period of at least forty-eight hours (Fig. 5). In all the dogs and in the cats that died, extrasystoles multiplied in frequency. At the time of death fibrillation of the ventricles occurred.

SUMMARY AND DISCUSSION.

The results of these experiments permit a comparison of the action of strophanthin in the normal and the infected animals. In cats the percentage of deaths following the injection of 0.1 mg. of strophanthin per kilo of body weight was the same in normal and in pneumonic cats. The number of recoveries in this series was larger than was expected. This result was due to the fact that the dose injected was not the lethal dose, but the average lethal dose (0.1 mg.) determined by Hatcher and Brody (7) and by Eggleston (8). The doses which the latter actually injected ranged from 0.085 mg. to 0.16.mg.; from these the average minimal lethal dose was calculated. In adopting the average dose as the standard one to inject, those of our cats that required more than 0.1 mg. naturally survived. The death rate was therefore low. The plan employed in dogs differed from that used in cats. A lethal dose was injected in each dog. Death occurred when an average of 0.12 mg. of strophanthin per kilo of body weight was injected. The same dose was required in normal and pneumonic dogs. The effect of strophanthin in both groups of infected and non-infected cats and dogs is, therefore, identical.

Whether the uniformity of strophanthin action in the two experimental groups may serve as the basis for assuming a like uniformity of action in normal individuals and in pneumonia patients is a subject which requires further analysis. The difficulty in transferring the experimental results to patients lies in the question of whether the type of pneumonia produced in animals is the same as that found in man. Clinically, the two diseases present both resemblances and differences. The animals become definitely ill and show the symptoms already described. The illness, however, is of short duration and apparently reaches its height in the majority of animals in twenty-four to seventy-two hours. Before the expiration of this time, the temperature frequently returns to normal. Many of the infected animals, when they survive, recover in three to five days. The mortality in dogs infected with pneumococci is given by Lamar and Meltzer as 16 per cent. In the present series of twelve infected cats, the mortality was also 16 per cent. These findings differ from human pneumonia in the following particulars: The infection is not so severe; the temperature, though elevated at first, soon falls; the duration of the disease is short; and convalescence is rapid. The mortality is slightly lower. Musser and Norris (12) give the human mortality at 21.06 per cent. Pathologically the two diseases also show differences. The gross appearance of the lungs is not dissimilar, but in the animals the consolidated portions are somewhat dry and they fail to show a stage of gray hepatization (5). The amount of fibrin present is small. There is comparatively slight congestion of the alveolar walls and of the walls of the bronchi.

The relation of experimental pneumonias to the human disease has been discussed by a number of investigators. Almost all believe that the two types are similar. if not identical. Among the first to express this opinion was Sternberg (13) ; and later Gamaléia (14). Prudden and Northrup (15), Kinyoun and Rosenau (16), Wadsworth (17), Lamar and Meltzer (5), Wollstein and Meltzer (9) coincided with his view. Lamar and Meltzer, especially, have insisted on the identity of the two processes. On the other hand, Welch (18) in his study of experimental pneumonia, says: "Many inoculations of cultures of virulent pneumococci into the trachea and lungs of dogs have been made in my laboratory by Dr. Canfield and myself, but in no instance were we able to produce an inflammation of the lungs which we were willing to identify with acute lobar pneumonia as found in human beings." But he adds that. in the majority of experiments, there was no demonstrable consolidation and that pleurisy and more or less extensive areas of pneumonia were produced only in a few animals.

The inference consequently cannot be drawn that an effect obtained with strophanthin in the experimental disease may be anticipated in man. The striking similarity in action in infected and uninfected animals renders it likely, however, that the usual action of the drug in man may be expected in the presence of pneumonia. We have accumulated evidence, to be published later, which shows that this action actually takes place in the human disease. As far as evidence obtained electrocardiographically is concerned, our experiments show that strophanthin causes the same electrical changes in the heart when the animals are infected as it does under normal conditions.

CONCLUSIONS.

1. When a like amount of strophanthin is injected intravenously, the mortality is the same in both normal cats and in cats suffering from experimental pneumonia.

2. The minimum lethal dose of strophanthin is the same in normal dogs and in dogs suffering from experimental pneumonia.

3. The presence of an acute infection in these animals does not interfere with the action of strophanthin on the heart.

4. Electrocardiographically the changes occurring in the heart's action when strophanthin is injected are found to be similar in normal and in infected animals.

5. The identity of strophanthin action in infected and normal animals renders it probable that a like similarity may be anticipated in man, under normal conditions and in pneumonia.

BIBLIOGRAPHY.

1. Mackenzie, J., Diseases of the Heart, 3d edition, London, 1913, 379.
2. Gibson, G. A., Acute Pneumonia: Its Prognosis and Treatment, *Glasgow Med. Jour.*, 1911, lxxv, 321.
3. Meyer, A. W., Die Digitalis-Therapie ihre Indikationen und Kontraindikationen, *Jena*, 1912.
4. Gunn, J. W. C., The Influence of Temperature on the Action of Strophanthin on the Mammalian Heart, *Jour. Pharmacol. and Exper. Therap.*, 1914, vi, 39.
5. Lamar, R. V., and Meltzer, S. J., Experimental Pneumonia by Intrabronchial Insufflation, *Jour. Exper. Med.*, 1912, xv, 133.
6. Dochez, A. R., and Gillespie, L. J., A Biologic Classification of Pneumococci by Means of Immunity Reactions, *Jour. Am. Med. Assn.*, 1913, lxi, 727.

7. Hatcher, R. A., and Brody, J. G., Biological Standardization of Drugs, *Am. Jour. Pharm.*, 1910, lxxxii, 360.

8. Eggleston, C., Biological Standardization of the Digitalis Bodies by the Cat Method of Hatcher, *Cornell Univ. Med. Coll. Bull.*, 1914, iii, pt. 3, article 10; reprinted from *Am. Jour. Pharm.*, 1913, lxxxv, 99.

9. Wollstein, M., and Meltzer, S. J., Experimental Bronchopneumonia by Intrabronchial Insufflation, *Jour. Exper. Med.*, 1912, xvi, 126.

10. Cohn, A. E., Fraser, F. R., and Jamieson, R. A., The Influence of Digitalis on the T Wave of the Human Electrocardiogram, *Jour. Exper. Med.*, 1915, xxi, 593.

11. Rothberger, C. J., and Winterberg, H., Studien über die Bestimmung des Ausgangspunktes ventrikulärer Extrasystolen mit Hilfe des Elektrokardiogramms, *Arch. f. d. ges. Physiol.*, 1913, cliv, 571.

12. Musser, J. H., and Norris, G. W., Lobar Pneumonia, in Osler, W., and McRae, T., A System of Medicine, Philadelphia and New York, 1907, ii, 537.

13. Sternberg, G. M., The Pneumonia Coccus of Friedländer (Micrococcus Pasteuri, Sternberg), *Am. Jour. Med. Sc.*, 1885, N. S., xc, 106.

14. Gamaléia, M. N., Sur l'étiologie de la pneumonie fibrineuse chez l'homme, *Ann. de l'Inst. Pasteur*, 1888, ii, 440.

15. Prudden, T. M., and Northrup, W. P., Studies on the Etiology of the Pneumonia Complicating Diphtheria in Children, *Am. Jour. Med. Sc.*, 1889, N. S., xcvii, 562.

16. Kinyoun, J. J., and Rosenau. M. J., Infections Caused by Pneumococcus, *Rep. Superv. Surg. Marine Hosp. 1896–97*, Washington, 1899, 762.

17. Wadsworth, A., Experimental Studies on the Etiology of Acute Pneumonitis. *Am. Jour. Med. Sc.*, 1904, N. S., cxxvii, 851.

18. Welch, W. H., The Micrococcus lanceolatus with Especial Reference to the Etiology of Acute Lobar Pneumonia, *Bull. Johns Hopkins Hosp.*, 1892, iii, 125.

EXPLANATION OF PLATES.

In the following electrocardiograms leads were taken from the right fore to the left hind leg. In all figures the divisions of the abscissæ equal 0.04 of a second, the divisions of the ordinates equal 10⁻⁴ millivolts.

PLATE 74.

Fig. 1, A and B. These curves were taken from Normal Dog B. In A. made immediately before the injection of strophanthin, the P-R interval is 0 07 to 0.08 of a second. The T waves are prominent and upwardly directed. In B. taken 25 minutes after the injection was begun and when 0.43 mg. of strophanthin had been injected, P-R time increased to 0 11 or 0.12 of a second. The T wave is negative. The time since the commencement of the injection and the amount of strophanthin injected are given except in Fig. 3. In the cat curves the time only is indicated.

Fig. 2, A to C. Curves taken from Normal Dog C. Curve A was taken before the injection of strophanthin. The rhythm is regular. The P-R interval is 0.08 to 0.09 of a second. The T waves are negative. Curve B was made 65 minutes after the injection was begun; strophanthin 0.76 mg. had been injected. Complete dissociation between auricles and ventricles is shown. The rate of the auricles is much reduced. Curve C was made 20 minutes later than B; strophanthin 0.85 mg. had been injected. It shows a succession of ectopic ventricular beats. Two distinct types are seen. In the first and last parts of the curve the impulses arose in the wall of the right ventricle; those between, in the wall of the left ventricle. The latter show considerable variation. Auricular complexes cannot be distinguished.

PLATE 75.

FIG. 3, A to D. These curves were taken from Infected Dog 1. Curve A was made before the injection of strophanthin. The P–R time is 0.10 of a second. The T wave in some of the complexes is flat, in others it is composed of two parts, a low upwardly directed portion and a small negative wave. Curve B was taken 65 minutes after the injection of strophanthin was begun. The P–R time is lengthened to 0.15 of a second. The T waves consist of three parts: first, a gradual upward rise following the S wave; second, a rather sharp downward deflection; and third, an upward rise. Every third auricular impulse is blocked. Curve C was taken 10 minutes after Curve B. The complexes are normal. The P–R time is 0.15 of a second. The ventricular complex is similar to that seen in Fig. 2 B, the principal change being in the T wave. Ventricular extrasystoles occur which are probably of left ventricular origin. The T wave accompanying these is positive. Curve D was taken 25 minutes later than C; strophanthin 1.09 mg. had been injected. No auricular waves appear. In each pair of ectopic ventricular beats the first member has its origin in the left ventricle, and the second in the right ventricle. The third pair forms an exception; here the order is reversed.

FIG. 4, A and B. These curves were taken from Infected Dog 11. Curve A was made before the injection of strophanthin. The P–R time is 0.08 of a second. The T wave is made up of three parts, a small positive wave following directly on the S wave, a small negative, and a small positive. Curve B was made 15 minutes after the injection of strophanthin was begun; 0.24 mg. had been injected. The P–R time is unchanged. The T wave is composed of two parts, a negative portion of considerable depth and a small upward part.

PLATE 76.

FIG. 5, A to D. These curves were taken from Infected Cat 63. The animal survived the injection of strophanthin. Curve A was made before the injection. In the earlier portion of the curve the P wave has a normal relation to the R wave; later it gradually merges into the R wave until it disappears. The excursion of R is now increased and S is decreased. The T wave consists of two portions, a flat or upwardly directed portion and a small negative wave. Curve B was taken 13 minutes after the injection of strophanthin was begun. The P–R time is 0.07 of a second. The T wave consists of two portions, a flat or upwardly directed portion and a negative part of much greater extent than in Curve A. Curves C and D were taken 24 and 48 hours after the injection of strophanthin, respectively. In both curves the P wave is absent and in this

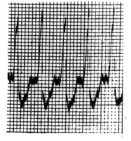

PLATE 74.

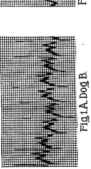

Fig.1A. Dog B.

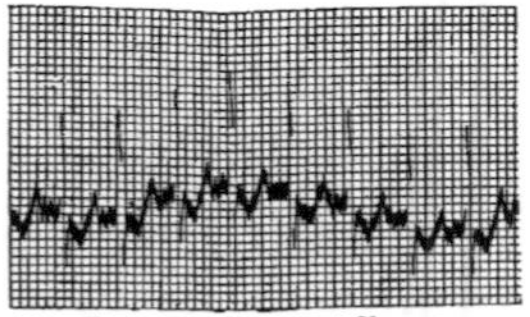

Fig.1B. 25min. 0.43 mgm.

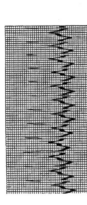

Fig.2A. Dog C.

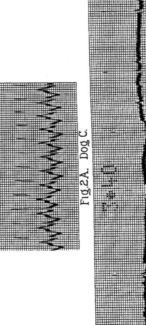

Fig.2B. 65 min. 0.76 mgm.

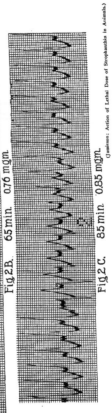

Fig.2 C. 85 min. 0.85 mgm.

(Jamieson: Action of Lethal Dose of Strophanthin in Animals.)

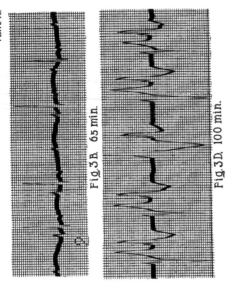

Fig.3B. 65 min.

Fig.3D. 100 min.

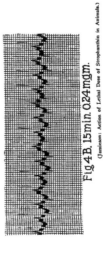

Fig.4B. 15 min. 0.24 mgm.

(Jamieson: Action of Lethal Dose of Strophanthin in Animals.)

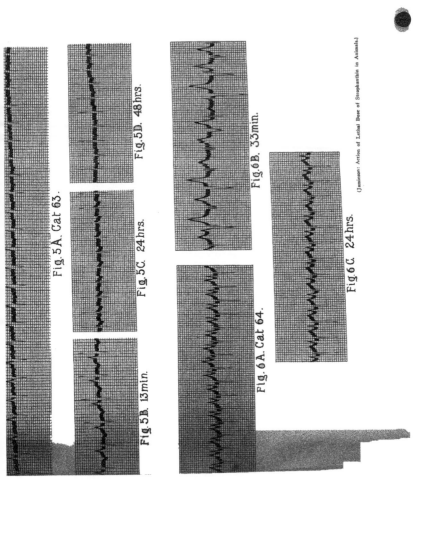

Fig. 5A. Cat 63.

Fig. 5B. 13 min.

Fig. 5C. 24 hrs.

Fig. 5D. 48 hrs.

Fig. 6A. Cat 64.

Fig. 6B. 33 min.

Fig. 6C. 24 hrs.

(Jamieson: Action of Lethal Dose of Strophanthin in Animals.)

respect they resemble the last portion of Curve A. A negative T wave has persisted in C and D, though in the latter T wave is more like the form seen in Curve A.

FIG. 6, A to C. These curves were taken from Infected Cat 64. This animal survived the injection of strophanthin. Curve A was made before the injection. The heart rhythm is normal. The P-R time is 0.07 of a second. The T wave is positive. Curve B was taken 33 minutes after the injection of strophanthin was begun. It shows a completely irregular heart action. No P waves are discernible. The ventricular complexes are ectopic, have various forms, and have their origin at several points in the walls of the right and left ventricles. Curve C was taken 24 hours later and shows the return of the heart to a normal rhythm. The P-R time is the same as in Curve A. The T waves are positive.

FURTHER INVESTIGATIONS ON THE ORIGIN OF TUMORS IN MICE.

I. Tumor Incidence and Tumor Age in Various Strains of Mice.

By A. E. C. LATHROP and LEO LOEB, M.D.

(*From the Department of Pathology of the Barnard Free Skin and Cancer Hospital, St. Louis.*)

(Received for publication, June 2, 1915.)

The starting point for our investigations was the observation of the endemic occurrence of tumors in animals.[1]

At that time, as well as on later occasions, we emphasized the fact that hereditary factors might be the cause of this endemic occurrence and that we had not been able to find any direct indication of infection; and we had furthermore occasion in a later publication to point out the importance of analyzing these conditions through breeding experiments. In 1907 we published some observations made at the mouse farm of Miss Lathrop, in Granby, Mass., which rendered it probable that the frequency of tumors in mice at certain places was in all probability due, not to infection, but to hereditary transmission in certain families.[2] Keeping mice in cages, where formerly tumor mice had lived, did not increase the ratio of cancer among those mice, while there was some indication that mice belonging to different families and kept on the same farm in Granby, Mass., showed a different tumor incidence, although the living conditions were approximately the same for the various families.

In 1910 we were enabled to resume these investigations on a larger scale on the same mouse farm in Granby, and in the following year we published the tree of one of the families of mice under our observation in which the hereditary transmission of tumors had been apparent.[3] Since our publication in 1907 there

---

[1] Loeb, L., On Carcinoma in Cattle, *Medicine*, 1900, vi, 286. Ueber das endemische Vorkommen des Krebses beim Tiere, *Centralbl. f. Bakteriol., 1te Abt., Orig.*, 1904, xxxvii, 235.

[2] Loeb, Further Observations on the Endemic Occurrence of Carcinoma, and on the Inoculability of Tumors, *Univ. Penn. Med. Bull.*, 1907–08, xx, 2.

[3] Loeb, Ueber einen Kontakt-Kombinationstumor bei einer weissen Maus, *Centralbl. f. allg. Path.*, 1911, xxii, 993.

appeared two investigations, one by Tyzzer[4] and a more recent one by Murray,[5] dealing with the heredity of tumors in mice. Both compared the frequency of tumors in a certain number of mice which were directly descended from tumor mice with those which had no tumor mice in their immediate ancestry. Tyzzer considers that his results suggest a possible influence of heredity, and Murray also considers his conclusions not yet as quite definite. Inasmuch as the results of Murray are more recent and are based on more extensive observations than those of Tyzzer, it might be of interest to analyze briefly the facts on which Murray's conclusions are based. Murray states that all the mice concerned in the experiments are descended ultimately from animals known to have suffered from cancer. "When new spontaneous cases of cancer in mice bred outside the laboratory are obtained, males bred in the laboratory from cancerous stock are mated with them, and as time goes on the process is repeated, so that a large number of animals with a composite known ancestry is obtained. This continual introduction of fresh cancerous stock to the parentage was decided on to eliminate, as far as possible, the influence of other indeterminate peculiarities which might conceivably influence the incidence of cancer indirectly in the nearly related animals derived by inbreeding from a single pair." Murray based his conclusions on the observation of 562 females which attained the age of 6 months or more. These 562 mice were divided into two unequal groups, one in which cancer occurred either in the mother, one or the other grandmother, or in all three, and another in which cancer was more remote in the ancestry. In the first lot of mice 18.2 per cent, in the latter 8.2 per cent tumor mice were found.

The principal criticism which can be made against conclusions based on this procedure is the following: Mice whose mothers or grandmothers had no tumors might belong to strains rich in tumors. In this case their offspring would in all likelihood develop many tumors, while mice whose direct ancestry had tumors might belong to strains very poor in tumors and their offspring would then develop less tumors than the offspring of non-tumor mice. For instance. if we should select non-cancerous female mice from our strains "English," "8½ plus English," "Michigan Wild plus 101," "Nov. 3d Strain," "Nov. 8th Strain," and compare the cancer rate of their offspring with the cancer rate of tumor mice from the strains "8½," "London," "Heitler," "Silver," "Cream," "German plus Carter," we would find that the offspring of the non-tumorous females is much richer in tumors than the offspring of tumorous animals. There is therefore a very large factor of chance in the results obtained by Murray. Accidentally the mice whose grandparents or parents had had tumors belonged to strains somewhat richer in tumors than the strains to which the other mice belonged. Definite conclusions can only be obtained if we use strains the tumor rate of which is known. These results we published in 1913.[6] We showed

[4] Tyzzer, E. E., A Series of 20 Spontaneous Tumors in Mice, with the Accompanying Pathological Changes and the Results of the Inoculation of Certain of These Tumors into Normal Mice, *Jour. Med. Research,* 1907-08, xvii, 155.

[5] Murray, J. A., Cancerous Ancestry and the Incidence of Cancer in Mice, *Fourth Scientific Report of the Imperial Cancer Research Fund,* 1911, 114.

[6] Lathrop, A. E. C., and Loeb, L., The Incidence of Cancer in Various Strains of Mice, *Proc. Soc. Exper. Biol. and Med.,* 1913, xi, 34. Loeb, L., Some Recent Results of Cancer Investigations, *Lancet-Clinic,* 1913, cx, 664.

among other facts on the basis of very extensive observations that different
strains kept at the same farm differed markedly in their tumor rate, that this
difference in tumor rate was a characteristic which was maintained approximately constant through several generations, and we gave definite figures for
the frequency of tumors in different strains. Thus the great importance of the
factor of heredity in the transmission of tumors in mice was definitely proven.

Since 1913 we have continued our investigations and we wish to
report (1) the extension of the work, (2) more detailed figures concerning the cancer rate and cancer age in the successive generations
of the various strains, and (3) some new facts especially regarding
the relation between cancer rate and cancer age. These data will
also form a basis for succeeding communications.

We found it best to divide the mice into three classes, according
to age: (1) in Class I were mice 7, 8, 9, 10, 11, and 12 months old;
(2) in Class II were mice 13, 14, 15, 16, and 17 months old; (3)
in Class III were mice 18 months old and older. Mice dead without tumor below the age of seven months are disregarded. In our
tables we give the generations of the mice as F_1, F_2, F_3, and so on,
designating as F_1 the first generation which we included in our new
records. On the left side of the table we state the number of mice
that die without tumors, on the right side those with tumors. First
we give the total figure of observed mice, then in brackets the
number of mice that die in each of the three periods of life. The
Roman figures after each Arabic figure indicate the period of age
during which a certain number of mice died. We give first the absolute figures and below in each case the percentage figures. In
the summary of the various strains we add the percentage of mice
dying with and without tumors in the respective periods expressed
in per cent of the total number of mice dying in this period. We
considered only female mice and included all the subcutaneous
tumors. The large majority of these are the common adenocarcinomata and carcinomata. There occur a few other tumors, as, for
instance, squamous cell carcinoma and sarcoma. The number of
these latter tumors is, however, in our material apparently so small
that the results cannot be affected by including all the subcutaneous
tumors in our calculations. Tumors of internal organs were disregarded at present. We shall report on this phase of our work on a

later occasion after the microscopic study has been concluded. Macroscopically visible tumors in internal organs of mice are of little significance for our present studies. While the material on which our conclusions are based is great, we intend to continue our work, and if there should be an error in our conclusions, it will in all probability be cleared up by our further experiments.

I. Tumor Incidence and Tumor Age of Non-Hybrid Strains.

In this class we consider strains, which, though of composite origin, were kept inbred in our experiments for a number of generations.

1. Strain English.

These are mice imported from England in 1905. Their offspring were brought to Granby, Mass., in the spring of 1906. The mice have a dull reddish color and their offspring have various colors, especially sable, agouti, and dull reddish. Later another female mouse of the same English strain with a bright red color is added. By breeding various red colored mice together an attempt is made to obtain red stock. However, among the offspring there are many black, sable, agouti, and chocolate colored animals, but rarely a mouse with a good red color. All the mice belonging to this strain were derived from a very small number of animals. The strain had been propagated for a number of years at the time when we started our experiments. The same holds good in the case of the majority of other strains. It may have been different in those strains which had been more recently imported, as the " European " and " German," " Heitler " and " London." But even in the Heitler and London the fact that the results even in small groups are, on the whole, concordant with each other does not make it probable that the various individuals within a strain which we used for propagation represented very different kinds of mice. Selection among the original gave, after breeding through several generations, a strain with a good proportion of black-eyed tans, and another with a small proportion of good black-eyed reds. There were also occasionally pink-eyed mice in these groups. One pink-eyed reddish tan colored male of English origin was mated to white females of Massachusetts origin. The offspring of this group again showed various colors, some of them white. These whites bred together for one generation gave again pure white offspring. The other offspring of the reddish tan male and white Massachusetts females were mostly black and sable in color. They all had black eyes. The latter when bred together without selection had generally dull colored offspring, but there were a very small number with pink eyes and a reddish or tan color, and two or three of a dull bluish smoke shade. These various pink-eyed animals had all pink-eyed offspring, usually of a non-characteristic color in the first generation, but in the succeeding generations the clear tan shade appeared; these tan mice, when bred together, bred true. Their offspring were the only mice which had exactly this color. These tan colored mice gave origin to the English substrains "101" and "Tan." The latter are mostly the offspring of Tumor Mice 121 and 122. They had therefore all tan color. Other mice of the same origin had, as mentioned above, a dull bluish color; there were two or three of those among the grandchildren of the original pink-eyed reddish tan male and the white Massachusetts female. These bluish mice bred together had some offspring of the same color and a few with distinct silver color. The silver colored were rare at first. Mice with somewhat similar, approximately silver color occurred also occasionally among the original English mice, which had no admixture of white Massachusetts stock. Through several generations these silver mice were added to the silver mice of the mixed origin. These silver mice selected and interbred gradually bred true. Occasionally, however, there occurred among them some mice of lighter, almost silver fawn color. The latter bred together also breed true and have only silver fawn offspring. Thus we have four distinct substrains of the English strains, which are characterized by their color which remains distinct: (1) the "Tan" mice with pink eyes, mainly offspring of Tumor Mice 121 and 122; (2) the substrain "101," also tan mice; (3) the "Silver"; and (4) the "Silver Fawn."

In addition to those we have two substrains in which mice of various colors occur: (1) "English Sable," members of the original English family. One pair of sable color was selected and gave origin to this substrain. This substrain has never had mice with the red and tan colors which some of the original English mice showed, but had many sable, some black, and many white mice. As white males were usually kept, the colored mice have now become eliminated and all "Sable" are at the present time white. The colored mice belonging to this substrain have never had pink eyes, but the white mice at present are pink-eyed, like ordinary white mice. (2) The last substrain is "English A," non-selected groups from the general stock. There were some groups all sable in color, some all red or orange; in other groups various colors were represented, sable, agouti, black, white, dull tans, and dull reds. In no case, however, did a real silver occur in such a mixed group. They occurred only a few times in earlier generations and were either discarded if of dull color, or added to the first two Silver generations, if of fairly clear color. Mice in this strain differed not only in their color, but also in their resistance. On the whole, the red and yellow pink-eyed varieties of English mice are not very rugged and not satisfactory as breeders. The females of the 101 strain were prolific, they very frequently had new litters, but they were usually small and died often at early ages. All the Silver as well as the 101 and Tan had pink eyes. None among the English Sable, with the exception of the white ones, had pink eyes. Of the substrain A some had pink eyes, but the majority were black-eyed. No. 481 a and her descendants had pink eyes, although they are mice of a relatively dark, dull agouti color.

The following substrains were thus distinguished:

1. English A, composed of various groups of English, not line bred.
2. Family 101, the offspring of pink-eyed tans.
3. English Tan E, the offspring of Tumor Mice 121 and 122.
4. English Sable.

5. English Silver.
6. English Silver Fawn.

It was of special interest to compare the tumor rates in the various groups which had a similar origin and to determine whether the various groups differed markedly in their tumor rates.

(a) Earlier groups.

I. Substrain English A.

Without tumors.   With tumors.

44 (33 I 4 II 7 III)   71 (25 I 39 II 7 III)
37% (75% I 9% II 16% III)   63% (35% I 55% II 10% III)

(b) Later groups.

12 (7 I 5 II)   21 (13 I 7 II 1 III)
36% (58% I 42% II)   64% (62% I 33% II 5% III)

(c) Additional records (winter, 1914-15).

8 (3 I 4 II 1 III)   19 (14 I 5 II)
30% (38% I 50% II 12% III)   70% (79% I 21% II)

We see that the tumor rate in the same group at different periods remained approximately the same (63, 64, and 70 per cent), that a large per cent of the tumors appear in the first period of life, and a large part of the remaining tumors in the second period. As a result of the high tumor incidence in young mice, the mortality is in early life so great that the number of mice attaining an old age is relatively small.

The figures for the total of this lot are:

Without tumors   With tumors.

64 (43 I 13 II 8 III)   111 (52 I 51 II 8 III)
37% (67½% I 20% II 12½% III)   63% (47% I 46% II 7% III)
II + III 26%   III 50%   I + II 35%   II + III 74%   III 50%   I + II 65%
 I 35%   I 65%

In two cases the offspring of tumor mice belonging to this group were followed separately. Through four generations the offspring of Tumor Mouse 481 a. (dull agouti with pink eyes) were recorded.

Without tumors.   With tumors.

3 (3 I)   11 (8 I 3 II)
21% (100% I)   79% (72% I 28% II)

In this case the result is in accordance with the behavior of the group as a whole.

In the offspring of Tumor Mouse 281, which were followed through five generations, the tumor rate is very much lower.

(a) Earlier generation.

Without tumors.	With tumors.
7 (7 I)	1 (1 II)

(b) Later generations.

5 (1 I 3 II 1 III)	2 (1 I 1 II)
Total: 12 (8 I 3 II 1 III)	3 (1 I 2 II)
80% (68% I 24% II 8% III)	20% (34% I 66% II)

It is possible that we accidentally isolated in this last family recessives with a low tumor rate. But as the mice died early in this case, and the number of animals we observed is small, we must leave this question at present undecided.

2. Substrain Family 101.

No. 101 was a pink-eyed English Tan tumor mouse from a family in which all three females had tumors.

Without tumors.	With tumors.
F₁ 3 (3 I)	
F₂ 6 (6 I)	10 (9 I 1 II)
37% (100% I)	63% (90% I 10% II)
F₃ 7 (3 I 4 II)	22 (17 I 5 II)
24% (43% I 57% II)	76% (77% I 23% II)
F₄ 9 (5 I 4 II)	27 (17 I 9 II 1 III)
25% (55% I 45% II)	75% (63% I 33% II 4% III)
F₅ 2 (1 I 1 II)	6 (1 I 5 II)
25% (50% I 50% II)	75% (17% I 83% II)
F₆ 2 (2 I)	5 (2 I 3 II)
28% (100% I)	72% (40% I 60% II)
Total: 29 (20 I 9 II)	70 (46 I 23 II 1 III)
29% (69% I 31% II)	71% (66% I 33% II 1% III)
II + III 28% I 30% I + II 29%	II + III 72% I 70% I + II 71%

We see that we have here to deal with a high tumor rate (71 per cent), that the majority of the tumors appear in the first period of life, and the rest in the second period, and furthermore that in the six generations which we observed in this family the tumor rate and the tumor age were about the same. If we disregard the mice dying in the first period of life, the tumor incidence remains about the same (72 per cent). In this family a tumor was observed to appear in a six months old mouse.

3. Substrain English Tan E.

Offspring (pink-eyed) of two early Tan Tumor Mice 121 and 122.

Without tumors.	With tumors.
3 (3 I)	8 (6 I 2 II)
27% (100% I)	73% (75% I 25% II)

This is a small group of similar character to the preceding family. Here tumor incidence and tumor age are similar to those in 101.

4. Substrain English Sable.

(a) Older records.

Without tumors.	With tumors.
F₁ 3 (2 or 3 I 1 uncertain II)	6 (3 I 3 II)
F₂ 12 (11 I (6 of these below 6 months) 1 III)	40 (33 I 5 II 2 III)
F₃ 11 (8 I 1 II 2 III)	25 (21 I 3 II 1 III)
F₄ 3 (1 I 2 II)	21 (17 I 3 II 1 III)
F₅ 0	5 (4 I 1 II)
Total: 29 (22 I 4 II 3 III)	97 (78 I 15 II 4 III)
23% (76% I 14% II 10% III)	77% (81% I 15% II 4% III)
II + III 29%	II + III 71%

(b) Newer records.

F₂ 2 (1 I 1 II)	5 (3 I 2 II)
F₃ 19 (11 I 7 II 1 III)	24 (14 I 8 II 2 III)
F₄ 4 (3 I 1 II)	4 (1 I 3 II)
Offspring of Tumor Mouse 217 1 (1 I)	6 (2 I 4 II)
Total: 26 (16 I 9 II 1 III)	39 (20 I 17 II 2 III)
40% (62% I 34% II 4% III)	60% (51% I 44% II 5% III)
II + III 34%	II + III 60%

Total of (a) and (b):

55 (38 I 13 II 4 III)	136 (98 I 32 II 6 III)
28% (69% I 24% II 7% III)	72% (73% I 23% II 4% III)
II + III 31%	II + III 69%

(c) Family of Tumor Mouse 437. (No. 437 raised young while she had a tumor.)

F₁ 1 (1 I)	7 (1 I 6 II)
F₂ 2 (2 I)	4 (2 I 2 II)
F₃ 2 (2 I)	5 (4 I 1 II)
F₄ 0	
Total: 5 (5 I)	16 (7 I 9 II)
24% (100% I)	76% (44% I 56% II)

(d) Additional records (winter, 1914–15).

F_3 9 (4 I   3 II   2 III)   15 (5 I   4 II   6 III)
F_5 7 (2 I   3 II   2 III)   8 (4 I   3 II   1 III)
F_6 0   1 (1 I)
Total: 16 (6 I   6 II   4 III)   24 (10 I   7 II   7 III)
40% (37½% I   37½% II   25% III)   60% (42% I   29% II   29% III)

Total of English Sable: 176 (115 I   48 II   13 III)
76 (49 I   19 II   8 III)   70% (66% I   27% II   7% III)
30% (65% I   25% II   10% III)   II + III 69%   III 62%   I 70%   I +
II + III 31%   III 38%   I 30%   I +   II 71%
II 29%

If we compare the total of English Sable with the total of English
101, we see that the figures are very similar as to the incidence of
cancer and the age at which cancer appears. In both cases (without
consideration of the mice alive in the various periods) the majority
of tumors appear in the first period of life. In English A the
tumor incidence is only slightly less, and here also the first period
of life is the one in which there appear more tumors than in any
other period; but here again the preponderance of the first period
over the other periods is not so great as in the case of the Sable and
101. English Tan and English Sable are similar. Within the
Sable the various generations and families of individual tumor mice
behave in a similar way as the total. Of course, certain variations
occur, but they are relatively slight and they are especially noticeable where the number of observed mice is relatively small; only in
one family a different tumor rate may prevail, but even here it is not
certain. The offspring of mice born while the mother had a tumor
are not more liable to have tumors than other mice of the same group
(family of Tumor Mouse 437). On the whole, the variations in
the incidence of tumors and the age at which they appear follow a
parallel course; if the incidence is somewhat lower in a group, the
tumors also appear at a somewhat later period of life. At least
this parallelism is noticeable in the records within the groups of the
English strain just mentioned. It remains to be seen whether or
not this parallelism is accidental.

· *5. Substrain English Silver.*

They were bred for color. All the English Silver mice are poor breeders.

(a) Older records.

Without tumors.   With tumors.

74 (29 I 39 II 6 III)   5 (1 I 4 II)
94% (39% I 53% II 8% III)   6% (20% I 80% II)
II + III 92%   II + III 8%

There were 9 groups. In 7 no tumors were observed, in 2 groups tumors appeared. In 1 of these 2 groups 2 out of 21 mice, in the other 3 out of 13, had tumors.

(b) Additional records (winter, 1914-15).

35 (15 I 10 II 10 III)   3 (3 I)
92% (43% I 28½% II 28½% III) 8% (100% I)

Here tumors appeared in 2 groups out of 6. In one there were 2 tumor mice in a group of 10, in the other 1 out of 7 had a tumor.

Total of all English Silver:

109 (44 I 49 II 16 III)   8 (4 I 4 II)
93% (41% I 45% II 14% III)   7% (50% I 50% II)

We see that tumor incidence is the same in both sets (a) and (b). The difference between the other English and the English Silver is most striking. In the latter the tumor rate is very low throughout. Furthermore, it is of interest that out of 15 groups tumors appeared in only 4, and in 3 out of these 4 groups there were more than 1 tumor mouse, while in 11 groups no tumor appeared.

Of great interest is the fact that while the tumor incidence in the English Silver is very low, the mice that had tumors had them at about as early an age as the strains with high tumor rate. The English Silver that did not have tumors in the first or second period of life were not more liable to acquire them in the third period than they were in the first and second periods. It seems therefore that incidence of tumor and tumor age are unit factors which are to some extent independent and transmitted independently of each other. Furthermore, we see here established the interesting fact that there is a linking in this case between the color of the mice and the rate of tumor incidence. A low rate goes with a silver color, while a high rate goes with the sable, tan, and white among the English strains.

6. Substrain English Silver Fawn.

The Silver Fawn have been more prolific than the Silver, but just as delicate. They rarely raise their young. Two groups were kept: (1) 1 to 9 months old, 2 to 11 months old, 1 to 13 months old, 1 to 15 months old, without tumors; (2) their offspring, 1 to 9 months old, 2 to 12 months old, 3 to 13 months old, 2 to 16 months old, without tumors. This record makes it probable that tumors are here as rare as among the Silver. We have therefore split off from the English strain a substrain Silver, in which there went hand in hand with the color a a very low tumor incidence. And from the Silver again another variety of color, Silver Fawn, was split off, in which again there was linked with the color a low tumor incidence.

2. Strain "Cream."

The mice that gave origin to this strain were obtained in the year 1903 from two different sources, Atlantic City and Springfield, Ohio. Up to the year 1911 only the cream colored mice were kept and used for breeding, while the black and sable colored mice which appeared more frequently were discarded. From 1911 on, females of all colors were kept. The females in this strain are poor breeders; the mice grow slowly, but ultimately reach the same size as the other mice. Many of these mice reach old age.

(a) Older records.

Without tumors.	With tumors.
127 (35 I 50 II 42 III)	3 (1 II 2 III)
98% (27% I 39% II 34% III)	2% (33% II 66% III)
II + III 97% III 97.6%	II + III 3% III 2.4%

There were 26 groups in this lot. In one group there appeared 2 tumor mice, and in another group 1 tumor mouse. In 24 groups there appeared no tumor mouse. As in the case of the English mice, the rate of tumor incidence is not essentially changed, if only mice in Periods II and III or mice in Period III of life are considered.

(b) Additional records (winter, 1914–15).

94 (28 I 41 II 25 III)	2 (1 II 1 III)
98% (29% I 44% II 27% III)	2% (50% II 50% III)

These 2 tumor mice appeared in 2 different groups. The tumor incidence is here the same as in Group (a). The low tumor rate in this strain is therefore a constant condition.

Total Cream strain:

221 (63 I 91 II 67 III)	5 (2 II 3 III)
98% (29% I 41% II 30% III)	2% (40% II 60% III)
II + III 97% III 96%	II + III 3% III 4%

The tumor rate is extremely low, 2 per cent. If we leave the mice dying in the first or first and second period of life out of consideration, the tumor rate is not essentially changed.

Cream X.

Seven mice of the Cream strain (with black, cream, and sable color) were separated from the others, and four generations of these mice were bred. They were called Cream X.

(a) Older records of Cream X.

Without tumors.	With tumors.
44 (20 I 13 II 11 III)	0
100% (45% I 30% II 25% III)	

(b) Recent records.

86 (28 I 32 II 26 III)	5 (2 I 3 II)
94½% (33% I 37% II 30% III)	5½% (40% I 60% II)

Total of Cream X:

130 (48 I 45 II 37 III)	5 (2 I 3 II)
96% (37% I 35% II 28% III)	4% (40% I 60% II)

Thus the tumor rate in the Cream X is of the same order as in the Cream, although they had been kept entirely separate. We may therefore conclude that in this case we have to deal with relatively pure strains as far as the tumor incidence is concerned. Of 14 groups of Cream X, 12 were free from tumors; in each of 2 groups (holding 9 and 7 mice respectively) 2 tumor mice were found. If we arrange the Cream X according to generations, we find the following:

F_1 34 (13 I 12 II 9 III)	
100% (38% I 35% II 27% III)	
F_2 39 (15 I 12 II 12 III)	
100% (38% I 31% II 31% III)	
F_3 39 (14 I 12 II 13 III)	5 (1 I 3 II)
89% (36% I 31% II 33% III)	11% (20% I 80% III)
F_4 18 (6 I 9 II 3 III)	1 (1 I)
95% (33% I 50% II 17% III)	5% (100% I)

Taking Cream and Cream X together we find that tumors occur here apparently later than in the English strain, also later than in the Silver; absolutely the majority of tumors appear here in the second and not in the third period of life.

3. Strain No. 8.

No. 8 is of mixed origin. They are in part descended from mice received from Springfield, Mass., and from Ohio; in part from a pair of plum-silver mice sometime imported from England, but not related to the English strain mentioned previously. From this mixed origin descended some plum-silver females which were in Dec., 1907, mated with the son of a tumor mouse found in Granby. Their descendants are the No. 8 strain. The male, the son of a tumor mouse, was later, when he had grown old, left to his daughters and gave origin to the strain designated 8½. No. 8 is a strain which breeds well and is healthy; the mice live longer than the English strain.

(a) Older records of No. 8.

Without tumors.	With tumors.
F₂ 1 (1 III)	4 (4 III)
F₃ 14 (2 I 3 II 9 unknown age)	8 (3 I 2 II 3 III)
F₄ 24 (22 I 1 II 1 III)	11 (3 I 6 II 2 III)
F₅ 18 (9 I 9 III)	9 (4 I 2 II 3 III)
F₆ 39 (8 I 15 II 16 III)	17 (7 I 8 II 2 III)
F₇ 29 (19 I 8 II 2 III)	3 (1 II 2 III)
F₉ 12 (7 I 3 II 2 III)	4 (1 II 3 III)
F₁₀ 12 (3 I 3 II 6 III)	8 (4 II 4 III)
Total:	
149 (70 I 33 II 37 III 9 unknown age)	64 (17 I 24 II 23 III)
70% (47% I 22% II 31% III)	
II + III 60% III 62%	30% (26% I 37% II 37% III)
	II +III 40% III 38%

In this group is maintained through the various generations a medium tumor rate, considerably lower than in the English strain, and the tumor age is higher; also higher than in the English Silver, which have a considerably lower tumor rate than the No. 8 strain.

(b) One tumor mouse, No. 100, from the fourth generation of the No. 8 strain was mated with a male of No. 8, F₄ and had the following offspring:

F₁ 0	3 (2 II 1 III)
F₂ 4 (1 I 3 III)	6 (1 I 4 II 1 III)
F₃ 5 (5 I)	0
F₄ 3 (2 I 1 II)	6 (6 I)
Total:	
12 (8 I 1 II 3 III)	15 (7 I 6 II 2 III)
44% (67% I 8% II 25% III)	56% (74% I 40% II 13% III)

The tumor rate is here apparently higher than in the No. 8 strain, and the tumor age somewhat lower.

(c) Additional records (winter, 1914–15).

F₃ 26 (7 I 13 II 6 III)   2 (2 III)

In the 8th generation the tumor rate is lower than usual. The two tumors appeared here in the last period of life. We have in this case to deal with a relatively small number of mice. Until further observations indicate a constancy in the general lowering of the tumor rate in this strain, we cannot attach too much importance to this observation.

Total of No. 8:
187 (85 I 47 II 46 III 9 unknown age)   81 (24 I 30 II 27 III)
70% (45% I 24% II 24% III)   30% (29% I 37% II 34% III)
II + III 62% III 63%   II + III 38% III 37%

In this case the tumor rate increases somewhat if we consider only the mice in the second and third periods of life.

4. Strain 8 1/2.

This is a strain closely related to Strain 8. The male (son of a tumor mouse) which gave origin to No. 8 was mated to his daughters, which latter belonged to the No. 8 strain. This strain is less rich in tumors than No. 8, and the tumors appear also at a rather late period of life. This fact may be explained by assuming that the male in this case belonged to a strain less rich in tumors than the females.

Without tumors.	With tumors.
$F_1 + F_2$ 24 (10 I 14 II)	1 (1 II)
F_3 14 (6 II 8 III)	4 (2 I 1 II 1 III)
F_4 51 (14 I 18 II 19 III)	1 (1 II)
F_5 40 (17 I 14 II 9 III)	6 (6 II)
F_6 2 (2 I)	0

Total:
131 (43 I 52 II 36 III)   27 (6 I 17 II 4 III)
83% (33% I 40% II 27% III)   17% (22% I 63% II 15% III)
II + III 81% III 90%   II + III 19% III 10%

Leaving out of consideration mice which did not reach the second or third period of life modifies the figures only slightly without essentially changing the results.

5. Strain Carter.

This strain is the offspring of a son of a pink-eyed English tumor mouse of orange color and of 9 females (7 of which were white, 2 brown with white spots) obtained in Utica, N. Y. The females of this strain are as good breeders as the average No. 8 or English, but do not grow up as fast as the latter.

Without tumors.	With tumors.
F_1 1 (1 III)	
F_2 11 (5 I 2 II 4 III)	3 (2 II 1 III)
F_3 19 (7 I 7 II 5 III)	16 (8 I 5 II 3 III)
F_4 10 (3 I 3 II 4 III)	7 (4 I 2 II 1 III)

Total:

41 (15 I 14 II 12 III)   26 (12 I 9 II 5 III)
61% (36% I 35% II 29% III)   39% (46% I 35% II 19% III)
II + III 65% III 71%   II + III 35% III 29%

Disregarding mice that died during the first period of life without tumors, but including those acquiring tumors during this period, the figures are: without tumors 50% (54% II 46% III); with tumors 50% (46% I 35% II 19% III). The tumor rate in this strain is higher than that of No. 8 and 8½ and lower than that of English. The tumor age is slightly higher than in the case of the English and lower than in Nos. 8 and 8½. Wherever the figures are sufficiently large in the various generations, the results are similar.

6. Strain European.

Thirteen mice of various sizes were imported from Europe, Feb., 1910. They are on the whole of smaller size than the Granby mice. They and their offspring suffer greatly from mouse typhus. In Oct., 1912, their number has not yet increased. Two of the offspring of the imported stock have tumors. Only a relatively small number of this lot lived long enough to be included in the records.

Without tumors.	With tumors.
F₄ 7 (5 I 2 II)	0
F₅ 9 (5 I 4 II)	2 (2 I)

Total:

16 (10 I 6 II)   3 (2 I 1 II)
84% (62% I 38% II)   16% (66% I 34% II)

Substrain European from Trio.

Two females and one male (all probably of one litter) were separated from the rest of the European mice; one of these two females had a tumor when not yet quite a year old (No. 228). The offspring of these mice were more healthy, but here also occasionally typhus appeared.

Without tumors.	With tumors.
F₂ 5 (1 I 3 II 1 III)	1 (1 II)
F₃ 18 (6 II 12 III)	2 (1 II 1 III)
F₄ 49 (23 I 14 II 12 III)	11 (5 I 4 II 2 III)
F₅ 7 (3 I 4 II)	1 (1 II)

Total:

79 (27 I 27 II 25 III)   15 (5 I 7 II 3 III)
84% (34% I 34% II 32% III)   16% (33% I 47% II 20% III)
II + III 84%   II + III 16%

The same tumor rate occurs here as in the other European group, notwithstanding the much larger number of mice in this group. The figures in both substrains are therefore confirmatory of each other. The tumor rate is similar to that in No. 8½. The tumor age is somewhat lower than that of No. 8 and No. 8½. The results in the various generations agree fairly well with each other.

7. Strain German.

These mice were imported from Germany in May, 1910. They are somewhat smaller than the average Granby mice, but a little larger than the European. They are a very prolific strain, but again die early on account of disease. Three just weaned young females and one male were the only ones to thrive. All the mice recorded are the offspring of these four mice.

Without tumors.	With tumors.
F₁ 0	3 (1 I 2 II or III)
F₂ 0	2 (1 I 1 II or III)
F₄ 6 (2 I 3 II 1 III)	3 (2 I 1 III)
F₅ 0	2 (2 II)
F₆ 4 (4 I)	

Total:

10 (6 I 3 II 1 III)	10 (3 I 5 II 2 III)
50% (60% I 30% II 10% III)	50% (30% I 50% II 20% III)
II + III 36%	II + III 64%

The number of mice is not large in this strain, but the indications are that we have to deal with a strain rich in tumors, which, however, do not appear as early as is the case of the English. In this lot the mortality in early life was great. The tumor rate increases, therefore, if we exclude the animals dying in the first period of life.

8. Strain Heitler.

This strain is the offspring of mice imported more recently from Germany.

Without tumors.	With tumors.
F₁ 13 (4 I 7 II 2 III)	3 (3 II)
81% (31% I 54% II 15% III)	19% (100% II)
F₂ 39 (24 I 5 II 10 III)	9 (1 I 8 II)
81% (62% I 13% II 25% III)	19% (11% I 89% II)
F₃ 13 (9 I 2 II 2 III)	11 (1 I 10 II)
54% (70% I 15% II 15% III)	46% (9% I 91% II)
F₄ 9 (5 I 4 II)	5 (2 I 3 II)
64% (56% I 44% II)	36% (40% I 60% II)

Total:

74 (42 I 18 II 14 III)	28 (4 I 24 II)
73% (56% I 24% II 20% III)	27% (14% I 86% II)
II + III 57%	II + III 43%

In this strain the tumor rate as a whole is similar to that of No. 8. The tumor age also is similar, inasmuch as the first period of life is poor in tumors, even more so than in Nos. 8 and 8½. It is an interesting fact that in this strain the large majority of tumors appear in the second period of life, while the third period is free from tumors. For this reason the tumor rate increases, if we omit from consideration the mice dying in the first period of life, just as we did in the case of No. 8. On the whole, the figures are within certain limits similar in the different generations.

9. *Strain London.*

This strain comprises about a dozen animals with various colors imported from London, England, late in the summer of 1911.
Of five imported females, four had tumors.

(a) Older records.

Without tumors.	With tumors.
F_1 22 (7 I 5 II 10 III)	9 (4 I 2 II 3 III)
F_2 19 (8 I 7 II 4 III)	7 (1 I 3 II 3 III)

Total:
41 (15 I 12 II 14 III)   16 (5 I 5 II 6 III)
72% (37% I 29% II 34% III)   28% (31% I 31% II 38% III)
II + III 73% III 70%   II + III 27% III 30%

(b) Family of an imported London mouse with tumor (481).

F_1 4 (3 I 1 III)	0
F_2 1 (1 II)	0
F_3 1 (1 I)	1 (1 I)

(c) Later records (winter, 1914-15).

F_1 9 (1 I 3 II 5 III)	3 (3 III)
F_2 16 (7 I 4 II 5 III)	2 (1 II 1 III)
F_3 15 (9 I 5 II 1 III)	11 (3 I 8 II)

Total:
40 (17 I 12 II 11 III)   16 (3 I 9 II 4 III)
54% (42% I 30% II 28% III)   26% (19% I 56% II 25% III)
II + III 64% III 54%   II + III 36% III 26%

Thus the later records agree very well with the former one, as to the rate of tumors and tumor age. The irregularities found in the individual generations increase in inverse ratio to the number of the animals observed; but on the whole the variations are not very great. The tumor rate and tumor age are similar to those of No. 8. Omitting animals in the first or in the first and second period of life does not greatly alter the result.

Total:

87 (36 I 25 II 26 III)   33 (9 I 14 II 10 III)

73% (41% I 29% II 30% III)   27% (27% I 43% II 30% III)

II. Tumor Incidence and Tumor Age of Some Hybrids.

These and a number of other strains, not yet mentioned, were used for hybridization and have thus produced a considerable number of additional strains which we observed through successive generations. Inasmuch as we shall report on the crosses later in connection with another problem, we shall here cite only a few of them.

1. German + Carter.

(a) Early records.

Without tumors.   With tumors.

195 (56 I 57 II 82 III)   15 (2 I 3 II 10 III)

93% (29% I 29% II 42% III)   7% (13% I 20% II 63% III)

(b) Later records.

131 (23 I 31 II 77 III)   17 (2 I 8 II 7 III)

88% (17% I 24% II 59% III)   12% (12% I 47% II 41% III)

Total:

326 (79 I 88 II 159 III)   32 (4 I 11 II 17 III)

91% (24% I 24% II 52% III)   9% (12% I 34% II 54% III)

II + III 90% III 89% I 95% I +   II + III 10% III 11% I 5% I + II

II 92%   8%

Hence the earlier and later records are similar in regard to the tumor rate as well as to the tumor age. The tumors appear very late. In the successive generations some variations occur; they are, however, on the whole not very great; the deviation from the mean is greatest where the number of observed mice is smallest. The tumors appear in all generations at a late period of life.

F_1 11 (6 I 5 III)   3 (3 III)

78% (55% I 45% III)   22% (100% III)

F_2 71 (15 I 18 II 38 III)   3 (3 III)

96% (21% I 25% II 54% III)   4% (100% III)

F₃ 205 (49 I 48 II 108 III)
  90% (24% I 23% II 53% III)
F₄ 39 (9 I 22 II 8 III)
  95.1% (20% I 60% II 20% III)

24 (3 I 10 II 11 III)
10% (13% I 42% II 45% III)
2 (1 I 1 II)
4.9% (50% I 50% II)

2. *Silver + 10.*

Without tumors.	With tumors.

F₁ 1 (1 III)
F₂ 40 (19 I 15 II 6 III)
  70% (48% I 37% II 15% III)
F₃ 95 (31 I 34 II 30 III)
  65% (33% I 35% II 32% III)
F₄ 13 (4 I 1 II 8 III)
62% (31% I 7% II 62% III)
F₅ 1 (1 III)

3 (1 II 2 III)
17 (3 I 7 II 7 III)
30% (18% I 41% II 41% III)
52 (20 I 17 II 15 III)
35% (38% I 33% II 29% III)
8 (3 I 4 II 1 III)
38% (38% I 49% II 13% III)
4 (4 III)

Total:

150 (54 I 50 II 46 III)
64% (35 2/3% I 33 1/3% II 31% III)
II + III 62% III 61%

84 (26 I 29 II 29 III)
36% (32% I 34% II 34% III)
II + III 38% III 39%

The various generations behave in a similar manner as far as tumor rate and tumor age are concerned. The tumors appear at a medium age.

3. *European + No. 10 (Nov. 3d Strain).*

The tumors appear in the successive generations at a similar rate. Variations occur especially if the number of mice is relatively small in a certain generation. The tumors appear on the average at a medium age in the various generations.

Without tumors.	With tumors.

F₁ 1 (1 III)
F₂ 5 (3 I 2 II)
  24% (60% I 40% II)
F₃ 21 (9 I 6 II 6 III)
  22% (43% I 28% II 29% III)
F₄ 42 (30 I 12 II)
  37% (71% I 29% II)
F₅ 2 (1 I 1 II)
  9% 50% I 50% II)

1 (1 III)
16 (6 I 10 II)
76% (37% I 63% II)
75 (16 I 36 II 25 III)
78% (21% I 48% II 31% III)
71 (33 I 30 II 8 III)
67% (46% I 43% II 11% III)
20 (8 I 5 II 7 III)
91% (40% I 25% II 35% III)

Total:

71 (43 I 21 II 7 III)
28% (60% I 30% II 10% III)
II + III 19% III 15%
I + II 31% I 41%

183 (63 I 81 II 39 III)
72% (34% I 43% II 23% III)
II + III 81% III 85%
I + II 69% I 59%

The tumors appear not so early as in the case of the English. The tumor age is similar to No. 8.

III. Tumor Age.

In discussing the various strains of mice, we have already had occasion to mention the age at which the tumors appeared. We did not, however, express the frequency of tumors of a certain strain in relation to the number of mice alive at a certain period. This is done in the following pages. Here those strains which have a sufficient number of mice for the purpose of such a calculation are divided into three classes according to their age. We include here furthermore strains which were not mentioned previously. In each case we state how many mice were alive in each of the three periods and what percentage of the mice alive had tumors at that period of life. Thus we come to a more accurate determination of the frequency of tumors at a certain period of life and we can decide whether the various strains have the same tumor age and whether this tumor age is correlated with other factors. The figures which we thus obtain are only an approximation to the real conditions: they are not absolutely correct, because in computing the figures we ignored the fact that during the various periods mice died at different times, while we assumed for the purpose of obtaining a relatively small number of comparable figures in each strain, that they lived to the end of that .period of life. However, we believe that this inaccuracy does not interfere seriously with a representation of the actual facts.

In defining the tumor age of the various strains we had further to consider the fact that the figures obtained stating the per cent of tumor mice living at a certain period depended also on the absolute frequency of tumors of a certain strain and not only on the distribution of tumors according to the age of the animals. In order to eliminate the former factor it was necessary to determine in all strains the relation between the percentage of tumor mice in each of the three periods of life.

On the following pages we state in each strain the number of tumor mice expressed in per cent of living mice in each period.

We then state the frequency of tumors in the strain as a whole, and then follows the relation between tumor rate in the first and second plus the third periods, and the relation between tumor rate in the first and the second periods, between tumor rate in the second and third, and lastly between the tumor rates in the first plus the second and third periods of life. Thus we obtain an insight into the relative frequency of tumors in the three different periods of life.

European + No. 10 (Nov. 3d strain).

I Period	254 mice	63 tumors	= 25%		72%	I : II+III=1 : 7 I : II=1 : 3
II "	148 "	120 "	= 81%			II : III=1 : 1 I+II : III=1.2 : 1
III "	46 "	39 "	= 88%			

English A.

I Period	175 mice	52 tumors	= 29%		63%	I : II+III=1 : 4 I : II=1 : 2.2
II "	80 "	51 "	= 64%			II : III=1.3 : 1 I+II : III=1.9 : 1
III "	16 "	8 "	= 50%			

English Sable.

I Period	252 mice	115 tumors	= 46%		70%	I : II+III=1 : 2.8 I : II=1 : 1.5
II "	88 "	61 "	= 69%			II : III=1 : 1 I+II : III=1.9 : 1
III "	21 "	13 "	= 62%			

English except Silver. Total:

I Period	537 mice	219 tumors	= 41%		68%	I : II+III=1 : 3.2 I : II=1 : 1.7
II "	203 "	146 "	= 72%			II : III=1.3 : 1 I+II : III=2 : 1
III "	38 "	22 "	= 58%			

English Silver.

I Period	117 mice	8 tumors	= 6.1%		7%	I : II+III=1.2 : 1 I : II=1.2 : 1
II "	69 "	4 "	= 5.1%			
III "	16 "	0 "	= 0 %			

No. 8.

I Period	268 mice	24 tumors	= 8%		30%	I : II+III= 1 : 9.3 I : II=1 : 4.7
II "	150 "	57 "	, = 38%			II : III=1 : 1 I+II : III=1.2 : 1
III· "	73 "	27 "	= 37%			

No. 8½.

I Period	158 mice	6 tumors	= 4%		17%	I : II+III=1 : 6.2 I : II=1 : 3.7
II "	109 "	17 "	= 15%			II : III=1.5 : 1 I+II : III=1.9 : 1
III "	40 "	4 "	= 10%			

Carter.

I Period	67 mice	12 tumors	= 17 %		39%	I : II+III=1 : 3 I : II=1 : 1.3
II "	40 "	9 "	= 22.5%			II : III=1 : 1.2 I+II : III=1.4 : 1
III "	17 "	5 "	= 28 %			

German.

I Period 20 mice 3 tumors = 15% ⎫
II " 11 " 5 " = 45% ⎬ 50% I : II+III=1 : 7.4 I : II=1 : 3
III " 3 " 2 " = 66% ⎭ II : III=1 : 1.5 I+II : III=1 : 1.1

European.

I Period 94 mice 5 tumors = 5% ⎫
II " 62 " 7 " = 11% ⎬ 16% I : II+III=1 : 4.2 I : II=1 : 2.2
III " 28 " 3 " = 10% ⎭ II : III=1.1 : 1 I+II : III=1.6 : 1

Heitler.

I Period 102 mice 4 tumors = 4% ⎫
II " 56 " 24 " = 43% ⎬ 27% I : II+III=1 : 10.7 I : II=1 : 10 7
III " 14 " 0 " = 0% ⎭ II : III∞ I+II : III∞

London.

I Period 120 mice 9 tumors = 7% ⎫
II " 75 " 14 " = 19% ⎬ 27% I : II+III=1 : 6.6 I : II=1 : 2.7
III " 36 " 10 " = 27% ⎭ II : III=1 : 1.4 I+II : III=1 : 1

European + No. 10 (Nov. 8 strain).

I Period 96 mice 20 tumors = 21% ⎫
II " 61 " 33 " = 54% ⎬ 65% I : II+III=1 : 6.1 I : II=1 : 2.6
III " 12 " 9 " = 75% ⎭ II : III=1 : 1.4 I+II : III=1 : 1

Cream + 10.

I Period 174 mice 4 tumors = 3% ⎫
II " 152 " 23 " = 15% ⎬ 36% I : II+III=1 : 17 I : II=1 : 5
III " 96 " 35 " = 36% ⎭ II : III=1 : 2.4 I+II : III=1 : 2

No. 8 + German.

I Period 244 mice 21 tumors = 9% ⎫
II " 183 " 51 " = 28% ⎬ 41% I : II+III=1 : 6.8 I : II=1 : 3.1
III " 86 " 28 " = 33% ⎭ II : III=1 : 1.2 I+II : III=1 : 1.1

101 + 103.

I Period 152 mice 6 tumors = 4% ⎫
II " 97 " 27 " = 29% ⎬ 34% I : II+III=1 : 17 I : II=1 : 7 2
III " 45 " 18 " = 40% ⎭ II : III=1 : 1.4 I+II : III=1 : 1.2

German + Carter.

I Period 358 mice 4 tumors = 1% ⎫
II " 275 " 11 " = 4% ⎬ 9% I : II+III=1 : 14 I : II=1 : 4
III " 176 " 17 " = 10% ⎭ II : III=1 : 2.5 I+II : III=1 : 2

European Hybrid + 8½ F₄.

I Period 473 mice 8 tumors = 2% ⎫
II " 364 " 19 " = 5% ⎬ 16% I : II+III=1 : 13 I : II=1 : 2.5
III " 213 " 50 " = 21% ⎭ II : III=1 : 4.2 I+II : III=1 : 3

Silver + 10.

I Period 234 mice 26 tumors = 11% ⎫
II " 154 " 29 " = 18% ⎬ 36% I : II+III=1 : 5.2 I : II=1 : 1.6
III " 75 " 29 " = 39% ⎭ II : III=1 : 2.2 I+II : III=1 : 1.4

(103 + European) F_1 + III. Daughter of
  No. 10.
I Period 168 mice   3 tumors = 2% ⎫
II   "   115   "   11   "   = 9% ⎬ 17% I : II+III=1 : 15   I : II=1 : 4.5
III "   67   "   14   "   = 21% ⎭ II : III=1 : 2.2 I+II : III=1 : 3

European 151 + 8 F_8
I Period 125 mice   8 tumors = 6% ⎫
II   "   86   "   20   "   = 23% ⎬ 30% I : II+III=1 : 7   I : II=1 : 3.8
III "   47   "   9   "   = 19% ⎭ II : III=1.2 : 1 I+II : III=1.5 : 1

European + 102, 103.
I Period 146 mice   2 tumors = 1.5% ⎫
II   "   109   "   9   "   = 8 % ⎬ 21% I : II+III=1 : 22   I : II=1 : 5.3
III "   76   "   19   "   = 25 % ⎭ II : III=1 : 3.1 I+II : III=1 : 2.6   —

London + (European + 103) F_3.
I Period   78 mice   0 tumors = 0% ⎫
II   "   66   "   2   "   = 3% ⎬ 5% II : III=1 : 2
III "   33   "   2   "   = 6% ⎭

Cream.
I Period 177 mice   0 tumors = 0% ⎫
II   "   120   "   2   "   = 2% ⎬ 2.3% II : III=1 : 2
III "   53   "   2   "   = 4% ⎭

8½ + English Sable.
I Period   48 mice   20 tumors = 41% ⎫
II   "   15   "   9   "   = 60% ⎬ 61% I : II+III=1 : 1.5   I : II=1 : 1.5
III "   1   "   0   "   = 0% ⎭

8½ + No. 10 F_1 (Nov. 8 strain).
I Period   82 mice   6 tumors = 7% ⎫
II   "   68   "   19   "   = 27% ⎬ 49% I : II+III=1 : 10   I : II=1 : 4
III "   35   "   15   "   = 43% ⎭ II : III=1 : 1.6 I+II : III=1 : 1.3

(8½ + No. 10) + No. 10 (Nov. 8).
I Period   72 mice   5 tumors = 7% ⎫
II   "   48   "   9   "   = 19% ⎬ 36% I : II+III=1 : 11   I : II=1 : 2.7
III "   20   "   12   "   = 60% ⎭ II : III=1 : 3.2 I+II : III=1 : 2.3

Michigan Wild + 101 F_2.
 I Period   50 mice   13 tumors = 26% ⎫
II   "   23   "   9   "   = 39% ⎬ 58% I : II+III=1 : 4   I : II=1 : 1.5
III "   10   "   7   "   = 70% ⎭ II : III=1 : 1.8 I+II : III=1 : 1.1

121 (English Tan) + Cream.
I Period   92 mice   16 tumors = 7% ⎫
II   "   56   "   18   "   = 31% ⎬ 42% I : II+III=1 : 8.8 I : II=1 : 4.4
III "   16   "   5   "   = 31% ⎭ II : III=1 : 1 I+II : III=1.2 : 1

European + 146 (English Tan).
I Period   74 mice   3 tumors = 4% ⎫
II   "   58   "   13   "   = 22% ⎬ 28% I : II+III=1 : 11   I : II=1 : 5.5
III "   22   "   5   "   = 23% ⎭ II : III=1 : 1 I+II : III=1.1 : 1

English 121 + German.

I Period 39 mice 10 tumors = 25%

II   "   18 "   7   "   = 39% $\Big\}$49% I : II+III=1 : 3.6 I : II=1 : 1.6

III "   4 "   2 "   = 50% II : III=1 : 1.3 I+II : III=1.3 : 1

White English + (8 + German) F$_4$.

I Period 82 mice 27 tumors = 33%

II   "   33 "   25 "   = 73% $\Big\}$63% I : II+III=1 : 2.2

III "   1 "   0 "   = 0%

We may arrange the various strains into four classes according to the relative frequency of tumors in the various periods of life. While there are some slight variations within each class, on the whole the various strains in each class behave in a similar manner.

Class I. Early Tumors.

I : II + III = 1 : 2.8 I : II = 1 : 1.6 II : III = 1 : 1.2 I + II : III = 1.5 : 1

English 68%.

English Silver 7%.

8½ + English Sable 61%.

English (121) + German 49%.

White English + (8 + German) F$_4$ 63%.

Michigan Wild + 101 F$_2$ (in Period III, more tumors) 58%.

Carter 39%.

European 16%.

Average tumor rate 45%.

Class II. Tumors of Medium Age.

I : II + III = 1 : 7.4 I : II = 1 : 3.5 II : III = 1 : 1.2 I + II : III =1.2 : 1.

Nov. 3 strain 72%.

Nov. 8 strain 65%.

No. 8 30%.

No. 8½ 17%.

Heitler (in II period very high tumor rate) 27%.

London 27%.

European + 146 (English Tan), (not quite as good. in Period I, less tumors) 28%.

No. 8 + German 41%.

Silver + 10 (less tumors in Period III, more in II, nearer Class I) 36%.

European Hybrids 151 + 8 F$_3$ (Period I low, Period II high. Period III lower, somewhat better than Class II) 30%.

121 + Cream 42%.

German 50%.

Average tumor rate 38%.

Class III. Great Frequency of Tumors in Period III.

I : II + III = 1 : 10.5　I : II = 1 : 3.3　II : III = 1 : 2.4　I + II : III = 1 : 1.8.

8½ + No. 10 F₁ (Nov. 8 strain) 49%.

(8½ + No. 10) + No. 10 (Nov. 8 strain) (in Period III many tumors) 36%.

Average tumor rate 42½%.

Class IV. Late Tumors.

I : II + III = 1 : 16.3　I : II = 1 : 4.7　II : III = 1 : 2.6　I + II : III = 1 : 2.3.

Cream + 10 36%.

Cream 2.3%.

101 + 103 = [101 + (103 + European)] 34%.

German + Carter 9%.

European Hybrid + 8½ F₄ = [(European + 102) + 8½ F.] 16%.

Hybrids (103 + European) F₂ + III, daughter of No. 10 17%.

European + 102, 103 21%.

London + (European + 103) F₃ 5%.

Average tumor rate 17.5%.

In Class I, in which the tumors appear relatively early, we find in the combined second and third periods of life 2.8 times as many tumors as in the first period; in the second period of life 1.6 times as many tumors as in the first period; in the third period 1.2 times as many tumors as in the second period; and in the combined first and second periods 1.5 times as many tumors as in the third period. While, therefore, in this class the tumors are more frequent in the second period than in the first, and in the third than in the second, the differences in the percentages are not very marked. The tumors appear here relatively early, the percentage of tumors in the first period of life being considerable. We include in this group eight strains, counting the various English substrains as one strain, with the exception of the Silver. We added to the name of each strain the number giving the frequency of tumor mice in per cent of the total number of mice, irrespective of the age at which they appear. If we designate as a high tumor rate 40 to 100 per cent, as a medium tumor rate 20 to 40 per cent, and as a low tumor rate 0 to 20 per cent, we find that five of the eight strains have a high tumor rate, one a medium tumor rate (with a figure near the lower border of the high tumor rate), and two a low tumor rate. Of these two, one strain (Silver) is related to the other English strain, which has a high tumor rate. We see therefore that the tumor age can vary independently of the tumor frequency. English Silver has like the other English strain an early tumor age, but in contradistinction to the other English mice a very low tumor rate. The average tumor rate in Class I is 45 per cent. It would be higher if we had counted the various English strains separately.

In Class II the tumors in the second period are 3.5 times more frequent than in the first period; the tumors in the second period are therefore here relatively (compared to the frequency in the first period) considerably more frequent than in the first class. In the third period the tumors are 1.2 times more frequent than in the second period; this is the same proportion between the frequency in the third and second periods as in the first class; but absolutely the tumors are here more frequent in the third period; 11 strains belong to this class; 4 of these have a high, 6 a medium, and 1 a low tumor rate. The average tumor rate is 38 per cent.

Class III consists of only two strains, which as far as the relation between the number of tumors in the first and second period of life is concerned, stand between the first and second class, but are nearer the second class, in which, however, the relative frequency of tumors in the third period is greater than in the second class. The average tumor rate is here 42.5 per cent.

Class IV contains the late tumors. The relative preponderance of the frequency of tumors in the second over that in the first period is here greater than in the first, second, or third class, and again the preponderance of the third period over the second period is greater than in the former classes. Eight strains are included in this class. None of these show a high tumor rate, 2 show a medium, and 6 a low tumor rate. The average tumor rate in this class is 17.5 per cent.

We may draw from this analysis the following conclusions:

1. There exists a certain relationship between tumor frequency and tumor age. On the whole, the more frequent the tumors, the earlier they appear in the various strains. It might be conceivable that the frequency of the tumors was independent of the tumor age;

that in strains in which the tumor frequency is greater, the tumors appear in the same percentage in the various periods of life, but this is evidently not the case. In strains in which the tumor frequency is greater, they appear on the whole also at an earlier period of life.

2. This parallelism between tumor frequency and tumor age is, however, not complete. The tumor age seems to be as characteristic for a strain as the tumor rate. In strains with a similar rate of frequency, the tumor age may be different. This difference is probably not accidental, because (1) if we have substrains related to each other, the tumor age is usually similar in all of them, and (2) the tumor age of the constituent strains seems to influence the tumor age of the crosses. How far this latter relation holds good, we shall discuss in another communication. On the other hand, we found in the case of the substrain Silver that it had a similar tumor age to the English strain, although the tumor rate of the Silver is . considerably lower than that of the other English strains.

3. The tumor age is transmitted from generation to generation in a similar manner to the tumor rate. We may therefore conclude that in all probability tumor rate and tumor age represent distinct · unit factors which frequently, but not in all cases, are in some way linked to each other.

4. We may furthermore conclude that the age where the maximum of tumors occurs varies in different strains. While in some it appears in the second period of life, in others it is in the third period. Here again the maximum is on the whole reached at an earlier period of life in strains with a high tumor rate. But here also peculiarities exist in different strains.

SUMMARY.

1. It is possible to split a strain of mice into certain substrains in which the tumor incidence is in some way linked to the color of the mice. Thus we could split off from the English strain, which as a whole and in various substrains with mixed colors (English A and Sable) has a high tumor rate, substrains with light tan color and pink eyes (101 and Tan) which have a high tumor rate like the large majority of the English mice, and two other apparently

recessive strains breeding true, which have a very low tumor incidence (Silver and Silver Fawn). Therefore, certain combinations of factors which determine certain colors of mice determine at the same time the tumor incidence of these strains or substrains. In the majority of cases isolated families bred through several generations separately from the majority of the other substrains give approximately the same tumor rates as the others; in some cases, however, it may perhaps be possible to separate from the main strain a family with a different rate.

2. The tumor incidence and the tumor age found in the earlier periods of our work are approximately the same as in the more recent period. On the whole, the results obtained in successive generations of the same strain also agree well with each other; the results are fairly constant; the deviations which occur are in most cases due to the small number of animals observed in the certain generations. Discarding all the mice dying in the first or in the first and second periods of life usually does not alter essentially the tumor ratio of a certain strain.

· 3. A certain relationship exists between tumor frequency and tumor age. On the whole, the more frequent the tumors, the earlier they appear in the various strains. This parallelism between tumor frequency and tumor age is, however, not complete. The tumor age seems to be as characteristic for a certain strain as the tumor rate. Certain substrains which differ in tumor frequency may show approximately the same tumor age. Strains with similar tumor frequency may show a different tumor age. We may therefore conclude that in all probability tumor rate and tumor age represent distinct unit factors, which are frequently, but not in all cases, linked in some manner to each other.

4. The age at which the maximum of tumors appears varies in different strains. The maximum may fall into the second or third period of life. On the whole, the maximum is reached at an earlier period of life in those strains which have a high tumor rate. But here also peculiarities exist in different strains.

THE RELATION OF THE SPLEEN TO BLOOD DE-STRUCTION AND REGENERATION AND TO HEMOLYTIC JAUNDICE.

XII. The Importance in the Production of Hemolytic Jaundice of the Path of Hemoglobin to the Liver.

By J. HAROLD AUSTIN, M.D., and O. H. PERRY PEPPER, M.D.

(*From the John Herr Musser Department of Research Medicine of the University of Pennsylvania, Philadelphia.*)

(Received for publication, June 15, 1915.)

In connection with our studies (1) on the influence of splenectomy in diminishing the tendency to jaundice after the administration of hemolytic agents, our attention has been called to the possible importance of a factor purely mechanical; namely, the relation of the spleen to the blood supply of the liver.

Ponfick (2) in a discussion of hemoglobinemia of whatever origin and its consequences called attention to certain facts. He recalls that the earliest observers had already recognized that when hemolysis is going on the spleen becomes greatly swollen in consequence of the accumulation of disintegrating erythrocytes; for this swollen spleen Ponfick uses the term spodogenous (σποδός, waste products). Ponfick notes that simultaneously the liver eliminates a bile very rich in pigments and suggests that this is derived from the hemoglobin set free in the spleen, carried by the portal circulation to the liver and removed by this organ. Ponfick further expressed the view, based on experiments not quoted in detail, that the liver could completely remove and transform into bile pigment liberated hemoglobin up to the extent of one-sixtieth of the total hemoglobin of the body, but that hemoglobin set free in excess of this amount passes through the liver and is eliminated by the kidneys, causing hemoglobinuria. One-sixtieth of the total hemoglobin in the dog is about 0.18 gm. per kilo. In an earlier paper of this series, Pearce, Austin, and Eisenbrey (3) have shown that the injection of from 0.14 to 0.35 gm. per kilo of hemoglobin as laked blood will cause the appearance of hemoglobinuria. but that a factor of great importance, apparently overlooked by Ponfick, is the rate at which the hemoglobin is liberated in the circulation. The more slowly it is introduced the larger is the quantity that the liver can take up without permitting the concentration in the blood to reach at any time that required for the production of hemoglobinuria. It was further shown in the same paper that while small amounts of injected hemoglobin are removed by the liver and presumably excreted as bile pigments in the bile without the occurrence of jaundice, if the injected hemoglobin be in excess of 0.30 to 0.40 gm. per kilo, the liver is unable to eliminate all the bile pigment formed from the excess of hemoglobin and some of the bile pigment is resorbed from the liver and under these circumstances appears in the urine. It was noted in this respect also that the rate of injection is of great importance in determining the amount of hemoglobin that the liver will tolerate without the appearance of bile pigments in the urine. Very slow but long continued liberation of hemoglobin can eventually overtax the hepatic excretory power and lead to the appearance of bile pigments in the urine, although the hemoglobin liberation may have been slow enough to permit of its continued adequate removal from the circulation by the liver with at no time the development of hemoglobinuria. Thus the first effect of hemoglobin liberation into the blood is an increased bile pigment content of the bile. This was shown experimentally by Tarchanoff (4). If the amount of hemoglobin be small enough and its liberation slow enough this is the only effect. A slightly larger amount rapidly liberated will produce hemoglobinuria. A larger amount extremely slowly liberated will produce bile in the urine. A larger amount liberated at an intermediate rate may produce both hemoglobinuria and bile pigments in the urine. Ponfick in his experiments noted bile pigments in the urine but follows Tarchanoff's explanation and attributes them to the now generally repudiated theory of extrahepatic conversion of hemoglobin into bile pigment.

Following Ponfick many other workers have attributed importance to the spleen as the site of disintegration of erythrocytes; among these may be mentioned Hunter (5), Gabbi (6), and Mya (7). Bottazzi (8) in his studies of the blood after splenectomy noted an increased resistance of the erythrocytes to hypotonic salt solutions and it was to this factor that Banti (9) attributed the greater resistance and diminished tendency to jaundice of splenectomized animals receiving hemolytic agents.

However, Pugliese and Luzzatti (10) failed to confirm Bottazzi's findings, but, noting with Banti and others the diminished tendency to jaundice after splenectomy, they made further studies along the lines suggested by Ponfick's observations and elaborated the following hypothesis. The spleen is the natural location for the disintegration of erythrocytes after the administration of hemolytic poisons, and the hemoglobin so liberated is carried directly by the portal system to the liver, there at least in large measure to be removed, converted into bile pigment, and excreted in the bile; or, if present in great quantity, to be resorbed and appear in the urine and tissues as bile pigments and thus produce jaundice. On the other hand, in the absence of the spleen, the blood cells undergo disintegration elsewhere, probably chiefly in the bone marrow, as noted by Martinotti and Barbacci (11). Hemoglobin liberated in the bone marrow could under these circumstances reach the liver only through the general circulation. It would therefore be diluted and moreover would reach the liver largely through the hepatic artery,—a vessel normally carrying blood for nutritive purposes, not for purposes of elaboration. They further note that the slow flow of blood through the bone marrow might delay the removal of the liberated hemoglobin. For these reasons in the splenectomized animal it is to be expected that the hemoglobin would reach the liver much more gradually and at a rate, indeed, which might well lie within the capacity of the liver for complete excretion as bile pigment; hence no reabsorption of bile pigments would occur and jaundice would not develop.

These authors were able to show by the aid of a bile fistula that while the other constituents of the bile are but little altered by splenectomy, the bile pigments are reduced to about one-half. Moreover, after the administration of hemolytic poisons while the bile of the splenectomized animal shows an increase in the bile pigments, this is not so pronounced as in the normal animal after hemolytic poisons; the increased pigmentation of the bile is, however, of longer duration in the splenectomized animal.

It may be noted in passing that these observers failed to find disintegrating erythrocytes in the lymph nodes after splenectomy; a phenomenon which Pearce and Austin (12) found conspicuous; nor do the findings of Karsner and Pearce (13) regarding the fragility of the corpuscles after splenectomy confirm those of Pugliese and Luzzatti, but on the contrary agree with those of Bottazzi, to the effect that there is an increased resistance of the erythrocytes to hypotonic salt solutions and to hemolytic immune serum after splenectomy.

In order to test the hypothesis of Pugliese and Luzzatti, we have injected a constant amount per kilo of hemoglobin solution in the form of laked blood into a series of dogs, injecting each at least twice, once into the general circulation by way of the femoral vein and once into the portal circulation by way of a mesenteric vein. The rate of injection has been always the same. In some instances, the femoral injection was given first, in other instances the mesenteric. At least five days were allowed to elapse between injections. We have thus been able to study the effect of the site of the injection upon the development of hemoglobinuria and of bile pigments in the urine, and from our results believe we may draw conclusions as to the fate of hemoglobin when liberated into the portal system, on the one hand, or into the general circulation, on the other.

In these experiments, a fasting, normal dog was bled, the blood defibrinated, the cells thrown down by centrifuging, and the supernatant serum removed. About four volumes of distilled water were then added to the cells and the mixture agitated for fifteen to twenty minutes to induce hemolysis. The solution was then centrifuged rapidly for twenty minutes to remove most of the cell stromata, was made isotonic by addition of sodium chloride, and centrifuged to remove any globulin thrown out of solution upon adding the salt. One cc. of this solution was then diluted with 99 cc. of distilled water and its hemoglobin strength determined in a Fleischl-Miescher hemoglobinometer. As much of this solution as should equal either 0.3 gm. or 0.4 gm. of hemoglobin per kilo of body weight was then injected intravenously into a normal dog. Injections were given at such a rate that the entire injection should occupy one minute per kilo of body weight. All bleedings and injections were made under ether anesthesia. Injections into the general circulation were made into one of the small veins of the leg. Portal injections were made by drawing a loop of intestine from the abdomen under aseptic precautions, and injecting through a needle into a small mesenteric vein. In some instances water was given by stomach tube at the close of the operation. The urine was then collected, the dog being kept in a metabolism cage, and if hemoglobin appeared the amount was estimated either directly in the Fleischl-Miescher hemoglobinometer or by comparison with a standardized acid hematin solution. In addition, we followed the jaundice by the persistence of bile pigments in the urine after hemoglobin injection into either the mesenteric or the femoral vein. The urine was examined for bile pigments by the Rosenbach test.

TABLE I.

Influence of Site of Injection on Amount of Hemoglobin Eliminated in the Urine.

Hemoglobin injections.	Date.	Gm. of hemoglobin per kilo eliminated after injection into	
		Femoral vein.	Mesenteric vein.
Dog 26 (0.4 gm. per kilo).............	Mar. 20	0.085	
	" 26		0.043
" 12 (0.3 " ").............	Feb. 25		0.029
	Apr. 15	0.043	
" 5 (0.3 " ").............	Jan. 9	0.024	
	" 15		None.
	Mar. 2		"
	" 26	0.026	
" 3 (0.3 " ").............	Jan. 7	0.017	
	" 15		Trace.
Splenectomized................	Feb. 19		
	Mar. 2		None.
	" 20	0.025	
" 49 (0.3 gm. per kilo).............	Feb. 25	0.014	
Three months after splenectomy..	Apr. 15		0.010

The results as regards hemoglobinuria are shown in Table I. In each of five dogs used, the output of hemoglobin by the kidney was much less when the hemoglobin was introduced into the mesenteric vein than when introduced into the femoral vein, and this is true regardless of which injection was performed first. This we attribute to the removal of the hemoglobin to a greater extent by the liver when the injection is made into a mesenteric vein with the result that the hemoglobin reaching the general circulation is less concentrated and is less likely to be eliminated by the kidneys and appear in the urine.

TABLE II.

Persistence of Bile Pigment in the Urine after Hemoglobin Injection as Determined by Point of Injection.

Hemoglobin injections.	Date.	Persistence after injection into	
		Femoral vein.	Mesenteric vein.
		days	*days*
Dog 4 (0.3 gm. per kilo)..............	Jan. 9	4	
	" 16	4	
" 5 (0.3 " " ")..............	" 15	3	
	" 15		12+
" 25 (0.4 " " ")..............	Mar. 21	4	
	" 26		7+
" 26 (0.4 " " ")......	" 20		
	" 26		7+
" 12 (0.3 " " ")	Feb. 25		5
	Apr. 15	4	
" 3 (0.3 " " ")..............	Jan. 7	0	
	" 15		
Splenectomized	Feb. 19		
	Mar. 2		
	" 20	0	
" 49 (0.3 gm per kilo)..............	Feb. 25	4	
Three months after splenectomy..	Apr. 15		

The results of the study of the degree and persistence of jaundice (as indicated by bile pigments in the urine) in the dogs after the two types of injection are shown in Table II. It will be seen that in six dogs studied the jaundice was distinctly more persistent after mesenteric than after femoral injection, and this was true regardless of which injection was made first. In Dog 4, two successive injections were made into the femoral vein to determine whether the second injection would give a result notably different from the first. Such was not the case, the duration of the bile pigments being the same after each injection when both were made into the femoral vein.

In our studies both of hemoglobinuria and of the persistence of jaundice after hemoglobin injections we have employed splenectomized as well as normal dogs, but have found that splenectomy has no influence upon the fate of hemoglobin injected into either the general or portal circulation.

SUMMARY.

These experiments indicate, therefore that when hemoglobin is set free in the portal circulation a larger amount is held by the liver and converted rapidly into bile pigment than is the case when it is set free in the general circulation, and that, under the former condition, over-loading of the liver with bile pigment more readily occurs and jaundice is more apt to develop.

This mechanical influence must, therefore, be a factor in the lessened tendency after splenctomy to the jaundice which follows blood destruction due to hemolytic agents, for whether the spleen be an active factor in destroying the erythrocytes or whether it plays merely a passive part as a place for the deposition of the disintegrating cells, there can be no question that in this organ, when it is present, a large number of cells undergo their final disintegration after the action of hemolytic poisons, and that the hemoglobin there liberated passes by the portal system directly to the liver. When the spleen is removed, this disintegration occurs in other organs, notably in the lymph nodes and bone marrow, and the hemoglobin from these organs passes not into the portal but into the general circulation, from which it reaches the liver more gradually and in a more dilute form.

BIBLIOGRAPHY.

1. Pearce, R. M., Austin, J. H., and Krumbhaar, E. B., The Relation of the Spleen to Blood Destruction and Regeneration and to Hemolytic Jaundice. I. Reactions to Hemolytic Serum at Various Intervals after Splenectomy, *Jour. Exper. Med.*, 1912, xvi, 363. Pearce, R. M., Austin, J. H., and Eisenbrey, A. B., II. The Relation of Hemoglobinemia to Hemoglobinuria and Jaundice in Normal and Splenectomized Animals, *Jour. Exper. Med.*, 1912, xvi, 375. Pearce, R. M., Austin, J. H., and Musser, J. H., Jr., III. The Changes in the Blood Following Splenectomy and Their Relation to the Production of Hemolytic Jaundice, *Jour. Exper. Med.*, 1912, xvi, 758; VII. The Effect of Hemolytic Serum in Splenectomized Dogs, *ibid.*, 1913, xviii, 494.

2. Ponfick, E., Ueber Haemoglobinemie und ihre Folgen, *Berl. klin. Wchnschr.*, 1883, xx, 389.

3. Pearce, Austin, and Eisenbrey, *loc. cit.*

4. Tarchanoff, J. F., Ueber die Bildung von Gallenpigment aus Blutfarbstoff im Thierkörper, *Arch. f. d. ges. Physiol.,* 1874, ix, 53.
5. Hunter, W., Physiology and Pathology of Blood Destruction, *Lancet,* 1892, ii, 1209, 1259, 1315, 1371.
6. Gabbi, U., Ueber die normale Hämatolyse mit besonderer Berücksichtigung der Hämatolyse in der Milz, *Beitr. z. path. Anat. u. z. allg. Path.,* 1893, xiv, 351.
7. Mya, G., Sur la régénération sanguine dans l'anémie par destruction globulaire, *Arch. ital. de biol.,* 1891–92, xvi, 108.
8. Bottazzi, F., La milza come organo emocatatonistico, *Sperimentale, Sez. biol.,* 1894, xlviii, 433.
9. Banti, G., La milza nelle itterizie pleiochromiche, *Gaz. d. Osp.,* 1895, xvi, 489.
10. Pugliese, A., and Luzzatti, T., Contributo alla fisiologia della milza: milza e veleni ematici, *Arch: p. le sc. med.,* 1900, xxiv, 1. Pugliese, A.. Rate et poisons hématiques, *Arch. ital. de biol.,* 1900. xxxiii, 349; Die Absonderung und Zusammensetzung der Galle nach Extirpation der Milz, *Arch. f. Anat. u. Physiol., Physiol. Abt.,* 1899, 60; La secrezione e la composizione della bile negli animali smilzati, *Polyclinico, Sez. med.,* 1899. vi. 121 : La secretion et la composition de la bile chez les animaux privés de la rate, *Arch. ital. de biol.,* 1900, xxxiii, 359.
11. Martinotti, G., and Barbacci, O., La tumefazione acuta della milza nelle malattie infettive, *Morgagni,* 1890, xxxii, 521. 593.
12. Pearce, R. M., and Austin, J. H., V. Changes in the Endothelial Cells of the Lymph Nodes and Liver in Splenectomized Animals Receiving Hemolytic Serum, *Jour. Exper. Med.,* 1912, xvi, 780.
13. Karsner, H. T., and Pearce, R. M., IV. A Study, by the Methods of Immunology, of the Increased Resistance of the Red Blood Corpuscles after Splenectomy, *Jour. Exper. Med.,* 1912, xvi, 769.

THE RELATION OF THE SPLEEN TO BLOOD DESTRUCTION AND REGENERATION AND TO HEMOLYTIC JAUNDICE.

XIII. The Influence of Diet upon the Anemia Following Splenectomy.

By RICHARD M. PEARCE, M.D., J. HAROLD AUSTIN, M.D., and O. H. PERRY PEPPER, M.D.

(*From the John Herr Musser Department of Research Medicine of the University of Pennsylvania, Philadelphia.*)

(Received for publication, June 15, 1915.)

In our various experimental studies (1) of the changes in the blood that follow, in the dog, the removal of the spleen, we have been impressed by the wide variation in the degree of anemia. In the severer forms of anemia the changes are marked, as, for example, a drop in red cells from 5,100,000 to 2,970,000 and in hemoglobin from 105 to 50 per cent in five weeks; in a milder form, the red cells may not fall below 4,000,000 to 4,500,000, with a decrease in hemoglobin to 65 or 75 per cent. Most of the animals show changes corresponding roughly to one or the other of these types. There remains, however, a relatively small group in which the changes are very slight, representing a decrease of only about 1,000,000 red cells and only 10 to 15 per cent of hemoglobin. These slighter changes, falling sometimes within the limit of error of the methods of blood counting, led us to ask ourselves the question: May there not be some factor other than the absence of the spleen that aids in producing the severer forms of anemia? That the absence of the spleen is the main factor there can be no doubt, for even in the milder forms the slight changes usually correspond to the period (three to six weeks) in which the severer anemias reached their lowest point; also, animals with the milder forms show the same tendency to slow repair when a hemolytic agent is given, as do the severer forms. There is, therefore, no

682

doubt about the removal of the spleen being responsible for a disturbance of the mechanism concerned in maintaining the normal blood picture. The question remains, however: Why does the degree of anemia vary? Our first thought was that diet might have some influence. Until the present investigation was undertaken, all animals, except those used in the study of iron metabolism (2), had been kept upon the same general diet; *i. c.,* a mixture of meat, bread, cereals, and vegetables, in all essentials the table scraps upon which dogs are usually fed. This was always supplied in abundance and each dog received all he would eat, and, as our records show that dogs splenectomized for periods of several months or a year or more gained in weight on this diet, we considered it perfectly satisfactory. However, we did not know the exact caloric value of this mixed diet and, moreover, as it was essentially a boiled diet, it might possibly be deficient in some substance essential to the proper function of the hemopoietic system.

In this connection we recalled (1) the observations of Asher and Vogel (3), that while an iron-poor diet (sugar, starch, and lard) has no effect upon the blood picture in a normal dog, in the splenectomized dog on the same diet a great decrease in number of red cells and amount of hemoglobin occurs, and further, that if under the latter circumstances an iron-rich (flesh) diet is given, the blood picture quickly returns to normal; (2) Richet's (4) observation that in order to maintain splenectomized dogs at the same weight as normal dogs, a much larger quantity of food is necessary; and (3) Paton's (5) conclusion that splenectomy in the dog has no influence either upon the blood picture or the general metabolism.

The various observations quoted, with the exception of Paton's studies of metabolism, are directly opposed to the results of our studies. In regard to Asher and Vogel's contention, we have not found by a direct quantitative study of the elimination of iron that splenectomy seriously influences iron metabolism (6). Moreover, the improvement in the anemia which they describe as the result of feeding iron-rich food corresponds in our opinion to the spontaneous repair of the anemia which usually begins about the end of the fourth week. In other words, the improvement they noticed was in part at least due, in our opinion, to the normal repair and not to the effect of the iron-rich food. Their conclusions would be more convincing if they had prevented entirely or lessened the severity of the anemia by beginning the feeding before or immediately after splenectomy instead of waiting nearly three weeks. This phase of the problem we have investigated in our present study. Richet's point, as also Paton's observations, demands carefully conducted metabolism studies and these, carried out simultaneously with our dietary studies, are presented in a separate communication (7). Here, however, it may be noted that in the normal animal we have found that the removal of the spleen has no influence upon metabolism. Moreover, we have not seen noteworthy changes in the weight of our splenectomized animals. For a few days after splenectomy, a slight loss may occur, but in all long time experiments an increase in weight has been observed.

The studies of Paton and his associates as to the changes in the blood after splenectomy are the most carefully conducted of any in the literature and for this reason we have been greatly disturbed that our results were so different. Their studies, however, were limited to two splenectomized animals and two controls and it may be that by chance the former correspond to the milder anemias which we observed. In this connection it is of interest to note that in the splenectomized dog they did obtain a decrease of 600,000 to 800,000 red cells, which, however, as it occurred also in a normal dog was not, by contrast, of apparent importance. Moreover, they used puppies, about two and a half months old, and it is possible that in such young animals the mechanism of blood regeneration may be more active than it is in older animals. Our observations were all upon full grown animals. In Paton's blood work diet is not mentioned, but in his metabolism work the dogs were, for part of the time at least, on a meat (high iron) diet, which, if used in the blood work also, might have been, if Asher and Vogel are correct, a factor in decreasing the anemia. Thus it is evident from our own experience and from this brief review of the discordant conclusions in the literature of the subject that information of value might be obtained from carefully controlled dietary studies. Such studies we present at this time.

In our first group of experiments, animals were placed on calorically sufficient diets, the protein being furnished in the form of beef heart, beef spleen, or commercial casein, and the fat and carbohydrate in the form of lard and bread crumbs. Beef spleen was introduced on account of its large iron content, in contrast with that

TABLE I.

Days.	Raw beef heart, lard, and bread.						Days.	Raw beef heart, lard, and bread.		
	Dog 79 (splenectomized).			Dog 83 (splenectomized).				Dog 81 (control).		
	Weight.	Red cell count.	Hemo-globin.	Weight.	Red cell count.	Hemo-globin.		Weight.	Red cell count.	Hemo-globin.
	Before splenectomy							kilos		per cent
							1	9.1	6,320,000	97
	kilos		per cent	kilos		per cent	4– 7		6,650,000	102
	9.3	7,930,000	110	8.7	7,000,000	107	12–18		6,880,000	96
		7,560,000	114		7,770,000	105	26–40	10.0	6,800,000	90
	After splenectomy						40–60	10.1	6,000,000	96
5– 7	8.8	7,420,000	97	8.6	7,910,000	105				
10–14		7,230,000	98		7,720,000	103				
18–23		6,250,000	98	8.6	6,500,000	96				
26–33		6,810,000	96	8.7	7,270,000	101				
38–40	9.8	7,360,000	104	8.3	6,880,000	97				
45–48	10.5	6,690,000	100	8.5	6,260,000	96				
52–61	10.8	6,640,000	95	8.6	6,240,000	93				

of the beef heart and the casein. Several blood examinations were made during a period of ten days to two weeks before splenectomy and at intervals seldom exceeding a week after operation. All operations were done under ether anesthesia and for most of these we are indebted to Dr. Max M. Peet of the Department of Surgical Research. The animals were kept under absolutely uniform conditions. In Tables I, II, and III, which show the results of these

TABLE II.

Days.	Casein, lard, and bread.						Days.	Casein, lard, and bread.		
	Dog 82 (splenectomized).			Dog 84 (splenectomized).				Dog 80 (control).		
	Weight	Red cell count.	Hemoglobin.	Weight	Red cell count.	Hemoglobin.		Weight	Red cell count	Hemoglobin.
Before splenectomy								kilos		per cent
	kilos		per cent	kilos		per cent	1	9.1	7,860,000	90
	16.7	5,500,000	98	8.7	7,780,000	98	4– 7		7,420,000	92
		5,551,000	85		7,710,000	97	12–18		7,020,000	96
After splenectomy							26–40	8.3	7,370,000	96
							40–60	9.2	7,180,000	103
5– 7	16.5	4,600,000	75	8.6	7,090,000	89				
10–14	17.2	4,688,000	65	8.8	7,140,000	92				
18–23		4,210,000	76	8.9	6,600,000	91				
26–33		5,160,000	82	9.1	7,010,000	92				
38–40		6,040,000	88	9.2	6,940,000	87				
45–48					6,630,000	91				
52–61	16.9	6,060,000	90	9.1						

studies, only the last two blood counts of the preliminary periods are given. These represent, usually, counts made respectively 1 to 2 and 5 to 7 days before splenectomy. As the blood of the several dogs was not always examined at exactly the same intervals after splenectomy, in order to shorten the table only enough blood counts are given to show the general trend of the blood picture. The estimation of iron and of nitrogen in the diet is based on the average of several estimations of the food materials used. These figures with a calculation of the caloric value of the food are given in Table IV.

Discussion of Tables I, II, and III.

By comparing the Tables I, II, and III it is at once evident that in no instance did the general nutrition of the animal suffer. A slight loss of weight occurred after operation. but this was soon re-

TABLE III.

Days	Dog 85 (splenectomized). Raw beef spleen, lard, and bread.			Dog 86 (splenectomized). Raw beef spleen, lard, and bread.			Dog 87 (splenectomized).			Dog 90 (control). Raw beef spleen, lard, and bread.			Days.
	Weight.	Red cell count.	Hemo-globin.	Weight.	Red cell count.	Hemo-globin.	Weight.	Red cell count.	Hemo-globin.	Weight.	Red cell count.	Hemo-globin.	
	kilos		*per cent*	*kilos*		*per cent*	*kilos*		*per cent*	*kilos*		*per cent*	
Before splenectomy													
	6.7	6,200,000	88	12.2	5,330,000	95	10.1	6,580,000	105	7.2	5,800,000	92	1
					5,450,000	95		5,900,000	100		5,940,000	94	4–7
											5,680,000	92	12–18
After splenectomy													
5–7	6.5	5,880,000	78	11.5	4,360,000	76	9.4	6,490,000	108	Discontinued because of development of distemper.			
10–14		4,720,000	75		4,780,000	85		6,250,000	98				
18–23		4,850,000	80		5,040,000	82		6,080,000	96				
26–33		4,820,000	80		5,170,000	85		6,360,000	100				
38–40	6.9	5,180,000	82	12.8	5,800,000	94	9.9	6,410,000	101				
45–48		5,740,000	82		5,820,000	96		5,820,000	96				
52–64	6.9	6,040,000	94	12.3			10.0	5,730,000	102				

TABLE IV.

Nitrogen and Iron Content and Caloric Value of Diets of Tables I, II, and III.

Dog No.	Actual total per day.			Per kilo of body weight per day.		
	Nitrogen.	Iron.	Calories.	Nitrogen.	Iron.	Calories.
	gm.	mg.		gm.	mg.	
79	6.5	11.8	709	0.72	1.31	79
83	9.8	18.0	1,043	1.17	2.15	124
81	6.5	11.8	665	0.68	1.23	69
82	13.6	11.3	1,194	0.81	0.68	71
84	10.6	9.1	956	1.18	1.01	106
80	7.4	6.2	664	0.82	0.69	74
85	4.6	352.0	517	0.69	53.00	77
86	8.3	653.4	893	0.69	54.00	74
87	6.8	543.0	752	0.68	54.00	75
90	4.6	352.0	517	0.64	49.00	72

gained. Also it is seen that in no instance does a splenectomized dog maintain the same constant level of red cell and hemoglobin content as do the non-splenectomized animals. The change, however, in Dogs 79, 83, 84, and 87 is so slight as to be within the limit of error of the methods of blood examination; in Dogs 82, 85, and 86 the change is more marked, but even here one can hardly refer to the condition present as a frank anemia. It is, however, of significance that in all instances the variations are more marked than in the controls and also that they usually occur after about four weeks. the period, in severe splenectomy anemia, marked by the lowest counts. On the other hand, the question arises: Are these favorable results due to the diet, that is, to the general character of the diet or to the presence in the animals on a meat diet of large amounts of iron? That iron in the diet is a factor seems doubtful in view of the fact that two of the three animals (85 and 86) fed with beef spleen showed the most marked changes of any in the group. Beef spleen was selected because it contains a large amount of iron, according to our analyses 235 mg. per 100 grams, presumably in large part in organic combination and therefore readily utilizable. Fresh beef heart and casein, on the other hand, contain only 4.6 mg. and 7.2 mg. per 100 grams respectively, and if iron is an important factor in preventing anemia after splenectomy, one would not expect animals fed with spleen to show the changes evident in the figures given for Dogs 85 and 86; rather, one would expect figures as in

Dog 87. That the administration of abundant organic iron in the form of beef spleen did not prevent the anemia is in accord with our studies (8) of iron metabolism in the absence of the spleen and opposed to the conclusion of Asher and his associates.

On the other hand, in view of the slight changes which occurred in many of the animals, it is impossible to avoid the question as to whether a diet, adequate for the normal dog, is in some way inadequate for the splenectomized dog; if so, the value of our views concerning the severer types of anemia following splenectomy based on our earlier experiments upon dogs fed on a general mixed diet would depend upon whether or not the inadequacy of diet held for all animals operated upon, or only for animals without a spleen. If anemia occurred in dogs fed on the mixed diet after other operations than splenectomy, it would be at once evident that the food, while sufficient for a normal dog, was not sufficient for a convalescent dog. On the other hand, if the anemia could be demonstrated only after splenectomy, there would be established a point of importance in regard to the spleen in its relation to metabolism, and our observations on the anemia after splenectomy would not only be substantiated, but would gain an added importance. To settle this point, it was essential, therefore, to study in animals on our routine mixed diet the effect of splenectomy, and as a control some other simple operation involving the removal of an organ. Nephrectomy was selected as an operation quite analogous, from the technical point of view, to splenectomy and accordingly two healthy dogs were placed upon ordinary kennel diet for seventeen days; upon each dog a nephrectomy was then performed and the animal kept on the same diet for twenty-three days longer; splenectomy was then performed upon each dog and the animals kept on the same diet for thirty-eight days more. Blood counts were made at frequent intervals throughout the experiment. Whereas during the seventeen days on the diet before operation and during the twenty-three days following nephrectomy no significant change (Table V) in the hemoglobin or red blood cells was observed in either animal, after the splenectomy both showed a well marked fall in hemoglobin and red blood cell count. It is noteworthy also that relatively slight changes in weight occurred.

TABLE V.

Effect of Splenectomy Controlled by Previous Nephrectomy. Mixed Diet.

Date.	Dog 23.			Dog 42.		
	Weight.	Hemo-globin.	Red blood count.	Weight.	Hemo-globin.	Red blood count.
	kilos	*per cent*		*kilos*	*per cent*	
May 18..	10.9	100	6,840,000	14.1	102	7,810,000
" 25..	10.9	98	7,240,000	14.0	102	7,850,000
June 1..	11.0	92	6,760,000	14.9	101	8,330,000
" 4..	Nephrectomy			Nephrectomy		
" 8..	11.1	93	6,280,000	14.0	98	7,210,000
" 15..	11.3	90	6,020,000	14.6	95	7,120,000
" 25..	11.4	97	6,690,000	14.8	97	7,010,000
" 27..	Splenectomy			Splenectomy		
July 3..	10.9	90	6,600,000	15.1	90	6,840,000
" 10..	10.8	83	5,890,000	14.9	92	6,830,000
" 18..	10.8	77	5,740,000	14.9	88	6,480,000
" 27..	10.4	66	5,080,000	14.3	75	6,060,000
Aug. 4..	10.4	68	4,540,000	14.2	76	6,350,000

From these observations four conclusions may be drawn: (1) that inasmuch as the animals did not lose greatly in weight, the routine table scrap diet is a satisfactory food for animals after surgical operations; (2) that on this diet operation involving the removal of an organ other than the spleen does not cause anemia; (3) that the anemia following splenectomy is not to be explained, in view of the fact that the splenectomized animals maintained their average weight, by insufficient nutrition; and (4) if the anemia is in any way related to the diet it is either (a) because some toxic substance operative in the absence of the spleen is present in this particular food, or (b) because some substance present in the diet and normally utilizable can not be utilized in the absence of the spleen. In connection with this last conclusion it occurred to us that as the routine kennel diet is essentially a cooked diet it was possible that in the cooking there occurred the destruction by heat of some vitamine-like substance normally utilized by the spleen. To control this point a new series of observations were made (Table VI).

TABLE VI.

The Influence upon the Anemia Following Splenectomy of a Raw and a Cooked Diet.

Dog No.	Diet.	Before splenectomy.				After splenectomy.				Food values per kilo of body weight.	
		Period.	Weight before operation.	Hemoglobin.	Red cell count.	Period.	Final weight.	Hemoglobin.	Red cell count.	Nitrogen.	Calories.
		days	*kilos*	*per cent*		*weeks*	*kilos*	*per cent*		*gm.*	
48	Raw	28	13.4	99	5,450,000	6th– 9th	14.0	96	5,590,000	0.41	69
57	"	25	10.9	100	6,220,000	6th–12th	12.4	83	4,499,000	0.46	75
53	"	24	8.0	99	6,140,000	8th–12th	8.9	75	4,920,000	0.74	69
50	"	41	8.5	104	6,910,000	10th–13th	8.6	83	5,551,000	0.15	72
52	Cooked	48	10.8	105	6,760,000	7th– 9th	11.4	77	5,130,000	0.40	69
56*	"	121	8.4	88	6,250,000	6th–10th	7.8	61	4,880,000	0.40	73

Controls.

Dog No.	Diet.	Period.	Weight before operation.	Hemoglobin.	Red cell count.						
9	Raw	108	11.4	102	6,620,000 (initial period)						
			12.9	95	6,590,000 (final period)						
56*	Cooked	121	8.2	100	6,460,000 (initial period)						
			8.4	88	6,250,000 (final period)						

* This animal was used first as a control for the cooked diet and later was splenectomized.

Six animals were placed upon a calorically sufficient diet, accurately determined; the only differences were that four received raw and two cooked meat. Examination of the blood was made at intervals of not longer than seven days. At the same time metabolism studies, the results of which are reported elsewhere (7), were made on some of the animals (Dogs 48, 52, 56, and 57). The diet in each of these experiments consisted of beef heart, lard, and sugar, a small amount of sodium chloride, and sufficient bone ash to ensure firm feces. Details as to nitrogen content and caloric value of the food are given in the table.

In connection with Table VI, it should be explained that, in order to place the figures covering all animals in one graphic table, the counts given represent averages of several examinations. The figures before splenectomy represent the averages of the last three counts before operation; the figures after splenectomy the average of the three lowest consecutive counts. The figures for the two control animals represent the average of the first three and last three counts respectively.

It is evident from a study of Table VI that there is a greater tendency for animals on the cooked diet to develop anemia than is the case with those receiving raw meat. Thus in the latter group no change in the blood picture was evident in one animal while in the other three with moderate anemia, the hemoglobin did not fall below 75 or the red cells much below 5,000,000. On the other hand in the group receiving cooked meat both showed a marked change in the blood picture and in one a hemoglobin content as low as 61. As in all our previous studies the hemoglobin decrease is relatively greater than the fall in red cells. That the amount of protein given in the raw food is not an important matter is seen by contrasting Dog 53 on a high nitrogen diet with Dog 50 on a low nitrogen diet. In these two animals the calories of the diet were maintained by varying the amount of fat. The difference in the degree of anemia is negligible. In connection with the problem of the influence of cooked diet, it is noteworthy that Dog 56, which served as a control to Dog 52 for four months before it was splenectomized, and was for all this time on a cooked diet, showed during this time a falling off in the hemoglobin content of its blood. Moreover this animal was the only one showing a persistent loss of weight after splenectomy. Definite conclusions cannot be drawn from such a small number of experiments, but the fact that splenectomized animals on cooked beef develop an anemia of a degree more closely approaching that of animals on the usual kennel diet, which is essentially a cooked diet, while animals on a raw diet have a less severe anemia, suggests that heat destroys something in the diet which in the absence of the spleen tends to cause anemia. In view, however, of the relatively slight differences which we have found, experiments on a large number of animals on diverse diets must be made before a final decision is reached. Such observations are now in progress and we hope later to amplify this statement of the influence of diet.

SUMMARY.

The anemia which develops after splenectomy is most marked in animals on a mixed table scrap diet of meat, bread, cereals, and vegetables, which is essentially a cooked diet. Control studies in which a unilateral nephrectomy precedes splenectomy demonstrate that the anemia is not due to operation, hemorrhage, or accidents of convalescence but develops only in the absence of the spleen. The results of studies of the influence of food containing a large amount of iron in presumably easily utilizable form, as in raw beef spleen, do not support the view that the anemia is due to lack of iron in the food. Observation on the influence of a diet of raw meat as contrasted with cooked meat shows a more severe anemia in animals on the cooked diet and suggests the possibility that heat alters some substance which, in the absence of the spleen, the body cannot utilize. A final conclusion in regard to this point must, however, await the results of more detailed studies now in progress.

BIBLIOGRAPHY.

1. Musser, J. H., Jr., An Experimental Study of the Changes in the Blood Following Splenectomy, *Arch. Int. Med.*, 1912, ix, 592. Pearce, R. M., Austin, J. H., and Musser, J. H., Jr., The Relation of the Spleen to Blood Destruction and Regeneration and to Hemolytic Jaundice. III. The Changes in the Blood Following Splenectomy and Their Relation to the Production of Hemolytic Jaundice, *Jour. Exper. Med.*, 1912, xvi, 758. Musser, J. H., Jr., and Krumbhaar, E. B., VI. The Blood Picture at Various Periods after Splenectomy, *Jour. Exper. Med.*, 1913, xviii, 487. Pearce, R. M., and Peet, M. M., VII. The Effect of Hemolytic Serum in Splenectomized Dogs, *Jour. Exper. Med.*, 1913, xviii, 494. Krumbhaar, Musser, and Pearce, VIII. Regeneration of the Blood of Splenectomized Dogs after the Administration of Hemolytic Agents, *ibid.*, 1913, xviii, 665. Pearce, R. M., and Pepper, O. H. P., IX. The Changes in the Bone Marrow after Splenectomy, *Jour. Exper. Med.*, 1914, xx, 19. Krumbhaar and Musser, X. Concerning the Supposed Regulatory Influence of the Spleen in the Formation and Destruction of Erythrocytes, *ibid.*, 1914, xx, 108.
2. Austin, J. H., and Pearce, R. M., XI. The Influence of the Spleen on Iron Metabolism, *Jour. Exper. Med.*, 1914, xx, 122.
3. Asher, L., and Vogel, H., Beiträge zur Physiologie der Drüsen. XVIII. Fortgesetzte Beiträge zur Funktion der Milz als Organ des Eisenstoffwechsels, *Biochem. Ztschr.*, 1912, xliii, 386.
4. Richet, C., Des effets de l'ablation de la rate sur la nutrition chez les chiens, *Jour. de physiol. et de path. gén.*, 1912, xiv, 689; 1913, xv, 579.

5. Paton, D. N., Studies of the Metabolism in the Dog before and after Removal of the Spleen, *Jour. Physiol.*, 1899–1900, xxv, 443. Paton, D. N., Gulland, G. L., and Fowler, J. S., The Relation of the Spleen to the Formation of the Blood Corpuscles, *Jour. Physiol.*, 1902, xxviii, 83.
6. Austin and Pearce, *loc. cit.*
7. Goldschmidt, S., and Pearce, R. M., Studies of Metabolism in the Dog before and after Removal of the Spleen, *Jour. Exper. Med.*, 1915, xxii, 319.
8. Austin and Pearce, *loc. cit.*

STUDIES ON THE CIRCULATION IN MAN.

XVI. A Study of the Development of the Collateral Circulation in the Right Hand after Ligation of the Innominate Artery for Subclavian Aneurysm.

By G. N. STEWART, M.D.

(From the H. K. Cushing Laboratory of Experimental Medicine of Western
Reserve University, Cleveland.)

(Received for publication, June 28, 1915.)

Through the kindness of my colleague, Dr. Carl A. Hamann, I have been enabled to study two cases in which he successfully ligated the innominate and common carotid arteries for subclavian aneurysm. I have employed the method of measuring the blood flow in the hands previously described by me.[1]

The results in the first case, that of Mrs. K., 68 years of age, have already been published,[2] and need only be briefly alluded to here for comparison with the second case. I did not have the opportunity of examining the blood flow before the operation on Mrs. K. The operation was performed on Feb. 26, 1913. On Mar. 20, the flow in the right hand was 1.50 gm. per 100 cc. of hand per minute, and in the left 5.32 gm. (ratio 1 : 3.54), with room temperature 26.7° C. On Mar. 21 the flows were 1.83 gm. and 6.38 gm. for the right and left hands, respectively (ratio 1 : 3.48), with room temperature 22.7° C. On July 9, 1913 (19 weeks after the operation), the flow in the right hand was 8.26 gm. per 100 cc. per minute and in the left 10.69 gm. (ratio 1 : 1.3), with room temperature 26.2° C. There was no pulse in the accessible arteries of the right arm. Yet it is obvious from the blood flow measurements that a very satisfactory collateral circulation had been established. At the present time the patient is still alive, and a pulse has returned.

· The second case was that of a colored man, Arthur B., aged 25 years, height 5 feet, 4 inches, weight 132 pounds. He was admitted to the City Hospital May 6, 1915, complaining of pain in the right shoulder and right arm. The pain began 3 weeks before he applied for admission coincidently with the appearance of a small lump below the right clavicle, which rapidly increased in size. The pain and muscular weakness in the right arm soon forced him to quit work. On admission the loss of power in the right forearm and hand was marked; in the upper arm and shoulder it was less marked, though evident. There is no atrophy

---

[1] Stewart, G. N., *Heart*, 1911, iii, 33.
[2] Stewart, *Arch. Int. Med.*, 1914, xiii, 1.

or edema of the right arm. The circumference of the right arm at the middle of the biceps is 30 cm., of the left 27 cm. The circumference of the right forearm is 28½ cm., of the left 27 cm. The man is right handed. The left radial pulse is of greater volume than the right. The pulse in the two radials is synchronous. Blood pressure in right arm, systolic 110, diastolic 88; in left arm, systolic 130, diastolic 70. The fingers of the right hand are markedly clubbed.

May 11, 1915. Blood: leucocytes 11,600, hemoglobin 80 per cent. Wassermann + + +.

On May 11 the innominate and common carotid arteries were ligated.

May 19, 1915. He is fairly well. The right hand is not cold. He has no pain in the right arm, but the arm and hand feel tired. The radial side of the palm of the hand feels numb. There is no pulse at the left wrist. Pulse rate 116 (sitting). The blood flow in the hands was measured May 8, that is. 3 days before the operation, and again on May 22, 11 days after the operation. Further examinations were made on May 28, June 4, and June 11.

The patient's friends prevailed on him to leave the hospital on May 24. and he subsequently returned from time to time for the blood flow examinations.

On May 8 the flow in the right hand was 12.52 grams per 100 cc. of hand per minute for the last nine minutes in the calorimeters. and that in the left hand 6.36 grams, with average room temperature of 22.5° C. It may appear puzzling at first thought that the flow in the right hand should be double that in the left. while the amplitude of the right radial pulse is so much smaller than that of the left. The pulse as felt by the finger, however, is only a rough criterion of the blood flow on the assumption that the anatomical conditions are normal. In the present case the pulse wave must be supposed to be greatly diminished and its form distorted in passing through the aneurysm. but that is no reason for expecting that the mass movement of the blood should be diminished as well. The systolic pressure in the right arm was 110, the diastolic 88 mm. of mercury. The pulse pressure, which can alone be detected by the finger, is only 22 mm. of mercury. In the left arm the systolic pressure was 130, the diastolic 70, and the pulse pressure 60 mm. of mercury. But while this makes it clear that there is no ground for expecting a smaller flow on the side of the smaller pulse. why should the flow be so much larger on that side? The explanation is probably twofold: first, there is evidence of pressure on constituents of the brachial plexus supplying the right hand. Now pressure sufficient to cause loss of power in the skeletal muscles may be assumed to cause also some loss of vasomotor tone. since the vasomotor fibers in the brachial plexus cannot conceivably be protected from the pressure. ' A loss of vasomotor tone in a hand will of course be accompanied by an increased blood flow. As a matter of fact, I have found that in early unilateral brachial neuritis the blood flow in the corresponding hand is decidedly greater than in the normal hand. Secondly, it is very likely that a dilated right subclavian artery offers a freer passage to the blood than the normal left subclavian does. That such reciprocal relations have an important influence on the distribution of the blood is indicated in an interesting manner by the results of the first blood flow examination after ligation of the innominate. On May 22 (eleven days after the operation) the flow in the right hand was 3.44 grams per 100 cc. per minute and that in the left hand 15.38 grams (ratio 1 : 4.47), with room temperature 25.0° C. The flow in the right hand has, of course, been greatly reduced by the ligation, but the interesting point is that the flow in the left hand has been correspondingly increased. Thus, 100 cc. of right hand and 100 cc. of left hand together received 18.88 grams of blood per minute before the ligation and 18.82 grams after ligation, exactly the same amount. But the distribution is totally different. Of course, this extremely exact correspondence is accidental, but it cannot be accidental that the flow in the left hand should have been so much smaller than that in the right before the operation and should have been so greatly increased after it. The cutting off of the path through the innominate and right common carotid obviously permitted more blood to enter the alternative route of the left subclavian and left carotid. That the flow in the left carotid was increased after the operation was indicated by the plainly visible throbbing of the left temporal artery. I have elsewhere discussed[3] the reciprocal effect of occlusion of one path upon the corresponding vascular path on the other side of the body.

The next blood flow examination was made on May 28 (seventeen days after the ligation). The flow in the right hand was 4.76 grams, and in the left 15.31 grams per 100 cc. per minute (for a period of five minutes when the flows were at the maximum for the two hands), with room temperature 26.0° C. The ratio of the

[3] Stewart, *Jour. Exper. Med.*, 1915, xxii, 1.

flows was 1 to 3.21, indicating a steady improvement in the collateral circulation. Including a period of vasoconstriction due to a psychical cause, which of course diminished the circulation more in the left hand than in the right, the flows (for ten minutes) were 4.15 grams per 100 cc. per minute for the right and 12.17 grams for the left hand (ratio 1 to 2.93). The reflex change in the flow elicited in the right hand by immersing the left hand in warm water was small, as is always the case in a part whose circulation is mechanically obstructed. For the three minutes immediately preceding the vasomotor test the flow in the right hand was 4.04 grams per 100 cc. per minute. For the first four minutes of immersion of the left hand in warm water the flow in the right sank to 3.05 grams per 100 cc. per minute, to rise to 4.86 grams per 100 cc. per minute for the remaining four minutes of the period, an insignificant reaction.

On June 4 (twenty-four days after the operation) the flow in the right hand was 4.86 grams and in the left 9.00 grams (ratio 1 to 1.85) per 100 cc. per minute for the last 18 minutes in the calorimeters, with room temperature 23.9° C. The patient came to the hospital for the examination on rather a cool morning, naturally with bare hands, and vasoconstriction due to this was probably responsible for cutting down the flow in the left hand. For the reason already given the effect on the right hand would be comparatively insignificant. The ratio is therefore probably to some extent artificial, and gives an unduly favorable view of the development of the collateral circulation at this time. Nevertheless the fact that in spite of the vasoconstriction the flow in the right hand is absolutely greater than at the last examination shows clearly enough that the collateral circulation is still opening up.

The last examination was made on June 11 (thirty-one days after the operation). The right hand was now being freely used, the only symptoms which troubled the patient being numbness along the palmar surface of the thumb and the radial surface of the index finger. The hand was fairly strong, although not of course as strong as the left hand. The flow in the right hand was 8.55 grams per 100 cc. per minute and in the left 14.24 grams (ratio 1 to 1.66). There was no pulse in the accessible arteries of the right anterior extremity.

The collateral circulation has therefore developed much more rapidly than in the other case. This is doubtless to be attributed in part at least to the youth of the patient and the consequent greater distensibility of his arteries and the greater driving power of his heart.

Protocols.

First Examination of Blood Flow.—Arthur B. May 8, 1915. Hands in bath at 10.27 a. m., in calorimeters at 10.38½, out of calorimeters at 10.51. Pulse 84.

Time.	Temperature of			Time.	Temperature of		
	Calorimeters.		Room.		Calorimeters.		Room.
	Right.	Left.			Right.	Left.	
10.38	31.31	31.21		10.46	31.73	31.39	
10.40	31.36	31.22		10.47	31.79	31.41	22.1
10.41	31.38	31.25	21.4	10.48	31.84	31.425	
10.42	31.42	31.26	21.9	10.49	31.92	31.45	22.5
10.43	31.49	31.28		10.50	31.99	31.47	
10.44	31.56	31.30		10.51	32.05	31.495	22.9
10.45	31.64	31.33		11.02	31.89	31.33	

Cooling of calorimeters in 11 minutes, right 0.16°, left 0.165°. Volume of right hand 482 cc., of left 427 cc. Water equivalent of calorimeters with contents, right 3,480, left 3,436. Rectal temperature 37.65° C.

Second Examination.—May 22, 1915. Hands in bath at 2.11 p. m., in calorimeters at 2.20, out of calorimeters at 2.31. Pulse 100.

Time.	Temperature of			Time.	Temperature of		
	Calorimeters.		Room.		Calorimeters.		Room.
	Right.	Left.			Right.	Left.	
2.19	31.98	31.98		2.27	32.015	32.41	
2.21	31.96	32.02		2.28	32.03	32.48	25.1
2.22	31.97	32.08	25.0	2.29	32.04	32.55	
2.23	31.97	32.14		2.30	32.055	32.62	25.0
2.24	31.975	32.20	25.0	2.31	32.07	32.68	
2.25	31.98	32.26		2.43	31.95	32.54	
2.26	32.00	32.34	25.0				

Cooling of calorimeters in 12 minutes, right 0.12°, left 0.14°. Volume of right hand 485 cc., of left 410 cc. The mark on the left wrist was inadvertently put somewhat lower than usual, so that a somewhat smaller volume of the left hand was in the calorimeter. Water equivalent of calorimeters with contents, right 3,483, left 3,423 cc. Mouth temperature 37.0° C.

Third Examination.—May 28, 1915. Hands in bath at 2.10½ p. m., in calorimeters at 2.22. At 2.36 the left hand was immersed in water at 44° C. At 2.44 the right hand was removed from the calorimeter.

Time.	Temperature of Calorimeters. Right.	Temperature of Calorimeters. Left.	Room.	Time.	Temperature of Calorimeters. Right.	Temperature of Calorimeters. Left.	Room.
2.21	31.95	31.95		2.34	32.09	32.58	25.9
2.23	31.945	31.99	25.9	2.35	32.105	32.63	
2.24	31.96	32.04		2.36	32.12	32.67	
2.25	31.97	32.08	26.0	2.37	32.125		25.7
2.26	31.975	32.12		2.38	32.135		
2.27	31.985	32.16	26.0	2.39	32.14		
2.28	31.99	32.21		2.40	32.15		25.9
2.29	32.005	32.27	26.1	2.41	32.165		
2.30	32.02	32.34		2.42	32.185		
2.31	32.04	32.40	26.1	2.43	32.205		25.9
2.32	32.055	32.47		2.44	32.22		
2.33*	32.08	32.54		2.52	32.14	32.48	

* Here he began to concern himself about the preparations being made for the warm water test, causing some psychical vasoconstriction.

Cooling of calorimeters, right 0.08° in 8 minutes, left 0.19° in 16 minutes. Volume of right hand 458 cc., of left 420 cc. Water equivalent of calorimeters with contents, right 3,461, left 3,431 cc. Rectal temperature 37.44° C.

Fourth Examination.—June 4, 1915. The day was rather cool and the examination was begun soon after his arrival at the hospital. Pulse 78. Hands in bath at 11.33 a. m., in calorimeters at 11.42½, out of calorimeters at 12.05.

Time.	Temperature of Calorimeters. Right.	Temperature of Calorimeters. Left.	Room.	Time.	Temperature of Calorimeters. Right.	Temperature of Calorimeters. Left.	Room.
11.42	31.70	31.65		11.56	31.805	31.975	
11.44	31.68	31.66	23.6	11.57	31.82	32.01	
11.45	31.69	31.67	23.8	11.58	31.83	32.04	23.9
11.46	31.695	31.675	23.8	11.59	31.845	32.07	
11.47	31.70	31.70		12.00	31.855	32.09	
11.48	31.71	31.74	24.0	12.01*	31.875	32.14	24.0
11.49	31.73	31.78		12.02	31.895	32.175	
11.50	31.74	31.80	23.9	12.03	31.91	32.19	24.0
11.51	31.75	31.83	24.0	12.04	31.925	32.23	
11.52	31.755	31.85		12.05	31.95	32.275	
11.53	31.76	31.88		12.13	31.84	32.16	
11.54	31.775	31.92	23.9				
11.55	31.795	31.95					

*Here he is beginning to fidget and says the right arm and hand are getting tired.

Cooling of calorimeters in 8 minutes, right 0.11°, left 0.115°. Volume of right hand 446 cc., of left 413 cc. Water equivalent of calorimeters with contents, right 3,452, left 3,425. Rectal temperature 37.19° C.

Fifth Examination.—June 11, 1915. Pulse 84. Hands in bath at 11.03 a. m., in calorimeters at 11.12¾, out of calorimeters at 11.30.

Time.	Temperature of Calorimeters.		Room.	Time.	Temperature of Calorimeters.		Room.
	Right.	Left.			Right.	Left.	
11.12	31.77	31.74		11.23	32.10	32.29	
11.14	31.80	31.79	25.0	11.24	32.13	32.35	25.1
11.15	31.84	31.85	25.1	11.25	32.165	32.41	
11.16	31.88	31.91		11.26	32.20	32.46	25.1
11.17	31.905	31.97	25.0	11.27	32.24	32.515	
11.18	31.93	32.02		11.28	32.26	32.56	
11.19	31.965	32.075		11.29	32.29	32.61	
11.20	32.00	32.135	25.0	11.30	32.32	32.66	
11.21	32.04	32.18		11.37	32.235	32.57	
11.22	32.07	32.24					

Cooling of calorimeters in 7 minutes, right 0.085°, left 0.09°. Volume of right hand 457 cc., of left 425 cc. Water equivalent of calorimeters with contents, right 3,460, left 3,435 cc. Rectal temperature 36.91° C.

SUMMARY.

The development of the collateral circulation after ligation of the innominate and right common carotid arteries for subclavian aneurysm was studied in two cases by measuring the rate of blood flow in the hands from time to time.

·In a woman, sixty-eight years old, the flow in the right hand three weeks after the operation was two-sevenths of that in the left. Nineteen weeks after the operation the flow in the right hand was more than three-fourths of that in the left, although no pulse returned until long afterwards.

In a man, twenty-five years old, the flow in the right hand eleven days after the operation was between one-fourth and one-fifth of that in the left. Seventeen days after the operation the flow in the right hand was nearly one-third of the flow in the left. Twenty-four days after the operation the flow in the right hand had increased to more than one-half of the left hand flow. Thirty-one days after the operation the flow in the right hand was three-fifths of that in the left, without return, as yet, of any pulsation.

Before the operation the flow in the right hand was markedly greater than in the left, notwithstanding the small size of the right radial pulse as compared with the left. The explanation of this fact is discussed.

THE ACCELERATION OF ESTERASE ACTION.

Studies on Ferment Action. XXVIII.

By JAMES W. JOBLING, M.D., A. A. EGGSTEIN, M.D., and
WILLIAM PETERSEN, M.D.

(From the Department of Pathology, Medical Department, Vanderbilt University, Nashville.)

(Received for publication, July 3, 1915.)

The importance ascribed by various authors to the ferments of the body in the processes of immunity makes it important to study not only the ferments normally present and their activity under normal conditions, but also to find means by which they can be increased in amount, or, failing in this, to bring about a greater activity of those present.

We have already shown (1) that the antiferments of the serum, of the cells of the body, and of bacterial cells, are composed of unsaturated fatty acids probably in the form of esters, and that oxidation of these acids renders them inactive. It is probable that dissociation of the esters might also render them inactive as antiferments, as the activity of the acids is dependent upon their degree of dispersion, and this probably would be lessened by dissociation owing to their relative insolubility. Bearing this in mind we should naturally attach considerable importance to any ferment, such as esterase, which can cause a dissociation of esters; for the removal of the protective esters from bacteria may render them more susceptible to the action of proteolytic ferments, which in turn, by hydrolysis, can render the toxic substances non-toxic. It is for this reason that we wish to report upon a group of substances, which without bearing a very definite relation to one another, have the ability to accelerate greatly the action of both serum and tissue esterases. A great deal of work has been devoted to the study of substances which accelerate the action of lipase and esterase, particularly by Biondi (2), Magnus (3), Loevenhart (4), von Hess (5).

701

Falk (6), Whipple (7), and Quinan (8), but so far as we are aware, none of these authors has worked with the agents which we have found to be the most active.

Technique.

The tissue esterase was prepared from dogs' livers. The livers were cut up in a meat machine and then extracted with glycerine for four days in the thermostat. The resulting mass was filtered first through cotton and then through filter paper, and the filtrate saturated with ammonium sulphate. The precipitate thus obtained was redissolved in water and reprecipitated several times with ammonium sulphate until the filtrate became colorless. The final precipitate was dissolved in a small amount of water and dialyzed against running water until it failed to react with either barium hydrate or Nessler's solution. The preparation was again filtered through filter paper, a small amount of toluol was added, and it was then stored in the ice box.   1 cc. of the esterase prepared in this manner when incubated for four hours with 1 cc. of ethyl butyrate usually gave an acidity equal to 7 cc. of N/100 sodium hydrate. We attempted to concentrate further the esterase by fractional precipitation with ammonium sulphate, but were not very successful. Esterases prepared in this manner and kept in the ice box were just as active after three months as they were originally.   Serum esterase from dogs, rabbits, and guinea pigs was also used.

The experiments were conducted in 150 cc. Erlenmeyer flasks. The agent to be tested was put into the flask with the esterase, the volume brought up to 10 cc. with 0.9 per cent sodium chloride solution, and 1 cc. of neutral ethyl butyrate added. Various control flasks containing the agent alone, the agent plus esterase but without ethyl butyrate, and the agent plus ethyl butyrate but without esterase were set up at the same time. The flasks after the addition of 0.5 cc. of toluol to each were tightly stoppered, shaken 100 times, and placed in the incubator for four hours. 25 cc. of neutral 95 per cent ethyl alcohol were then added to each flask, and the acidity was determined by titrating against N/50 sodium hydrate, with phenolphthalein as the indicator. In the tables, however, the results are expressed in the number of cc. of N/100 sodium hydrate required to neutralize the mixtures after four hours' incubation. These figures show the increase in acidity obtained after deducting that found in the control flasks, and in each instance represent the average of several experiments.

It was noted that esterases prepared from different livers were not always activated to the same degree by the various reagents used. In several instances esterases which were considered weak preparations when acting alone gave much higher figures in the presence of certain reagents than ones which were considered to be much stronger. Thus one preparation which gave an acidity equal to 8 cc. N/100 sodium hydrate when incubated for four hours with 1 cc. of ethyl butyrate gave 70 cc. in the presence of sodium citrate, while another preparation which gave 7 cc. of acidity with ethyl butyrate alone, gave only 40 cc. when incubated with sodium citrate solution.

EXPERIMENTS.

Many substances have been tested during the progress of the work, but we shall discuss only those which exert a definite accelerating action.

The most active agent with which we have worked is sodium citrate. As far as we are aware, there are no statements in the literature concerning the accelerating action this salt exerts on the activity of esterase. Falk (6) discovered that sodium acetate accelerated the action of vegetable lipase, and Quinan (8) observed that it also accelerated the action of tissue esterases. Quinan found that the specific action is a progressive one which increases with increasing concentrations of the agent. Table I indicates that the accelerating of sodium citrate also increases with increasing concentrations of the salt, and that it is far more active than any accelerating agent yet reported.

Table I shows the accelerating action of different amounts of sodium citrate on liver esterase. The figures represent the excess of acidity over that obtained with esterase alone.

As shown in Table I, sodium citrate both in large and small amounts markedly increases esterase action. Potassium citrate is also very active as an accelerating agent, but the sodium salt is more so. As both potassium and sodium citrate accelerate the action of the ferment, experiments were made to determine if the citric acid was the activating part of the salt, but it was found to have little if any stimulating action.

TABLE I.

Amount used. gm.	Increased acidity. cc. N/100 NaOH
1.0	110
0.8	106
0.6	103
0.4	98
0.2	90
0.1	74
0.01	18
0.001	2
0.0001	0

The fact that sodium and potassium citrate inhibit the coagulation of blood led us to test other substances which influence coagulation. Of those which prevent coagulation we used potassium oxalate, hirudin, and cobra venom, while calcium lactate represented those which accelerate coagulation. Of course, we are aware that these substances probably do not act in the same manner in preventing coagulation.

Cobra venom and hirudin were found to be slightly active in accelerating esterase action, while calcium lactate and potassium oxalate were very active. It will be seen that these reagents accelerate esterase action regardless of their influence on coagulation. Table II shows the activating action of calcium lactate and potassium oxalate. The figures show the increase in acidity above that obtained with the esterase alone, which was 6.7 cc.

TABLE II.

Action of Calcium Lactate and Potassium Oxalate on Esterase Action.

Amount used. gm.	Potassium oxalate. cc. N/100 NaOH	Calcium lactate. cc. N/100 NaOH
1.0	40.5	33
0.8	55.0	35
0.6	52.5	38
0.4	40.5	35
0.2	28.5	20
0.1	22.5	10
0.01	6.8	6
0.001	1.0	1
0.0001	0	0

In view of the activating action of sodium citrate *in vitro,* we made some experiments to see if it would act in a similar manner *in vivo.*

Dog 1.—Weight 5 kilos. Was given 2 gm. of sodium citrate intravenously. Animal immediately developed dyspnea, the pupils became dilated, and it was evidently in distress. Complete recovery within 1 hour.

Table III shows the activity of the serum before and after injection. The figures show the acidity remaining after subtracting the controls.

TABLE III.

Amount of serum. cc.	Time.	Results. cc. $N/100$ $NaOH$
1	Before injection	0.4
	15 min. after injection	0.8
	1 hour " "	0.1
	5 hours " "	0.8

The changes are not marked. As the citrate is very toxic when given intravenously, another experiment was conducted in which 5 gm. were injected into the peritoneal cavity of a dog.

Dog 2.—Weight 5 kilos. Injected with 5 gm. of sodium citrate dissolved in sterile salt solution into the peritoneal cavity. After the injection the dog immediately became very sick and restless, the abdomen became distended, and apparently there was considerable pain. The animal was killed two hours after the injection, as it was almost dead. Blood was obtained from the hepatic vein and from the heart. Considerable peritoneal exudate was present, and this was also tested. Table IV shows the activity of the serum esterase before and after the injection of sodium citrate. The controls have been subtracted.

TABLE IV.

Amount of serum. cc.	Time	Results cc. $N/100$ $NaOH$
1	Before injection	0.8
1	(Heart) 2 hours after injection (dead)	1.0
	(Liver) " " " " "	1.4
	Peritoneal exudate	0.4

In both instances toxic symptoms were observed and a definite though slight rise in serum esterase followed. This rise may have been due either to the presence of the citrate in the serum. or to mobilization of the ferment as a result of the injury to the cells of the body.

CONCLUSIONS.

1. Potassium citrate, potassium oxalate, and calcium lactate accelerate the action of tissue and serum esterases.

2. Intravenous and intraperitoneal injections of large amounts of sodium citrate do not cause a definite increase in the activity of the serum esterase.

BIBLIOGRAPHY.

1. Jobling, J. W., and Petersen, W., *Jour. Exper. Med.*, 1914, xix, 459; xx, 452.
2. Biondi, C., *Virchows Arch. f. path. Anat.*, 1896, cxliv, 373.
3. Magnus, R., *Ztschr. f. physiol. Chem.*, 1904, xlii, 150.
4. Loevenhart, A. S., and Souder, C. G., *Jour. Biol. Chem.*, 1906–07, ii, 415.
   Loevenhart, A. S., *ibid.*, p. 427.
5. von Hess, C. L., *Jour. Biol. Chem.*, 1911–12, x, 381.
6. Falk, K. G., *Jour. Am. Chem. Soc.*, 1913, xxxv, 601.
7. Whipple, G. H., Peightal, T. C., and Clark, A. H., *Bull. Johns Hopkins Hosp.*, 1913, xxiv, 343.
8. Quinan, C., *Jour. Med. Research*, 1915, xxxii, 73.

THE RELATION OF SERUM ESTERASE TO LIVER DESTRUCTION.

Studies on Ferment Action. XXIX.

By JAMES W. JOBLING, M.D., A. A. EGGSTEIN, M.D., and
WILLIAM PETERSEN, M.D.

(*From the Department of Pathology, Medical Department, Vanderbilt University, Nashville.*)

(Received for publication, July 3, 1915.)

The variations of esterase in the blood of animals which have been poisoned by phosphorus or chloroform have been studied by Whipple (1), von Hess (2), and Loevenhart (3). Whipple observed that the esterase is increased in strength in destructive conditions of the liver, the degree of increase bearing a definite relation to the extensiveness of the process. He concludes that the increase is due to the liberation of the ferment from the liver cells which are undergoing disintegration. In certain of his experiments he tied the larger vessels of the liver, but still observed the increase in the general circulation. He explains these results by assuming the presence of collateral vessels which carry the ferment to the general circulation. Just about the time our experiments were completed an article appeared by Quinan (4) in which he reports the results of his study of tissues of healthy dogs and dogs poisoned with phosphorus and chloroform. He found that in the poisoned animals the esterase content of the liver cells was decreased 27 to 38 per cent, while kidney and muscle tissue showed a corresponding increase. He concludes that the tissues containing the largest amount of esterase must be the ones which furnish the excess ferment in the serum in chloroform and phosphorus poisoning.

Technique.

The work was conducted with blood obtained from various sources from animals poisoned with chloroform and phosphorus. In each instance blood was drawn from the ear of the dog about one hour before the chloroform or phosphorus was administered. This blood was assumed to represent the normal esterase strength of that particular animal. In some animals other tests were made on the same day, but more frequently a test was made on each subsequent day until the animal died or was killed. The animals were anesthetized on the day the serum esterase reached its maximum strength, and

707

blood was drawn from the portal and hepatic veins, the inferior vena cava below and above the entrance of the hepatic vein, and from the heart. This procedure was adopted in order to get the blood from the liver when presumably it would contain the most esterase. At this time the hepatic blood should contain an amount much larger than that present in the general circulation owing to its subsequent dilution.

The chloroform was given subcutaneously in olive oil to dogs in the proportion of 2 cc. of chloroform for each 1,000 gm. The phosphorus oil was given in 2 cc. doses on two successive days. The degree of fatty degeneration of the livers varied; in some it was very marked, in others slight.

The livers were ground up in a meat machine, weighed, and extracted with glycerine for seventy-two hours in the incubator. The mass was then diluted with two volumes of water and filtered until clear. The clear filtrate was saturated with ammonium sulphate, and the precipitate redissolved and reprecipitated several times. It was finally dissolved in a small amount of water, dialyzed free from ammonium sulphate, and its esterase activity determined. The esterase activity of the solutions thus prepared is represented as the activity of 1 gram of the original tissue. No doubt some of the ferment is lost by these procedures, but the loss is probably about the same in all. In any case the differences are so striking that our interpretation can hardly be wrong. For purposes of comparison extracts from normal livers were prepared in the same manner. The technique of determining the esterase strength has been described in a previous paper (5).

Esterase Content of Normal and Injured Liver Tissue.

Dog 1.—Normal. Killed, the liver removed, and treated as already described. The esterase activity of 1 gm. of liver tissue gave an acidity equal to 104.4 cc. of N/100 NaOH when incubated for 4 hours with neutral ethyl butyrate. A number of normal livers have been examined in the same manner, and while the results were not the same in all cases, the esterase activity per gm. of weight was always greatly in excess of that obtained with the livers of poisoned animals.

Dog 2.—Received 3 cc. of phosphorus oil on Apr. 3. Animal anesthetized Apr. 6. The liver showed marked fatty degeneration. The esterase activity per gm. of weight of the liver was equal to 53.4 cc. N/100 NaOH.

Dog 3.—Received 10 cc. of chloroform in olive oil subcutaneously. Animal anesthetized the following day. The liver showed marked fatty degeneration. The esterase activity per gm. of weight was equal to 39 cc. N/100 NaOH.

As will be seen from the above experiments, which are merely examples of others, the esterase activity of normal liver tissue gave an acidity equal to 104.4 cc. N/100 NaOH, the phosphorus liver 53.4 cc., and the chloroform liver 39 cc. In other words, the esterase activity of normal liver tissue is two to three times greater than that from liver tissue obtained from animals poisoned with phosphorus or chloroform. This confirms the observations of Quinan.

Changes in Serum Esterase.

Dog 29.—Weight 6 kilos. Inoculated subcutaneously daily for three days beginning Mar. 18, 1915, with 1 cc. of oil of phosphorus. It was anesthetized the day following the last dose, and blood drawn from the portal vein, the hepatic vein, the inferior vena cava, and from the heart. Table I gives the esterase activity of the specimens of blood.

TABLE I.

Source of blood.	Esterase activity. cc. N/100 NaOH
Heart	3.7
Portal vein	4.8
Inferior vena cava	4.7
Hepatic vein	3.3

In this dog the destruction of liver tissue was not very marked, and there was but little increase in the activity of the serum esterase. However, even here, with a slight degree of liver destruction, the esterase action of the blood from the hepatic vein was less than that obtained from other sources. The blood from the portal vein and inferior vena cava was most active.

Dog 49.—Weight 7.3 kilos. Inoculated daily for three days beginning Mar. 29, 1915, with 1 cc. of phosphorus oil. It was anesthetized the day after the last dose was given, and blood was obtained from the portal vein, hepatic vein, inferior vena cava, and the heart. The liver showed a mild degree of fatty degeneration. Table II gives the esterase activity of the blood.

TABLE II.

Source of blood.	Esterase activity cc. N 100 NaOH
Heart	2.4
Portal vein	Lost.
Inferior vena cava	2.0
Hepatic vein	2.0

Here, also, the esterase activity of the blood obtained from the hepatic vein is less than that present in the blood from other sources.

Dog 59.—Weight 6.3 kilos. Inoculated subcutaneously daily for three days beginning Apr. 3, 1915, with 1 cc. of phosphorus oil. The animal was anesthetized the day following the last dose. The liver showed marked fatty degeneration. Table III gives the esterase activity of the different specimens of blood.

TABLE III.

Source of blood.	Esterase activity. cc. $N/100$ $NaOH$
Heart	19.0
Portal vein	21.0
Inferior vena cava, below liver	16.5
" " " above "	15.5
Hepatic vein	14.0

This animal, with extensive liver changes, shows a very high esterase action of the serum. This is particularly evident in the portal blood, while the least active is the blood from the hepatic vein. This decrease in the activity of the blood from the hepatic vein is interesting as it is less than either the portal or arterial blood, and suggests the possibility of the esterase being adsorbed or destroyed as it passes through the liver.

Dog 60.—Weight 6.2 kilos. Inoculated subcutaneously on Apr. 5, 1915, with 2 cc. of phosphorus oil, and on Apr. 6 with 1 cc. The animal was anesthetized on the day following the last dose. The liver presented a high degree of fatty degeneration. Table IV gives the esterase activity of the blood from different sources.

TABLE IV.

Source of blood.	Esterase activity. cc. $N/100$ $NaOH$
Heart	7.3
Portal vein	7.3
Inferior vena cava	8.1
Hepatic vein	6.1

Here, also, the blood from the hepatic vein was less active. In this case the blood from the inferior vena cava was more active than that from the portal vein.

Dog 64.—Weight 3.5 kilos. Inoculated subcutaneously Apr. 15, 1915 at 9 a. m., with 10 cc. of chloroform in olive oil. The animal was anesthetized the following day. The serum of the animal obtained before the inoculation gave an acidity with ethyl butyrate of 0.4 cc. N/100 NaOH after 4 hours' incubation, while that obtained at 4 p. m. of the same day gave an acidity equal to 12.4 cc. Table V gives the esterase activity of the blood.

TABLE V.

Source of blood.	Esterase activity. cc $N/100$ $NaOH$
Heart	18.6
Portal vein	16.3
Inferior vena cava, below liver	19.7
" " " above "	17.3
Hepatic vein	16.3

This animal had extensive fatty degeneration of the liver and the serum had a high esterase content. There was a general rise in esterase action until the animal was killed. In seven hours it increased from 0.3 to 12.4 cc., with a further increase of 6.6 cc., during the next sixteen hours. Here again the esterase activity of the blood from the hepatic vein is low, but it is just as strong as that from the portal vein. The blood from the inferior vena cava is the most active.

DISCUSSION.

Quinan's results (4) show that in chloroform and phosphorus poisoning there is a decrease in the esterase strength of the liver cells, while it is increased in the kidney and muscle tissue. He does not prove, however, that the increase in serum esterase may not be due to a mobilization of the ferment from the disintegrating liver cells. The decrease of ferment in the liver might be due to its rapid mobilization.

It is evident, however, from the results of the experiments just described, that the increase in serum esterase in animals poisoned with chloroform or phosphorus cannot be due to a mobilization of the ferment owing to the destruction of the liver cells. If this were true we should expect to find the esterase strength of the blood from the hepatic vein higher than that from other portions of the body. but the reverse holds true. In fact. it would seem that the injured liver cells actually adsorb or destroy some of the excess esterase. as the blood from the hepatic vein contains less than is found elsewhere. The fact that the excess serum esterase in these conditions is not derived from the degenerating liver cells affords a rational explanation of the results obtained by Whipple (1) in Eck fistula dogs.

where the hepatic artery, with all its branches, has been ligated. This author found that the serum esterases were increased in all conditions accompanied with destruction of liver cells. If this holds true, we must assume either that the agent causing the destruction of the liver cells causes also a mobilization of the ferment from other sources, or that the products of the disintegration of the liver cells enter the general circulation and bring about this mobilization. This view is confirmed by the results of some of the experiments reported in our work on the effects of the intravenous injection of protein cleavage products, where we found that the serum esterase activity was greatly increased. In Whipple's experiments it is probable that the products of cell disintegration are removed by the lymphatics, and in this way get into the general circulation. Quinan has shown that in these conditions the esterases are increased in the kidneys and muscles, and so it is possible that they are derived from these tissues, but in several of our experiments we observed the highest esterase action in the portal blood, so this source must also be considered.

CONCLUSIONS.

1. Liver tissue showing fatty degeneration obtained from animals poisoned with phosphorus or chloroform contains a decreased amount of esterase.

2. The serum of animals poisoned with phosphorus or chloroform has a high esterase activity.

3. The increased amount of esterase in the serum is not derived from the disintegrating liver cells as the esterase in the blood of the hepatic vein is less than that found elsewhere.

BIBLIOGRAPHY.

1. Whipple, G. H., *Bull. Johns Hopkins Hosp.*, 1913, xxiv, 357.
2. von Hess, C. L., *Jour. Biol. Chem.*, 1911–12, x, 381.
3. Loevenhart, A. S., *Am. Jour. Physiol.*, 1901–02, vi, 331.
4. Quinan, C., *Jour. Med. Research*, 1915, xxxii, 73.
5. Jobling, J. W., Eggstein, A. A., and Petersen, W., *Jour. Exper. Med.*, 1915, xxii, 701.

FURTHER INVESTIGATIONS ON THE ORIGIN OF TUMORS IN MICE.

II. Tumor Incidence and Tumor Age in Hybrids.

By A. E. C. LATHROP and LEO LOEB, M.D.

(*From the Department of Pathology of the Barnard Free Skin and Cancer Hospital, St. Louis.*)

(Received for publication, August 2, 1915.)

We wish to report the results that we obtained in crossing the strains described in our preceding paper.[1] We used some of the hybrids thus obtained for further hybridization. Soon after the beginning of our studies of heredity in cancer in mice, we started experiments in hybridization in order to determine whether tendency to a high or a low tumor rate prevails in the offspring of two parents, one of which had a high, while the other had a low tumor rate; or, to state it differently, whether a tendency to develop cancer is a dominant or recessive character.[2] We referred to these experiments briefly in a former publication.[3] We stated that in mating a certain strain rich in tumors with a strain very poor in tumors we obtained hybrids which have so far been rich in tumors. In this case the male, which is itself not liable to have tumors, transmits the liability to have tumors to the daughters. We mentioned another strain rich in tumors which if mated to a strain poor in tumors produced hybrids rich in tumors. We stated that we intended to collect further data before arriving at a definite conclusion. We can now on the basis of a large number of experiments confirm our previous preliminary conclusion that the higher tumor rate may be dominant in crosses. But we found that the result of hybridization varies in

---

[1] Lathrop, A. E. C., and Loeb, L., *Jour. Exper. Med.*, 1915, xxii, 646.

[2] We use the terms "dominant," "recessive," and "unit factor" at present merely as convenient means of expression, without implying by the use of these terms the adoption of further going hypotheses concerning the mechanism of heredity in cancer.

[3] Lathrop, A. E. C., and Loeb, L., *Proc. Soc. Exper. Biol. and Med.*, 1913, xi, 34.

different crosses. We again gave special attention to the inheritance of the tumor age. Yet while the material on which our further conclusions are based is large, we repeat what we said in the preceding paper, that we are still continuing our experiments and that we are ready to modify our conclusions whenever our further work should fail to confirm our first results. We lay special stress on communicating the observations as such and regard some of our conclusions as only tentative. We take this attitude especially as we were often obliged, through lack of sufficient funds in this work, to direct the breeding in such a way that simultaneously with our scientific aims economic needs were satisfied. Otherwise the experiments could not have been completed.

Hybrids Produced.

European + 102 and 103: 16% (16%) + 30% (38%). I age class + II age class.[4]

Hybrids: total 21% (25%). Mixture of 102 and 103: 22% (26%), 29% (32%), 29% (33%). European + 102: 14% (18%), 17% (21%). European + 103: 10%. IV age class.

The European mice were imported. 102 and 103 were the offspring of an old male, the son of a tumor mouse, and plum-silver females which had mixed origin and gave origin to the No. 8 strain. The old male was later mated to his daughters (among them 102 and 103) and their offspring constituted the 8½ strain. 102 and 103 after having been mated to the European had young ones at the same time and their offspring could only be partly distinguished as to whether 102 or 103 was the mother of a certain mouse. 102 and 103 are there-

[4] Before discussing in each case the result of the hybridization, we mention the two parent strains which entered into the cross; the male parent is named first, then the female parent. In the case of the first group mentioned the European was the male and 102 or 103 the female parent. Then follows the tumor rate of the father, 16% (including all age classes), and the tumor rate of the II and III age classes only, in brackets (16%) of the father strain; in the case of the mother strain the figures are 30% for all age classes and 38% for the II and III age class. Then follows the age class of the father strain: I age class; and of the mother strain: II age class. In giving the tumor rate of the hybrids the figure for the second and third age class follows again in brackets. The age class of the hybrids is given last. In the case of the first hybrid group several figures are given for the tumor rate of the hybrids; in some of the offspring 102, in others 103 was the mother, in still others the offspring of 102 and 103 were mixed. The lists of the mice with and without tumors in the different generations and in different age classes are arranged in the same way as in the preceding paper (Lathrop and Loeb, *Jour. Exper. Med.,* loc. cit.). On the left side of the page are the mice without, on the right side the mice with tumors; the Roman figures signify the age classes.

fore essentially of a similar character to the No. 8 and No. 8½ strain. Altogether a record was kept of 146 mice of this strain. The variations among the different strains and different families are relatively small. The tumor rate throughout is low, on the whole similar to European, or intermediate between the European and No. 8. The tumors appear late, and belong to the IV age class. They herein differ from their parents, where the tumors appeared early; namely, in the I and II age class. Both parents of these hybrids had a low or medium tumor rate, and the tumor rate of the hybrids agrees with that of the parents. In several cases these hybrids were used for further hybridizations.

(a) (European + 102) $F_1 + 8½$ F_4 : 14% (18%), 17% (21%) + 17% (19%). IV + II age class.

In this case both parents had a similar, low tumor rate; the tumor rate of the hybrids is the same as that of the parents, 16% (18%). The tumors appear late, the fourth is dominant over the second age class. Five generations of these hybrids were observed, and the results in all agreed with each other in regard to tumor rate as well as to tumor age.

Without tumors.	With tumors.
F_1 26 (3 I 8 II 15 III)	5 (2 I 3 III)
84% (12% I 31% II 57% III)	16% (40% I 60% III)
F_2 96 (18 I 26 II 52 III)	26 (2 I 7 II 17 III)
79% (17% I 27% II 56% III)	21% (8% I 27% II 65% III)
F_3 210 (49 I 66 II 95 III)	40 (4 I 11 II 25 III)
84% (23% I 31% II 46% III)	16% (10% I 27% II 63% III)
F_4 47 (19 I 9 II 19 III)	3 (3 III)
94% (40½% I 19% II 40½% III)	6% (100% III)
F_5 16 (11 I 3 II 2 III)	3 (1 II 2 III)
84% (70% I 18% II 12% III)	16% (34% II 66% III)
Total: 396 (101 I 112 II 183 III)	77 (8 I 19 II 50 III)
84% (25% I 28% II 47% III)	16% (10% I 25% II 65% III)
II + III 82%	II + III 18%

(b) 101 (English) + (European + 103) : 71% (72%) + 10% or total 21% (25%). I age class + IV age class.

In this case males from a strain rich in tumors (101 English) were mated to females from a strain poor in tumors (European + 103). In the strain of the father the tumors appeared early, in the strain of the mother late. In the hybrids the tumor rate is intermediate. 34% (46%). The tumors in the offspring belong to the IV age class; the late appearance of the tumors seems again to be dominant. The hybrid mice were observed through three generations and the results in the various generations agreed well, on the whole. The figures for the sum of all the generations are as follows:

Without tumors.	With tumors.
101 (49 I 25 II 27 III)	51 (61 I 27 II 18 III)
66% (49% I 25% II 26% III)	34% (12% I 52% II 36% III)
II + III 54% III 60%	II + III 46% III 40%

(c) London + (European + 103) F_2: 28% (27%) + 10% (European + 103) or + 21% (25%) European + 102 and 103 (total). II age class + IV age class.

The parents have here a low or medium low tumor rate. In the father the tumors belonged to the II, in the mother to the IV age class. In the hybrids the tumor rate is low (5%), even slightly lower than in the mother. The tumors appear late.

The total figures are:

Without tumors.	With tumors.
74 (12 I 31 II 31 III)	4 (2 II 2 III)
95% (16% I 42% II 42% III)	5% (50% II 50% III)

The IV age class is again dominant.

(d) (European + 103) F, + III daughter of No. 10 : 10% or total 21% (25%) + ?30% (38%). IV age class + ?II age class.

In order to understand the character of these and the following hybrids, it will be necessary to state the character of the family of No. 10. No. 10 was a female mouse which was found to have a tumor on Sept. 8, 1910. She belonged probably to the No. 8 strain. She was in all probability mated to a male from the No. 8 strain. Sept. 10, 1910, she gave birth to three daughters, which are designated as follows: I (Nov. 3d), II (Nov. 8th), and III. The third daughter failed to grow for a considerable period of time, but later grew to normal size. Daughter I (Nov. 3d) was first mated to a European male 151, afterwards to an English Silver male, and last to a Cream male. Daughter II (Nov. 8th) was first mated to the same European male 151, and later to a male from the No. 8½ strain. The III daughter was mated to a son of European + 103. The European 151 was later crossed with females of No. 8 F₆. These various crosses are represented in Table I.

TABLE I.

No. 10 ♀ + No. ?8 ♂

151 European ♂
mated to:

♀ I Nov. 3d	♀ II Nov. 8th	♀ III
(a) + I Nov. 3d ♀	(a) + 151 European ♂	(a) + 151 European ♂
(b) + II Nov. 8th ♀	(b) + English Silver ♂	(b) + 8½ ♂
(c) + No. 8 F₆ ♀	(c) + Cream ♂	(a) + (European + 103) F₁ ♂

The offspring from the cross (European + 103) F₁ + III daughter of No. 10 gave a tumor rate of 17% (22%). The tumors belonged to the IV age class. The tumor rate and tumor age of the father were dominant in this case. About the tumor rate of the No. 10 family we have no definite knowledge, although we must assume that the I and II daughters had a very high tumor rate.

Without tumors.	With tumors.
F₁ 8 (2 II 6 III)	0
F₂ 94 (35 I 22 II 37 III)	11 (4 II 7 III)
90% (37% I 23% II 40% III)	10% (36% II 64% III)
F₃ 38 (15 I 13 II 10 III)	17 (3 I 7 II 7 III)
69% (39% I 34% II 27% III)	31% (18% I 41% II 41% III)
Total: 140 (50 I 37 II 53 III)	28 (3 I 11 II 14 III)
83% (36% I 27% II 37% III)	17% (11% I 39% II 50% III)
II + III 78% III 79%	II + III 22% III 21%

(e) English Sable (son of Tumor Mouse 198) + [(European + 103) F_1 + III daughter of No. 10] F_2: 70% (69%) + 17% (22%). I age class + IV age class.

In this case only a few of the offspring were obtained, but the tumor rate is high, the tumors appear late, and the hybrids belong to the IV age class.

Without tumors.	With tumors.
F_1 I (I I)	5 (3 II 2 III)
83% tumor rate, IV age class.	

Here the high tumor rate of the English Sable male was dominant over the low tumor rate of the two females of the [(European + 103) F_1 + III daughter of No. 10] F_2 to whom they were mated. The latter had a very low tumor rate. On the other hand, it is probable that the age class (IV) of the females prevailed over the age class of the English Sable (I).

Summary.—In considering these six crosses, we shall distinguish tumor rate and age class. In two hybrids we find marked divergence of the tumor rate in the two parents; namely, 101 + (European + 103), where the tumor rate was intermediate, and English Sable + [(European + 103) F_1 + III daughter of No. 10] F_2. In this case the higher tumor rate seemed to dominate. In the other four crosses the tumor rates of the parents were either very similar to each other or differed not very markedly. In two of these cases perhaps the lower tumor rate dominated; in the two other the tumor rate of the offspring was similar to that of the parents. In the case of the age classes, on the other hand, the higher age class (later tumors) dominated in all cases. Every time the presence of the combination European + 102 or 103 called forth the IV age class, independently of the behavior of the tumor rate in the hybrids. This was already noticeable in the first cross: European + 102, 103. The age class was here higher than in either of the parent strains. It appears, therefore, that the age classes are determined by unit factors different from those determining tumor rate, and that dominance of the higher tumor rate may be combined with dominance of the higher age class (lateness of tumors).

(a) 151 (European) + No. 8 F_3: 16% (16%) (?) + 30" (38°) I age class (?) + II age class.

The hybrids had a tumor rate similar to No. 8: 30% (34°). The age class is likewise that of No. 8; *viz.*, II. The tumors appear, however, slightly earlier than corresponds to Age Class II. Four generations were observed which agreed fairly well.

Without tumors.	With tumors.

Total: 88 (31 I   19 II   38 III)   37 (8 I   20 II   9 III)
      70% (35% I   22% II   43% III)   30% (22% I   54% II   24% III)
II + III 66%   III 87%   II + III 34%   III 19%

(b) 151 (European) + I daughter of No. 10 (Nov. 3d) : 16% (16%) (?) + high
tumor rate. I age class (?) + II or III age class (?).

The hybrids show a tumor rate of 72% (81%); they belong to the II age
class. We may assume that daughters I and II of No. 10 had the tendency to
a high tumor rate and that it was dominant over the lower tumor rate of European. ·However, there is not combined with the high tumor rate an early
appearance of the tumors; they belong to the II age class. The figures for these
hybrids were given in our preceding paper.[5]

(c) [151 (European) + I daughter of No. 10 Nov. 3d] + 101 : F_2 72% (81%)
+ 71% (72%). II age class + I age class.

In this class both parents have a high tumor rate, and the hybrids have also
a high tumor rate (55%), although not quite as high as the parents. This slight
decrease in the tumor rate of the hybrids may be accidental, due to the relatively small number of observed mice. The tumors belong to the I age class.

Without tumors.	With tumors.

12 (8 I   2 II   2 III)   15 (10 I   5 II)
45% (66% I   17% II   17% III)   55% (66⅔% I   33⅓% II)
Four generations were observed. The results in the various generations agreed.

(d) 151 (European) + II daughter of No. 10 (Nov. 8th) : 16% (16%) (?) +
high tumor rate. I age class (?) + II or III age class (?).

The II daughter of No. 10 had probably a similar tendency to a high tumor
rate as the I daughter. The hybrids have here again a high tumor rate, but
their rate is not quite as high as that of the preceding strain of hybrids, the
offspring of the same European father and the I sister. The tumor rate of the
hybrids was 65% (69%), the tumors belonged to the II age class.

Without tumors.	With tumors.

F_1 3 (3 I)   0
F_2 4 (2 I   2 II)   3 (2 I   1 III)
F_3 24 (8 I   14 II   2 III)   44 (17 I   23 II   4 III)
   35% (33⅓% I   58⅓% II   8⅓% III)   65% (39% I   52% II   9% III)
F_4 3 (2 I   1 III)   15 (1 I   10 II   4 III)
   16% (66⅔% I   33⅓% III)   84% (7% I   67% II   26% III)
Total: 34 (15 I   16 II   3 III)   62 (20 I   33 II   9 III)
   35% (44% I   47% II   9% III)   65% (32% I   54% II   14% III)
II + III 31%   III 25%   II + III 69%   III 75%

The same European male and two sisters produced therefore hybrids with a
similar tumor rate and a similar age class.

(e) English Silver + I daughter of No. 10 (Nov. 3d) : 7% + ? high tumor rate.
I age class + ?II age class.

[5] Lathrop and Loeb, *Jour. Exper. Med., loc. cit.*

After the I daughter of No. 10 had been mated to European 151, she was mated to an English Silver male with a very low tumor rate, but belonging to an early age class. The hybrids have a tumor rate of 36% (38%) which is much higher than the tumor rate of English Silver and intermediate between the tumor rate of the English Silver and the tumor rate of the descendants of this female and the European 151 male. The tumor mice belong to the II and not to the I age class, as do the tumor mice of the English Silver. The detailed figures for the various generations of these hybrids are given in the preceding paper.[6]

(f) English Sable + (English Silver + I daughter of No. 10, Nov. 3d) : 70% (69%) + 36% (38%). I age class + II age class.

Here a strain with high tumor rate is crossed with one with medium tumor rate. The descendants have a high tumor rate, 69%, similar to the strain of the father. The tumors stand probably somewhat between the I and II age class.

Without tumors.	With tumors.
5 (4 I 1 II)	11 (5 I 6 II)
31% (80% I 20% II)	69% (45% I 55% II)

The high tumor rate is here dominant.

(g) Cream + I daughter of No. 10 (Nov. 3d) : 2% (3%) + high tumor rate. IV age class + ?II or III age class.

The hybrids have a tumor rate of 36% (38%). The tumors belong to the IV age class. Tumor rate of the hybrids is similar to the tumor rate of the English Silver + I daughter of No. 10 (Nov. 3d). English Silver and Cream have a similar tumor rate, and the hybrids also have a similar tumor rate. But while the tumors of the English Silver belonged to the first age class, the tumors of the Cream belong to the IV age class and correspondingly the tumors of the English Silver hybrids belong to the II age class, and the tumors of the Cream hybrids belong to the IV age class. As far as the age of the tumors is concerned, the higher class (lateness of the tumors) is dominant; while as far as the tumor incidence is concerned, the hybrids show an intermediate condition between both parents. Neither in this case nor in the case of the English Silver hybrids with the I daughter of No. 10 does the lower tumor rate dominate. The various generations of this hybrid strain agree very well with each other.

Without tumors.	With tumors.
F_1 13 (1 I 2 II 10 III)	5 (5 III)
There may be a few mice of the F_2 generation included.	
72% (8% I 15% II 77% III)	28% (100% III)
F_2 27 (2 I 17 II 8 III)	13 (3 I 7 II 3 III)
68% (7% I 64% II 29% III)	32% (23% I 56% II 23% III)
F_3 54 (11 I 6 II 37 III)	40 (1 I 14 II 25 III)
58% (20% I 11% II 69% III)	42% (2½% I 35½% II 62% III)
F_4 18 (4 I 8 II 6 III)	4 (2 II 2 III)
82% (22% I 44% II 34% III)	18% (50% II 50% III)
Total: 112 (18 I 33 II 61 III)	62 (4 I 23 II 35 III)
64% (16% I 29% II 55% III)	36% (6% I 37% II 57% III)
II + III 62% III 64%	II + III 38% III 36%

[6] Lathrop and Loeb, *Jour. Exper. Med., loc. cit.*

(h) No. $8\frac{1}{2}$ + II daughter of No. 10 (Nov. 8th) : 17 % (19 %) + ? high tumor rate. II age class + ?II or III age class.

The hybrids show here a high tumor rate, similar to that of the mother, possibly slightly lower: 49% (50%). The tumors belong to the III age class. Here again it seems that greater tumor incidence and lateness of the tumors are dominant. The various generations were as follows:

Without tumors.	With tumors.
F_1 2 (1 I 1 II)	0
F_2 5 (1 II 4 III)	2 (1 II 1 III)
F_3 27 (4 I 10 II 13 III)	28 (6 I 13 II 9 III)
49% (15% I 36% II 49% III)	51% (21% I 47% II 32% III)
F_4 8 (3 I 2 II 3 III)	10 (5 II 5 III)
44% (37% I 26% II 37% III)	56% (50% II 50% III)
Total: 42 (8 I 14 II 20 III)	40 (6 I 19 II 15 III)
51% (19% I 33% II 48% III)	49% (15% I 47½% II 37½% III)
II + III 50% III 57%	II + III 50% III 43%

(i) A subgroup of this strain was formed. After the aged $8\frac{1}{2}$ male of the preceding group had died, the II daughter of No. 10 (Nov. 8th) was mated with one of her sons. [$8\frac{1}{2}$ + II daughter of No. 10 (Nov. 8th)] + II daughter of No. 10 (Nov. 8th) : 49% (50%) + high tumor rate. III age class + ?II or III age class.

Here the tumor rate is somewhat lower than in the preceding strain; but this is mainly due to the greater mortality in earlier periods of life in mice of this strain.

Without tumors.	With tumors.
Total: 46 (19 I 19 II 8 III)	26 (5 I 9 II 12 III)
64% (41% I 41% II 18% III)	36% (19% I 35% II 46% III)
II + III 56% III 40%	II + III 44% III 60%

Summary.—In all probability there was in no instance a low tumor incidence dominant, while in several cases a high tumor rate was dominant, and in others an intermediate tumor rate prevailed. It is almost certain that the I and II daughters of No. 10 had a tendency to a high tumor rate, and we may assume that the European male 151 had a tendency to a tumor rate perhaps somewhat higher than the average European. It is interesting that the two daughters of No. 10 gave with the English Silver and Cream, which had both very low tumor rates, a lower tumor rate than with European, which had apparently a somewhat higher tumor rate. The high tumor rate is again dominant in the hybrids between English Sable and English Silver + I daughter of No. 10. The high tumor rate is also dominant or almost dominant in the crosses between 8½ + II daughter of No. 10. The dominance of the high tumor rate in this group is, however, not complete in each case. It is of interest that the crosses with the Cream and English Silver males, in which the males had in both cases very low and similar tumor rates, showed also similar tumor rates, and that these tumor rates were lower than in the hybrids between either 8½ or European with the I or II daughter of No. 10. Again the age class of tumors appears to be transmitted to hybrids as a distinct unit factor at least partly independent of the tumor rate, and the lateness of tumors may dominate, when at the same time a high tumor rate is dominant. We found in this class undoubted instances in which a low tumor incidence did not prevail.

Hybridizations in Which the Tumor Incidence of the Two Parent Strains Differed Markedly.

(a) Grandson of Tumor Mouse 121 (English Tan) + 3 Cream females: 73% + 2% (3%). I age class + IV age class.

The Cream mothers were without tumors. The hybrids had a tumor rate intermediate between those of the parent strains; *viz.*, 42% (41%). The age class of the hybrids was also intermediate; *viz.*, II.

Without tumors.	With tumors.
F_1 1 (1 I)	5 (3 I 2 II)
F_2 27 (12 I 9 II 6 III)	17 (5 I 7 II 5 III)
61% (45% I 33% II 22% III)	39% (29% I 42% II 29% III)
F_3 18 (2 I 11 II 5 III)	13 (6 I 7 II)
58% (11% I 61% II 28% III)	42% (46% I 54% II)
F_4 7 (5 I 2 II)	4 (2 I 2 II)
64% (71 % I 29% II)	36% (50% I 50% II)
Total: 53 (20 I 22 II 11 III)	39 (16 I 18 II 5 III)
58% (37% I 42% II 21% III)	42% (41% I 46% II 13% III)
II + III 59% III 69%	II + III 41% III 31%

In this case the father that does not have a tendency to tumors influences the tumor rate of the female children.

(b) In another case an English Sable male (No. 4.444) was mated to Cream females. English Sable + Cream: 70% (69%) + 2% (3"). I age class + IV age class.

In this case the hybrids, three generations of which were observed, were without tumors. Here the low tumor incidence of the Cream was apparently dominant over the high tumor incidence of the English Sable.

Without tumors.	With tumors.
25 (7 I 11 II 7 III)	0
100% (28% I 44% II 28% III)	0%

(c) No. 8½ + English Sable: 17% (19%) + 70% (69%). II age class + I age class.

The English Sable females used in this experiment were the daughters of the male No. 4,444 used in the preceding experiment, in which the English Sable was mated to Cream females. The hybrids in Experiment (c) had a tumor rate of 61% (60%); the tumors belonged to the I age class. The high tumor incidence and early tumor age of the English are here dominant.

(d) European + English Tan (daughter of Tumor Mouse 146, granddaughter of Tumor Mouse 121) : 16% (16%) + 73%. I age class + I age class.

In this case the hybrids have a tumor rate of 28% (31%). The tumors belong to the II age class. The low tumor incidence of the European seems to be dominant. The tumors appear later than in the English strain. In this case a remarkable decrease in the tumor rate is found in the third generation.

Without tumors.	With tumors.
F₁ 2 (1 II 1 III)	6 (1 I 3 II 2 III)
25% (50% II 50% III)	75% (17% I 50% II 33% III)
F₂ 21 (5 I 11 II 5 III)	12 (2 I 8 II 2 III)
64% (24% I 52% II 24% III)	36% (17% I 66% II 17% III)
F₃ 30 (8 I 11 II 11 III)	3 (2 II 1 III)
91% (26% I 37% II 37% III)	9% (66% I 34% II)
Total: 53 (13 I 23 II 17 III)	21 (3 I 13 II 5 III)
72% (24% I 43% II 33% III)	28% (14% I 62% II 24% III)
II + III 69%	II + III 31%

(e) Michigan Wild + English 101 F₂: (?) + 71% (72%). (?) age class + I age class.

In this experiment a male gray wild mouse, which had been caught in Michigan, was mated with English 101 females. Nothing definite is known about the tumor incidence of the strain to which the gray mouse belonged; there are, however, indications that the tumor incidence of wild gray mice is much lower than that of English 101. The offspring have here the tumor incidence and tumor age of the English 101. Hybrids 58% (70%). I age class.

Without tumors.	With tumors.
F₁ 4 (1 I 3 III)	6 (1 I 5 III)
F₂ 9 (7 I 2 II)	5 (1 I 2 II 2 III)
F₃ 6 (4 I 2 II)	14 (7 I 7 II)
F₄ 2 (2 I)	4 (4 I)
Total: 21 (14 I 4 II 3 III)	29 (13 I 9 II 7 III)
42% (67% I 19% II 14% III)	58% (45% I 31% II 24% III)
II + III 30%	II + III 70%

It is possible that as a result of some accident a few mice of the first generation of hybrids were the offspring of hybrid males (Michigan Wild and English

101 F₂) F₁ and the English mother: (Michigan Wild + English 101 F₂) F₁ + English 101 F₂. They may, therefore, have been ¾ instead of ½ English. But inasmuch as these particular females had no offspring, all the other generations were entirely ½ English.

(f) Cream + European 2% (3%) + 16% (16%). IV age class + I age class. A black colored Cream male was mated to a pure European female (which became later Tumor Mouse No. 428). While No. 428 had tumors she raised three young mice, all females and very wild. These three daughters all developed tumors. In this case it is improbable that the low tumor incidence of the Cream was dominant. The higher rate of the European mother which herself had a tumor was evidently dominant.

Summary.—By combining Cream and two English strains, the tumor rate and age class in one case are intermediate, in the other case the low tumor incidence of Cream is dominant. The same English strain mated to 8½ is dominant. In a similar way the I daughter of No. 10 if mated to Cream gave a tumor rate similar to the intermediate one of the hybrids (English Tan + Cream), while the cross of the I daughter of No. 10 with a European, or of the II daughter of No. 10 with 8½, gave a high tumor rate. In a cross between European and English Tan the lower rate of the European is dominant, the tumors appear also later, while in a cross between Cream and European and probably also between a wild gray mouse and 101 English, the higher tumor incidence prevailed.

Waltzer Group.

(a) Waltzer + No. 8. In the autumn of 1909 a pure waltzing mouse was obtained from the Harvard Laboratory of Experimental Psychology. Tumors among these waltzing mice were extremely rare. This waltzing mouse was mated to a female of a Cream strain, with yellow color.[7] The offspring of this pair were black and did not waltz, a fact in accordance with the recessive character of the waltzing. These F₁ females were again mated to a pure male Waltzer. These offspring F₂ (¾ Waltzer) were in part black, some were spotted, none were waltzing. The females of F₂ were again mated to a pure male Waltzer. The offspring F₃ (⅞ Waltzer) were spotted black and white: there were no yellow mice among them; they were not waltzing.

An F₃ male (⅞ Waltzer and ⅛ Cream) was mated to white females of the No. 8 strain. Their offspring, which comprised mainly black mice or black mice with very little white are the first generation of the Waltzer + No. 8 hybrids. Mice with yellow color appeared only in the II generation of the hybrids. These mice were propagated through six generations; they were sensitive and died early and the number observed is therefore not great.

[7] These Creams were not identical with the Cream strain mentioned in the preceding paper or in other crosses, but they were equally poor in tumors.

724 *Further Investigations on the Origin of Tumors in Mice.*

⅞ Waltzer ⅛ Cream + No. 8 : 1 to 5% (?) + 30% (38%). (?) age class + II age class.
The hybrids have a tumor rate of 31% (33%). The tumors appear early. The tumor rate of No. 8 is here apparently dominant.

Without tumors.	With tumors.
22 (12 I 4 II 6 III)	10 (5 I 4 II 1 III)
69% (55% I 18% II 27% III)	31% (50% I 40% II 10% III)
II + III 67%	II + III 33%

(b) (Waltzer + 8) + English 101 F_2 : 31% (33%) + 71% (72%). Early tumor + I age class.
A male of the preceding group was mated to a female of the English 101 strain. Only a small number of the offspring were observed, but the indications are that there is a high tumor rate, probably a higher rate than in the Waltzer + No. 8.

Without tumors.	With tumors.
5 (2 I 3 II)	4 (1 II 3 III)

German Group.

In a number of crosses the German strain was used as one of the parent strains. Owing to the prevailing infection, the number of German mice observed was not very large, but the indications are that the German strain is fairly rich in tumors and that the tumors belong to the II age class.

(a) No. 8 + German : 30% (38%) + 50% (60%). II age class + II age class.
The hybrids show a tumor rate of 41% (41%). The tumors belong to the II age class. The tumor rate is therefore perhaps somewhat higher than that of No. 8 and not quite as high as that of the German mice. But the difference between the rates of the two parent strains is not great enough to make these results decisive. The age class is the same as that of the parents. The tumor rate in the second generation was in this case lower than that of the following generations.

Without tumors.	With tumors.
F_2 31 (2 I 14 II 15 III)	6 (4 II 2 III)
84% (6% I 45% II 49% III)	16% (66% II 34% III)
F_3 83 (27 I 24 II 32 III)	64 (12 I 32 II 20 III)
56% (32% I 29% II 39% III)	44% (19% I 50% II 31% III)
F_4 19 (5 I 5 II 9 III)	20 (4 I 10 II 6 III)
49% (26% I 26% II 48% III)	51% (20% I 50% II 30% III)
F_5-F_7 11 (6 I 3 II 2 III)	10 (5 I 5 II)
52% (55% I 27% II 18% III)	48% (50% I 50% II)
Total : 144 (40 I 46 II 58 III)	100 (21 I 51 II 28 III)
59% (28% I 32% II 40% III)	41% (21% I 51% II 28% III)
II + III 59% III 68%	II + III 41% III 32%

(b) German + Carter: 50% (60%) + 39 % (35%). II age class + I age class.
In our preceding paper we gave the tumor rate and tumor age of the various generations of these hybrids. This cross represents the only case in which the tumor rate and tumor age of the hybrids did not resemble that of either both or one of the parents. The hybrids had a tumor rate of 9% (10%). The tumors belonged to the IV age class.

Without tumors.	With tumors.
326 (79 I 88 II 159 III)	32 (4 I 11 II 17 III)
91% (24% I 27% II 49% III)	9% (12½% I 39% II 48½% III)

The result in this case is difficult to explain. We may have had accidentally to deal with recessives on one or both sides.

(c) English Tan (son of Tumor Mouse 121) + German 73% + 50% (60%).
I age class + II age class.
Both parents have here a high tumor rate. •The hybrids have also a high tumor rate, but apparently a somewhat lower rate than the English Tan. They belong to the age class of the English Tan (I age class).

Without tumors.	With tumors.
20 (11 I 7 II 2 III)	19 (10 I 7 II 2 III)
51% (55% I 35% II 10% III)	49% (52% I 36% II 12% III)

(d) German + (German + English Tan 143) F_1. The German father was mated to his daughters.
The English female which was mated to the German was Tumor Mouse 143:
50% (60%) + [50% (60%) + 73%]. II age class + I age class.

Without tumors.	With tumors.
2 (2 I) One of these two mice died at	5 (3 I 2 II)
the age of 7 mos.	

A strain probably rich in tumors and with early tumors.

(e) English + (8 + German) F_4. In this cross two English males were used, one belonging to the substrain English A, the other to the substrain English Sable: 77% (71%) and 63% (74%) + 41% (41%). Total (8 + German) or 51% (53%) : (8 + German) F_4. I age class + II age class.
The hybrids have a tumor rate similar to that of the English: 63% (76%).
The tumors belong to the I age class. The crosses with both males gave similar results.

Without tumors.	With tumors.
Total: 30 (22 I 7 II 1 III)	52 (27 I 25 II)
37% (74% I 23% II 3% III)	63% (52% I 48% II)
II + III 24%	II + III 76%

In this strain the animals died young.

(f) German + 108. 108 was a white mouse of old Granby stock. She was related to the mice from which the father of the No. 8 strain was derived.
50% (60%) + ? low tumor rate. II age class + (?) age class.
In this case the hybrids had a low tumor rate (8%).

Without tumors.	With tumors.
22 (7 I 10 II 5 III)	2 (1 I 1 (?))
92% (32% I 45% II 23% III)	8%

It is probable that the low tumor rate of 108 prevailed.

(g) German + 6 F₄. No. 6 was a strain in which tumors occurred, but were apparently not very frequent, probably considerably lower than in the German strain. 50% (60%) + ?. II age class + (?) age class.

In this case the F₃ and F₄ generations of the hybrids had less tumors than the F₁ and F₂ generations. In the hybrids the tumor rate was 37% (41%) and the tumors stood somewhere between the II and III age class.

Without tumors.	With tumors.
33 (11 I 10 II 12 III)	19 (4 I 6 II 9 III)
63% (33% I 30% II 37% III)	37% (21% I 32% II 47% III)
II + III 59% III 57%	II + III 41% III 43%

Summary.—If German and English mice are crossed, there is a high tumor rate in the hybrids; both parents have in this case a high tumor rate and the tumors appear on the whole early. In crossing No. 8 with German, the crosses seem to have an intermediate tumor rate. In crossing the (No. 8 + German) with English the high tumor rate of the English prevails and the tumors are early. In crossing German with 108 a very low tumor rate results. It is probable that 108 belonged to a family poor in tumors and that a low tumor rate was dominant. The tumor rate in the crosses of German and 6 F₄ is apparently intermediate between that of the parents. It seems then that in this group there may occur a dominance of the high as well as of the low tumor rate and an intermediate tumor rate may also occur. An unusual result was obtained in the case of the German + Carter in which the tumor rate was considerably lower than in either of the parents, and in this case the tumors appeared also much later than in the parents. This is the only case in which such an abnormality was observed in our experiments.

Summary of Table II.—In hybrids in which the English are one of the parent strains, the high tumor rate of the English is dominant in the majority of cases. An intermediate tumor rate prevails in hybrids in which the other parent has a very marked tendency to low tumor rate (Cream, European + 103 European) or where the other parent has a medium rate (⅞ Waltzer ⅛ Cream + No. 8). When the second parent has a very low tumor rate (Cream), the

TABLE II.

Behavior of Various Strains in Hybridization Experiments.

English.	Tumor rate.	Age class.
101 + (European + 103) d_1	i !	L !
English Sable (198) + [(103 + European) F_1 + III daughter of No. 10] d_1	h !	L !
(European 151 + I daughter of No. 10) F_2 + 101 F_2 d_1	c !	h !
English Sable d + (Silver + I daughter of No. 10)	h !	i (?)
English Tan (121) + Cream (3 females)	i !	i !
English Sable (4,444) + Cream d	L l	
8½ + English Sable (4,444) d d_1	h !	h !
European d d_1 + English Tan (daughter of 146, granddaughter of 121)	L (i) !	< L !
Michigan Wild + 101 F_2 d d_1	h (?)	h (?)
(⅞ Waltzer ⅛ Cream + No. 8) d_1 + 101 F_2	i	L ?
English Tan (121) d_1 + German d	L	h
German + (German + English Tan 143) F_1 d d_1.	h	h
English d d_1 + (8 + German) F_4	h l	h !

d indicates dominance of tumor rate of a parent strain.

d_1 indicates dominance of age class of a parent strain.

c indicates tumor rate or age class of parents which is similar.

i indicates a tumor rate or age class intermediate between both parents.

L indicates the tumor rate or age class of the parent with lower tumor rate or late tumor dominant.

h indicates the tumor rate or age class of the parent with the higher tumor rate or early tumors dominant.

l indicates fairly definite results.

? indicates doubtful results.

< indicates that tumors in a cross are later than in either of the parent strains.

low tumor rate may prevail. The age class of the English strain is not to the same extent dominant as their tumor rate.

TABLE III.

European + 103 or 102 (L) (10% to 25%). IV age class	Tumor rate.	Age class.
1. London + (European + 103) F_3 d d_1	L (c)	L !
2. (European + 102) F_1 d_1 + 8½ F_4	c !	L !
3. 101 + (European + 103) d_1	i !	L !
4. (103 + European) F_1 d d_1 + III daughter of No. 10	L (c) ?	L
5. English Sable (198) d_1 + [(103 + European) F_1 + III daughter of No. 10] d_1	h !	L !

Summary of Table III.—European + 103 (102) is a strain in which the rather low rate did not become entirely recessive in the offspring; the tumor rate of this parent strain was either dominant or the result was intermediate. The tendency of this strain to very late tumors prevailed in all cases, even when its low tumor rate did not entirely prevail.

TABLE IV.

European.	Tumor rate.	Age class.
European d d_1 + 103, 102	L (?)	L
European d d_1 + English Tan (daughter of 146)	L (i) !	< L !
Cream + European	h (?) !	

Summary of Table IV.—The low tumor rate of the European had a tendency to be dominant with those strains which did not have a very high tumor rate. With the English it was dominant or intermediate; but the higher tumor rate of the European dominated over the lower tumor rate of the Cream.

TABLE V.

European 151	Tumor rate.	Age class.
European 151 d_1 + 8 F_8 d	h (c) ?	i (L)
European 151 + I daughter of No. 10 d	h !	L
(European 151 + I daughter of No. 10) F_3 + 101		
F_2 d_1	c !	h !
European 151 + II daughter of No. 10 d	h !	L ?

Summary of Table V.—The lower tumor rate of European 151 was recessive in crosses whose mother was the first or second daughter of No. 10.

TABLE VI.

No. 8.	Tumor age.	Age class.
European 151 + 8 F_8 d	h (c) ?	i (L)
⅞ Waltzer ⅛ Cream + No. 8 d d_1	h !	h
(⅞ Waltzer ⅛ Cream + No. 8) d_1 + 101 F_2	i !	< L (?)
No. 8 F_3 + German	i !	c
English d d_1 + (8 + German) F_4	h !	h !

Summary of Table VI.—The higher tumor rate of No. 8 prevailed probably over the lower tumor rate of European 151; it also prevailed over ⅞ Waltzer ⅛ Cream.

TABLE VII.

I and II daughters of No. 10.	Tumor rate	Age class.
European 151 + I daughter of No. 10 d d_1	h !	L (?)
(European 151 + I daughter of No. 10) + 101		
F_2 d_1	c !	h !
European 151 + II daughter of No. 10 d d_1	h !	L (?)
Silver d + I daughter of No. 10 d_1	i !	L !
Cream d_1 + I daughter of No. 10	i !	L !
8½ + II daughter of No. 10 d	h (i) !	c (L) ?
(8½ + II daughter of No. 10) d + II daughter		
of No. 10 d_1	L (c)	L

Summary of Table VII.—The high tumor rate of the I and II daughters of No. 10 was dominant over the lower rate of European 151 and No. 8½. It was intermediate with the very low tumor rates of Silver and Cream. Here we find again that various strains with a very low tumor rate depress the dominance of the parent with tendency to high tumor rate more than strains with a tendency to not as low a tumor rate. The independence of the inheritance of the age class from that of the tumor rate is here apparently quite definite.

TABLE VIII.

Cream.	Tumor rate.	Tumor age.
Cream d_1 + I daughter of No. 10	i !	L !
English Tan (121) + Cream (3 females)	i !	i !
English Sable (4,444) + Cream d	L !	
Cream + European d	h (?) !	
⅞ Waltzer ⅛ Cream + No. 8 d	h !	h
(⅞ Waltzer ⅛ Cream + No. 8) d_1 + 101 F_2	i	L (?)

Summary of Table VIII.—The low tumor rate of the Cream is apparently only in one case quite recessive, *viz.*, with European; in other cases it is dominant or the result is intermediate. The age class can be intermediate, when the tumor rate is intermediate, but the lateness of the tumors in the Cream mice can prevail with an intermediate tumor rate.

TABLE IX.

Silver.	Tumor rate.	Tumor age
Silver + I daughter of No. 10 d_1	i !	L !
English Sable + (Silver + I daughter of No. 10) d_1	h !	i (?)

Summary of Table IX.—The Silver strain resembles in its behavior the Cream. With a strain with a high tumor rate it produces crosses with an intermediate tumor rate. The greater lateness again prevails. In a further cross between such an intermediate strain and a strain with a high tumor rate, the high tumor rate of the English dominates.

TABLE X.

No. 8½.	Tumor rate.	Age Class.
8½ + II daughter of No. 10 d	h (i) !	c (I.) ?
(European + 102) F_1 d_1 + 8½ F_4	c !	I. !
(8½ + II daughter of No. 10) d d_1 + II daughter of No. 10	I. (c) !	I.
8½ + English Sable (4.444) d d_1	h !	h !

Summary of Table X.—The tumor rate of the No. 8½ strain is recessive in combination with a strain with high tumor rate or intermediate crosses result.   In these crosses the I age class can also prevail over the later tumors of No. 8½.   In combination with (European + 102), however, the late tumor age of the (European + 102) prevails.

TABLE XI.

German.	Tumor rate.	Age class.
No. 8 F_3 + German	i (?)	c !
German + Carter	< L !	< L !
English Tan (121) d_1 + German d	L	h !
German + (German + English Tan 143) F_1	h	h
English d d_1 + (8 + German) F_4	h !	h !
German + 108 (= No. 8 ?) d	L	
German + 6 F_4	i (?)	c (?)

SUMMARY AND CONCLUSIONS.

1. In crossing strains known to differ in their tumor rates, the hybrids show in a considerable number of cases a tumor rate corresponding to the parent with a high tumor incidence; in some cases the offspring have the tumor rate of the parent with the low tumor incidence; in certain cases the tumor rate of the offspring is intermediate between those of the parents.

That these results are not accidental follows from the fact that we could show in some cases that two sisters crossed with the same strains or with the same male give similar offspring, and in other cases we could show that the same individual crossed successively with two strains that behave similarly produces hybrids with a similar tumor incidence.

2. There exists some evidence for the conclusion that different strains in being crossed with other strains differ in their power to impress their tumor rate upon the crosses.   Thus the English strain and the I and II daughters of No. 10 have the tendency to transmit to the offspring a high tumor rate, while Cream, Silver, and some European other than 151 have a tendency to transmit a low tumor rate.   While crosses of these daughters of No. 10 with European 151 or with No. 8½ show the high tumor rate of the mothers, the crosses of one of the same females with Cream or Silver show an intermediate tumor rate.

3. We find further evidence for our conclusion previously stated

that age class of the tumors and tumor rate are not dependent on the same factor. The age class enters into the crosses as a factor independent of the tumor rate. Thus we find in the crosses between the first daughter of No. 10 and Cream, and in the crosses between the same female and English Silver a similar tumor rate, but the age classes differ in conformity with the difference in the age classes of the parents.

We find, furthermore, that while in some cases a tumor rate and an age class that correspond to each other (high tumor rate, early tumors—low tumor rate, late tumors) are transmitted to the offspring, in other cases tumor rate and age class transmitted to the crosses diverge.

4. It seems that certain strains with very late tumors if mated with strains with earlier tumors have a tendency to transmit to the offspring their own tendency to very late tumors. With a certain strain lateness of the tumors seems to be dominant, while a low tumor rate is not necessarily dominant in the same crosses. This was noticeable in the crosses into which the strain European + 102 or 103 entered as one of the parents.

5. If both parents have a similar tumor rate the offspring have usually a similar tumor rate. There was, however, one exception to this rule in the case of the German + Carter mice, in which the offspring showed a much lower tumor rate and higher age class than either of the parent strains.

THE INFLUENCE UPON THE SPLEEN AND THE THYROID OF THE COMPLETE REMOVAL OF THE EXTERNAL FUNCTION OF THE PANCREAS.

By J. E. SWEET, M.D., and JAMES W. ELLIS, M.D.

(*From the Department of Surgical Research of the University of Pennsylvania, Philadelphia.*)

PLATE 77.

(Received for publication, July 2, 1915.)

In the course of studies upon the pancreas in which the external function of the gland was completely removed, either by double ligation of both ducts, cutting and interposing omentum, or by the complete removal of the duodenal portion of the gland, two findings were encountered which seem worthy of a brief communication. The first is that a striking simple atrophy of the spleen rapidly follows such an operation. The second is that the thyroid apparatus of these animals shows a constant change, evidenced macroscopically by a translucency which may amount to an actual transparency, microscopically by an evident increase in the amount of colloid, chemically by a marked increase of the iodine content of the gland, and physiologically by a greatly delayed appearance of tetany, after the complete operative removal of the thyroids and parathyroids.

The details of the operation are shown in Fig. 1. The pancreas of the dog, in our experience, always possesses two ducts, a smaller one draining a definite island of tissue and opening at or near the ampulla of the bile duct, a, and a larger one draining the far greater portion of the pancreas, b. Several authors have described instances in which more than two ducts have been found, so that we have usually tied off the pancreas at A and B, and completely removed the duodenal, duct-bearing portion of the pancreas between the lines A and B.

The results as regards the effects on the spleen and the thyroid

732

are the same with either method, provided the blocking of the ducts by the interposed omentum has been successful.

Such animals lose weight at first rather rapidly; after several months they reach a level at which they continue or even gain slightly. None of our animals developed sugar in the urine; all showed the voluminous fatty stools typical of loss of pancreatic juice.

Both of the observations here recorded were purely accidental. The operations were undertaken for a definite purpose, and the changes in the spleen and thyroids would doubtless not have attracted our attention had it not been for the extreme changes noted in the first animal of the series. The outline of this animal's spleen (one-half actual size) is shown in Text-fig. 1 (Dog 1). We did not have the measurements of this spleen at the time of the operation upon the pancreas, but if we may judge from the routine measurements of spleens made since, it is fair to assume that this spleen measured at operation at least 18 cm. in length, while at autopsy it measured 6.75 cm.

Following this first observation we repeated the experiment, carefully measuring the spleen at the time of the operation upon the pancreas. The results are seen in Table I and in Text-fig. 1, which shows the simple outline drawings of the spleens of two animals, the outer outline being the spleen at operation, the inner outline the same spleen at autopsy. Under the microscope this atrophy appears as a simple atrophy.

That the splenic atrophy is something more than the expression of the share taken by the spleen in the general loss of body weight is shown, we believe, in the comparative study of some of these animals.

Dog 1 lost 45.6 per cent of its original body weight. While we have not the actual figures, it seems fair to assume that the spleen was at autopsy at least 66 per cent smaller than at the time of the operation. The actual figures for Dog 66 show a loss of weight of 37.9 per cent, with a loss of spleen, as measured by the length of the organ, of 61.9 per cent. Dog 1 reached a constant weight after 150 days, having lost 45.6 per cent of its original weight. Dog 48 was killed at the end of 131 days, having lost 40 per cent of its original weight, but the spleen of Dog 1, the spleen which first

TABLE I.

No. of dog.	Weight at operation.	Weight at autopsy.	Measurements of spleen at operation.	Measurements of spleen at autopsy.	Length of life after operation.	Thyroparathyroidectomy.	Tetany.
	gm.	*gm.*	*cm.*	*cm.*			
I.....	11,400	6,200	-	Length...... 6.75 / Width of head. 1.75 / " tail. 0.75	Killed after 10 mos.		
A.....	10,080		Length......18 / Width of head. 5 / " tail . 5	Length......12 / Width of head. 3.5 / " tail . 2.25	" 21 days		
B.....	8,190		Length......16 / With of head. 6 / " tail . 4	Length......10 / With of tail . 3.5 / " tail . 2.25	" 12 "		
D.....	10,220		Length......16 / Width of head. 5.5 / " tail . 2.75	Length...... 9.3 / Width of head. 3 / " tail . 1.5	Died after 3 days of acute pancreatitis		
48.....	9,680	5,805	Length......18.5 / Width of head. 6 / " tail . 4.5	Length...... 9 / Width of head. 3 / " tail . 2	Killed after 131 days		
24.....	26,122	12,300	Length......24 / Width of head 13 / " tail . 7	Length......12.25 / Width of head. 6 / " tail . 3	" 76 "		
49.....	7,410	4,850	Length......19 / Width of head. 8 / " tail . 3.5	Length......11.5 / Width of head. 4.5 / " tail . 2.5	" 62 "		
66.....	12,580	7,800	Length......21 / Width of head. 8.5 / " tail . 3.5	Length...... 8 / Width of head. 4.5 / " tail . 1.15	" 123 "	96 days after pancreatectomy	19 days after removal of thyroids and parathyroids. Recovered.
C.....	9,220	6, 60	Length......18 / Width of head. 5 / " tail . 4	Length......11.5 / Width of head. 3 / " tail . 2	Died " 49 "	35 days after pancreatectomy	None.
E 25..	13,000	11,630	Length......20 / Width of head. 7.25 / " tail . 3	Length......13.5 / Width of head. 3.2 / " tail . 2	Killed " 19 "	7 days after pancreatectomy	Slight tetany. Recovered.

attracted our attention, was only about one-half the size of the spleen of Dog 48, and showed clearly from its slate-colored, shrunken appearance a greater degree of atrophy than did the spleen of Dog 48. This we are inclined to interpret as meaning that the observed atro-

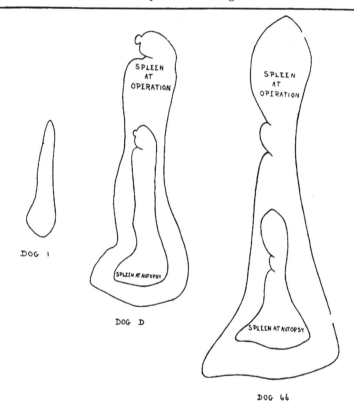

Text-Fig. 1. Outline drawings of the spleen before and after the complete removal of the external function of the pancreas.

phy of the spleen had continued in Dog 1 after the body weight had become stationary. That this atrophy is due to something other than the general loss of weight we think is further shown in the dog designated D, which died of an acute pancreatitis three days after operation. The atrophy of the spleen is here clearly marked (Text-fig. 1), although the subcutaneous and intra-abdominal fat had not appreciably decreased in amount in the short space of three days. The actual weight of the dog at autopsy was not determined.

The observations regarding the thyroid apparatus can be supported in such a variety of ways that their accuracy is hardly to be doubted. The macroscopical appearance of the gland at autopsy is marked by a peculiar translucency which often amounts to an actual transparency. If such a lobe is held towards the light both parathyroids can be plainly seen; in the dog the lower parathyroid body is imbedded completely within the thyroid tissue, and can usually be found only by cutting serial sections; the outline of an opaque object held between the thyroid and the light can be clearly seen through the translucent lobe.

Microscopically the thyroid is seen to contain somewhat more than the normal amount of colloid, with a consequent flattening of the alveolar cells. The parathyroids, as far as we have discovered, are normal; in some instances they have given us the impression of hypertrophy.

In view of the well known variations in the macroscopic and microscopic appearance of the thyroid, one is inclined to speak with reserve concerning conditions based upon such variations. We are therefore fortunate in the case of the thyroid apparatus in having two definite proofs of change, the physiological test and the chemical determination of the specific content of the thyroid.

The physiological results are shown in the three dogs, 66, C, and E 25. Dog 66 developed mild tetanic attacks 19 days after the complete thyroparathyroidectomy. Dog C developed no tetany during the 14 days after the thyroid operation. Dog E 25 developed tetany 7 days after thyroparathyroidectomy.

We have already stated that the parathyroids appear entirely normal, unless, indeed, they are somewhat enlarged. We offer these findings concerning tetany without attempting an explanation. It has been our experience, in several fairly large series of parathyroid tetany experiments, that a normal dog will develop a severe and fatal tetany on the third or fourth day following the complete extirpation of the thyroid apparatus; in a very few instances, not over 5 per cent of our cases, the animal has failed to show any symptoms whatever of tetany.

The iodine determinations in eighteen normal thyroids and in seven thyroids from operated animals are shown in Table II, in which the result is expressed in milligrams of iodine per gram of dry thyroid tissue.

TABLE II.

Source of thyroid.	Weight of dry gland. *mg.*	Iodine per gm. of dry gland. *mg.*
Normal dog	0.2460	0.723
" "	0.1530	0.953
" "	0.4530	1.036
	0.5710	1.205
	1.0000	1.402
	0.3550	1.068
	0.1620	0.960
	0.7128	2.676
	0.4248	1.220
	0.1621	1.200
	0.42	0.400
	0.4	1.050
	0.3	0.405
	0.7	0.228
	0.25	0.800
	0.30	0.993
	0.2	0.500
" "	0.13	Negative.
Pancreatectomy		
Dog 124	0.5	3.151
" 228	0.25	6.960
" 279	0.4456	1.434
' 257	0.1430	9.025
' 67 P	0.6934	1.134
' 267	0.3502	1.316
" 19 P	0.3448	1.827

Average of 18 normal determinations = 0.9393 mg.
Average of 7 pancreatectomies   = 3.678 mg.

We take pleasure in expressing our thanks to Mr. Robert B. Krauss of the Laboratory of the Henry Phipps Institute, for the iodine determinations which were made by the method elaborated by him.[1]

We have found nothing in the literature concerning the thyroid, the pancreas, or the spleen which aids in the explanation of the

[1] Krauss, R. B., The Determination of Iodine in the Presence of Organic Matter, *Jour. Biol. Chem.*, 1915, xxii, 151.

above findings.   Any discussion of their meaning at present would therefore be mere speculation, and should, at best, await the confirmation of our results.

SUMMARY.

The complete removal of the function of the pancreas concerned in digestion is followed by marked changes in the spleen and in the thyroid apparatus.   Second, the spleen shows an extreme simple atrophy.   Third, the thyroid apparatus exhibits a constant change shown by the macroscopic transparency of the gland, by the microscopic increase in the amount of colloid, by the chemical increase of the iodine content of the gland, and by the functional test of the delayed appearance of tetany after the complete removal of the thyroid apparatus.

EXPLANATION OF PLATE 77.

Fig. 1.   Diagrammatic outline of the pancreas of the dog (posterior aspect) showing the points at which the operation is carried out.   P = pylorus; d = bile duct.   For further description, see page 732.

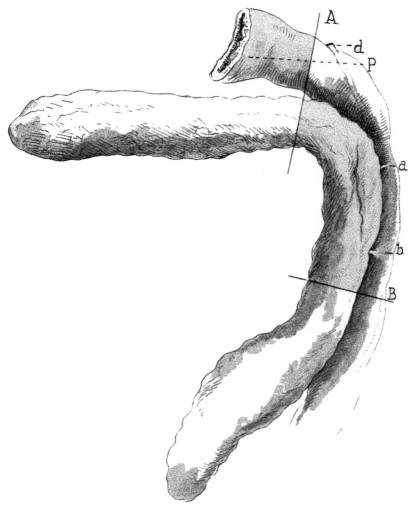

Fig. 1.

(Sweet and Ellis: External Function of Pancreas

THE INFLUENCE UPON TADPOLES OF FEEDING DESICCATED THYROID GLAND IN VARIABLE AMOUNTS AND OF VARIABLE IODINE CONTENTS.

By C. H. LENHART, M.D.

(*From the H. K. Cushing Laboratory of Experimental Medicine of Western Reserve University, Cleveland.*)

PLATE 78.

(Received for publication, July 2, 1915.)

In 1912 and 1914 Gudernatsch[1] reported his studies of the variable effects upon the growth and differentiation of tadpoles produced by feeding different kinds of animal tissues, such as thyroid, thymus, muscle, pancreas, liver, testicle, etc. His most striking findings were that thyroid feeding hastened the differentiation of the tadpoles, at the same time inhibiting their growth, so that he was able to obtain pigmy frogs; and that thymus feeding prevented or delayed their differentiation but favored their growth, so that giant tadpoles resulted. He used fresh tissues, and, in the case of the thyroid, without determining the amount of iodine present.

In view of the known relations of iodine to thyroid activity, it seemed probable that the iodine content of the thyroid fed might also modify its effect on tadpoles. With this in view a supply of tadpoles was brought to the Laboratory on May 9, 1914. These tadpoles were of uniform size, and their age was estimated at about one week.

The stock tadpoles were kept in large granite baking dishes. Those used for experimental observations were kept in small granite basins of about 200 cc. capacity, in which were placed a few small stones. The water in all the basins was completely changed twice

[1] Gudernatsch, J. F., Feeding Experiments on Tadpoles. I. The Influence of Specific Organs Given as Food on Growth and Differentiation. *Arch. f. Entwcklngsmechn. d. Organ.*, 1912–13. xxxv. 457: Feeding Experiments on Tadpoles. II. A Further Contribution to the Knowledge of Organs with Internal Secretion. *Am. Jour. Anat.*, 1913–14. xv. 431.

daily. The experimental basins were kept on tables in the middle of a large room, so that all would be exposed to similar light and temperature conditions. The room temperature was recorded every afternoon. After a few changes by way of trial, it was decided to feed the stock with fresh hog's liver every day, while the experimental animals were fed liver and thyroid on alternate days. The liver was cut up into small pieces, but not crushed. In the earlier experiments the liver was put into the basins in the forenoon and left till late in the afternoon, but this was abandoned because on hot days there was evidence of fermentation or putrefaction which led to the death of some of the tadpoles. The plan of allowing the liver to remain in the basins for one hour and then changing the water eliminated this danger even in the hottest weather. The thyroid was fed in the form of dried powder, in each case the iodine content having been previously determined by Dr. Marine. In the earlier experiments ten tadpoles were placed in each dish, with about 200 cc. of city tap water. Later but five tadpoles were placed in this quantity of water. Beyond taking photographs of the several series, no objective measurements were made of the changes produced.

The questions particularly studied may be summarized as follows: First, the effect upon both growth and differentiation of non-thyroid iodine; *e. g.*, potassium iodide and iodalbin (Parke, Davis & Co.). Second, the effect of thyroid feeding; (a) feeding constant amounts of a series of desiccated thyroids containing progressively increasing percentages of iodine; (b) feeding different quantities of one particular thyroid; (c) feeding thyroid obtained from different species of animals. Third, an attempt to counteract the effect of the thyroid feeding by keeping the tadpoles exposed to cold, by the use of cracker dust, quinine, egg white, egg yolk (both cooked and uncooked), and egg yolk extracted with acetone.

It may at once be stated that the effect of the potassium iodide solutions was negative. As to iodalbin, the results were indefinite. The animals showed early tail absorption, and most of them showed some emaciation at the time of their death; but they all developed some disease resembling a general body edema. Iodalbin contains about 21 per cent of iodine so loosely bound that the toxic effects from free iodine had to be considered, and hence these results cannot be accepted as suggesting a thyroid-like action for iodalbin. Since these observations were made Morse[2] has published positive results from artificially iodized proteins, and states that the effect is comparable to that produced by thyroid iodine. Further observations must be made before this can be accepted, since there is no conclusive evidence that artificially iodized proteins exhibit an iodothyreoglobulin-like action.

The effect of thyroid feeding was very marked and closely associated with both the iodine content and the amount fed. The details will be exhibited later.

As to factors protecting against the effect of thyroid feeding, only two were found which were certain in their action; namely, exposure to cold and feeding carbohydrate in the form of cracker dust.

We may now examine more in detail the positive results.

The Effect of Thyroid Feeding.

Preparations of desiccated thyroid from human, canine, sheep, and ox glands were used. The human thyroids were obtained from Dr. Crile's clinic and include simple and exophthalmic goitres. All acted alike qualitatively. The sheep and ox glands available were too few to furnish an extended series of experiments, and may be dismissed from present consideration with the statement that there is no reason to believe that, with the material available, one could not get as gradated a series of results as we shall show can be gotten by the use of desiccated dog thyroid. With human glands a graduated series of effects was obtained, but it was not so sharp as with dog's thyroid. This is to be expected because of so many unknown factors in the life history, treatment, etc. Then, too, some of the thyroids had been in 10 per cent formalin for a day or two before desiccation. As regards the effect of formalin one can only state that it does not destroy the thyroid effect.

As examples of the experimental findings, the following protocols are exhibited.

Series I, Dog Thyroid.—Dishes 1, 2, and 3. The tadpoles in this experiment received 50 mg. respectively of three thyroids whose iodine contents were 0.05.

[2] Morse, M., The Effective Principle in Thyroid Accelerating Involutions in Frog Larvae, *Jour. Biol. Chem.*, 1914, xix, 421.

1.40, and 2.92 mg. of iodine per gm. of dried thyroid. This dosage was given every other day, as in all other experiments, unless otherwise indicated. The feeding was started May 16. No liver was fed in this case after the thyroid was started. The tadpoles in Dish 3 were all dead as early as 15 days, and in Dish 2 in 11 days. These were instances of high iodine contents. Four of the tadpoles in Dish 1, receiving thyroid of low iodine content, were living and active as late as Aug. 3—79 days—when the experiment was terminated. These tadpoles were about the size of the controls, but more differentiated, presenting formed (jointed) hind legs. Those in Dishes 2 and 3 died early, were much emaciated, and only slightly differentiated as compared with the controls.

Series I, Dog Thyroid.—Dishes 4, 5, and 6. This experiment is the exact duplicate of the previous one, except that liver was fed on alternate days. The tadpoles in Dish 6 were all dead in 10 days, and in Dish 5 in 2 days,—instances again of early death after feeding with thyroid of high iodine content. Those in Dish 4, getting low iodine thyroid, showed one tadpole still alive and active after 79 days. The tadpoles in this dish developed functional hind legs, were of large size, and had long, well preserved tails. Compared with the controls they showed no emaciation, but marked differentiation. They were larger than those in Dish 1, perhaps due to the liver feeding. In contrast, the tadpoles in Dishes 5 and 6, getting high iodine thyroid, died early, with much emaciation and before there was time for much differentiation. The emaciation was extreme. They literally melted down, the tails rapidly disappearing.

Series I, Dog Thyroid.—Dishes 7, 8, and 9. Here conditions were the same as in the first and second experiments above, except that the thyroid was given only twice. The tadpoles in Dishes 8 and 9, receiving high iodine thyroid, were all dead in 16 and 10 days, respectively, while those in Dish 7, getting low iodine thyroid, were not all dead till 57 days had passed. Those in Dishes 8 and 9 were the more emaciated. Differentiation was not especially affected in any. This experiment shows that only two doses of thyroid of a certain iodine strength will initiate emaciation and lead to early death, the effect being more marked in the case of thyroids with higher iodine contents. Gudernatsch also observed that one feeding with thyroid was sufficient to induce the emaciation and death.

Series II, Dog Thyroid.—In this experiment a series of thyroids containing respectively 0 05, 0.08, 0.18, 0.54, 0.71, and 1.40 mg. (Figs. 1 to 6) of iodine were fed in 50 mg. doses every other day, beginning May 23. Liver was given on alternate days. As early as four days the series as a whole showed a progressive decrease in size and activity in proportion as the iodine percentage increased. Within five days a most remarkable difference was seen, from large active tadpoles in Dish 1, getting the thyroid of lowest iodine content, to markedly emaciated, inert, and highly metamorphosed tadpoles in the dish getting the highest iodine thyroid. At the end of 72 days there was one tadpole living in each of the first three dishes. All were dead in Dishes 4 and 5 within 19 days, and in Dish 6 all were dead within 11 days. The number of days that intervened before the first tadpole died in each dish of the series ran as follows: 8 (accidental), 52, 33, 17, 9, and 5 days. For the second dead in each dish the figures ran: 54, 54, 35, 19, 11, and 7 days. For the third dead: 59, 68, 41, 15, and 11 days. This clearly shows that the death rate parallels the iodine contents. As to differentiation the notes cannot be given in detail, but by way of summary it may be stated that the tadpoles getting high iodine thyroid (Dishes 4, 5, and 6) emaciated so rapidly and died so soon that little differentiation took place. In 41 days the tadpoles in Dish 1 had formed hind legs and were larger than the controls; *i. e.,* they showed marked differentiation together with growth instead of emaciation. On the same date the tadpoles in Dish 2 showed formed hind legs, but were smaller than those in Dish 1, while those in Dish 3 compared in every way with the controls. The final result then seems to be a balance between the tendency to emaciation and to hastened differentiation, and all degrees of differentiation may be associated with all degrees of size. Gudernatsch's more uniform results were undoubtedly due to using thyroid of more constant or high iodine content.

As previously stated, experiments were also made in which the quantity of a particular thyroid fed was varied; *e. g.,* feeding in 10, 20, 30, 40, and 50 mg. doses of some one particular thyroid. The first experiment of this kind to be reported is Series IV, where a dog thyroid containing 1.40 mg. of iodine per gm. of dried gland was fed beginning June 1. The number of days when all were dead ran, respective to the increasing amounts given, 20. 10, 8, 7, 8. Emaciation was very rapid and marked in all, so much so that in the larger doses there was little time for any differentiation. The tadpoles in Dish 1, on the other hand, getting only 10 mg. of thyroid, proceeded within 14 days to the formation of front and hind legs, large frog mouth, and prominent eyes.

Series VII, Dog Thyroid.—Here as in the previous experiment increasing doses were given of a thyroid containing only 0.54 mg. of iodine. Here the number of days within which all were dead were 32, 37, 37, 39, and 42, not very strikingly different. Emaciation was of little consequence in this series, and death was probably largely due to advanced differentiation. which was hastened in all as compared with the controls.

On the whole, the experiments of varying the quantity of thyroid fed are not nearly as clear cut as those where the iodine percentage was varied. I feel, however, that it is only a matter of obtaining a thyroid of suitable iodine content and arranging the quantities fed in a suitably gradated series, in order to get a well gradated series of effects.

The Protective Effect of Feeding Cracker Dust.

Series VIII, Human Thyroid.—In conjunction with some experiments on the possible inhibiting effect of quinine on metabolism, the following experiment was made. One group was fed cracker dust in addition to the regular liver feeding, and the other liver alone. Both groups received 50 mg. of human thyroid with an iodine content of 2.58 mg. per gm. of dried gland. every second day. beginning June 16. Dates of death were of little importance in this case, as they were mostly due to drowning on account of the high degree of differentiation reached. The tadpoles receiving cracker dust became large and acquired functioning hind legs, front legs, and frog-shaped bodies. Those receiving no cracker dust were smaller, had formed hind legs, but no front legs : on the whole they were more nearly like tadpoles, the first more like frogs. Both sets were larger than the controls.

Series XI and XII received every other day dog thyroid with an iodine content of 1.40 mg. Each series was divided into 5 dishes getting 10, 20, 30, 40, and 50 mg. of thyroid, respectively, beginning June 30. ˙ Series XI got cracker dust every second day alternately to the thyroid feeding. Within 8 days the cracker series showed a very distinct progressively increasing emaciation, proportional to the increasing amounts of thyroid fed. This bears out the previous experiments on the effect of variable quantity of thyroid. As early as the sixth day the other group, not getting cracker, showed a marked absorption of tails with decreased activity, while on the seventh day the disparity between the two groups was exceedingly well marked. All the non-cracker group were like small round balls with short conical tails. Owing to the severity of the reaction there was not much difference between the different members of this group. The death dates show a marked shortening of life in the group not receiving cracker. We may conclude then that the feeding of cracker dust delays the tendency of thyroid, when sufficiently active, to hasten death and also tends to prevent emaciation.

The Protective Effect of Exposure to Cold.

Believing that the effects thus far observed were largely due to the well known pharmacological action of thyroid of increasing metabolism, it was thought this action might be lessened by exposing the tadpoles to a lower temperature. Being cold-blooded animals, this would tend to lower their metabolism. With this in mind the following experiments were made.

Series V, Dog Thyroid.—To ten tadpoles kept in a refrigerator were given every second day 50 mg. of a thyroid (dog) containing 1.40 mg. of iodine, which had by previous experiments been shown to have a marked effect. The first dose was given on June 1. All the tadpoles were dead in 27 days, while of the controls kept at room temperature all were dead in 9 days. The controls became markedly emaciated. Those on ice became emaciated toward the end; but earlier, while the controls were still living, they were distinctly larger and less emaciated than the controls. Gradually their tails became absorbed, their bodies smaller, hind and front legs developed along with a frog facies, so that shortly before death they were small but well differentiated; *i. e.*, really pigmy frogs. The controls emaciated so rapidly that there was little differentiation.

Series VIII, Human Thyroid.—These tadpoles were fed cracker dust every second day in addition to the regular liver feeding, and were given every second day 50 mg. of a thyroid containing 2.58 mg. of iodine. Thyroid was started on June 16. One dish was kept on ice, one at room temperature. Of those on ice four were still living at the end of 49 days, when the experiment was terminated, while all of those kept at room temperature were dead in 28 days. Those on ice were of good size, with well preserved tails and slight hind leg buds at the last date of observation, Aug. 3. Those kept at room temperature, compared on the same dates with those on ice, were always larger. Also their differentiation went much further, in that they developed functioning hind legs, front legs, and frog-shaped body and head. And while they died earlier, death was not due to emaciation but to drowning, owing to their complete differentiation.

We may interpret this experiment as follows: At room temperature the stimulation of metabolism by this particular thyroid was not sufficient, in the presence of a more sufficient food supply (cracker), to lead to emaciation, but on the contrary the animals grew large and practically completely differentiated, meeting death by drowning. In the tadpoles kept on ice, metabolism was lowered by the cold so that the tadpoles grew only a little and differentiated only slightly; that is, the stimulating effect of thyroid on metabolism was to some extent counteracted. In the first experiment (Series V) cold protected against the extreme emaciation produced by a certain thyroid at room temperature. In the second experiment (Series VIII) cold tended to counteract the mild stimulation of a certain thyroid which at room temperature led to a high degree of differentiation.

DISCUSSION AND CONCLUSIONS.

We may conclude that the feeding of dried thyroid gland to tadpoles causes an early differentiation in proportion to the quantity fed or the percentage of iodine content of the gland used. With the larger doses and the higher iodine percentages, metabolism is stimulated to such an extent that the animals emaciate rapidly and die early, before there is time for much differentiation. With smaller amounts and lower iodine percentages the size of the animals is roughly inversely proportional to the amount or percentage, so that a close association of differentiation with pigmy size is not characteristic of thyroid feeding as such, as Gudernatsch seems to conclude. One may see early and marked differentiation along with large size. It all seems a question of dosage. The larger sizes are associated with slower differentiation, the smaller sizes with more rapid differentiation, and the smallest sizes may show no differentiation at all, due to the extremely rapid and marked emaciation, and early death. Non-thyroid iodine does not have this effect. The thyroid effect is inhibited by exposure to cold and by cracker feeding. Exposure to cold probably acts by lowering metabolism; cracker feeding, by substituting food other than the animal's own tissues to meet the increased demands caused by the stimulating effect of the thyroid feeding.

Gudernatsch in his earlier paper speaks of the thyroid as stimulating metabolism, which leads to early differentiation and suppresses growth. Later he seems to lean to the view that the thyroid possesses some specific influence on differentiation. It may all be a matter of words, but our present conception is that we are simply dealing with the well known action of thyroid on metabolism. As the iodine content increases, the thyroid increasingly stimulates the metabolism of the tadpole, which undergoes changes in size, increased growth or rapid emaciation, according to the strength of the action. The tadpole being a larval form, the tissues first to be stimulated to increased metabolism, and later the first to be consumed, are naturally those tissues whose normal function is approaching a normal end, and which, in the normal course of events, are about to undergo metamorphosis. Hastening of differentiation seems then to ensue not as a specific stimulation of differentiation, but only to be the normal result of the stimulation of general metabolism. The seeming specificity of the result lies not in a new action of thyroid, but in its application to a living organism at a specific time in its development.

Most important, of course, is the confirmation of what we may be justified in regarding as an established fact; namely, that the activity and potency of the physiologically active substance of the thyroid is measurable in terms of its percentage iodine content.

Finally, it may be pointed out that the reaction of tadpoles to thyroid feeding is so sensitive that the procedure might well serve as a biological test for the activity of thyroid tissue, superior even to chemical methods.

EXPLANATION OF PLATE 78.

FIGS. 1 TO 6. Photographs of Series II, dog thyroid experiments. Experiment begun May 23 and photographed 7 days later. All were fed 50 mg. of thyroid every other day.

No. 1 received thyroid containing 0.05 mg. of iodine per gm. dried.
" 2 " " " 0.08 " " " " " "
" 3 " " " 0.18 " " " " " "
" 4 " " " 0.54 " " " " " "
" 5 " " " 0.71 " " " " " "
" 6 " " " 1.40 " " " " " "

No change is seen in Nos. 1, 2, and 3 because of the short time interval and the low iodine content of thyroid used, while Nos. 4, 5, and 6 show the characteristic increasing effect of thyroid paralleling the iodine content.

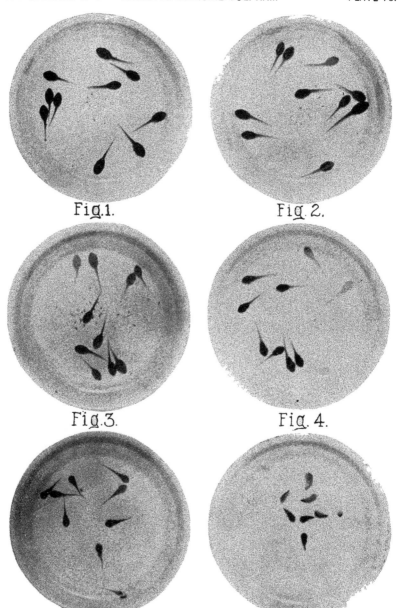

Fig.1.   Fig. 2.

Fig.3.   Fig. 4.

EXPERIMENTAL PNEUMONIA (FRIEDLÄNDER TYPE).

By WARREN R. SISSON, M.D., and I. CHANDLER WALKER, M.D.

(*From the Departments of Medicine and Pathology of the Harvard Medical School and the Peter Bent Brigham Hospital, Boston.*)

PLATES 79 TO 81.

(Received for publication, July 2, 1915.)

INTRODUCTION.

Certain cases of lobar pneumonia in man have for a long time been considered to be caused by *Bacillus mucosus capsulatus* (Friedländer's bacillus). Although this view has been repeatedly questioned, the clinical and pathological studies offer fairly convincing evidence in its support.[1] Few experimental studies on this subject have been undertaken, and little attempt has been made either to reproduce the disease in animals or to study the pathology of the early stages of the condition.

Previous Experimental Work.

Kokawa,[2] working with guinea pigs, mice, and rabbits, was unable to reproduce a definite lobar pneumonia with Friedländer's bacillus. In two instances lesions corresponding more nearly to bronchopneumonia were found. Intratracheal and direct injections of bouillon cultures into the lung were the methods employed. Based upon a very limited number of experiments, Kokawa concluded that the lung could be infected with Friedländer's bacillus and that the infection was aerogenous in origin. Furthermore, he considered that the bacilli were able to cause a pneumonia only when very virulent or when present in great numbers. Certain predisposing factors, according to this author, such as trauma of the lung parenchyma, are necessary for the production of the pathological process. Lamar and Meltzer,[3] working more recently on the subject of experimental pneumonia, have reproduced the disease successfully in two dogs after intrabronchial insufflation of the organism. The pathological findings in this work are not described in detail.

Technique Employed and Results of Experiments.

The following experimental studies are based upon findings in thirty-nine cats. The cats employed for the most part were excep-

[1] Sisson, W. R., and Thompson, C. B., *Am. Jour. Med. Sc.*, 1915, cl, 713.
[2] Kokawa, J., *Ztschr. f. Hyg. u. Infectionskrankh.*, 1905, l, 364.
[3] Lamar, R. V., and Meltzer, S. J., *Jour. Exper. Med.*, 1912, xv, 133.

tionally large, well nourished, well developed animals. They had
been kept in confinement for but a very short time and showed no
evidence of snuffles or other disease. Each cat was treated as
follows:

2 gm. of urethane were given subcutaneously 2 hours before the operation.
During insufflation the cat was subjected to a very light ether anesthesia which
lasted from 5 to 10 minutes.

In the first 13 experiments, the operation consisted of the following procedures. With the animal on its back, the mouth was held open, the tongue
pulled forward and downward. A small filiform catheter was now inserted
between the vocal cords and carefully pushed in until fairly definite resistance
was encountered, indicating that the tip of the catheter had entered a bronchus.
5 to 8 cc. of a 12 hour culture of *B. mucosus capsulatus* were now insufflated
slowly through the catheter with a syringe.

The culture used in this first series of experiments was recovered
from the lung exudate of a patient that died 31½ hours after the
onset of the first symptoms. The organism had the characteristics
of *Bacillus mucosus capsulatus*. 1 cc. of the culture employed, injected intraperitoneally into a medium sized guinea pig, killed the
animal in about 18 hours. 38.4 per cent of the cats insufflated with
the organism in this manner developed definite lobar pneumonic
processes which involved entire lobes or the larger portions of lobes.
These cats all yielded positive cultures of *Bacillus mucosus capsulatus* from the heart's blood at autopsy. The remaining animals
showed pneumonic processes of various appearances. In some of
these, entire lobes were fairly firm, dark red in color, and microscopically showed almost complete atelectasis of the tissue. This was
sometimes associated with marked congestion and often with some
alveolar exudation. Often, small hemorrhagic areas were found; in
others, lesions typical of bronchopneumonia occurred. One of the
cats in this series of experiments developed a pericarditis, otherwise
no lesions other than the pulmonary ones were found at autopsy.
All the animals that showed the lobar type of involvement became
extremely prostrated within 10 hours after the insufflation. After
24 hours the animal would lie in a semicomatose condition. The cat
was usually killed as shortly before a probable exitus as could be
estimated. In no instance did the animal live more than 72 hours.
In two instances the cats were found dead at the end of 48 hours.
The following experiments[4] will serve to illustrate the course of the

[4] For complete series of experiments see Table I.

disease and the pathological findings in the cats that developed lobar pneumonia after insufflation of cultures of Friedländer's bacillus.

Experiment 3.—Large, well developed, well nourished cat. 7 cc. of culture were insufflated intrabronchially. 24 hours later, the animal was found in a markedly prostrated condition, semicomatose, with respiration definitely accelerated, and gray, slimy, viscid material exuding from the mouth. The cat was killed at this time with chloroform.

Autopsy Findings.—The right inferior lobe is enormously enlarged, being approximately three to four times the size of the corresponding lobe on the left. The visceral pleura over the enlarged lobe is cloudy, but smooth. The left lung weighs 9.5 gm., the right 46 gm. (The normal weights are about 9 for 13 gm. respectively.) The right lower lobe is firm and consolidated except for narrow marginal areas where alveolar structure can be made out. The color is in general dark red and the lobular markings are accentuated because of injection of the entire lobular lymphatics. On cross-section the cut surface is covered with a large amount of stringy, mucoid, slightly sanguinous, grayish material. On pressure, this can be easily expressed from numerous foci. The surface, on removal of the exudate, presents a very striking, mottled, marble-like appearance. Poorly defined, gray areas are surrounded by hemorrhagic areas. No clearly defined lobular distribution of the process can be made out. Pieces of the lung sink in water. The tissue does not present a granular appearance. Over the surface of the left superior lobe, about the hilus, are raised (2 to 5 mm. in diameter), injected, nodular areas. These, on cross-section, represent small, grayish, circumscribed foci of consolidation with hemorrhagic margins. The findings elsewhere show nothing remarkable.

Cultures taken from the heart's blood, lung exudate, and spleen all showed a pure culture of *Bacillus mucosus capsulatus.*

Microscopic Findings.—Sections were taken from various portions of the right inferior lobe and fixed in formalin and Zenker's fluid. Stained in hematoxylin and eosin, methylene blue, and also by Gram's method. All the sections show alveolar spaces completely filled with exudation except in tissue taken from the margin of the lung. Here the demarcation between air-containing and infiltrated lung tissue often is very sharply shown. In most of the sections, consolidation is complete and does not suggest a focal or bronchopneumonia. The alveolar exudate consists chiefly of dense blue-staining masses of extracellular. Gram-negative, capsulated bacilli and polymorphonuclear cells (Fig. 3). Scattered throughout the field are spaces almost completely filled with red blood cells. Large vacuolated mononuclear cells are rarely seen. Some of the bronchi are completely filled with exudate, others are air-containing.

Sections taken from the focal lesions in the left lung show alveolar spaces about bronchi that are filled with exudate similar to that described above.

Experiment 5.—Large adult, male cat. Technique the same as in the previous experiment. 24 hours after insufflation the animal is markedly prostrated, refuses food, is dyspneic, and coughs. Killed 28 hours after insufflation.

Autopsy Findings.—The pleural cavities contain no free fluid; the visceral pleura of the right inferior lobe is somewhat opaque. The left lung weighs 21 gm., the right 51 gm. The right inferior lobe is greatly enlarged, also the adjoining middle lobe of this lung. Both are very much redder than normal and in places suggest infarcted lung tissue. The consistence is firm, non-crepitant, except for marginal areas. On cross-section the surface is homogeneous in appearance, and moist. A small amount of seropurulent, slightly sanguinous exudate can be expressed from the bronchi. This is not viscid in character. The scraped surface has a slightly granular appearance.

Cultures taken from the heart's blood and from the lung exudate show pure growth of characteristic capsulated bacilli.

Microscopical Findings.—Sections taken from various portions of the consolidated lung were fixed in Zenker's fluid and formalin and stained with methylene blue and eosin and by Gram's method. All the alveoli are completely filled with polymorphonuclear cells. Scattered through the exudate are rare vacuolated, large mononuclear cells, few capsulated bacilli, a small amount of fibrin, and few red blood cells. The capillaries of septal walls are slightly dilated. The lumina of the bronchi contain granular coagulum and polymorphonuclear leucocytes.

Experiment 6.—Large, well nourished, adult cat. The technique employed was the same as in the last two experiments. The animal 24 hours after insufflation is inactive, refuses food, and seems somewhat dyspneic. 48 hours after injection dyspnea is more marked; found in semicomatose condition. Killed at this time.

Autopsy Findings.—The left lung weighs 12 gm., the right 42 gm. The pleural cavities contain no free fluid. The visceral pleura of the right inferior lobe is somewhat opaque. This lobe is greatly enlarged. The entire lobe is apparently completely consolidated and is moderately firm in consistence. On cross-section the surface presents an almost homogeneous appearance. It is pink to light red in color. A small amount of mucopurulent exudate can be scraped off with the edge of the knife and expressed from the bronchi. Cultures from the heart's blood and the lung exudate show a pure growth of capsulated bacilli.

Microscopical Findings.—Sections were taken from the consolidated lobe, fixed, and stained as in the previous experiments. They show alveolar spaces uniformly and completely filled with exudate. This everywhere has the same characteristics, and consists chiefly of polymorphonuclear cells and few large phagocytic mononuclear cells, often containing capsulated Gram-negative bacilli. There are a few red blood cells and little fibrin present. The bronchi often contain a similar exudate.

The second method employed to reproduce the Friedländer type of pneumonia was the direct injection of bouillon cultures into the femoral vein. Four animals received 5 cc. of a culture similar to that used in the preceding series. These showed evidence of slight toxemia within twenty-four hours after the injection. The clinical picture was quite different from that in the preceding experiments and it is believed that all the animals would have survived. They were killed from one to five days after the injection. The autopsies failed to show any pulmonary lesions.

A third series of ten animals were treated as follows: The ether anesthesia was followed by an intrabronchial insufflation of strong ammonia fumes for a period of thirty seconds to one minute. In this connection it should be stated that the injury to the lung parenchyma produced by this procedure was not extensive. In the majority of animals no lesions or only slight hemorrhagic foci were found at autopsy after the insufflation of ammonia fumes. The fumes were forced in through a catheter by blowing through a series of bottles arranged with two-hole stoppers. In most instances the fumes elicited definite coughing. Immediately after the insufflation of the ammonia, 5 cc. of a 12 hour bouillon culture of *Bacillus mucosus capsulatus* were injected into the femoral vein. 30 per cent of these cats developed pulmonary lesions that had the characteristics of lobar pneumonia. In each of these cases the animals developed symptoms quite similar to those in the first series that showed lobar pneumonic processes. The extreme prostration was striking. The animals in this last series that developed no pulmonary lesions acted like those injected intravenously without intrabronchial insufflation. The autopsy findings were also similar.

Experiment 30 will serve to illustrate the course of the disease and the pathological findings in a cat developing a lobar pneumonia 56 hours after the intravenous injection of the organism and intrabronchial insufflation of ammonia fumes.

Experiment 26.—Large, well nourished, and well developed cat. 24 hours after injection the animal is markedly prostrated, refuses food, and is slightly dyspneic. After 48 hours it is in a semicomatose condition with marked dyspnea. It attempts to hold up its head as if to facilitate respiration. It was killed at this time.

Autopsy Findings.—The pleural cavities contain no free fluid. The visceral pleura of the right inferior lobe is distinctly opaque and is covered over limited areas by a thin deposit of fibrin. There is extreme enlargement of this lobe; it is firm, non-crepitant, and apparently completely consolidated. The right lung together with the heart was preserved in Kaiserling's solution as a permanent specimen (Fig. 1). This weighs 110 gm. (normal weight 30 to 35 gm.). The inferior lobe is dark red in color and shows poor definition of lobules with absence of alveolar markings except along the margins. The cut surface of the consolidated lobe (Kaiserling preparation) presents a non-homogeneous appearance, but apparently is diffusely infiltrated (Fig. 2). Large hemorrhagic, poorly defined areas in the central portion are everywhere surrounded by gray, slightly hemorrhagic tissue. The process does not suggest a lobular distribution.

Culture taken from the heart's blood shows no growth after 48 hours.

Microscopical Examination.—(Kaiserling material.) Although the sections are not entirely satisfactory, the alveolar spaces are found to be filled with an exudate made up of polymorphonuclear cells, many capsulated bacilli, considerable coarse fibrin, many red blood cells, and a few large vacuolated mononuclear phagocytic cells. The amount of exudate varies somewhat, however; nowhere are air-containing alveoli seen. The bronchi often contain an exudate similar to that in the alveoli. The bronchial epithelium is everywhere intact.

TABLE I.

Series.	Experiment No.	Insufflation.	Intravenous injection.	Pneumonia, lobar.	Blood culture.	Duration.
						hrs.
I	1	B. mucosus capsulatus		0	0	
	2	"		0	0	
	3	"		+	+	24
	4			0	Streptococci	
	5			+	+	28
	6			+	+	48
	7			+	+	27
	8			+	+	36
	9			0	0	
	10			0	Streptococci	
	11			0		
	12			0	"	
	13			0	"	
II	14		B. mucosus capsulatus	0	0	
	15		"	0	0	
	16		"	0	0	
	17			0	0	
III	18	Ammonia fumes		+	+	36
	19	"		0	0	
	20	"		0	0	
	21		"	+	+	72
	22			0		
	23			0	0	
	24			0	+after 30 hrs.	
	25			0	0	
	26			++	0	56
	27		"	0	0	
IV	28		Pneumococci (Group 1)	0	0	
	29		"	+	0	72
	30		"	0		
	31		Streptococci	0	0	
	32		"	0	0	
	33		"	0	0	
	34		"	0	+	
	35		Pneumococci (Group 4)	0	0	
	36		"	0	0	
	37	"	"	0	0	
	38	"	"	0	0	

In a fourth series of experiments, various strains of streptococci and pneumococci were injected intravenously after intrabronchial insufflation of ammonia fumes, as in the previous series. The organisms were obtained either from lung punctures or from the lung tissue of patients dying of lobar pneumonia. Eleven cats were each given 5 cc. of a 24 hour bouillon or hydrocele fluid culture of the various organisms. But one of the animals developed pulmonary lesions. In Experiment 29 after the injection of a culture of pneumococci (Group 1) a very early lobar pneumonia was found.

SUMMARY.

The foregoing experiments show that in cats a definite lobar pneumonia may be caused by *Bacillus mucosus capsulatus.* Judging both from the clinical course and from the pathological findings, this form of pulmonary infection differs from the usual pneumococcus types of pneumonia and closely resembles the so called Friedländer's bacillus or *Bacillus pneumoniæ* in man. In all instances in which a lobar pneumonia was found after the injection of the bacillus, a similar organism was recovered from the lung, and in no case was this associated with other organisms. The course of the disease in cats is very short, the animals developing early symptoms of profound toxemia. In 87 per cent of the animals showing a lobar pneumonia positive blood cultures were obtained. The pathological findings, judging from the early stages of the disease, are subject to considerable variation. In some instances the process may suggest a pseudolobar or confluent lobular distribution. In these cases the lung has a mottled, marble-like appearance. In the majority of cases, however, the process gave a more homogeneous appearance, suggesting a diffuse and uniform distribution. Foci of hemorrhage were not uncommon in both. Such areas cause the mottled appearance sometimes found. In all instances the consolidated lung presents a greater infiltration of tissue than is usually seen in other types of experimental pneumonia. Although the exudate as seen on the cut surface may be abundant and especially viscid in character, this is not present in most cases. The cut surface of the consolidated lung does not present a granular appearance. The histological findings are also subject to considerable variation. In most instances the infundibular and alveolar spaces are completely filled with an exudate made up chiefly of polymorphonuclear cells. Associated with

these are the capsulated bacilli, large vacuolated mononuclear phagocytic cells, and red blood cells, and occasionally small amounts of
fibrin. The organisms may vary greatly in numbers. Some sections
show spaces almost completely filled with bacilli. The contrast between spaces containing an exudate consisting chiefly of polymorphonuclear forms and an adjoining one filled with organisms is often
striking (Fig. 4). The bacilli found are both intra- and extracellular.

The large vacuolated cells are numerous in this type of pneumonia.
They apparently are the first cells to become phagocytic. Often
they are seen to contain as many as 10 to 15 capsulated bacilli, while
polymorphonuclear cells in the same exudate contain no organisms.
The histogenesis of these cells seems to be somewhat clearer from
the study of these early stages of pneumonia. In many instances
one sees swollen, partially desquamated epithelial cells along the
alveolar wall. These closely resemble the large vacuolated forms.
Various types of these vacuolated mononuclear cells were observed.
These may well represent stages of development from the desquamating epithelial cell to the large vacuolated form. Although similar cells may arise elsewhere, we have been led to regard them in
our studies as epithelial in origin (Fig. 5).

The number of red blood cells and the amount of fibrin present
in the exudate vary greatly. Small foci consisting of alveolar
spaces filled with erythrocytes are not uncommon. The fibrin is very
much less abundant than in most types of pneumonia.

From the above experiments it is seen that a lobar pneumonia in
cats can be produced at least by two methods, either by intrabronchial
insufflation of the organism or by direct injection into the veins,
provided that in the latter case an irritant is introduced into the
lungs. In each case there is little doubt but that a local injury of
the lung parenchyma was produced. Without this injury (that is,
by intravenous injection of the organism alone), no pulmonary lesion
was obtained.

Further studies with both these methods must be undertaken to
ascertain more exactly the sequence of the pathological process. It
seems probable that they are identical in each case.

The results obtained from the second method employed to produce a lobar pneumonia offer suggestive evidence in support of a hematogenous causation of this disease in at least certain instances. It is not proposed to discuss the aerogenous versus hematogenous theories at this time. Kidd[5] has recently reviewed the subject and states that the aerogenous theory for the causation of pneumonia is most widely held. This view has gained credence especially since the work of Meltzer and Lamar. In spite of this, Kidd emphasizes the fact that based on our knowledge of pulmonary infections in man and upon theoretical grounds and upon certain experimental facts, the hematogenous theory seems more plausible.

No definite conclusions can be drawn from the last series of experiments. From this limited study it seems probable that lobar pneumonic processes are produced less easily after intravenous injection of various cocci and insufflation of irritating substances than by similar treatment with Friedländer's bacillus.

CONCLUSIONS.

1. A typical lobar pneumonia can be produced in cats either by intrabronchial insufflation of *Bacillus mucosus capsulatus* (Friedländer's bacillus) or by its intravenous injection, provided that a local injury of the lungs coexists.

2. Pathologically, this form of experimental pneumonia. although subject to considerable variation, has distinctive characteristics.

The authors wish to take this opportunity to thank Dr. W. T. Councilman and Dr. H. A. Christian for valuable suggestions in connection with this work; and Mr. H. S. Aiken for the accompanying illustrations.

EXPLANATION OF PLATES.

PLATE 79.

FIG. 1.   Photograph of right lung (Experiment 26), lateroposterior view, showing markedly enlarged consolidated inferior lobe (about actual size).

FIG. 2.   Cross-section of lung and heart (Experiment 26), as seen in Fig. 1, showing complete consolidation of inferior right lobe (about actual size).

PLATE 80.

FIG. 3.   Section from consolidated lung (Experiment 3), showing the character of the exudate (low power). A. capsulated bacilli; B. polymorphonuclear cells; C, fibrin; D, dilated capillary of alveolar wall.

[5] Kidd, P., *Lancet*, 1912, i, 1590.

756   *Experimental Pneumonia.*

Fig. 4. Section from consolidated lung (Experiment 3), showing the difference in the character of the exudate in adjoining alveoli (high dry). A, numerous capsulated bacilli; B, polymorphonuclear forms.

PLATE 81.

Fig. 5. Section from consolidated lung (schematic reproduction) (Experiment 8), showing vacuolated type of cells (oil immersion). A, epithelial cell of alveolar wall; B, early forms of desquamated epithelial cells, showing vacuoles; C, typical large vacuolated cells containing capsulated bacilli and pigment; D, polymorphonuclear cells containing capsulated bacilli.

FIG. 1.

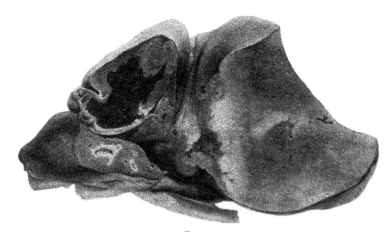

FIG. 2.

(Sisson and Walker: Experimental Pneumonia.)

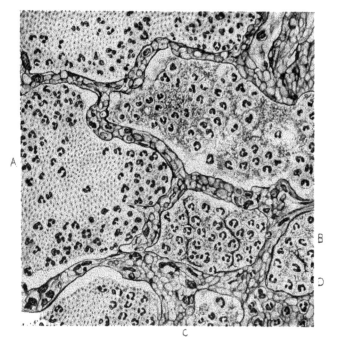

A

B

D

C

FIG. 3.

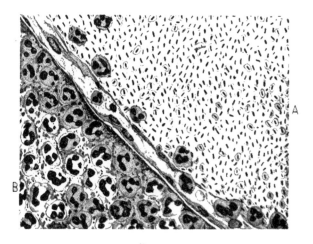

A

B

FIG.

(Sisson and Walker: Experimental Pye...

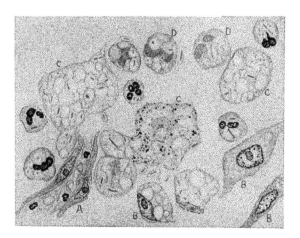

FIG. 5.

(Sisson and Walker: Experimental Pneumonia.)

AN EXPERIMENTAL STUDY OF BLOOD GLYCOLYSIS. THE EFFECTS OF THYROID AND ADRENAL EXTRACTS AND PHLORHIZIN ON GLY- COLYSIS IN VITRO.

By GEORGE M. MACKENZIE, M.D.

(*From the Department of Pathology of the College of Physicians and Surgeons, Columbia University, New York.*)

(Received for publication, June 17, 1915.)

Efforts to localize and measure glycolysis have been very un- satisfactory, especially since the theory of Cohnheim[1] concerning glucose destruction by the combined action of extracts of muscle and pancreas has been shattered by Levene and Meyer.[2] Levene and Meyer showed that the sugar was not really oxidized in such mix- tures, but converted into other forms which fail to reduce copper. Hydrolysis of the resulting fluid restores the glucose with its original power of reduction. Nevertheless, Levene and Meyer admitted that glycolysis could be brought about by leucocytes, and many others have shown that if blood be incubated for several hours the amount of glucose remaining is distinctly less than in the freshly shed blood. This oxidation of glucose in the blood is probably, however, of very slight physiological importance.

It is known that glycosuria appears after the injection of certain materials into the body, and it was thought desirable to determine, if it could be demonstrated by using the ordinary blood glycolysis as a test, whether these substances have any effect upon the action of a sugar-splitting enzyme. It was realized that other explanations of their action are generally accepted and that their application to a somewhat artificial process of glycolysis outside of the body might be criticized. Nevertheless, the experiments were made to learn whether, in any degree, they could act to inhibit a glycolytic process.

[1] Cohnheim, O., *Ztschr. f. physiol. Chem.*, 1904, xlii, 401 ; 1906, xlvii, 253.
[2] Levene, P. A. and Meyer, G. M., *Jour. Biol. Chem.*, 1911, ix, 97.

Study of Blood Glycolysis. Effects of Thyroid and Adrenal Extracts and Phlorhizin on Glycolysis in Vitro.

Method.

Dogs were used in all the experiments. The animals were etherized and either a carotid or femoral artery exposed. A cannula, with rubber tube attached, was inserted in the artery and the blood drawn directly into a sterile bottle containing glass beads. The blood was defibrinated and 20 cc. portions pipetted off and placed in sterile 150 cc. Erlenmeyer flasks which were plugged with sterile cotton. The amount of reducing substance in the blood drawn was determined immediately; in most of the experiments duplicate portions were used. Table III shows the difference found in the duplicate portions of the same blood. The other portions, after the substances being studied for their effect on glycolysis had been added, were placed in the incubator at 38° C. for varying periods of time and then the reducing substance determined. For comparison of the rate of glycolysis an untreated portion of the same blood was, in each case, incubated along with the treated portions.

In all the determinations the protein was precipitated with colloidal ferric hydroxide and sodium sulphate[3] and the reducing substance found by the method of Bertrand.[4]

Inasmuch as only comparative values were sought rather than an accurate determination of the normal blood sugar level of the dog, no effort was made to avoid the effects of excitement, struggling, and ether. The figures, therefore, are somewhat higher than those of Shaffer,[5] Embden, Lüthje, and Liefmann,[6] who eliminated these factors as completely as possible. Shaffer's results varied between 0.02 and 0.065 per cent, Embden, Lüthje, and Liefmann's between 0.057 and 0.088 per cent. Macleod[7] gives the figures of several observers which vary between 0.086 and 0.207 per cent. His own vary between 0.079 and 0.205 per cent.

[3] Rona, P., and Michaelis, L., *Biochem. Ztschr.*, 1908, vii, 329. Macleod, J. J. R., Diabetes, London, 1913, 28.

[4] Bertrand, G., *Bull. de la Soc. chim. de Paris*, 1906, xxxv, 1285.

[5] Shaffer, P. A., *Jour. Biol. Chem.*, 1914, xix, 297.

[6] Embden, G., Lüthje, H., and Liefmann, E., *Beitr. z. chem. Phys. u. Path.*, 1907, x, 265.

[7] Macleod, *loc. cit.; Jour. Biol. Chem.*, 1913, xv, 497.

In all the experiments sterile conditions were maintained as far as possible. Cultures were taken from the incubated blood in the first few experiments. A few of these were contaminated but most of them were sterile. To determine the effect of contamination upon the rate of disappearance of reducing substance, duplicate portions of the same blood were incubated for the same period of time, one under aseptic conditions, the other without sterile glassware and without any effort to avoid contamination. The results are shown in Table I. The loss of reducing substance is tabulated here, and in other tables that follow, as percentile glycolysis, which expresses the ratio of the amount of sugar lost during incubation to the amount originally present; *i.e.*, $\dfrac{\text{sugar lost per 100 cc.}}{\text{sugar originally present per 100 cc.}} =$ percentile glycolysis.

TABLE I.

Comparative Values for Blood Glycolysis under Sterile and Non-Sterile Conditions.

Experiment.	Glucose before incubation.	Condition of blood during incubation.	Duration of Incubation.	Glucose after incubation.	Percentile glycolysis.
	per cent		*hrs.*	*per cent*	
I	0.130	Sterile	3	0.073	43.8
		Non-sterile	3	0.076	41.5
II	0.130	Sterile	3	0.075	42.3
		Non-sterile	3	0.073	43.8
III	0.100	Sterile	3	0.074	26.0
		Non-sterile	3	0.075	25.0

From this it appears that the difference in the amount of glycolysis taking place in sterile and contaminated blood during a period of three hours' incubation is within the limits of experimental error, and, therefore, that failure to maintain sterile conditions probably has little or no effect on the amount of glycolysis occurring within a period of three hours.

In order to determine how much variation there might be in the rate of glycolysis in several portions of the same blood taken at the same time and kept under the same conditions, three 20 cc. portions of the same blood were incubated in separate flasks and the percentage of sugar then determined. The results are shown in Table II.

TABLE II.

Variations in the Amount of Glycolysis in Several Portions of the Same Blood.

Experiment.	Glucose before incubation.	Portions.	Duration of incubation.	Glucose after incubation.	Percentile glycolysis.
	per cent		*hrs.*	*per cent*	
		A	3	0.069	31.0
I	0.100	B	3	0.074	26.0
		C	3	0.073	27.0
		A	3	0.070	46.1
II	0.130	B	3	0.073	43.8
		C	3	0.075	42.3

This indicates that different portions of the same blood under identical conditions show very slight variation in the rate of glycolysis, and this factor, therefore, does not explain irregularities in the results after adrenalin, thyroid extract, and phlorhizin have been added to the blood.

Table III is given to show the amount of experimental error by the methods used. The greatest difference found was about 3 per cent.

TABLE III.

Percentage of Blood Sugar in Two Portions of the Same Blood.

	Portions.	1	2	3	4	5	6	7	8
Percentage of blood sugar	A	0.110	0.099	0.174	0.176	0.100	0.175	0.131	0.133
	B	0.112	0.096	0.172	0.175	0.101	0.173	0.129	0.135

Effect of Fresh Thyroid Extract upon Blood Glycolysis.

The results of experimental work upon the relation of the thyroid secretion to carbohydrate metabolism are somewhat conflicting. Macleod[8] after reviewing the work that has been done concludes that the thyroid glands, in so far as carbohydrate metabolism is concerned, have no direct influence whatever. The results which some observers[9] have obtained, tending to show that the thyroid furnishes a substance which lowers the assimilation limit for dextrose, he ascribes to parathyroid effect.

Clinical observations on the blood sugar, glycosuria, and dextrose assimilation limit in cases of hyperthyroidism and hypothyroidism

[8] Macleod, Diabetes, London, 1913, 104.

[9] Eppinger, H., Falta, W., and Rudinger, C., *Wien. klin. Wchnschr.*, 1908, xxi, 241; *Ztschr. f. klin. Med.*, 1908, lxvi, 1; 1909, lxvii, 380.

seem to indicate, however, that the gland does have some effect upon carbohydrate metabolism. When the thyroid is supplying the blood with abnormally large quantities of its secretion, the blood sugar rises, glycosuria often appears, and the dextrose assimilation limit is lowered. The reverse is the case in myxedema. If the thyroid secretion could be shown to contain an antiferment it might act simply by preventing the sugar-splitting ferments of the blood and tissues from oxidizing the carbohydrate. This would result in hyperglycemia, glycosuria, and lowered tolerance for glucose. The following four experiments were done to determine whether such an antiferment could be extracted from the gland.

Thyroid extract was prepared by extirpating the entire gland of a dog, and grinding it up in a sterile mortar with sterile normal salt solution or sterile Ringer solution, care being taken to exclude the parathyroids in so far as possible. The amount of extract used, whether 5 or 10 cc., represented one entire gland. The extract was added to the blood taken from another animal and the mixture incubated. Controls with the amounts of Ringer solution or saline used in the extract were incubated at the same time. For comparison one portion of untreated blood was incubated for the same period. The results are shown in Table IV.

TABLE IV.

Experiment.	Glucose before incuation.	Mixtures incubated.	Duration of incubation.	Blood sugar after incubation.	Percentile glycolysis.
	per ent		*hrs.*	*per cent*	
IX	0.098	Blood 20 cc. + 10 cc. fresh thyroid extract	3	0.060	39.7
		" " " " " " Ringer solution	3	0.054	44.8
		" " " untreated	3	0.080	18.4
X	0.173	" " " + 5 cc. fresh thyroid extract	3½	0.096	44.5
		" " " " " " Ringer solution	3½	0.106	38.7
		" " " untreated	3½	0.119	31.1
XI	0.175	" " " + 5 cc. fresh thyroid extract	2½	0.132	24.6
		" " " " " " Ringer solution	2½	0.138	21.1
		" " " untreated	2½	0.145	17.1
XII	0.152	" " " + 5 cc. fresh thyroid extract	3	0.094	38.1
		" " " " " " 0.9 per cent sodium chloride	3	0.088	42.1
		Blood 20 cc. untreated	3	0.103	32.2

From these results one cannot conclude that the thyroid secretion contains any substance which materially affects the rate of glycolysis *in vitro*. In each case there is apparently some acceleration of glycolysis both in the flask containing the extract and in the flask containing the Ringer or physiological salt solution. In Experiment IX the acceleration is well marked, in the others, slight. Possibly this is due simply to the dilution, because in the experiment in which the glycolysis was most marked the dilution was greatest. The experiments do not, of course, settle the question of whether or not the thyroid gland furnishes an antiferment which inhibits glycolysis; it is possible that the gland contains such a substance but that it is not extracted by the methods used. To the extent, however, that the conditions of glycolysis *in vitro* are similar to glycolysis in circulating blood, they furnish an argument against the hypothesis that the thyroid gland affects carbohydrate metabolism by inhibiting ferment action.

Effect of Adrenalin on Blood Glycolysis.

Blum,[10] fourteen years ago, showed that subcutaneous or intravenous administration of adrenal extract causes glycosuria. Since then this work has been abundantly confirmed, and it has been shown that adrenalin regularly causes a hyperglycemia. Experiments of Pollak[11] showed that adrenalin produces glycosuria even after the resection of the splanchnic nerve, while the work of Claude Bernard[12] and Eckhard demonstrated that, after such resection, puncture of the floor of the fourth ventricle is not followed by glycosuria. The adrenalin is, therefore, thought to act by stimulating the peripheral sympathetic nerve endings in the liver. Moreover, it causes the hyperglycemia even in animals whose livers have been rendered glycogen-free by starvation. This is thought to be due to the production of sugar from protein. Despite these explanations of the action of adrenalin in the production of hyperglycemia and glycosuria, it seemed to be worth while to determine whether or not the adrenal secretions contain a substance acting as an antiferment. With this possibility in mind the following experiments were done.

[10] Blum, F., *Deutsch. Arch. f. klin. Med.*, 1901, lxxi, 146.
[11] Pollak, L., *Arch. f. exper. Path. u. Pharmakol.*, 1909, lxi, 149.
[12] Bernard, C., Leçons sur la physiologie et la pathologie du système nerveux, Paris, 1858.

Parke, Davis and Co. adrenalin was used. Quantities varying from 1 to 5 cc. were added to 20 cc. portions of blood and the mixtures incubated for varying periods. The sugar content of the blood was determined immediately after drawing the blood, and in the incubated portions immediately after the expiration of the period of incubation. A portion of untreated blood was also incubated for comparison. The results are shown in Table V.

TABLE V.

Experiment No.	Adrenalin 1:1,000.	Glucose before incubation.	Duration of incubation at 38° C.	Glucose after incubation.	Percentile glycolysis.	Glucose in untreated blood after incubation.	Percentile glycolysis in untreated blood.
	cc.	per cent	hrs.	per cent		per cent	
IX	5	0.098	3	0.039	60.2	0.080	18.3
X	5	0.173	3⅓	0.102	41.0	0.119	31.1
XI	5	0.175	2½	0.150	14.2	0.145	17.1
XIII	5	0.152	3	0.064	57.8	0.103	32.2
XIV	1	0.100	2½	0.049	51.0	0.037	63.0
XIV	2	0.100	2½	0.043	57.0	0.037	63.0
XIV	3	0.100	2½	0.044	56.0	0.037	63.0
XV	2	0.174	5	0.124	28.8	0.120	31.0

In Experiments IX and XIII the blood with added adrenalin showed a definite acceleration of glycolysis as compared with the untreated blood. In Experiment IX about three times as much reducing substance disappeared from the adrenalin blood as from the untreated blood, in Experiment XIII a little less than twice as much. In the other experiments the difference in the percentile glycolysis between the treated and the untreated blood was so slight as to be almost within the limits of experimental error.

The results are so irregular that one is not justified in concluding that adrenalin contains any substance which either accelerates or inhibits blood glycolysis. The explanation of the apparent acceleration in two experiments is not clear.

Effect of Phlorhizin on Blood Glycolysis.

It is believed that the glycosuria after administration of phlorhizin[13] results from the power of phlorhizin to render the kidney more permeable to sugar in the blood, thus lowering the threshold for sugar excretion. There is usually a hypoglycemia. Beyond the fact that phlorhizin renders the kidney more permeable to

[13] For a full discussion and complete bibliography of phlorhizin, see Lusk, G., *Ergebn. d. Physiol.*, 1912, xii, 315.

glucose, little is known of its action. The possibility of its acting as an antiferment suggested itself.

Phlorhizin in varying amounts was dissolved in sodium bicarbonate solution, warm Ringer solution, and blood serum, and then added to blood which was incubated for varying periods. Dry phlorhizin also was added to blood in one experiment. The amount of reducing substance lost during the incubation of these mixtures was compared with the loss of reducing substance from an untreated portion of the same blood taken at the same time and incubated for the same period. The results are shown in Table VI.

TABLE VI.

Experiment No.	Phlorhizin.	Solvent.	Glucose before incubation.	Duration of incubation at 38° C.	Glucose after incubation.	Percentile glycolysis.	Glucose in untreated blood after incubation.	Percentile glycolysis in untreated blood.
	mg.		*per cent*	*hrs.*	*per cent*		*per cent*	
III	20	1 cc. 2 per cent sodium bicarbonate	0.102	2	0.077	24.5	0.083	18.6
IV	100	5 cc. 2 per cent sodium bicarbonate	0.152	3	0.103	32.2	0.93	38.8
V	100	5 cc. 2 per cent sodium bicarbonate	0.151	3	0.107	29.1	0.102	32.4
VIII	100	5 cc. 2 per cent sodium bicarbonate	0.144	2½	0.095	34.2	0.126	12.5
IX	50	5 cc. Ringer solution	0.098	3	0.162	39.6	0.80	18.3
X	50	5 " " "	0.174	2½	0.132	24.7	0.119	31.2
XIII*	60	10 cc. serum	0.152	3	0.185	26.5	0.103	32.2
XV	100	Dry	0.174	5	0.142	18.4	0.120	31.0
XV	50	5 cc. Ringer solution	0.174	5	0.109	37.0	0.120	31.0

* In this experiment deduction was made for the amount of sugar contained in the 10 cc. of serum used to dissolve the phlorhizin. The phlorhizin was partly in solution and partly in suspension.

In six of the nine experiments the difference in the percentile glycolysis between the phlorhizin and the untreated blood was not more than 7 per cent. This difference is too slight to justify an inference that the phlorhizin affected the rate of glycolysis. In Experiment VIII there was apparently an acceleration of the glycolytic process; in Experiment XV apparently a retardation; in Experiment IX instead of a loss during the period of incubation there was an increase of 39 per cent in the amount of reducing substance. I cannot give any satisfactory explanation for this. From these results one cannot conclude that phlorhizin has any effect upon the rate of glycolysis.

CONCLUSIONS.

1. Blood glycolysis *in vitro* during a period of three hours' incubation proceeds at practically the same rate under sterile conditions and when no effort is made to prevent contamination.

2. Fresh thyroid extract, adrenalin, and phlorhizin do not contain any substances which have a constant effect upon the rate of blood glycolysis outside of the body. No evidence of the presence of an antiferment was found.

It is a great pleasure to express my indebtedness to Professor W. G. MacCallum, under whose direction this work was done.

DIET AND TUMOR GROWTH.

By WILLIAM H. WOGLOM, M.D.

(From Columbia University, George Crocker Special Research Fund, New York.)

(Received for publication, July 7, 1915.)

One of the most important among many papers dealing with the relations between food and tumor growth was recently published by Van Alstyne and Beebe,[1] who found that a transplantable neoplasm known as the Buffalo rat sarcoma grew better when the hosts were fed on a diet containing carbohydrates (especially lactose) than when they were kept on a non-carbohydrate regimen, provided that these substances were administered continuously for several weeks before inoculation. The enhanced suitability of the lactose-fed animals was sometimes shown by both increased inoculation percentage and more vigorous proliferation of the tumors, at other times by rapid growth alone, the percentage in the latter case being similar to that obtaining for the controls. If these findings were applicable to all propagable neoplasms, the authors point out, it might be possible to rescue a poorly growing tumor, or to obtain large amounts of tissue for experimental purposes; it is evident, also, that the labor and expense incidental to adapting spontaneous new growths for transplantation could be materially curtailed by the simple expedient of keeping on hand a number of lactose-fed animals, a possibility which demanded the fullest investigation.

Unfortunately, the various carcinomata employed in the following experiments do not appear to be stimulated by the administration of lactose, although this was given in some instances throughout a period much longer than that regarded by Van Alstyne and Beebe as sufficient; hence, transplantable neoplasms must vary in regard to their sensitiveness toward qualitative differences in diet, as Rous[2]

---

[1] Van Alstyne, E. V. N., and Beebe, S. P., *Jour. Med. Research*, 1913-14, N. S., xxiv, 217.

[2] Rous, P., *Proc. N. Y. Path. Soc.*, 1914, N. S., xiv, 126; *Jour. Exper. Med.*, 1914, xx, 433.

has shown they differ in respect of their reaction toward quantitative differences, or else the findings of these authors must be referred to the intervention of some uncontrolled factor.

Two objections may be urged against their experiments. These are, first, that their rats may not have been of the same breed, and consequently may not have been equally susceptible to the tumor; and, secondly, that the Buffalo rat sarcoma is one of the most erratic of all the transplantable neoplasms now under investigation. Thus, as Wood and McLean[3] have recently pointed out in a communication from this laboratory, it frequently undergoes spontaneous ab-

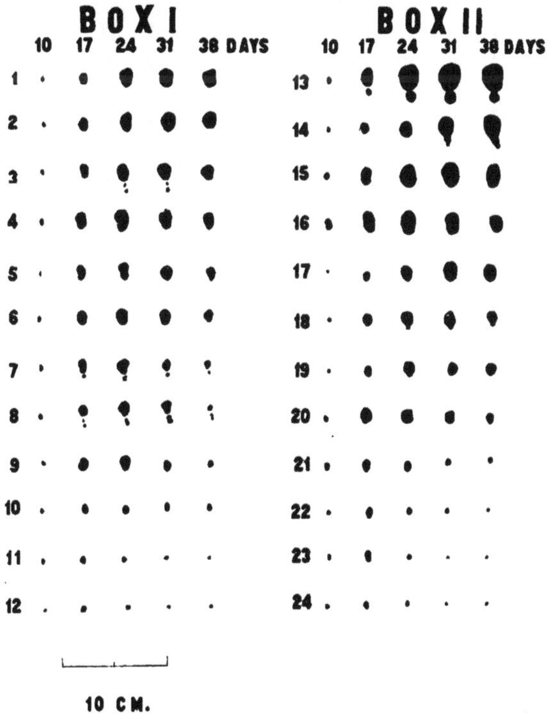

BOX I

	10	17	24	31	38 DAYS
1					
2					
3					
4					
5					
6					
7					
8					
9					
10					
11					
12					

BOX II

	10	17	24	31	38 DAYS
13					
14					
15					
16					
17					
18					
19					
20					
21					
22					
23					
24					

10 CM.

TEXT-FIG. I. Experiment I. Normal untreated rats bearing the Buffalo rat sarcoma and divided into two groups of twelve each immediately after inoculation.

[3] Wood, F. C., and McLean, E. H., *Proc. Soc. Exper. Biol. and Med.*, 1014-15, xii, 135.

sorption; again, its growth is irregular to the highest degree, as Text-figs. 1, 2, and 3 will demonstrate. These three figures, reproducing pages from the book in which the routine transplantations of this tumor are charted, show how variable is its growth even in normal untreated rats of the same age and breed kept under similar conditions and fed on the same diet, and, at the same time, suggest the employment of a more stable neoplasm for cancer research.

TEXT-FIG. 2. Experiment 2. Normal untreated rats bearing the Buffalo rat sarcoma and divided into two groups immediately after inoculation.

To rule out as far as possible the two variables just cited, the experiments were repeated on mice of the same breed, and with transplantable carcinomata, since these are generally somewhat more uniform than the sarcomata in their behavior. Various growths were employed, possessing this characteristic in common, that their proliferative capacity was not of the highest; any stimulation, there-fore, howsoever slight, would become immediately apparent.   Among the mouse carcinomata, Crocker Fund tumors Nos. 15, 46, and 58 have an average inoculation percentage of about 60, grow rather slowly, and sometimes undergo spontaneous absorption, while the

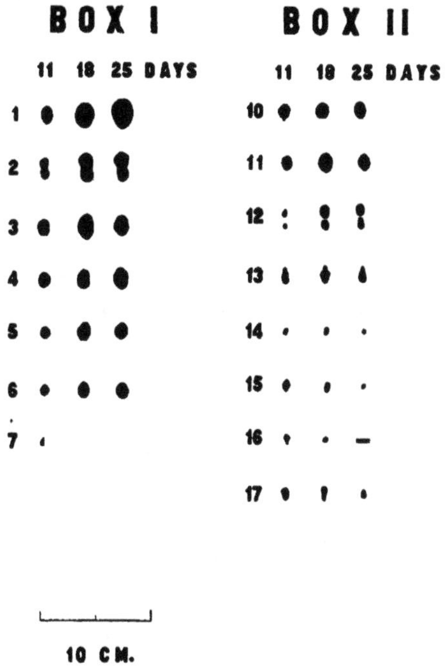

10 CM.

Text-Fig. 3.   Experiment 3.   Normal untreated rats bearing the Buffalo rat sarcoma and divided into two groups immediately after inoculation.

Ehrlich carcinoma (Ehrlich's *"Stamm 33"*) grows in from 70 to 100 per cent of engrafted animals, and does not often retrogress. Two spontaneous mammary carcinomata, Nos. 36 and 46, were also used, and, finally, the Flexner-Jobling rat adenocarcinoma, with an inoculation percentage of from 50 to 100, moderate growth energy, and a rather pronounced tendency to regress.

The lactose added in these experiments to the regular laboratory regimen of bread and vegetables was given in lard, sufficient of this

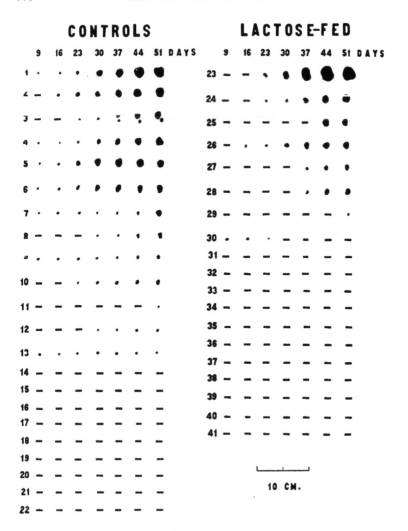

TEXT-FIG. 4. Experiment 4. $\frac{15}{8E}$. Nos. 1 to 22, normal control mice; average weight 15 gm. Nos. 23 to 41, mice fed with lactose for about six months; average weight 22 gm. Both groups inoculated in the right axilla with 0.02 gm. of tumor by the needle method.

Experiment 4.—(Text-fig. 4.) Only one mouse (No. 23) among the lactose-fed had a larger tumor than the controls, while, totally cancelling this almost negligible difference, the inoculation percentage was a little higher in the control series. If the lactose exerted any effect at all in this group, and in Experiment 6, it was merely to make older mice almost or quite as susceptible as their younger controls, a receptivity far below that described by Van Alstyne and Beebe. The tumor in No. 25 was intra-abdominal, a fact which accounts for its late discovery.

Experiment 5.—(Text-fig. 5.) Here the tumors in the control mice were larger than those in the lactose-fed, the inoculation percentage being identical in the two series.

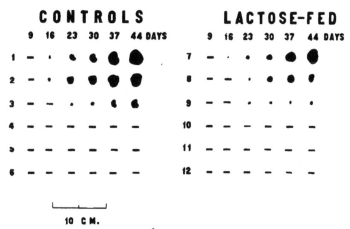

TEXT-FIG. 5. Experiment 5. $\frac{46}{9D}$. Nos. 1 to 6, normal control mice; average weight 27 gm. Nos. 7 to 12, mice fed with lactose for about six months; average weight 25 gm. Both groups inoculated in the right axilla with 0.02 gm. of tumor by the needle method.

mixture being put into the cages to have some still remaining at the next feeding, and it was continued in each case until the termination of the experiment; the controls received only bread and vegetables. Save for this difference in food, the lactose-fed animals and their controls were kept under identical conditions throughout. Ingestion of the lactose was proved, for the rats at least, by examination of the stomach contents (Experiment 12), those of the group fed with this substance reducing Fehling's solution in every case, a reaction which did not take place in any of the controls. Except in Experi-

LACTOSE-FED

CONTROLS

Experiment 6.—(Text-fig. 6.) Between the two groups there was but slight difference, either in the rate at which the tumor had grown or in the inoculation percentage.

Experiment 7.—(Text-fig. 7.) In this experiment the tumors in the lactose-fed were a little larger than those in the controls, but the inoculation percentage was about the same.

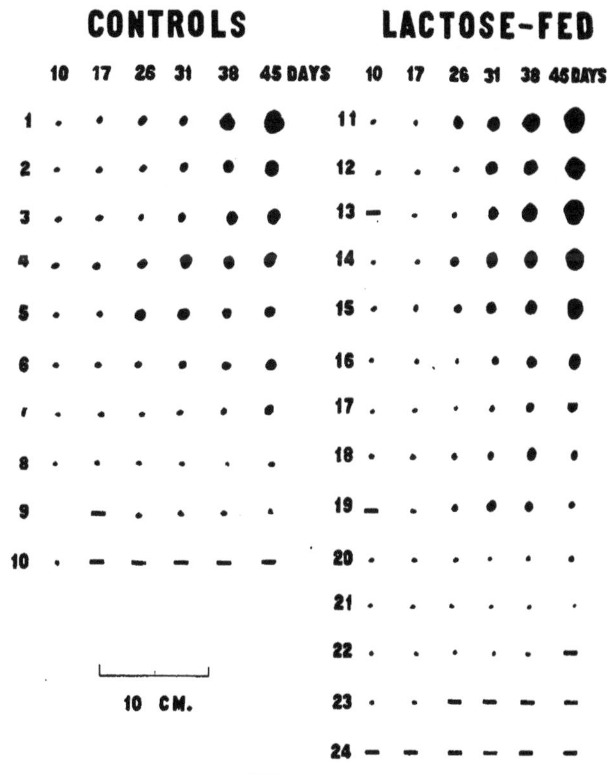

TEXT-FIG. 7. Experiment 7. $\frac{EC}{99B}$. Nos. 1 to 10, normal control mice. Nos. 11 to 24, mice fed with lactose for about two months. Both groups inoculated in the right axilla with 0.01 gm. of tumor by the needle method. On account of an oversight, the mice of this experiment were not weighed.

ments 4, 6, and 12, the weights of the animals were approximately equal. In 4 and 6, however, the controls set aside at the beginning of the experiment had to be replaced by younger animals of the same breed, as they had all died when the time came for inoculation. For

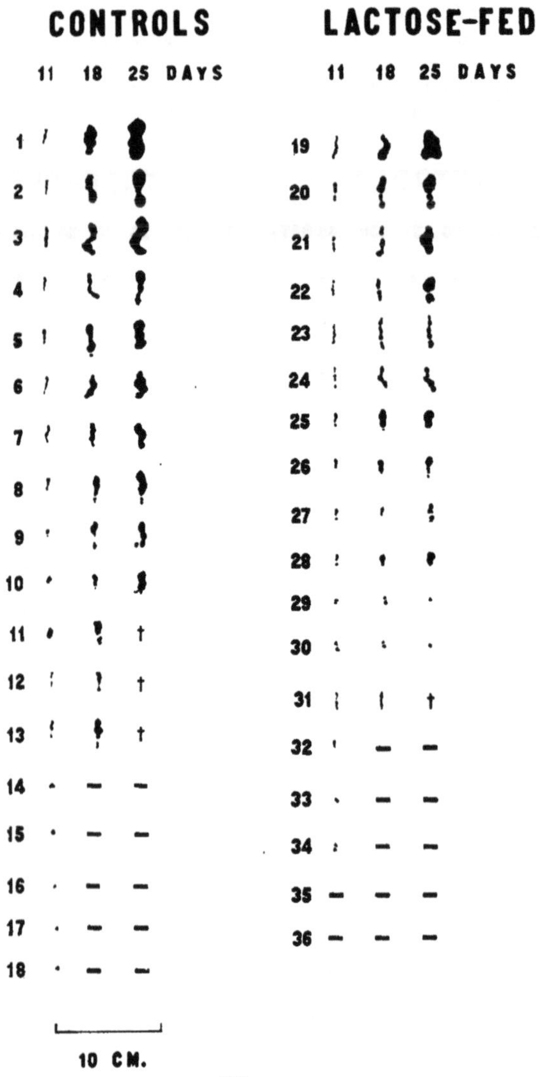

CONTROLS LACTOSE-FED

| | 11 | 18 | 25 DAYS | | | 11 | 18 | 25 DAYS |

Text-Fig. 8. Experiment 8. $\frac{EC}{98A}$. Nos. 1 to 18, normal control mice; average weight 15 gm. Nos. 19 to 36, mice fed with lactose for about one month. Average weight 17 gm. Both groups inoculated in the right axilla with 0.02 cc. of a tumor emulsion by the syringe method.

Experiment 8.—(Text-fig. 8.) The tumors were somewhat larger in the control mice, and the inoculation percentage was practically similar in the two groups.

Experiment 9.—(Text-fig. 9.) Here the tumors were a trifle larger in the lactose-fed series, though the inoculation percentages were approximately equal.

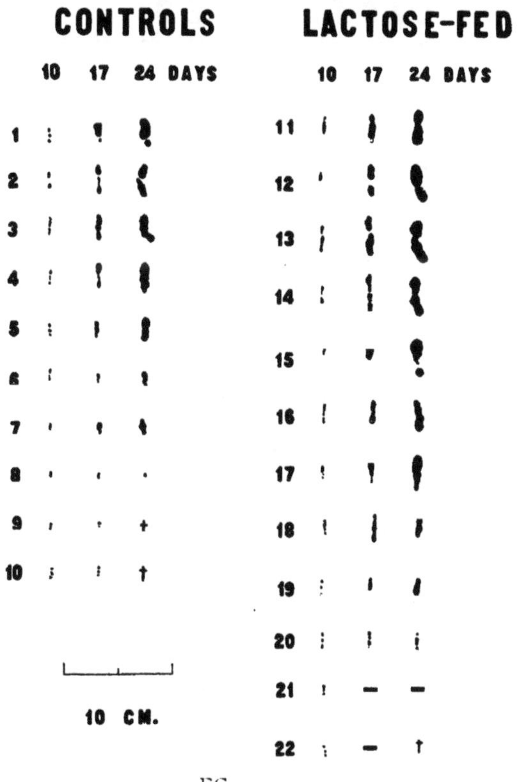

TEXT-FIG. 9. Experiment 9. $\frac{EC}{4B}$. Nos. 1 to 10, normal control mice; average weight 16 gm. Nos. 11 to 22, mice fed with lactose for about six weeks; average weight 17 gm. Both groups inoculated in the right axilla with 0.02 cc. of a tumor emulsion by the syringe method.

Experiment 12 it was not possible to get rats of exactly the same size; here, however, the lactose-fed group was the younger.

Summing up, it appears that in a total of nine experiments comprising 123 lactose-fed animals and 109 controls, the tumors were

Experiment 10.—(Text-fig. 10.) This is the only one of the entire series which would even suggest that a diet containing lactose may render an animal more suitable for tumor growth. Of seventeen lactose-fed mice which lived long enough to appear in the final reckoning, three had fairly large neoplasms, while not one of the twelve controls developed a tumor.

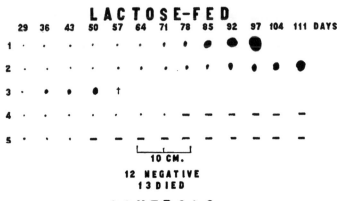

LACTOSE-FED

12 NEGATIVE
13 DIED

CONTROLS

12 NEGATIVE
17 DIED

TEXT-FIG. 10. Experiment 10. $\frac{46}{1}$. Nos. 1 to 5, mice fed with lactose for about six weeks; average weight 16 gm. The average weight of the control mice was also 16 gm. Both groups inoculated in the right axilla with 0.01 gm. of tumor by the needle method.

slightly larger or the inoculation percentage a trifle higher in the lactose-fed mice in three, in the control mice in three, and in three no difference could be discerned between the two groups. Such an even distribution makes it almost certain that the addition of lactose to the diet did not in the slightest affect the growth of the tumors employed, and intimates, though of course it does not prove, that the findings published by Van Alstyne and Beebe may have been the outcome of chance.

Other substances beside lactose have been brought recently into connection with tumor growth. Robertson and Burnett[4] have as-

[4] Robertson, T. B., and Burnett, T. C., *Proc. Soc. Exper. Biol. and Med.*, 1912–13, x, 59, 140. Burnett, T. C., *ibid.*, 1913–14, xi, 42; 1914–15, xii, 33. Robertson and Burnett, *Jour. Exper. Med.*, 1913, xvii, 344.

Experiment 11.—(Text-fig. 11.) This experiment reverses the findings of the previous experiment, since the only tumors which were found occurred in the controls. Both, however, were much smaller than the three mentioned in Experiment 10.

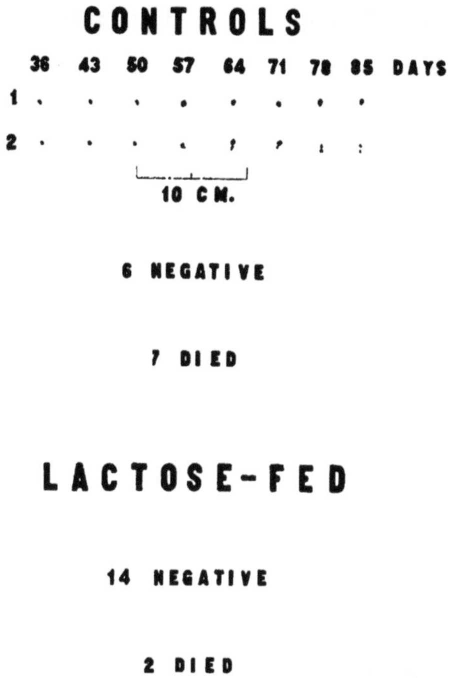

CONTROLS

36 43 50 57 64 71 78 85 DAYS

1 .   .   .   .

2 .   .   .   .   *   *   ,   :

|____.___.____|
10 CM.

6 NEGATIVE

7 DIED

LACTOSE-FED

14 NEGATIVE

2 DIED

TEXT-FIG. 11. Experiment 11. $\frac{36}{1}$. Nos. 1 and 2, normal control mice; average weight 22 gm. The lactose mice, which averaged 24 gm. in weight, had been fed with lactose for about six weeks. Both groups inoculated in the right axilla with 0.02 gm. of tumor by the needle method.

serted, for example, that cholesterol acts as a stimulus when injected into the tumor or even when introduced subcutaneously at a remote point. Now, while this may or may not be true, the authors have fallen into an error which Van Alstyne and Beebe carefully avoided, the only too common error of confusing tumor growth with tumor genesis. In other words, they have extended their findings to explain the constantly increasing tendency toward neoplasia which is associated with advancing age, and have ascribed it to an accumula-

Experiment 12.—(Text-fig. 12.) There is no difference between the two groups, either as regards the size of the tumors, or the inoculation percentage, notwithstanding the fact that the lactose-fed rats were younger than the controls.

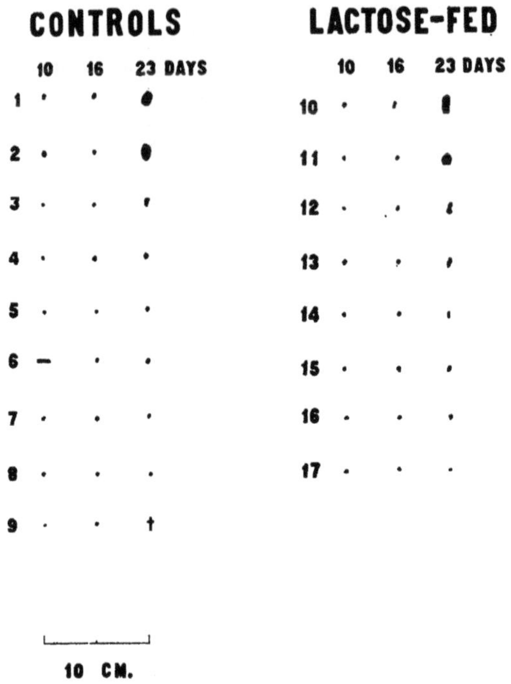

CONTROLS   **LACTOSE-FED**

	10	16	23 DAYS			10	16	23 DAYS
1	.	.	●		10	.	,	▮
2	.	.	●		11	.	.	●
3	.	.	▮		12	.	.	▮
4	.	.	.		13	.	,	▮
5	.	.	.		14	.	.	▮
6	—	.	.		15	.	.	▮
7	.	.	.		16	.	.	.
8	.	.	.		17	.	.	.
9	.	.	†					

|—————————|
10 CM.

Text-Fig. 12. Experiment 12. $\dfrac{FRC}{3E}$.   Nos. 1 to 9, normal control rats; average weight 65 gm. Nos. 10 to 17, rats fed with lactose for about six weeks; average weight 52 gm. Both groups inoculated in the right axilla with 0.02 gm. of tumor by the needle method.

tion of cholesterol within the organism. Such a broad assumption is scarcely warranted by any of the data which it has been possible so far to accumulate; indeed, all the evidence points to an opposite conception—that the conditions responsible for the origin of a tumor differ widely from those necessary for its continued growth, once it has developed. This hypothesis, accepted by both Bashford and Ehrlich, rests on the complementary observations that spontaneous tumors sometimes arise in mice known to be refractory to transplantable growths, and that grafts succeed much more readily in young than in old animals, although it is in the latter that spontaneous neoplasms originate in the great majority of instances.

SUMMARY.

Lactose does not increase the receptivity of mice and rats for the transplantable carcinomata of their species.

OBSERVATIONS ON ANTITYPHOID VACCINATION.[1]

By HENRY J. NICHOLS, M.D.,

Captain, Medical Corps, U. S. Army.

(From the Department of Pathology of the Army Medical School, Washington,
and the Laboratory Service of the Letterman General Hospital,
San Francisco.)

(Received for publication, June 24, 1915.)

In this paper, some evidence, experimental or clinical, is presented
on the following subjects: (1) The infectious power of a living,
sensitized vaccine; (2) the absence of immunity in vaccinated
rabbits to direct gall bladder infections; (3) the relative pathogenicity, virulence, and toxicity of the strain used in the Army
vaccine; (4) local vaccine reactions after typhoid fever and after
immunization; and (5) skin reactions as an index of immunity.

The Infectious Power of a Living Sensitized Vaccine.

I have recently reported (1) some work on this subject in which
it was shown that Metchnikoff and Besredka's sensitized living vaccine does not produce gall bladder lesions in the rabbit after subcutaneous and intravenous injections, and the inference was drawn
that this vaccine could safely be used subcutaneously in man. The
first transplant of the vaccine, however, produced a gall bladder
lesion in the rabbit after intravenous injection, and it was inferred
that this vaccine would produce the disease in man if accidentally
taken by mouth. Its general use, especially in the military service,
was, therefore, concluded to be dangerous.

A little further work has been done. On Oct. 28, 1914, three rabbits were
inoculated directly into the gall bladder with 1 cc. of Metchnikoff and Besredka's
vaccine. The vaccine was then more than one year old, but still gave a growth
of typhoid bacilli in pure culture. On Nov. 10 the stools of the three animals
were examined by Endo plates and Russell's double sugar media method, and
typhoid bacilli were easily recovered from each of the three animals. On autopsy

[1] Read at the Annual Meeting of the American Society of Tropical Medicine,
San Francisco, June 15, 1915.

a purulent cholecystitis was found, and pure cultures of the typhoid bacillus were cultivated from the pus in all three animals.

This experiment confirms the former one done with the first transplant of the vaccine because the bile in this instance was simply a good medium for multiplication. Apparently, therefore, this strain is infectious, unless it is exposed to both amboceptor and complement. If it is not treated with amboceptor it is infectious, as when introduced into the blood stream; if it is sensitized but not exposed to complement it is also infectious, as when introduced directly into the gall bladder. Since neither amboceptor nor complement are present in the alimentary canal, any food or water accidentally contaminated with this vaccine would be apt to spread the disease among non-immunes. From this point of view, this vaccine is simply a pure culture of pathogenic typhoid bacilli and should be treated accordingly. The toxicity of a living sensitized vaccine for rabbits and guinea pigs is considerably less than that of an unsensitized vaccine, as has been shown by Besredka (2) and by Cecil (3); but its infectious power is unimpaired except when complement is present.

Absence of Immunity in Vaccinated Rabbits to Direct Gall Bladder Infections.

The claim for the superiority of a living antityphoid vaccine rests on analogy and on experimental work done with the higher monkeys, which cannot be easily repeated on account of the difficulty of obtaining the animals. The rabbit is not a suitable animal for testing immunity following vaccination in my experience, at least, because regular infections of the gall bladder cannot be secured by intravenous injections among controls, and because direct gall bladder injections result in lesions in spite of any method of immunization. This last point is illustrated by the following experiment.

Three rabbits were immunized with Besredka's vaccine as follows:
1............Dec. 24, 1914, 0.5 cc. subcutaneously.
2............ " 31, " 1.0 " intravenously.
3............Jan. 7, 1915, 1.0 " "

On Jan. 24 the gall bladders were exposed, and after withdrawing about 0.5 cc. for culture, 0.5 cc. of the living vaccine was injected. Preliminary cultures of the bile were sterile. On Feb. 8, 12, and 22 the animals were autopsied and all showed a marked purulent cholecystitis, with pure cultures of typhoid bacilli.

Hence in this experiment no immunity could be demonstrated with a living vaccine when the test injection was given directly into the gall bladder. Similar results have been obtained by others using other forms of vaccine, and it is evident that the rabbit cannot be used for testing immunity experimentally under these conditions.

Living cultures, unsensitized but heated to 50° C. for one hour, have been used without bad results by Castellani (4). No personal work has been done with this form of vaccine, but it seems likely that this vaccine would also be infectious if accidentally taken by mouth, as bacilli are readily cultivated from it on agar. The attenuation by heat apparently makes it safe for subcutaneous injections and probably small quantities of even a virulent culture could be safely injected subcutaneously. But in view of the excellent results obtained with an absolutely non-infectious, killed, and cresolized vaccine, and in view of the evident infectiousness by mouth of a sensitized living vaccine, and in view of the added danger of accidental contamination of a living vaccine itself such as occurred in the Philippines and in India with living cholera and plague vaccines, the field of usefulness of a living antityphoid vaccine does not seem to be very wide.

The Relative Pathogenicity, Virulence, and Toxicity of the Army (Rawling's) Strain.

The most convincing clinical evidence of the protective value of antityphoid immunization has been furnished by the recent experiences of the English and the American armies. The same single strain has been used in the preparation of the vaccine in both services, and substantial protection has been secured in spite of exposures to typhoid in all parts of the world. In our service, the latest statistics (5) show no lowering of our previous remarkable records, but failure in immunization has been reported in several civil communities (6). In view of the results, some interest attaches to the Army strain and some observations have been made on its relative pathogenicity, virulence, and toxicity, as compared with nine other strains. The strain was originally obtained from the spleen of a soldier who died of typhoid in England in 1900. It was selected originally by Leishman for experimental use in preparing vaccines, not on account of its low toxicity or superior immunizing properties, but because it gave a remarkably even emulsion when washed off agar with salt solution (7).

Culturally no peculiarities have been noted as far as I am aware. *Pathogenicity.*—The strain is still pathogenic, as may be seen in the following experiments.

On Oct. 8, 1914, three rabbits were inoculated directly into the gall bladder with ½₀ of a fresh agar slant growth; a month later, autopsy showed a definite cholecystitis with pure cultures of the typhoid bacillus in two out of three animals. The strain is therefore still definitely pathogenic.

Virulence.—As gall bladder inoculations with many other strains give 100 per cent of infections, the result just stated suggests that the strain is somewhat avirulent. Further evidence on this point is brought out by intravenous injections in the rabbit. Six animals were inoculated intravenously with one-half the fatal dose of living bacilli, and after a month no gall bladder lesions were found in any of them. With more invasive strains a certain percentage of infection is usually secured in this way, about 30 per cent in my experience.

TABLE I.

Strain No.	Age.	1/10	1/7	1/5	1/4	1/2	1
1	1 week	—	×*	×	×		
2	1 month						
3	" "	—	×	×	×		
4	2 months	—	—	×	×		
5	7 "	—	—	×			
6	8 years	—	—	×	×		
7	" "	—	—	—	—		
8	?	—	—	—	×		
9	14 years	—	—	—	—	—	—
Army	" "	—	—	—	—	—	—

* In all the tables × indicates death in 24 hours.

Guinea pigs are not quite as suitable as rabbits for the determination of virulence, because it is not so easy to verify the multiplication of the bacilli, but guinea pigs have usually been used in this work, and the results here also point to the low virulence of this strain. In Table I the comparative results are given of intraperitoneal injections in guinea pigs of the Army strain and nine other strains. The figures indicate fractions of a standard agar slant growth.

It is seen that the more recently isolated strains are more virulent in general, but some of the older strains are still virulent. The Army strain fails to kill after injection of a whole culture. In one series the growth from two and one-half agar slants was required to kill a guinea pig.

Toxicity.—Virulence is often spoken of as identical with toxicity in work of this sort, but in my experience real virulence or invasiveness seems to have no definite relation to toxicity. The strains which, when living, are more virulent for rabbits and guinea pigs have, when killed, no more toxic effects on rabbits and guinea pigs, than the strain in question. No consistent toxic action can be obtained with the more virulent strains in smaller quantities than with the Army strain. This subject is of some clinical importance, as local and general reactions seem to depend directly on toxicity rather than on virulence.

The toxicity of a vaccine can be experimentally determined by the results of intravenous injections in rabbits. For example, the effect of a colon vaccine can be cited. As is well known clinically, a colon vaccine gives severe local and general reactions unless administered in small quantities, such as tens of millons rather than in hundreds of millions, as in the case of typhoid vaccines. And a colon vaccine will kill a rabbit in a much smaller dose than is required in the case of a typhoid vaccine. A number of observations have been made on the toxicity of the Army vaccine for rabbits, after intravenous injection. The results obtained by using different lots kept for different times are shown in Table II. The doses are in cc. per kilo of body weight.

TABLE II.

Period kept.	3 cc.	4 cc.	5 cc.	Period kept.	3 cc.	4 cc.	5 cc.
1 month	X(X)*	X		8 months	(X)—	—	
2 months	(X)			9 "			
3 "	—	X		10 "	—	—	—
5 "	(X)			11 "			
6 "	(X)—	—	—	12 "	—	—	—
7 "			—				

* In the tables (X) indicates death later from cachexia.

A certain amount of allowance must, of course, be made for differences in individual rabbits, but this result shows that the vaccine is distinctly toxic, especially during the first three months.

As was said before, more virulent strains tested in the same way have not proved any more toxic. The results of the use on rabbits of fresh vaccines made from three most virulent strains are as follows (Table III).

TABLE III.

Strain No.	3 cc.	4 cc.	5 cc.
1	—	—	—
2	—	—	×
3	×	—	—

As can be seen, no consistent increase in toxic action of more virulent strains can be demonstrated.

Period of Maximum Toxicity.—One lot of the regular vaccine was tested every month for six months; the vaccine was kept on ice except during shipment across the country. The results are shown in Table IV.

TABLE IV.

Period kept.	3 cc.	4 cc.	5 cc.	Period kept.	3 cc.	4 cc.	5 cc.
1 week	—	—	—	4 months	—	—	—
1 month	—	—	—	5 "		—	—
2 months	—	—	—	6 "		—	—
3 "	(×)	×	×				

This result seems to indicate that the toxicity increases with age and is at its maximum at three months. At present the bulk of the Army vaccine is used only in the first four months after manufacture. Apparently there is a gradual disintegration of the bacilli with greatest liberation of endotoxins at three months.

Increased Toxicity by Killing Bacilli with Tricresol.—The method used in preparing the Army vaccine consists of killing at 53° to 54° C. for one hour and then adding 0.25 tricresol as a precautionary antiseptic. The heating kills nearly all the bacilli, although a few colonies usually develop on plating 1 cc. before adding tricresol. It has been proposed to kill the bacilli directly by the tricresol: this procedure increases the toxicity of the vaccine as may be seen from Table V.

TABLE V.

Method of preparation.	Dose per kilo intravenously.		
	2 CC.	3 CC.	5 CC.
1. Fresh vaccine, heated at 53°–54° C. for 1 hr.......	—	—	
2. Fresh vaccine, heated and tricresolized	—	—	
3. Fresh vaccine, tricresolized but not heated.........	×	×	
Control: tricresolized sodium chloride....	—	—	—

Both rabbits given the tricresolized unheated vaccine died in twenty-four hours, while the others were not seriously affected. When the bacilli are treated directly with tricresol, the endotoxins are apparently liberated more rapidly than when the bacilli are first killed with heat.

DISCUSSION.

From the results given above I believe that a rationale of the action of the Army vaccine can be formulated as based on its toxicity. The strain used, while avirulent, is pathogenic and distinctly toxic. In preparing the vaccine, no effort is made to reduce its toxicity, but at the same time procedures which exaggerate the toxic action are avoided. The doses given are the maximum ones which can be borne without serious disturbance by the majority of individuals. The doses are spaced at seven or ten days in order to secure a response to the gradually liberated endotoxins. Finally, the vaccine is not used after four months, when its toxicity is on the wane. I do not claim that the toxic action of the vaccine has been the leading consideration in all the steps of its development, but I do hold that the net result conforms to the theory that the efficiency of the Army vaccine depends on its toxicity.

.Various workers have attempted to devise atoxic vaccines by sensitization, fractionation, etc. From the point of view outlined above, such efforts belong to the "something for nothing" class. The most reasonable theory of the action of the vaccine seems to me to be that of Pfeiffer and Bessau (8) who hold that the toxic and immunizing fractions are identical, and that efforts to separate them are wrong in principle. Fortunately, from this point of view, it is not easy to prepare a really non-toxic antityphoid vaccine. The toxic action may be masked or slight, but it is demonstrable in most vaccines called atoxic. In the opinion of those who have been most responsible for the introduction of this method of protection, something must be paid for immunity in terms of local or general reactions (Leishman (9)).

The theory of endotoxins received a set-back during the recent split products and ferment period, but as some of this latter work has still more recently been shaken to its foundations by the work of Jobling (10), Bronfenbrenner (11), and others, the endotoxin theory seems to still be the best working hypothesis for immunization and for keeping qualities of the vaccine.

Local Vaccine Reactions after Typhoid Fever and after Artificial Immunization.

A careful study of a very large number of temperature reactions following vaccination against typhoid has been made by Major Russell (12), and the figures show a very small percentage of reactions of 100° F. or over. The temperature reaction was selected as the most objective evidence which could be accurately obtained. Some misapprehension has apparently been created by these figures on account of a failure to recognize what they really are, namely, temperature reactions, which are not necessarily an index 'of a patient's general feeling nor of the local reaction. From what has been said about the toxicity of the vaccine it is only natural to expect some local and general reaction, and in fact this price is usually paid for protection, although the temperature is not much affected.

A similar study of temperature reactions as a result of a vaccination in those who have had typhoid fever, or have been previously artificially immunized, has also been made by Major Russell, and no notable changes in the percentages have been forthcoming. On the other hand the local reactions under the skin seem to be more marked in some cases after typhoid fever and after vaccination. I have no exact statistics on this point but the consensus of opinion is that the local reactions may be more severe than on primary vaccination. In the more marked cases the arm may become intensely red and swollen from the elbow nearly to the shoulder. I have recently seen a soldier who volunteered for a course of injections whenever the subject was mentioned, and who had taken ten injections in the last five years. Through an oversight he was

tions and only fourteen, or 78 per cent, gave positive reactions, of which 60 per cent were strong. Among those immunized from one to four years ago there were no strong reactions. If the typhoidin test is an index of immunity according to these results nearly one-fourth of those vaccinated a month previously are not protected and over one-third of those vaccinated up to four years are not protected. But our experience in the Army has not borne out this view, as our protection is very much greater for the periods mentioned.

The most important reason why the test is not an index of immunity, in my opinion, is the occurrence of reactions with a paratyphoid A control (Table VII). Twenty-four cases have been tested; five gave a history of typhoid and nineteen had been vaccinated against typhoid. Sixteen, or 66 per cent, gave reactions, eight, or 33 per cent, of which were strong.

TABLE VII.

Reactions.	Previous typhoid.	Vaccinated against typhoid.	Total.	Per cent.
Para A control				
Positive...................	3	13	16	66.6
Negative..................	2	6	8	
Typhoidin				
Positive...................	2	16	18	75
Negative..................	3	3	6	

Apparently we are dealing here with group reactions. The percentage of positives with the paratyphoidin was not quite as high as that of the typhoidin reactions, neither were they as strong, but the close parallelism is striking. If this skin reaction is an index of immunity, there should be considerable immunity following antityphoid vaccination against paratyphoid infections. But one of the clear cut results of typhoid immunization is that there is no group immunity to paratyphoid infections and where these diseases are prevalent a combined vaccine has been urged by Castellani and others. Both in the English service and in our own the occurrence of paratyphoid among those immunized against typhoid has been emphasized and in the present war Dreyer (14) and others have just reported that paratyphoid is of the same frequency among those vaccinated against typhoid as among those not vaccinated.

The nature of the proposed typhoidin test is still an open question. For other clinical skin tests we have some approximate explanation. The tuberculin and luetin reactions are supposed to depend in some way on the continued presence in the body of the living virus of the diseases; the Schick test apparently depends on an antitoxin; but in the typhoidin test we have to deal, not with an antitoxin or with a living virus, but with the complex effect of the previous presence of typhoid bacilli in the body. The same questions are raised as in the case of the local subcutaneous reaction already mentioned. The reaction may well be an evidence of proteid immunity without being an evidence of true immunity to the disease.

From my experience, the following reasons may be advanced for a different interpretation of the typhoidin skin reaction:

1. The test was negative in 25 per cent of cases who have had typhoid and most of whom had also been vaccinated, while an attack of typhoid fever is supposed to give permanent immunity in over 90 per cent of cases.

2. It was negative in 36 per cent of those who have been vaccinated within four years, while experience has shown that vaccination protects in much more than 64 per cent of cases.

3. Paratyphoid A controls reacted in nearly as high a percentage (66) as typhoidin (75), while clinical experience is unanimous that there is no immunity to paratyphoid infections after typhoid immunization.

Personally it seems to me that a positive typhoidin reaction is of considerable theoretical interest and indicates that the individual reacting has had typhoid fever or possibly paratyphoid or has been injected with typhoid bacilli. A negative reaction does not have the same negative value and is not so conclusive.

SUMMARY.

1. Metchnikoff and Besredka's living sensitized vaccine produces a typhoid cholecystitis when injected directly into the gall bladder of rabbits. It is therefore infectious.

2. Rabbits cannot be successfully immunized with this vaccine against direct gall bladder infections. Accordingly, rabbits cannot be used to test immunity in this way.

3. The strain used in the Army vaccine is pathogenic, relatively avirulent, and distinctly toxic. Its efficacy is believed to depend on its toxicity.

4. Vaccinations in those who have had typhoid and revaccinations produce more severe local reactions than original vaccinations in some instances.

5. The typhoidin skin test is not believed to be an index of true immunity, but rather an indication of typhoid proteid sensitization, which is not so complete, so permanent, or so specific as true immunity.

BIBLIOGRAPHY.

1. Nichols, H. J., *Jour. Exper. Med.*, 1914, xx, 573.
2. Besredka, A., *Ann. de l'Inst. Pasteur*, 1913, xxvii, 607.
3. Cecil, R. L., *Jour. Infect. Dis.*, 1915, xvi, 26.
4. Castellani, A., *Lancet*, 1909, ii, 528.
5. Lyster, W., *Jour. Am. Med. Assn.*, 1915, lxv, 510.
6. Sawyer, W. A., *Jour. Am. Med. Assn.*, 1915, lxv, 1413.
7. Leishman, W. B., Harrison, W. S., Smallman, A. B., and Tulloch, M. G., *Jour. Hyg.*, 1905, v, 381.
8. Pfeiffer, R., and Bessau, G., *Centralbl. f. Bakteriol., 1te Abt., Orig.*, 1912, lxiv, 172.
9. Leishman, W. B., *Glasgow Med. Jour.*, 1912, lxxvii, 401.
10. Jobling, J. W., and Petersen, W., *Jour. Exper. Med.*, 1914, xx, 37. Jobling, J. W., Petersen, W., and Eggstein, A. A., *Jour. Exper. Med.*, 1915, xxi, 239.
11. Bronfenbrenner, J., *Jour. Exper. Med.*, 1915, xxi, 221.
12. Russell, F. F., *Am. Jour. Med. Sc.*, 1913, cxlvi, 803.
13. Gay, F. P., and Force, J. N., *Arch. Int. Med.*, 1914, xiii, 471. Gay, F. P., and Claypole, E. J., *Arch. Int. Med.*, 1914, xiv, 671.
14. Dreyer, G., Walker, W. A., and Gibson, A. G., *Lancet*, 1915, i, 324.

THE RELATIONSHIP OF CHRONIC PROTEIN INTOXI-
CATION IN ANIMALS TO ANAPHYLAXIS.[1]

By WARFIELD T. LONGCOPE, M.D.

(*From the Medical Clinic of the Presbyterian Hospital, Columbia University,
New York.*)

(Received for publication, June 25, 1915.)

The experiments which furnish the subject of this paper represent
a continuation of some observations made upon the effect of repeated injections of foreign protein in animals sensitized to these
specific proteins.[2]   Under these conditions areas of degeneration
with extensive inflammatory reaction are produced in the myocardium, the liver, and the kidneys.   In different species of animals the
effects vary somewhat, since the kidney is involved in the dog, cat,
rabbit, and guinea pig, the liver in the rabbit and cat,[3] and the
myocardium[4] in the rabbit and guinea pig.   A detailed description
of these changes has already been published, and it is, therefore,
necessary only to state that they consist in focal areas of degeneration, infiltrated with small round cells, which in the heart are scattered throughout the myocardium of both ventricles, in the liver
are usually situated about the portal spaces, and in the kidney are
seen most often through the midzone, whence they extend into the
cortex.   In their most advanced stages these alterations produce the
picture of an extensive myocarditis, a periportal cirrhosis of the
liver, or a wide-spread subacute nephritis with involvement, at times,
of the glomeruli.

At first it was thought that sensitization to some foreign protein, such as horse serum or egg-white, was necessary before subsequent injections of the same protein could bring about the development of such changes, but further study showed that similar, though

---

[1] Read before the Association of American Physicians, Washington, May
10, 1915.

[2] Longcope, W. T., *Jour. Exper. Med.*, 1913, xviii, 678.

[3] Longcope, *Tr. Assn. Am. Phys.*, 1913, xxviii, 407.

[4] Longcope, *Arch. Int. Med.*, 1915, xv, 1079.

less marked, alterations might occur in animals receiving but a single large dose of horse serum or egg-white. The question, therefore, arose as to whether the effects produced were the result of introducing repeatedly into the bodies of these animals small quantities of a substance which was in itself primarily toxic for the cells of the kidneys, liver, and myocardium, or whether the horse serum and egg albumen became injurious only after the animals were sensitized to these substances. In the latter case the subacute inflammatory reactions might readily be dependent upon the repeated anaphylactic shocks to which the animals were subjected.

Though a number of observers interested principally in the fact that egg-white, when introduced parenterally into animals, is eliminated as such in the urine, have studied casually the kidneys of animals soon after an injection of egg-white, very few have described definite alterations in the kidneys. Adams[5] and Maschke[6] state that the kidneys of rabbits after single injections of egg-white are normal. Sollmann and Brown[7] observed under the same conditions occasional swelling and degeneration of the tubular epithelium. Chiray[8] could not discover any changes in the kidneys of rabbits after a single intravenous, intraperitoneal, or subcutaneous injection of egg albumen, but noted after repeated injections, which were limited to a few animals, degeneration of the tubular epithelium with cellular reaction and scar formation that led to chronic nephritis.

The following experiments were undertaken with the view of determining, if possible, whether the toxic action of horse serum and egg-white was a primary property of these proteins or developed only when the animals were sensitive to such foreign proteins. The latter view would express the idea which has long been held by Vaughan.[9]

It was thought that if the protein itself were toxic, it might be possible to demonstrate this toxicity by injecting repeatedly at weekly intervals proteins always of a different type, so that the animal might receive large quantities of foreign protein but never the same protein twice, thereby avoiding anaphylactic shock.

The following experiment was, therefore, performed:

[5] Adams, B., Inaugural Dissertation, Leipzig, 1880.

[6] Hirsch, C., and Maschke, W., *Berl. klin. Wchnschr.*, 1912, xlix, 145.

[7] Sollmann, T., and Brown, E. D., *Jour. Exper. Med.*, 1902, vi, 207.

[8] Chiray, M. M., Thèse de Paris, 1906.

[9] Vaughan, V. C., Vaughan, V. C., Jr., and Vaughan, J. W., Protein Split Products in Relation to Immunity and Disease, Philadelphia and New York, 1913.

Four lots of rabbits were selected from a single stock. Lot A, consisting of 11 rabbits, was used as control. Lot B, consisting of 11 rabbits, was sensitized to horse serum or egg-white by intraperitoneal inoculation of 2 to 4 cc. of these proteins on 3 consecutive days. After an interval of several weeks they were reinoculated intravenously at weekly intervals with horse serum in amounts varying from 0.5 to 6 cc., or with egg-white in amounts varying from 0.4 to 0.8 cc. Lot C, consisting of 9 rabbits, was injected intravenously at weekly intervals with horse serum, egg-white, beef serum, sheep serum, dog serum, cat serum, human serum, pig serum, edestin, and casein. The edestin and casein were made up in 2 per cent solution in sodium carbonate. The proteins were given in doses of 1 cc. per kilo of body weight, the actual dose varying from 1 to 2.7 cc. It was so arranged that the total amount of serum, etc., which each rabbit in this group received was as great or greater than that given to the animals in Group B.

Finally, in Lot D, a series of 17 rabbits was given each a single dose of horse serum or egg-white, either intravenously or intraperitoneally, and killed at intervals varying from 3 to 29 weeks. 5 of these received 2 cc. of egg-white intraperitoneally. 4 received 2 cc. of horse serum intravenously, and 1 rabbit 1 cc. of horse serum intraperitoneally; 2 rabbits received 2 cc. intraperitoneally, and 1 rabbit received 1 cc. intraperitoneally; 1 received 3 cc. and 1 received 5 cc. of sheep serum intravenously; 1 rabbit received 6 cc. of dog serum intravenously and 1 received 6 of cat serum intravenously.

In order to eliminate any possible effect which fresh serum might have, the sera for all these experiments were heated for one-half hour at 56° C.

Table I gives the results of these experiments.

TABLE I.

Animals and tissues examined.	Lot A. Normal control rabbits. Un-inoculated.	Lot B. Rabbits sensitized to egg-white or horse serum and after 3 weeks reinjected with egg-white or horse serum.	Lot C. Rabbits receiving repeated inoculations of different proteins (horse serum, egg-white, beef, sheep, dog, cat, human, and pig serum, edestin, and casein).	Lot D. Rabbits receiving a single dose of foreign protein.
No. of rabbits	11	9	9	17
No. of intoxicating doses......	0	2–10	1–10	1
Involvement of myocardium ..	0	9	3	–
Involvement of liver..........	0	9	1	
Involvement of kidneys.......	1	9	.	12
Total No. of animals	11	9		17
Total No. positive results......	1	9		12

The uninoculated control rabbits (Lot A) all showed normal organs except one. In this one rabbit a few areas of round cell infiltration were seen in the pelvis of the kidneys.

Of the eleven sensitized rabbits in Lot B, two died before intoxicating doses of protein were administered.   They showed normal organs.   The nine sensitized rabbits receiving subsequent injections of serum all showed extensive and advanced changes in the heart muscle, liver, and kidneys.

In Lot C, only one-third of the animals developed changes in one or the other of these organs, and these were much less extensive than the lesions appearing in Lot B.

In Lot D, where the animals received but a single inoculation of egg-white or horse serum, the results corresponded quite well to those observed in Lot C.

It is, therefore, evident that though previous sensitization increases the incidence and severity of the changes, it is not necessary for their development.

During further experiments it was noted that when alterations occurred in the internal organs after a single inoculation they did not make their appearance for several days after inoculation.   Experiments were, therefore, designed to determine more accurately the time of appearance of these lesions and to determine why this should be so long delayed.   If the lesions were dependent upon the introduction of a toxic substance, one would expect, especially when the toxin was given in large amounts, that it would act much more promptly.

It is well known that the skin of an animal sensitized to a foreign protein such as horse serum will react specifically to the intracutaneous injection of that protein.   In rabbits this specific sensitiveness of the skin usually develops within two to twenty days after the primary inoculation of horse serum, and, as Knox, Moss, and Brown[10] have shown, manifests itself after the intracutaneous injection of minute quantities of serum by the appearance in twelve to thirty-six hours of a large area of reddening, swelling, and edema, which persists at the site of inoculation for twelve to twenty-four hours.   According to the view which generally prevails, the reaction is associated with the union of the antigen (horse serum) with antibodies formed against this specific protein.

This method was, therefore, utilized to determine whether or

[10] Knox, J. H. M., Moss, W. L., and Brown, G. L., *Jour. Exper. Med.,* 1910, xii, 562.

not the lesions in the internal organs following a single inoculation of horse serum were associated with the appearance of skin sensitiveness and, therefore, probably with the development of antibodies towards horse serum in the body of the animal.

It was at first•hoped that repeated intracutaneous injections of horse serum in amounts varying from 0.02 to 0.1 cc. might be made, in order that the exact time at which sensitization developed could be fixed, but in normal rabbits used as controls the repeated intracutaneous injections, at four to six day intervals, of horse serum even in such small amounts led eventually to a positive reaction, which was interpreted as a true Arthus phenomenon. In such animals, however, general sensitiveness did not develop, since the injection of large quantities of horse serum intravenously in animals giving a positive skin test did not produce symptoms of anaphylaxis. In view of this fact it was possible to use the intracutaneous test only once.

For the final experiment 15 white rabbits were injected intravenously with from 5.5 to 9.0 cc. of horse serum in divided doses during 3 consecutive days.

TABLE II.

Association between the Development of Lesions in the Heart, Liver, and Kidneys and the Appearance of Skin Sensiveness in 15 Rabbits Receiving Single Injections of Horse Serum.

Rabbit No.	Amount of serum.	Lived after injection.	Skin reaction	Heart.	Liver.	Kidney.
	cc.	*days*				
15	5.5	7	Not made	0		
9	8.5	8	" "			
17	5.5	8	+			
10	8.5	10	+			
7	8.5	12	−			
8	8.5	12	+			
11	5.5	13	Not made	"	..	
5	8.5	14	−		"	
6	8.5	14	+		.. ?	
13	5.5	17	+		"	
16	5.5	21	−			
18	5.5	22	− 20 hrs.	.+	...	.. ..
12	5.5	27	+ +	.. ..	...	"
1	8.0	30	+ +	...	...	.. ..
2	9.0	30	+	..	.. ..	...
19 (control)	0					
20 "	0					
4 "	0		+ pure Arthus	"		

At varying intervals after the first dose an intracutaneous injection of from 0 02 to 0.05 cc. of horse serum was made on one side of the shaved abdomen, while an equal amount of sheep or beef serum was injected in the skin of the opposite side as a control. The animals, except in one instance, were observed for from 36 to 48 hours and after the result of the skin test had been noted, killed, carefully autopsied, and their organs examined histologically. The result of this experiment is given in Table II.

It will be seen that from the seventh to the twenty-first day after inoculation there were occasionally observed scattered and slight changes in the liver or kidney. When such alterations were found during this period they always occurred in rabbits that gave a positive skin test. All animals that gave a positive skin test, however, did not show changes in the internal organs.

After the twenty-second day the incidence and extent of the changes in the internal organs increased greatly and the skin reaction, except in one animal, was very intense. Uninoculated controls and one animal subjected to intracutaneous inoculation alone showed no changes in the internal organs.

Table II shows very well that the lesions in the internal organs appear either at the time the skin sensitiveness has made its appearance or directly afterwards, and therefore, after the animal has, hypothetically, produced antibodies to the foreign protein. One may liken the condition to serum disease in man during which time profound and very obvious disturbances may take place. In the rabbit the signs and symptoms so characteristic of the disease in man are lacking, though in one or two of the rabbits in this series moderate subcutaneous edema and generalized enlargement of the lymph nodes was observed at autopsy. Von Pirquet and Schick[11] have noted under the same conditions in rabbits alterations in the total and differential leucocyte counts that are analogous to those that are observed in children during serum sickness. The effect of anaphylactic shock is, however, so unlike in different species of animals that it is not permissible to draw general conclusions from the results obtained in any one species.

[11] von Pirquet, C. F., and Schick, B., Die Serumkrankheit, Leipzig and Vienna, 1905.

CONCLUSIONS.

Such foreign proteins as horse serum and egg-white in the amounts employed in these experiments do not produce evidences of intoxication immediately after injection into rabbits. Single large injections do, however, produce changes in the parenchymatous organs after a period of ten to twenty-one days. These develop at the time or immediately after the animal has formed antibodies for the foreign proteins.

The mechanism of the development of the lesions in the myocardium, liver, and kidneys of rabbits is thus the same, whether a single inoculation is given or whether repeated inoculations are made in sensitized animals. By the latter method, however, much more marked and extensive changes may be produced.

THE EFFECT OF ROENTGEN RAYS ON THE RATE OF GROWTH OF SPONTANEOUS TUMORS IN MICE.

By JAMES B. MURPHY, M.D., AND JOHN J. MORTON, M.D.

(*From the Laboratories of The Rockefeller Institute for Medical Research.*)

(Received for publication, September 1, 1915.)

In a previous communication we reported the existence of a relationship between the resistant state to transplantable tumors in mice and a lymphocytic crisis in the circulating blood. We further demonstrated that by a previous destruction of the lymphoid tissue of these animals with x-ray a potentially resistant animal was rendered susceptible to cancer inoculation.[1]

In the course of some experiments on x-ray we have noted that the lymphoid elements, after extensive depletion by x-ray,[2] will soon start to regenerate actively. This process will continue, as has been noted before, to a period of overproduction of the lymphoid elements. The rapidity with which this occurs depends somewhat on the amount of original destruction and somewhat on the general condition of the animal. We have further noted that by one small dose of x-ray we could obtain in a certain proportion of animals a stimulation of the lymphoid elements, preceded by a comparatively short period in which the lymphocytes were below normal. This suggested an explanation of certain therapeutic effects of x-ray.

Our first problem was to determine whether or not x-rays in a small dose administered to an animal as a whole would produce an effect on the subsequent growth of a cancer, different from that produced by a similar dose applied directly to the cancer outside the body. For this purpose it was necessary in one set of animals to confine the x-ray effect to the animal alone, ruling out any possible action on the cancer, and, in a second set to confine the x-ray effect to the cancer, preventing an indirect effect on the animal. Spontaneous tumors of the mouse were selected for this work as a more

---

[1] Murphy, Jas. B., and Morton, J. J., *Jour. Exper. Med.*, 1915, xxii, 204.
[2] Heineke, H., *Mitt. a. d. Grenzgeb. d. Med. u. Chir.*, 1905, xiv, 21.

suitable material than the transplanted tumors, for reasons that will be explained later. The results are given here in a brief summary. A complete analysis of the size and characters of the cancers with autopsy findings will be given in a subsequent report.

Series I.—There were 52 mice with various stages and types of spontaneous cancers in this group. The tumors were removed as completely as possible by operation, and, with the cancer out, the whole animal was exposed to a stimulating dose of x-ray (Coolidge tube). Immediately afterwards a graft of the original cancer was replaced in the groin of the animal. In 26 of the 52 animals treated in this fashion, there resulted a complete immunity to the recurrence of the disease. Only those animals were included in this number that lived and remained in good physical condition for at least five weeks after the treatment. The majority lived from two to four months, some to eight months, and some are still living. There has been no evidence of a local recurrence at the site of operation, nor where the graft was implanted, or of metastasis in those that have died. Among the remaining 26 animals of the series the average time for the appearance of the graft was five weeks and four days, a figure which contrasts strongly with the figure for the control animals. The number of recurrences at the original location of the tumor was 11 among the 52 animals, all occurring in the latter 26.

Series II.—For a control series we had 29 mice with spontaneous tumors of various sorts. These were operated on in the same manner as the animals in the first series, but a graft of the cancer was returned without treatment to either the animal or the cancer. The tumors were kept outside the body for the same length of time as in the first series. In 28 of the 29 the grafts grew progressively. In one the graft grew for a period and then retrogressed to complete absorption. The average time for the grafts to become palpable was one week and five days. This is about the same figure obtained by Rous.[3] Local recurrences of the cancer occurred in 14 of the 29 animals.

Series III.—The cancers in these animals were removed in the same manner as in the first two series, but in this group the cancers were subjected to the same amount of treatment that the animals had

[3] Rous, P., *Jour. Exper. Med.*, 1914, xx, 433.

received in the first series. A graft from the cancer, after this treatment outside of the body, was returned to the groin of the original host, as in the other experiments. 10 mice with spontaneous tumors were used for this series, and in all 10 the returned grafts grew. The average time for these to become palpable was one week and three days. There was a local recurrence of the tumor in 4 of the 10 animals.

A tabulation of the figures for the three experiments is given for comparison in Table I.

TABLE I.

	Immune.	Susceptible.	Local recurrence of tumor.	Average time for appearance of graft.
	per cent	*per cent*	*per cent*	
Series I*............	50.0	50.0	21.2	5 wks., 4 days.
" II............	3.4	96.6	48.3	1 wk., 5 "
" III............	0.0	100	40.0	1 " 3 "

* Series I, animals treated by x-ray, while cancer was out. Later a graft of the tumor was returned. Series II, control animals in which cancer was removed and a graft returned without treatment to either animal or tumor. Series III, cancers removed and subjected direct to x-ray treatment and a graft returned to the original host.

It will be seen from the figures in Table I that x-ray administered directly to the cancer outside of the body is insufficient to prevent the growth of a graft returned to the original host. On the other hand, the same small dose of x-ray given to the animal with the cancer removed was sufficient to render 50 per cent of the mice so treated immune to returned untreated grafts of their own tumors and greatly retard the growth in the other 50 per cent.

The contrast between the control animals with 1 immune out of 29, and the x-rayed series with 26 immune out of 52 is striking, as is also the comparison of local recurrence in the two series. The delayed appearance of the graft in the treated series is important and this period could perhaps have been prolonged or a recurrence prevented altogether by a second exposure to x-ray after a suitable interval.

Total white counts and differentials were done on all of these mice before operation[4] or treatment and on part of them at intervals after. So far in the limited number counted systematically afterward, our treatment has given in the successful cases a definite increase in the lymphocytes. Whether or not this increase is vitally concerned in the immunity process to spontaneous tumors is a point to which at present we are unwilling to commit ourselves. The results in the light of our previous experiments are strongly suggestive of this, however.

We have demonstrated a direct effect of x-rays on the animal, which renders it more highly resistant to replants of its own cancer than would normally be the case. There is also an absence of any demonstrable effect of this small dose of x-ray when administered to the cancer direct. Grafts of such tumors when returned to their original host grow as well as do the controls.[5]

---

[4] Counts made on over 100 untreated mice with spontaneous tumors have failed to show an abnormally low lymphocytic content in the circulating blood. This result is contrary to that obtained by Baeslack (Baeslack, F. W., *Ztschr. f. Immunitätsforsch., Orig.*, 1913–14, xx, 421) in counts on two mice with spontaneous tumors. We were also unable to confirm his reported decline in the numbers of lymphocytes in animals with growing transplantable cancers (Murphy and Morton, *loc. cit.*). He also gives differential counts on four mice with natural immunity to transplanted tumors, but as no total white cell counts are given it is impossible to tell whether his percentage variations are due to fluctuations in the polymorphonuclear cell or in the lymphocyte.

[5] We have avoided in this communication any discussion of the massive and contradictory literature on direct x-ray effects. We are unaware of any experiments that bring out the above points.

A FURTHER STUDY ON THE BIOLOGIC CLASSIFI-CATION OF PNEUMOCOCCI.

By OSWALD T. AVERY, M.D.

(*From the Hospital of The Rockefeller Institute for Medical Research.*)

(Received for publication, September 17, 1915.)

The biologic classification of pneumococci, according to Dochez and Gillespie,[1] divides these organisms into four main groups. These groups represent well defined types which are distinct and readily differentiated by immunological reactions. In previous communications[2] the varieties of pneumococci, their relation to disease, and their significance in problems of specific therapy have been discussed in detail. The present work is confined to a study of a limited number of strains of pneumococci which, because of certain serological reactions, are closely allied to the second group, and appear to represent distinct subgroups of Pneumococcus Type II.

In classifying pneumococci by serological methods, agglutination and protection experiments have been employed. Agglutination of typical pneumococci of the second group in an immune serum of the homologous type is prompt and characteristic. There is occasionally encountered, however, an organism with which the agglutination reaction in Antipneumococcus Serum II is incomplete and less prompt, often being delayed several hours. The occurrence of strains showing these peculiarities has been frequent enough to direct attention to them, since in the presence of a positive, but atypical agglutination, some confusion may arise in diagnosis. Theoretically, too, it is of interest to determine, if possible, whether or not these variations in agglutinability indicate essential biologic differences.

The facts presented are a result of a study of ten strains of pneumococci, all of which agglutinated in Antipneumococcus Serum II, atypically as described above. Cross-immunity reactions and

---

[1] Dochez, A. R., and Gillespie, L. J., *Jour. Am. Med. Assn.*, 1913, lxi, 727.

[2] Cole, Rufus, *Arch. Int. Med.*, 1914, xiv, 56; *New York Med. Jour.*, 1915, ci, 1, 59. Dochez, A. R., and Avery, O. T., *Jour. Exper. Med.*, 1915, xxi, 114.

absorption tests indicate that these organisms possess characteristic group relationships among themselves and a common antigenic relation to the original and typical Type II pneumococcus.

Description of Cultures.—Ten strains of pneumococci were chosen, five of which were isolated from disease and five from the sputum of normal individuals. Table I gives the source and culture designation of the strains studied.

TABLE I.

Pneumococcus.	Source.	Obtained from	Remarks.
J	Acute maxillary antrum	Pus	Recovery.
L	" lobar pneumonia	Lung puncture	"
M	" " "	Blood	Septicemia, fatal.
Jn	" " "	Sputum	Recovery.
W	Primary pneumonia (child)	Lung puncture	Fatal.
As	Normal individual	Sputum	No infection.
Ar	" "	"	" "
F C B	" "	"	" "
S 13	Diabetic patient	"	" "
H	" "	"	" "

These ten strains possess the common characteristic of partial agglutination in antipneumococcus serum of Type II. Culturally and biologically they present all the usual characters of typical pneumococci: such as inulin fermentation, bile solubility, and more or less distinct capsule formation. Their virulence on isolation was distinctly lower than that of the typical Type II organism.

Agglutination as illustrated in Table II establishes a definite relationship between the strains studied and the type organism of Pneumococcus Group II. The reaction of all ten strains in Serum II is distinct, but never as prompt or complete as in the case of the typical Group II organism. With all typical Type II pneumococci the agglutination begins almost immediately and is complete in half an hour, while the reaction of these ten strains is always delayed and often incomplete at the end of the period of observation. It is further evident that these minor agglutinins tend to disappear in the higher dilution and are completely absent in dilution of 1 to 80, at which titer the reaction of the type organism still persists.

This variation in agglutination was at first attributed to the possibility that there might exist, among pneumococci of Group II.

Agglutination Experiments.

TABLE II.

Determination of the Titer of Agglutination of Strains in Antipneumococcus Serum II, Using Antipneumococcus Serum I and Normal Horse Serum (N) as Controls.

Dilution.	1:1			1:10			1:20			1:40			1:80		
Sera.	N	I	II	N	I	II	N	I	II	N	I	II	N	I	II
Culture															
As 11^5	–	–	++	–	–	++	–	–	++	–	–	–	–	–	–
L 6^8	–	–	++	–	–	++	–	–	++	–	–	+	–	–	–
Jn 8^9	–	±	++	–	–	++	–	–	++	–	–	++	–	–	–
M 5^{10}	–	–	++	–	–	++	–	–	++	–	–	+	–	–	–
Ar 9^5	–	±	++	–	–	++	–	–	++	–	–	+	–	–	–
J 7^8	–	–	++	–	–	++	–	–	++	–	–	+	–	–	–
W 5^9	–	–	++	–	–	++	–	–	++	–	–	±	–	–	–
F C B 9^5	–	+	++	–	±	+	–	–	–	–	–	–	–	–	–
S 13 12^3	–	+±	++	–	–	++	–	–	+	–	–	–	–	–	–
H 7^3	–	+	++	±	–	+	–	–	+	–	–	±	–	–	–
II 46^{11}	–	–	++	–	–	++	–	–	++	–	–	++	–	–	++
I 115^3	–	+	–	–	+	–	–	+	–	–	+	–	–	±	–

The reactions were read after 2 hrs. at 37° C. and over night on ice.

The numerals following the culture indicate the animal passage; the exponent indicates the number of generations removed from the last passage.

TABLE III.

The Effect of Increased Virulence upon the Agglutinability of the Atypical Group II Pneumococcus in Antipneumococcus Serum II.

Antipneumococcus Serum II.	Dilution.			
	1:10	1:20	1:40	1:80
Pneumococcus				
M 0^9	++	++	±	–
M 5^{11}	++	++	±	–
L 0^9	++	++	–	–
L 6^{11}	++	++	±	–
Jn 0^9	+	±	–	–
Jn 10^3	++	+	–	–
F C B 1^{16}	++	–	–	–
F C B 11^3	++	–	–	–
II 46^{11}	++	++	++	++

The numerals after the culture indicate number of mouse passages.

The exponent indicates the number of generations removed from the last passage.

strains of poor agglutinability, analogous to similar conditions among other bacterial groups. It was also thought that, since these organisms were of lower virulence than the type organism, this fact might bear some relation to their agglutinability. Agglutination tests were carried out with certain atypical strains, the virulence of which had been enhanced by repeated animal passage. The effect of the animal passage on the agglutination titer was then determined. Four strains were chosen, the virulence of which was raised by mouse passage until it had attained a point comparable to that of the typical Type II pneumococcus, with a minimum lethal dose of 0.000001 cc. of broth culture.

It is evident from Table III that enhancing the virulence of these strains did not affect their agglutinability. The variations in agglutinations between these strains and the type pneumococcus appear to be not merely differences in agglutinability, but suggest rather the possibility that actual differences in agglutinogenic properties may characterize these organisms. To determine this a univalent immune serum was prepared by immunization of rabbits to each of the ten strains.

Agglutination by Immune Rabbit Sera of the Ten Strains of Atypical Group II Pneumococci and the Results Obtained by Cross-Agglutination.

TABLE IV.

Agglutination of the Ten Strains by Homologous Sera and the Effect of Such Sera on Stock Cultures of Pneumococcus of Type II.

	Immune rabbit sera.									
Sera.	As		Ar		S 13		F C B			
Culture.	As	II	Ar	II	S 13	II	F C B	II	II	II
1 : 1	+ +	—	+ +	—	+ +	—	+ +	—	+ +	—
1 : 10	+ +	—	+ +	—	+ +	—	+ +	—	+ +	—
1 : 20	+ +	—	+	—	+ +	—	+	—		—
1 : 40	+ +	—	±	—	+ +	—		—		—

Sera.	L		J							
Culture.	L	II	J	II	Jn	II	W	II	M	II
1 : 1	+ +	—	+ +	—	+ +	—	+ +	—		
1 : 10	+ +	—	+ +	—	+ +	—	+ +	—		
1 : 20	+ +	—	+ +	—	+ +	—	+ +	—		
1 : 40	+ +	—	+ +	—	+ +	—	+ +	—		

Table IV shows that the rabbit immune atypical sera each agglutinated strongly the homologous strain of pneumococcus, but that none of these sera had any effect on the type culture of Group II. This failure of the antisera of the ten strains to agglutinate the Type II pneumococcus shows differences in the antigenic properties of these organisms and is in striking contrast to the positive reactions of agglutination with these same strains by Antipneumococcus Serum II (Table II). This failure of reversibility of the agglutination reaction would seem, therefore, to be due to actual differences in the agglutinogenic groups of these various organisms.

TABLE V.

Cross-Agglutination.

Immune sera.	Pneumococcus cultures.										
	As	S 13	F C B	Ar	L	J	Jn	W	M	H	II
As	++	−	−	−	++	−	++	−	++	−	−
S 13	−	++	−	−	−	−	−	−	−	−	−
F C B	−	−	++	−	−	−	−	−	−	−	−
Ar	−	−	−	++	−	+	−	+	−	−	−
L	++	−	−	−	++	−	++	−	++	−	−
J	−	−	−	++	−	++	−	++	−	−	−
Jn	++	−	−	−	++	−	++	−	++	−	−
W	−	−	−	++	−	+	−	++	−	−	−
M	+	−	−	−	+	−	+	−	+	−	−
H	−	−	−	−	−	−	−	−	−	++	−
Normal rabbit	−	−	−	−	−	−	−	−	−	−	−
II	+	++	+	+	++	+	++	++	++	++	++

Table V reveals the striking fact that in a series of organisms wholly contained within Group II certain separate relationships are demonstrable. An analysis of this table indicates clearly the existence of three distinct groups which appear to represent subdivisions of the main Group II, in the antiserum of which all members of each of these subdivisions are agglutinated. For purposes of convenience these subdivisions will be referred to as Subgroups, II A, II B, and II X, and are shown more clearly in Table VI.

Subgroup II A consists of four strains, the immune reactions of which are specific within the group, being identical with those of all other strains of the group, but bearing no relation to those of Subgroups II B, or II X. Similarly, Subgroup II B consists of three other strains characterized by the possession of specific immunity

TABLE VI.

Subgroup II A.

Immune sera.	Pneumococcus cultures.				
	As	L	Jn	M	II
As	++	++	++	++	−
L	++	++	++	++	−
Jn	++	++	++	++	−
M	++	·+	+	+	−
II	+	++	++	++	++

Subgroup II B.

Immune sera.	Pneumococcus cultures.			
	Ar	J	W	II
Ar	++	+	+	−
J	++	++	++	
W	++	+	++	−
II	+	+	++	++

Subgroup II X.

Immune sera.	Pneumococcus cultures.			
	F C B	S 13	H	II
F C B	++	−	−	−
S 13	−	++	−	
H	−	−	++	−
II	+	++	++	++

reactions, identical for members of this subgroup alone. The remaining three strains have been placed in Subgroup II X which, like the larger Group IV of the original biologic classification, is peculiar in that it seems to consist of a heterogeneous series of independent strains which do not cross in their immunity reactions with members of the other two subgroups or with each other. All, however, possess the common character of atypical agglutination in Antipneumococcus Serum II. This subgroup, like its prototype Group IV, seems to be infinitely variable, and to be characterized by the absence of cross-immunity reactions and by lower virulence.

Protection Experiments.—The protection of animals against infection is generally conceded to be one of the most specific of immunological reactions and hence one of the most satisfactory methods of classification. White mice were given intraperitoneally graduated doses of pneumococci and at the same time a fixed quantity of immune serum. All animals except the virulence controls received 0.2

cc. of immune serum intraperitoneally. This quantity of Immune Serum II as a rule protects mice against 0.01 cc. of broth culture of the homologous organism which, given alone, kills mice regularly in doses of 0.000001 cc. All animals surviving for five days were considered effectively protected.

In experimental pneumococcal infection the specificity of the protective power of an immune serum is evident only when the culture employed is fully virulent. Of the ten strains of pneumococci it was found possible to raise the virulence of seven by animal passage. The virulence of the other three strains could not be increased sufficiently for use in protection experiments, although they were passed successively through 11, 14, and 18 animals, respectively. They were even then of such low virulence that it was impossible to kill mice with the moderate doses necessary for the successful carrying out of the test. Of these three strains, two belonged to Subgroup II X and one to Subgroup II A.

Further Evidence of the Specificity of Group Relationships by Protection Experiments.

TABLE VII.

Protective Action of Antipneumococcus Serum II.
The Relation between the Specificity of the Protective Action of Immune Serum and the Virulence of the Infecting Organism.
Pneumococcus L (Subgroup II A) after One Animal Passage.

Pneumococcus Lr¹.	Virulence controls.	Serum II.	Serum I.	Normal serum.
cc.				
0.01		S.*	D. 96	S.
0.001	D. 72	"	S.	"
0.0001	S.	"	"	D. 96
0.00001	"	"	"	S.
0.000001	D. 72	"	D. 96	D. 96

Pneumococcus L (Subgroup II A) after Seven Animal Passages.

Pneumococcus L 7¹.	Virulence controls.	Serum II.	Serum I.
cc.			
0.01		S.	D. 18
0.001		"	" "
0.0001	D. 36	"	" 24
0.00001	" "	"	" 36
0.000001	" "	"	" "

* In the tables D. stands for. died; S. for survived. The figures represent the number of hours before the death of the animal.

Table VII shows that after increasing the virulence of pneumococcus L (Subgroup II A) to a degree sufficient to apply the test, definite protection was afforded by stock Antipneumococcus Serum II against 10,000 times the minimal lethal dose of culture. The odd survivals and non-specific reactions with the same strain before its virulence was raised by animal passage is evidence of the futility of attempting specific reactions of a protective nature with avirulent organisms. That this phenomenon does not represent a reversion to type brought about by animal passage is evidenced by the fact that the antigenic properties of these organisms remained unaffected by such treatment. Of the seven strains, the virulence of which was increased by mouse passage, Antipneumococcus Serum Type II protected against six, three of which belonged to Subgroup II A and three to Subgroup II B. The seventh strain, an organism of Subgroup II X, although made equally virulent by nine mouse passages, was not protected against by immune serum of Type II.

Cross-Protection Tests with Homologous and Heterologous Sera of the Three Subgroups of Pneumococcus II.

TABLE VIII.

Protective Action of Sera of Subgroup II A against a Pneumococcus of the Same Group, and Failure of Sera of Subgroups II B and II X to Protect against the Same Organism.

Pneumococcus Subgroup II A.	Immune sera.					
	Subgroup II A.		Subgroup II B.	Subgroup II X.	Antipneumococcus Type II.	Virulence controls.
Jn 9[1]	Jn	L	W	F C B		
cc.						
0.01	D. 72	S.	D. 18	D. 18	D. 96	
0.001	S.	"	" "	" "	S.	
0.0001	"	"	" "	" "	"	D. 18
0.00001	"	"	" "	" 24	"	" "
0.000001	"	"	" "	" "	"	"

Tables VIII and IX demonstrate that an immune serum of Subgroups II A and II B protects against any organism of the homologous subgroup, but fails to protect against any strain of the other two subgroups. Table X emphasizes the individual character of organ-

TABLE IX.

Protective Action of Sera of Subgroup II B against a Pneumococcus of the Same Subgroup and Failure of Sera of Subgroups II A and II X to Protect against the Same Organism.

Pneumococcus Subgroup II B.	Immune sera.					Virulence controls.
	Subgroup II B.		Subgroup II A.	Subgroup II X.§	Antipneumo-coccus Type II.	
W 6¹	W	J	Jn	F C B		
cc.						
0.01	D. 26	S.	D. 18	D. 18	D. 96	
0.001	S.	"	" "	" "	S.	
0.0001	"	"	" "	" "	"	D. 18
0.00001	"	"	" "	" "	"	" "
0.000001	"	"	" "	" "	"	" "

TABLE X.

Protective Action of Serum of Subgroup II X against the Homologous Organisms Only and Failure of Sera of Subgroups II A and II B and Antipneumococcus Serum II To Protect against the Same.

Pneumococcus Subgroup II X.	Immune sera.					Virulence controls.
	Subgroup II X.		Subgroup II A.	Subgroup II B.	Antipneumo-coccus Type II.	
F C B II¹	F C B	S 13	Jn	W		
cc.						
0.01	D. 18	D. 18	D. 18	D. 18	D. 18	
0.001	S.	" "	" "	" "	" "	
0.0001	"	" "	" "	" "	" "	D. 18
0.00001	"	" "	" 22	" "	" "	" "
0.000001	"	" "	" "	" 36	" "	" "

TABLE XI.

Lack of Protective Power of Immune Sera of Subgroups II A, II B, and II X against Typical Pneumococcus II.

Pneumococcus Type II.	Immune sera.			Virulence controls.
	Subgroup II A.	Subgroup II B.	Subgroup II X.	
II 46¹	Jn	W	F C B	
cc.				
0.01	D. 18	D. 18	D. 18	
0.001	" "	" "	" "	
0.0001	" "	" "	" "	D. 18
0.00001	" "	" "	" "	" "
0.000001	" "	" "	" 26	" "

isms of Subgroup II X. A serum produced by immunization with any given strain of this type protects against that particular organism and against no other. As previously noted, there is a complete lack of crossing in the immunity reactions of the individual members of Subgroup II X. Table XI shows that in protective action, as in agglutination (Table IV), the immune sera of Subgroups II A, II B, and II X have no effect on the typical pneumococcus of Type II. That these immunologic reactions between the original Type II pneumococcus and organisms of the subgroups are not reversible seems to indicate degrees of difference in antigenic characters. While the serological specificity of the subgroups definitely separates one from the other, nevertheless their immune reactions with the antisera of the typical Type II organism indicate that they are all biologically related. The correlation of these subgroups is further proven by absorption tests.

Absorption Experiments.—The phenomenon of specific absorption of agglutinins from an immune serum by the homologous organism was first described by Castellani.[3] This investigator found that from a polyvalent serum produced by immunization with two microorganisms of different species, the agglutinins for either one could be removed by fractional absorption with the homologous strain, while in a serum thus exhausted, the antibodies for the second organism remained intact. It has been shown also that in a univalent serum against *Bacillus typhosus* not only are agglutinins present for that organism alone, but that in the same serum there also exist partial or minor agglutinins for bacilli which are biologically similar, and which fall within the same general group. Absorption of a typhoid immune serum by *Bacillus typhosus* removes not only the agglutinins for that organism, but completely exhausts the serum of its minor antibodies for the closely allied organisms. This reaction, however, is not reversible, for removal of the partial agglutinius by absorption with a member of the intermediary species leaves the antibodies for *Bacillus typhosus* practically undiminished. The significance of this phenomenon in bacterial classification is obvious, and its applicability to the present study is evident in the following protocols.

[3] Castellani, A., *Ztschr. f. Hyg. u. Infectionskrankh.*, 1902, xl. 1.

Specificity of Absorption Reaction. Absorption of Antipneumococcus Serum I with Pneumococci of Groups I and II.

Technique.—Specific absorption. Antipneumococcus Serum I. 2 cc. of serum were diluted with 3 cc. of salt solution. To the 5 cc. of diluted serum the live, washed bacterial residue of 150 cc. of a twenty-four hour broth culture of Pneumococcus I was added, allowed to stand in contact over night in the ice box, then centrifuged, the clear supernatant serum pipetted off, and passed through a Berkefeld filter.

Non-specific absorption. Antipneumococcus Serum I was absorbed with Pneumococcus II. The technique was the same as above.

Control. Antipneumococcus Serum I diluted, and, without the addition of any bacteria, filtered by the same technique.

TABLE XII.

Specific and Non-Specific Absorption of Agglutinins from Antipneumococcus Serum I by Pneumococci of Groups I and II.
Agglutination.

Culture pneumococcus, Group I.	Antipneumococcus Serum I.		
	Specific absorption with Pneumococcus I.	Non-specific absorption with Pneumococcus II.	Control. Serum unabsorbed.
	−	++	++

TABLE XIII.

Protective Power of Antipneumococcus Serum I after Absorption with Pneumococci of Groups I and II.
Protection.

Culture pneumococcus, Group I.	Virulence controls.	Antipneumococcus Serum I.		
		Specific absorption with Pneumococcus I.	Non-specific absorption with Pneumococcus II.	Control. Serum unabsorbed.
cc.				
0.01		D. 18	S.	S.
0.001		" 20	"	"
0.0001	D. 18	" 24	"	"
0.00001	" "	" "	"	"
0.000001	" 20	" 48	"	"

Tables XII and XIII demonstrate the specificity of the absorption reaction with pneumococci of the fixed Types I and II. Saturating an immune serum of Group I with pneumococcus of the same type completely exhausts that serum of all its agglutinins and protective antibodies, while absorption of the same serum with organisms of Group II does not appreciably diminish these immune substances for pneumococci of Group I.

Absorption of Antipneumococcus Serum II with Pneumococcus II and Organisms of Its Subgroups II A, II B, and II X.

Technique.—To 10 cc. of undiluted Antipneumococcus Serum II was added the washed bacterial residue from 150 cc. of an eighteen hour broth culture of the given strain of pneumococcus. Before being added to the serum the bacteria were killed by heating at 56° C. for forty-five minutes. The serum mixtures were incubated in the water bath for two hours at 37° C. and allowed to remain in contact over night in the ice box. The clumps of agglutinated bacteria were whirled out by centrifugation, and the clear supernatant serum was pipetted off. This serum was absorbed a second time by the same technique, tested for the absence of agglutinins, and called exhausted serum.

TABLE XIV.

Cross-Agglutination Reactions with Antipneumococcus Serum II Exhausted by Absorption with Type Strains of Pneumococcus II and Its Subgroups II A, II B, and II X.

Antipneumococcus Serum II. Absorbed by	Pneumococcus.										
	Subgroup II A.				Subgroup II B.			Subgroup II X.		Group II.	
	Jn	As	L	M	W	Ar	J	S 13	FCB	H	Type II.
Subgroup A, Jn	−	−	−	−	++	+±	++	++	+	++	++
" B, W	++	++	++	+±	−	−	−	++	+	++	++
" X, S 13	++	++	++	+±	++	++	++	−	+	++	++
Group II	−	−	−	−	−	−	−	−	−	−	−

Table XIV shows that specific absorption of Antipneumococcus Serum II with the typical Type II pneumococcus removes all the agglutinins, not only for the homologous organism, but also all the partial agglutinins for its subgroups, II A, II B, and II X. In other words, specific absorption of Antipneumococcus Serum II completely exhausts it of both major and minor agglutinins. Conversely, however, absorption of the same immune serum with a representative strain of Subgroup II A removes the minor agglutinins for members of that subgroup only, leaving intact the antibodies for the Type II pneumococcus and its other subgroups, II B and II X. Similarly, absorption of Antipneumococcus Serum II with any member of Subgroup II B takes out the agglutinins for all the Subgroup II B organisms, but leaves unaffected the antibodies for the Type II pneumococcus and its subgroups, II A and II X. The lack of cross-immunological reactions among the heterogeneous organisms within

Subgroup II X already noted in the previous agglutination and protective experiments (Tables VI and X) is again evident in the absorption tests. Saturation of Immune Serum II with a pneumococcus of Subgroup II X robs the serum of its agglutinins for that individual strain only, and for no other.

The results obtained by absorption experiments with Antipneumococcus Serum II, a serum produced by intensive immunization of the horse with a single strain of the typical Type II pneumococcus, corroborate the same specific groupings obtained by the cross-immunological reactions of agglutination and protection with immune rabbit sera of the individual strains.

DISCUSSION.

The biologic classification of the pneumococcus distinguishes four distinct groups. These types are based upon well defined immunologic differences. The accuracy with which these groups may be differentiated and the constancy of their relative frequence in disease and health emphasize the importance of their recognition in clinical and epidemiological studies. The exactness with which the large number of strains studied have conformed to type indicates the extraordinary uniformity and comparative fixity of the specific groups. These distinctive differences in antigenic properties not only offer a reliable method for the more exact determination of the varieties of pneumococcus, but afford the only rational basis for the study of immunotherapy in pneumococcal infection.

The second group of pneumococci of the original classification consists of highly virulent organisms which are responsible for about one-third of all cases of lobar pneumonia of pneumococcus origin. Organisms of this group produce infections which are clinically severe, and the mortality of which is about 35 per cent. The serological reactions by means of which Group II was originally identified are sharply defined. So characteristic and prompt is the agglutination of any strain of this type in Antipneumococcus Serum II that any deviation from the normal reaction is quickly recognized. The isolation of an occasional strain of pneumococcus which agglutinates atypically in Antipneumococcus Serum II led to an attempt to determine the nature of these organisms and their relation to the type pneumococcus of Group II. Of the ten strains studied five were isolated from disease and five from the sputum of normal individuals. These organisms all exhibited the usual cultural and biochemical characteristics of the pneumococcus; namely, inulin fermentation, bile solubility, and varying degrees of capsular development. The facts developed by this study indicate the existence of pneumococci which are biologically similar and closely allied to the typical organism of Group II. These organisms possess partial antigenic characters common to the Type II pneumococcus, but they vary from the typical representative of this group by a diversity of relationships among themselves, and by a lack of the reversibility of their immune reactions with the type organism. Because of these variations these organisms may be classified as subvarieties of Pneumococcus Group II. All strains of the three subgroups thus far recognized are agglutinated by Antipneumococcus Serum II, but the diminished intensity of the reaction serves to distinguish them from typical II pneumococci. The incomplete reaction of agglutination of these subvarieties in the immune serum of Type II is apparently similar to the diminished reactions which occur in many immune sera with other organisms closely allied to the type used in producing the serum. Such reactions occur only in the higher concentrations of immune sera and may be attributed to the so called minor agglutinius. In such cases, also, a non-reversibility of the immune reaction has been noted. For instance, a potent typhoid immune serum may agglutinate *Bacillus coli* in the higher concentrations, but an anticolon serum may not affect *Bacillus typhosus*. All strains of these subgroups of pneumococci are partially agglutinated in the higher concentrations of Antipneumococcus Serum II, while conversely the immune serum produced by any strain of these subgroups fails to react with the typical II pneumococcus.

In addition to the partial agglutination of these strains in Antipneumococcus Serum II and the absence of reverse immunity reactions, these subvarieties are further characterized by certain interrelationships of a definite antigenic nature, by virtue of which they may be classified into at least three subgroups, which have been called Subgroups II A, II B, and II X. It has been shown in the preceding protocols that a given member of either Subgroup II A or

II B is characterized by the possession of immunity reactions identical with those of all other strains of the homologous subgroup, and that these reactions are specific only within the group. Subgroup II X is peculiar in that it seems to consist of a heterogeneous series of independent strains which do not cross in their immunity reactions with members of the other two subgroups or with each other. This subgroup, like its prototype, Group IV, of the original biologic classification, is of lower virulence, infinitely variable in its composition, and lacking in cross-immunity reactions. As has been noted, serum Type II fails to protect against organisms of this subgroup, and inasmuch as specific protection is regarded as the ultimate criterion for classification, it is doubtful whether organisms of Subgroup II X are of sufficiently close relationship to be included within Group II.

The specificity of these subgroups, II A and II B, as tested by the immunity reactions of agglutination and protection is further confirmed by the phenomenon of absorption. Saturation of Antipneumococcus Serum II with a typical Group II pneumococcus removes all the agglutinins both for the type organism and its three subgroups. Absorption of the same serum, however, with a member of either Subgroup II A or II B removes only the partial agglutinins for the homologous subgroup, but leaves intact the antibodies for the typical II pneumococcus and the other subgroup. Absorption of Antipneumococcus Serum II, on the other hand, by any member of Subgroup II X takes out the antibodies for that particular strain only, and for no other.

In the present discussion no attempt is made to interpret the experimental data in terms of their phylogenetic significance. Whether the subvarieties of the second group of pneumococci represent strains which have acquired independently certain adaptive characters, or whether they are related to each other and to the fixed type by the lineage of common descent is interesting. However, the limited nature of the present study precludes the formulation of any hypothesis as to origin.

SUMMARY.

1. At least three subgroups of Pneumococcus Type II may be recognized by specific immune reactions. T' ¬y have been called Subgroups II A, II B, and II X.

2. That the organisms of these three subgroups are biologically related to Pneumococcus Type II is shown by the following facts: (a) Agglutination with Antipneumococcus Serum II. (b) Protection with Antipneumococcus Serum II, except Subgroup II X. (c) Absorption of Antipneumococcus Serum II with typical Type II pneumococcus removes the antibodies for all subgroups. (d) Absorption of Antipneumococcus Serum II with a member of Subgroups II A or II B removes only the antibodies for the homologous subgroup. Absorption of Antipneumococcus Serum II with any given member of Subgroup II X removes the antibodies for that particular strain only.

3. That the three subgroups, although biologically related to Pneumococcus Type II, possess, nevertheless, specific differential characters which separate them one from another, is evidenced by the following facts: (a) The organisms of any subgroup are not agglutinated by the antisera of the other two subgroups. (b) They are not protected against by the sera of the other subgroups. (c) They do not absorb from Antipneumococcus Serum II the specific immune bodies of the other subgroups.

4. Subgroups II A and II B are characterized by immunity reactions identical within the respective group.

5. Subgroup II X consists of heterogeneous strains which do not cross in their immunity reaction with each other or with Subgroups II A or II B.

The author acknowledges his indebtedness to Dr. A. L. Bloomfield of the Johns Hopkins Hospital for five of the cultures of pneumococcus used in this study.

SPONTANEOUS AND EXPERIMENTAL LEUKEMIA OF THE FOWL.

By HARRY C. SCHMEISSER, M.D.

(*From the Department of Pathology of Johns Hopkins University, Baltimore.*)

(Received for publication, July 23, 1915.)

During the past three years a transmissible leukemia of the fowl has come under observation, and, in the following paper, a summary of the literature, together with a brief presentation of the clinical and anatomical aspects of the disease will be given.[1]

HISTORICAL.

Moore (1) reported infectious leukemia in fowls, but this has since been regarded as not true leukemia. Butterfield (2) and Mohler[2] recognized the condition from the postmortem findings, and the first careful study, including the clinical course and anatomical changes, was made by Warthin (3) in 1907. The predominating cell in the blood and tissues was the large lymphocyte,[3] and the condition was diagnosed leukemic lymphocystostoma. Other cases described by both Butterfield and Warthin lacked the blood changes but showed the same tissue picture and were considered by these authors as examples of aleukemic leukemia.

Kon (4) and later Soshestrenski (5) each report a similar case of leukemia. They consider their case as true splenic leukemia.

Ellermann and Bang (6, 7), in 1908, were the first to transmit the disease successfully from a spontaneous case to other healthy fowls. They report two typical spontaneous cases in every respect similar to Warthin's first case. They transmitted the disease by inoculation of organic emulsion through three generations, producing blood picture and organic findings identical with the spontaneous cases. They also found by inoculation of organic extract from a pseudoleukemic fowl (used in the same sense as aleukemia of Warthin) that a picture of true leukemia resulted, and concluded that the two conditions are etiologically identical.

In several subsequent reports (8, 9) Ellermann and Bang call attention to the following points: (1) that in transmitted cases the disease may appear as

[1] A more detailed and fully illustrated study will appear in the *Rep. Johns Hopkins Hosp.*, 1915, xvii (in press).

[2] Cited by Butterfield (2). Mohler personally never reported his cases.

[3] Obviously the large mononuclear of the classification presented in this paper.

typical leukemia, pseudoleukemia, or as an anemia with changes in the bone marrow; (2) that mitoses in the blood are pathological and always present in leukemia; (3) that the blood of leukemic fowls contains the virus; and (4) that the disease can be produced by a cell-free Berkefeld filtrate. They conclude (a) that leukemia must be an infectious disease, and (b) that it is to be placed among the diseases due to a filterable virus.

Schridde (10) questioned the infectious etiology of leukemia and claimed that chickens injected with extracts of entirely normal organs present the same changes as Ellermann and Bang reported for leukemia.

Hirschfeld and Jacoby (11, 12) in 1909 observed a typical case of leukemia, and the following year they reported (13) the transmission of the disease from a leukemic animal[4] into the fifth generation.

Burckhardt's claim (14, 15), supported by Friedberger (16), that leukemia could be produced in the fowl by inoculation with pure culture of fowl tubercle bacilli is only of passing interest.

Ellermann (17, 18) in answer to the above objection of Schridde points out that in the latter's experiments only the blood picture was produced, and in the absence of the characteristic organic changes of which Schridde makes no mention, his experiments are of no importance. The injection of emulsion of normal organs Ellermann never found to cause any change in the blood.

Ellermann again reports the successful transmission of the disease with Berkefeld filtrates. Hirschfeld and Jacoby (19) and Burckhardt (15), on the other hand, report unsuccessful results with Berkefeld filtrate.

Ellermann claims to have shown that the leukemic virus can be separated from the virus of tuberculosis by filtration, and that, therefore, the two diseases are distinct. He further claims to be able to separate the spontaneous and transmitted leukemia into the types (a) myeloid and (b) lymphatic. He says that a myeloid type may occur in one generation and a lymphatic in the next, or both types in the same generation and that this is highly suggestive that both forms in man are due to one and the same infective agent.

The Normal Fowl.

The following data of the normal fowl are confined to those portions of the body which are involved in leukemia and have been compiled from a large number of young adult Plymouth Rock hens.

Blood.—The blood is readily obtained from the vein under the wing. From a small needle puncture the blood flows under pressure. It is thick, dark red, and clots quickly.

The number of red blood cells averages 3,000,000 to 4,000,000 per cmm. The number of white blood cells[5] varies between 20,000 to 80,000 per cmm. Actually the proportion of white blood cells to red blood cells varies between 1 to 50 and 1 to 150. The hemoglobin (Sahli) averages 60 to 70 per cent.

[4] This animal was given to them by Ellermann and Bang.

[5] These were determined by the indirect method.

The blood cells stained by Wilson's method may be classified as follows:

Erythrocytes.—1. Normocyte: elliptical disk. Nucleus same shape as cell, deep blue, slightly pyknotic. Cytoplasm yellow and glassy. (Both cell and nucleus are uniform in size, shape, and staining.)

Blood Platelets.—Length of normocyte; width less than that of normocyte. Nucleus round, purple; chromatin diffuse; diameter equal to width of its cell. Cytoplasm pale gray with vacuoles about nucleus, frequently containing small circumscribed red structures. They may vary in size and shape.

Leucocytes.—1. Polymorphonuclear leucocyte with eosinophilic rods: round, diameter about length of normocyte. Nucleus, of two or more lobes, pale blue; chromatin diffuse. Cytoplasm colorless with bright red, spindle-shaped rods.

2. Polymorphonuclear leucocyte with eosinophilic granules: about the same in shape and size. Nucleus of two or more lobes, purple, slightly pyknotic. Cytoplasm faintly blue with dull red granules.

3. Lymphocyte: round, diameter about width of normocyte. Nucleus round, purple; chromatin diffuse. Cytoplasm small in amount, to one side of nucleus, pale blue. Same cell may be slightly larger. Thus a division into small and large lymphocyte may be made.

4. Large mononuclear cell: round or oval, diameter about length of normocyte (at times more or less). Nucleus round, oval, or slightly irregular, and larger; otherwise similar to nucleus of lymphocyte. Cytoplasm abundant, completely surrounds nucleus; pale blue. (A suggestion of fine granules.)

5. Mast cell: about same size and shape. Nucleus round or oval, very pale blue. Cytoplasm abundant, colorless, mostly to one side of nucleus with purple granules; some scattered over nucleus.

Differential count, 300 cells.	*Per cent.*
Polymorphonuclears with eosinophilic rods	29.6
" " " granules	4.3
Lymphocytes	42.3
Large mononuclear cells	19.4
Mast cells	2.2
Unclassified cells	2.2
	100.0

Gross Anatomy.—A fowl of average size and weight (1,760 grams) has abundant subcutaneous fat and large muscles. The inner surface of the skin is slightly yellow.

The cervical lymph glands[6] are sometimes difficult to find. They are present as two chains of six to ten glands, one on each side of the neck, lying in the fat upon the internal jugular vein, and extend-

[6] Kon (4) considers these structures the thymus. He (4) and Ellermann (17, 18) state that the fowl has no lymph glands. Soshestrenski (5) asserts their existence by stating that, in his case, they were not enlarged.

ing from the middle of the neck to the base of the heart. They are elliptical, flat, lobulated bodies, averaging 1 × 0.5 × 0.2 cm. The lobules are pink and separated by narrow septa of fat. No other lymph glands were found in the entire body.

The peritoneal cavity contains an omentum rich in fat, averaging 1 cm. in thickness. It takes its origin from the anterior surface and lower margin of the gizzard, extending down the intestines, etc.

The liver is a bilobed organ, weighing 50 grams, or 2.8 per cent of the body weight. It extends down between the omentum and sternum, the margin of each lobe reaching as far as 3 cm. above the tip of the xiphoid. It is uniformly reddish brown, moderately soft, and friable. The lobulation is usually distinct, at times difficult to see. The cut surface shows small blood vessels in long and cross-section.

The spleen lies just behind the liver. It measures 2 × 1.5 × 1 cm. and weighs 1 gram, or 0.05 per cent of the body weight. It is small, soft, and reddish brown. Beneath the capsule bluish white Malpighian bodies, slightly larger than a pin point, may be indistinctly seen. On section the capsule is very delicate, the trabeculæ are few, but usually definite, containing small blood vessels. The Malpighian bodies at times are prominent. The pulp is not raised above the edge of the capsule. The kidneys weigh 12 grams, or 0.7 per cent of the body weight. They are uniformly reddish brown with a slightly nodular surface.

The bone marrow is mottled, bright red and yellow. It is very soft, at times semifluid. It is always rich in fat.

Histology.—The cervical lymph glands are divided into lobules of parenchyma, separated by fatty tissue. Each lobule is surrounded by a delicate fibrous capsule and has a very fine reticulum, in which the lymphocytes are diffusely scattered. These are small round cells with round, deeply staining, pyknotic nuclei and a narrow rim of pink cytoplasm, usually incompletely surrounding the latter.

Sometimes red blood cells are associated with the lymphocyte; but these are mostly confined to the capillaries. They appear as elongated yellowish pink cells with solid black nuclei, which are seen as rods or dots, according to whether they are in long or cross-section.   Scattered through the lobule are small, sharply outlined, hyaline structures, which Kon (4) considered as corresponding to Hassal's corpuscles of the thymus.

In the liver the lobule is difficult to limit.   Periportal spaces are not easily made out.   They contain the usual vessels, surrounded by very little acellular fibrous tissue, although lymphocytes may be present diffusely or in small follicles.   The liver cells are arranged in trabeculæ, separated by capillaries.   All blood vessels and capillaries are filled almost exclusively with red blood cells.

In the spleen the Malpighian bodies are numerous and indistinct. They are composed of the usual lymphocytes surrounding very small arteries.   Red blood cells are limited to the pulp where they occur more or less in clusters, although sinuses or inclosures of any kind cannot be definitely demonstrated.   The pulp is also diffusely infiltrated with lymphocytes.

The kidneys for all practical purposes have a structure similar to the human kidney.

The bone marrow under low power shows a framework of fatty tissue enclosing nests of marrow cells.   Stained with hematoxylin and eosin these may be grouped as follows:

Erythrocytes.—1. Normocyte (a): elongated, varying in shape due to pressure. Nucleus a solid black rod or dot (long or cross-section). Cytoplasm yellowish pink and glassy.

2. Normoblast (b): round; diameter about width of normocyte. Nucleus uniformly black. Cytoplasm pink or faintly blue, and glassy. Frequently a narrow clear zone is seen about the nucleus.

3. Megaloblast (b): the same, larger, about the length of a normocyte. Nucleus slightly pyknotic.

Leucocytes.—1. Polymorphonuclear myelocyte with eosinophilic rods[7] (c): round. Nucleus, two or more solid, black lobes. Cytoplasm colorless with bright red, spindle-shaped rods.

2. Polymorphonuclear myelocyte with eosinophilic granules[7] (b): round; about same size. Nucleus two or more slightly vesicular lobes. Cytoplasm colorless with bright red granules.

3. Mononuclear myelocyte with eosinophilic granules (a): round or oval, varying in size, mostly larger than 1 and 2. Nucleus round, oval, or horseshoe-shaped, eccentric, slightly or very vesicular. Cytoplasm colorless with bright red granules.

7 These are obviously the polymorphonuclear with eosinophilic rods, the polymorphonuclear with eosinophilic granules, and the large mononuclear cell of the normal blood.

4. Large mononuclear myelocyte[7] (d) : round, slightly larger. Nucleus round, very vesicular, one or more nucleoli. Cytoplasm basophilic, moderate in amount. *Reticular Cells.*—Branched; nucleus elongated, vesicular; several nucleoli. Cytoplasm pink, giving off delicate fibers to form reticulum.

Lymphocytes, mast cells, platelets, mitoses of red and white cells were not seen. The letters in parenthesis indicate the order of predominance.

Spontaneous Leukemia.

On October 31, 1912, a typical case of leukemia of the fowl was brought to the pathological laboratory. The animal was a Plymouth Rock hen and had just been killed. Nothing is known of its clinical history.

Autopsy Findings.—Blood smears taken from the heart were stained by Wilson's method. A detailed description of the cells of leukemic blood will be reserved for the experimental leukemia where it was possible to study these more carefully. Suffice it here to comment briefly upon the most striking features.

There was an enormous increase in white blood cells, the proportion of white to red cells being 1 to 1.3. The predominating cells were the large mononuclear and the mononuclear myelocyte with eosinophilic granules. The latter is abnormal in the blood, normal in the bone marrow, and when found outside of the bone marrow is typical of leukemia. Lymphocytes and polymorphonuclears were strikingly decreased. Of the latter those with red granules and the mast cell were rare. Mitoses of the large mononuclears were common. The red blood cells appeared poor in hemoglobin, showed anisocytosis, poikilocytosis, and polychromatophilia. These cytoplasmatic changes were usually associated with an increase in the size of the nucleus. Premature red blood cells, normoblasts, and megaloblasts were present in large numbers.

Differential count, 300 cells.

	per ct.
Polymorphonuclears with eosinophilic rods	8
" " " granules	0
Lymphocytes	2
Large mononuclear cells	30
Mast cells	0
Mononuclear myelocytes with eosinophilic granules	52
Unclassified cells	8
	100

Gross Examination of Tissues.—The comb, the featherless area about the eyes, the wattles, the buccal mucous membrane, and the conjunctivæ were very pale. The anterior chamber of the left eye was filled with an old blood clot. A moderately firm, slightly nodular tumor 3.7 × 2 × 1.5 cm. occupied the triangular space on the left side of the head between the angle of the mouth, the ear, and the angle of the lower jaw, extending slightly below the ramus of the latter. Emaciation was extreme. Subcutaneous fat was practically absent. The muscles were greatly atrophied. A second, slightly smaller tumor 1.7 × 1 × 1 cm. was found just inside of the ramus and below the orbit, apparently communicating over the ramus with the first growth and continuous below with a slightly larger third mass.[8]

The cervical lymph glands were somewhat enlarged and showed indistinct lobulation.

In the peritoneal cavity the omental fat was entirely absent and the lower margin of the right and left lobes of the liver extended almost to the pubis.

The liver was enormous. Its surface was extremely mottled. It was reddish brown, and specked with innumerable gray or slightly yellow spots from pin point to a few mm. in diameter, frequently closely packed to form irregular areas, the largest being 1 cm. in diameter. In addition there were scattered gray or slightly yellow nodules averaging about 2 mm. in diameter. The sectioned surface was similar in appearance. The blood vessels were surrounded by a gray zone.

The spleen was enormous. It was about the size of a small hen's egg and diffusely gray.

At the apex of the heart there were several gray spots, pin point to pin head in size.

The kidneys were greatly enlarged and both showed nodules similar to those in the liver.

The bone marrow was gray and poor in fat.

The remaining organs appeared normal.

Microscopical Examination of Tissues.—The normal structure of the cervical lymph glands was changed. The interlobular fat had

[8] A more detailed description and study of these tumors will be reserved for a future communication.

entirely disappeared with approximation of the greatly swollen parenchymatous lobules. Here and there were intra- and interlobular foci of myeloid tissue. These frequently enclosed an artery or vein. There was a more diffuse infiltration of the parenchyma by myeloid cells, filling capillaries, veins, arteries, and in places breaking through the lobular capsule into the interlobular tissue. Of the infiltrating cells two predominated: (1) A large mononuclear, usually round, at times slightly polygonal, with a single round, oval or indented, vesicular, at times multiple nucleus with one or more nucleoli. Its cytoplasm was granular and slightly basophilic. (2) The mononuclear myelocyte with eosinophilic granules, described under normal bone marrow. Mitoses of both cells were common. Normoblasts and megaloblasts were also present.

In the liver the process was most extreme. Very little liver tissue remained. Everywhere were closely packed masses of myeloid cells, in which the liver trabeculæ had completely disappeared. These masses of cells frequently surrounded blood vessels, both arteries and veins, infiltrating the walls of the latter and filling the lumina of both. Within the vessels, the white blood cells were present in about equal proportions with the normocytes. These cells also occurred everywhere between the liver columns, spreading them apart. It was usually difficult to demonstrate the capillary wall. The hepatic cells in these areas had not suffered so much. Some of the circumscribed infiltrations showed a coarse sclerosis. The infiltrating cells were the same as in the cervical lymph glands. A few polymorphonuclear cells of both types were present.

The pulp of the spleen was diffusely infiltrated with closely packed leukemic cells, crowding the reticulum and distending blood vessels, separating and compressing the Malpighian bodies. Mononuclear myelocytes with eosinophilic granules although present were somewhat scarce. The large mononuclear with mitoses was the common cell.

The bone marrow was greatly changed. It consisted of closely packed white marrow cells, with complete atrophy of the fat, and a great rarity of normocytes. One or two small areas of sclerosis were present. The marrow cells, both red and white, answered the same description as normal bone marrow, with the exception of the large mononuclear myelocyte, which showed considerable variation in its nucleus. This was single or multiple, round, oval, or horse-shoe-shaped. The order of predominance had changed. The large mononuclear myelocyte was present in far greater numbers than any other cell, and showed extensive mitoses. Normoblasts and megaloblasts with mitoses followed next in frequency, then normo-cytes. Polymorphonuclear myelocytes, both with eosinophilic rods and granules, had disappeared entirely. The mononuclear myelocyte with eosinophilic granules could not be demonstrated with certainty. As in the normal marrow, lymphocytes, mast cells, and platelets were seen.

In the remaining organs, the blood vessels were filled with the characteristic blood. Cell infiltrations, both diffuse and focal, occurred in the heart, lungs, and kidneys.

Summary. Blood.—(a) Although a total count was not made, a great increase in the total number of white blood cells was evident from the appearance of the blood smears, and blood vessels in sections. The actual proportion of white to red cells (1 to 1.3) substantiated this fact. The differential count showed a marked increase of the large mononuclear cell at the expense of the other white blood cells of the normal blood. In addition, a true myelocyte appeared in the circulation. Mitoses of the large mononuclear in the circulation were common. (b) There was a corresponding decrease in the total number of red blood cells. These were poor in hemo-globin and showed variation in size and shape, basophilic staining of the cytoplasm, and swelling of their nuclei. Premature red cells also occurred.

Organs.—(a) Many organs contained diffuse or circumscribed infiltrations of myeloid cells. The large mononuclear and mononu-clear myelocyte with eosinophilic granules predominated in these in-filtrations although the other cells of the normal bone marrow were also present. This myelosis involved especially the liver, spleen, kidneys, and bone marrow, resulting in an extensive enlargement of the first three organs. Almost all the remaining organs show infiltration, but to a less degree. (b) The proportion of white to red cells was greatly increased in the blood vessels. The predom-inating cells were the same as in the infiltration.

Conclusion.—If we consider that the cells characteristic of 'both the leukemic infiltration and blood are the same, and that under normal conditions the mononuclear monocyte with eosinophilic granules occurs only in the bone marrow and the large mononuclear only in the marrow and blood, it is evident from the summarized facts that the above case must be considered a typical case of myeloid leukemia.

Experimental Leukemia.

From the above animal, the disease was transmitted into the 5th generation. A total of 105 animals was used in conducting many different kinds of experiments. Of this number, 22 in all developed leukemia. In 4 additional animals, a definite diagnosis could not be established, although they were highly suggestive.

This paper will be confined to a report of the simple transmission of the disease by injection of an organic emulsion.

Five series of experiments were conducted, each consisting of 5. 10, or 15 fowls, with 20 to 40 per cent of positive animals. Of the total number of 40 chickens injected, 13 developed leukemia; *i. e.,* 32.5 per cent. In addition, a definite diagnosis could not be made in 3, which were very suspicious.

For transmission pieces of liver, sometimes also spleen. were thoroughly macerated in 0.9 per cent salt solution and filtered through a single layer of fine linen or a small amount of raw cotton. This emulsion could be used for intraperitoneal injection. When used intravenously, however. instant death resulted. and it was necessary to filter it also through two layers of filter paper with the aid of a suction pump for successful intravenous inoculations. This filtrate still contained blood and parenchymatous cells. A 15 per cent emulsion gave the best results. 10 cc. of the emulsion were injected either intravenously or intraperitoneally. Both methods were sometimes combined when a total of 20 cc. was administered. The vein under the wing was selected for injection.

Animals used for transmission were young adult hens of the same breed as the spontaneous case and exemplified by the normal control.

Clinical History.—The incubation period is usually from 5 to 6 weeks, with a maximum of 16 weeks.

The onset is as a rule abrupt, beginning with a slight but progressive, at times a sudden and intense, pallor, affecting comb, featherless area about eyes, and wattles. This is almost invariably associated with jaundice, which is likewise progressive and usually reaches an extreme grade. A characteristic yellowish pink color results which gives the fowl a ghastly appearance. Comb, etc., may be surprisingly red or extremely pale without jaundice. The animal emaciates rapidly and to an extreme degree. The actual loss in weight during the disease is striking. No. 28 at the onset of the disease weighed 1.485 gm. The duration of the disease was 3 weeks and 4 days. Weight at death was 897 gm., a loss, therefore, of 588 gm. in 25 days, or at the rate of 24 gm. per day.

At first the fowl acts normally, or only slightly ill. In a few days, however, it is very sick, stops eating and drinking, and stands about with head retracted, eyes closed, and tail drooping. It prefers to assume a squatting position. If made to move, it does so very slowly and carefully. Weakness becomes more and more marked, until on the last day it lies on the floor, wings drooping, eyes closed, and is at times dyspneic. Fever has never been noticed.

After the animal is observed to be sick the illness lasts from 1 to 2 weeks. One case, exceptionally acute, lasted only 33 hours. The longest course was 4 weeks.

Only in one case, out of 23 leukemic animals, did a spontaneous cure result.

Blood.—With the onset of anemia, the vein under the wing collapses and the blood undergoes a change. It soon flows with the greatest ease, is pale yellow and watery, and shows no tendency to clot. The animal apparently would bleed to death from a pin point wound if hemorrhage were not artificially arrested.

In most cases there is a progressive and extreme decrease in the total number of red blood cells. At the onset the count may be but slightly below the normal, 2,224,000, while just before death it at times reaches 630,000 per cmm. In one case the count remained normal. The white blood cells are invariably high, 131,200 to 210,-000 per cmm. The ratio of white blood cells to red blood cells varies between 1 to 3 and 1 to 9.

The hemoglobin usually falls steadily and reaches a very low point. From slightly below normal, 40 to 50 per cent at onset, it frequently drops to 10 to 15 per cent just before death. In one case there was no change at all.

The morphology of the leukemic blood is exceptionally interesting. In addition to the enormous increase in the number of white cells, all the cellular elements show marked changes and many new forms make their appearance.

The blood cells stained by Wilson's method may be classified as follows:

Erythrocytes.—1. Normocyte: (a) normal, except for variation in the amount of hemoglobin; (b) showing anisocytosis, poikilocytosis, and polychromatophilia, mostly associated with a swelling of the nucleus and a separating of its chromatin.

2. Normoblast: round; diameter less than length of normal normocyte. Nucleus of the same color, or slightly purple, with more scattered chromatin than nucleus of normal normocyte. Cytoplasm greenish blue and glassy, frequently with a clear zone about the nucleus.

3. Megaloblast: the same, except that the diameter is equal to or greater than the length of the normal normocyte.

4. Mitotic cells: all stages, from monaster to complete division of nucleus. (a) Round or elliptical; diameter about the length of the normal normocyte. Dense, deep blue chromosomes, massed in center of cell. Cytoplasm greenish blue and glassy. (b) The same, with two masses of chromosomes in opposite extremes of cell. (c) The same (at times with a slight constriction in the middle of the cell) with irregular dense, deep blue or slightly purple chromatin masses in place of individual chromosomes. (d) The same, with two nuclei, similar in appearance to those of the normoblast. (e) The same as the normal normocyte with two nuclei, similar in appearance to those of the normoblast, but only slightly larger than nucleus of normal normocyte. a, b, c, and d may show polychromatophilic cytoplasm.

Blood Platelets.—1. The same as normal, except that both cell and nucleus are larger.

2. Larger than normal with two nuclei.

Leucocytes.—1. Polymorphonuclear leucocyte with eosinophilic rods: the same as in normal blood, at times possibly a little smaller.

2. Polymorphonuclear leucocyte with eosinophilic granules: the same as in normal blood, at times possibly a little smaller.

3. Lymphocyte: the same as in normal blood.

4. Large mononuclear cell: (a) as described in normal blood with a little less cytoplasm. (b) The same size to one-half times larger than (a), with both nucleus and cytoplasm paler. (a) and (b) cannot be separated absolutely. Every gradation. (c) Mitoses: all stages from monaster to complete separation of nucleus: (a') elliptical; diameter about 1.5 times the length of the normal normocyte. Dense, purple chromosomes centrally massed. Cytoplasm pale bluish gray, granular. (b') The same, with two masses of chromosomes in opposite extremes of the cell. (c') The same, with two purple nuclei. (d) The same, with three purple nuclei.

5. Mast cell: as described in normal blood.

6. Mononuclear myelocyte with eosinophilic granules: round or slightly elliptical, diameter about length of normal normocyte or longer. Nucleus oval, pale blue with dense chromatin, eccentric. Cytoplasm colorless with small and large bright red granules; some scattered over nucleus.

Cytoplasmatic masses without nuclei: round or slightly oval; diameter varies, usually about width of normocyte. No nucleus. Cytoplasm grayish blue, sometimes with vacuoles.

The above is the blood picture common to all leukemic animals. The order of predominance of the white cells is typical of leukemia, differing from the normal, but agreeing with that of the spontaneous case. The large mononuclear is present in by far the greatest number. The other cells of the normal blood are decreased. Polymorphonuclears with eosinophilic granules and mast cells are very scarce. The mononuclear myelocyte with eosinophilic granules, although not as common as in the spontaneous case, can usually be demonstrated in every leukemic animal. The platelets are greatly increased in number.

Differential count, 300 cells.

	Per cent.
Polymorphonuclears with eosinophilic rods	6
" " " granules	0
Lymphocytes	4
Large mononuclear cells	86
. Mast cells	1
Mononuclear myelocytes with eosinophilic granules	1
Unclassified cells	2
	100

Autopsy Findings. Gross and Microscopical.—The weight of the animal in every case was far below its weight before injection.[9] The two lowest weights, at autopsy, were 675 and 897 grams.

The jaundice, at times, involved the skin of the entire body.[10] The conjunctivæ and buccal mucous membrane were always very pale. Emaciation was extreme. There was a great scarcity or entire absence of subcutaneous fat. Muscular atrophy was usually marked; at times practically only the skeleton remained.

The cervical lymph glands, macroscopically, were, as a rule, uninvolved. In two cases, they were definitely enlarged, in the one very much so. They measured 1.5 × 8 × 4 cm. and 2 × 1 × 5 cm., respectively. They appeared uniformly gray with absence of lobulation due to parenchymatous swelling and atrophy of interlobular fat. In a third case they were of normal size and appearance, but showed several gray nodules 1 mm. in diameter. Microscopically, those glands which appeared normal in gross were usually free

[9] Animals which remained negative invariably gained weight.
[10] Probably this was more frequent but not recognized, except when very grave, because of the normal yellow tint of the skin.

of myeloid infiltration, although their blood vessels contained leukemic blood. The diffusely enlarged glands showed a marked swelling of their parenchyma with complete atrophy of the interlobular fat. Myeloid cells, both the large mononuclear and the mononuclear myelocyte with eosinophilic granules, were scattered through the lobules. Foci, entirely of the first, or exclusively of the second, were localized principally in the interlobular connective tissue. They were rich in mitotic figures. The gray nodules proved to be a mass of proliferating myeloid tissue.

In the peritoneal cavity, the omental fat was greatly decreased. At best, it was present only in moderate amount. As a rule, it was replaced by a thin yellow membrane. Ascites occurred in 6 out of 13 cases, at times in sufficient amount to distend the abdomen. It was always associated with a serofibrinous mass which covered the liver and most of the other abdominal viscera and the outer surface of the pericardium. With the exception of Nos. 31 and 70 every animal of the thirteen had an enlarged liver. The margins of the right and left lobes of the smallest of the enlarged livers were respectively 1 and 3 cm. above the xiphoid, while the margins of the largest liver extended 3.5 and 3 cm. respectively below the xiphoid. In fact, the latter organ *in situ* was so enormous that it filled the entire peritoneal cavity and was the only viscus visible on opening the abdomen.

The liver was frequently enormous. Its weight ranged from 80 to 265 grams; *i. e.,* over five times normal, or 6.2 to 10.3 per cent of body weight, an increase of 3.6 times normal.

The external appearance of the liver was usually very characteristic. It frequently appeared diffusely gray, due to very closely packed subcapsular spots, pin point to 2 mm. in diameter. These were usually more scattered and translucent, at times slightly yellow and opaque. They were often fused to form larger areas with a diameter of 0.5 to 1 cm., or they were arranged in a delicate gray network. In addition, numerous gray or slightly yellow nodules ranging from 0.2 to 0.5 cm. in diameter were often present. The whole gave the surface an extremely mottled appearance. Lobulation at times was visible, usually indistinct. The liver was either slightly firm or friable. The sectioned surface appeared very similar to the external surface. Dots and lines often formed gray borders along the blood vessels. The latter also occupied the centers of some of the larger, circumscribed gray areas. The perivascular connective tissue in rare instances was bile-stained. Microscopically, there was usually a diffuse infiltration of myeloid cells. They were closely packed within and without the intralobular capillaries. The liver trabeculæ showed fatty degeneration and atrophy. The large mononuclear, rich in mitotic figures, was by far the predominating cell. The mononuclear myelocyte with eosinophilic granules was very scarce. Scattered through the section were foci of large mononuclears, or of mononuclear myelocytes with eosinophilic granules. Some foci contained both types of cells. They abounded in mitotic figures. The liver cells in these foci had entirely disappeared. The small, gray nodules consisted of circumscribed masses of myeloid tissue, usually occupying the periportal spaces and composed of a central portion of large mononuclears surrounded by the mononuclear myelocyte with eosinophilic granules. In the blood vessels and capillaries the large mononuclear at times seemed to exceed greatly the normocyte in number.

The spleen with the exception of Nos. 31 and 70 was, in every case, enlarged, usually extremely. The smallest of the enlarged organs measured 3 × 2 × 1.7 cm., the largest 4 × 3 × 2.3 cm. The latter was just twice normal in every dimension. The lowest weight was 6 grams, the highest 18 grams; *i. e.*, 18 times normal, or 0.6 to 1.6 per cent of body weight; *i. e.*, 32 times normal.

The surface of the spleen was sometimes normal. More frequently, it was studded with scattered gray or slightly yellow spots and nodules. The first was from 1 to 3 mm., and the largest of the second from 0.5 to 1 in diameter. Again, the organ appeared diffusely gray. On section it was usually similar to the surface. When diffusely gray, the pulp was somewhat granular and in excess. Microscopically the entire pulp was often diffusely infiltrated with the large mononuclear cell separating and compressing the Malpighian bodies. In addition, circumscribed closely packed masses of pure large mononuclears or mononuclear myelocytes with eosinophilic granules were scattered through the section. Occasionally, the myelosis was only present in foci, with practically no disturbance of normal splenic structure. Mitoses were abundant wherever the large mononuclear occurred.

The kidneys of 5 of the 13 cases were enlarged. They weighed from 12 to 26 grams; or 0.8 to 1.4 per cent of the body weight.

The surface and section of the kidney were entirely normal, but usually there were below its capsule scattered, gray, pin point dots, delicate lines, or even one or two nodules, the largest 0.5 cm. in diameter. Again, they appeared diffusely gray. Microscopically, the kidneys always showed more or less distention of their blood vessels and intertubular capillaries with the characteristic leukemic blood in which white blood cells, at times, even seem to exceed the normocytes, and in which the large mononuclears predominated. When capillary distention was moderate, the tubules appeared normal, but at times they were so enormously overfilled as to cause extreme atrophy and degeneration of the renal epithelium, resulting frequently in circumscribed areas devoid of any parenchyma. These areas were more pronounced about the larger blood vessels. The walls and the perivascular tissue of some of the vessels were infiltrated with actively generating myeloid cells, especially the large mononuclear.

The bone marrow was always involved, usually more or less characteristically. It was increased in amount, moderately soft. and red. with countless gray dots to slightly larger areas: or it was diffusely gray. Microscopically, the normal structure of the marrow was greatly changed. The fatty tissue had completely disappeared and the marrow cells formed a solid mass. The description of the red and white marrow cells agreed with that given under normal marrow. Possibly the large mononuclear was more commonly polygonal, due to pressure. Mitoses of both the large mononuclear and the erythrocytes were common. The order of predominance was changed from normal. The large mononuclear myelocyte was by far the most abundant; the normoblast. the megaloblast. and the mononuclear myelocyte with eosinophilic granules followed in about equal numbers; the normocyte was least numerous. Both types of polymorphonuclear as well as the lymphocyte. mast cells. and platelet were not seen at all. Mitoses of the mononuclear myelocyte with eosinophilic granules were not demonstrated.

In the remaining organs and tissues the blood vessels were filled with the characteristic blood; and in many, infiltrations of myeloid cells occurred in moderate degree.

Summary.—The essential points deduced from the simple transmission of leukemia by injection of an organic emulsion may be stated as follows:

Clinical History.—After an incubation period, usually from 5 to 6 weeks, the fowl suddenly becomes pale, jaundiced, emaciates rapidly, loses weight, and shows signs of extreme weakness, followed almost without exception by death in 1 to 2 weeks.

Blood.—(a) The total number of white blood cells is greatly increased, resulting in a proportion of one white to three red blood cells. The differential count showed a marked increase and predominance of the large mononuclear over the other white cells of the normal blood, which have decreased. The mononuclear myelocyte with eosinophilic granules is present in more or less numbers in practically every case. Besides the normal appearing large mononuclear, one sees many larger pale forms. In these cells, typical and atypical, mitoses in all stages are common. (b) The total number of red blood cells is correspondingly decreased, with a marked fall in hemoglobin. They present varieties in size, shape, and staining, associated with swelling of the nucleus. Normoblasts and megaloblasts, with mitoses in all stages, make their appearance. (c) There is an increase in the number of blood platelets, associated with an increase, both in size of the cell and its nucleus. The cells frequently contain more than one nucleus. (d) The clotting power of blood is greatly decreased.

Organs.—(a) A diffuse or focal infiltration of marrow cells occurs in many organs. The large mononuclear and mononuclear myelocyte with eosinophilic granules predominate and may occur separately or together in the same nodule. Extensive mitoses occur in both these cell types. This myelosis affects more particularly the liver, spleen, kidneys, and bone marrow, causing a great increase in the size of the first three organs. Rarely the cervical lymph glands are also very much enlarged. Most of the remaining organs and neighboring tissues may contain infiltrations, but not so extensive. (b) The proportion of white to red blood cells is greatly increased in the blood vessels. The same cells predominate as in the infiltration. (c) The atrophic and degenerative changes of the parenchyma and of the adipose tissue are marked; ascites occurs.

Conclusion.—From the above experimental study the following conclusions may be drawn: The injection of an organic emulsion causes a picture of myeloid leukemia in every respect similar to spontaneous leukemia as it occurs in the fowl. The clinical picture and changes produced in blood and organs are analogous to those which occur in human leukemia.

CONCLUSIONS.

1. The spontaneous occurrence of myeloid leukemia of the fowl is confirmed.

2. Myeloid leukemia of the fowl is transmissible by intravenous or intraperitoneal injection of an organic emulsion.

The latter is in confirmation of the work of Ellermann and Bang (6, 7), who first successfully transmitted the disease. They were followed by Hirschfeld and Jacoby (13), whose successful transmissions, however, seem to be limited to a strain which had its origin in a fowl presented to them by Ellermann and Bang. Burckhardt (14, 15) likewise transmitted the disease, but here again the stock animal came from Hirschfeld and Jacoby and therefore indirectly from Ellermann and Bang.

The transmission reported above is of special interest, because it originated in an animal absolutely unrelated to that of the previous investigators.

BIBLIOGRAPHY.

1. Moore, V. A., Infectious Leukaemia in Fowls; A Bacterial Disease Frequently Mistaken for Fowl Cholera, *Twelfth and Thirteenth Annual Reports, U. S. Bureau of Animal Industry,* 1897, 185.
2. Butterfield, E. E., Aleukaemic Lymphadenoid Tumors of the Hen. *Folia Haematol.,* 1905, ii, 649.
3. Warthin, A. S., Leukemia of the Common Fowl. *Jour. Infect. Dis.,* 1007, iv, 369.
4. Kon, J., Über Leukämie beim Huhn, *Virchows Arch. f. path. Anat.,* 1907, cxc, 338.
5. Soshestrenski, N. A., Leukaemia of the Chicken. *Uchen. zapiski Kazan. Vet. Inst.,* 1908, xxv, 215.
6. Ellermann, V., and Bang, O., Experimentelle Leukämie bei Hühnern. Vorläufige Mitteilung. *Centralbl. f. Bakteriol., 11e Abt., Orig.,* 1008, xlvi, 4.

7. Ellermann and Bang, Experimentelle Leukämie bei Hühnern, *ibid.*, p. 596.
8. Ellermann, V., Experimentelle Leukämie bei Hühnern, *Verhandl. d. deutsch. path. Gesellsch.*, 1908, xii, 224.
9. Ellermann and Bang, Experimentelle Leukämie bei Hühnern. II, *Ztschr. f. Hyg. u. Infectionskrankh.*, 1909, lxiii, 231.
10. Schridde, H., Gibt es eine infektiöse Aetiologie der Leukämie?, *Deutsch. med. Wchnschr.*, 1909, xxxv, 280.
11. Hirschfeld, H., and Jacoby, M., Zur Kenntniss der übertragbaren Hühnerleukämie, *Berl. klin. Wchnschr.*, 1909, xlvi, 159.
12. Jacoby, M., Uebertragbare Hühnerleukämie, *Berl. klin. Wchnschr.*, 1909, xlvi, 314.
13. Hirschfeld, H., and Jacoby, M., Uebertragungsversuche mit Hühnerleukämie, *Ztschr. f. klin. Med.*, 1910, lxix, 107.
14. Burckhardt, J. L., Referat über die Sitzungen der achtundzwanzigsten Abteilung (Hygiene und Bakteriologie) der 82. Versammlung deutscher Naturforscher und Aerzte zu Königsberg, *Ztschr. f. Immunitätsforsch., Ref.*, 1910, ii, 810.
15. Burckhardt, Ueber das Blutbild bei Hühnertuberkulose und dessen Beziehungen zur sogenannten Hühnerleukämie nebst Bemerkungen über das normale Hühnerblut, *Ztschr. f. Immunitätsforsch., Orig.*, 1912, xiv, 544.
16. Friedberger, Referat über die Sitzungen der achtundzwanzigsten Abteilung (Hygiene und Bakteriologie) der 82. Versammlung deutscher Naturforscher und Aerzte zu Königsberg, *Ztschr. f. Immunitätsforsch., Ref.*, 1910, ii, 811.
17. Ellermann, [Virus of leukæmia in poultry], *Ugesk. f. Læger*, 1913, lxxv, 1685.
18. Ellermann, Untersuchungen über das Virus der Hühnerleukämie, *Ztschr. f. klin. Med.*, 1914, lxxix, 43.
19. Hirschfeld and Jacoby, Uebertragbare Hühnerleukämie und ihre Unabhängigkeit von der Hühnertuberkulose, *Ztschr. f. klin. Med.*, 1912, lxxv, 501.

INDEX TO VOLUME XXII.

Lightning Source UK Ltd.
Milton Keynes UK
UKHW021306210119
335934UK00013B/754/P